MW01069554

Reinforced Concrete Design

NINTH EDITION

Abi Aghayere

Professor

Department of Civil, Architectural, and Environmental Engineering

Drexel University

330 Hudson Street, NY NY 10013

Vice President, Portfolio Management: Andrew Gilfillan
Editorial Assistant: Lara Dimmick
Senior Vice President, Marketing: David Gesell
Marketing Coordinator: Elizabeth MacKenzie-Lamb
Director, Digital Studio and Content Production: Brian Hyland
Managing Producer: Jennifer Sargunar
Content Producer (Team Lead): Faraz Sharique Ali
Manager, Rights Management: Johanna Burke
Operations Specialist: Deidra Smith
Cover Design: Cenveo Publisher Services
Cover Photo: William Tao & Associates, Inc.
Full-Service Project Management: Integra Software Services Pvt. Ltd.
Composition: Integra Software Services Pvt. Ltd.
Printer/Binder: LSC Communications, Inc.
Cover Printer: Phoenix Color/Hagerstown
Text Font: PalatinoLTPro

Library of Congress Cataloging-in-Publication Data

Names: Aghayere, Abi O., author.
Title: Reinforced concrete design / Abi Aghayere, Professor, Department of Civil, Architectural, and Environmental Engineering, Drexel University.
Description: Ninth edition. | Upper Saddle River, New Jersey : Pearson Education, Inc., 2018. | Includes bibliographical references and index.
Identifiers: LCCN 2017056582| ISBN 9780134715353 | ISBN 0134715357
Subjects: LCSH: Reinforced concrete.
Classification: LCC TA444 .L44 2018 | DDC 624.1/8341--dc23 LC record available at https://lccn.loc.gov/2017056582

1 18

ISBN 10: 0-13-471535-7
ISBN 13: 978-0-13-471535-3

NOTICE TO THE READER

The information contained in this book has been prepared in accordance with recognized engineering principles and is for general information only. Although it is believed to be accurate, this information should not be used for any specific application without competent professional examination and verification of its accuracy, suitability, and applicability by a licensed professional engineer, architect, or designer. The authors and publisher of this book make no warranty of any kind, expressed or implied, with regard to the material contained in this book nor shall they be liable for any special, consequential, or exemplary damages resulting, in whole or in part, from the reader's use of or reliance on this material.

To the cherished memory of my mother, Regina Ekeneza-Obasogie, and
my great-grand mother, Aghayubini Osawe

PREFACE

The primary objective of *Reinforced Concrete Design*, ninth edition, remains the same as that of the previous editions that were co-authored with George F. Limbrunner, who is now retired: to provide a basic and thorough understanding of the strength and behavior of reinforced concrete members and reinforced concrete structural systems.

With the recent changes in the ACI 318 Code, and relevant reinforced concrete research and literature continuing to become available at a rapid rate, it is the intent of this book to translate this vast amount of information and data into an integrated source that reflects the latest information available. This book is intended to help the reader understand the fundamentals of reinforced concrete design and behavior, and good practices in the industry. It will be useful to students in undergraduate civil and architectural engineering programs, and engineering technology and architecture programs. In addition, practicing structural engineers and engineers preparing for the licensure exams will find this text to be a helpful and practical resource.

This ninth edition has been prepared with the primary objective of updating its contents to conform to the latest *Building Code Requirements for Structural Concrete* (ACI 318-14) of the American Concrete Institute. The ACI 318-14 is a complete reorganization of the Code compared to previous editions of the Code, in addition to incorporating a number of technical changes. Throughout the text, frequent references are made to the pertinent sections of the ACI Code. Because the ACI Code serves as the design standard in the United States, it is strongly recommended that the Code be used as a companion publication to this book.

In addition to the necessary changes to conform to the new code, some sections have been edited, new sections have been added, and the student design project problems have been enhanced and several drawings updated. Working on the design project helps the student appreciate how and where the individual reinforced concrete elements covered in the different chapters fit within the context of a real life building project. Thus, they can see how what they have learned is readily applicable to, and usable in, the real world of engineering and construction. Answers to selected problems are furnished at the back of the text.

This text is suitable for any undergraduate concrete design course which would typically include topics from Chapters 1, 2, 3, 4, 5, 6, 7, and parts of Chapters 9 and 10. The remaining topics and chapters could make up a significant portion of a second undergraduate concrete design course. This text covers more topics than those required in the Concrete I course, and many of the topics required in the Foundation Design course, of the Basic Education for Structural Engineers Curriculum published by the Structural Engineering Institute (SEI) of the American Society of Civil Engineers (ASCE), the National Council of Structural Engineering Associations (NCSEA), and the Council of American Structural Engineers (CASE).

Throughout the nine editions, the text content has maintained primarily a fundamental and practice-oriented approach to the design and analysis of reinforced concrete structural members using numerous examples and a step-by-step solution format. In addition, there are chapters that provide a conceptual approach on such topics as prestressed concrete and detailing of reinforced concrete structures. The metric system (SI) is introduced in Appendix C with several example problems.

Form design is an important consideration in most structural design problems involving concrete members, and Chapter 12 illustrates procedures for the design of job-built forms for slabs, beams, and columns. Appropriate tables are included that will expedite the design process. In Chapter 14, we introduce the reader to several practical considerations and rules of thumb for the design of reinforced concrete beams, girders, columns and one-way slabs, and methods for strengthening existing reinforced concrete structures.

NEW TO THIS EDITION

- The entire text has been revised to conform to the latest ACI Code: ACI 318-14.
- The design of concrete mixes and admixtures is discussed in Chapter 1 and an introduction to gravity load distribution (tributary areas and tributary width) and concrete slab systems is presented.
- The design of slab-on-grade is included in Chapter 2.
- The design of corbels and brackets is included in Chapter 4.
- Structural integrity reinforcement is presented in Chapter 5 and more examples on the development of reinforcement have been added to the chapter.
- Two-way slab design using the direct design method is covered in Chapter 6 together with an introduction to the equivalent frame method.
- In Chapter 7, the procedures for calculating the long-term deflections of continuous beams and girders are included, in addition to a new example on the long-term deflection of a continuous T-beam. Deflection control measures are discussed and the design of concrete floor systems for vibrations is also included in Chapter 7.
- The design of columns subject to axial load plus bi-axial bending, the moment magnification factor, and the design of slender columns in non-sway or braced frames are introduced in Chapter 9.
- A discussion of the types of information contained in a geotechnical report and the design of eccentrically loaded spread footings with a new example to illustrate the design procedure are presented in Chapter 10.
- The design of pile caps and deep beams using the strut-and-tie method is covered in Chapter 10.
- A new section on diaphragms, chords, and drag struts is included in Chapter 14. Additional sections added to this chapter include one-way slabs subjected to concentrated loads, fire resistance of structural concrete, concrete pour strips to mitigate shrinkage cracks, concrete specifications, and load testing of existing structures.
- A second student design project problem has been added in Chapter 14.

This book has been thoroughly tested over the years in engineering, architecture, and engineering technology programs, and should serve as a valuable design guide and resource for engineering and architectural students, technologists, and design engineers. In addition, it will aid engineers and architects preparing for state licensing examinations for professional registration.

Acknowledgments

Thanks are due to George Limbrunner—the founding coauthor of this text since 1977, and with whom I was coauthor for the sixth through the eighth editions—for the rich legacy and practical imprint he left on this text.

I would also like to thank the peer reviewers for the previous editions and particularly those who did the review for this edition for their many valued contributions. We are excited about continuing the practice-oriented and easy-to-understand emphasis of this textbook while introducing some new topics of interest based on the feedback we have received on the previous editions.

As in the past, appreciation is extended to our students, past and present, and our colleagues, whose constructive feedback and enthusiasm have provided encouragement for this edition. Special thanks to Jason Vigil, S.E., P.E., for his invaluable input and his immense help in preparing the figures for this edition. Thanks are due as well to Ben Okorosobo, P.Eng., consulting structural engineer, for his helpful insights during the preparation of this text. I am also indebted to the late University Professor James G. MacGregor for his mentorship during and after my doctoral studies at the University of Alberta.

I'm grateful to my forever-bride, Josephine, for her enduring support and patience, and to my children—Osa, Ito, Odosa, and Eghosa—for their continuous encouragement. Finally, I am most grateful to Almighty God for His amazing grace and strength that made this project possible.

Download Instructor Resources from the Instructor Resource Center

To access supplementary materials online, instructors need to request an instructor access code. Go to www.pearsonhighered.com/irc to register for an instructor access code. Within 48 hours of registering, you will receive a confirming e-mail including an instructor access code. Once you have received your code, locate your text in the online catalog and click on the Instructor Resources button on the left side of the catalog product page. Select a supplement, and a login page will appear. Once you have logged in, you can access instructor material for all Pearson textbooks. If you have any difficulties accessing the site or downloading a supplement, please contact Customer Service at http://support.pearson.com/getsupport

Abi Aghayere
Philadelphia, PA

CONTENTS

Chapter 1

MATERIALS AND MECHANICS OF BENDING, AND CONCRETE SLAB SYSTEMS 1

1-1 Concrete 1
1-2 The ACI Building Code 1
1-3 Cement and Water 1
1-4 Aggregates 1
1-5 Concrete Mixes 2
1-6 Concrete in Compression 3
1-7 Concrete in Tension 5
1-8 Reinforcing Steel 5
1-9 Concrete Cover 8
1-10 Beams: Mechanics of Bending Review 8
1-11 Concrete Slab Systems 13
1-12 Gravity Load Distribution in Concrete Slab Systems 14
References 16
Problems 17

Chapter 2

RECTANGULAR REINFORCED CONCRETE BEAMS AND SLABS: TENSION STEEL ONLY 21

2-1 Introduction 21
2-2 Analysis and Design Method 21
2-3 Behavior Under Load 22
2-4 Strength Design Method Assumptions 23
2-5 Flexural Strength of Rectangular Beams 24
2-6 Equivalent Stress Distribution 25
2-7 Balanced, Brittle, and Ductile Failure Modes 27
2-8 Ductility Requirements 28
2-9 Strength Requirements 30
2-10 Rectangular Beam Analysis for Moment (Tension Reinforcement Only) 31
2-11 Summary of Procedure for Rectangular Beam Analysis for ϕM_n (Tension Reinforcement Only) 34
2-12 Slabs: Introduction 34
2-13 One-Way Slabs: Analysis for Moment 34
2-14 Rectangular Beam Design for Moment (Tension Reinforcement Only) 37
2-15 Summary Of Procedure for Rectangular Reinforced Concrete Beam Design for Moment (Tension Reinforcement Only) 41
2-16 Design of One-Way Slabs for Moment (Tension Reinforcement Only) 41
2-17 Summary of Procedure for Design of One-Way Slabs for Moment (To Satisfy ACI Minimum Thickness, h) 43
2-18 Slabs-On-Grade 43
References 46
Problems 46

Chapter 3

REINFORCED CONCRETE BEAMS: T-BEAMS, L-BEAMS, AND DOUBLY REINFORCED BEAMS 51

3-1 T-Beams and L-Beams: Introduction 51

3-2 T-Beam and L-Beam Analysis 53

3-3 Analysis of Beams Having Irregular Cross Sections 56

3-4 T-Beam and L-Beam Design (for Moment) 57

3-5 Summary of Procedure for Analysis of T-Beams and L-Beams (for Moment) 60

3-6 Summary of Procedure for Design of T-Beams and L-Beams (for Moment) 61

3-7 Doubly Reinforced Beams: Introduction 62

3-8 Doubly Reinforced Beam Analysis for Moment (Condition I) 62

3-9 Doubly Reinforced Beam Analysis for Moment (Condition II) 65

3-10 Summary of Procedure for Analysis of Doubly Reinforced Beams (for Moment) 67

3-11 Doubly Reinforced Beam Design for Moment 68

3-12 Summary of Procedure for Design of Doubly Reinforced Beams (for Moment) 69

3-13 Additional Code Requirements for Doubly Reinforced Beams 70

Problems 71

Chapter 4

SHEAR AND TORSION 75

4-1 Introduction 75

4-2 Shear Reinforcement Design Requirements 76

4-3 Shear Analysis Procedure 78

4-4 Stirrup Design Procedure 79

4-5 Torsion of Reinforced Concrete Members 86

4-6 Corbels and Brackets 94

References 99

Problems 99

Chapter 5

DEVELOPMENT, SPLICES, AND SIMPLE-SPAN BAR CUTOFFS 104

5-1 Bond Stress and Development Length: Introduction 104

5-2 Development Length: Tension Bars 106

5-3 Development Length: Compression Bars 111

5-4 Development Length: Standard Hooks in Tension 112

5-5 Development of Web Reinforcement 115

5-6 Splices 117

5-7 Tension Splices 117

5-8 Compression Splices 117

5-9 Simple-Span Bar Cutoffs and Bends 118

5-10 Code Requirements for Development of Positive Moment Steel at Simple Supports 122

5-11 Structural Integrity Reinforcement-Beams 125

References 126

Problems 126

Chapter 6

CONTINUOUS ONE-WAY AND TWO-WAY FLOOR SYSTEMS 130

6-1 Introduction 130

6-2 Continuous-Span Bar Cutoffs 132

6-3 Design of Continuous One-Way Floor Systems 133

6-4 Analysis and Design of Continuous Two-Way Slabs 145

References 180

Problems 180

Chapter 7

SERVICEABILITY 183

7-1 Introduction 183

7-2 Deflections 183

7-3 Calculation of I_{cr} 184

7-4 Immediate Deflection 186

7-5 Long-Term Deflection 186

7-6 Procedure for Calculating the Deflection of Simply Supported and Continuous Beams and Slabs 189

7-7 Procedure for Calculating the Deflection of Continuous Girders 190

7-8 Deflection Control Measures in Reinforced Concrete Structures 193

7-9 Crack Control 194

7-10 Floor Vibrations 195

7-11 Gross and Cracked Section Properties of Concrete Sections 197

References 198

Problems 198

Chapter 8

WALLS 200

8-1 Introduction 200

8-2 Lateral Forces on Retaining Walls 201

8-3 Design of Reinforced Concrete Cantilever Retaining Walls 204

8-4 Design Considerations for Bearing Walls 219

8-5 Design Considerations for Basement Walls 221

8-6 Lateral Load Resisting Systems in Concrete Buildings 221

8-7 Concrete Moment Frames 222

8-8 Shear Walls 223

References 231

Problems 231

Chapter 9

COLUMNS 234

9-1 Introduction 234

9-2 Strength of Reinforced Concrete Columns: Small Eccentricity 235

9-3 Code Requirements Concerning Column Details 236

9-4 Analysis of Short Columns: Small Eccentricity 238

9-5 Design of Short Columns: Small Eccentricity 239

9-6 Summary of Procedure for Analysis and Design of Short Columns with Small Eccentricities 241

9-7 The Load-Moment Relationship 241

9-8 Columns Subjected to Axial Load at Large Eccentricity 242

9-9 ϕ Factor Considerations 242

9-10 Analysis of Short Columns: Large Eccentricity 243

9-11 Biaxial Bending 250

9-12 The Slender Column 252

9-13 Concrete Column Schedule 256

References 257

Problems 257

Chapter 10

FOUNDATIONS 260

10-1 Introduction 260

10-2 The Geotechnical Report 261

10-3 Wall Footings 262

10-4 Wall Footings Under Light Loads 267

10-5 Individual Reinforced Concrete Footings for Columns 267

10-6 Square Reinforced Concrete Footings 270

10-7 Rectangular Reinforced Concrete Footings 273

10-8 Eccentrically Loaded Footings 277

10-9 Combined Footings 282

10-10 Cantilever or Strap Footings 284

10-11 Analysis and Design of Mat Foundations 286

10-12 Deep Foundations-Piles, Drilled Shaft (Caissons), and Pile Caps 287

10-13 Strut-and-Tie Models for Pile Caps and Deep Beams 292

References 300

Problems 301

Chapter 11

PRESTRESSED CONCRETE FUNDAMENTALS 303

11-1 Introduction 303

11-2 Design Approach and Basic Concepts 303

11-3 Stress Patterns in Prestressed Concrete Beams 305

11-4 Prestressed Concrete Materials 306

11-5 Analysis of Rectangular Prestressed Concrete Beams 307

11-6 Alternative Methods of Elastic Analysis: Load Balancing Method 310

11-7 Flexural Strength Analysis 313

11-8 Notes on Prestressed Concrete Design 315

References 315

Problems 315

Chapter 12

CONCRETE FORMWORK 317

12-1 Introduction 317

12-2 Formwork Requirements 317

12-3 Formwork Materials and Accessories 318

12-4 Loads and Pressures on Forms 319

12-5 The Design Approach 321

12-6 Design of Formwork for Slabs 326

12-7 Design of Formwork for Beams 331

12-8 Wall Form Design 335

12-9 Forms for Columns 339

References 342

Problems 342

Chapter 13

DETAILING REINFORCED CONCRETE STRUCTURES 344

13-1 Introduction 344

13-2 Placing or Shop Drawings 345

13-3 Marking Systems and Bar Marks 345

13-4 Schedules 352

13-5 Fabricating Standards 352

13-6 Bar Lists 353

13-7 Extras 354

13-8 Bar Supports and Bar Placement 355

13-9 Computer Detailing 356

References 359

Chapter 14

PRACTICAL CONSIDERATIONS IN THE DESIGN OF REINFORCED CONCRETE BUILDINGS 360

14-1 Introduction 360

14-2 Rules of Thumb and Practical Considerations for Reinforced Concrete Design 360

14-3 Approximate Moments and Shears in Continuous Girders 362

14-4 Strengthening and Rehabilitation of Existing Reinforced Concrete Structures 364

14-5 Diaphragms, Drag Struts, and Chords 369

14-6 One-Way Slabs Subjected to Concentrated Loads 371

14-7 Load Testing of Structures 373

14-8 Closure or Pour Strips in Reinforced Concrete Floors 375

14-9 Fire Resistance of Concrete Structural Elements 376

14-10 Analysis and Design of Edge-Supported Two-Way Slabs on Stiff Supports 377

14-11 Cast-In Place Concrete Specifications 379

14-12 Student Design Projects 381

References 385

APPENDIX A TABLES AND DIAGRAMS 386

APPENDIX B SUPPLEMENTARY AIDS AND GUIDELINES 403

B-1 ACCURACY FOR COMPUTATIONS FOR REINFORCED CONCRETE 403

B-2 FLOW DIAGRAMS 403

APPENDIX C METRICATION 408

C-1 THE INTERNATIONAL SYSTEM OF UNITS (SI) 408

C-2 SI STYLE AND USAGE 410

C-3 CONVERSION FACTORS 411

REFERENCES 416

APPENDIX D ANSWERS TO SELECTED PROBLEMS 417

INDEX 419

CHAPTER **ONE**

MATERIALS AND MECHANICS OF BENDING, AND CONCRETE SLAB SYSTEMS

1-1 Concrete
1-2 The ACI Building Code
1-3 Cement and Water
1-4 Aggregates
1-5 Concrete Mixes
1-6 Concrete in Compression
1-7 Concrete in Tension
1-8 Reinforcing Steel
1-9 Concrete Cover
1-10 Beams: Mechanics of Bending Review
1-11 Concrete Slab Systems
1-12 Gravity Load Distribution in Concrete Slab Systems

1-1 CONCRETE

Concrete, the most commonly used construction material worldwide, is a composite material that consists primarily of a mixture of cement and fine and coarse aggregates (sand, gravel, crushed rock, and/or other materials) to which water has been added as a necessary ingredient for the chemical reaction during the curing process. The bulk of the mixture consists of the fine and coarse aggregates. The resulting concrete strength and durability are a function of the proportions of the mix as well as other factors, such as the concrete placing, finishing, and curing history.

The compressive strength of concrete is relatively high. Yet, it is a relatively brittle material, the tensile strength of which is small compared with its compressive strength. Hence steel reinforcing rods (which have high tensile and compressive strength) are used in combination with the concrete; the steel will resist the tension and the concrete the compression. *Reinforced concrete* is the result of this combination of steel and concrete. In many instances, steel and concrete are positioned in members so that they both resist compression.

1-2 THE ACI BUILDING CODE

The design and construction of reinforced concrete buildings is controlled by the *Building Code Requirements for Structural Concrete* (ACI 318-14) of the American Concrete Institute (ACI) [1]. The use of the term *code* in this text refers to the ACI Code unless otherwise stipulated. The code is revised, updated, and currently currently reissued on a 3-year cycle. It has been incorporated into the building codes of almost all states and municipalities throughout the United States, however. When so incorporated, it has official sanction, becomes a legal document, and is part of the law controlling reinforced concrete design and construction in a particular area.

1-3 CEMENT AND WATER

Structural concrete uses, almost exclusively, hydraulic cement. With this cement, water is necessary for the chemical reaction of *hydration*. In the process of hydration, the cement sets and bonds the fresh concrete into one mass. *Portland cement*, which originated in England, is undoubtedly the most common form of cement. Portland cement consists chiefly of calcium and aluminum silicates. The raw materials are limestones, which provide calcium oxide (CaO), and clays or shales, which furnish silicon dioxide (SiO_2) and aluminum oxide (Al_2O_3). Following processing, cement is marketed in bulk or in 94-lb (1-ft^3) bags.

In fresh concrete, the ratio of the amount of water to the amount of cement, by weight, is termed the *water-cementitious material ratio*. This ratio can also be expressed in terms of gallons of water per bag of cement. For complete hydration of the cement in a mix, a water-cementitious material ratio of 0.35 to 0.40 (4 to $4\frac{1}{2}$ gal of water/bag of cement) is required. To increase the *workability* of the concrete (the ease with which it can be mixed, handled, and placed), higher water-cementitious material ratios are normally used or superplasticizers are added to the mix.

1-4 AGGREGATES

In ordinary structural concretes, the aggregates occupy approximately 70% to 75% of the volume of the hardened mass. Gradation of aggregate size to produce close packing is desirable because, in general, the more densely the aggregate can be packed, the better are the strength and durability because of the inter-locking of the aggregates.

Aggregates are typically classified as fine or coarse. *Fine aggregate* is generally sand and may be categorized as consisting of particles that will pass a No. 4 sieve (four openings per linear inch). *Coarse aggregate* consists of particles that would be retained on a No. 4 sieve. The maximum size of coarse aggregate in reinforced concrete is governed by various ACI Code requirements. These requirements are established primarily to ensure that the concrete can be placed with ease into the forms without any danger of jam up between adjacent bars or between bars and the sides of the forms. Section 26.4.2.1 of ACI 318 states that maximum size of coarse aggregate should not exceed the least of the following:

(i) one-fifth the narrowest dimension between sides of forms
(ii) one-third the depth of slabs
(iii) three-fourths the minimum specified clear spacing between individual reinforcing bars or wires, bundles of bars, prestressed reinforcement, individual tendons, bundled tendons, or ducts

Note that smaller aggregate sizes have a relatively larger total surface area and thus require more cement paste to coat the surfaces of the aggregate which results in higher drying shrinkage. In typical building construction, the nominal maximum coarse aggregate size of $^3/_4''$ is commonly specified, though maximum aggregate size of up to $1^1/_2''$ is also sometimes specified to achieve reduced shrinkage.

1-5 CONCRETE MIXES

There are numerous factors to consider when determining all of the components of a concrete mix. The most basic mix consists of water, cement, coarse aggregate, and fine aggregate (sand). Beyond these basic components, one or more chemical admixtures can be used depending on the required performance of the concrete. Since the admixtures are a proprietary product, the dosage is based on the manufacturer's recommendations. Admixtures are often dosed in units of fluid ounces per 100 pounds of cementitious material (cwt). The strength of the concrete could be adversely impacted by certain admixtures, and there could also be compatibility issues with multiple admixtures. To control these issues and to also ensure the performance of the concrete, concrete suppliers will typically go through the process of creating batches of concrete and testing them. Once a certain mix is validated, it can then be used in the future using the same components and proportions. A sample mix design is shown in the below table.

The most critical parameter in the design and selection of a concrete mix is the design strength f'_c, which is defined as the compressive strength of the concrete at 28 days after placement, but there are other factors to consider in order to satisfy the concrete durability requirements of the ACI Code. For each group of structural members, the exposure classes must first be assigned based on ACI 318–14, Table 19.3.1.1 depending on the degree of exposure of the concrete to freeze/thaw (F), sulfate (S), water (W), and corrosion (C). Once the exposure classes are known, ACI 318–14, Table 19.3.2.1 is used to determine the minimum design strength and water-cementitious material ratio. This table is also used to determine whether or not air entrainment is needed. If air entrainment is needed, then ACI 318–14, Table 19.3.3.1 is used. The amount of air entrainment needed is a function of the exposure class and the aggregate size. Air entraining typically ranges between 4.5% and 7%. The design data for a sample mix is shown below:

Sample Mix Design Data

Mix Design Number: **4001-A** (this number is specific to the concrete supplier)
Minimum compressive strength at 28 days
$f'_c = 4000\,\text{psi}$
Used for: Footings, piers, and foundation walls
Slump: 4 in.; 8 in. with HRWR[1]
Air content: 6.0% +/− 1.5%
water-to-cementitious material ratio: 0.40

Materials	Weight per Cubic Yard
Cement, Type I/II, ASTM C150	540 lbs
Fly Ash, ASTM C618	95 lbs
Natural Sand, ASTM C33	1230 lbs
#1 and #2 crushed stone, ASTM C33	1780 lbs
Potable Water, ASTM C1602	255 lbs
Air Entrainment, ASTM C260	1.5 oz./cwt
Water Reducer, ASTM C494 Type A	3.0 oz./cwt
High-Range Water Reducer, ASTM C494 Type F*	7 oz. to 10 oz./cwt
Non-Chloride Accelerator, ASTM C494*	0 oz. to 26 oz./cwt

*Added at the site upon request

The most common admixtures are outlined as follows:

Air Entrainment

Air entrainment provides small air bubbles in the concrete to act as a buffer against volumetric changes during freeze/thaw cycles and thus makes the concrete more durable. It also makes the concrete more workable. A rule of thumb is that the compressive strength of concrete will decrease by 5% for every 1% increase in entrained air.

Water Reducers

When it is desired to make the concrete more workable, a common cost-effective solution is to simply add water, but this will have a negative impact on

[1]High range water reducer (HRWR)

the strength of the concrete. Water-reducing admixtures will increase the slump and thus the workability of concrete while not adversely impacting the water-cementitious material ratio.

When a much higher degree of workability is needed, a superplasticizer (high-range water reducer) can be used which will yield slumps around 8 in. and higher. There are also products available that are called mid-range water reducers that provide a more moderate amount of slump increase.

Accelerators and Retarders

It is sometimes desired to decrease or increase the setting time of the concrete, and so an accelerator or retarder product could be used. Accelerators can be used to increase the rate at which the concrete reaches its design strength. This may be desired due to weather conditions or to meet a certain construction schedule. Calcium chloride is a common accelerator component, but it cannot be used in reinforced concrete since the chloride will promote corrosion in the steel reinforcement.

Set retarders are used to delay the chemical reaction with the cement that allows the concrete to set. This might be used to offset the effect of high temperatures since high temperatures can produce a faster setting time. They might also be used, for example, during a paving operation to allow more time between concrete batches and reduce the possibility of a cold joint in the *pavement*.

Corrosion-Inhibitor

Corrosion-inhibiting admixtures are used to deter corrosion of reinforcing steel in concrete that is exposed to water or salt. This will increase the durability and life span of the concrete structure.

Other chemical admixtures include shrinkage compensating and permeability reducing admixtures. For more information on chemical admixtures, the reader should refer to ACI 212.3R-16-Report on Chemical Admixtures for Concrete.

Concrete that is placed between 50° F and 85° F do not require any special mix design or temperature considerations. When concrete is placed during extreme weather temperatures outside of these range, special considerations for cold and hot weather concrete must be accounted for [13, 14].

When concrete is poured in cold temperatures without any precautionary measures taken, the concrete will freeze and thus destory the bond between the concrete and the rebar. Pouring concrete at excessively high temperatures causes the concrete to set faster than anticipated, leading to the formation of cold joints. High temperatures also lead to uncontrolled and signigicant cracking of the concrete which negatively impacts strength and durability. Detailed information on the temperature requirements for concrete can be found in ACI 305 - Guide to Hot Weather Concreting and ACI 306 - Guide to Cold Weather Concreting. Special mix considerations also apply to mass concrete pours, such as thick mat foundations, where the high temperatures generated from the hydration process, if not adequately controlled, can lead to cracking of the concrete.

1-6 CONCRETE IN COMPRESSION

The theory and techniques relative to the design and proportioning of concrete mixes, as well as the placing, finishing, and curing of concrete, are outside the scope of this book and are adequately discussed in many other publications [2–5]. Field testing, quality control, and inspection are also adequately covered elsewhere. This is not to imply that these are of less importance in overall concrete construction technology but only to reiterate that the objective of this book is to deal with the design and analysis of reinforced concrete members.

We are concerned primarily with how a reinforced concrete member behaves when subjected to load. It is generally accepted that the behavior of a reinforced concrete member under load depends on the stress–strain relationship of the materials, as well as the type of stress to which it is subjected. With concrete used principally in compression, the compressive stress–strain curve is of primary interest.

The compressive strength of concrete is denoted as f'_c and is assigned the unit of *pounds per square inch* (psi). This is the unit for f'_c used in the ACI Code equations. For calculations, f'_c is frequently used with the unit *kips per square inch* (ksi).

A test that has been standardized by the American Society for Testing and Materials (ASTM C39) [6] is used to determine the compressive strength (f'_c) of concrete. The test involves compression loading to failure of a specimen cylinder of concrete. The compressive strength so determined is the highest compressive stress to which the specimen is subjected. Note in Figure 1-1 that f'_c is not the stress that exists in the specimen at failure but that which occurs at a strain of about 0.002 (though the concrete strain at f'_c may vary between approximately 0.0015 and 0.0025). Currently, 28-day concrete strengths (f'_c) range from 2500 to upwards of 10,000 psi, with 3000 to 4000 psi being common for reinforced concrete structures and 5000 to 6000 psi being common for prestressed concrete members, and higher strengths used for columns in high-rise buildings. High strength concrete with 12,000 psi compressive strength was used recently for the columns, shear walls, and drilled caissons for a project at the Hudson Yards in New York City. For normal weight concrete, the compressive strength at 28 days is specified in the ACI Code

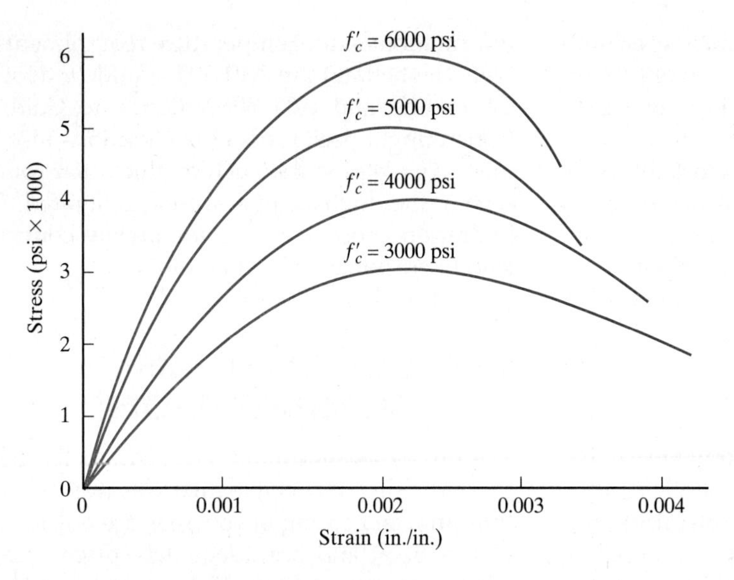

FIGURE 1-1 Typical stress–strain curves for concrete.

as the design strength. However, for higher strength concrete, the compressive strength at 56 or 90 days is commonly used [15]. The curves shown in Figure 1-1 represent the result of compression tests on 28-day standard cylinders for varying design mixes. The strain at rupture of concrete varies as indicated by the plots in Figure 1-1, but in developing the ACI Code equations for flexure, a limiting crushing strain of 0.003 is assumed (ACI 318-14, Section 22.2.2.1).

A review of the stress–strain curves for different-strength concretes reveals that the maximum compressive strength is generally achieved at a unit strain of approximately 0.002 in./in. Stress then decreases, accompanied by additional strain. Higher-strength concretes are more brittle and will fracture at a lower maximum strain than will the lower-strength concretes. The initial slope of the curve varies, unlike that of steel, and only approximates a straight line. For steel, where stresses are below the yield point and the material behaves elastically, the stress–strain plot will be a straight line. The slope of the straight line is the modulus of elasticity. For concrete, however, we observe that the straight-line portion of the plot is very short, if it exists at all. Therefore, there exists no constant value of modulus of elasticity for a given concrete because the stress–strain ratio is not constant. It may also be observed that the slope of the initial portion of the curve (if it approximates a straight line) varies with concretes of different strengths. Even if we assume a straight-line portion, the modulus of elasticity is different for concretes of different strengths. At low and moderate stresses (up to about $0.5f'_c$), concrete is commonly assumed to behave elastically.

The ACI Code, Section 19.2.2, provides the accepted empirical expression for *modulus of elasticity*:

$$E_c = w_c^{1.5} 33 \sqrt{f'_c}$$

where

E_c = modulus of elasticity of concrete in compression (psi)

w_c = unit weight of concrete (lb/ft^3)

f'_c = compressive strength of concrete (psi)

This expression is valid for concretes having w_c between 90 and 160 lb/ft^3. For normal-weight concrete, the unit weight w_c will vary with the mix proportions and with the character and size of the aggregates. If the unit weight is taken as 144 lb/ft^3, the resulting expression for modulus of elasticity is

$$E_c = 57{,}000\sqrt{f'_c} \quad \text{(see Table A-6 for values of } E_c\text{)}$$

It should also be noted that the stress–strain curve for the same-strength concrete may be of different shapes if the condition of loading varies appreciably. With different *rates of strain* (loading), we will have different-shape curves. Generally, the maximum strength of a given concrete is smaller at slower rates of strain.

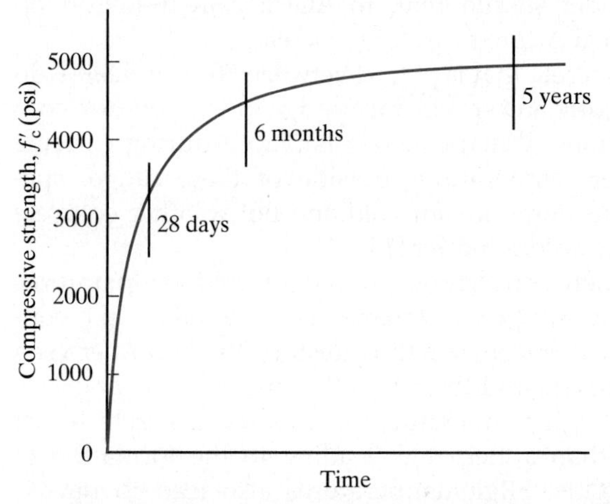

FIGURE 1-2 Strength–time relationship for concrete.

Concrete strength varies with time, and the specified concrete strength is usually that strength that occurs 28 days after the placing of concrete. A typical strength–time curve for normal stone concrete is shown in Figure 1-2. Generally, concrete attains approximately 70% of its 28-day strength in 7 days and approximately 85% to 90% in 14 days.

Concrete, under load, exhibits a phenomenon termed *creep*. This is the property by which concrete continues to deform (or strain) over long periods of time while under constant load. Creep occurs at a decreasing rate over a period of time and may cease after several years. Generally, high-strength concretes exhibit less creep than do lower-strength concretes. The magnitude of the creep deformations is proportional to the magnitude of the applied load as well as to the length of time of load application.

1-7 CONCRETE IN TENSION

The tensile and compressive strengths of concrete are not proportional, and an increase in compressive strength is accompanied by an appreciably smaller percentage increase in tensile strength. According to the ACI Code Commentary, the tensile strength of normal-weight concrete in flexure is about 10% to 15% of the compressive strength.

The true tensile strength of concrete is difficult to determine. The *split-cylinder test* (ASTM C496) [6] has been used to determine the tensile strength of lightweight aggregate concrete and is generally accepted as a good measure of the true tensile strength. The split-cylinder test uses a standard 6-in.-diameter, 12-in.-long cylinder placed on its side in a testing machine. A compressive line load is applied uniformly along the length of the cylinder, with support furnished along the full length of the bottom of the cylinder. The compressive load produces a transverse tensile stress, and the cylinder will split in half along a diameter when its tensile strength is reached.

The tensile stress at which splitting occurs is referred to as the *splitting tensile strength*, f_{ct}, and may be calculated by the following expression derived from the theory of elasticity:

$$f_{ct} = \frac{2P}{\pi LD}$$

where

f_{ct} = splitting tensile strength of lightweight aggregate concrete (psi)
P = applied load at splitting (lb)
L = length of cylinder (in.)
D = diameter of cylinder (in.)

Another common approach has been to use the *modulus of rupture*, f_r (which is the maximum tensile bending stress in a plain concrete test beam at failure), as a measure of tensile strength (ASTM C78) [6]. The moment that produces a tensile stress just equal to the modulus of rupture is termed the *cracking moment*, M_{cr}, and may be calculated using methods discussed in Section 1-9. The ACI Code recommends that the modulus of rupture f_r be taken as $7.5\lambda\sqrt{f'_c}$, where f'_c is in psi (ACI 318 Equation 19.2.3.1). Greek lowercase lambda (λ) is a modification factor reflecting the lower tensile strength of lightweight concrete relative to normal-weight concrete. The values for λ are as follows:

Normal-weight concrete—1.0
Sand-lightweight concrete—0.85
All-lightweight concrete—0.75

Interpolation between these values is permitted. See ACI Code Table 19.2.4.2 for details. If the average splitting tensile strength f_{ct} is specified, then $\lambda = f_{ct}/(6.7\sqrt{f_{cm}}) \leq 1.0$, where f_{cm} is the average measured compressive strength, in psi.

1-8 REINFORCING STEEL

Concrete cannot withstand very much tensile stress without cracking; therefore, tensile reinforcement must be embedded in the concrete to overcome this deficiency. In the United States, this reinforcement is in the form of steel reinforcing bars or welded wire reinforcing composed of steel wire. In addition, reinforcing in the form of structural steel shapes, steel pipe, steel tubing, and high-strength steel tendons is permitted by the ACI Code. Many other approaches have been taken in the search for an economical reinforcement for concrete. Principal among these are the fiber-reinforced concretes, where the reinforcement is obtained through the use of short fibers of steel or other materials, such as fiberglass. For the purpose of this book, our discussion will primarily include steel reinforcing bars and welded wire reinforcing. High-strength steel tendons are used mainly in prestressed concrete construction (see Chapter 11).

The specifications for steel reinforcement published by the ASTM are generally accepted for the steel used in reinforced concrete construction in the United States and are identified in ACI 318 Section 20.2.

The ACI Code states that reinforcing bars should be secure and in place prior to the placement of concrete, thus the practice of *wet-setting* of rebar is not permitted. While it is not uncommon in practice to see some contractors wanting to place rebar in wet concrete, this should not be permitted because the rebar displaces the aggregates, and proper bond between the concrete and rebar cannot easily be assured.

The steel bars used for reinforcing are, almost exclusively, round deformed bars with some form of patterned ribbed projections rolled onto their surfaces. The patterns vary depending on the producer, but

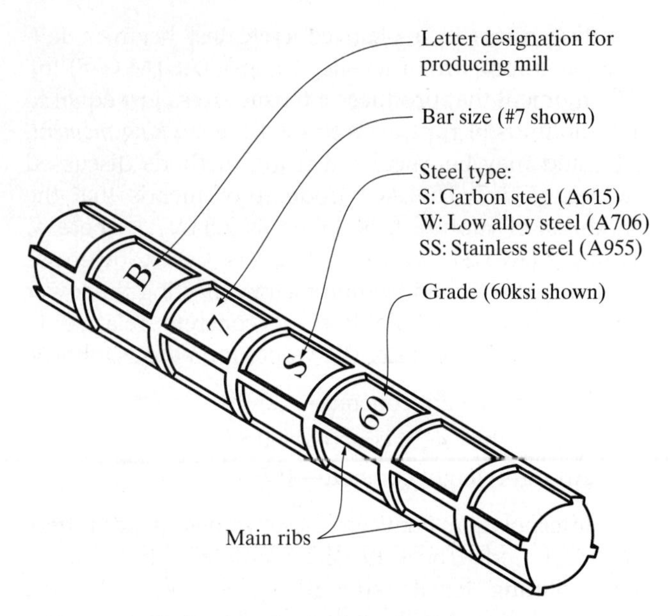

FIGURE 1-3 Rebar marks and designation.

all patterns should conform to ASTM specifications. Steel reinforcing bars are readily available in straight lengths of 60 ft. Smaller sizes are also available in coil stock for use in automatic bending machines. The bars vary in designation from No. 3 through No. 11, with two additional bars, No. 14 and No. 18. See Figure 1-3 for a sample rebar designation.

For bars No. 3 through No. 8, the designation represents the bar diameter in eighths of an inch. The No. 9, No. 10, and No. 11 bars have diameters that provide areas equal to 1-in.-square bars, $1^1/_8$-in.-square bars, and $1^1/_4$-in.-square bars, respectively. The No. 14 and No. 18 bars correspond to or have the same cross-sectional areas as $1^1/_2$-in.-square bars and 2-in.-square bars, respectively, and are commonly available only by special order. Round, plain reinforcing bars are permitted for spirals (lateral reinforcing) in concrete compression members.

ASTM specifications require that identification marks be rolled onto the bar to provide the following information: a letter or symbol indicating the producer's mill, a number indicating the size of the bar, a symbol or letter indicating the type of steel from which the bar was rolled, and for grade 60 bars, either the number 60 or a single continuous longitudinal line (called a *grade line*) through at least five deformation spaces. The *grade* indicates the minimum specified yield stress in ksi. For instance, a grade 60 steel bar has a minimum specified yield stress of 60 ksi. No symbol indicating grade is rolled onto grade 40 or 50 steel bars (see Figure 1-3). Grade 75 bars can have either two grade lines through at least five deformation spaces or the grade mark 75. Reference [7] is an excellent resource covering the various aspects of bar identification.

Reinforcing bars are usually made from newly manufactured steel (billet steel). Steel types and ASTM specification numbers for bars are tabulated in Table A-1. Note that ASTM A615, which is billet steel, is available in grades 40, 60, 75, and 80. Grade 80 steel is allowed for non-seismic applications per ASTM 615 and ASTM 706 [8]. (The full range of bar sizes is not available in grades 40, 75, and 80, however.) Grade 75 steel is approximately 20% stronger than grade 60 steel, requiring a corresponding reduction in the required area of reinforcement, though the installed cost of grade 75 steel reinforcement is slightly higher than the cost for grade 60 steel. ASTM A706, low-alloy steel, which was developed to satisfy the requirement for reinforcing bars with controlled tensile properties and controlled chemical composition for weldability, is available in only one grade. Tables A-2 and A-3 contain useful information on cross-sectional areas of bars.

The most useful physical properties of reinforcing steel for reinforced concrete design calculations are yield stress (f_y) and modulus of elasticity. A typical stress–strain diagram for reinforcing steel is shown in Figure 1-4a. The idealized stress–strain diagram of Figure 1-4b is discussed in Chapter 2.

The yield stress (or yield point) of steel is determined through procedures governed by ASTM standards. For practical purposes, the yield stress may be thought of as that stress at which the steel

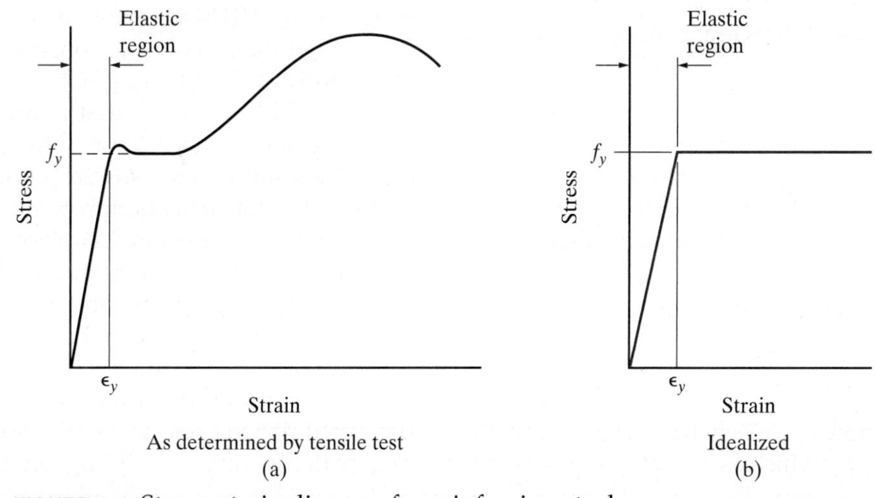

FIGURE 1-4 Stress–strain diagram for reinforcing steel.

exhibits increasing strain with no increase in stress. For reinforcement without a sharply defined yield point, ACI 318-14, similar to the ASTM standards, defines the yield strength as the 0.2% proof stress (that is, the offset stress at a 0.2% strain). The yield stress of the steel will usually be one of the known (or given) quantities in a reinforced concrete design or analysis problem. See Table A-1 for the range of f_y.

The modulus of elasticity of carbon reinforcing steel (the slope of the stress–strain curve in the elastic region) varies over a very small range and has been adopted as 29,000,000 psi (ACI Code Section 20.2.2.2).

Unhindered corrosion of reinforcing steel will lead to cracking and spalling of the concrete in which it is embedded. Quality concrete, under normal conditions, provides good protection against corrosion for steel embedded in the concrete with adequate cover (minimum requirements are discussed in Chapter 2). This protection is attributed to, among other factors, the high alkalinity of the concrete. Where reinforced concrete structures (or parts of structures) are subjected to corrosive conditions, however, some type of corrosion protection system should be used to prevent deterioration. Examples of such structures are bridge decks, parking garage decks, wastewater treatment plants, and industrial and chemical processing facilities.

One method used to minimize the corrosion of the reinforcing steel is to coat the bars with a suitable protective coating. The protective coating can be a nonmetallic material such as epoxy or a metallic material such as zinc (galvanizing). The ACI Code requires epoxy-coated reinforcing bars to comply with ASTM A775 or ASTM A934 and galvanized (zinc-coated) bars to comply with ASTM A767 (ACI 318 Section 20.6.2). The bars to be epoxy-coated or zinc-coated (galvanized) must meet the code requirements for uncoated bars as tabulated in Table A-1.

Welded wire reinforcing (WWR) (commonly called *mesh*) is another type of reinforcement. It consists of cold-drawn wire in orthogonal patterns, square or rectangular, resistance welded at all intersections. It may be supplied in either rolls or sheets, depending on wire size. WWR with wire diameters larger than about ¼ in. is usually available only in sheets.

Both plain and deformed WWR products are available. According to ACI 318 Section 20.2.1.7, deformed wire, plain wire, welded deformed wire reinforcement, and welded plain wire reinforcement shall conform to ASTM A 1064 for carbon steel and ASTM A 1022 for stainless steel. Depending on the application, both materials have the maximum f_y permitted for design that varies from 60,000 psi to 100,000 psi (see ACI 318 Tables 20.2.2.4a and 20.2.2.4b). The deformed wire is usually more expensive, but it can be expected to have an improved bond with the concrete.

A rational method of designating wire sizes to replace the formerly used gauge system has been adopted by the wire industry. Plain wires are described by the letter W followed by a number equal to 100 times the cross-sectional area of the wire in square inches. Deformed wire sizes are similarly described, but the letter D is used. Thus a W9 wire has an area of 0.090 in.^{2} and a D8 wire has an area of 0.080 in.^{2} A W8 wire has the same cross-sectional area as the D8 but is plain rather than deformed. Sizes between full numbers are given by decimals, such as W9.5.

Generally, the material is indicated by the symbol WWR, followed by spacings first of longitudinal wires, then of transverse wires, and last by the sizes of longitudinal and transverse wires. Thus WWR6 × 12-W16 × W8 indicates a plain WWR with 6-in. longitudinal spacing, 12-in. transverse spacing, and a cross-sectional area equal to 0.16 in.^{2} for the longitudinal wires and 0.08 in.^{2} for the transverse wires.

Additional information about WWR, as well as tables relating size number with wire diameter, area, and weight, may be obtained through the Wire Reinforcement Institute [9] or the Concrete Reinforcing Steel Institute [9 and 10]. Table 1-1 contains common WWR sizes with the area of steel in in.^{2}/ft width.

Most concrete is reinforced in some way to resist tensile forces (Figure 1-5). Some structural elements, particularly footings, are sometimes made of *plain concrete*, however. Plain concrete is defined as structural

TABLE 1-1 Welded Wire Reinforcement (Meets ASTM A185, f_y = 60 ksi)

Wire gauge	W-number	Area of steel in.^{2}/ft (each direction)	Wire diameter (in.)
6 × 6-10 × 10	6 × 6-W1.4 × W1.4	0.029	0.135 (10ga)
6 × 6-8 × 8	6 × 6-W2.0 × W2.0	0.041	0.162 (8ga)
6 × 6-6 × 6	6 × 6-W2.9 × W2.9	0.058	0.192 (6ga)
6 × 6-4 × 4	6 × 6-W4.0 × W4.0	0.080	0.225 (4ga)
4 × 4-10 × 10	4 × 4-W1.4 × W1.4	0.043	0.135 (10ga)
4 × 4-8 × 8	4 × 4-W2.0 × W2.0	0.062	0.162 (8ga)
4 × 4-6 × 6	4 × 4-W2.9 × W2.9	0.087	0.192 (6ga)
4 × 4-4 × 4	4 × 4-W4.0 × W4.0	0.120	0.225 (4ga)

FIGURE 1-5 Concrete construction in progress. Note formwork, reinforcing bars, and pumping of concrete.
(George Limbrunner)

concrete with no reinforcement or with less reinforcement than the minimum amount specified for reinforced concrete. Plain concrete is discussed further in Chapter 10.

1-9 CONCRETE COVER

The clear distance between the concrete surface and the face of the rebar is called the cover. The reinforcement in concrete needs to have this cover for several reasons:

- To protect the reinforcement against corrosion
- To provide adequate surface area for bond between the concrete and the reinforcement
- To protect the reinforcement against loss of strength in a fire

For protection against corrosion and to ensure adequate bond between the reinforcement and the surrounding concrete, ACI 318-14 Table 20.6.1.3.1 specifies the concrete cover requirements for cast-in-place reinforced concrete members. For slabs and walls not exposed to weather or in contact with the ground, the minimum clear cover to the outermost reinforcement surface is $^3/_4$″ for No. 11 and smaller bars, and 1.5″ for No. 14 and larger bars. For beams and columns not exposed to weather or in contact with the ground, the minimum clear cover to the outermost reinforcement surface is 1.5″. For all structural elements cast against and permanently in contact with soil or ground (e.g., footings, caissons, grade beams, foundation and basement walls), the specified concrete cover is 3″. These members are typically placed without the use of forms. For concrete elements not cast against soil or ground, but exposed to the weather or in contact with ground, the specified concrete cover is 2″ for No. 6 through No. 18 bars, and 1.5″ for No. 5 and smaller bars. The surfaces of these members would typically be formed. The presence of forms allows for greater accuracy in establishing the proper clear cover distance.

The fire protection requirements of the Code may sometimes necessitate a higher concrete cover than the above-specified concrete covers, depending on the required fire rating. See Chapter 14 for discussions on concrete cover requirements as a function of the fire ratings.

1-10 BEAMS: MECHANICS OF BENDING REVIEW

The concept of bending stresses in homogeneous elastic beams is generally discussed at great length in all strength of materials textbooks and courses. Beams composed of material such as steel or timber are categorized as homogeneous, with each exhibiting elastic behavior up to some limiting point. Within the limits of elastic behavior, the internal bending stress distribution developed at any cross section is linear (straight line), varying from zero at the neutral axis to a maximum at the outer fibers.

The accepted expression for the maximum bending stress in a beam is termed the *flexure formula*,

$$f_b = \frac{Mc}{I}$$

where

f_b = calculated bending stress at the outer fiber of the cross section

M = the applied moment

c = distance from the neutral axis to the outside tension or compression fiber of the beam

I = moment of inertia of the cross section about the neutral axis

The flexure formula represents the relationship between bending stress, bending moment, and the geometric properties of the beam cross section. By rearranging the flexure formula, the maximum moment that may be applied to the beam cross section, called the *resisting moment*, M_R, may be found:

$$M_R = \frac{F_b I}{c}$$

where F_b = the allowable bending stress.

This procedure is straightforward for a beam of known cross section for which the moment of inertia can easily be found. For a reinforced concrete beam, however, the use of the flexure formula presents some complications, because the beam is not homogeneous and concrete does not behave elastically over its full range of strength. As a result, a somewhat different approach that uses the beam's internal bending stress distribution is recommended. This approach is termed the *internal couple method*.

Recall from strength of materials that a couple is a pure moment composed of two equal, opposite, and parallel forces separated by a distance called the *moment arm*, which is commonly denoted Z. In the internal couple method, the couple represents an internal resisting moment and is composed of a compressive force C above the neutral axis (assuming a single-span, simply supported beam that develops compressive stress above the neutral axis) and a parallel internal tensile force T below the neutral axis.

As with all couples, and because the forces acting on any cross section of the beam must be in equilibrium, C must equal T. The internal couple must be equal and opposite to the bending moment at the same location, which is computed from the external loads. It represents a couple developed by the bending action of the beam.

The internal couple method of determining beam strength is more general and may be applied to homogeneous or nonhomogeneous beams having linear (straight-line) or nonlinear stress distributions. For reinforced concrete beams, it has the advantage of using the basic resistance pattern found in the beam.

The following three analysis examples dealing with plain (unreinforced) concrete beams provide an introduction to the internal couple method. Note that the unreinforced beams are considered homogeneous and elastic. This is valid if the moment is small and tensile bending stresses in the concrete are low (less than the tensile bending strength of the concrete) with no cracking of the concrete developing. For this condition, the entire beam cross section carries bending stresses. Therefore, the analysis for bending stresses in the uncracked beam can be based on the properties of the gross cross-sectional area using the elastic-based flexure formula. The use of the flexure formula is valid as long as the maximum tensile stress in the concrete does not exceed the modulus of rupture f_r. If a moment is applied that causes the maximum tensile stress just to reach the modulus of rupture, the cross section will be on the verge of cracking. This moment is called the *cracking moment*, M_{cr}.

These examples use both the internal couple approach and the flexure formula approach so that the results may be compared.

Example 1-1

A normal-weight plain concrete beam is 6 in. × 12 in. in cross section, as shown in Figure 1-6. The beam is simply supported on a span of 4 ft and is subjected to a midspan concentrated load of 4500 lb. Assume $f'_c = 3000$ psi.

a. Calculate the maximum concrete tensile stress using the internal couple method.
b. Repeat part (a) using the flexure formula approach.
c. Compare the maximum concrete tensile stress with the value for modulus of rupture f_r using the ACI-recommended value based on f'_c.

Solution:

Calculate the weight of the beam (weight per unit length):

$$\begin{aligned}\text{weight of beam} &= \text{volume per unit length} \times \text{unit weight}\\ &= \frac{6 \text{ in.}(12 \text{ in.})}{144 \text{ in.}^2/\text{ft}^2}(150 \text{ lb/ft}^3)\\ &= 75 \text{ lb/ft}\end{aligned}$$

Calculate the maximum applied moment:

$$\begin{aligned}M_{max} &= \frac{PL}{4} + \frac{wL^2}{8}\\ &= \frac{4500 \text{ lb}(4 \text{ ft})}{4} + \frac{75 \text{ lb/ft}(4 \text{ ft})^2}{8}\\ &= 4650 \text{ ft-lb}\end{aligned}$$

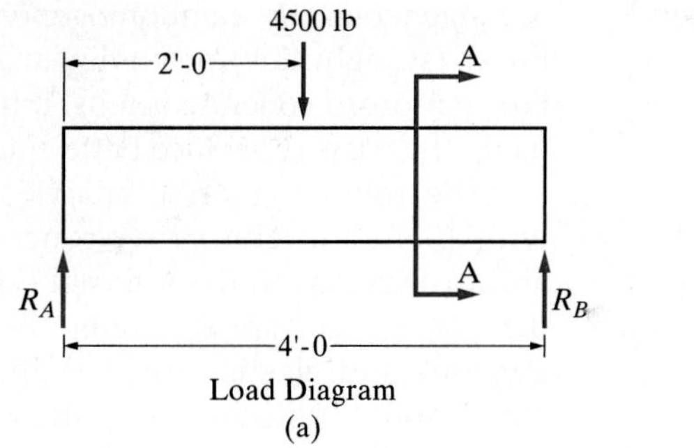

FIGURE 1-6 Loading diagram and section for Example 1-1.

a. Internal couple method

1. Because the beam is homogeneous, elastic, and symmetrical with respect to both the X–X and Y–Y axes, the neutral axis (N.A.) is at midheight. Stresses and strains vary linearly from zero at the neutral axis (which is also the centroidal axis) to a maximum at the outer fiber. As the member is subjected to positive moment, the area above the N.A. is stressed in compression and the area below the N.A. is stressed in tension. These stresses result from the bending behavior of the member and are shown in Figure 1-7.
2. C represents the resultant compressive force above the N.A. T represents the resultant tensile force below the N.A. C and T each act at the centroid of their respective triangles of stress distribution. Therefore, $Z = 8$ in. C and T must be equal (since $\Sigma H_F = 0$). The two forces act together to form the internal couple (or internal resisting moment) of magnitude CZ or TZ.
3. The internal resisting moment must equal the bending moment due to external loads at any section. Therefore,

$$M = CZ = TZ$$

$$4650 \text{ ft-lb } (12 \text{ in./ft}) = C \text{ (8 in.)}$$

from which

$$C = 6975 \, lb = T$$

4. C = average stress × area of beam on which stress acts

$$C = \tfrac{1}{2} f_{top}(y_{top})(b) = \tfrac{1}{2} f_{top}(6 \text{ in.})(6 \text{ in.}) = 6975 \text{ lb}$$

Solving for f_{top} yields

$$f_{top} = 388 \text{ psi} = f_{bott}$$

The modulus of elasticity of the concrete,

$$E_c = 57{,}000\sqrt{f'_c} = 57{,}000\sqrt{3000 \text{ psi}}$$
$$= 3{,}122{,}019 \text{ psi}$$

The concrete strain at the top of the beam is

$$\varepsilon_{top} = \frac{f_{top}}{E_c} = \frac{388 \text{ psi}}{3{,}122{,}019 \text{ psi}} = 0.000124 \text{ in./in.}$$

b. Flexure formula approach

$$I = \frac{bh^3}{12} = \frac{6(12^3)}{12} = 864 \text{ in.}^4$$

$$f_{top} = f_{bott} = \frac{Mc}{I} = \frac{4650(12)(6)}{864} = 388 \text{ psi}$$

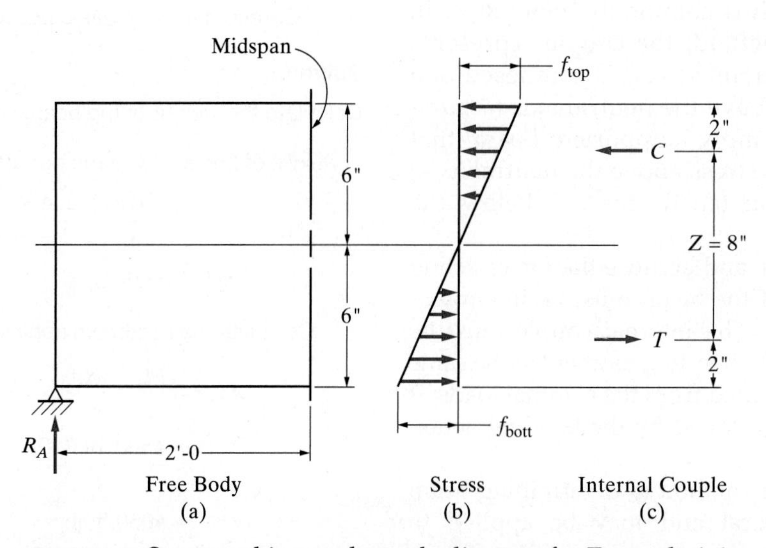

FIGURE 1-7 Stress and internal couple diagram for Example 1-1.

c. The ACI-recommended value for the modulus of rupture (based on f'_c) is

$$f_r = 7.5\lambda\sqrt{f'_c} = 7.5(1.0)\sqrt{3000}$$
$$f_r = 411 \text{ psi}$$

The calculated tensile stress (f_{bott}) of 388 psi is about 6% below the modulus of rupture, the stress at which flexural cracking would be expected.

Example 1-1 is based on elastic theory and assumes the following: (1) a plane section before bending remains a plane section after bending (the variation in strain throughout the depth of the member is linear from zero at the neutral axis), and (2) the modulus of elasticity is constant; therefore, stress is proportional to strain and the stress distribution throughout the depth of the beam is also linear from zero at the neutral axis to a maximum at the outer fibers.

The internal couple approach may also be used to find the moment strength (resisting moment) of a beam.

Example 1-2

Calculate the cracking moment M_{cr} for the plain concrete beam shown in Figure 1-8. Assume normal-weight concrete and $f'_c = 4000$ psi.

a. Use the internal couple method.
b. Check using the flexure formula.

Solution:

The moment that produces a tensile stress just equal to the modulus of rupture f_r is called the cracking moment, M_{cr}. The modulus of rupture for normal-weight concrete is calculated from ACI Equation 19.2.3.1:

$$f_r = 7.5\sqrt{f'_c} = 7.5\sqrt{4000} = 474 \text{ psi}$$

For convenience, we will use force units of kips (1 kip = 1000 lb). Therefore, $f_r = 0.474$ ksi.

a. Using the internal couple method

$$Z = 14 - 2(2.33) = 9.34 \text{ in.}$$
$$C = T = \tfrac{1}{2}(0.474)(8)(7) = 13.27 \text{ kips}$$
$$M_{cr} = CZ = TZ = \frac{13.27(9.34)}{12} = 10.33 \text{ ft.-kips}$$

b. Check using the flexure formula

$$f = \frac{Mc}{I}$$
$$M_R = M_{cr} = \frac{f_r I}{c}$$
$$I = \frac{bh^3}{12} = \frac{8(14)^3}{12} = 1829 \text{ in.}^4$$
$$M_{cr} = \frac{f_r I}{c} = \frac{0.474(1829)}{7(12)} = 10.32 \text{ ft.-kips}$$

The internal couple method may also be used to analyze irregularly shaped cross sections, although for homogeneous beams it is more cumbersome than the use of the flexure formula.

Example 1-3

Calculate the cracking moment (resisting moment) for the T-shaped unreinforced concrete beam shown in Figure 1-9. Use $f'_c = 4000$ psi. Assume positive moment (compression in the top). Use the internal couple method and check using the flexure formula.

Solution:

The neutral axis must be located so that the strain and stress diagrams may be defined. The location of the neutral axis with respect to the noted reference axis is calculated from

$$\bar{y} = \frac{\Sigma(Ay)}{\Sigma A} = \frac{4(20)(22) + 5(20)(10)}{4(20) + 5(20)} = 15.33 \text{ in.}$$

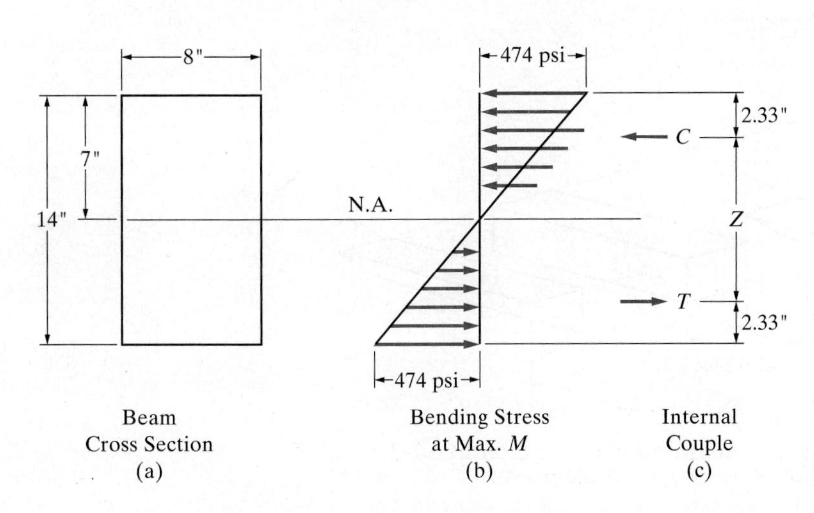

FIGURE 1-8 Sketch for Example 1-2.

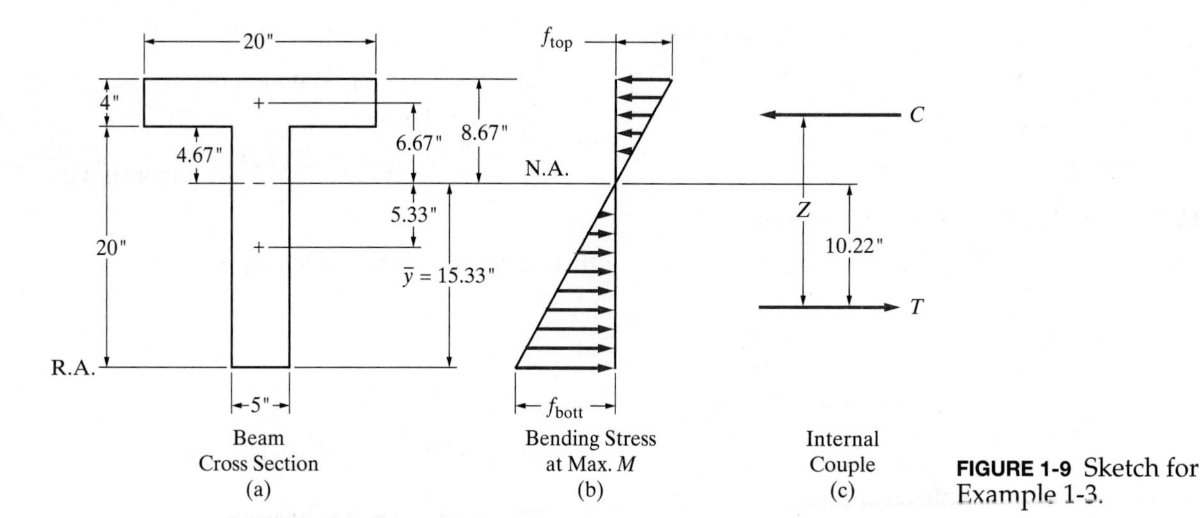

FIGURE 1-9 Sketch for Example 1-3.

The bottom of the cross section is stressed in tension. Note that the stress at the bottom will be numerically larger than at the top because of the relative distances from the N.A. The stress at the bottom of the cross section will be set equal to the modulus of rupture ($\lambda = 1.0$ for normal-weight concrete):

$$f_{bott} = f_r = 7.5\lambda\sqrt{f'_c} = 7.5(1.0)\sqrt{4000} = 474 \text{ psi} = 0.474 \text{ ksi}$$

Using similar triangles in Figure 1-9b, the stress at the top of the flange is

$$f_{top} = \frac{8.67}{15.33}(0.474) = 0.268 \text{ ksi}$$

The modulus of elasticity of the concrete,

$$E_c = 57{,}000\sqrt{f'_c} = 57{,}000\sqrt{4000 \text{ psi}} = 3{,}605{,}000 \text{ psi}$$

The concrete strain at the top of the beam is

$$\varepsilon_{top} = \frac{f_{top}}{E_c} = \frac{268 \text{ psi}}{3{,}605{,}000 \text{ psi}} = 0.000074 \text{ in./in.}$$

Similarly, the stress at the bottom of the flange is

$$f_{\text{bott of flange}} = \frac{4.67}{15.33}(0.474) = 0.1444 \text{ ksi}$$

The total tensile force can be evaluated as follows:

$$T = \text{average stress} \times \text{area}$$
$$= \tfrac{1}{2}(0.474)(15.33)(5) = 18.17 \text{ kips}$$

and its location below the N.A. is calculated from

$$\tfrac{2}{3}(15.33) = 10.22 \text{ in. (below the N.A.)}$$

The compressive force will be broken up into components because of the irregular area, as shown in Figure 1-10. Referring to both Figures 1-9 and 1-10, the component internal compressive forces, component internal couples, and M_R may now be evaluated. The component forces are first calculated:

$$C_1 = 0.1444(20)(4) = 11.55 \text{ kips}$$
$$C_2 = \tfrac{1}{2}(0.1236)(20)(4) = 4.94 \text{ kips}$$
$$C_3 = \tfrac{1}{2}(0.1444)(5)(4.67) = 1.686 \text{ kips}$$
$$\text{total } C = C_1 + C_2 + C_3 = 18.18 \text{ kips}$$
$$C \approx T \quad \text{(O.K.)}$$

Next we calculate the moment arm distance from each component compressive force to the tensile force T:

$$Z_1 = 10.22 + 4.67 + \tfrac{1}{2}(4.00) = 16.89 \text{ in.}$$
$$Z_2 = 10.22 + 4.67 + \tfrac{2}{3}(4.00) = 17.56 \text{ in.}$$
$$Z_3 = 10.22 + \tfrac{2}{3}(4.67) = 13.33 \text{ in.}$$

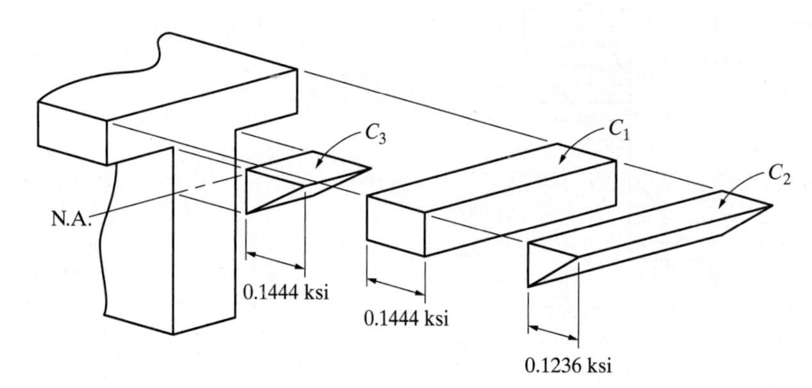

FIGURE 1-10 Component compression forces for Example 1-3.

The magnitudes of the component internal couples are then calculated from force × moment arm as follows:

$$M_{R_1} = 11.55(16.89) = 195.1 \text{ in.-kips}$$
$$M_{R_2} = 4.94(17.56) = 86.7 \text{ in.-kips}$$
$$M_{R_3} = 1.686(13.33) = 22.5 \text{ in.-kips}$$
$$M_{cr} = M_R = M_{R_1} + M_{R_2} + M_{R_3} = 304 \text{ in.-kips}$$

Check using the flexure formula. The moment of inertia is calculated using the transfer formula for moment of inertia from statics:

$$I = \Sigma I_o + \Sigma Ad^2$$
$$I = \frac{1}{12}(20)(4^3) + \tfrac{1}{12}(5)(20^3) + 4(20)(6.67^2) + 5(20)(5.33^2)$$
$$= 9840 \text{ in.}^4$$
$$M_{cr} = M_R = \frac{f_r I}{c} = \frac{0.474(9840)}{15.33} = 304 \text{ in. - kips}$$

(Checks O.K.)

As mentioned previously, the three examples are for plain, unreinforced, and uncracked concrete beams that are considered homogeneous and elastic within the bending stress limit of the modulus of rupture. The internal couple method is also applicable to nonhomogeneous beams with nonlinear stress distributions of any shape, however. Because reinforced concrete beams are nonhomogeneous, the flexure formula is not directly applicable. Therefore, the basic approach used for reinforced concrete beams is the internal couple method (see Chapters 2 and 3).

1-11 CONCRETE SLAB SYSTEMS

The two types of floor systems used in reinforced and prestressed concrete structures are one way and two-way slab systems.

One-Way Slab Systems

One-way concrete floor systems are usually supported by stiff beams or walls and the slab spans or bends predominantly in one direction (usually in the shorter direction of the rectangular slab panel) and the load transfer to the members supporting the slab occurs predominantly in the shorter direction (see Figure 1-11). This is the case when the clear span of the longer side of the rectangular slab panel (ℓ_1) is greater than or equal to twice the clear span of the shorter side of the slab (ℓ_2); that is, $\ell_1/\ell_2 \geq 2$. In a reinforced or prestressed concrete one-way slab system, the curvature of the slab is predominantly in *the shorter* direction (i.e., parallel to the ℓ_2 dimension). The design of one-way reinforced concrete slab systems is covered in Chapter 2.

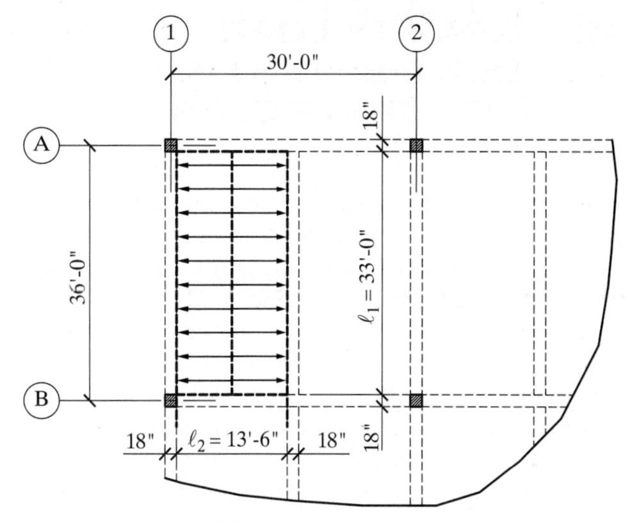

FIGURE 1-11 Load distribution in one-way slab systems.

Two-Way Slab Systems

Two-way concrete slab systems are supported on columns and span in two orthogonal directions, or the slab is supported on all four sides by beams or walls and the curvature of the slab, and the load transfer occurs in both orthogonal directions (see Figure 1-12). A two-way slab system occurs when the clear span of the longer side of the rectangular slab panel (ℓ_1) is less than twice the clear span of the shorter side of the slab panel (ℓ_2); that is, $\ell_1/\ell_2 < 2$. Examples of two-way slabs include flat plates, flat slabs, flat slabs with beams, and slabs supported on stiff beams or walls on all four sides of a rectangular slab panel. Note that ℓ_1 is the larger dimension of the rectangular slab panel bounded on all four sides by columns or beams or walls and ℓ_2 is the smaller dimension of the rectangular slab panel. Further treatment of two-way slab systems is covered in Chapter 6.

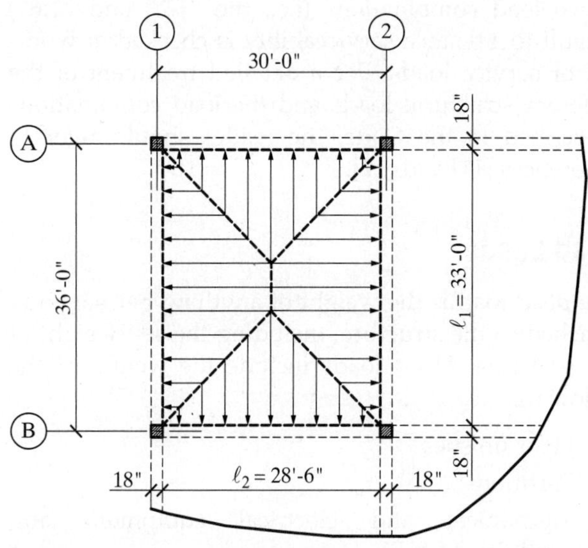

FIGURE 1-12 Load distribution in two-way slab systems.

1-12 GRAVITY LOAD DISTRIBUTION IN CONCRETE SLAB SYSTEMS

Reinforced concrete structures are subjected to a variety of loads which includes dead loads, floor live loads, roof live loads, snow loads, hydrostatic pressures, lateral soil pressures, wind loads, and seismic or earthquake loads. In this section, we introduce the reader to the distribution of gravity loads in concrete slab systems.

Gravity loads that act on structures include dead loads, floor live loads, snow loads, and roof live loads. In concrete floor and roof systems, the pertinent load combinations that usually govern for ultimate strength design are as follows:

Floor Slab:

$1.4D$
$1.2D + 1.6L$

Roof Slab:

$1.4D$
$1.2D + 1.6\,(L_r \text{ or } S \text{ or } R) + 0.5W$
$1.2D + 1.0W + 0.5(L_r \text{ or } S \text{ or } R)$

where,

D = dead load
L = floor live load
L_r = roof live load
S = snow load
R = rain load
W = wind load

When designing for serviceability limit state (e.g., deflections and vibrations), all the load factors in the above load combinations (i.e., the "1.2" and "1.6") default to 1.0 since serviceability is checked at working or service loads. For a detailed treatment of the different structural loads and the load combinations prescribed in the Code, the reader should refer to References [11] and [12].

Dead Loads

The dead load is the weight of anything permanently attached to the structure, including the self-weight of the structure. Dead loads include the weight of the following:

- Floor finishes
- Partitions
- Mechanical and electrical equipment and conduits (M & E)
- Glazing
- Cladding

The weights of common building materials are given in Section C3 of ASCE 7-16. The density of unreinforced concrete is approximately 145 lb/ft^3 and the density of reinforced concrete is approximately 150 lb/ft^3.

A sample dead load calculation for a building is as follows:

Components of Roof Dead Load in a Reinforced Concrete Building:

5 ply + gravel = 6.5 psf
Insulation and membrane = 3.5 psf
4″ concrete slab ≅ 50 psf
(slab weight = $4''/12 \times 150\ lb/ft^3$ = 50 psf)
Mechanical/Electrical (M & E) ≅ 10 psf (20 psf for industrial buildings)
Suspended ceiling ≅ 2 psf

Total roof dead load, D ≅ 72 psf (unfactored)

Components of Floor Dead Load in a Reinforced Concrete Building:

6″ concrete slab[1] = 75 psf
Floor finishes ≅ 8 psf (estimate weight of actual finishes specified by the architect)
M & E ≅ 10 psf (20 psf for industrial building)
Suspended ceiling ≅ 2 psf
Partitions[2, 3] ≅ 15 psf

Total floor dead load, D = 115 psf (unfactored)

Tributary Width and Areas

In this section, we introduce the concept of tributary area and tributary width. Beams in reinforced concrete floor or roof systems share the uniformly distributed loads in proportion to their distance from adjacent parallel beams. The tributary width (TW) of a beam is a measure of the total width of floor or roof supported by the beam. The tributary area (A_T) is the area of the floor supported by a structural element. For a beam, it is the product of

[1]Weight of 6″ slab = $6''/12 \times 150\ lb/ft^3$ = 75 psf
[2]The minimum value per ASCE 7-16, Section 4.3.2, is 15 psf. Actual weight of partitions may be higher.
[3]The Code allows partition loads to be neglected whenever the floor live load, L, is greater than or equal to 80 psf (ASCE 7-16, Section 4.3.2)

the tributary width and the span of the beam. For a column, it is the area bounded by lines mid-way to adjacent columns above, below, to the right, and to the left of the column under consideration. In a roof or floor system, the load path for gravity loads involves the beams supporting uniform loads from the floor and the beams transferring their reactions to the girders, which then in turn transfer their reactions to the columns.

Tributary width width (TW) = Width of floor or roof supported by a **beam** or **girder**
= ($\frac{1}{2}$ × the distance to the adjacent beam on the right + $\frac{1}{2}$ × distance to the adjacent beam on the left)

Tributary area (A_T) = Area of roof or floor supported by a **beam, girder**, or **column**

The tributary area of a beam, $A_{T\,(\text{beam})}$ = TW of beam × span of beam

The tributary area of a column,
$A_{T(\text{column})}$ = Floor or roof area supported by the column
= the rectangular area bounded by one-half the distance to adjacent columns on all sides of the column under consideration

Example 1-4

Load Calculations for One-Way Slabs and Supporting Beams

The elevation and section through a reinforced concrete building is shown in Figure 1-13. Determine the factored design loads for the floor slab, interior beam (A), and spandrel beam (B). Assume the following loads:

- Office building live load (i.e., floor live load = 50 psf)
- Block wall weighs 60 psf of vertical surface area
- Brick veneer weighs 40 psf of vertical surface area

Solution:

Floor Dead Load:

Finishes	8 psf
6″ concrete slab	75 psf
(6″/12 × 150 pcf)	
Suspended ceiling	= 2 psf
M & E	10 psf
Partitions	20 psf

Total dead load, ≅ 115 psf

Live load, L = 50 psf (for office buildings)

Slab Loads:

$$\text{Factored load, } W_u = 1.2\,D + 1.6\,L = 1.2\,(115) + 1.6\,(50) = \underline{218 \text{ psf}}$$

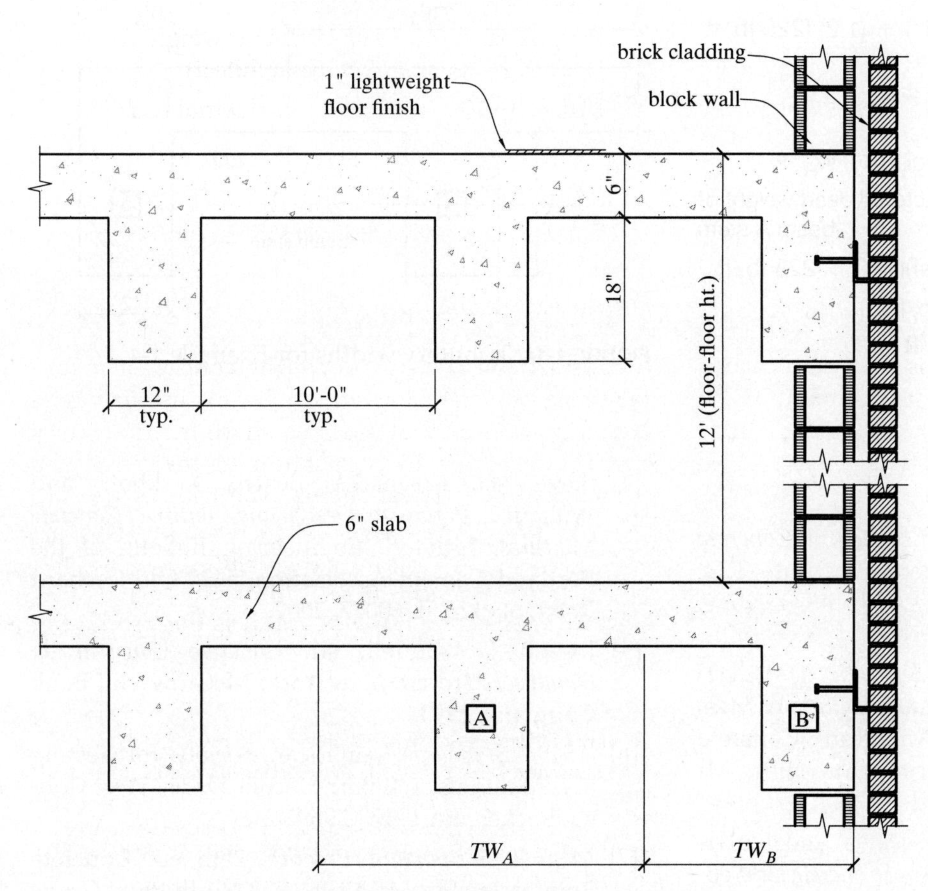

FIGURE 1-13 Building section for Example 1-4.

In the design of slabs, a 12-inch-wide strip is usually considered. Therefore, the load on the slab in pounds per linear foot (lb/ft) per foot width of slab is calculated as

$$w_u = (218 \text{ psf})(1 \text{ ft}) = 218 \text{ lb/ft per foot width of slab}$$

This load is used in the ultimate strength design of the slab.

$$\text{Service or unfactored load, } w_s = 1.0\,D + 1.0\,L = 115 + 50 = 165 \text{ psf}$$

$$w_s = (165 \text{ psf})(1 \text{ ft width}) = 165 \text{ lb/ft per foot width of slap}$$

This unfactored load will be used for the design of the slab for serviceability limit state such as deflections.

The loads to the interior and spandrel beams are based on the tributary widths of each beam as shown in Figure 1-14.

Typical Interior Beam Loads

$$\text{Tributary width of beam A, } TW_A = \frac{10' - 0''}{2} + 1' - 0'' + \frac{10'-0''}{2} = 11'-0''$$

$$\text{Unfactored weight of beam stem} = \left(\frac{12''}{12}\right)\left(\frac{18''}{12}\right)(150 \text{ pcf}) = \mathbf{225 \text{ lb/ft}}$$

$$\text{Factored load on beam A, } w_u = w_u(\text{slab}) \times TW_A + \text{factored dead weight of beam A stem}$$
$$= (218 \text{ psf})(11') + (1.2)(225 \text{ lb/ft}) = 2668 \text{ lb/ft} = \mathbf{2.67 \text{ kip/ft}}$$

$$\text{Service or unfactoredload on beam, } w_s = w_s(\text{slab}) \times TW_A + \text{unfactored dead weight of beam A stem}$$
$$= (165 \text{ psf})(11') + 225 \text{ lb/ft} = 2040 \text{ lb/ft} \equiv \mathbf{2.04 \text{ k/ft}}$$

Spandrel Beam Loads

Tributary width of spandrel beam B,

$$TW_B = \frac{10'-0''}{2} + 1'-0'' = 6'-0''$$

$$\text{Unfactored weight of beam stem} = \left(\frac{12''}{12}\right)\left(\frac{18''}{12}\right)(150 \text{ pcf}) = 225 \text{ lb/ft}$$

Height of block back-up wall = 12 ft − 2 ft = 10 ft

Height of exterior brick veneer/cladding = 12 ft

Unfactored weight of exterior block and brick wall

$$= (60 \text{ psf})(10') + (40 \text{ psf})(12') = 1080 \text{ lb/ft}$$

Factored load on beam,

$$w_u = W_u, \text{slab} \times TW_B + \text{factored weight of beam stem} + \text{factored weight of wall}$$
$$= (218 \text{ psf})(6') + (1.2)(225 \text{ lb/ft}) + (1.2)(1080 \text{ lb/ft}) = 2874 \text{ lb/ft} = \mathbf{2.87 \text{ kip/ft}}$$

Service load on beam, w_S

$$w_{S,\text{ slab}} \times TW_B + \text{unfactored weight of beam stem} + \text{unfactored weight of wall}$$
$$= (165 \text{ psf}) \times (6') + (225 \text{ lb/ft}) + (1080 \text{ lb/ft}) = 2295 \text{ lb/ft} \cong \mathbf{2.30 \text{ kip/ft}}$$

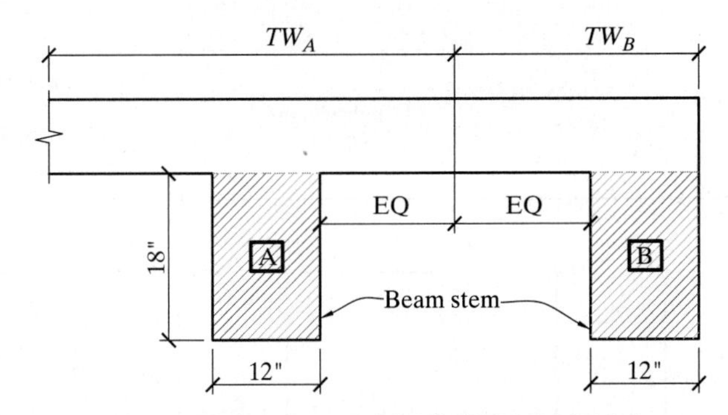

FIGURE 1-14 Tributary widths for Example 1-4.

References

[1] *Building Code Requirements for Structural Concrete* (ACI 318-14). American Concrete Institute, P.O. Box 9094, Farmington Hills, MI 48333-9094, 2014.

[2] ACI Committee 211. *Standard Practice for Selecting Proportions for Normal, Heavyweight, and Mass Concrete* (ACI 211.1-91). American Concrete Institute, P.O. Box 9094, Farmington Hills, MI 48333-9094, 1991. (Reapproved 2002.)

[3] George E. Troxell, Harmer E. Davis, and Joe W. Kelly. *Composition and Properties of Concrete*, 2nd ed. New York: McGraw-Hill Book Company, 1968.

[4] Steven H. Kosmatka, Beatrix Kerkhoff, and William C. Panarese. *Design and Control of Concrete Mixtures*, 14th ed. Engineering Bulletin of the Portland Cement Association, 5420 Old Orchard Road, Skokie, IL 60077, 2002.

[5] Joseph J. Waddell, ed. *Concrete Construction Handbook*, 3rd ed. New York: McGraw-Hill Book Company, 1993.

[6] *ASTM Standards*. American Society for Testing and Materials, 100 Barr Harbor Drive, West Conshohocken, PA 19428-2959.

[7] *Manual of Standard Practice*, 28th ed. Concrete Reinforcing Steel Institute, 933 North Plum Grove Road, Schaumburg, IL 60173, 2009.

[8] Clifford W. Schwinger. "ASTM A615 Grade 75 Reinforcing Steel–When, Why & How to Use It," *STRUCTURE Magazine*, pp. 34–35, August 2011.

[9] *Manual of Standard Practice* (WWR-500-R-16) Wire Reinforcement Institute, 942 Main Street, Suite 300, Hartford, CT 06103, 2016.

[10] Concrete Reinforcing Steel Institute, 933 North Plum Grove Road, Schaumburg, IL 60173.

[11] Abi Aghayere and Jason Vigil. *Structural Steel Design: A Practice-Oriented Approach*, 2nd ed. New York: Pearson, 2015.

[12] David A. Fanella. *Structural Loads – 2012 IBC and ASCE/SEI 7-10*, International Code Council, 2012.

[13] Cawsie Jijina and J. Benjamin Alper. "Cold and Hot Weather Concrete," STRUCTURE Magazine, pp. 14-16, September 2017.

[14] S. H. Kosmatka and M. L. Wilson. "Design and Control of Concrete Mixtures," 16th Edition, Portland Cement Association (PCA), 2016.

[15] Jeffrey Smilow and Ahmad Rahimian. "55 Hudson Yards," STRUCTURE Magazine, pp. 36–39, September 2017.

Problems

Note: In the following problems, assume plain concrete to have a weight of 145 pcf (conservative) unless otherwise noted.

1-1. The unit weight of normal-weight reinforced concrete is commonly assumed to be 150 lb/ft^3. Find the weight per lineal foot (lb/ft) for a normal-weight reinforced concrete beam that
a. Has a rectangular cross section 16 in. wide and 28 in. deep.
b. Has a cross section as shown in the accompanying diagram.

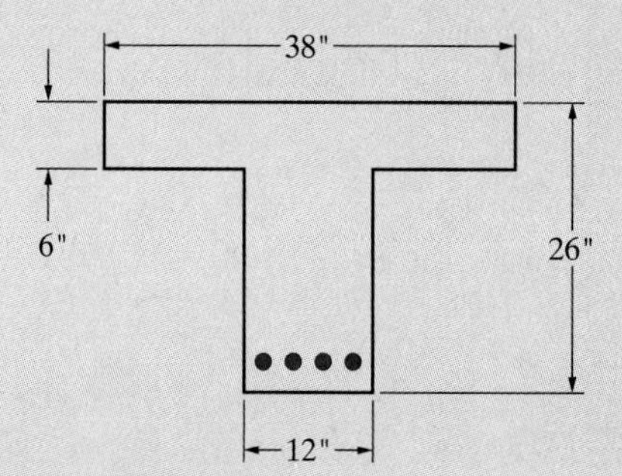

PROBLEM 1-1

1-2. Develop a spreadsheet application that will display in a table the values of modulus of elasticity E_c for concrete having unit weight ranging from 95 pcf to 155 pcf (in steps of 5 pcf) and compressive strength ranging from 3500 to 7000 psi (in steps of 500 psi). Display the modulus of elasticity rounded to the nearest 1000 psi.

1-3. A normal-weight concrete test beam 6 in. by 6 in. in cross section and supported on a simple span of 24 in. was loaded with a point load at midspan. The beam failed at a load of 2100 lb. Using this information, determine the modulus of rupture f_r of the concrete and compare with the ACI-recommended value based on an assumed concrete strength f'_c of 3000 psi.

1-4. A plain concrete beam has cross-sectional dimensions of 10 in. by 10 in. The concrete is known to have a modulus of rupture f_r of 350 psi. The beam spans between simple supports. Determine the span length at which this beam will fail due to its own weight. Assume a unit weight of 145 pcf.

1-5. The normal-weight plain concrete beam shown is on a simple span of 10 ft. It carries a dead load (which includes the weight of the beam) of 0.5 kip/ft. There is a concentrated load of 2 kips located at midspan. Use $f'_c = 4000$ psi. Compute the maximum bending stress. Use the internal couple method and check with the flexure formula.

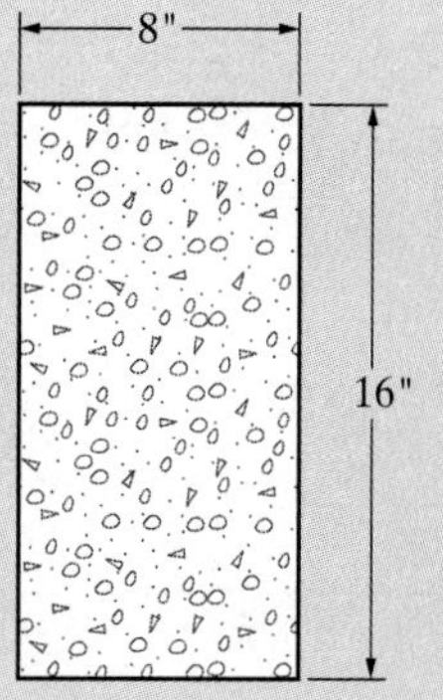

PROBLEM 1-5

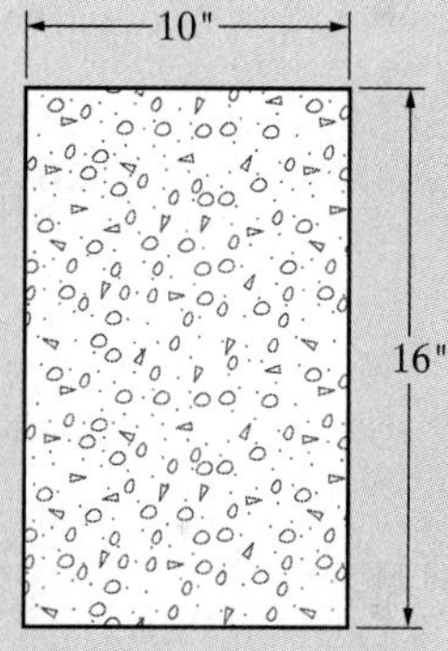

PROBLEM 1-6

1-6. Calculate the cracking moment (resisting moment) for the unreinforced concrete beam shown. Assume normal-weight concrete with $f'_c = 3000$ psi. Use the internal couple method and check with the flexure formula.

1-7. Develop a spreadsheet application to solve Problem 1-6. Set up the spreadsheet so that a table will be generated in which the width of the beam varies from 8 in. to 16 in. (1-in. increments) and the depth varies from 12 in. to 24 in. (1-in. increments.) The spreadsheet should allow the user to input any value for f'_c between 3000 psi and 8000 psi.

1-8. Rework Example 1-3 but invert the beam so that the flange is on the bottom and the web extends vertically upward. Calculate the cracking moment using the internal couple method and check using the flexure formula. Assume positive moment.

1-9. Calculate the cracking moment (resisting moment) for the U-shaped unreinforced concrete beam shown. Assume normal-weight concrete with $f'_c = 3500$ psi. Use the internal couple method and check with the flexure formula. Assume positive moment.

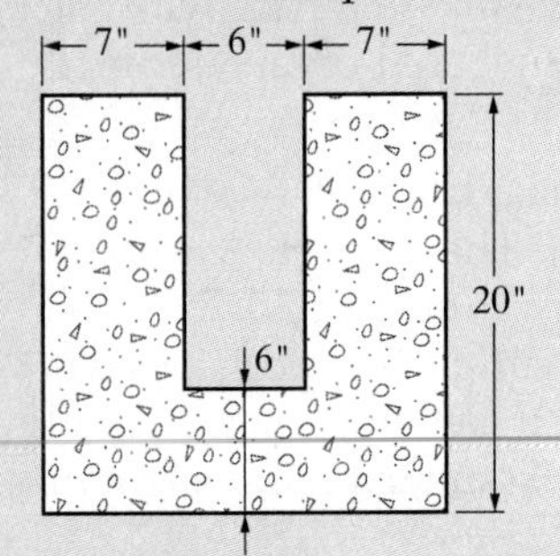

PROBLEM 1-9

1-10. The plain concrete beam shown is used on a 12-ft simple span. The concrete is normal weight with $f'_c = 3000$ psi. Assume positive moment.

a. Calculate the cracking moment.

b. Calculate the value of the concentrated load P at midspan that would cause the concrete beam to crack. (Be sure to include the weight of the beam.)

1-11. Using the section below, determine the following:

a. The service dead load in psf supported by the slab.

b. The factored total uniform load in psf supported by the slab. Assume a floor live load of 100 psf.

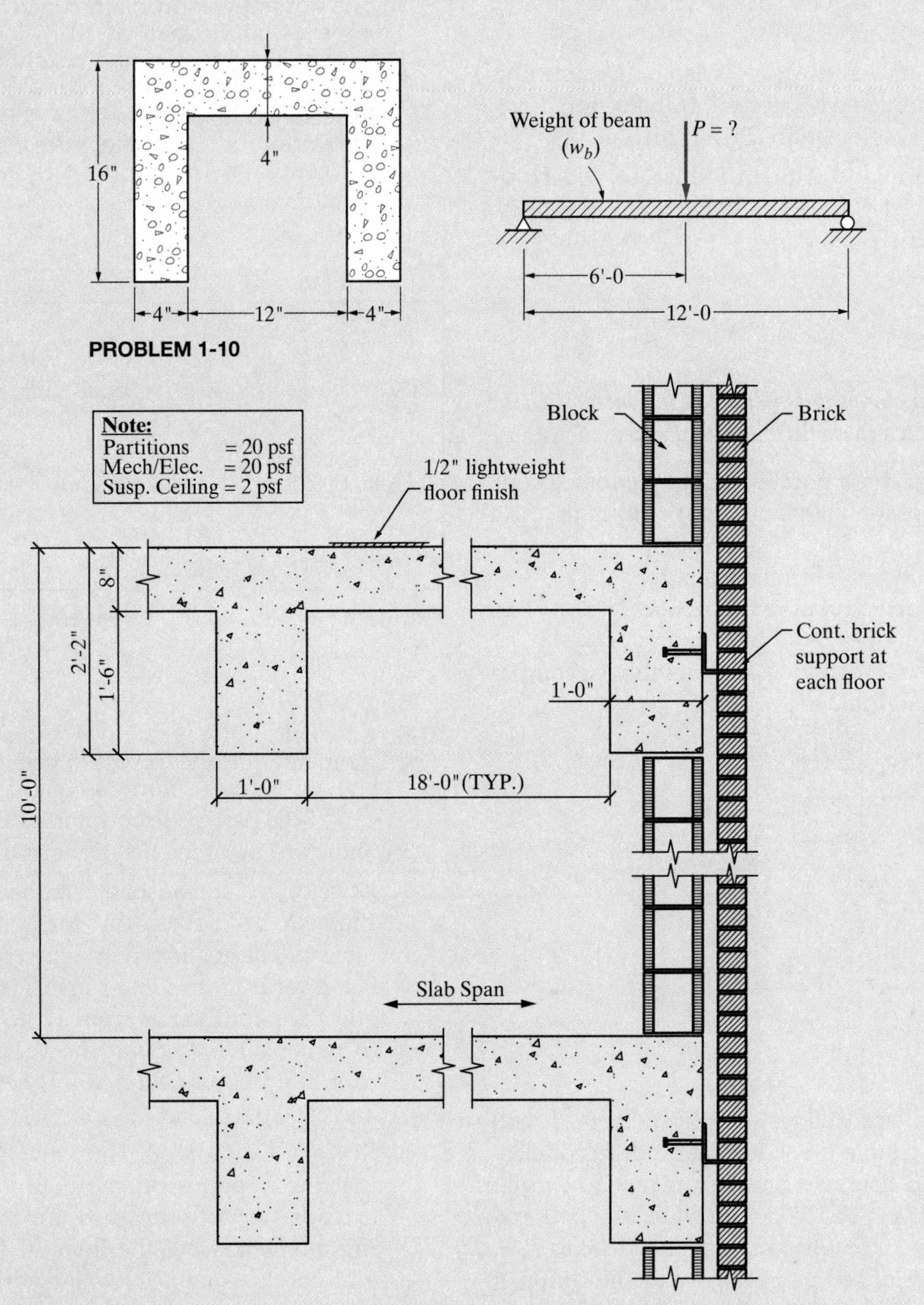

PROBLEM 1-10

PROBLEM 1-11

c. The service dead and service live load in kip/ft supported by the spandrel or edge beam. Assume the block backup wall weighs 55 psf and the brick cladding weighs 40 psf of the vertical surface are of the wall.

d. The factored total uniform load in kip/ft supported by the spandrel beam.

e. The factored total uniform load in kip/ft supported by a typical interior beam.

1-12. A partial floor plan for a three-story reinforced concrete office building with 10 ft floor-to-floor heights, located in Rochester, New York, is shown in the figure below. Assume the following:

- 7″ thick slab on the roof and floors
- 5-ply roofing plus gravel on the roof slab plus 3.5 psf for insulation
- 1/2 in. lightweight finish on the floor slabs
- 20 psf for the partitions
- 10 psf for mechanical and electrical equipment
- 2 psf for the suspended ceiling
- Exterior cladding:
- 55 psf backup block wall—7 ft 10 in. high
- 40 psf brick veneer—10 ft high
- Beams: 16″ wide × 26″ deep; Girders: 18″ wide × 26″ deep
- Ignore live load reduction

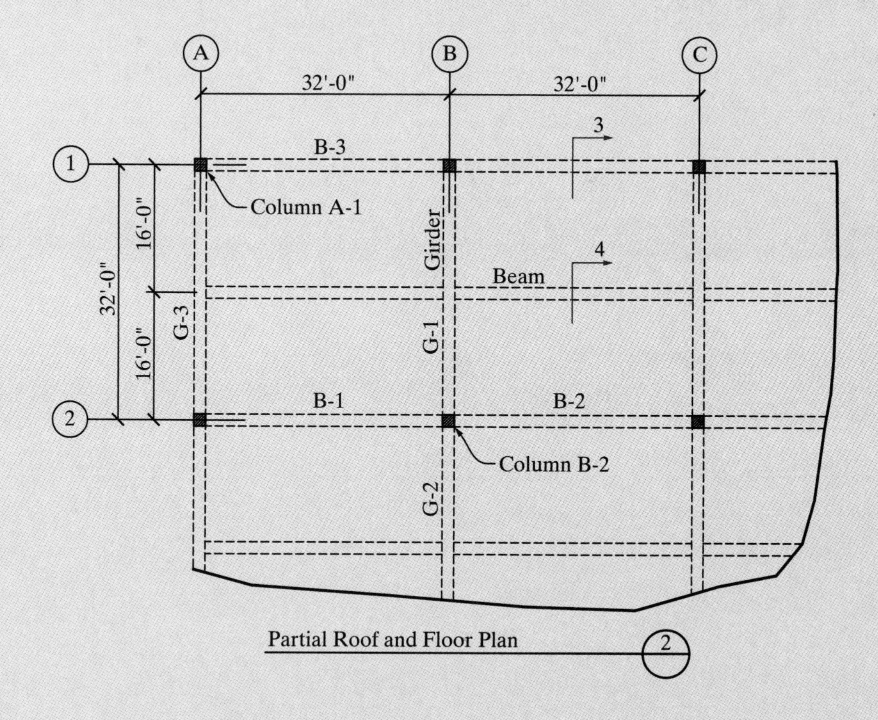

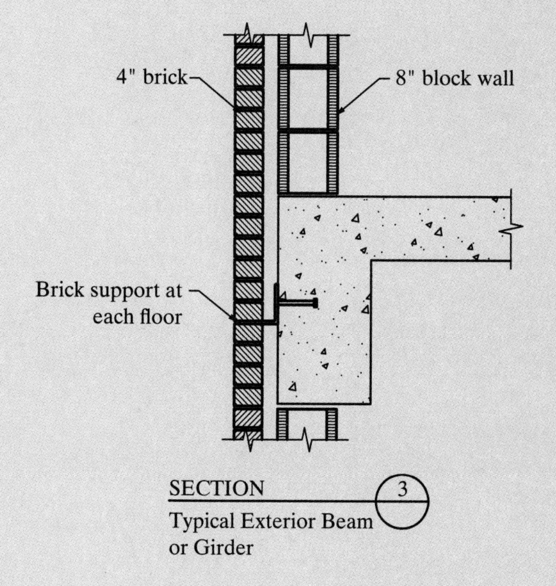

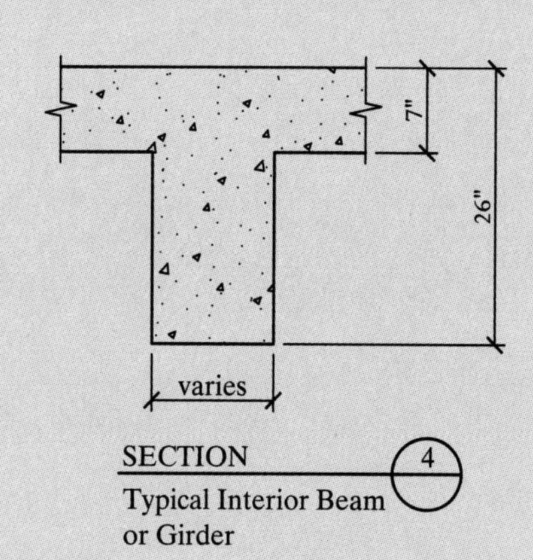

PROBLEM 1-12

Determine the following:

a. The factored total uniform load in psf on the roof and floor slabs

b. The factored total uniform load in kip/ft supported by the

i. Typical interior roof beam
ii. Typical interior floor beam
iii. Typical roof spandrel beam
iv. Typical floor spandrel beam

c. The factored concentrated load from the beam reactions in kip on the following girders and the factored self-weight of the girder stem in kip/ft. For the purposes of calculating the beam reactions, assume the beams are simply supported:

i. Typical interior roof girder
ii. Typical interior floor girder
iii. Typical roof spandrel girder
iv. Typical floor spandrel girder

d. The total factored load and the total service load in kip on Column B-2. Assume 16″ × 16″ columns. For the purposes of calculating the beam and girder reactions, assume the beams and girders are simply supported. Present your results in a tabular format.

e. The total factored load and the total service load in kip on Column A-1. Assume 16″ × 16″ columns. For the purposes of calculating the beam and girder reactions, assume the beams and girders are simply supported. Present your results in a tabular format.

CHAPTER **TWO**

RECTANGULAR REINFORCED CONCRETE BEAMS AND SLABS: TENSION STEEL ONLY

2-1 Introduction
2-2 Analysis and Design Method
2-3 Behavior under Load
2-4 Strength Design Method Assumptions
2-5 Flexural Strength of Rectangular Beams
2-6 Equivalent Stress Distribution
2-7 Balanced, Brittle, and Ductile Failure Modes
2-8 Ductility Requirements
2-9 Strength Requirements
2-10 Rectangular Beam Analysis for Moment (Tension Reinforcement Only)
2-11 Summary of Procedure for Rectangular Beam Analysis for ϕM_n (Tension Reinforcement Only)
2-12 Slabs: Introduction
2-13 One-Way Slabs: Analysis for Moment
2-14 Rectangular Beam Design for Moment (Tension Reinforcement Only)
2-15 Summary of Procedure for Rectangular Reinforced Concrete Beam Design for Moment (Tension Reinforcement Only)
2-16 Design of One-Way Slabs for Moment (Tension Reinforcement Only)
2-17 Summary of Procedure for Design of One-Way Slabs for Moment (To Satisfy ACI Minimum Thickness, *h*)
2-18 Slabs-on-Grade

2-1 INTRODUCTION

In the ACI Code, two types of limit states are considered: the ultimate limit state (i.e. strength design) where the failure of the structure due to bending, shear, crushing or buckling, e.t.c., is imminent and this occurs at the factored loads, and the serviceability limit state which results from excessive deflection, cracking and vibrations at service loads.

When a beam is subjected to bending moments (also termed *flexure*), bending strains are produced. Under positive moment (as normally defined), compressive strains are produced in the top of the beam and tensile strains are produced in the bottom. These *strains* produce *stresses* in the beam, compression in the top, and tension in the bottom. Bending members must therefore be able to resist both tensile and compressive stresses.

For a concrete flexural member (beam, wall, slab, and so on) to have any significant load-carrying capacity, its basic inability to resist tensile stresses must be overcome. By embedding reinforcement (usually deformed steel bars) in the tension zones, a *reinforced concrete* member is created. When properly designed and constructed, members composed of these materials perform very adequately when subjected to flexure.

Initially, we will consider simply supported single-span beams that, as they carry only positive moment (tension in the bottom), will be reinforced with steel bars placed near the bottom of the beam.

2-2 ANALYSIS AND DESIGN METHOD

In the beam examples in Chapter 1, we assumed both a straight-line strain distribution and straight-line stress distribution from the neutral axis to the outer fibers. This, in effect, stated that stress was proportional to strain. This analysis is sometimes called *elastic design*.

As stated in Chapter 1, elastic design is considered valid for the homogeneous plain concrete beam as long as the tensile stress does not exceed the modulus of rupture, that stress at which tensile cracking commences. With homogeneous materials used in construction, such as structural steel and timber, the limit of stress–strain proportionality is generally termed the *proportional limit*. Note that the modulus of rupture for the plain concrete beam may be considered analogous to the proportional limit for structural steel and timber with respect to the limit of stress–strain proportionality.

With structural steel, the proportional limit and yield stress have nearly the same value, and when using the allowable stress design (ASD) method, an allowable bending stress is determined by applying a factor of safety to the yield stress.

With timber, the determination of an allowable bending stress is less straightforward, but it may be thought of as some fraction of the bending stress that causes failure. Using the allowable bending stress and the assumed linear stress–strain relationship, both the analysis and design of timber members and structural steel members (using the ASD method) uses a simplified method that is similar to that used in the Chapter 1 examples.

Even though a reinforced concrete beam was known to be a nonhomogeneous member, for many years the elastic behavior approach was considered valid for concrete design, and it was known as the *working stress design* (WSD) method. The basic assumptions for the WSD method were as follows: (1) A plane section before bending remains a plane section after bending; (2) Hooke's law (stress is proportional to strain) applies to both the steel and the concrete; (3) the tensile strength of concrete is zero and the reinforcing steel carries all the tension; and (4) the bond between the concrete and the steel is perfect, so no slip occurs.

Based on these assumptions, the flexure formula was still used even though the beam was nonhomogeneous. This was accomplished by theoretically transforming one material into another based on the ratio of the concrete and steel moduli of elasticity.

Although the WSD method was convenient and was used for many years, it has been replaced with a more modern and realistic approach for the analysis and design of reinforced concrete. One basis for this approach is that at some point in the loading, the proportional stress–strain relationship for the compressive concrete ceases to exist. When first developed, this method was called the *ultimate strength design* (USD) method. Since then, the name has been changed to the *strength design method.*

The assumptions for the strength design method are similar to those itemized for the WSD method, with one notable exception. Research has indicated that the compressive concrete stress is approximately proportional to strain up to only moderate loads. With an increase in load, the approximate proportionality ceases to exist, and the compressive stress diagram takes a shape similar to the concrete compressive stress–strain curve of Figure 1-1. Additional assumptions for strength design are discussed in Section 2-4.

A major difference between ASD and strength design lies in the way the applied loads (i.e., *service loads*—the loads that are specified in the general building code) are handled and in the determination of the capacity (strength) of the reinforced concrete members. In the strength design method, service loads are amplified using load factors. Members are then designed so that their practical strength at failure, which is somewhat less than the true strength at failure, is sufficient to resist the amplified loads. The strength at failure is still commonly called the *ultimate strength*, and the load at or near failure is commonly called the *ultimate load.* The stress pattern assumed for strength design is such that predicted strengths are in substantial agreement with test results.

2-3 BEHAVIOR UNDER LOAD

Before discussing the strength design method, let us review the behavior of a long-span, rectangular reinforced concrete beam as the load on the beam increases from zero to the magnitude that would cause failure. The reinforced concrete simple beam of Figure 2-1 is assumed subjected to downward loading, which will cause positive moment in the beam. Steel reinforcing, three bars in this example, is located near the bottom of the beam, which is the tension side. Note that the overall depth of the beam is designated h, whereas

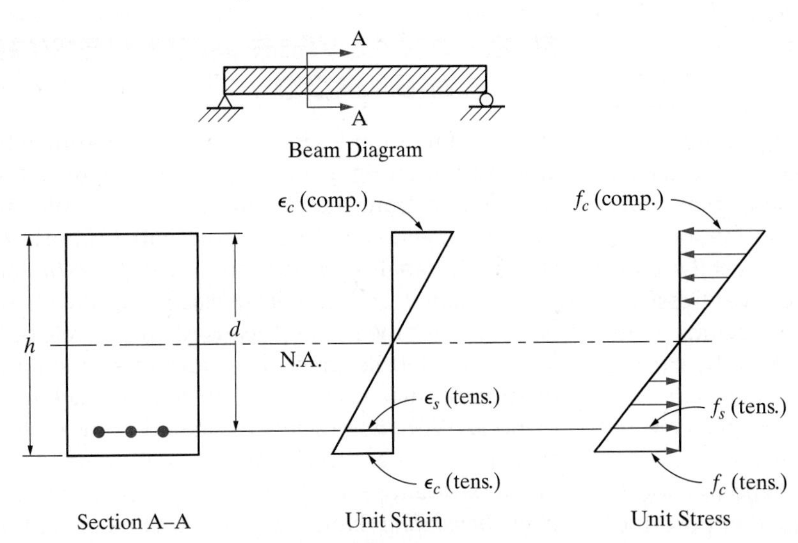

FIGURE 2-1 Flexural behavior at very small loads.

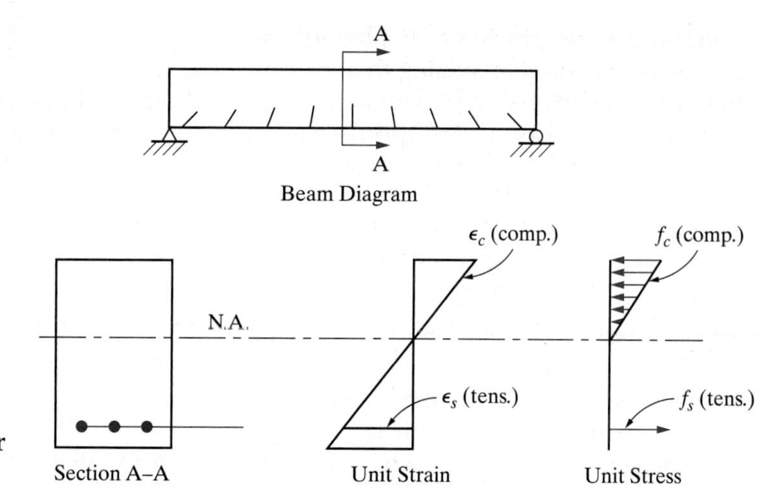

FIGURE 2-2 Flexural behavior at moderate loads.

the location of the steel, referenced to the compression face, is defined by the *effective depth, d.* The effective depth is measured to the centroid of the reinforcing steel. In this example, the centroid is at the center of the single layer of bars. If there are multiple layers of bars, then the effective depth is measured from the compression face to the centroid of the bar group.

At very small loads, assuming that the concrete has not cracked, both concrete and steel will resist the tension, and concrete alone will resist the compression. The stress distribution will be as shown in Figure 2-1. The strain variation will be linear from the neutral axis to the outer fiber. Note that stresses also vary linearly from zero at the neutral axis and are, for all practical purposes, proportional to strains. This will be the case when stresses are low (below the modulus of rupture).

At moderate loads, the tensile strength of the concrete will be exceeded, and the concrete will crack (hairline cracks) in the manner shown in Figure 2-2. Because the concrete cannot transmit any tension across a crack, the steel bars will then resist the entire tension. The stress distribution at or near a cracked section then becomes as shown in Figure 2-2. This stress pattern exists up to approximately a concrete stress f_c of about $f'_c/2$. The concrete compressive stress is still assumed to be proportional to the concrete strain.

With further load increase, the compressive strains and stresses will increase; they will cease to be proportional, however, and some nonlinear stress curve will result on the compression side of the beam. This stress curve above the neutral axis will be essentially the same shape as the concrete stress–strain curve (see Figure 1-1). The stress and strain distribution that exists at or near the ultimate load is shown in Figure 2-3. Eventually, the ultimate capacity of the beam will be reached and the beam will fail. The actual mechanism of the failure is discussed later in this chapter.

At this point the reader may well recognize that the process of attaining the ultimate capacity of a member is irreversible. The member has cracked and deflected significantly; the steel has yielded and will not return to its original length. If other members in the structure have similarly reached their ultimate capacities, the structure itself is probably crumbling and in a state of distress or partial ruin, even though it may not have completely collapsed. Naturally, although we cannot ensure that this state will never be reached, factors are introduced to create the commonly accepted margins of safety. Nevertheless, the ultimate capacities of members are, at present, the basis for reinforced concrete analysis and design. In this text, it is in such a context that we will speak of failures of members.

2-4 STRENGTH DESIGN METHOD ASSUMPTIONS

The development of the strength design approach depends on the following basic assumptions:

1. A plane section before bending remains a plane section after bending. That is, the strain throughout the depth of the member varies linearly from zero at the neutral axis. Tests have shown this assumption to be essentially correct.

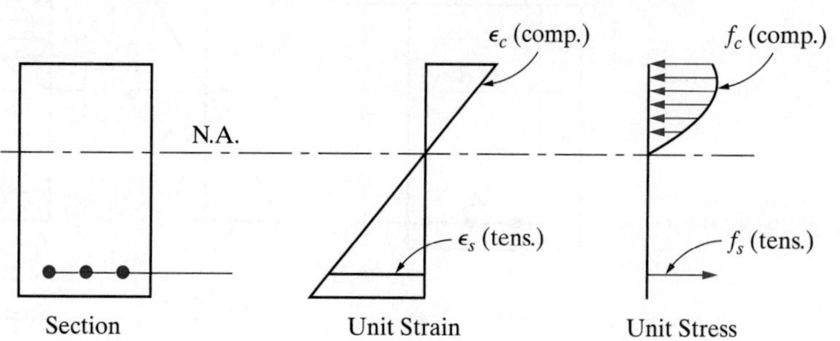

FIGURE 2-3 Flexural behavior near ultimate load.

2. Stresses and strains are approximately proportional only up to moderate loads (assuming that the concrete stress does not exceed approximately $f'_c/2$). When the load is increased and approaches an ultimate load, stresses and strains are no longer proportional. Hence, the variation in concrete stress is no longer linear.
3. In calculating the ultimate moment capacity of a beam, the tensile strength of the concrete is neglected.
4. The maximum usable concrete compressive strain at the extreme fiber is assumed equal to 0.003. This value is based on extensive testing, which indicated that the flexural concrete strain at failure for rectangular beams generally ranges from 0.003 to 0.004 in./in. Hence, the assumption that the concrete is about to crush when the maximum strain reaches 0.003 is slightly conservative.
5. The steel is assumed to be uniformly strained to the strain that exists at the level of the centroid of the steel. Also, if the strain in the steel (ϵ_s) is less than the yield strain of the steel (ϵ_y), the stress in the steel is $E_s\epsilon_s$. This assumes that for stresses less than f_y, the steel stress is proportional to strain. For strains equal to or greater than ϵ_y, the stress in the reinforcement will be considered independent of strain and equal to f_y. See the idealized stress–strain diagram for steel shown in Figure 1-4b.
6. The bond between the steel and concrete is perfect and no slip occurs.

Assumptions 4 and 5 constitute what may be termed *code criteria* with respect to failure. The true ultimate strength of a member will be somewhat greater than that computed using these assumptions. The strength method of design and analysis of the ACI Code is based on these criteria, however, and consequently so is our basis for bending member design and analysis. Note that "cracking in reinforced concrete is not a defect. The very basis of reinforced concrete design is that concrete has no tensile strength and that sufficient reinforcement is provided to control crack widths." Therefore, "cracking is an inherent part of reinforced concrete, and if controlled, will not be detrimental to the performance of the structure."[10]

2-5 FLEXURAL STRENGTH OF RECTANGULAR BEAMS

Based on the assumptions previously stated, we can now examine the strains, stresses, and forces that exist in a reinforced concrete beam subjected to its *ultimate moment*, that is, the moment that exists just prior to the failure of the beam. In Figure 2-4, the assumed beam has a width b and an effective depth d and it is reinforced with a steel area of A_s. (A_s is the total *cross-sectional area* of tension steel present.)

Based on the preceding assumptions, it is possible that a beam may be loaded to the point where the maximum tensile steel unit stress equals its yield stress (as a limit) and the concrete compressive strain is less than 0.003 in./in. It is also possible that in another beam, the maximum concrete compressive strain will equal 0.003 in./in. and the tensile steel unit stress will be less than its yield stress f_y. When either condition occurs, it implies a specific mode of failure, which will be discussed later.

As stated previously, the compressive stress distribution above the neutral axis for a flexural member is similar to the concrete compressive stress–strain curve as depicted in Figure 1-1. As may be observed in Figure 2-4, the ultimate compressive stress f'_c does not occur at the outer fiber, neither is the shape of the curve the same for different-strength concretes.

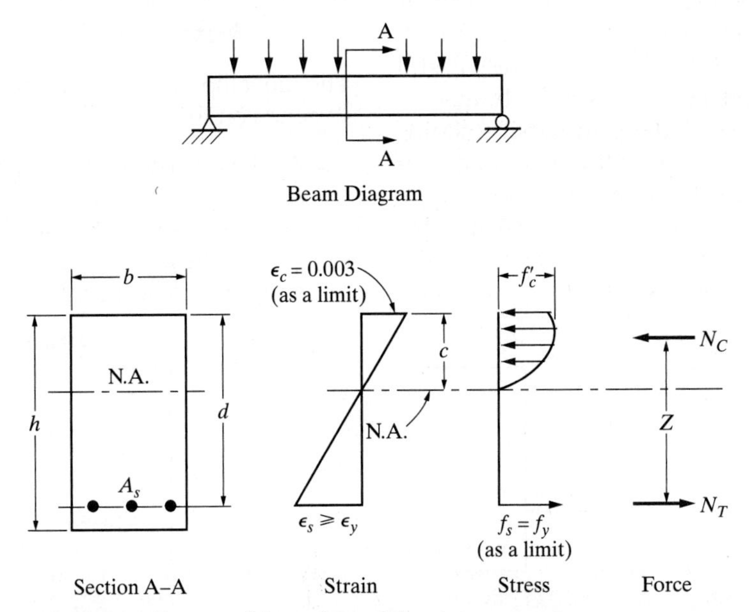

FIGURE 2-4 Beam subjected to ultimate moment.

Actually, the magnitudes of the compressive concrete stresses are defined by some irregular curve, which could vary not only from concrete to concrete but also from beam to beam. Present theories accept that, at ultimate moment, compressive stresses and strains in concrete are not proportional. Although strains are assumed linear, with maximum strain of 0.003 in./in. at the extreme outer compressive fiber, the maximum concrete compressive stress f'_c develops at some intermediate level near, but not at, the extreme outer fiber.

The flexural strength or resisting moment of a rectangular beam is created by the development of these internal stresses that, in turn, may be represented as internal forces. As observed in Figure 2-4, N_C represents a theoretical internal resultant compressive force that in effect constitutes the total internal compression above the neutral axis. N_T represents a theoretical internal resultant tensile force that in effect constitutes the total internal tension below the neutral axis.

These two forces, which are parallel, equal, and opposite and separated by a distance Z, constitute an internal resisting couple whose maximum value may be termed the *nominal moment strength* of the bending member. As a limit, this nominal moment strength must be capable of resisting the design bending moment induced by the applied loads. Consequently, if we wish to design a beam for a prescribed loading condition, we must arrange its concrete dimensions and the steel reinforcements so that it is capable of developing a moment strength at least equal to the maximum bending moment induced by the loads.

The determination of the moment strength is complex because of the shape of the compressive stress diagram above the neutral axis. Not only is N_C difficult to evaluate but its location relative to the tensile steel is difficult to establish. Because the moment strength is actually a function of the magnitude of N_C and Z, however, it is not really necessary to know the exact shape of the compressive stress distribution above the neutral axis. To determine the moment strength, it is necessary to know only (1) the total resultant compressive force N_C in the concrete and (2) its location from the outer compressive fiber (from which the distance Z may be established). These two values may easily be established by replacing the unknown complex compressive stress distribution by a fictitious one of simple geometrical shape, provided the fictitious distribution results in the same total compressive force N_C applied at the same location as in the actual distribution when it is at the point of failure.

2-6 EQUIVALENT STRESS DISTRIBUTION

For purposes of simplification and practical application, a fictitious but equivalent rectangular concrete stress distribution was proposed by Whitney [1] and subsequently adopted by the ACI Code (Sections 22.2.2.3 and 22.2.2.4). The ACI Code also stipulates that other compressive stress distribution shapes may be used, provided results are in substantial agreement with comprehensive test results. Because of the simplicity of the rectangular shape, however, it has become the more widely used fictitious stress distribution for design purposes.

With respect to this equivalent stress distribution as shown in Figure 2-5, the average stress intensity is taken as 0.85 f'_c and is assumed to act over the upper area of the beam cross section defined by the width b and a depth of a. The magnitude of a may be determined by

$$a = \beta_1 c$$

where

c = distance from the outer compressive fiber to the neutral axis

β_1 = a factor that is a function of the strength of the concrete as follows and as shown in Figure 2-6:

For 2500 psi $\leq f'_c \leq$ 4000 psi: $\beta_1 = 0.85$.
For $f'_c >$ 4000 psi:

$$\beta_1 = 0.85 - \frac{0.05(f'_c - 4000)}{1000} \geq 0.65$$

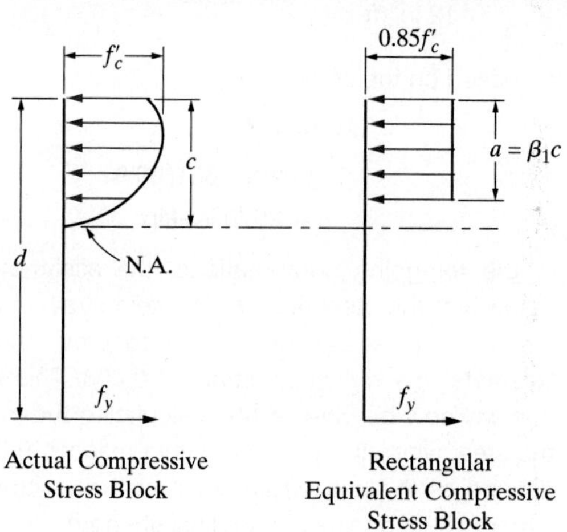

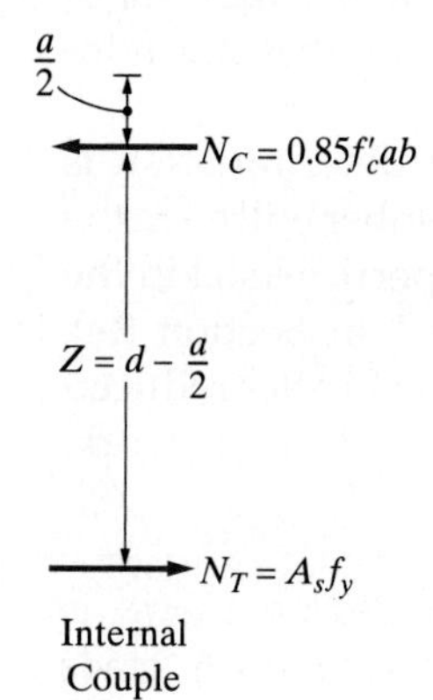

FIGURE 2-5 Equivalent stress block for strength design and analysis.

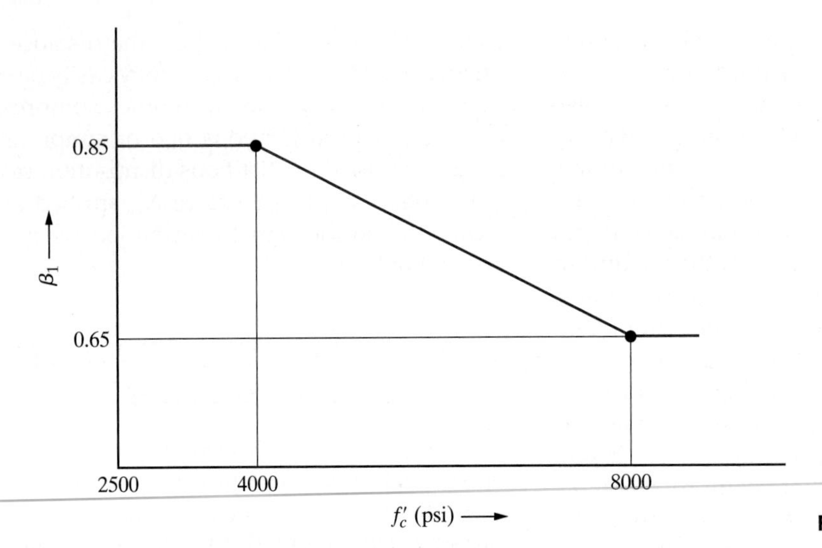

FIGURE 2-6 β_1 v. f'_c.

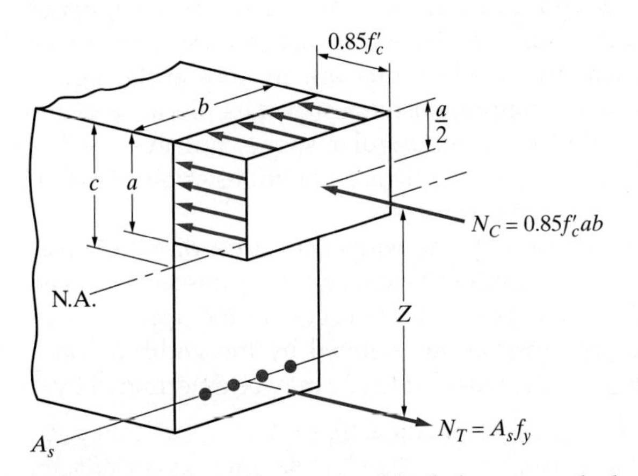

FIGURE 2-7 Equivalent stress block for strength design and analysis.

It is in no way maintained that the compressive stresses are actually distributed in this most unlikely manner. It is maintained, however, that this equivalent rectangular distribution gives results close to those of the complex actual stress distribution. An isometric view of the accepted internal relationships is shown in Figure 2-7.

Using the equivalent stress distribution in combination with the strength design assumptions, we may now determine the nominal moment strength M_n of rectangular reinforced concrete beams that are reinforced for tension only.

The nominal moment strength determination is based on the assumption that the member will have the exact dimensions and material properties used in the design computations. As discussed in Section 2-9, the nominal moment strength will be further reduced when it is used in practical analysis and design work.

Example 2-1

Determine M_n for a beam of cross section shown in Figure 2-8, where $f'_c = 4000$ psi. Assume A615 grade 60 steel.

Solution:

1. We will assume that f_y exists in the steel, subject to later check. By $\Sigma H = 0$,

$$N_C = N_T$$

$$(0.85\, f'_c)ab = A_s f_y$$

$$a = \frac{A_s f_y}{0.85 f'_c b} = \frac{2.37(60)}{0.85(4)(10)} = 4.18 \text{ in.}$$

This is the depth of the stress block that must exist if there is to be horizontal equilibrium.

2. Calculate the length of the lever arm, Z:

$$Z = d - \frac{a}{2} = 23 - \frac{4.18}{2} = 20.9 \text{ in.}$$

3. Calculate M_n:

$$M_n = N_C Z \text{ or } N_T Z$$

Based on the concrete,

$$M_n = N_C Z = (0.85 f'_c)ab\,(20.9)$$
$$= 0.85(4.0)(4.18)(10)(20.9)$$
$$= 2970 \text{ in.-kip}$$

$$\frac{2970 \text{ in.-kip}}{12 \text{ in./ft}} = 248 \text{ ft.-kip}$$

or, based on the steel,

$$M_n = A_s f_y Z$$
$$= 2.37(60)(20.9)$$
$$= 2970 \text{ in.-kip}$$

4. In the foregoing computations, the assumption was made that the steel reached its yield strain (and therefore its yield stress) before the concrete reached its "ultimate" (by definition) strain of 0.003. This assumption will now be checked by calculating the strain ϵ_t in the steel when the concrete strain reaches 0.003. ϵ_t is defined as the net tensile strain at the centroid of the extreme tension steel at nominal strength.

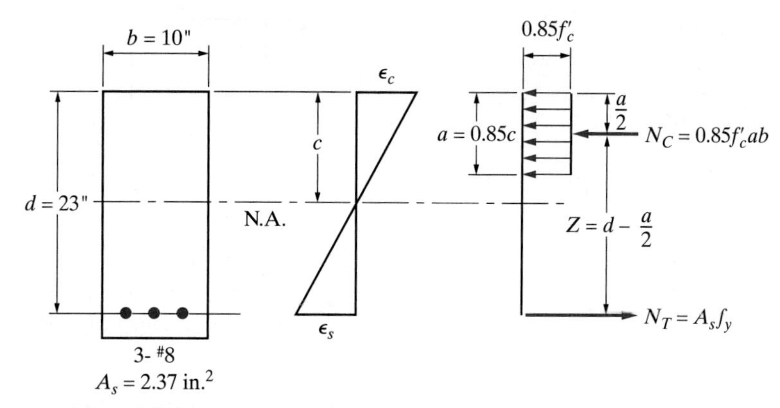

FIGURE 2-8 Sketch for Example 2-1

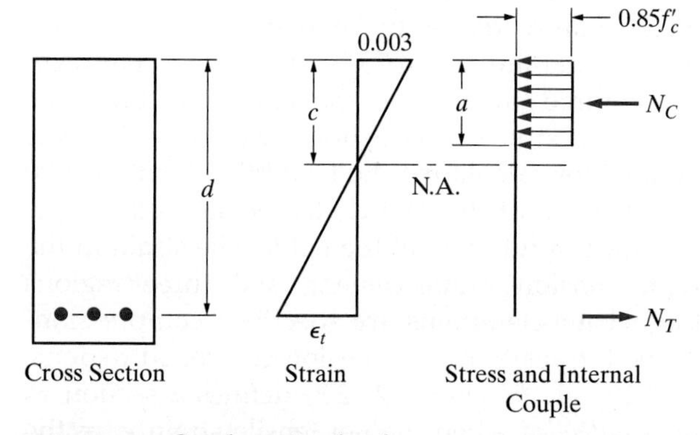

FIGURE 2-9 Steel strain check.

Referring to Figure 2-9, we may locate the neutral axis as follows:

$$a = \beta_1 c \quad \text{(ACI Code, Section 22.2.2.4.1)}$$

$$\beta_1 = 0.85 \quad \text{because } f'_c = 4000 \text{ psi}$$

Therefore

$$c = \frac{a}{0.85} = \frac{4.18}{0.85} = 4.92 \text{ in.}$$

By similar triangles in the strain diagram, we may find the strain in the steel when the concrete strain is 0.003:

$$\frac{0.003}{c} = \frac{\epsilon_t}{d - c}$$

Then

$$\epsilon_t = \frac{d - c}{c}(0.003) = \frac{23 - 4.92}{4.92}(0.003)$$

$$= 0.011 \text{ in./in.}$$

The strain at which the steel yields (ϵ_y) may be determined from the basic definition of the modulus of elasticity, E = stress/strain:

$$\epsilon_y = \frac{f_y}{E_s} = \frac{60{,}000}{29{,}000{,}000} = 0.00207 \text{ in./in.}$$

(see Table A-1)

This represents the strain in the steel when the stress first reaches 60,000 psi.

Because the computed strain in the steel (0.011) is greater than the yield strain (0.00207), the steel reaches its yield stress before the concrete reaches its strain of 0.003, and the assumption that the stress in the steel is equal to the yield stress was correct. (See assumption 5 in Section 2-4 and the idealized stress–strain diagram for steel shown in Figure 1-4b.)

2-7 BALANCED, BRITTLE, AND DUCTILE FAILURE MODES

In Figure 2-9, the *ratio* between the steel strain and the maximum concrete strain is fixed once the neutral axis is established. The location of the neutral axis will vary based on the amount of tension steel in the cross section, as the stress block is just deep enough to ensure that the resultant compressive force is equal to the resultant tensile force ($\Sigma H = 0$). If more tension bars are added to the bottom of a reinforced concrete cross section, the depth of the compressive stress block will be greater, and therefore the neutral axis will be lower. Referring again to Figure 2-9, if there were just enough steel to put the neutral axis at a location where the yield strain in the steel and the maximum concrete strain of 0.003 existed at the same time, the cross section would be said to be *balanced* (see the ACI Code, Section R21.2.2). The amount of steel required to create this condition is relatively large. However, the balanced condition is the dividing line between two distinct types of reinforced concrete beams that are characterized by their failure modes. If a beam has more steel than is required to create the balanced condition, the beam will fail in a *brittle mode.* The additional steel will cause the neutral axis to be low (see Figure 2-10). This will in turn cause the concrete to reach a strain of 0.003 *before* the steel yields. Should more moment (and therefore strain) be applied to the beam cross section, failure will be initiated by a sudden crushing of the concrete. This brittle failure mode is undesirable.

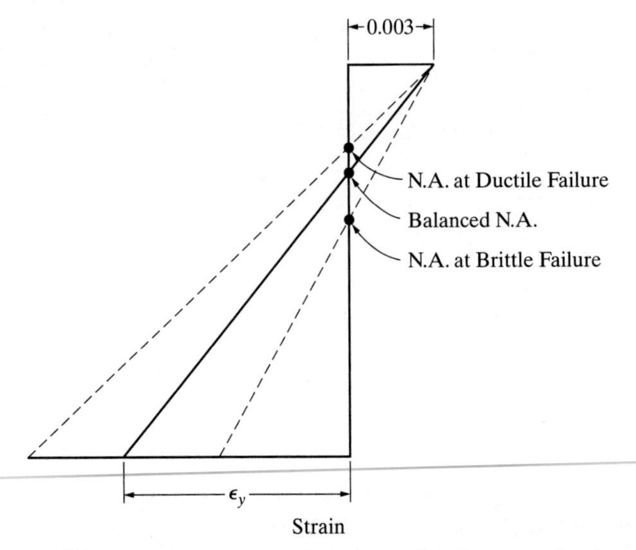

FIGURE 2-10 Strain distribution and failure modes in flexural members.

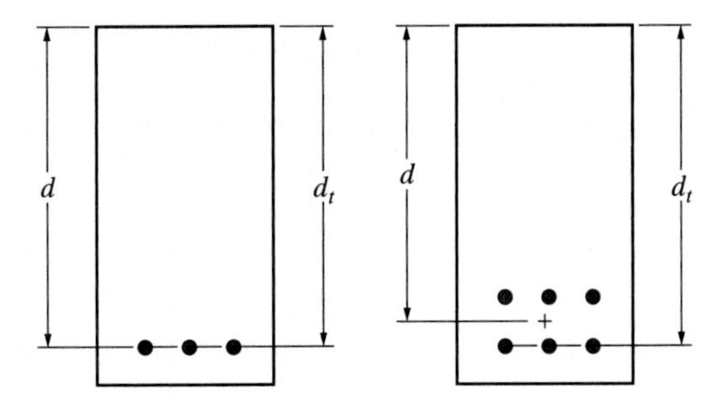

FIGURE 2-11 Definition of extreme tension steel.

If a beam has less steel than is required to create the balanced condition, the beam will fail in a *ductile mode*. The neutral axis will be higher than the balanced neutral axis, and the steel will reach its yield strain (and therefore its yield stress) before the concrete reaches a strain of 0.003. Figure 2-10 shows these variations in neutral axis location for beams that are on the verge of failure. In each case, the concrete has been strained to 0.003. Following the ductile failure mode case through to failure, we see that a slight additional load will cause the steel to stretch a considerable amount. The strains in the concrete and the steel continue to increase. The tensile force is not increasing since the steel stress has reached the maximum stress, f_y. Because the compressive force cannot increase ($\Sigma H = 0$), and because the concrete strain and therefore its stress are increasing, the area under compression must be decreasing and the neutral axis must rise. This process continues until the reduced area fails in compression as a secondary effect. This failure due to yielding is a gradual one, with the beam showing greatly increased deflection after the steel reaches the yield point; hence, there is adequate warning of impending failure. The steel, being ductile, will not actually pull apart even at failure of the beam. The ductile failure mode is desirable and is required by the ACI Code, as discussed in the following section.

2-8 DUCTILITY REQUIREMENTS

Although failure due to yielding of steel is gradual, with adequate warning of collapse, failure due to crushing of concrete is sudden and without warning. We have seen that flexural members with less steel than is required to produce the balanced condition will fail by yielding of the steel due to the strain in the steel exceeding the yield strain. In previous editions of the ACI Code prior to 2005, in order to assure a ductile failure mode, the maximum tension reinforcement was limited to 75 % of the steel reinforcement required to achieve the balanced condition, and uniform strength reduction factor ($\phi = 0.9$) was prescribed for calculating the flexural capacity. In the current version of the ACI Code, the strength reduction factor is a function of the net tensile strain in the extreme tension reinforcement, and three regions of the net tensile strains are specified: compression-controlled, transition, and tension-controlled regions. The ACI Code (Section R21.2.2) defines a section as *tension-controlled* when the net tensile strain ϵ_t in the extreme tension steel is equal to or greater than 0.005 when the concrete has reached its assumed strain limit of 0.003. The minimum net tensile strain, ϵ_t, in non-prestressed slabs is 0.004 (ACI 318 Section 7.3.3.1). With reference to Figure 2-11, note that the extreme tension steel is located at d_t from the extreme compression face. For a single layer of steel, $d_t = d$, the effective depth. For multiple layers of steel, $d_t > d$.

Further, ACI Code (Section R21.2.2) defines a section as *compression-controlled* when the net tensile strain ϵ_t in the extreme tension steel is equal to or less than the yield strain ϵ_y of the steel just as the concrete in compression reaches its assumed strain limit of 0.003. See Table A-1 for values of ϵ_y. (The code permits ϵ_y to be taken as 0.002 for grade 60 steel [see ACI 21.2.2.1].)

For non-prestressed flexural members subjected to little or no axial load, the net tensile strain ϵ_t at nominal strength shall not be less than 0.004. Because all values of ϵ_y for current reinforcing steel bars are less than 0.004, this ensures a tension-controlled flexural member, one that will exhibit ductility and fail by yielding of the steel.

Sections in which ϵ_t falls between the compression-controlled strain limit ϵ_y and the tension-controlled limit 0.005 constitute a transition region between compression-controlled and tension-controlled sections. Whether a section is tension-controlled, compression-controlled, or in the transition region has implications, which are discussed shortly. See Figure 2-12 for a graphical representation of the strain limit definitions.

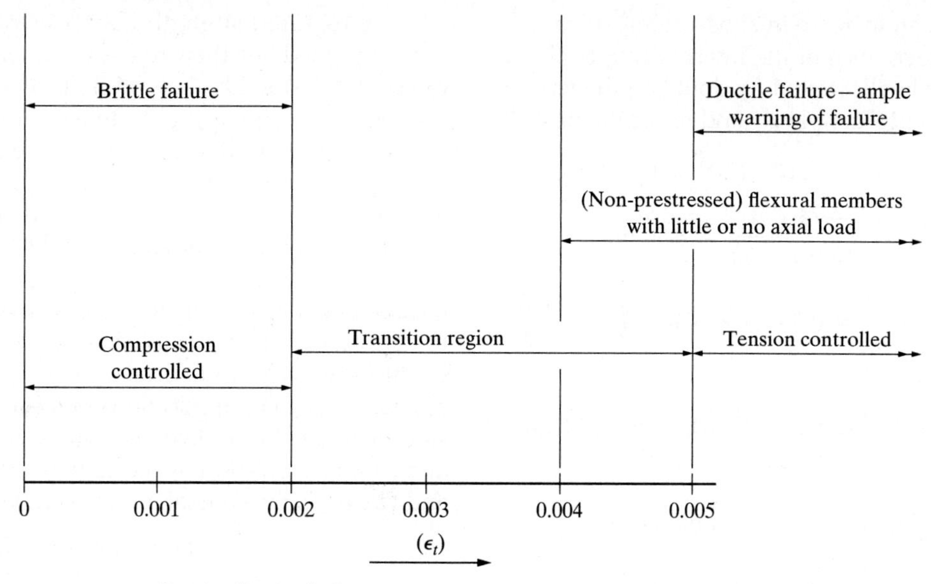

FIGURE 2-12 Strain–limit definitions.

Example 2-2

For the beam cross section of Example 2-1, determine the amount of steel A_s required to cause the strain in the tension steel ϵ_t to be 0.005 just as the maximum strain in the concrete reaches 0.003. Use $f'_c = 3000$ psi and $f_y = 60{,}000$ psi.

Solution:

The cross section, strain, stress, and internal couple are shown in Figure 2-13. Note that $d_t = d$.

1. Determine the location of the neutral axis. By similar triangles:

$$\frac{c}{0.003} = \frac{23 - c}{0.005}$$

$$0.005c = 0.003(23 - c)$$

$$0.005c + 0.003c = 0.003(23) = 0.0690$$

from which

$$c = \frac{0.0690}{0.008} = 8.63 \text{ in.}$$

2. Determine N_C:

$$a = \beta_1 c = 0.85(8.63) = 7.34 \text{ in.}$$

$$N_C = 0.85\, f'_c\, ab$$
$$= 0.85(3 \text{ kip/in.}^2)(7.34 \text{ in.})(10 \text{ in.}) = 187.2 \text{ kip}$$

3. Determine A_s:

$$N_C = N_T = 187.2 \text{ kip}$$

$$N_T = A_s f_y$$

$$A_s = \frac{N_T}{f_y} = \frac{187.2 \text{ kip}}{60 \text{ kip/in.}^2} = 3.12 \text{ in.}^2$$

For ϵ_t to reach 0.005 just as the maximum concrete strain reaches 0.003, A_s must be 3.12 in.^{2}. This is the maximum amount of steel for the section to be a tension-controlled section. Any steel in excess of 3.12 in.^{2} will cause ϵ_t to be less than 0.005. Recall that the code does not allow $\epsilon_t < 0.004$ in a non-prestressed flexural member with little or no axial load.

At this point, note again that heavily reinforced beams are less efficient than their more lightly reinforced counterparts. One structural reason for this is that, for a

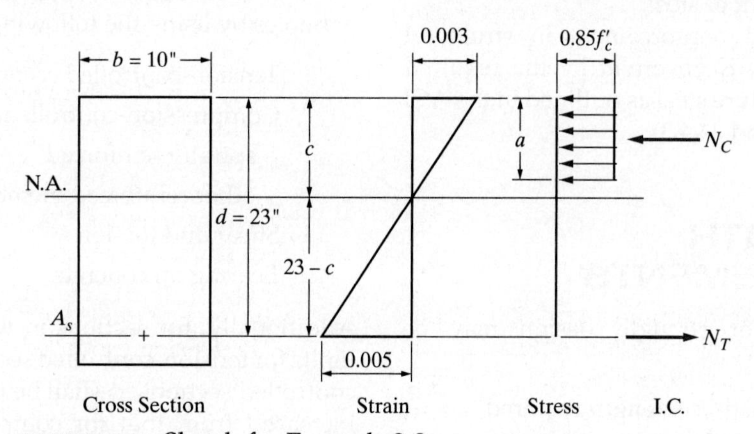

FIGURE 2-13 Sketch for Example 2-2.

given beam size, an increase in A_s is accompanied by a decrease in the lever arm of the internal couple ($Z = d - a/2$). This may be illustrated by doubling the tension steel for the beam of Example 2-1 and recalculating M_n:

$$A_s = 2(2.37) = 4.74 \text{ in.}^2 \quad (100\% \text{ increase})$$

$$a = \frac{A_s f_y}{0.85 f'_c b} = \frac{4.74(60)}{0.85(4)(10)} = 8.36 \text{ in.}$$

$$M_n = 0.85 f'_c ab\left(d - \frac{a}{2}\right) = 0.85(4)(8.36)(10)\left(23 - \frac{8.36}{2}\right)$$

$$= 446 \text{ ft.-kip}$$

For the beam of Example 2-1, M_n was 248 ft.-kip; therefore,

$$\frac{446 - 248}{248} \times 100 = 80\% \text{ increase}$$

Sections 9.6.1.1 and 9.6.1.2 of the ACI Code also establish a lower limit on the amount of tension reinforcement for flexural members. The code states that where tensile reinforcement is required by analysis, the steel area A_s shall not be less than that given by

$$A_{s,\min} = \frac{3\sqrt{f'_c}}{f_y} b_w d \geq \frac{200}{f_y} b_w d$$

(Note that for rectangular beams, $b_w = b$.)

$A_{s,\min}$ is conveniently calculated using Table A-5, where the larger of $3\sqrt{f'_c}/f_y$ and $200/f_y$ is tabulated. Note that $A_{s,\min}$ is the product of the tabulated value and $b_w d$.

The lower limit guards against sudden failure essentially by ensuring that a beam with a very small amount of tensile reinforcement has a greater moment strength as a reinforced concrete section than that of the corresponding plain concrete section computed from its modulus of rupture. Alternatively, it is satisfactory to provide an area of tensile reinforcement that is one-third greater than that required by analysis (ACI Code, Section 9.6.1.3). This requirement applies especially to grade beams, wall beams, and other deep flexural members where the minimum reinforcement requirement specified in ACI 9.6.1.2 would result in an excessively large amount of steel.

Minimum required reinforcement in structural slabs (see Section 2-13) is governed by the required shrinkage and temperature steel as outlined in the ACI Code (Sections 24.4.1 and 24.4.3).

2-9 STRENGTH REQUIREMENTS

The basic criterion for strength design may be expressed as

strength furnished ≥ strength required

All members and all sections of members must be proportioned to meet this criterion.

The required strength may be expressed in terms of design loads or their related moments, shears, and forces. Design loads may be defined as service loads multiplied by the appropriate *load factors*. (When the word *design* is used as an adjective [e.g., design load] throughout the ACI Code, it indicates that load factors are included.) The subscript u is used to indicate design loads, moments, shears, and forces.

The ACI Code, Section 5.3, specifies load factors to be used and load combinations to be investigated. Loads to be considered are dead loads, live loads, fluid loads, loads due to weight and pressure of soil, and snow loads, among others. For applications in this text, we will consider only dead load, live load, and loads due to weight and pressure of soil. For the combination of dead load and live load, the general representation is as follows:

$$U = 1.2D + 1.6L \geq 1.4D$$

where U is defined as the required strength to resist factored loads or related internal moments and forces, D is the service dead load, and L is the service live load. (The term *service load* generally refers to the load specified in applicable building codes as representing minimum requirements.) The factors 1.2 and 1.6 (and 1.4) represent load factors. The load factors are part of the overall safety provision in reinforced concrete structures and are meant to reflect the variability in load effects. Dead loads can be more accurately estimated than live loads, and hence a lower load factor is used for dead load. Live loads have a greater variability than dead loads, and hence a higher load factor is used.

A second part of the overall safety provision, provided in the ACI Code, Section 21.2, is the reduction of the theoretical capacity of a structural element by a *strength-reduction factor* ϕ. This provides for the possibility that small adverse variations in material strengths, workmanship, and dimensions, although within acceptable tolerances and limits of good practice, may combine to result in undercapacity. In effect, the nominal strength of a member, when multiplied by the ϕ factor, will furnish us with a practical strength that is obviously less than the nominal strength.

The ACI Code, Section 21.2, provides for these variables by using the following ϕ factors:

Tension-controlled sections	0.90
Compression-controlled sections	
spirally reinforced	0.75
other reinforced members	0.65
Shear and torsion	0.75
Bearing on concrete	0.65

Additionally, for sections in which ϵ_t is between the limits for tension-controlled sections and compression-controlled sections, ϕ shall be permitted to be linearly increased from that for compression-controlled sections to 0.90 as ϵ_t increases from the compression-controlled strain limit to 0.005.

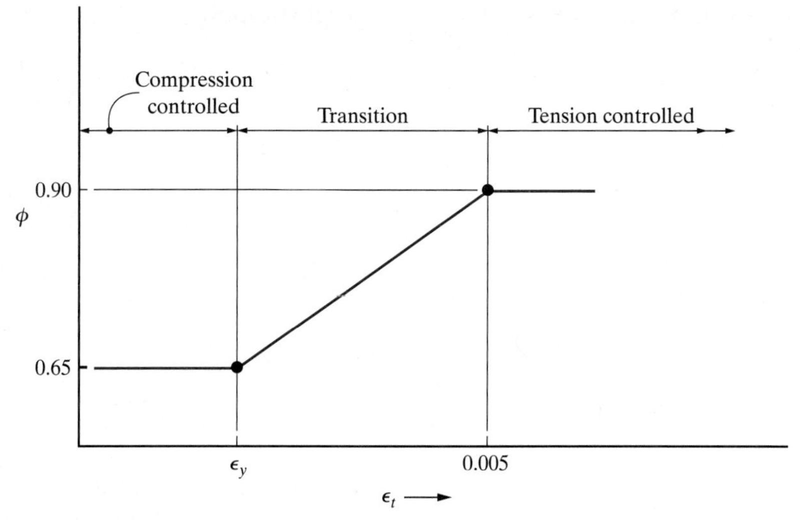

FIGURE 2-14 ϕ v. ϵ_t.

In each case, the practical strength of a reinforced concrete member will be the product of the nominal strength and the ϕ factor. Therefore, in terms of moment, we can say that

$$\text{practical moment strength} = \phi M_n$$

Therefore, the general design equation for the ultimate limit state is

$$\phi R_n > U$$

Where, ϕR_n = the design strength or capacity

U = required strength or the factored load effects or the demand

The individual values assigned to ϕ factors are determined using statistical studies of structural resistances and loading variabilities tempered by subjective factors reflecting the consequences of a structural failure.

The variation in ϕ for tension-controlled sections, transition sections, and compression-controlled (nonspirally reinforced) sections is shown in Figure 2-14. For the transition section, an expression for ϕ as a function of ϵ_t can be developed:

$$\phi = 0.65 + \left(\frac{0.90 - 0.65}{0.005 - \epsilon_y}\right)(\epsilon_t - \epsilon_y)$$

This is valid for the range of $\epsilon_y \leq \epsilon_t \leq 0.005$. Assuming $\epsilon_y = 0.002$, this expression becomes:

$$\phi = 0.65 + (\epsilon_t - 0.002)\left(\frac{250}{3}\right)$$

The calculation of ϵ_t can also be simplified using the depth of the stress block a, which is commonly calculated in analysis problems. Refer to Figure 2-8 and set $d_t = d$.

$$\epsilon_t = \frac{d_t - c}{c}(0.003)$$

Substitute $c = a/\beta_1$:

$$\epsilon_t = \frac{d_t - \dfrac{a}{\beta_1}}{\dfrac{a}{\beta_1}}(0.003) = \frac{0.003\,\beta_1 d_t}{a} - 0.003 \qquad \textbf{(2-1)}$$

2-10 RECTANGULAR BEAM ANALYSIS FOR MOMENT (TENSION REINFORCEMENT ONLY)

The *flexural analysis problem* is characterized by knowing precisely what comprises the cross section of a beam. That is, the following data are *known*: tension bar size and number (or A_s), beam width (b), effective depth (d), or total depth (h), f'_c, and f_y. To be found, basically, is the beam strength, although this may be manifested in various ways: Find ϕM_n, check the adequacy of the given beam, or find an allowable load that the beam can carry. The *flexural design problem*, on the other hand, requires the determination of one or more of the dimensions of the cross section or the determination of the main tension steel to use. It will be important to recognize the differences between these two types of problems because the methods of solution are different.

To expedite reinforced concrete analysis and design calculations, use is frequently made of tables of pertinent quantities. Tables find their greatest use in the design process, but because they are also useful for analysis, they are developed here.

In Example 2-1, we calculated the nominal moment strength M_n for a rectangular reinforced concrete section (tension steel only). Based on concrete:

$$M_n = 0.85 f'_c ab\left(d - \frac{a}{2}\right)$$

and

$$a = \frac{A_s f_y}{0.85 f'_c b}$$

We will now develop a modified expression for M_n (and ϕM_n), which will then be adapted for table use. It is convenient to use the concept of *reinforcement ratio* ρ

(lowercase Greek "rho"), the ratio of tension steel area to effective concrete area:

$$\rho = \frac{A_s}{bd}$$

from which

$$A_s = \rho bd$$

Substituting this into the expression for a:

$$a = \frac{A_s f_y}{(0.85f_c')b} = \frac{\rho bdf_y}{(0.85f_c')b} = \frac{\rho df_y}{0.85f_c'}$$

Arbitrarily define ω (omega):

$$\omega = \rho\frac{f_y}{f_c'}$$

Then

$$a = \frac{\omega d}{0.85}$$

Substitute into the ϕM_n expression:

$$\phi M_n = \phi(0.85f_c')(b)\frac{\omega d}{0.85}\left[d - \frac{\omega d}{2(0.85)}\right]$$

Simplify and rearrange:

$$\phi M_n = \phi bd^2 f_c'\omega(1 - 0.59\omega)$$

Arbitrarily define $\bar{k}$:

$$\bar{k} = f_c'\omega(1 - 0.59\omega)$$

The term $\bar{k}$ is sometimes called the *coefficient of resistance*. It varies with ρ, f_c', and f_y. The general expression for ϕM_n can now be written as

$$\phi M_n = \phi bd^2\bar{k}$$

Before introducing the tables, we discuss another important quantity that depends on these same three factors and that is included in the tables.

In Example 2-1, we calculated the strain ϵ_t in the extreme tension steel at nominal strength (when the maximum concrete strain reaches 0.003). Assuming one layer of steel ($d = d_t$), it can be shown that ϵ_t is fixed if ρ, f_c', and f_y are known. Referring to Figure 2-8 and Example 2-1:

$$\epsilon_t = \frac{d - c}{c}(0.003)$$

Also,

$$a = \beta_1 c$$

From which

$$c = \frac{a}{\beta_1}$$

We previously developed

$$a = \frac{\rho df_y}{0.85f_c'}$$

By substitution,

$$c = \frac{\rho df_y}{0.85f_c'\beta_1}$$

Again by substitution,

$$\epsilon_t = \frac{d - c}{c}(0.003) = \frac{\left(d - \dfrac{\rho df_y}{0.85\,f_c'\,\beta_1}\right)}{\left(\dfrac{\rho df_y}{0.85\,f_c'\,\beta_1}\right)}(0.003)$$

From which

$$\epsilon_t = \frac{0.00255f_c'\beta_1}{\rho f_y} - 0.003 \qquad \textbf{(2-2)}$$

Note, again, that this is based on the assumption that $d = d_t$. For multiple layers of steel, use basic principles or Equation (2-1).

Tables A-7 through A-11 give the values of the coefficient of resistance $\bar{k}$ for values of ρ and various combinations of f'_c and f_y. The maximum tabulated value of ρ is that which corresponds to an ϵ_t value of 0.004, the minimum value allowed by code for nonprestressed flexural members with little or no axial load (ACI Code, Sections 7.3.3.1, 8.3.3.1, and 9.3.3.1). Values of ϵ_t are tabulated between 0.004 and 0.005 because in this transition region ϕ must be determined (ACI Code, Section 21.2.2). When $\epsilon_t > 0.005$, sections are tension-controlled and $\phi = 0.90$. Note that for the tables, it is assumed that $d = d_t$, as it would be for one layer of steel. If there are multiple layers of steel (see Figure 2-11), then $d_t > d$ and ϵ_t will be larger than the tabulated ϵ_t. Therefore, the tables are conservative with regard to the determination of ϕ based on ϵ_t.

Example 2-3

Determine if the beam shown in Figure 2-15 is adequate as governed by the ACI Code. The loads shown are service loads. The uniformly distributed load is DL = 0.65 kip/ft, LL = 0.80 kip/ft. The dead load excludes the beam weight. The point load is a live load; $f_c' = 4000$ psi, $f_y = 60{,}000$ psi.

Solution:

A logical approach to this type of problem is to compare the practical moment strength (ϕM_n) with the applied design moment resulting from the *factored* loads. This latter moment will be noted M_u. If the beam is adequate for the moment, $\phi M_n \geq M_u$. The procedure outlined here for ϕM_n is summarized in Section 2-11.

Determination of ϕM_n

1. Given:

$$f_y = 60{,}000 \text{ psi}$$
$$b = 12 \text{ in.}$$
$$d = 17.5 \text{ in.}$$
$$A_s = 3.16 \text{ in.}^2 \quad \text{(Table A-2)}$$
$$f_c' = 4000 \text{ psi}$$

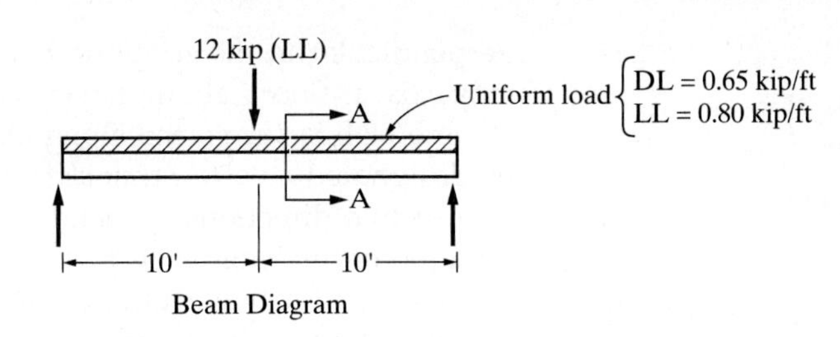

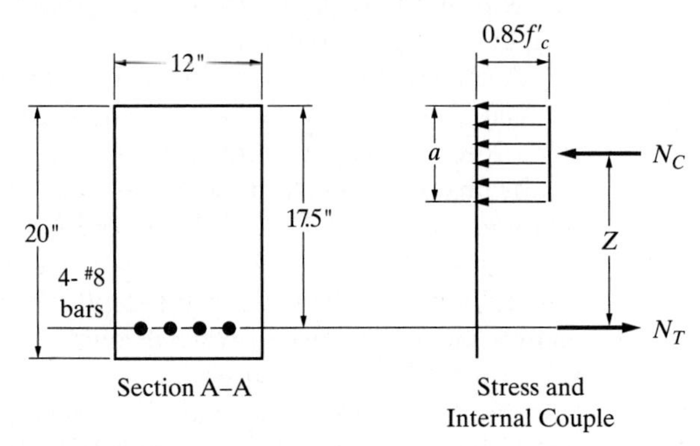

FIGURE 2-15 Sketch for Example 2-3.

2. To be found: ϕM_n and M_u.
3. $$\rho = \frac{A_s}{bd} = \frac{3.16}{12(17.5)} = 0.01505$$
4. From Table A-5:
$$A_{s,\min} = 0.0033\, b_w d$$
$$= 0.0033(12)(17.5) = 0.69 \text{ in.}^2 \qquad \text{(O.K.)}$$
$$3.16 \text{ in.}^2 > 0.69 \text{ in.}^2$$
5. From Table A-10, ϵ_t is not tabulated; therefore $\epsilon_t > 0.005$, the section is tension-controlled, and $\phi = 0.90$. Also from Table A-10, $\bar{k} = 0.7809$. (It is conservative to use the lower tabulated value. An interpolation between 0.7809 and 0.7853 could be done, but it is not warranted.)
6. Calculate ϕM_n from $\phi N_C Z$ or $\phi N_T Z$:
$$a = \frac{A_s f_y}{0.85 f'_c b} = \frac{3.16(60)}{0.85(4)(12)} = 4.65 \text{ in.}$$
$$Z = d - \frac{a}{2} = 17.5 - \frac{4.65}{2} = 15.18 \text{ in.}$$

Based on steel:
$$M_n = A_s f_y Z$$
$$= 3.16(60)(15.18) = 2880 \text{ in.-kip}$$
$$\frac{2880}{12} = 240 \text{ ft.-kip}$$
$$\phi M_n = 0.90(240) = 216 \text{ ft.-kip}$$

or

Calculate ϕM_n using the coefficient of resistance $\bar{k}$:
$$\phi M_n = \phi b d^2 \bar{k}$$
$$= \frac{0.90(12 \text{ in.})(17.5 \text{ in.})^2(0.7809 \text{ kip/in.}^2)}{12 \text{ in./ft}}$$
$$= 215 \text{ ft.-kip}$$

Find M_u: Because the given service dead load excluded the beam weight, it will now be calculated. As the beam weight is a uniformly distributed load, it will be found in terms of weight per linear foot (kip/ft):

$$\text{beam weight} = \text{beam volume per foot of length} \times 0.150 \text{ kip/ft}^3$$
$$= \frac{20 \text{ in.}(12 \text{ in.})}{144 \frac{\text{in.}^2}{\text{ft}^2}} \times 1 \text{ ft} \times 0.150 \text{ kip/ft}^3$$
$$= 0.250 \text{ kip (per linear foot)}$$
$$= 0.250 \text{ kip/ft}$$

Then summarizing the loads,

$$\text{superimposed service uniform dead load} = 0.65 \text{ kip/ft}$$
$$\text{total service uniform dead load} = 0.250 + 0.65$$
$$= 0.90 \text{ kip/ft} = w_{DL}$$
$$\text{total service uniform live load} = 0.80 \text{ kip/ft} = w_{LL}$$

Total factored uniform load:
$$w_u = 1.2 w_{DL} + 1.6 w_{LL}$$
$$= 1.2(0.90) + 1.6(0.80) = 2.36 \text{ kip/ft}$$

Superimposed concentrated live load = 12 kip = P_{LL}.
Total factored concentrated load:
$$P_u = 1.6 P_{LL} = 1.6(12) = 19.2 \text{ kip}$$
$$M_u = \frac{w_u \ell^2}{8} + \frac{P_u \ell}{4}$$
$$= \frac{2.36(20)^2}{8} + \frac{19.2(20)}{4}$$
$$= 214 \text{ ft.-kip} < 215 \text{ ft.-kip}$$

Therefore, the beam is adequate.

2-11 SUMMARY OF PROCEDURE FOR RECTANGULAR BEAM ANALYSIS FOR ϕM_n (TENSION REINFORCEMENT ONLY)

1. List the known quantities. Use a sketch.
2. Determine what is to be found. (An "analysis" may require any of the following to be found: ϕM_n, allowable service live load or dead load, maximum allowable span.)
3. Calculate the reinforcement ratio:

$$\rho = \frac{A_s}{bd}$$

4. Calculate $A_{s,\min}$ and compare with A_s (use Table A-5).
5. Determine ϵ_t by calculation or table ($\epsilon_t \geq 0.004$). Determine ϕ. Select $\overline{k}$ if it is to be used.
6. Calculate ϕM_n from $\phi N_C Z$ or $\phi N_T Z$, where

$$N_C = 0.85 f'_c ab$$

$$N_T = A_s f_y$$

$$a = \frac{A_s f_y}{0.85 f'_c b}$$

$$Z = d - \frac{a}{2}$$

or

calculate ϕM_n from $\phi M_n = \phi b d^2 \overline{k}$.

A flow diagram of this procedure is presented in Appendix B.

2-12 SLABS: INTRODUCTION

Slabs constitute a specialized category of bending members and are used in both structural steel and reinforced concrete structures. As discussed in Chapter 1, slabs can be classified as one-way or two-way slabs. Probably the most basic and common type of slab is the *one-way slab*. A one-way slab may be described as a structural reinforced concrete slab supported on two opposite sides so that the bending occurs in one direction only—that is, perpendicular to the supported edges. One-way slabs may exist as floor slabs and in concrete stairs.

If a slab is supported along all four edges, it may be designated as a *two-way slab*, with bending occurring in two directions perpendicular to each other. If the ratio of the lengths of the two perpendicular sides is in excess of 2, however, the slab may be assumed to act as a one-way slab with bending primarily occurring in the short direction.

A specific type of two-way slab is categorized as a *flat slab*. A flat slab may be defined as a concrete slab reinforced in two or more directions, generally without beams or girders to transfer the loads to supporting members. The slab, then, could be considered to be supported on a grid of shallow beams, which are themselves integral with and have the same depth as the slab. The columns tend to punch upward through the slab, resulting in a high shearing stress along with inclined slab cracking. (This "punching shear" is considered later in the discussion on column footings.) Thus, it is common to both thicken the slab in the vicinity of the column, utilizing a *drop panel*, and at the same time enlarge the top of the column in the shape of an inverted frustum called a *column capital* (see Figure 2-16). Another type of two-way slab is a *flat plate*. This is similar to the flat slab, without the drop panels and column capitals; hence, it is a slab of constant thickness supported directly on columns. Generally, the flat plate is used where spans are smaller and loads lighter than those requiring a flat slab design (see Figure 2-16).

2-13 ONE-WAY SLABS: ANALYSIS FOR MOMENT

In our discussion of slabs in this chapter, we will be primarily concerned with one-way slabs. Examples of such slabs are shown in Figures 2-17A and 2-17B. Such a slab is assumed to be a rectangular beam with a width $b = 12$ in. When loaded with a uniformly distributed load, the slab deflects so that it has curvature, and therefore bending moment, in only one direction. Hence, the slab is analyzed and designed as though it were composed of 12-in.-wide segments placed side by side with a total depth equal to the slab thickness. With a width of 12 in., the uniformly distributed load,

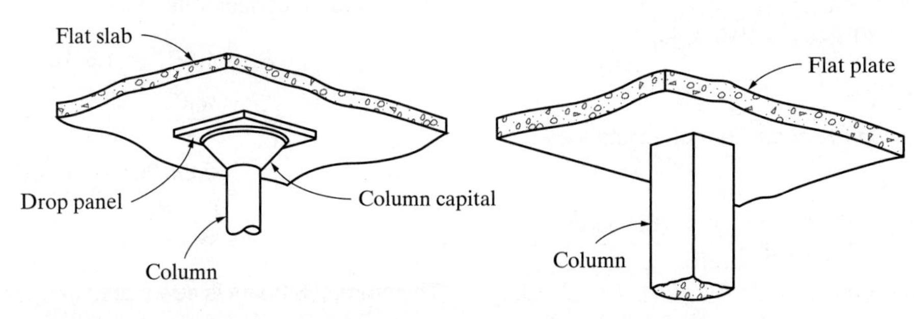

FIGURE 2-16 Reinforced concrete slabs.

generally specified in pounds per square foot (psf) for buildings, automatically becomes the load per linear foot (lb/ft) for the design of the slab (see Figure 2-17A).

In the one-way slab, the main reinforcement for bending is placed perpendicular to the supports. Because analysis and design will be done for a typical 12-in.-wide segment, it will be necessary to specify the amount of steel in that segment. Reinforcing steel in slabs is normally specified by bar size and center-to-center spacing, and the amount of steel considered to exist in the 12-in.-wide typical segment is an *average amount*. Table A-4 is provided to facilitate this determination. For example, if a slab is reinforced with No. 7 bars spaced 15 in. apart (center to center), a typical 12-in.-wide segment of the slab would contain an average steel area of 0.48 in.^{2}. This is denoted 0.48 in.^{2}/ft.

In addition, the ACI Code stipulates that reinforcement for shrinkage and temperature stresses perpendicular to the principal reinforcement must be provided in structural floor and roof slabs, where the principal reinforcement extends in one direction only. The area of temperature and shrinkage reinforcement is 0.0018bh, and the spacing should not exceed 5h or 18 in. (see ACI Code, Section 24.4.3.3). The ACI Code (Table 24.4.3.2) further states that for grade 40 or 50 deformed bars, the minimum area of the principal or main reinforcement in one-way slabs must be $A_s = 0.0020bh$, and for grade 60 deformed bars, the minimum area must be $A_s = 0.0018bh$, where b = width of member (12 in. for slabs) and h = total slab thickness. For reinforcing with yield strength, f_y, greater than 60,000 psi, the minimum reinforcement is $A_s = 0.0018bh(60{,}000/f_y)$, but not less than $A_s = 0.0014bh$, where f_y is in psi. The ACI Code also stipulates that in structural slabs of uniform thickness, the minimum amount of reinforcement in the direction of the span (principal reinforcement) must not be less than that required for shrinkage and temperature reinforcement (ACI Code, Section 7.6.1). Further, according to the ACI Code, Section 7.7.2.3, the principal reinforcement shall not be spaced farther apart than three times the slab thickness nor more than 18 in.

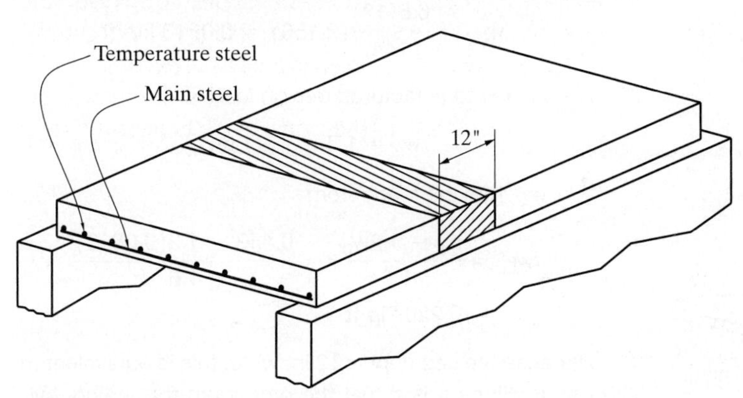

FIGURE 2-17A Floor or roof slab.
(Note: top reinforcement at slab supports not shown)

The required thickness of a one-way slab may depend on the bending, deflection, or shear strength requirements. The ACI Code imposes span/depth criteria in an effort to prevent excessive deflection, which might adversely affect the strength or performance of the structure at service loads. Table 7.3.1.1 (see also Chapter 6) of the ACI Code establishes minimum thicknesses for beams and one-way slabs in terms of fractions of the span length. According to the code, these may be used for members not supporting or attached to construction likely to be damaged by large deflections. Given that most any construction cannot tolerate large deflections, it is advisable to always check deflections and use the values obtained from Table 7.3.1.1 as an absolute minimum. For members of lesser thickness than that indicated in the table, deflection should be computed, and if satisfactory, the member may be used. The tabulated values are for use with non-prestressed reinforced concrete members made with normal-weight concrete and grade 60 reinforcement. If a different grade of reinforcement is used, the tabulated values must be multiplied by a factor equal to

$$0.4 + \frac{f_y}{100{,}000}$$

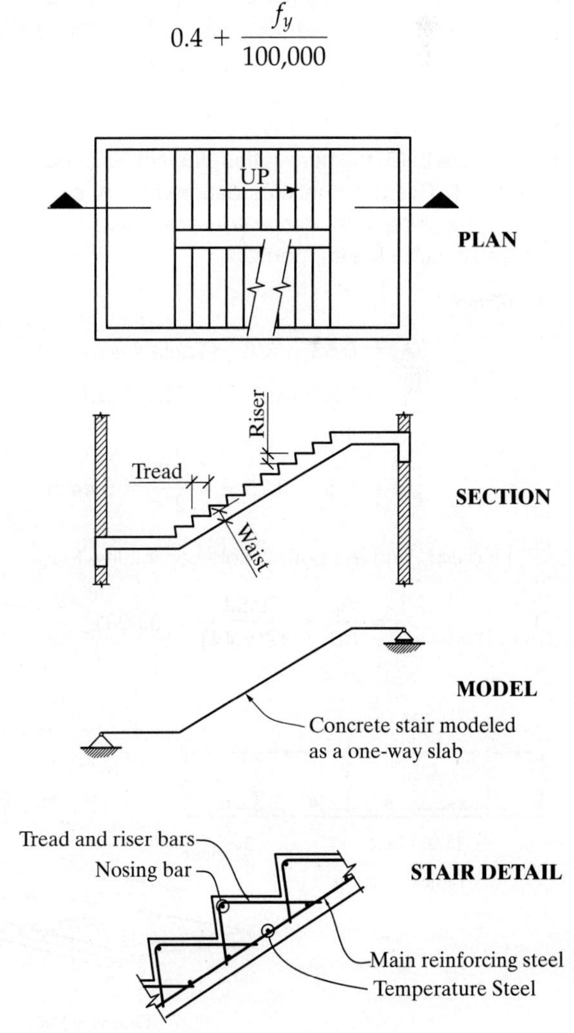

FIGURE 2-17B Reinforced concrete stair.

where f_y is in units of psi. As an example, for simply supported solid one-way slabs of normal-weight concrete and grade 60 steel, the minimum thickness required when deflections are not computed equals $\ell/20$, where ℓ is the span length of the slab. Deflections are discussed further in Chapter 7.

The ACI Code, Section 20.6.1.3.1, discusses *cover*, concrete protection for the reinforcement against weather and other effects, which is measured from the surface of the steel to the nearest surface of the concrete. The cover for reinforcement in slabs must be not less than 3/4 in. for surfaces not exposed directly to the weather or in contact with the ground. This is applicable for No. 11 and smaller bars. For surfaces exposed to the weather or in contact with the ground, the minimum cover for reinforcement is 2 in. for No. 6 through No. 18 bars and 1½ in. for No. 5 and smaller bars. If a slab is cast against and permanently in contact with the ground, the minimum concrete cover for all reinforcement is 3 in.

Example 2-4

A one-way structural interior slab having the cross section shown spans 12 ft. The steel is A615 grade 40. The concrete strength is 3000 psi, and the cover is 3/4 in. Determine the service live load (psf) that the slab can support.

Solution:

From Figure 2-18, the bars in this slab are No. 5 bars spaced 7 in. apart. This is sometimes denoted "7 in. o.c." (7 in. on center), meaning 7 in. center-to-center distance. They are perpendicular to the supports.

1. Given:

$$A_s = 0.53 \text{ in.}^2/\text{ft} \quad \text{(Table A-4)}$$

$$f'_c = 3000 \text{ psi}, \quad f_y = 40{,}000 \text{ psi}$$

$$b = 12 \text{ in.}$$

$$d = 6.5 - 0.75 - \frac{0.625}{2} = 5.44 \text{ in.}$$

2. Find ϕM_n and the permissible service live load.

3. $$\rho = \frac{A_s}{bd} = \frac{0.53}{12(5.44)} = 0.0081$$

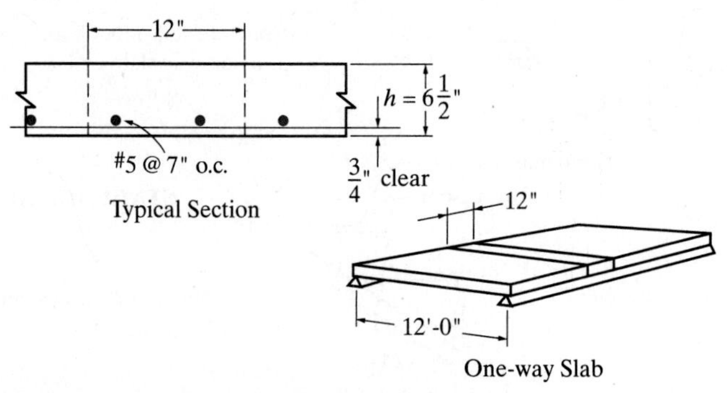

FIGURE 2-18 Sketch for Example 2-4.

4. The minimum flexural reinforcement for slabs is equal to what is required for shrinkage and temperature steel:

$$A_{s,\min} = 0.0020bh = 0.0020(12)(6.5) = 0.16 \text{ in.}^2/\text{ft}$$

$$0.53 \text{ in.}^2 > 0.16 \text{ in.}^2 \qquad \text{(O.K.)}$$

5. From Table A-7, for $\rho = 0.0081$, ϵ_t is not tabulated. Therefore, $\epsilon_t > 0.005$, this section is tension-controlled, and $\phi = 0.90$. Also from Table A-7, $\bar{k} = 0.3034$ ksi.

6. Calculate a, Z, and ϕM_n:

$$a = \frac{A_s f_y}{0.85 f'_c b} = \frac{0.53(40)}{0.85(3)(12)} = 0.693 \text{ in.}$$

$$Z = d - \frac{a}{2} = 5.44 - \frac{0.693}{2} = 5.09 \text{ in.}$$

$$\phi M_n = \phi A_s f_y Z = 0.9(0.53)(40)(5.09)$$
$$= 97.1 \text{ in.-kip (per foot of slab width)}$$

$$\phi M_n = \frac{97.1 \text{ in.-kip}}{12 \text{ in./ft}} = 8.09 \text{ ft.-kip}$$

or

from Table A-7, $\bar{k} = 0.3034$ ksi, from which

$$\phi M_n = \phi b d^2 \bar{k}$$
$$= \frac{0.90(12 \text{ in.})(5.44 \text{ in.})^2(0.3034 \text{ kip/in.}^2)}{12 \text{ in./ft}}$$
$$= 8.08 \text{ ft.-kip}$$

The service live load that the slab can support will be found next. (The notation M_u is used to denote moment resulting from *factored* applied loads.) As

$$M_u = \frac{w_u \ell^2}{8}$$

the total factored design load that can be supported by the slab is

$$w_u = \frac{8M_u}{\ell^2}$$

Because, as a limit, $M_u = \phi M_n$,

$$w_u = \frac{8\phi M_n}{\ell^2} = \frac{8(8.09)}{12^2} = 0.449 \text{ kip/ft}$$

The slab weight is

$$w_{DL} = \frac{6.5(12)}{144}(0.150) = 0.0813 \text{ kip/ft}$$

As the total factored design load is

$$w_u = 1.2w_{DL} + 1.6w_{LL}$$

w_{LL} may be found from

$$w_{LL} = \frac{w_u - 1.2w_{DL}}{1.6} = \frac{0.449 - 1.2(0.0813)}{1.6}$$
$$= 0.220 \text{ kip/ft}$$

Because the segment is 12 in. wide, this is equivalent to 220 psf. It will be noted that the procedure for finding ϕM_n for a one-way slab is almost identical to that for a beam. See Section 2-11 for a summary of this procedure.

2-14 RECTANGULAR BEAM DESIGN FOR MOMENT (TENSION REINFORCEMENT ONLY)

In the design of rectangular sections for moment, with f'_c and f_y usually prescribed, *three basic quantities* are to be determined: beam width, beam depth, and steel area. It should be recognized that there is a large multitude of combinations of these three quantities that will satisfy the moment strength required in a particular application. Theoretically, a wide, shallow beam may have the same ϕM_n as a narrow, deep beam. It must also be recognized that practical considerations and code restraints will affect the final choices of these quantities. There is no easy way to determine the *best* cross section, because economy depends on much more than simply the volume of concrete and amount of steel in a beam.

We have previously developed the analysis expression for the resisting moment of a rectangular beam with tension reinforcement only:

$$\phi M_n = \phi N_C Z = \phi N_T Z$$

We subsequently modified the equation for ϕM_n for the use of tables (Tables A-7 through A-11):

$$\phi M_n = \phi b d^2 \overline{k}$$

This equation will now be used for the design of rectangular reinforced concrete sections. The first example is one where the cross-section width b and overall depth h are known by either practical or architectural considerations, leaving the selection of the reinforcing bars as the only unknown.

Material properties, sizes, and availability of reinforcing steel in the form of bars and welded wire fabric were discussed in Section 1-7. In our discussion we will be concerned with bars only and the ACI Code recommendations as to details governing minimum clearance and cover requirements for steel reinforcing bars. Clearance details are governed by the requirement for concrete to pass through a layer of bars without undue segregation of the aggregates. Cover details are governed by the necessity that the concrete must provide protection against corrosion for the bars. Required minimums of both spacing and cover also play a role in preventing the splitting of the concrete in the proximity of highly stressed tension bars.

Spacing requirements in the ACI Code indicate that the clear space between bars in a single layer shall be not less than the following:

1. The bar diameter, but not less than 1 in. (ACI Code, Section 25.2.1).
2. $\frac{4}{3} \times$ maximum aggregate size (ACI Code, Sections 25.2.1 and 26.4.2.1).

Also, should multiple layers of bars be necessary, a 1-in. minimum clear distance is required between layers (ACI Code, Section 25.2.2) and bars in the upper layers shall be placed directly above bars in the bottom layer. When multiple layers of steel are required, short transverse *spacer bars* may be used to separate the layers and support the upper layers. A No. 8 spacer bar will provide the minimum 1-in. separation between layers. This detail is illustrated in Figure 3-6. In general, in this text, clear distance between layers of steel is noted with a dimension.

Cover requirements for cast-in-place concrete are stated in the ACI Code, Section 20.6.1.3.1. This listing is extensive. For beams, girders, and columns not exposed to the weather or in contact with the ground, however, the minimum concrete cover on any steel is $1\frac{1}{2}$ in. Cover requirements for slabs were discussed in Section 2-13 of this text.

Table A-3 combines spacing and cover requirements into a tabulation of minimum beam widths for multiples of various bars. The assumptions are stated. It should be noted that No. 3 stirrups are assumed. Stirrups are a special form of reinforcement that primarily resist shear forces and will be discussed in Chapter 4. Stirrups are common in rectangular beams and will be assumed to exist in all further rectangular beam examples and problems in this text. One type of stirrup, called a *loop stirrup*, can be observed in Figure 2-19, in Example 2-5.

Example 2-5

Design a rectangular reinforced concrete beam to carry a service dead load moment of 100 ft.-kip (which includes the moment due to the weight of the beam) and a service live load moment of 75 ft.-kip. Architectural considerations require the beam width to be 10 in. and the total depth (h) to be 25 in. Use $f'_c = 3000$ psi and $f_y = 60{,}000$ psi.

Solution:

Of the three basic quantities to be found, two are specified in this example, and the solution for the required steel

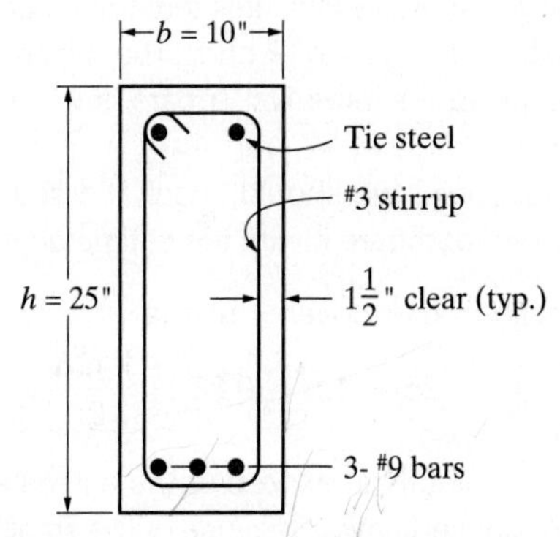

FIGURE 2-19 Design sketch for Example 2-5.

area is direct. The procedure outlined here is summarized in Section 2-15.

1. The total design moment is

$$M_u = 1.2M_{DL} + 1.6M_{LL}$$
$$= 1.2(100) + 1.6(75)$$
$$= 240 \text{ ft.-kip}$$

2. Estimate d to be equal to h – 3 in. This is conservative for a single layer of bars. The effective depth d in the resulting section will be a bit larger.

$$d = 25 - 3 = 22 \text{ in.}$$

Because $\phi M_n = \phi bd^2\bar{k}$ and because ϕM_n must equal M_u as a lower limit, the expression may be written as $M_u = \phi bd^2\bar{k}$ and, assuming $\phi = 0.90$, subject to later check,

$$\text{required } \bar{k} = \frac{M_u}{\phi bd^2} = \frac{240(12)}{0.9(10)(22)^2} = 0.6612 \text{ ksi}$$

3. From Table A-8, $\bar{k}$ of 0.6649 ksi will be provided if the steel ratio $\rho = 0.0131$. Therefore, required $\rho = 0.0131$. Also from Table A-8, because ϵ_t is not tabulated, $\epsilon_t >$ 0.005, this is a tension-controlled section, and $\phi = 0.90$.

4.
$$\text{required } A_s = \rho bd$$
$$= 0.0131(10)(22) = 2.88 \text{ in.}^2$$

Check $A_{s,min}$. From Table A-5,

$$A_{s,min} = 0.0033b_wd$$
$$= 0.0033(10)(22) = 0.73 \text{ in.}^2$$

5. Select the bars. Theoretically, any bar or combination of bars that provides at least 2.88 in.^{2} of steel area will satisfy the design requirements. Preferably, no fewer than two bars should be used. The bars should be of the same diameter and placed in one layer whenever possible. We next consider the bar selection based on the foregoing required A_s (2.88 in.^{2}). The following combinations may be considered:

2 No. 11 bars: $A_s = 3.12$ in.^{2}

3 No. 9 bars: $A_s = 3.00$ in.^{2}

4 No. 8 bars: $A_s = 3.16$ in.^{2}

5 No. 7 bars: $A_s = 3.00$ in.^{2}

A review of Table A-3 indicates that the most acceptable combination is three No. 9 bars. The minimum width of beam required for three No. 9 bars is $9^1/_2$ in., which is satisfactory.

At this point we should check the actual effective depth d and compare it with the estimated d:

$$\text{actual } d = h - \text{cover} - \text{stirrup} - d_b/2$$
$$= 25 - 1.5 - 0.38 - \frac{1.128}{2} = 22.6 \text{ in.}$$

This is slightly in excess of the estimated d and is therefore conservative. Because of the small difference, no revision is either suggested or required.

6. Generally, concrete dimensions should be to an increment no smaller than $\frac{1}{2}$ in. In this example, whole inch dimensions are given. The final design sketch should show the following (see Figure 2-19):
   a. Beam width
   b. Total beam depth
   c. Main reinforcement size and number of bars
   d. Cover on reinforcement
   e. Stirrup size

As a final note for Example 2-5, the reader has probably recognized that there is a direct solution that avoids the use of Tables A-7 through A-11. A quadratic equation results when the design expression is written as

$$M_u = \phi M_n = \phi A_s f_y\left(d - \frac{a}{2}\right)$$

where
$$a = \frac{A_s f_y}{0.85f'_c b}$$

With known quantities of M_u, f'_c, f_y, b, and d, the quadratic equation can be solved for the required steel area A_s. Generally, the use of the tables results in a much faster solution when performing hand calculations.

Rearranging the equation above yields,

$M_u = \phi A_s f_y\left(d - \dfrac{A_s f_y}{1.7 f'_c b}\right)$, from which we obtain the following quadratic equation:

$$A_s^2 - \left(\frac{1.7f'_c bd}{f_y}\right)A_s + \left(\frac{1.7f'_c b}{\phi f_y^2}\right)M_u = 0$$

The quadratic equation can be represented as $ax^2 + bx + c = 0$, for which the two solutions are:

$$x = \frac{-b \pm \sqrt{b^2 - 4ac}}{2a}$$

where

$$x = A_s \text{ (in.}^2\text{)}$$
$$a = 1$$
$$b = -\left(\frac{1.7f'_c bd}{f_y}\right) \text{ (in.-kip)}$$
$$c = \left(\frac{1.7f'_c b}{\phi f_y^2}\right)M_u \text{ (in.-kip)}$$

f'_c and f_y are in ksi and M_u is in units of in.-kip.

Where there are two positive solutions to the quadratic equation, the smaller of the two solutions is the correct area of steel. In the case where one solution is positive and the second solution is negative, the positive solution is the correct area of steel.

A rule of thumb or approximate equation commonly used in design practice to determine the approximate area of flexural reinforcement is [2]

$$A_s = \frac{M_u}{4d}$$

where

M_u = the factored moment in ft.-kip
d = the effective depth in inches
A_s = reinforcement area, in.^{2}

This equation is derived from the equation

$$M_u = \phi M_n = \phi A_s f_y\left(d - \frac{a}{2}\right)$$

by substituting f_y of 60 ksi, assuming tension-controlled section or $\phi = 0.9$, and assuming the internal moment arm

$$\left(d - \frac{a}{2}\right) \approx 0.9d$$

Both the quadratic equation and the rule of thumb methods can be used to determine the required area of flexural reinforcement for rectangular beams, T-beams, L-beams, and one-way slabs. Both methods will now be used to determine the required area of steel for the rectangular beam in Example 2-5.

Quadratic Equation Method:

$a = 1$

$$b = -\left(\frac{1.7f'_c bd}{f_y^2}\right) = -\frac{1.7(3\text{ ksi})(10\text{ in.})(22\text{ in.})}{60\text{ ksi}}$$

$$= -18.7\text{ in.-kip}$$

$$c = \left(\frac{1.7f'_c b}{\phi f_y^2}\right)M_u = \frac{1.7(3\text{ ksi})(10\text{ in.})}{0.9(60\text{ ksi})^2}(240\text{ ft.-kip})(12\text{ in./ft})$$

$$= 45.33\text{ in.-kip}$$

Therefore, $A_s = 2.86$ in.^{2} (the smaller of the two positive solutions). This required area of flexural steel compares well with 2.88 in.^{2} obtained using the Tables.

Rule of Thumb Method:

$$A_s(\text{in.}^2) = \frac{M_u(\text{ft.-kip})}{4d(\text{in.})} = \frac{240(\text{ft.-kip})}{4(22\text{ in.})} = 2.73\text{ in.}^2$$

This required area of flexural steel compares well with 2.88 in.^{2} obtained using the Tables.

A second type of design problem is presented in Example 2-6. This may be categorized as a *free design* because of the *three* unknown variables: beam width, beam depth, and area of reinforcing steel. There are, therefore, a large number of combinations of these variables that will theoretically solve the problem. We do have some idea as to the required or desired relationships between these unknowns, however.

Whenever possible, flexural members should be proportioned so they are tension-controlled sections ($\epsilon_t \geq 0.005$.) This allows the strength-reduction factor ϕ to be taken at its maximum value of 0.90. The reinforcement ratio ρ corresponding to $\epsilon_t = 0.005$ may be determined from Tables A-7 through A-11. Additionally, we know that the steel area must not be less than $A_{s,min}$ (from Table A-5). This in effect establishes the range of the acceptable amount of steel. Table A-5 contains recommended values of ρ and associated values of $\bar{k}$ to use for design purposes. These are recommended maximum values. Try not to use larger ρ values. If larger values are used, a smaller concrete section will result, with potential deflection problems.

These ρ values are based on a previous ACI Code, which stipulated that

$$\text{where } \rho > \frac{0.18f_c}{f_y}, \text{ deflection must be checked}$$

$$\text{where } \rho < \frac{0.18f_c}{f_y}, \text{ deflection need not be checked}$$

This stipulation was deleted in more current ACI Codes. Nevertheless, it remains a valid guide for selecting a preliminary value for the reinforcement ratio. The assumption of a value for reinforcement ratio will reduce from three to two the number of unknown quantities to be determined (yet to be found are b and d).

Experience and judgment developed over the years have also established a range of acceptable and economical depth/width ratios for rectangular beams. Although there is no code requirement for the d/b ratio to be within a given range, rectangular beams commonly have d/b ratios between 1 and 3. *Desirable d/b ratios lie between 1.5 and 2.2.* There are situations in which d/b ratios outside this range have applications, however; therefore, it is the designer's choice.

Example 2-6

Design a simply supported rectangular reinforced concrete beam with tension steel only to carry a service dead load of 1.35 kip/ft and a service live load of 1.90 kip/ft. (The dead load does not include the weight of the beam.) The span is 18 ft. Assume No. 3 stirrups. Use $f'_c = 4000$ psi and $f_y = 60{,}000$ psi. See Figure 2-20.

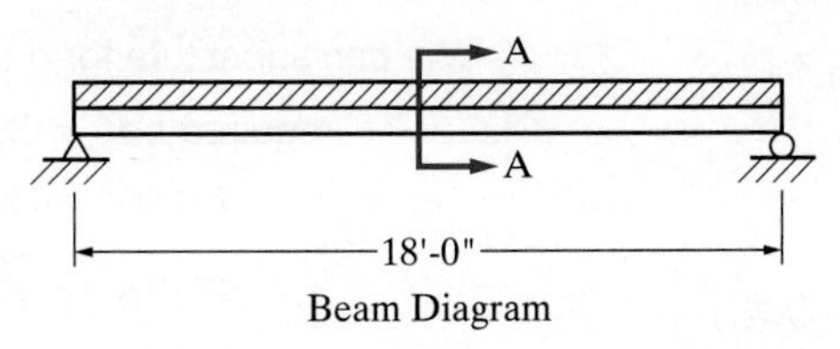

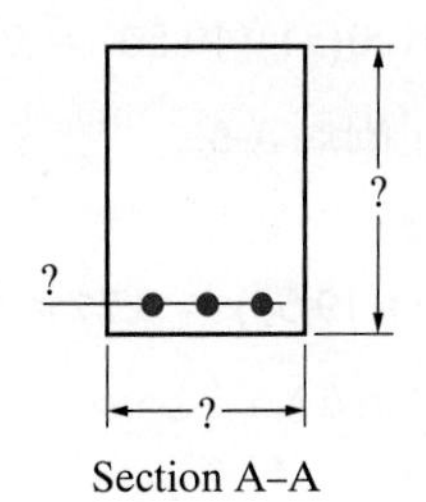

FIGURE 2-20 Sketch for Example 2-6.

Solution:

1. Find the applied design moment M_u, temporarily neglecting the beam weight, which will be included at a later step:

$$w_u = 1.2w_{DL} + 1.6w_{LL}$$
$$= 1.2(1.35) + 1.6(1.90) = 4.66 \text{ kip/ft}$$

Note that this is the factored design load,

$$M_u = \frac{w_u \ell^2}{8} = \frac{4.66(18)^2}{8} = 189 \text{ ft.-kip}$$

2. Assume a value for ρ. Use $\rho = 0.0120$ (see Table A-5).
3. From Table A-5, the associated $\bar{k}$ value is 0.6438 ksi (alternatively, Table A-10 can be used).
4. At this point we have two unknowns, b and d, which can be established using two different approaches. One approach is to assume b and then solve for d. This is logical as practical and architectural considerations often establish b within narrow limits. Assume $\phi = 0.90$, subject to later check.

   Assuming that $b = 11$ in. and utilizing the relationship $M_u = \phi b d^2 \bar{k}$,

$$\text{required } d = \sqrt{\frac{M_u}{\phi b \bar{k}}} = \sqrt{\frac{189(12)}{0.9(11)(0.6438)}} = 18.9 \text{ in.}$$

   A check of the d/b ratio gives $18.9/11 = 1.72$, which is a reasonable value.
5. At this point, the beam weight may be estimated. Realizing that the total design moment will increase, the final beam size may be estimated to be about 11 in. × 23 in. for purposes of calculating its weight. Thus

$$\text{beam dead load} = \frac{11(23)}{144}(0.150) = 0.264 \text{ kip/ft}$$

6. The additional M_u due to beam weight is

$$M_u = 1.2\left(\frac{0.264(18)^2}{8}\right) = 12.8 \text{ ft.-kip}$$
$$\text{total } M_u = 189 + 12.8 = 202 \text{ ft.-kip}$$

7. Using the same ρ, $\bar{k}$, and b as previously, compute the new required d:

$$\text{required } d = \sqrt{\frac{M_u}{\phi b \bar{k}}} = \sqrt{\frac{202(12)}{0.9(11)(0.6438)}} = 19.5 \text{ in.}$$

   A check of the d/b ratio gives $19.50/11 = 1.77$, which is reasonable.
8. $$\text{required } A_s = \rho b d$$
$$= 0.0120(11)(19.50) = 2.57 \text{ in.}^2$$

   Check $A_{s,min}$. From Table A-5,

$$A_{s,min} = 0.0033 b_w d$$
$$= 0.0033(11)(19.50) = 0.71 \text{ in.}^2 \quad \text{(O.K.)}$$

9. From Table A-2, select three No. 9 bars. Therefore, $A_s = 3.00$ in.^{2}. From Table A-3,

$$\text{minimum required } b = 9.5 \text{ in.} \quad \text{(O.K.)}$$

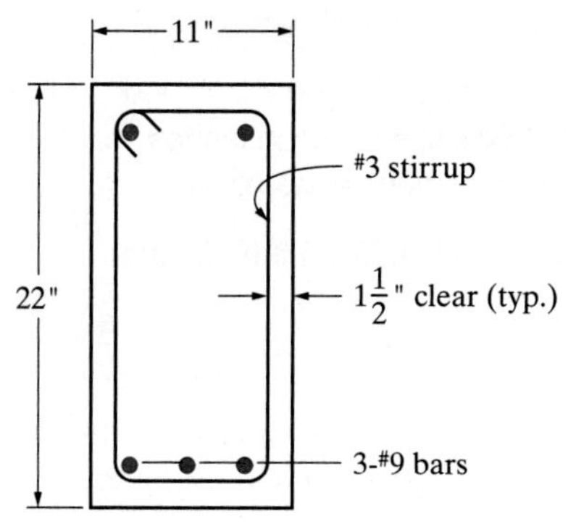

FIGURE 2-21 Design sketch for Example 2-6.

10. Determine the total beam depth h:

$$\text{required } h = 19.50 + \frac{1.13}{2} + 0.38 + 1.5 = 21.95 \text{ in.}$$

   Rounding up by a $\frac{1}{2}$ in. increment, we will use $h = 22.0$ in. The actual effective depth d may now be checked:

$$d = 22.0 - 1.5 - 0.38 - \frac{1.13}{2} = 19.55 \text{ in.} > 19.50 \text{ in.} \quad \text{(O.K.)}$$

11. A design sketch is shown in Figure 2-21. This design is not necessarily the best. The final ρ is

$$\frac{3.0}{11(19.55)} = 0.01395$$

   Check ϵ_t: from Table A-10, $\epsilon_t > 0.005$. Therefore, the assumed ϕ is O.K. The final d/b ratio is

$$\frac{19.55}{11} = 1.78 \quad \text{(O.K.)}$$

   The design moment was based on a section 11 in. × 23 in., where the designed beam turned out to be 11 in. × 22 in. Small modifications in proportions could very well make more efficient use of the steel and concrete provided.

   The percentage of overdesign could be determined (not necessary, however) by analyzing the designed cross section and comparing moment strength with applied moment.

   There are several alternative approaches to this problem. At step 4, we could establish a desired d/b ratio and then mathematically solve for b and d. For example, if we establish a desirable d/b ratio of 2.0, then $d = 2b$. Again, using the relationship $M_u = \phi b d^2 \bar{k}$,

$$\text{required } bd^2 = \frac{M_u}{\phi \bar{k}} = \frac{189(12)}{0.9(0.6438)} = 3914 \text{ in.}^3$$

   We can substitute for d as follows:

$$\text{required } bd^2 = 3914 \text{ in.}^3$$
$$b(2b)^2 = 3914 \text{ in.}^3$$
$$b^3 = \frac{3914}{4} = 979 \text{ in.}^3$$
$$\text{required } b = \sqrt[3]{979} = 9.93 \text{ in.}$$

Hence, assuming that $b = 10$ in., the required effective depth d can then be calculated in a manner similar to that previously used. Also, at step 7, with a width b and a depth h established, along with the related design moment M_u, we could revert to the procedure for design of a rectangular section where the area is known (see Example 2-5, step 2).

2-15 SUMMARY OF PROCEDURE FOR RECTANGULAR REINFORCED CONCRETE BEAM DESIGN FOR MOMENT (TENSION REINFORCEMENT ONLY)

A. Cross Section (b and h) Known; Find the Required A_s

1. Convert the service loads or moments to design M_u (include the beam weight which must first be assumed, and then checked later, since the beam size is not yet determined).
2. Based on knowing h, estimate d by using the relationship $d = h - 3$ in. (conservative for bars in a single layer). Calculate the required $\bar{k}$ using an assumed ϕ value of 0.90, subject to later check.

$$\text{required } \bar{k} = \frac{M_u}{\phi b d^2}$$

3. From Tables A-7 through A-11, find the required steel ratio ρ and ensure that $\epsilon_t \geq 0.005$. If ϵ_t is within the range $0.004 \leq \epsilon_t \leq 0.005$, then ϕ will have to be reduced.
4. Compute the required A_s:

$$\text{required } A_s = \rho b d$$

Check $A_{s,\text{min}}$. Use Table A-5.

5. Select the bars. Check to see if the bars can fit into the beam in one layer (preferable). Check the actual effective depth and compare with the assumed effective depth. If the actual effective depth is slightly in excess of the assumed effective depth, the design will be slightly conservative (on the safe side). If the actual effective depth is less than the assumed effective depth, the design is on the unconservative side and should be revised.
6. Sketch the design. (See Example 2-5, step 6, for a discussion.)

B. Design for Cross Section and Required A_s (i.e. b, h, and A_s are unknown)

1. Convert the service loads or moments to design moment M_u. An estimated beam weight may be included in the dead load if desired. Be sure to apply the load factor to this additional dead load.
2. Select a desired steel ratio ρ. (See Table A-5 for recommended values. Use the ρ values from Table A-5 unless a small cross section or decreased steel is desired.)
3. From Table A-5 (or from Tables A-7 through A-11), find $\bar{k}$.
4. Assume b and compute the d required:

$$\text{required } d = \sqrt{\frac{M_u}{\phi b \bar{k}}}$$

If the d/b ratio is reasonable (1.5 to 2.2), use these values for the beam. If the d/b ratio is not reasonable, increase or decrease b and compute the new required d.

5. Estimate h and compute the beam weight. Compare this with the estimated beam weight if an estimated beam weight was included.
6. Revise design M_u to include the moment due to the beam's own weight using the latest weight determined. Note that at this point, one could go to step 2 in design procedure A, where the cross section is known.
7. Using b and $\bar{k}$ previously determined along with the new total design M_u, find the new required d:

$$\text{required } d = \sqrt{\frac{M_u}{\phi b \bar{k}}}$$

Check to see if the d/b ratio is reasonable.

8. Find the required A_s:

$$\text{required } A_s = \rho b d$$

Check $A_{s,\text{min}}$. Use Table A-5.

9. Select the bars and check to see if the bars can fit into a beam of width b in one layer (preferable).
10. Establish the final h, rounding this upward to the next $\frac{1}{2}$ in. This will make the actual effective depth greater than the design effective depth, and the design will be slightly conservative (on the safe side).
11. Check ϵ_t. Check the ϕ assumption. Sketch the design. (See Example 2-5, step 6, for a discussion.)

A flow diagram of this procedure is presented in Appendix B.

2-16 DESIGN OF ONE-WAY SLABS FOR MOMENT (TENSION REINFORCEMENT ONLY)

As higher-strength steel and concrete have become available for use in reinforced concrete members, the sizes of the members have decreased. *Deflections* of members are affected very little by material strength but are affected greatly by the size of a cross section and its related moment of inertia. Therefore, deflections will be larger for a member of high-strength materials than would be the deflections for the same member fabricated from lower-strength materials, because the latter member will be essentially larger in cross-sectional area. Deflections are discussed in detail in Chapter 7. As discussed previously, one method that the ACI Code allows for limiting adverse deflections is by the use of a minimum slab thickness (see Section 2-13). A slab that meets the minimum thickness requirement from ACI 318-14, Table 7.3.1.1. must still be designed for flexure. Deflections need not be calculated or checked unless the slab supports or is attached to non-structural elements likely to be damaged by large deflections.

Example 2-7 illustrates the use of the ACI minimum thickness for one-way slabs. With respect to the span length to be used for design, a rational approach found in previous versions of ACI 318 [9] recommends the following for beams and slabs not integral with supports

$$\text{span length} = \text{clear span} + \text{depth of member}$$

but not to exceed the distance between centers of supports. For design purposes, we will use the distance between centers of supports, as the slab thickness is not yet determined.

Example 2-7

Design a simple-span one-way slab to carry a uniformly distributed live load of 400 psf. The span is 10 ft (center-to-center of supports). Use $f'_c = 4000$ psi and $f_y = 60{,}000$ psi. Select the thickness to be not less than the ACI minimum thickness requirement.

Solution:

Determine the required minimum h and use this to estimate the slab dead weight.

1. From ACI Table 7.3.1.1, for a simply supported, solid, one-way slab,

$$\text{minimum } h = \frac{\ell}{20} = \frac{10(12)}{20} = 6.0 \text{ in.}$$

   Try $h = 6$ in. and design a 12-in.-wide segment.
2. Determine the slab weight dead load:

$$\frac{6(12)}{144}(0.150) = 0.075 \text{ kip/ft}$$

   The total design load is

$$\begin{aligned} w_u &= 1.2w_{DL} + 1.6w_{LL} \\ &= 1.2(0.075) + 1.6(0.400) \\ &= 0.730 \text{ kip/ft} \end{aligned}$$

3. Determine the design moment:

$$M_u = \frac{w_u\ell^2}{8} = \frac{0.730(10)^2}{8} = 9.13 \text{ ft.-kip}$$

4. Establish the approximate d. Assuming No. 6 bars and minimum concrete cover on the bars of $^3/_4$ in.,

$$\text{assumed } d = 6.0 - 0.75 - 0.375 = 4.88 \text{ in.}$$

5. Determine the required $\bar{k}$ assuming $\phi = 0.90$:

$$\begin{aligned} \text{required } \bar{k} &= \frac{M_u}{\phi b d^2} \\ &= \frac{9.13(12)}{0.90(12)(4.88)^2} = 0.4260 \text{ ksi} \end{aligned}$$

6. From Table A-10, for a required $\bar{k} = 0.4260$, the required $\rho = 0.0077$. (Note that the required ρ selected is the next *higher* value from Table A-10.) Also note that ϵ_t is not tabulated. Therefore, $\epsilon_t > 0.005$, this is a tension-controlled section, and $\phi = 0.90$.
   Use $\rho = 0.0077$.
7. required $A_s = \rho b d = 0.0077(12)(4.88) = 0.45$ in.^{2}/ft
8. Select the main steel (from Table A-4). Select No. 5 bars at 8 in. o.c. ($A_s = 0.46$ in.^{2}). The assumption on bar size was satisfactory (actual $d >$ assumed d). The code requirements for maximum spacing have been discussed in Section 2-13. Minimum spacing of bars in slabs, practically, should not be less than 4 in., although the ACI Code allows bars to be placed closer together, as discussed in Example 2-5. Check the maximum spacing (ACI Code, Section 7.7.2.3):

$$\begin{aligned} \text{maximum spacing} &= 3h \text{ or } 18 \text{ in.} \\ 3h &= 3(6) = 18 \text{ in.} \quad \text{(O.K.)} \\ 8 \text{ in.} &< 18 \text{ in.} \end{aligned}$$

   Therefore, use No. 5 bars at 8 in. o.c.
9. Select shrinkage and temperature reinforcement (ACI Code, Table 24.4.3.2):

$$\begin{aligned} \text{required } A_s &= \textit{the greater of } 0.0018bh \text{ and } 0.0014bh \\ &= 0.0018(12)(6) = 0.13 \text{ in.}^2\text{/ft} \end{aligned}$$

   Select No. 3 bars at 10 in. o.c. ($A_s = 0.13$ in.^{2}) or No. 4 bars at 18 in. o.c. ($A_s = 0.13$ in.^{2}):

$$\text{maximum spacing} = 5h \text{ or } 18 \text{ in.}$$

   Use No. 3 bars at 10 in. o.c.
10. The main steel area must exceed the area required for shrinkage and temperature steel (ACI Code, Section 7.6.1.1):

$$0.46 \text{ in.}^2 > 0.13 \text{ in.}^2 \quad \text{(O.K.)}$$

11. A design sketch is shown in Figure 2-22.

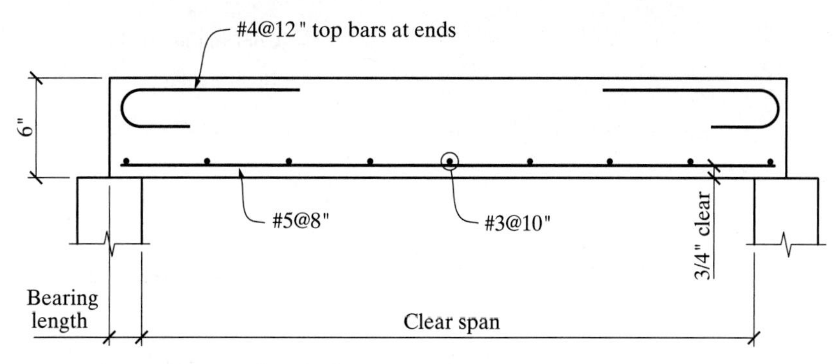

FIGURE 2-22 Design sketch for Example 2-7.

2-17 SUMMARY OF PROCEDURE FOR DESIGN OF ONE-WAY SLABS FOR MOMENT (TO SATISFY ACI MINIMUM THICKNESS, *h*)

1. Compute the minimum h based on the ACI Code, Table 7.3.1.1. For practical purposes, the slab thickness h can be rounded to the next higher $\frac{1}{4}$ in. for slabs up to 6 in. thickness and to the next higher $\frac{1}{2}$ in. for slabs thicker than 6 in.
2. Compute the slab weight and compute w_u (total design load).
3. Compute the design moment M_u.
4. Calculate an assumed effective depth d (assuming No. 6 bars and $\frac{3}{4}$ in. cover) by using the relationship
$$d = h - 1.12 \text{ in.}$$
5. Calculate the required $\bar{k}$ assuming $\phi = 0.90$:
$$\text{required } \bar{k} = \frac{M_u}{\phi b d^2}$$
6. From Tables A-7 through A-11, find the required steel ratio ρ. Check ϵ_t to verify the ϕ assumption. If $\epsilon_t < 0.005$, the slab must be made thicker.
7. Compute the required A_s:
$$\text{required } A_s = \rho b d$$
8. Select the main steel (Table A-4). Check with maximum spacing of $3h$ or 18 in. Check the assumption of step 4.
9. Select shrinkage and temperature steel as per the ACI Code:
$$\text{required } A_s = 0.0020bh \text{ (grade 40 and 50 steel)}$$
$$\text{required } A_s = 0.0018bh \text{ (grade 60 steel)}$$
Check with maximum spacing of $5h$ or 18 in.
10. The main steel area *cannot* be less than the area of steel required for shrinkage and temperature.
11. Sketch the design.

2-18 SLABS-ON-GRADE

The previous discussion primarily focused on framed structural slab. Another category of slabs, generally used as a floor, are called *slab-on-grade*. As the name implies, it is a slab that is supported throughout its entire area by some form of subgrade.

Concrete placed directly on soil is used as a floor surface in almost every structure and in other areas outside of a building such as sidewalks, equipment pads, and roads. The design of slabs-on-grade is significantly different from the design of a framed structural slab; there are numerous factors to consider in the design of slabs-on-grade, but we will cover the most basic and most common aspects of their design here.

The subgrade is the natural ground, graded and compacted, on which the floor slab is built. The pressure on the subgrade is generally low due to the rigidity of the concrete floor slab. The floors do not necessarily require strong support from the subgrade. It is important that the subgrade support be reasonably uniform without abrupt changes, horizontally, from hard to soft, however. The upper portion of the subgrade should be of uniform material and density.

The subbase is usually a thinner layer of material placed on top of the prepared subgrade. It is generally used when a uniform subgrade cannot be developed by grading and compaction. It will serve to equalize minor surface defects as well as provide a capillary break and a working platform for construction activities. A 4-in. thickness, of the sub-base, usually of compacted granular material, is sometimes used.

A typical slab-on-grade detail is shown in Figure 2-23. Each of the components shown can play a role in the performance and load-carrying capacity

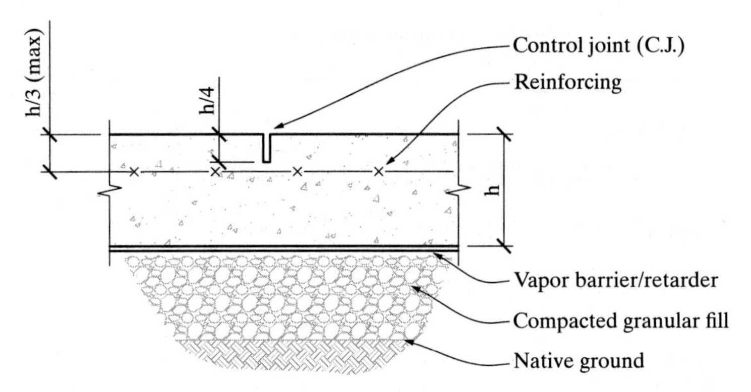

FIGURE 2-23 Slab-on-grade detail.

of the slab-on-grade. In a typical project, the existing native soil at the ground level is removed down to a suitable bearing surface. The slab-on-grade could be placed directly on native soil, but the amount removed during the construction process is typically greater than the proposed concrete slab thickness. It is also usually desirable to have a free-draining and well-compacted base directly under the slab-on-grade, and once the existing soil is removed down to the suitable bearing layer of native soil, a layer of compacted granular fill is placed next. This layer can vary from a few inches up to perhaps 24 inches.

The next item typically used in a slab-on-grade is a vapor barrier or vapor retarder placed directly beneath the concrete slab. ACI 302 identifies three possible cases for a slab-on-grade construction with regard to the use of the vapor barrier as indicated in Table 2.1. A vapor barrier would often be used where moisture migration through the slab would be detrimental to floor finishes or to other items such as products that are sensitive to water infiltration. The IBC (Ref. [11]) states that a vapor retarder is required as follows: "*For buildings of other occupancies where migration of moisture through the slab from below will not be detrimental to the intended occupancy of the building.*" It is often very difficult to establish the exact occupancy and conditions that may be present in a structure throughout its life, so the use of a vapor barrier is typically recommended as good construction practice. Regarding the location of the vapor barrier, it is required to be directly under the slab when moisture-sensitive finishes are used. Furthermore, the only condition where a layer of dry granular material (i.e., sand) can be placed on the vapor barrier is when a watertight roof is in place. This would be required to prevent moisture from accumulating in this layer of sand which could be detrimental to the slab curing and the occupancy. In the experience of the author, it is therefore best to place the vapor barrier directly below the slab, which allows for greater latitude in the construction sequence and for all possible occupancy conditions. The location of the vapor barrier needs to be considered in the layout of control joints, and that will be discussed later. Table 2.1 is provided to summarize the possible locations for a vapor barrier.

TABLE 2-1 Vapor Barrier Location

Vapor barrier location	Occupancy	Construction conditions
Vapor barrier not used	Moisture infiltration through the slab is not detrimental to the occupancy	N/A
Vapor barrier directly below the concrete slab	Moisture infiltration through the slab is detrimental to the occupancy and any occupancy with a moisture-sensitive floor covering	Any
Vapor barrier is above the subbase and below a layer of dry granular material (e.g., sand)	Moisture infiltration through the slab is detrimental to the occupancy	Watertight roof must be in place

The concrete slab thickness is a function of several factors:

1. Strength of subgrade and subbase
2. Strength of concrete
3. Magnitude and type of load (including contact area of loads)

To determine the slab thickness, several methods can be used, most of which are empirical or based on rules of thumb. It is also possible to analyze the slab by isolating a typical section. The slab-on-grade can be modeled as a beam element supported by an infinite number of supports each with a spring constant. Under a perfectly uniform load, the soil under the slab would compress uniformly and the slab would only see direct compression stress. The typical loading is often quite varied, and thus differential displacement

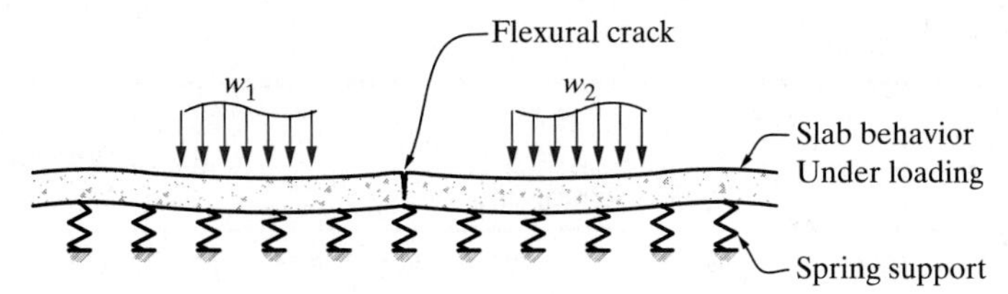

FIGURE 2-24 Slab-on-grade detail.

will occur across the slab, and flexural stresses would occur in the tension zones of the slab. Figure 2-24 shows the loading and displacement that a slab could experience. In this case, the unloaded zone between applied loads w_1 and w_2 would experience flexural tension stress, and if the slab were not sufficiently thick or if this tension zone was not sufficiently reinforced, a flexural crack would occur. In most typical structures, the slab-on-grade design does not involve any significant structural analysis, rather rules of thumb and historical practices are used. For most structures, the slab-on-grade thickness varies from 4 to 8 inches. For slabs-on-grade with heavy loading, and specifically heavy point loads, the reader is referred to ACI 360 for several empirical methods for determining the slab thickness and reinforcing.

The concrete strength for industrial and commercial slabs-on-grade should not be less than 4000 psi at 28 days. This furnishes satisfactory wear resistance in addition to strength. Generally, in residential construction the slabs on ground will have concrete strengths of 2500 or 3000 psi. A minimum of 3000 psi is recommended.

The reinforcement in the slab-on-grade is often welded wire fabric (WWF), but reinforcing bars in both directions could also be used for more heavily loaded slabs. The reinforcing is mainly used to mitigate flexural cracking from loading when the slab spans soft spots in the subgrade, and also to control and minimize shrinkage cracks. Despite several other relatively minor advantages that result from the use of steel, questions have been raised as to whether reinforcement is always necessary, particularly with uniform support of the slab and short joint spacings. The steel does not always prevent cracking nor does it add significantly to the load-carrying capacity of the slab. It is also usually more economical to obtain increased strength in concrete slabs-on-grade by increasing the thickness of the slab.

While there are methods for determining the exact amount of reinforcing to use in certain slabs, the reinforcing typically used is based on rules of thumb and historical practices (see Table 2-2). The reinforcing should be placed within the upper one-third of the slab and it also should be secured and in place prior to pouring the concrete. The reinforcing is typically not higher than 1″ below the finished concrete surface to allow for the proper clear cover and also construction tolerances for finishing.

Control joints are critical to the performance of the slab-on-grade. The timing of the joint placement and the location of the joints need to be properly addressed to avoid undesirable cracking and curling. Joints are typically placed within the first 12 to 24 hours of the concrete pour, and they typically have to be placed within 24 hours in order to be effective. As the concrete sets, moisture is released and the concrete undergoes volumetric changes such that it shrinks. This shrinkage will cause the concrete to crack, and so in order to mitigate or control this cracking, control joints are placed to force a crack to occur in a deliberately weakened section of the concrete. The concrete slab will also have a tendency to curl, which is when one side of the concrete dries faster than the other side. Concrete slabs are often cured by placing a vapor barrier on top of the slab and water is regularly applied to this concrete surface for a few weeks after the slab is poured to allow the slab to maintain a constant moisture content through the entire slab thickness. Note also that with the presence of a vapor barrier directly under the slab, this constant moisture content is more easily achieved than with the vapor barrier under a layer of sand or if the vapor barrier is not present at all.

The joints placed in a slab are usually made with a tool or they are cut with a saw. The depth of the joint should be one-quarter of the slab thickness. The joints should be placed between 24 and 36 times the slab thickness, and they should also be placed such that each slab panel that is created has an aspect ratio not greater than 1.5 (ratio of the long side to the short side). Slabs should also be allowed to float away from other building elements to avoid cracking. Where a slab abuts a wall, there should be an expansion material to prevent the concrete from bonding to the wall and to allow for differential movement. At building columns, joints are typically placed in a diamond pattern with an expansion joint to isolate the concrete that might be bonded to the column that will likely settle from the main floor slab. Figure 2-25 shows the common placement of control joints and details.

The table below summarizes typical types of slabs-on-grade and their uses.

Design aids and procedures (based primarily on research done for highway and airport pavements) have been developed and are available in specialized publications. (See references [3] through [8].)

TABLE 2-2 Typical Slab-on-Grade Types

Slab thickness	Common occupancies	Typical reinforcing	Load capacity (Ref. [12])	Recommended joint spacing
4″	Residential slab-on-grade; lightly loaded sidewalks; Residential garages	None or 6 × 6-W1.4 × W1.4 WWF	100 psf	8 ft to 12 ft
5″	Commercial, institutional slab-on-grade; public garages	6 × 6-W2.1 × W2.1 WWF or 6 × 6-W2.9 × W2.9 WWF	100 psf–200 psf	10 ft to 15 ft
6″	Industrial slab-on-grade with moderate vehicle traffic (passenger vehicles, fork trucks)	6 × 6-W2.9 × W2.9 WWF or 6 × 6-W4.0 × W4.0	400 psf– 500 psf	12 ft to 18 ft
7″	Industrial slab-on-grade with heavier vehicle traffic	Two layers of 6 × 6-W2.9 × W2.9 WWF or 6 × 6-W4.0 × W4.0	600 psf–800 psf	14 ft to 21 ft
8″	Heavy vehicle loading for loading docks around a building	#4@12″ or #4@16″	1500 psf	16 ft to 24 ft
10″	Heavy Industrial	Two layers of #5@12" each way	3000 psf – 3500 psf	20 ft–30 ft

References

[1] Charles S. Whitney, "Plastic Theory of Reinforced Concrete Design." *Trans. ASCE*, V. 68, 1942.

[2] JoAnn Browning, "Rule-of-Thumb for Flexural Steel Area," *Concrete Q & A, Concrete International*, October 2007, p. 91.

[3] *Guide for Concrete Floor and Slab Construction.* ACI 302.1R-15, Farmington Hills, MI: American Concrete Institute, P.O. Box 9094, 48333-9094, 2004.

[4] *Design of Slabs on Grade.* ACI 360R-10, Farmington Hills, MI: American Concrete Institute, P.O. Box 9094, 48333-9094, 2006.

[5] *Slabs on Grade.* Concrete Craftsman Series 1, Farmington Hills, MI: American Concrete Institute, P.O. Box 9094, 48333-9094, 1994.

[6] Robert G. Packard, *Slab Thickness Design for Industrial Concrete Floors on Grade.* Skokie, IL: Portland Cement Association, 5420 Old Orchard Road, 60077, 1996.

[7] J. A. Farny and S. M. Tarr, *Concrete Floors on Ground*. Skokie, IL: Portland Cement Association, 5420 Old Orchard Road, 60077, 2008.

[8] Robert B. Anderson, *Innovative Ways to Reinforce Slabs-on-Ground*. Hartford, CT: Tech Facts, Wire Reinforcement Institute, Inc., 942 Main Street, Suite 300, 06103, 1996.

[9] *Building Code Requirements for Structural Concrete.* ACI 318-14, Farmington Hills, MI: American Concrete Institute, P.O. Box 9094, 48333-9094, 2014.

[10] Ronald L. Kozikowski and Bruce A. Suprenant, "Controlling Early-Age Cracking in Mass Concrete," *Concrete International*, March 2015, pp. 59–62.

[11] International Codes Council, International Building Code (IBC)—2015, Falls Church, VA: ICC, 2015.

[12] *CRSI Handbook 1980*, 4th ed., Concrete Reinforcing Steel Institute, 1980.

[13] American Society of Civil Engineers, "Design Loads on Structures During Construction" (ASCE/SEI 37-14), Reston, VA, 2014

Problems

In the following problems, consider moment only and tension reinforcing only. For beams, assume $1\frac{1}{2}$-in. cover and No. 3 stirrups. For slabs, assume $\frac{3}{4}$-in. cover. Unless noted otherwise, given loads are superimposed service loads and do not include weights of the members.

2-1. **a.** The beam of the cross section shown has $f'_c = 3000$ psi and $f_y = 60{,}000$ psi. Neglect the tensile strain check and $A_{s,\min}$ check. Calculate M_n.

b. Same cross section as part (a) but the steel is changed to four No. 10 bars. Calculate M_n. Calculate the percent increase in M_n and A_s.

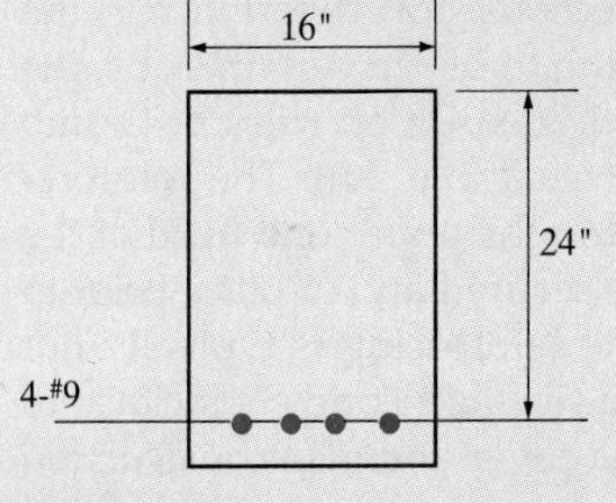

PROBLEM 2-1

c. Same cross section as part (a) but the depth d is increased to 28 in. Calculate M_n. Calculate the percent increase in M_n and d.

d. Same cross section as part (a) but f'_c is increased to 4000 psi. Calculate M_n. Calculate the percent increase in M_n and f'_c.

2-2. For the cross section of Problem 2-1(a), verify that the tension steel yields.

2-3. **a.** Calculate the practical moment strength ϕM_n for a rectangular reinforced concrete cross section having a width b of 13 in. and an effective depth d of 24 in. The tension reinforcing steel is four No. 8 bars. $f'_c = 4000$ psi and $f_y = 40{,}000$ psi.

b. Same as part (a) but $f_y = 60{,}000$ psi. Also calculate the percent increase in ϕM_n and f_y.

2-4. Determine ϕM_n for a reinforced concrete beam 16 in. wide by 32 in. deep reinforced with seven No. 10 bars (placed in two layers: five in the bottom layer, two in the top layer, with 1 in. clear between layers). Use $f'_c = 4000$ psi and $f_y = 60{,}000$ psi.

2-5. A reinforced concrete beam having the cross section shown is on a simple span of 28 ft. It carries uniform service loads of 3.60 kip/ft live load and 2.20 kip/ft dead load. Check the adequacy of the beam with respect to moment. Use $f'_c = 3000$ psi and $f_y = 40{,}000$ psi.

a. Reinforcing is six No. 10 bars.

b. Reinforcing is six No. 11 bars.

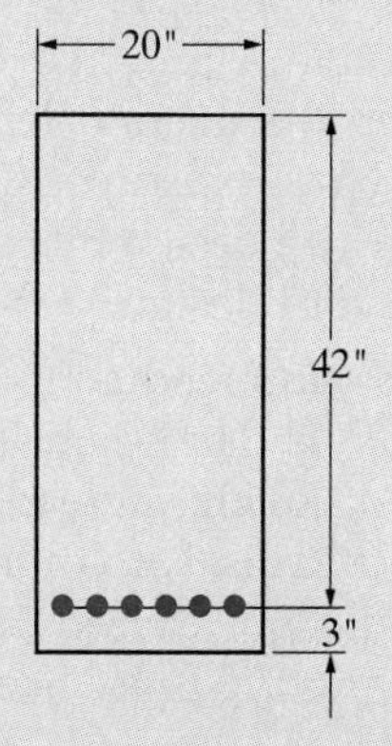

PROBLEM 2-5

2-6. Develop a spreadsheet application that will allow a user to input the basic information for the analysis of a rectangular reinforced concrete cross section and that will then calculate the practical moment strength ϕM_n. Set up the spreadsheet to be "user friendly" and fully label the output.

2-7. A 12-in.-wide by 20-in.-deep concrete beam is reinforced with three No. 8 bars. The beam supports a service live load of 2.5 kip/ft and a service dead load of 0.7 kip/ft on a simple span of 16 ft. Use $f'_c = 4000$ psi and $f_y = 60{,}000$ psi. Check the adequacy of the beam with respect to moment.

2-8. A reinforced concrete beam having the cross section shown is on a simple span of 26.5 ft. It supports uniformly distributed service loads of 3.20 kip/ft live load and 1.80 kip/ft dead load (excluding the beam weight). Reinforcing is as shown. Check the adequacy of the beam with respect to moment. Use $f'_c = 3000$ psi and $f_y = 60{,}000$ psi.

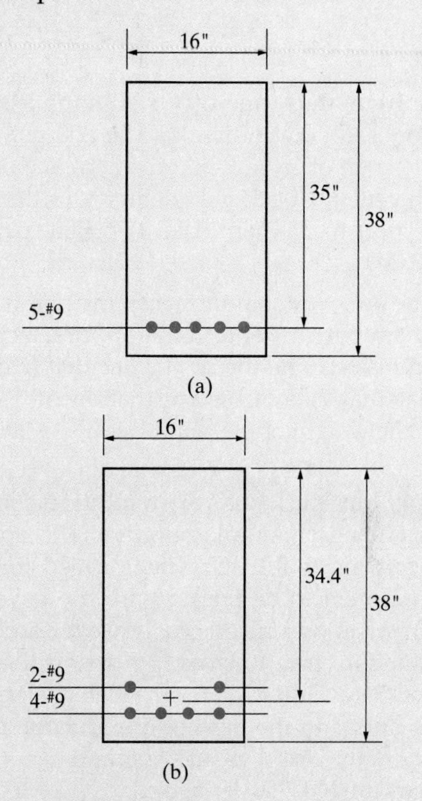

PROBLEM 2-8

2-9. A rectangular reinforced concrete beam carries *service loads* on a span of 20 ft as shown. Use $f'_c = 3000$ psi and $f_y = 60{,}000$ psi; $b = 14.5$ in., $h = 26$ in., and reinforcing is three No. 10 bars. Determine whether the beam is adequate with respect to moment.

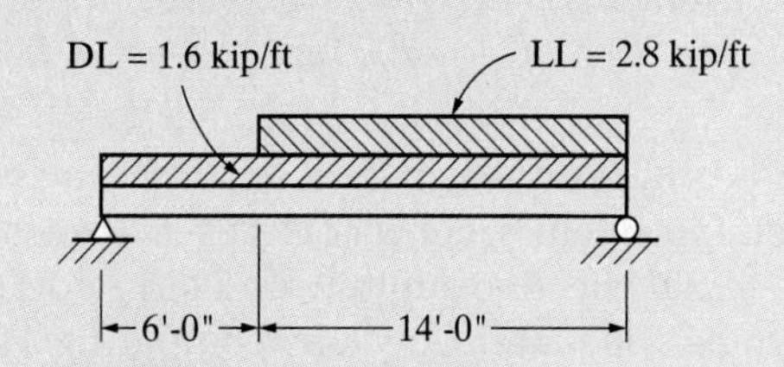

PROBLEM 2-9

2-10. A rectangular reinforced concrete beam 14 in. wide by 24 in. deep is to support a service dead load of 0.6 kip/ft and a service live load of 1.4 kip/ft. Reinforcing is four A615 grade 60 No. 9 bars. Use $f'_c = 4000$ psi. Determine the maximum simple span length on which this beam may be utilized.

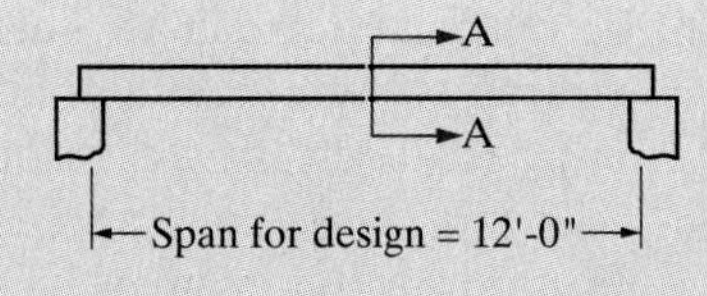

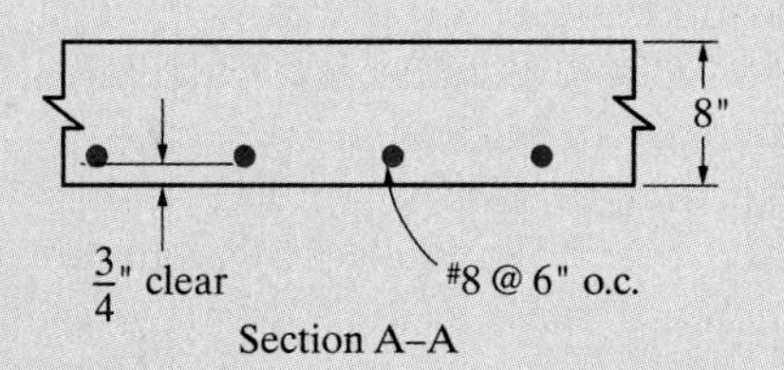

PROBLEM 2-12

2-11. A 10-in.-thick one-way slab supports a superimposed service live load of 600 psf on a simple span of 16 ft. Reinforcement is No. 7 at 6 in. on center. Check the adequacy of the slab with respect to moment. Use $f'_c = 3000$ psi and $f_y =$ 60,000 psi.

2-12. The one-way slab shown spans 12 ft from center of support to center of support. Calculate ϕM_n and determine the *service live load* (psf) that the slab may carry. (Assume that the only dead load is the weight of the slab.) Use $f'_c = 3000$ psi and $f_y = 40{,}000$ psi.

2-13. An $8\frac{1}{2}$-in.-thick one-way reinforced concrete slab overhangs a simple support. The span of the overhang is 8 ft. Drawings called for the reinforcement to be placed with top cover of 1 in. The steel was misplaced, however, and later was found to be as much as $3\frac{1}{2}$ in. below the top of the concrete. Find ϕM_n for the slab as designed and as built and the percent of reduction in flexural strength. Use $f'_c = 4000$ psi and $f_y = 60{,}000$ psi. Bars are No. 7 at 11 in. o.c.

2-14. Design a rectangular reinforced concrete beam to resist a total design moment M_u of 133 ft.-kip. (This includes the moment due to beam weight.) Architectural considerations require that the width (b) be $11\frac{1}{2}$ in. and the overall depth (h) be 23 in. Use $f'_c = 3000$ psi and $f_y = 60{,}000$ psi. Sketch your design.

2-15. Rework Problem 2-14 with $M_u = 400$ ft.-kip, $b = 16$ in., $h = 28$ in., $f'_c = 4000$ psi, and $f_y = 60{,}000$ psi.

2-16. For the beam designed in Problem 2-15, if the main reinforcement were incorrectly placed so that the actual effective depth were 24 in., would the beam be adequate? Check by comparing M_u with the ϕM_n resulting from the beam using actual steel, actual b, and $d = 24$ in.

2-17. Design a rectangular reinforced concrete beam (tension steel only) for a simple span of 32 ft. Uniform service loads are 0.85 kip/ft dead load and 1.0 kip/ft live load. The beam is to be $11\frac{1}{2}$ in. wide and 26 in. deep overall (form reuse consideration). Use $f'_c = 4000$ psi and $f_y = 60{,}000$ psi. Calculate ϕM_n for the beam designed.

2-18. Design a rectangular reinforced concrete beam (tension steel only) for a simple span of 30 ft. There is no superimposed dead load (other than the weight of the beam) and the superimposed live load is 1.35 kip/ft. The beam is to be 12 in. wide and 27 in. deep overall. Use $f'_c = 5000$ psi and A615 grade 60 steel. As a check, calculate ϕM_n for the beam *designed*.

2-19. Rework Problem 2-18 assuming that the superimposed live load has increased to 1.75 kip/ft and there is now a 1.0 kip/ft superimposed dead load (in addition to the beam weight). As before, check ϕM_n for the beam designed.

2-20. Design a simply supported rectangular reinforced concrete beam to span 22 ft and to carry uniform service loads of 1.6 kip/ft dead load and 1.4 kip/ft live load. The assumed dead load includes an estimated beam weight. Use A615 grade 60 steel and $f'_c = 3000$ psi. Use the recommended ρ from Table A-5. Make the beam width 15 in. and keep the overall depth (h) to full inches. Assume No. 3 stirrups. Check the adequacy of the beam you design by comparing M_u with ϕM_n. Sketch your design.

2-21. Design a rectangular reinforced concrete beam for a simple span of 30 ft. The beam is to carry uniform *service* loads of 1.0 kip/ft dead load and 2.0 kip/ft live load. Because of column sizes, the beam width should not exceed 16 in. Use $f'_c = 3000$ psi and $f_y = 60{,}000$ psi. Sketch your design.

2-22. Rework Problem 2-21 assuming that the total depth h is not to exceed 30 in. and that there is no limitation on the width b.

2-23. Design a rectangular reinforced concrete beam for a simple span of 32 ft. Uniform service loads are 1.5 kip/ft dead load and 2.0 kip/ft live load. The width of the beam is limited to 18 in. Use $f'_c = 3000$ psi and $f_y = 60{,}000$ psi. Sketch your design.

2-24. Rework Problem 2-23 assuming that the total depth h is not to exceed 32 in. and that there is no limitation on the width b.

2-25. Design a rectangular reinforced concrete beam for a simple span of 40 ft. Uniform service loads are 0.8 kip/ft dead load and 1.4 kip/ft live load. Use $f'_c = 4000$ psi and $f_y = 60{,}000$ psi. Sketch your design.

2-26. Design a rectangular reinforced concrete beam (tension steel only) for the span and *superimposed* service loads shown. Use $f'_c = 5000$ psi and A615 grade 60 steel.

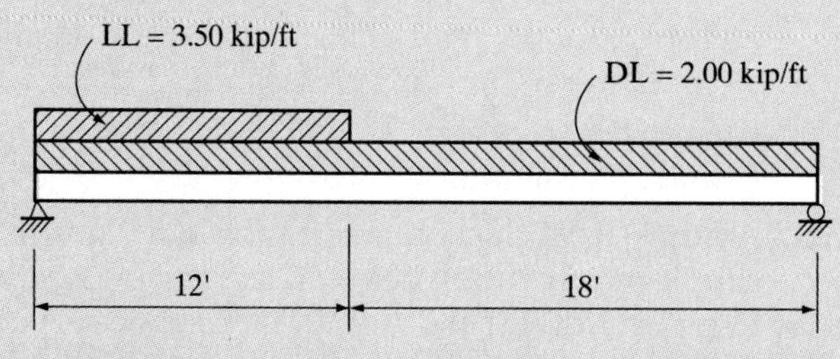

PROBLEM 2-26

2-27. Design a 28-ft, simple-span, rectangular reinforced concrete beam to support a uniform live load of 0.8 kip/ft and concentrated loads at midspan of 10 kip dead load and 14 kip live load. Use $f'_c = 5000$ psi and $f_y = 60{,}000$ psi.

2-28. Design a simply supported rectangular reinforced concrete beam for the span and *service loads* shown. Use $f'_c = 3000$ psi and $f_y = 60{,}000$ psi.

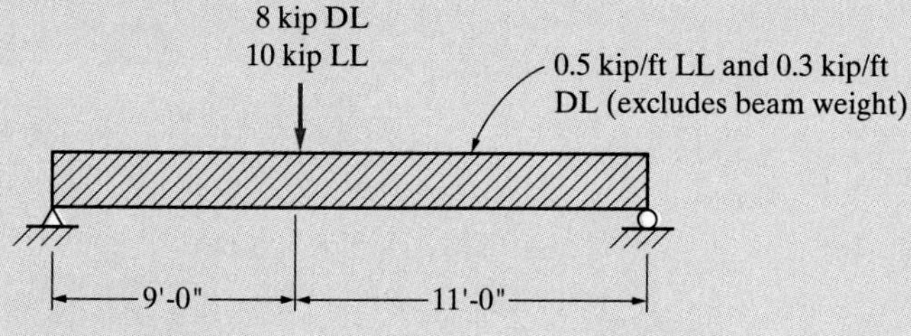

PROBLEM 2-28

2-29. Rework Problem 2-28 assuming that the beam is to be extended to overhang the right support by 10 ft. The uniform load also extends to the end of the overhang.

2-30. Design a simply supported one-way reinforced concrete floor slab to span 8 ft and carry a service live load of 300 psf and the dead load is the self-weight of the slab only. Use $f'_c = 3000$ psi and $f_y = 60{,}000$ psi. Sketch your design.

2-31. Design a simply supported one-way reinforced concrete floor slab to span 10 ft and carry a service live load of 175 psf and a service dead load of 25 psf. Use $f'_c = 3000$ psi and $f_y = 60{,}000$ psi. Make the slab thickness to $\frac{1}{2}$-in. increments.

a. Design the slab for the ACI Code minimum thickness.

b. Design the thinnest possible slab allowed by the ACI Code.

2-32. Design the simply supported one-way reinforced concrete slab as shown. The service live load is 200 psf and the dead load is the self-weight of the slab only. Use $f'_c = 3000$ psi and $f_y = 60{,}000$ psi. Sketch your design.

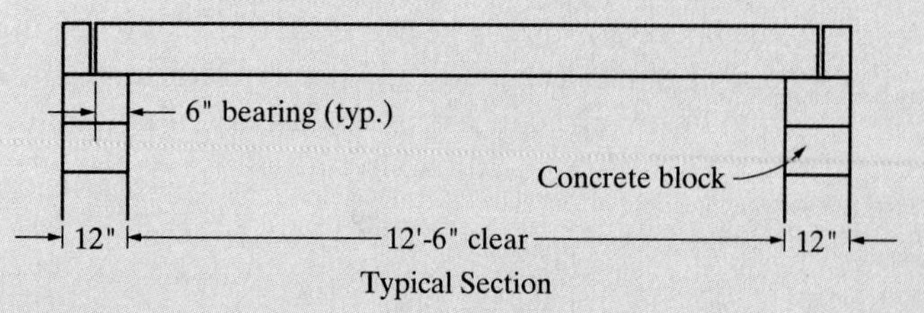

PROBLEM 2-32

2-33. a. Determine the service live load in psf that the slab shown can support. Ignore the design checks for $A_{s,min}$ and spacing for the main reinforcing. Determine the required temperature reinforcing (select bar size and spacing)

- Assume simple span of 14 ft. for the slab
- $f'_c = 3000$ psi, $\gamma_c = 150$ pcf
- $f_y = 60{,}000$ psi
- Service dead load is 50 psf plus the self-weight of the slab
- Use the load combination for dead plus live loads only (ignore live load reduction)

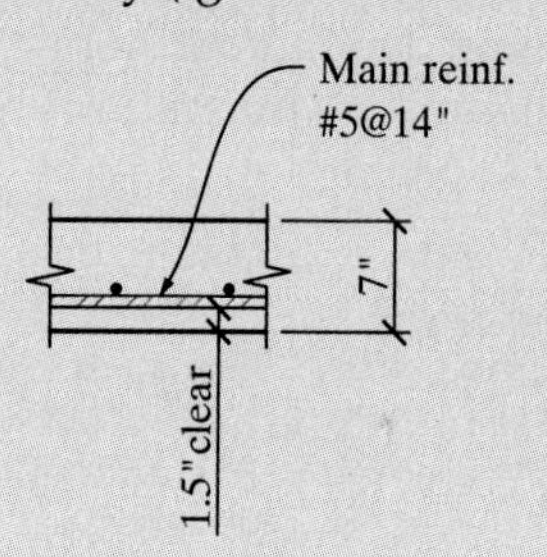

PROBLEM 2-33

2-34. Determine the required area of steel for bending (bar size and number) for the beam shown.

- Self-weight is included as shown
- $f'_c = 4000$ psi
- $f_y = 60{,}000$ psi
- Use the load combination for dead plus live loads only (ignore live load reduction)

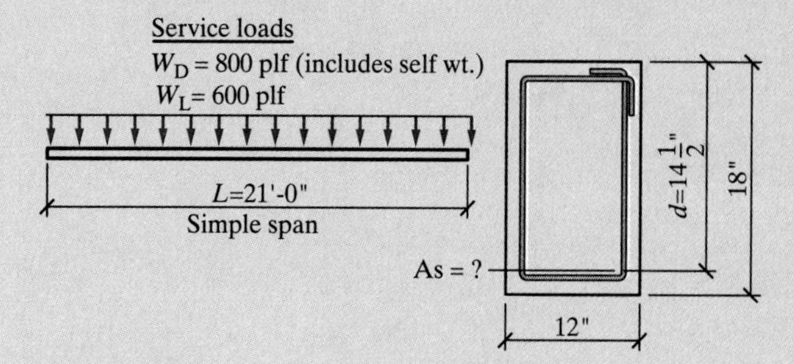

PROBLEM 2-34

2-35. The detail shown is a cross-section of two precast concrete lintels used to support gravity loads. Each lintel is 4″ × 8″ and has a single #4 for flexural reinforcing. The total factored uniform load on the assembly, including self-weight, is w_u = 500 plf. Use f'_c = 5000 psi and f_y = 60,000 psi. Use the load combination for dead plus live loads only (ignore live load reduction). Determine the maximum span in feet that this assembly could span based on bending strength only.

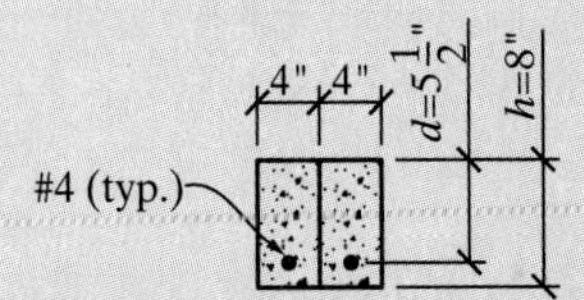

PROBLEM 2-35

2-36. For the floor plan and details shown below, assume simple spans for the slab. Use f'_c = 3,500 psi and f_y = 60,000 psi. Use a service dead load of 110 psf (includes self-weight)

The service live load is L = 175 psf. Use the load combination for dead plus live loads only (ignore live load reduction). Consider the design of a single 32″ wide slab section

Determine the following:

a. The required reinforcing for flexural loads only

b. The required temperature reinforcing (bar size and spacing)

c. Sketch the design showing a cross-section through the 32″ width and the actual number of bars

2-37. Assume a rectangular concrete beam with b = 11″ and h = 20″. Use f'_c = 4000 psi and f_y = 60,000 psi. The factored moment is Mu = 175 *k-ft* (includes beam self-weight)

Determine the following:

a. The required reinforcing for flexural loads; check minimum steel provisions

b. Sketch the design showing the reinforcing and all dimensions

2-38. Assume a 9″ thick concrete slab with d = 7″; simple span of 18 ft. between supporting beam. Use f'_c = 3000 psi and f_y = 60,000 psi. The service live load is 125 psf and the service dead load is 20 psf plus the self-weight of the slab. Use the load combination for dead plus live loads only (ignore live load reduction)

Determine the following:

a. The required reinforcing for flexural loads; check minimum steel provisions

b. Sketch the design showing the reinforcing and all dimensions through the slab section and sketch a floor plan showing the reinforcing

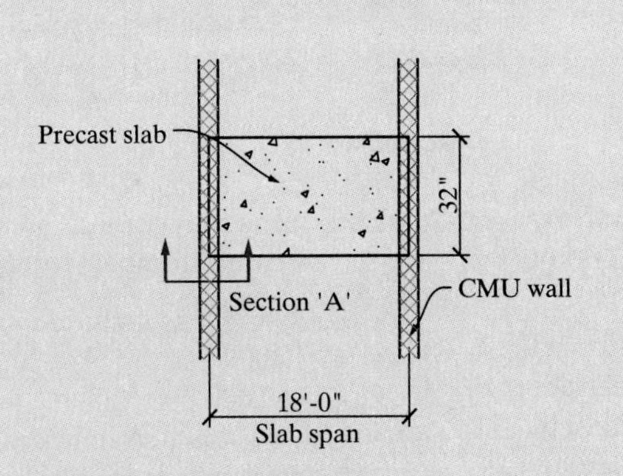

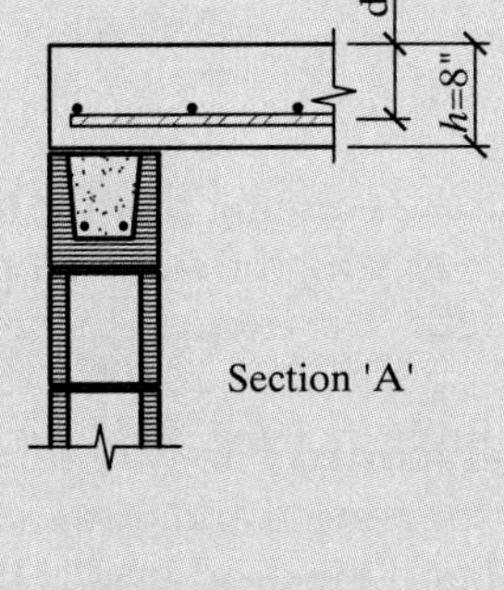

PROBLEM 2-36

REINFORCED CONCRETE BEAMS: T-BEAMS, L-BEAMS, AND DOUBLY REINFORCED BEAMS

3-1 T-Beams and L-Beams: Introduction

3-2 T-Beam and L-Beam Analysis

3-3 Analysis of Beams Having Irregular Cross Sections

3-4 T-Beam and L-Beam Design (For Moment)

3-5 Summary of Procedure for Analysis of T-Beams and L-Beams (For Moment)

3-6 Summary of Procedure for Design of T-Beams and L-Beams (For Moment)

3-7 Doubly Reinforced Beams: Introduction

3-8 Doubly Reinforced Beam Analysis for Moment (Condition I)

3-9 Doubly Reinforced Beam Analysis for Moment (Condition II)

3-10 Summary of Procedure for Analysis of Doubly Reinforced Beams (For Moment)

3-11 Doubly Reinforced Beam Design for Moment

3-12 Summary of Procedure for Design of Doubly Reinforced Beams (For Moment)

3-13 Additional Code Requirements for Doubly Reinforced Beams

3-1 T-BEAMS AND L-BEAMS: INTRODUCTION

Floors and roofs in reinforced concrete buildings may be composed of slabs that are supported so that loads are carried to columns and then to the building foundation. As previously discussed, these are termed *flat slabs* or *flat plates*. The span of such a slab cannot become very large before its own dead weight causes it to become uneconomical. Many types of systems have been devised to allow greater spans without the problem of excessive weight.

One such system, called a *beam and girder system*, is composed of a slab on supporting reinforced concrete beams and girders. The beam and girder framework, in turn, is supported by columns. In such a system, the beams and girders are commonly placed monolithically with the slab. Systems other than the monolithic system do exist, and these may make use of some precast and some cast-in-place concrete. These are generally of a proprietary nature. The typical monolithic system is shown in Figure 3-1. The beams are commonly spaced so that they intersect the girders at the midpoint, third points, or quarter points, as shown in Figure 3-2. To minimize the cost of concrete formwork, the same depth is often used for the beams and girders.

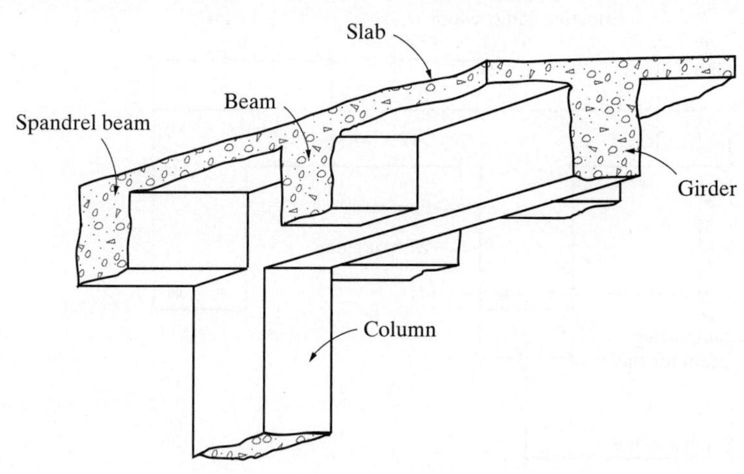

FIGURE 3-1 Beam and girder floor system.

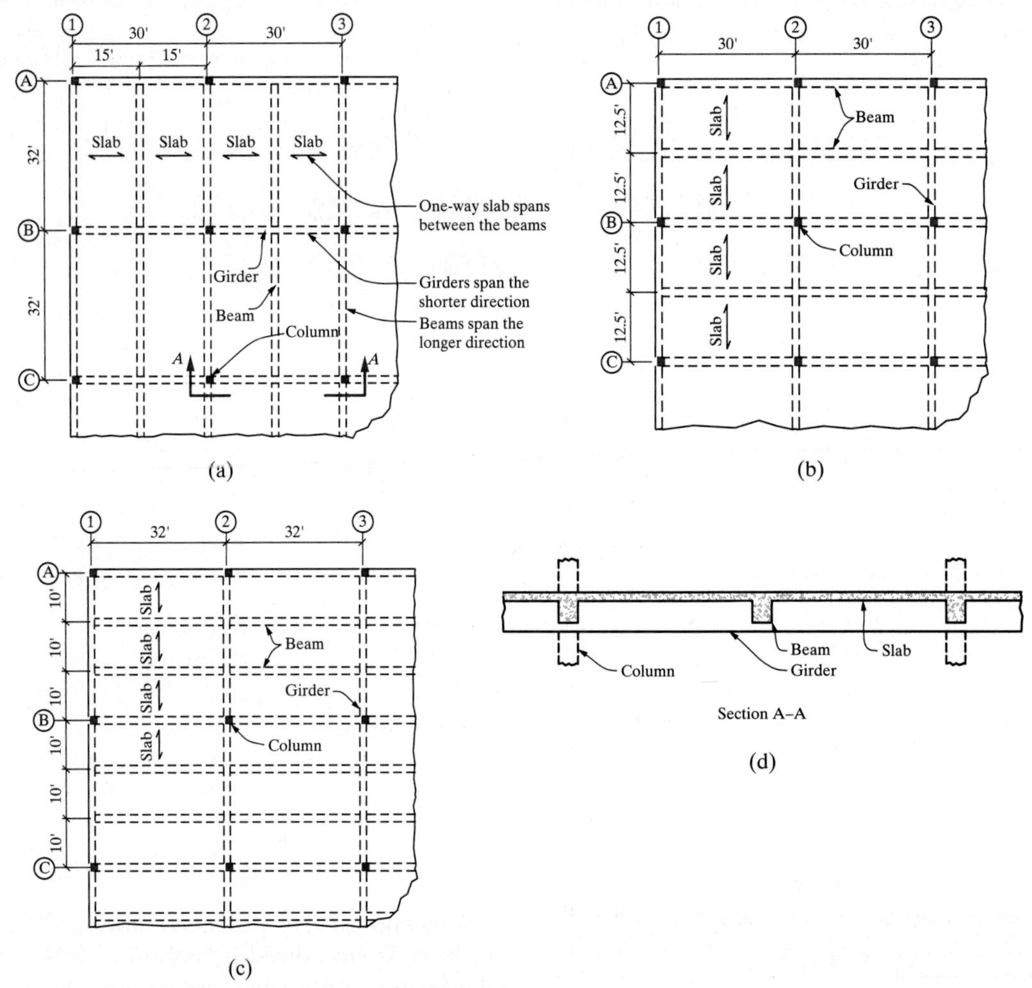

FIGURE 3-2 Common beam and girder layouts.

In the analysis and design of such floor and roof systems, it is common practice to assume that the monolithically placed slab and supporting beam interact as a unit in resisting *positive bending moment*. As shown in Figure 3-3, the slab becomes the compression flange, and the supporting beam becomes the web or stem. For the typical interior beams or girders (see Figure 3-2), the interacting flange and stem produce the cross section having the typical T-shape from which the T-beam gets its name (see Figure 3-3). Similarly, for a typical edge or spandrel beam/girder (e.g. the beams on grid line 1 in Figure 3-2a, and grid line A in Figures 3-2b and 3-2c, and the girders on grid line A in Figure 3-2a, and grid line 1 in Figures 3-2b and 3-2c), the interacting flange and stem produce the cross section having an inverted L-shape as shown in Figure 3-3. It should be noted that

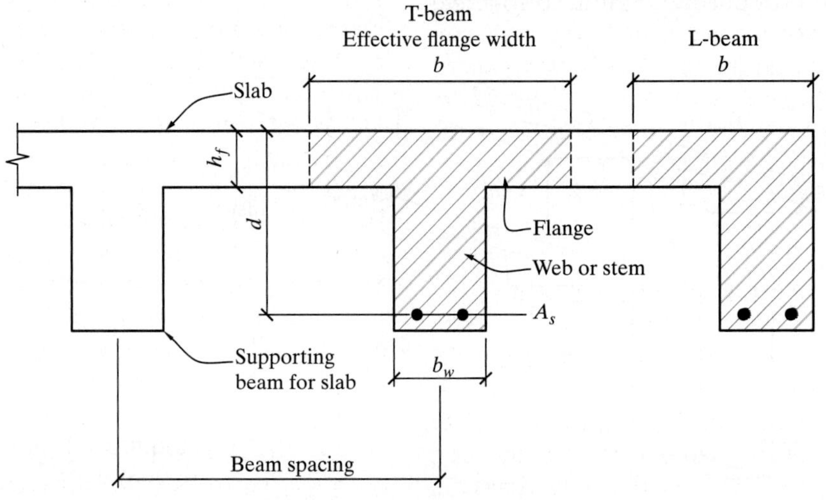

FIGURE 3-3 T-beam and L-beam as part of a floor system.

the slab, which comprises the T-beam flange, must itself be designed to span across the supporting beams. Therefore, the slab behaves as a bending member acting in two directions. It should also be noted that should the T-beam cross section be subjected to *negative bending moment*, the slab at the top of the stem will be in tension while the bottom of the stem is in compression. It will be seen that this situation will occur at interior supports of continuous beams, which are discussed later.

To simplify the complex two-way behavior of the flange, the ACI Code, for design and analysis purposes, has established criteria whereby the flange, when acting together with the web, will have a limited width that may be considered effective in resisting applied moment. This effective flange width for symmetrical shapes will always be equal to or less than the beam spacing (see Figure 3-3).

3-2 T-BEAM AND L-BEAM ANALYSIS

For purposes of analysis and design, the ACI Code, Section 6.3.2.1, has established limits on the effective flange width in the compression zone for T-beams (and inverted L-beams) in the positive moment regions as follows:

1. For T-beams (i.e., interior beams or girders), the effective flange width must not exceed the width of the beam web plus one-fourth of the clear span length of the beam, and the effective overhanging flange width on each side of the web must not exceed eight times the thickness of the slab nor one-half of the clear distance between adjacent beam webs. In other words, the effective flange width, b, must not exceed
   a. b_w + One-fourth of the clear span length.
   b. $b_w + 16h_f$.
   c. Center-to-center spacing of beams.
   The smallest of the three values will control.
2. For beams having a flange on one side only (i.e., spandrel or perimeter beams which are also referred to as inverted L-beams), the effective overhanging flange width must not exceed one-twelfth of the clear span length of the beam, nor six times the slab thickness, nor one-half of the clear distance to the next beam. Thus, the effective width, b, for inverted L-beams or girders is the smallest of the following:
   a. b_w + One-twelfth of the clear span length of the beam.
   b. $b_w + 6h_f$.
   c. $b_w + \frac{1}{2}$ (clear spacing between the webs of the L-beam and the adjacent interior beam).
3. For isolated beams in which the T-shape is used only for the purpose of providing additional compressive area, the flange thickness must not be less than one-half of the width of the web, and the total flange width must not be more than four times the web width.

In the positive moment regions of a beam (i.e., where the bottom of the beam is in tension and the top flange is in compression), such as the mid-span of the beam, the width of the compression zone is the effective width calculated above (i.e., b). In the negative moment regions of a beam (i.e., where the bottom of the beam is in compression and the top flange is in tension), such as the beam support locations, the width of the compression zone is the width of the beam stem (i.e., $b = b_w$).

The ductility requirements for T-beams are similar to those for rectangular beams. To ensure ductile behavior, the ACI Code, Section 9.3.3.1, requires a net tensile strain $\epsilon_t \geq 0.004$ for flexural members. A section is tension-controlled, that is, completely ductile when the net tensile strain $\epsilon_t \geq 0.005$. It is always desirable and more efficient in the design of flexural members to strive for a tension-controlled section. The T-shape can be a factor in the determination of net tensile strain for a T-beam.

The procedure for determining the minimum steel for a T-beam is the same as for a rectangular beam when the T-beam flange is in compression (positive moment). Where tensile reinforcement is required by analysis, the steel area, A_s, shall not be less than that given by

$$A_{s,\,\min} = \frac{3\sqrt{f'_c}}{f_y}b_w d \geq \frac{200}{f_y}b_w d$$

Note that for T-beams, b_w represents the width of the web. Also note that the first expression controls only if $f'_c > 4440$ psi. The above expressions for minimum steel also apply to continuous T-beams.

For negative moment (flange in tension) in statically determinate members,

$$A_{s,\,\min} = \text{the smaller of } \frac{6\sqrt{f'_c}}{f_y}b_w d \quad \text{or} \quad \frac{3\sqrt{f'_c}}{f_y}bd$$

The minimum steel requirements need not be applied if, at every section along the member, at least 33% more steel than is required by analysis is provided. Where multiple layers of reinforcement are used in a beam, the clear spacing between horizontal layers of reinforcement should be a minimum of 1 inch (ACI 318-14, Section 25.2.2). This spacing can be provided by using at least a #8 spacer bar.

Because of the relatively large compression area available in the flange of the T-beam, the moment strength is usually limited by the yielding of the tensile steel. Therefore, it is usual to assume that the tensile steel will yield before the concrete reaches its ultimate strain and crushes. The total tensile force, N_T, at the ultimate condition may then be found by

$$N_T = A_s f_y$$

To proceed with the analysis, the shape of the compressive stress block must be defined. As in our previous analyses, the total compressive force N_C must be equal to the total tensile force N_T. The shape of the stress block must be compatible with the area in compression. Two conditions may exist: The stress block may be completely within the flange, or it may cover the flange and extend into the web. These two conditions will result in what we will term, respectively, a *rectangular T-beam* and a *true T-beam*. In addition to the shape of the stress block, the basic difference between the two is that the rectangular T-beam with effective flange width b is analyzed in the same way as is a rectangular beam of width b, whereas the analysis of the true T-beam must consider the T-shaped stress block.

Example 3-1

The T-beam shown in Figure 3-4 is part of a floor system. Determine the practical moment strength ϕM_n if $f_y = 60{,}000$ psi (A615 grade 60) and $f'_c = 3000$ psi.

Solution:

1. Because the span length is not given, determine the effective flange width in terms of the flange thickness and beam spacing:

$$b_w + 16h_f = 10 + 16(2) = 42 \text{ in.}$$

$$\text{beam spacing} = 32 \text{ in. o.c.}$$

$$\text{Use } b = 32 \text{ in.}$$

2. Check $A_{s,\min}$. From Table A-5:

$$A_{s,\min} = 0.0033 b_w d = 0.0033(10)(12) = 0.40 \text{ in.}^2$$

$$0.40 \text{ in.}^2 < 3.0 \text{ in.}^2 \quad \text{(O.K)}$$

3. Assume that the steel yields and find N_T:

$$N_T = A_s f_y = 3.00(60{,}000) = 180{,}000 \text{ lb}$$

4. The flange alone, if fully stressed to $0.85f'_c$, would produce a total compressive force of

$$N_{Cf} = (0.85f'_c)h_f b = 0.85(3000)(2)(32) = 163{,}200 \text{ lb}$$

5. Because $180{,}000 > 163{,}200$, the stress block must extend below the flange far enough to provide the remaining compression:

$$180{,}000 - 163{,}200 = 16{,}800 \text{ lb}$$

Hence, the stress block extends below the flange and the analysis is one for a true T-beam.

6. The remaining compression $(N_T - N_{Cf})$ may be obtained by the additional web area:

$$N_T - N_{Cf} = (0.85f'_c)b_w(a - h_f)$$

Solving for a, we obtain

$$a = \frac{N_T - N_{Cf}}{(0.85f'_c)b_w} + h_f = \frac{16{,}800}{0.85(3000)(10)} + 2 = 2.66 \text{ in.}$$

7. Determine net tensile strain ϵ_t (check ductility). Using the relationship $a = \beta_1 c$, which is approximate for T-beams:

$$c = \frac{a}{\beta_1} = \frac{2.66 \text{ in.}}{0.85} = 3.13 \text{ in. (see Figure 3-4)}$$

The distance d_t of the extreme tensile reinforcement from the compression face is 12 in. Therefore, the net tensile strain in the extreme tensile reinforcement is

$$\epsilon_t = 0.003\frac{(d_t - c)}{c} = 0.003\frac{(12 - 3.13)}{3.13} = 0.0085$$

8. Determine the strength-reduction factor ϕ: because $0.0085 > 0.005$, this is a tension-controlled section and $\phi = 0.90$ (see Section 2-9).

9. To calculate the magnitude of the internal couple, it is necessary to know the lever-arm distance between N_C and N_T. The location of N_T is assumed at the centroid of the steel area, and we will locate N_C at the centroid of the T-shaped compression area (see Figure 3-5a). Using a reference axis at the top of the section, the centroid may be located a distance $\bar{y}$ below the reference axis, as follows:

$$\bar{y} = \frac{\Sigma(Ay)}{\Sigma A}$$

$$A_1 = 32(2) = 64 \text{ in.}^2, \quad A_2 = 10(0.66) = 6.6 \text{ in.}^2$$

$$\bar{y} = \frac{64(1) + 6.6(2 + 0.33)}{64 + 6.6} = 1.12 \text{ in.}$$

This locates N_C. Therefore,

$$Z = d - \bar{y} = 12 - 1.12 = 10.88 \text{ in.}$$

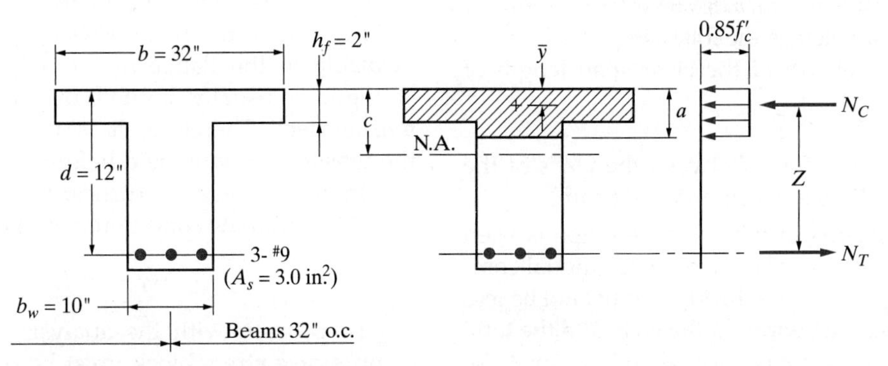

FIGURE 3-4 Sketch for Example 3-1.

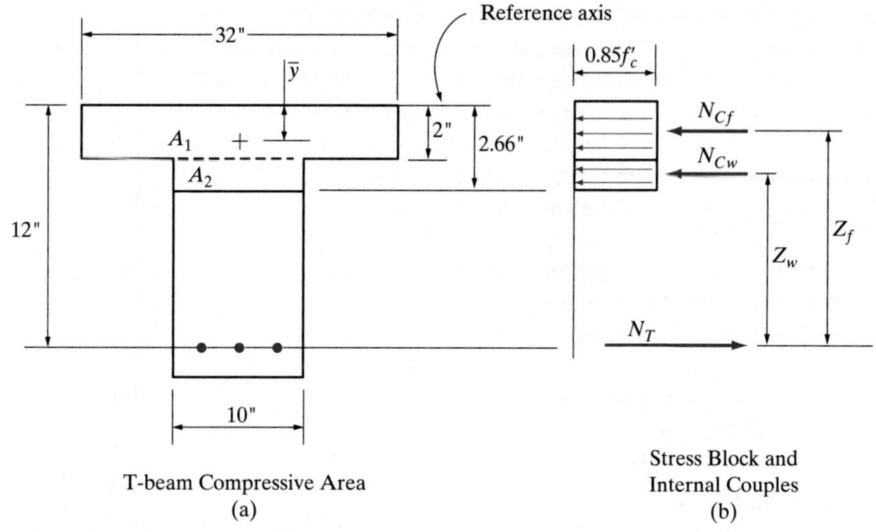

FIGURE 3-5 Sketch for Example 3-1.

The nominal (or ideal) moment strength may be found:

$$M_n = N_T Z = \frac{180{,}000(10.88)}{12{,}000} = 163 \text{ ft.-kip}$$

from which the practical moment strength is

$$\phi M_n = 0.9(163) = 147 \text{ ft.-kip}$$

In the solution of Example 3-1, step 9 may be accomplished in a slightly different way. If the total internal couple M_n is assumed to be composed of two component couples, a flange couple (using compressive force N_{Cf} in Figure 3-5b) and a web couple (using compressive force N_{Cw} in Figure 3-5b), then its magnitude can be calculated from

$$\begin{aligned} M_n &= \text{flange couple} + \text{web couple} \\ &= N_{Cf}Z_f + N_{Cw}Z_w \\ &= N_{Cf}\left(d - \frac{h_f}{2}\right) + (N_T - N_{Cf})\left[d - h_f - \left(\frac{a - h_f}{2}\right)\right] \end{aligned}$$

This avoids the calculation of the centroid location and results in the same moment strength. The concept of two component couples is used again in the design of true T-beams.

Example 3-2

For the T-beam shown in Figure 3-6, determine the practical moment strength ϕM_n if $f'_c = 3000$ psi and $f_y = 60{,}000$ psi. The beam span length is 24 ft.

Solution:

1. Find the effective flange width:

$$b_w + \tfrac{1}{4}\text{ span length} = 10 + \frac{24(12)}{4} = 82 \text{ in.}$$

$$b_w + 16h_f = 10 + 16(4) = 74 \text{ in.}$$

$$\text{beam spacing} = 60 \text{ in.}$$

$$\text{Use } b = 60 \text{ in.}$$

2. Check $A_{s,\min}$. From Table A-5:

$$\begin{aligned} A_{s,\min} &= 0.0033b_w d \\ &= 0.0033(10)(24) = 0.79 \text{ in.}^2 \end{aligned}$$

$$0.79 \text{ in.}^2 < 6.00 \text{ in.}^2 \qquad \text{(O.K.)}$$

3. Assume the steel yields and find N_T.

$$N_T = A_s f_y = 6.0(60{,}000) = 360{,}000 \text{ lb}$$

4. The flange itself is capable of furnishing a compression force of

$$N_{Cf} = (0.85f'_c)bh_f = 0.85(3000)(60.0)(4) = 612{,}000 \text{ lb}$$

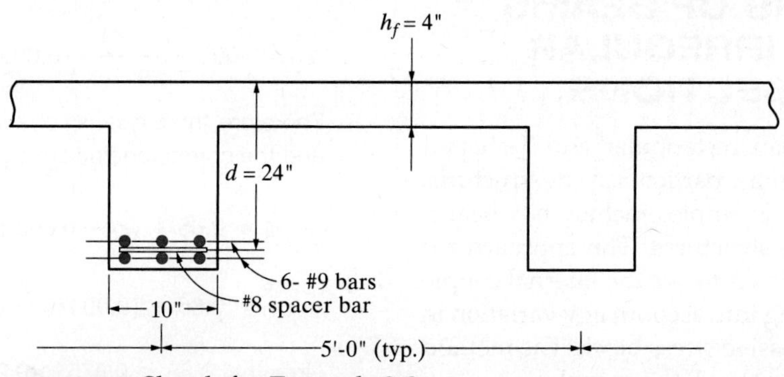

FIGURE 3-6 Sketch for Example 3-2.

5. Because 612,000 > 360,000, the flange furnishes sufficient compression area and the stress block lies entirely in the flange. Therefore, analyze the T-beam as a rectangular T-beam of width $b = 60$ in.
6. Solve for the depth of the compressive stress block.

$$a = \frac{A_s f_y}{0.85 f'_c b} = \frac{6.00(60{,}000)}{0.85(3000)(60)} = 2.35 \text{ in.}$$

7. Determine the net tensile strain (check ductility). The depth to the extreme tension steel d_t is determined using a #8 spacer bar between the two layers of #9 bars, as shown in Figure 3-6.

$$d_t = 24 + \frac{1.00}{2} + \frac{1.125}{2} = 25.1 \text{ in.}$$

$$c = \frac{a}{\beta_1} = \frac{2.35}{0.85} = 2.76 \text{ in.}$$

$$\epsilon_t = \frac{0.003(d_t - c)}{c} = \frac{0.003(25.1 - 2.76)}{2.76} = 0.0243$$

(Or use Equation [2-1] from Section 2-9.)

Because 0.0243 > 0.005, ductility is ensured, and this is a tension-controlled section.

8. Find ϕ. Because $\epsilon_t > 0.005$, $\phi = 0.90$.
9. Find ϕM_n.

$$\phi M_n = \phi A_s f_y\left(d - \frac{a}{2}\right)$$

$$= \frac{0.90(6.00)(60)\left(24 - \frac{2.35}{2}\right)}{12} = 616\text{ft.-kip}$$

This T-beam behaves like a rectangular beam having a width b of 60 in., so the expression for ϕM_n developed in Section 2-10 could be used at step 9:

$$\phi M_n = \phi b d^2 \overline{k}$$

$$\rho = \frac{A_s}{bd} = \frac{6.00}{60(24)} = 0.0042$$

From Table A-8, $\overline{k} = 0.2396$ ksi and $\epsilon_t > 0.005$. Therefore, $\phi = 0.90$.

Then,

$$\phi M_n = \frac{0.90(60)(24)^2(0.2396)}{12} = 621 \text{ ft.-kip}$$

3-3 ANALYSIS OF BEAMS HAVING IRREGULAR CROSS SECTIONS

Beams having other than rectangular and T-shaped cross sections are common, particularly in structures using precast elements. Examples include box beams that are used in bridge structures. The approach for the analysis of such beams is to use the internal couple in the normal way, taking into account any variation in the shape of the compressive stress block. The method is similar to that used for true T-beam analysis.

Example 3-3

The cross section shown in Figure 3-7 is sometimes referred to as an inverted T-girder. Find the practical moment strength ϕM_n. (The ledges in the beam cross section will possibly be used for support of precast slabs.) Use $f_y = 60{,}000$ psi (A615 grade 60) and $f'_c = 3000$ psi.

Solution:

1. The effective flange width may be considered to be 7 in.
2. Check $A_{s,\min}$. From Table A-5:

$$A_{s,\min} = 0.0033 b_w d$$
$$= 0.0033(17)(24) = 1.35 \text{ in.}^2$$
$$1.35 \text{ in.}^2 < 4.00 \text{ in.}^2 \qquad \text{(O.K.)}$$

3. Find N_T:

$$N_T = A_s f_y = 4.00(60{,}000) = 240{,}000 \text{ lb}$$

4. Determine the amount of compression that the 7 in × 4 in. area is capable of furnishing. (This is the area we are considering to be the flange.)

$$N_{Cf} = (0.85 f'_c) h_f b$$
$$= 0.85(3000)(4)(7) = 71{,}400 \text{ lb}$$

5. Because 240,000 lb > 71,400 lb, the compressive stress block must extend below the ledges to provide the remaining compression ($a > h_f$):

$$N_T - N_{Cf} = 240{,}000 - 71{,}400 = 168{,}600 \text{ lb}$$

6. The remaining compression will be furnished by additional beam area below the ledges. Referring to Figure 3-7,

$$a = \frac{N_T - N_{Cf}}{(0.8 f'_c) b_w} + h_f$$
$$= \frac{168{,}600}{0.85(3000)(17)} + 4$$
$$= 7.89 \text{ in. from top of beam}$$

7. Check ductility.

$$c = \frac{a}{\beta_1} = \frac{7.89 \text{ in.}}{0.85} = 9.28 \text{ in.}$$

The distance of the extreme tensile reinforcement from the compression face, d_t, is 24 in. Therefore, the net tensile strain in the extreme tensile reinforcement is

$$\epsilon_t = 0.003\frac{(d_t - c)}{c} = 0.003\frac{(24 - 9.28)}{9.28} = 0.00476$$

8. Therefore, this is a transition section ($0.004 < \epsilon_t < 0.005$), and the corresponding strength-reduction factor is

$$\phi = 0.65 + (\epsilon_t - 0.002)\left(\frac{250}{3}\right)$$
$$= 0.65 + (0.00476 - 0.002)\left(\frac{250}{3}\right) = 0.875$$

Note that $0.65 < 0.875 < 0.90$.

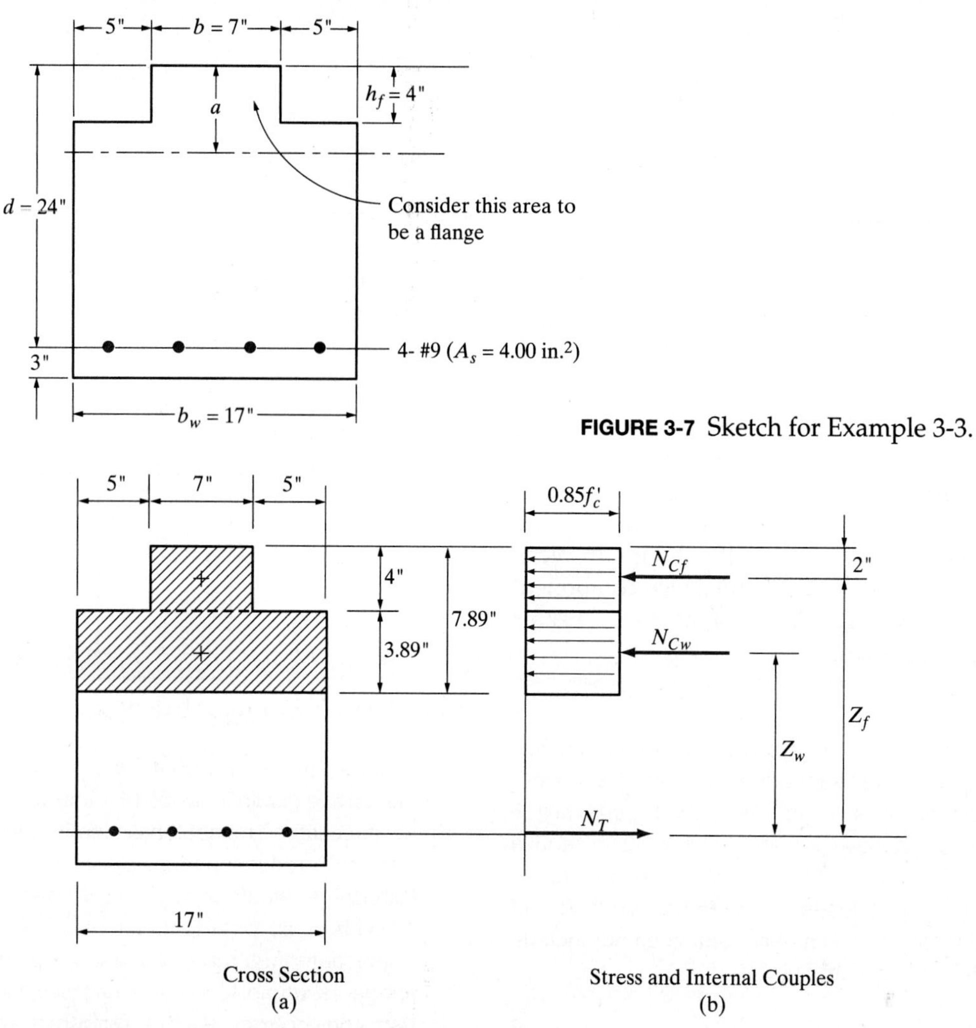

FIGURE 3-7 Sketch for Example 3-3.

FIGURE 3-8 Inverted T-girder for Example 3-3.

9. ϕM_n will be calculated considering two component internal couples, a flange couple and a web couple. Refer to Figure 3-8.

$$\phi M_n = \phi(N_{Cf}Z_f + N_{Cw}Z_w)$$

$$= \phi\left\{N_{Cf}\left(d - \frac{h_f}{2}\right) + (N_T - N_{Cf})\left[d - h_f - \left(\frac{a - h_f}{2}\right)\right]\right\}$$

$$= \frac{0.875\left\{71.4\left(24 - \frac{4}{2}\right) + 168.6\left[24 - 4 - \left(\frac{3.89}{2}\right)\right]\right\}}{12}$$

$$= 337 \text{ ft.-kip}$$

3-4 T-BEAM AND L-BEAM DESIGN (FOR MOMENT)

The design of the T-sections (and inverted L-sections) involves the dimensions of the flange and web and the area of the tension steel, a total of five unknowns. In the normal progression of a design, the flange thickness is determined by the design of the slab, and the web size is determined by the shear and moment requirements at the end supports of a beam in continuous construction. Practical considerations, such as column sizes and forming, may also dictate web width. Therefore, when the T-section is designed for positive moment, most of the five unknowns have been previously determined.

As indicated previously, the ACI Code dictates permissible effective flange width, b, when the flange is in compression (e.g. at the midspan). The flange itself generally provides more than sufficient compression area; therefore, the compressive stress block usually lies completely in the flange in the positive moment region of the beam. Thus, most T-beams (and inverted L-beams) are wide rectangular beams with respect to flexural behavior when the top of the beam is in compression (e.g. at the midspan). Note that when the flange of the beam is in tension as would occur in the negative moment regions (e.g. at the beam or girder supports), the web of the beam or girder will be in compression and the width of the beam or girder in the compression zone will be b_w (i.e. $b = b_w$).

The recommended method for the design of T-beams in the positive moment regions will depend on whether the T-beam behaves as a rectangular T-beam

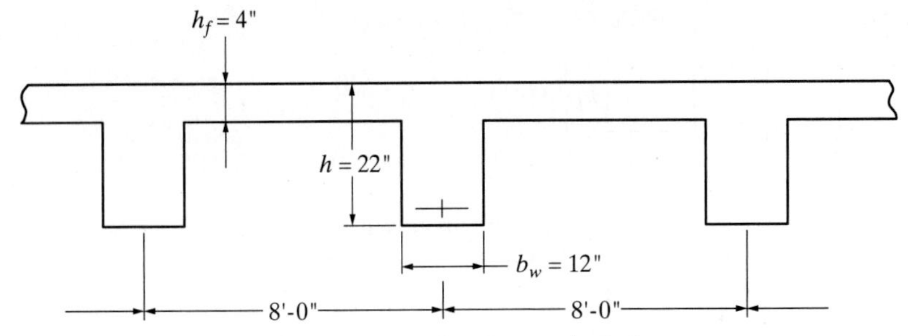

FIGURE 3-9 Typical floor section for Example 3-4.

or a true T-beam. The first step will be to answer this question. If the T-beam is determined to be a rectangular T-beam, the design procedure is the same as for the tensile reinforced rectangular beam, where the size of the cross section is known (see Section 2-15). If the T-beam is determined to be a true T-beam, the design proceeds by designing a flange component and a web component and combining the two. The design of inverted L-beams follows a similar approach.

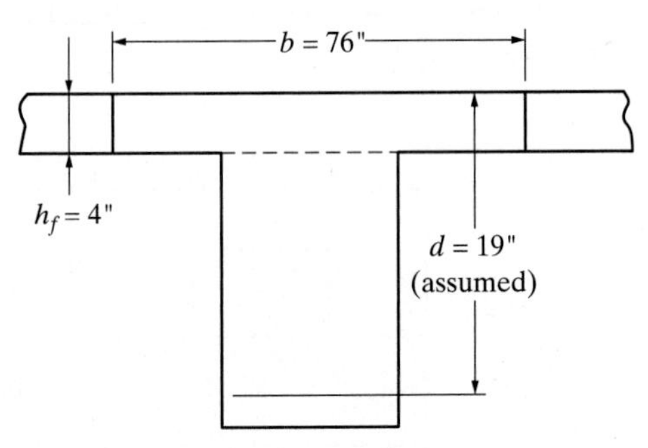

FIGURE 3-10 T-beam for Example 3-4.

Example 3-4

Design the T-beam for the floor system shown in Figure 3-9. The floor has a 4-in. slab supported by 22-ft-span-length beams cast monolithically with the slab. Beams are 8 ft-0 in. on center and have a web width of 12 in. and a total depth $= 22$ in.; $f'_c = 3000$ psi and $f_y = 60{,}000$ psi (A615 grade 60). Service loads are 0.125-ksf live load and 0.200-ksf dead load. The given dead load does not include the weight of the floor system.

Solution:

1. Establish the design moment:

$$\text{slab weight} = \frac{96(4)}{144}(0.150) = 0.400 \text{ kip/ft}$$

$$\text{stem weight} = \frac{12(18)}{144}(0.150) = 0.225 \text{ kip/ft}$$

$$\text{total weight} = 0.625 \text{ kip/ft}$$

$$\text{service DL} = 8(0.200) = 1.60 \text{ kip/ft}$$

$$\text{service LL} = 8(0.125) = 1.00 \text{ kip/ft}$$

Calculate the factored load and moment:

$$w_u = 1.2(0.625 + 1.60) + 1.6(1.00) = 4.27 \text{ kip/ft}$$

$$M_u = \frac{w_u \ell^2}{8} = \frac{4.27(22)^2}{8} = 258 \text{ ft.-kip}$$

2. Assume an effective depth $d = h - 3$ in.:

$$d = 22 - 3 = 19 \text{ in.}$$

3. Determine the effective flange width:

$$b_w + \tfrac{1}{4}\text{span length} = 12 + 0.25(22)(12) = 78 \text{ in.}$$

$$b_w + 16h_f = 12 + 16(4) = 76 \text{ in.}$$

$$\text{beam spacing} = 96 \text{ in.}$$

Use an effective flange width $b = 76$ in.

4. Assume a tension-controlled section—that is, the net tensile strain $\epsilon_t \geq 0.005$; this assumption will be checked later. The net tensile strain value of 0.005 gives a strength-reduction factor $\phi = 0.90$.
5. Determine whether the beam behaves as a true T-beam or as a rectangular beam by computing the practical moment strength ϕM_{nf} with the full effective flange assumed to be in compression. This assumes that the bottom of the compressive stress block coincides with the bottom of the flange, as shown in Figure 3-10. Thus,

$$\phi M_{nf} = \phi(0.85 f'_c) b h_f \left(d - \frac{h_f}{2}\right)$$

$$= \frac{0.9(0.85)(3)(76)(4)(19 - 4/2)}{12} = 988 \text{ ft.-kip}$$

6. Because 988 ft.-kip $>$ 258 ft.-kip, the total effective flange need not be completely utilized in compression (i.e., $a < h_f$), and the T-beam behaves as a wide rectangular beam with a width b of 76 in.
7. Design a rectangular beam with b and d as known values (see Section 2-15):

$$\text{required } \bar{k} = \frac{M_u}{\phi b d^2} = \frac{258(12)}{0.9(76)(19)^2} = 0.1254 \text{ ksi}$$

8. From Table A-8, select the required steel ratio to provide a $\bar{k}$ of 0.1254 ksi:

$$\text{required } \rho = 0.0021$$

9. Calculate the required steel area:

$$\text{required } A_s = \rho b d$$

$$= 0.0021(76)(19) = 3.03 \text{ in.}^2$$

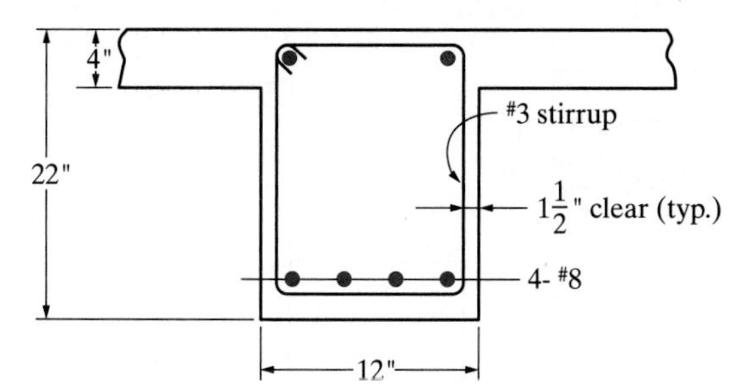

FIGURE 3-11 Design sketch for Example 3-4.

10. Select the steel bars. Use four No. 8 bars ($A_s = 3.16\text{ in.}^2$):
 From Table A-3:

$$\text{minimum } b_w = 11 \text{ in.} \qquad \text{(O.K.)}$$

 Check the effective depth d. Assume a No. 3 stirrup and $1\frac{1}{2}$-in. cover, as shown in Figure 3-11.

$$d = 22 - 1.5 - 0.38 - \frac{1.00}{2} = 19.62 \text{ in.}$$

$$19.62 \text{ in.} > 19 \text{ in.} \qquad \text{(O.K.)}$$

11. Check $A_{s,\min}$. From Table A-5:

$$A_{s,\min} = 0.0033 b_w d$$
$$= 0.0033(12)(19.62) = 0.78 \text{ in.}^2$$
$$0.78 \text{ in.}^2 < 3.16 \text{ in.}^2 \qquad \text{(O.K.)}$$

12. Check ϵ_t to ensure a tension-controlled section ($\epsilon_t \geq 0.005$). From Section 2-10, for a rectangular section:

$$\epsilon_t = \frac{0.00255 f'_c \beta_1}{\rho f_y} - 0.003$$
$$= \frac{0.00255(3)(0.85)}{0.0025(60)} - 0.003 = 0.0404$$

 Therefore, the net tensile strain is much larger than 0.005; this is a tension-controlled section, and $\phi = 0.90$, as assumed. (Note that $\epsilon_t > 0.005$ could also be confirmed from Table A-8.)
13. Sketch the design (see Figure 3-11).

Example 3-5

Design a T-beam having a cross section as shown in Figure 3-12. Assume that the effective flange width given is acceptable. The T-beam will carry a total design moment M_u of 340 ft.-kip. Use $f'_c = 3000$ psi and $f_y = 60{,}000$ psi. Use $1\frac{1}{2}$-in. cover and No. 3 stirrups.

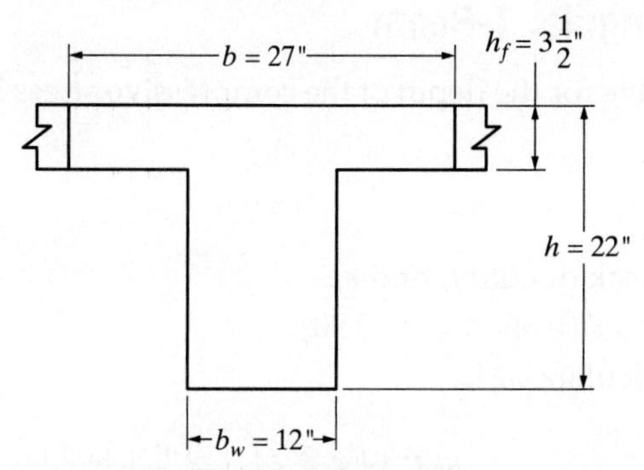

FIGURE 3-12 Sketch for Example 3-5.

Solution:

1. The design moment $M_u = 340$ ft.-kip (given).
2. Assume an effective depth of

$$d = 22 - 3 = 19 \text{ in.}$$

3. The effective flange width = 27 in. (given).
4. Assume a tension-controlled section—that is, the net tensile strain $\epsilon_t \geq 0.005$; this assumption will be checked later. The net tensile strain value of 0.005 gives a strength-reduction factor $\phi = 0.90$.
5. Determine ϕM_{nf} assuming the effective flange to be in compression over its full depth:

$$\phi M_{nf} = \phi(0.85 f'_c) b h_f \left(d - \frac{h_f}{2}\right)$$
$$= \frac{0.9(0.85)(3)(27)(3.5)(19 - 3.5/2)}{12} = 312 \text{ ft.-kip}$$

6. $\phi M_{nf} < M_u$; therefore, the beam must behave as a true T-beam.
7. Two component couples will be designed, a flange couple (subscript f) and a web couple (subscript w). Refer to Figure 3-13.

 Calculate the required steel area A_{sf} for the flange couple:

$$\text{estimated } d_f = h - 3 \text{ in.} = 22 - 3 = 19 \text{ in.}$$
$$\text{estimated } Z_f = d_f - h_f/2 = 19 - 3.5/2 = 17.25 \text{ in.}$$
$$\text{required } A_{sf} = \frac{\phi M_{nf}}{\phi f_y Z_f}$$
$$= \frac{312(12)}{0.9(60)(17.25)}$$
$$= 4.02 \text{ in.}^2$$

8. The web couple will be designed for the remaining applied moment ($M_u - \phi M_{nf}$). The design is for a rectangular reinforced concrete beam having a depth $h_w = h - h_f$ and a width of b_w.

$$h_w = 22 - 3.5 = 18.5 \text{ in.}$$
$$\text{estimated } d_w = h_w - 3 \text{ in.} = 18.5 - 3 = 15.5 \text{ in.}$$
$$\text{required } \bar{k} = \frac{M_u - M_{nf}}{\phi b_w d_w^2} = \frac{(340 - 312)(12)}{0.9(12)(15.5)^2}$$
$$= 0.1295 \text{ ksi}$$

 From Table A-8, the required $\rho = 0.0023$, from which we calculate

$$\text{required } A_{sw} = \rho b_w d_w = 0.0023(12)(15.5) = 0.43 \text{ in.}^2$$

9. Total required

$$A_s = A_{sf} + A_{sw} = 4.02 + 0.43 = 4.45 \text{ in.}^2$$

10. From Table A-2, select three No. 11 bars. $A_s = 4.68 \text{ in.}^2$ and minimum b_w is 11.0 in. Check d assuming No. 3 stirrups and $1\frac{1}{2}$ in. cover:

$$d = 22 - 1.5 - 0.38 - \frac{1.41}{2} = 19.42 \text{ in.}$$

$$19.42 \text{ in.} > \text{estimated } d_f = 19.0 \text{ in.} \qquad \text{(O.K.)}$$

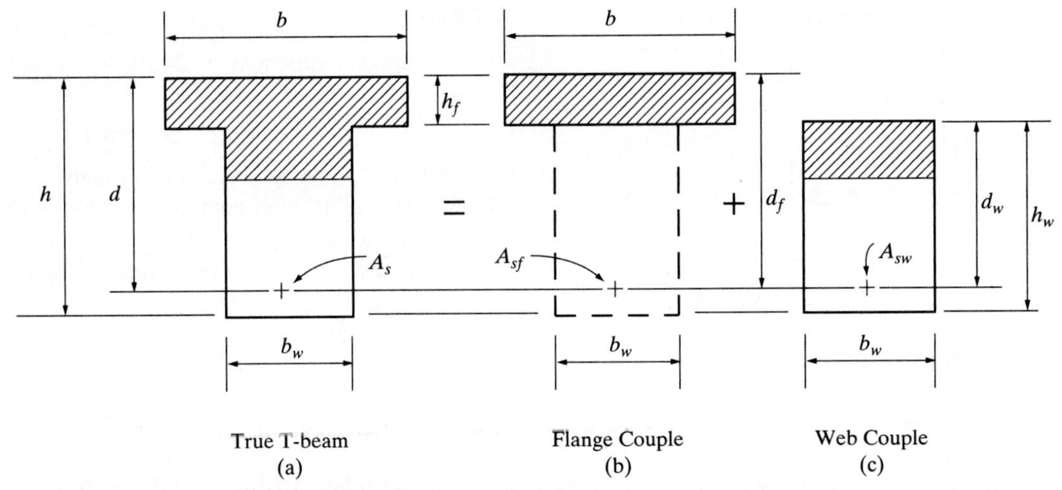

FIGURE 3-13 True T-beam design for Example 3-5.

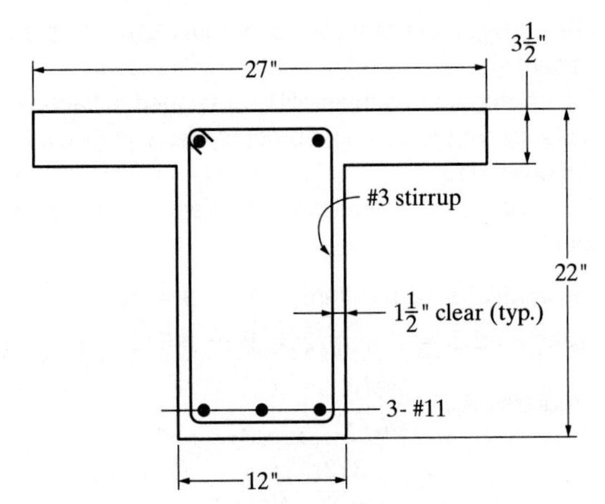

FIGURE 3-14 Design sketch for Example 3-5.

11. Check $A_{s,\min}$. From Table A-5:

$$A_{s,\min} = 0.0033 b_w d$$
$$= 0.0033(12)(19.42) = 0.77 \text{ in.}^2$$

$0.77 \text{ in.}^2 < 4.68 \text{ in.}^2$ (O.K.)

12. Check ϵ_t to ensure a tension-controlled section ($\epsilon_t \geq 0.005$):

$$d_t = d = 19.42 \text{ in.}$$

$$N_T = A_s f_y = 4.68(60) = 281 \text{ kip}$$

$$N_{Cf} = 0.85 f_c' b h_f = 0.85(3)(27)(3.5) = 241 \text{ kip}$$

$$a = \frac{N_T - N_{Cf}}{0.85 f_c' b_w} + h_f = \frac{281 - 241}{0.85(3)(12)} + 3.5 = 4.81 \text{ in.}$$

From Section 2-9, Equation (2-1):

$$\epsilon_t = \frac{0.003 \beta_1 d_t}{a} - 0.003$$

$$= \frac{0.003(0.85)(19.42)}{4.81} - 0.003 = 0.00730$$

$0.00730 > 0.005$; therefore, this is a tension-controlled section and $\phi = 0.90$, as assumed. (O.K.)

13. Sketch the design. See Figure 3-14.

3-5 SUMMARY OF PROCEDURE FOR ANALYSIS OF T-BEAMS AND L-BEAMS (FOR MOMENT)

1. Establish the effective flange width (i.e. the width of the beam in the compression zone) in the positive moment regions based on ACI criteria. In the negative moment region, the width of the beam in the compression zone is equal to the width of the beam stem, b_w.
2. Check $A_{s,\min}$. Use Table A-5.
3. To ensure ductility, assume a net tensile strain greater than or equal to 0.004; this assumption will be checked later. Compute the total tension in the steel:

$$N_T = A_s f_y$$

4. Compute the magnitude of the compression that the flange itself is capable of furnishing:

$$N_{Cf} = 0.85 f_c' b h_f$$

5. If $N_T > N_{Cf}$, the beam will behave as a true T-beam and the remaining compression, which equals $N_T - N_{Cf}$, will be furnished by additional web area. If $N_T < N_{Cf}$, the beam will behave as a rectangular beam of width b.

Rectangular T-Beam

6. Solve for the depth of the compressive stress block:

$$a = \frac{A_s f_y}{0.85 f_c' b}$$

7. Check ductility; find ϵ_t.
8. Find $\phi (0.65 \leq \phi \leq 0.90)$.
9. Calculate ϕM_n.

$$\phi M_n = \phi A_s f_y \left(d - \frac{a}{2}\right)$$

Or, in place of steps 6–9, calculate ρ, obtain $\bar{k}$, check ϵ_t, determine ϕ, and use

$$\phi M_n = \phi b d^2 \bar{k}$$

A flow diagram of this procedure is presented in Appendix B.

True T-Beam

6. Determine the depth of the compressive stress block:

$$a = \frac{N_T - N_{Cf}}{0.85 f'_c b_w} + h_f$$

7. Check ductility; find ϵ_t.
8. Find $\phi (0.65 \leq \phi \leq 0.90)$.
9. **a.** Locate the centroid of the total compressive area referenced to the top of the flange using the relationship

$$\bar{y} = \frac{\sum (Ay)}{\sum A}$$

from which

$$Z = d - \bar{y}$$

Compute the practical moment strength ϕM_n:

$$\phi M_n = \phi N_C Z \quad \text{or} \quad \phi N_T Z$$

or

b. Calculate ϕM_n using a summation of internal couples contributed by the flange and the web:

$$\phi M_n = \phi \left\{ N_{Cf}\left(d - \frac{h_f}{2}\right) + (N_T - N_{Cf})\left[d - h_f - \left(\frac{a - h_f}{2}\right)\right]\right\}$$

The procedures presented above apply also to edge or spandrel beams which are shaped like inverted L-beams (see Figure 3-3). The main difference between a T-beam and an L-beam in the positive moment region is in the value of the effective flange width, b (see Section 3-2). In the negative moment region, the width of the T-beam and the L-beam in the compression zone is equal to the width of the beam stem, b_w.

3-6 SUMMARY OF PROCEDURE FOR DESIGN OF T-BEAMS AND L-BEAMS (FOR MOMENT)

1. Compute the design moment M_u. In many cases, the size of the beam is unknown and an estimate of the beam depth and width of the beam stem has to be made before proceeding with the calculation of the factored load on the beam and the factored moment, M_u. The reader should refer to Chapter 6 for the calculation of the factored moments at the positive and negative moment regions of continuous beams.
2. Assume that the effective depth $d = h - 3$ in.

   The "3 in." above is an average value. For beams with one layer of tension reinforcement, an effective depth of $h - 2.5$ in. can be used, while for beams with two layers of tension reinforcement, an effective depth of $h - 3.5$ in. can be used.
3. Establish the effective flange width based on ACI criteria (see Section 3-2).
4. Assume a net tensile strain $\epsilon_t \geq 0.005$; this will give a strength-reduction factor $\phi = 0.90$. This assumption will be checked later.
5. Compute the practical moment strength ϕM_{nf} assuming that the total effective flange is in compression:

$$\phi M_{nf} = \phi (0.85 f'_c) b h_f \left(d - \frac{h_f}{2}\right)$$

6. If $\phi M_{nf} > M_u$, the beam will behave as a rectangular T-beam of width b. If $\phi M_{nf} < M_u$, the beam will behave as a true T-beam.

Rectangular T-Beam

7. Design a rectangular beam with b and d as known values. Compute the required $\bar{k}$:

$$\text{required } \bar{k} = \frac{M_u}{\phi b d^2}$$

8. From the tables in Appendix A, determine the required ρ for the required $\bar{k}$ of step 7.
9. Compute the required A_s:

$$\text{required } A_s = \rho b d$$

10. Select bars and check the beam width. Check the actual d and compare it with the assumed d. If the actual d is slightly in excess of the assumed d, the design will be slightly conservative (on the safe side). If the actual d is less than the assumed d, the design may be on the nonconservative side (depending on the steel provided) and should be more closely investigated for possible revision.
11. Check $A_{s,\min}$. Use Table A-5.
12. Check ductility. Find ϵ_t. If $d = d_t$, Tables A-7 through A-11 can be used, or use Equation (2-2) from Section 2-10. If $d \neq d_t$, use basic principles or Equation (2-1) from Section 2-9. Check the assumed value of ϕ

$$\epsilon_t = \frac{0.00255 f'_c \beta_1}{\rho f_y} - 0.003 \qquad (2\text{-}2)$$

13. Sketch the design.

True T-Beam

7. Using an estimated $d_f = h - 3''$ and $Z_f = d_f - h_f/2$, determine the steel area A_{sf} required for the flange couple:

$$\text{required } A_{sf} = \frac{\phi M_{nf}}{\phi f_y Z_f}$$

8. Design the web couple as a rectangular reinforced concrete beam having a total depth $h_w = h - h_f$, using an estimated $d_w = h_w - 3''$ and a beam width of b_w. Design for an applied moment of $M_u - \phi M_{nf}$. Determine required $\overline{k}$, required ρ, and required A_{sw}.
9. Total required $A_s = A_{sf} + A_{sw}$.
10. Select the bars. (Bars must fit into beam width b_w.) Check d as in step 10 of the rectangular T-beam design.
11. Check $A_{s,\min}$. Use Table A-5.
12. Check ductility. Find ϵ_t. Determine the stress block depth a and use the expression for ϵ_t at the end of Section 2-9.
13. Sketch the design.

Flowcharts of these procedures are presented in Appendix B.

The procedures presented above for T-beams or girders (i.e., interior beams or girders) also apply to inverted L-beams or girders (i.e., spandrel beams or girders)

3-7 DOUBLY REINFORCED BEAMS: INTRODUCTION

As indicated in Chapter 2, the practical moment strength of a rectangular reinforced concrete beam reinforced with only tensile steel may be determined by the expression $\phi M_n = \phi b d^2 \overline{k}$, where $\overline{k}$ is a function of the steel ratio ρ, f'_c, and f_y.

If we assume a given rectangular section with tension-only reinforcing that is tension-controlled, the upper limit of tension steel area can be established using the reinforcement ratio ρ associated with net tensile strain ϵ_t of 0.005. The maximum practical moment strength ϕM_n for the section may then be calculated using the associated value of $\overline{k}$.

Occasionally, practical and architectural considerations may dictate and limit beam sizes, whereby it becomes necessary to develop more moment strength from a given cross section. When this situation occurs, the ACI Code, Section 22.2.3.1, permits the addition of tensile steel over and above the code maximum provided that compression steel is also added in the compression zone of the cross section. The result constitutes a combined tensile and compressive reinforced beam commonly called a *doubly reinforced beam.*

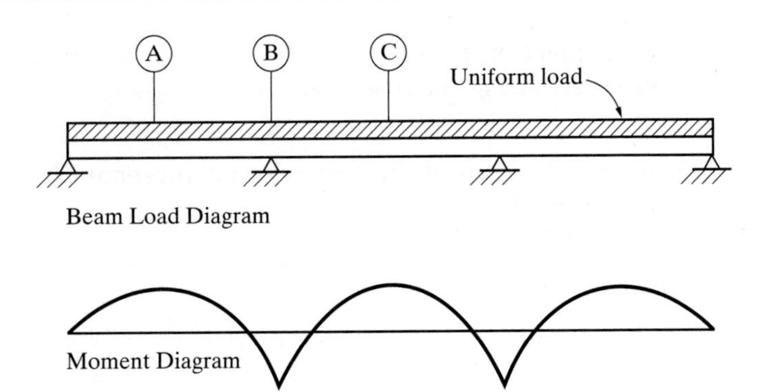

FIGURE 3-15 Continuous beam.

Where beams span more than two supports (continuous construction), practical considerations are sometimes the reason for the existence of main steel in compression zones. In Figure 3-15, positive moments exist at A and C; therefore, the main tensile reinforcement would be placed in the bottom of the beam. At B, however, a negative moment exists and the bottom of the beam is in compression. The tensile reinforcement must be placed near the top of the beam. It is general practice that at least some of the tension steel in each of these cases will be extended to the length of the beam and will pass through compression zones. In this case the compression steel may sometimes be used for additional strength. In Chapter 7, we will see that compression reinforcement aids significantly in reducing long-term deflections. In fact, the use of compression steel to increase the bending strength of a reinforced concrete beam is an inefficient way to utilize steel; however, in shallow concrete beams, the extra moment capacity might be needed. More commonly, deflection control will be the reason for using compression steel.

In general, the advantages of compression steel include reduction of long-term deflections due to creep, and an increase in beam ductility which is advantageous in seismic regions; they can also serve as stirrup support bars which eases fabrication of the reinforcement cage. When compression steel is not needed for design, two bars (one at each top corner) are provided to help with the fabrication of the stirrups.

3-8 DOUBLY REINFORCED BEAM ANALYSIS FOR MOMENT (CONDITION I)

The basic assumptions for the analysis of doubly reinforced beams are similar to those for tensile reinforced beams. One additional significant assumption is that the *compression steel stress* (f'_s) is a function of the strain at the level of the centroid of the compression steel. As discussed previously, the steel will behave elastically up to the point where the strain exceeds the yield strain ϵ_y. In other words, as a limit, $f'_s = f_y$

when the compression steel strain $\epsilon_s' \geq \epsilon_y$. If $\epsilon_s' < \epsilon_y$, the compression steel stress $f_s' = \epsilon_s' E_s$, where E_s is the modulus of elasticity of the steel.

With two different materials, concrete and steel, resisting the compressive force N_C, the total compression will now consist of two forces: NC_1, the compression resisted by the concrete, and NC_2, the compression resisted by the compressive steel. For analysis, the total resisting moment of the beam will be assumed to consist of two parts or two internal couples: the part due to the resistance of the compressive concrete and tensile steel, and the part due to the compressive steel and additional tensile steel. The two internal couples are illustrated in Figure 3-16.

Notation for doubly reinforced beams is as follows:

A_s' = total compression steel cross-sectional area

d = effective depth of tension steel

d' = depth to centroid of compression steel from compression face of beam

A_{s1} = amount of tension steel used by the concrete–steel couple

A_{s2} = amount of tension steel used by the steel–steel couple

A_s = total tension steel cross-sectional area ($A_s = A_{s1} + A_{s2}$)

M_{n1} = nominal moment strength of the concrete–steel couple

M_{n2} = nominal moment strength of the steel–steel couple

M_n = nominal moment strength of the (doubly reinforced) beam

ϵ_s = unit strain at the centroid of the tension steel

ϵ_s' = unit strain at the centroid of the compression steel

The total nominal moment strength may be developed as the sum of the two internal couples, neglecting the concrete displaced by the compression steel.

The strength of the steel–steel couple is evaluated as follows:

$$M_{n2} = N_{T2}Z_2$$

Assuming that $f_s = f_y$ (tensile steel yields),

$$M_{n2} = A_{s2}f_y(d - d')$$

Also, because $\sum H_F = 0$ and $NC_2 = NT_2$,

$$A_s'f_s' = A_{s2}f_y$$

If we assume that the compression steel yields and that $f_s' = f_y$, then

$$A_s'f_y = A_{s2}f_y$$

from which

$$A_s' = A_{s2}$$

Therefore,

$$M_{n2} = A_s'f_y(d - d')$$

The strength of the concrete–steel couple is evaluated as follows:

$$M_{n1} = N_{T1}Z_1$$

Assuming that the tensile steel yields and $f_s = f_y$,

$$M_{n1} = A_{s1}f_y\left(d - \frac{a}{2}\right)$$

Also, because $A_s = A_{s1} + A_{s2}$, then

$$A_{s1} = A_s - A_{s2}$$

and because $A_{s2} = A_s'$, then

$$A_{s1} = A_s - A_s'$$

Therefore,

$$M_{n1} = (A_s - A_s')f_y\left(d - \frac{a}{2}\right)$$

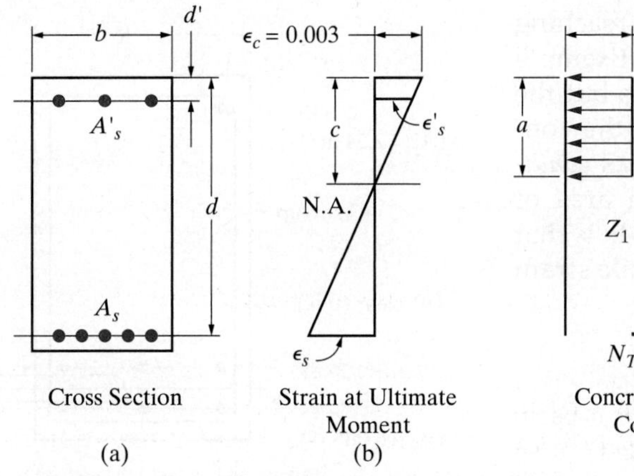
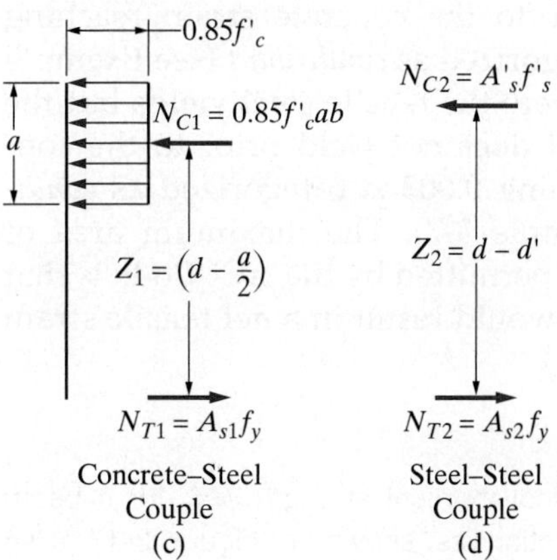

FIGURE 3-16 Doubly reinforced beam analysis.

Summing the two couples, we arrive at the nominal moment strength of a doubly reinforced beam:

$$M_n = M_{n1} + M_{n2}$$

$$= (A_s - A_s')f_y\left(d - \frac{a}{2}\right) + A_s'f_y(d - d')$$

The practical moment strength of ϕM_n may then be calculated.

The foregoing expressions are based on the assumption that both tension and compression steels yield prior to concrete strain reaching 0.003. This may be checked by determining the strains that exist at the nominal moment, which depend on the location of the neutral axis. The neutral axis may be located, as previously, by the depth of the compressive stress block and the relationship $a = \beta_1 c$. Thus,

$$N_T = N_{C1} + N_{C2}$$

$$A_s f_y = (0.85f_c')ab + A_s'f_y$$

$$a = \frac{(A_s - A_s')f_y}{(0.85f_c')b}$$

which may also be expressed as

$$a = \frac{A_{s1}f_y}{(0.85f_c')b}$$

With the calculation of the a distance, the neutral axis location c may be determined and the assumptions checked.

Check the net tensile strain ϵ_t in the extreme tensile reinforcement and ensure that $\epsilon_t \geq 0.004$ to satisfy the ACI Code, Section 9.3.3.1.

$$\epsilon_t = 0.003\frac{(d_t - c)}{c}$$

where d_t = the distance of the extreme tensile reinforcement from the compression face. The corresponding strength-reduction factor is then calculated as described in Section 2-9.

The case where both tensile and compressive steels yield prior to the concrete strain reaching 0.003 will be categorized as *condition I* (see Example 3-6). The case where the tensile steel yields but the compressive steel does not yield prior to the concrete strain reaching 0.003 is categorized as *condition II* (see Example 3-7). The maximum area of steel in the beam permitted by the ACI Code is that area of steel that would result in a net tensile strain of 0.004.

Example 3-6

Compute the practical moment strength ϕM_n for a beam having a cross section as shown in Figure 3-17. Use $f_c' = 3000$ psi and $f_y = 60{,}000$ psi.

Solution:

1. Assume that all the steel yields:

$$f_s' = f_y \quad \text{and} \quad f_s = f_y$$

Therefore, $A_{s2} = A_s'$ (see Figure 3-16d).

2.
$$A_s = A_{s1} + A_{s2}$$
$$A_{s1} = A_s - A_s'$$
$$= 6.00 - 2.54 = 3.46 \text{ in.}^2$$

From the concrete–steel couple, the stress block depth can be found:

$$a = \frac{A_{s1}f_y}{(0.85f_c')b} = \frac{3.46(60)}{0.85(3.0)(11)} = 7.40 \text{ in.}$$

3. Assuming that the same relationship ($a = \beta_1 c$) exists between the depth of the stress block and the beam's neutral axis as existed in singly reinforced beams, the neutral axis may now be located for purposes of checking the steel strains. From Figure 3-16 at the nominal moment,

$$c = \frac{a}{\beta_1} = \frac{a}{0.85} = \frac{7.40}{0.85} = 8.71 \text{ in.}$$

This value of c is based on the assumption in step 1 and will be verified in step 4.

4. Check the strains to determine whether the assumptions are valid and that both steels yield before the concrete crushes (see Figure 3-16b). The strains calculated exist at the nominal moment:

$$\epsilon_s' = \frac{c - d'}{c}(0.003)$$

$$= \frac{8.71 - 2.5}{8.71}(0.003) = 0.00214$$

Therefore, $\epsilon_s' > \epsilon_y = 0.00207$ (from Table A-1).

Check the ductility of the beam by calculating the net tensile strain in the extreme tensile reinforcement:

$$\epsilon_t = 0.003\frac{(d_t - c)}{c} = 0.003\frac{(21.2 - 8.71)}{8.71} = 0.0043$$

$$0.0043 > 0.004 \qquad \text{(O.K.)}$$

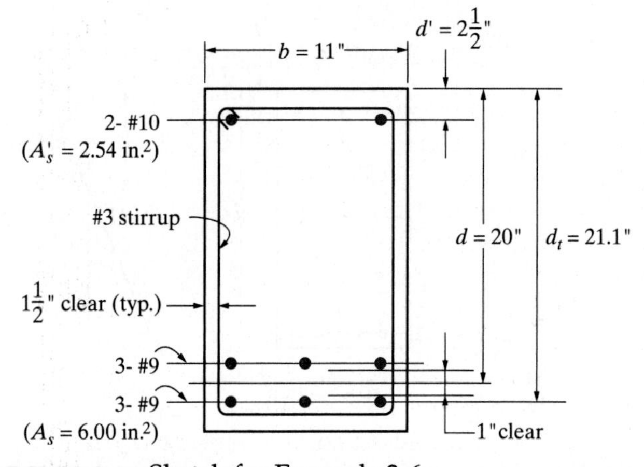

FIGURE 3-17 Sketch for Example 3-6.

Therefore, the tensile steel yields, and ductility is assured. Because $0.004 \leq \epsilon_t \leq 0.005$, this is a transition section, and the strength-reduction factor ϕ is calculated as follows:

$$\phi = 0.65 + (\epsilon_t - 0.002)\left(\frac{250}{3}\right)$$

$$= 0.65 + (0.0043 - 0.002)\left(\frac{250}{3}\right) = 0.84$$

$$0.65 < 0.84 < 0.90 \qquad \text{(O.K.)}$$

Because $\epsilon_s' > \epsilon_y$, the compression steel will yield before the concrete strain reaches 0.003, and $f_s' = f_y$. Therefore, the assumption concerning the compression steel stress is O.K.

5. From the concrete–steel couple:

$$M_{n1} = A_{s1}f_y\left(d - \frac{a}{2}\right)$$

$$= 3.46(60)\left(20 - \frac{7.40}{2}\right) = 3384 \text{ in.-kip}$$

$$\frac{3384}{12} = 282 \text{ ft.-kips}$$

From the steel–steel couple:

$$M_{n2} = A_s'f_y(d - d') = 2.54(60)(20 - 2.5)$$

$$= 2667 \text{ in.-kip}$$

$$\frac{2667}{12} = 222 \text{ ft.-kip}$$

$$M_n = M_{n1} + M_{n2}$$

$$= 282 + 222 = 504 \text{ ft.-kip}$$

6. $$\phi M_n = 0.84(504)$$

$$= 423 \text{ ft.-kip}$$

In this example, because both the compressive steel and tensile steel yield prior to the concrete reaching a compressive strain of 0.003 and because the net tensile strain ϵ_t is greater than 0.004, a ductile failure mode is assured.

3-9 DOUBLY REINFORCED BEAM ANALYSIS FOR MOMENT (CONDITION II)

As has been pointed out, usually the compression steel (A_s') will reach its yield stress before the concrete reaches a strain of 0.003. This may not occur in shallow beams reinforced with the higher-strength steels, however. Referring to Figure 3-16b, if the neutral axis is located relatively high in the cross section, it is possible that $\epsilon_s' < \epsilon_y$ at the nominal moment. The magnitude of ϵ_s' (and therefore f_s') depends on the location of the neutral axis. The depth of the compressive stress block a also depends on c, because $a = \beta_1 c$.

The total compressive force must be equal to the total tensile force $A_s f_y$, and an equilibrium equation can be written to solve for the exact required value of c. This turns out to be a quadratic equation. This situation and its solution may be observed in the following example.

Example 3-7

Compute the practical moment strength ϕM_n for a beam having a cross section as shown in Figure 3-18. Use $f_c' = 5000$ psi and $f_y = 60{,}000$ psi.

Solution:

1. Assume that all the steel yields. This results in

$$A_{s2} = A_s'$$

2. With reference to Figure 3-19b,

$$a = \frac{(A_s - A_s')f_y}{(0.85f_c')b}$$

$$= \frac{3.10(60)}{0.85(5)(11)}$$

$$= 3.98 \text{ in.}$$

3. Locate the neutral axis:

$$a = \beta_1 c, \quad \beta_1 = 0.80 \quad \text{(reference, ACI Code, Section 22.2.2.4.3)}$$

$$c = \frac{a}{\beta_1} = \frac{3.98}{0.80} = 4.98 \text{ in.}$$

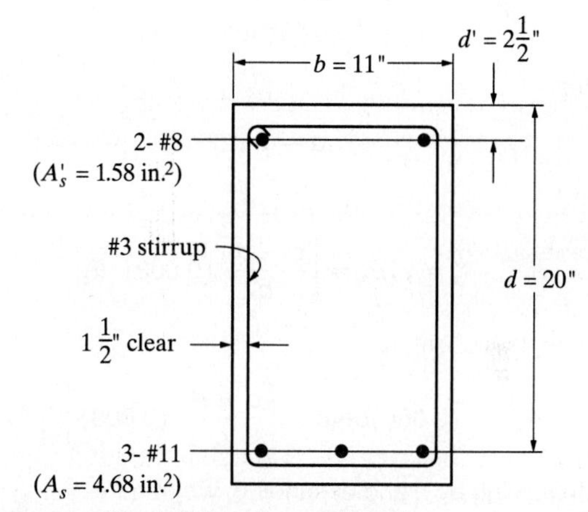

FIGURE 3-18 Sketch for Example 3-7.

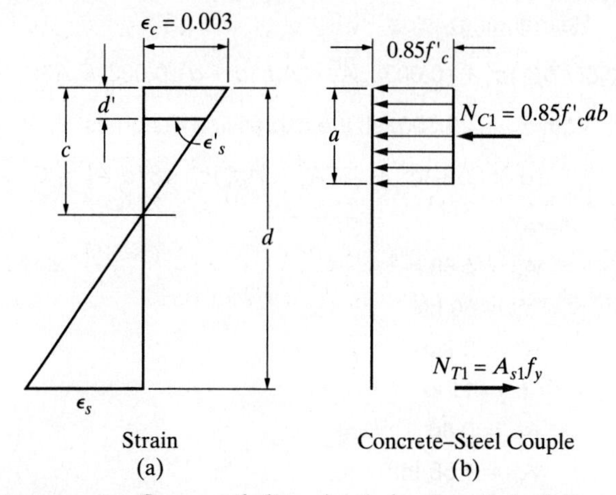

FIGURE 3-19 Compatibility check for Example 3-7.

This value of c is based on the assumption in step 1 and will be verified in step 4.

4. By similar triangles of Figure 3-19a, check the steel strains:

$$\text{compressive steel: } \epsilon'_s = \frac{0.003(c - d')}{c}$$

$$\epsilon_s = \frac{2.48}{4.98}(0.003) = 0.0015$$

Calculate the net tensile strain in the extreme reinforcement based on the depth of the neutral axis obtained in step 3. Note that $d_t = d$ and, therefore, $\epsilon_s = \epsilon_t$.

$$\epsilon_t = 0.003\frac{(d_t - c)}{c} = 0.003\frac{(20 - 4.98)}{4.98} = 0.009$$

$$0.009 > 0.004 \qquad \text{(O.K.)}$$

For grade 60 steel, $\epsilon_y = 0.00207$ (from Table A-1). Because $\epsilon_s > \epsilon_y < \epsilon'_s$, the tensile steel *has* yielded and the compression steel *has not* yielded. Therefore, the assumptions of step 1 are incorrect.

5. With the original assumptions incorrect, a solution for the location of the neutral axis must be established. With reference to Figure 3-16, c will be determined by using the condition that horizontal equilibrium exists. That is, $\sum H_F = 0$. Thus,

$$N_T = N_{C1} + N_{C2}$$
$$A_s f_y = (0.85f'_c)ba + f'_s A'_s$$

But

$$a = \beta_1 c$$

and

$$f'_s = \epsilon'_e E_s = \left[\frac{c - d'}{c}(0.003)\right]E_s$$

Then, by substitution,

$$A_s f_y = (0.85f'_c)b\beta_1 c + \left[\frac{c - d'}{c}(0.003)\right]E_s A'_s$$

Multiplying by c and expanding, we obtain

$$A_s f_y c = (0.85f'_c)b\beta_1 c^2 + c(0.003)E_s A'_s - d'(0.003)E_s A'_s$$

Rearranging yields

$$(0.85f'_c b\beta_1)c^2 + (0.003E_s A'_s - A_s f_y)c - d'(0.003)E_s A'_s = 0$$

With $E_s = 29{,}000$ ksi, the expression becomes

$$(0.85f'_c b\beta_1)c^2 + (87A'_s - A_s f_y)c - 87d'A'_s = 0$$

where

$$A_s = 4.68 \text{ in.}^2$$
$$f_y = 60 \text{ ksi}$$
$$f'_c = 5 \text{ ksi}$$
$$b = 11 \text{ in.}$$
$$\beta_1 = 0.80$$
$$A'_s = 1.58 \text{ in.}^2$$
$$d' = 2.5 \text{ in.}$$

Substitution yields

$$[0.85(5)(11)(0.80)]c^2 + [87(1.58) - 4.68(60)]c - 87(2.5)(1.58) = 0$$

$$37.4c^2 - 143.34c - 343.65 = 0$$
$$c^2 - 3.83c - 9.19 = 0$$

This may be solved using the usual formula for the roots of a quadratic equation:

$$\frac{-b \pm \sqrt{b^2 - 4ac}}{2a}$$

where the coefficients are

$$a = 1.0$$
$$b = -3.83$$
$$c = -9.19$$

Or the square may be completed as follows:

$$c^2 - 3.83c = 9.19$$

$$c^2 - 3.83c + \left(\frac{-3.83}{2}\right)^2 = 9.19 + \left(\frac{-3.83}{2}\right)^2$$

$$c^2 - 3.83c + 3.67 = 9.19 + 3.67 = 12.86$$
$$(c - 1.92)^2 = 12.86$$
$$c - 1.92 = \sqrt{12.86} = 3.59$$
$$c = 5.51 \text{ in.}$$

The solution of the quadratic equation for c may be simplified as follows:

$$c = \pm\sqrt{Q + R^2} - R$$

where

$$R = \frac{87A'_s - A_s f_y}{1.7f'_c b\beta_1}$$

$$Q = \frac{87d'A'_s}{0.85f'_c b\beta_1}$$

Note that basic units are kip and inches, so the value of f_y, for instance, must be in ksi, not in psi.

6. With this value of c, all the remaining unknowns may be found:

$$f'_s = \frac{c - d'}{c}(87) = \frac{5.51 - 2.50}{5.51}(87)$$

$$= 47.5 \text{ ksi} < 60 \text{ ksi} \quad \text{(as expected)}$$

7. $$a = \beta_1 c = 0.80(5.51) = 4.41 \text{ in.}$$

The actual net tensile strain is calculated as

$$\epsilon_t = 0.003\frac{(d_t - c)}{c} = 0.003\frac{(20 - 5.51)}{5.51} = 0.0079$$

$$0.0079 > 0.004 \qquad \text{(O.K.)}$$

Thus, the beam is ductile as assumed and because $\epsilon_t > 0.005$, the strength-reduction factor $\phi = 0.90$, as discussed in Section 2-9.

8. $N_{C1} = (0.85f_c')ab = 0.85(5)(4.41)(11.0) = 206.2 \text{ kip}$ (O.K.)

$$N_{C2} = A_s'f_s' = 47.5(1.58) = 75.1 \text{ kip}$$

$$N_C = 281.3 \text{ kip}$$

$$\text{Check: } N_T = A_sf_y = 4.68(60) = 281 \text{ kip}$$

$$N_T \approx N_C$$

9. $M_{n1} = N_{C1}Z_1 = N_{C1}\left(d - \frac{a}{2}\right) = 206.2\left(20 - \frac{4.41}{2}\right)$
$= 3670$ in.-kip

$M_{n2} = N_{C2}Z_2 = N_{C2}(d - d') = 75.1(20 - 2.5)$
$= 1314$ in.-kip

$M_n = 4984$ in.-kip $= 415.3$ ft.-kip

10. $\phi M_n = 0.9(415.3)$
$= 373.8$ ft.-kip

3-10 SUMMARY OF PROCEDURE FOR ANALYSIS OF DOUBLY REINFORCED BEAMS (FOR MOMENT)

1. Assume that all the steel yields, $f_s = f_s' = f_y$. Therefore,

$$A_{s2} = A_s'$$

2. Using the concrete–steel couple and $A_{s1} = A_s - A_s'$, compute the depth of the compression stress block:

$$a = \frac{A_{s1}f_y}{(0.85f_c')b} = \frac{(A_s - A_s')f_y}{(0.85f_c')b}$$

3. Compute the location of the neutral axis:

$$c = \frac{a}{\beta_1}$$

This value of c is based on the assumption in step 1 and will be verified in step 4.

4. Using the strain diagram, check the strain in the compression reinforcement and the net tensile strain in the extreme tensile reinforcement to determine whether the assumption in step 1 is valid:

$$\epsilon_s' = \frac{0.003(c - d')}{c}$$

$$\epsilon_t = \frac{0.003(d_t - c)}{c}$$

It is required that $\epsilon_t \geq 0.004$. Therefore, the tensile steel has yielded ($0.004 > \epsilon_y$). The following two conditions may exist. In each of the two cases, the strength-reduction factor ϕ must be determined as discussed in Section 2-9.

a. Condition I: $\epsilon_s' \geq \epsilon_y$. This indicates that the assumption of step 1 is correct and the compression steel has yielded.

b. Condition II: $\epsilon_s' < \epsilon_y$. This indicates that the assumption of step 1 is incorrect and the compression steel has not yielded.

Condition I

5. If ϵ_s' and ϵ_s both exceed ϵ_y, compute the nominal moment strengths Mn_1 and Mn_2. For a steel–steel couple:

$$M_{n2} = A_s'f_y(d - d')$$

For a concrete–steel couple:

$$M_{n1} = A_{s1}f_y\left(d - \frac{a}{2}\right)$$

and

$$M_n = M_{n1} + M_{n2}$$

6. Practical moment strength = ϕM_n.

Condition II

5. If ϵ_s' is less than ϵ_y and $\epsilon_s \geq \epsilon_y$, compute c using the following formula:

$$(0.85f_c'b\beta_1)c^2 + (87A_s' - A_sf_y)c - 87d'A_s' = 0$$

and solve the quadratic equation for c, or use the simplified formula approach from Example 3-7, step 5. Note that the basic units are kip and inches.

6. Compute the compressive steel stress (to be less than f_y):

$$f_s' = \frac{c - d'}{c}(87)$$

7. Solve for a using

$$a = \beta_1 c$$

To check ductility, recalculate the net tensile strain,

$$\epsilon_t = 0.003\frac{(d_t - c)}{c}$$

The strength-reduction factor ϕ is determined as discussed in Section 2-9.

8. Compute the compressive forces:

$$N_{C1} = (0.85f_c')ba$$

$$N_{C2} = A_s'f_s'$$

Check these by computing the tensile force:

$$N_T = A_sf_y$$

Note that N_T should equal $N_{C1} + N_{C2}$.

9. Compute the ideal resisting moment strengths of the individual couples:

$$M_{n1} = N_{C1}\left(d - \frac{a}{2}\right)$$

and

$$M_{n2} = N_{C2}(d - d')$$
$$M_n = M_{n1} + M_{n2}$$

10. Practical moment strength $= \phi M_n$.

3-11 DOUBLY REINFORCED BEAM DESIGN FOR MOMENT

If a check shows that a singly reinforced rectangular section is inadequate and the size of the beam cannot be increased, a doubly reinforced section may be designed using a procedure that consists of the separate design of the two component couples such that their summation will result in a beam of the required strength.

Example 3-8

Design a rectangular reinforced concrete beam to carry a design moment M_u of 697 ft.-kip. Physical limitations require that $b = 14$ in. and $h = 30$ in. If compression steel is needed, $d' = 3$ in. Use $f'_c = 3000$ psi and $f_y = 60{,}000$ psi. The beam cross section is shown in Figure 3-20.

Solution:

1. Assume that $d = h - 4 = 26$ in. (because of the probability of two rows of steel).
2. The design moment M_u is given; $M_u = 697$ ft.-kip.
3. Determine if a singly reinforced beam will work. From Table A-8, maximum ρ for $\epsilon_t = 0.005$ is 0.01355. This is if $d = d_t$. For $d_t > d$, the maximum ρ for moment-strength calculation can be found by proportion. Assuming a #8 bar, #3 stirrups, and $1\frac{1}{2}$-in. cover,

$$d_t = 30 - 1.5 - 0.38 - \frac{1.00}{2} = 27.6 \text{ in.}$$

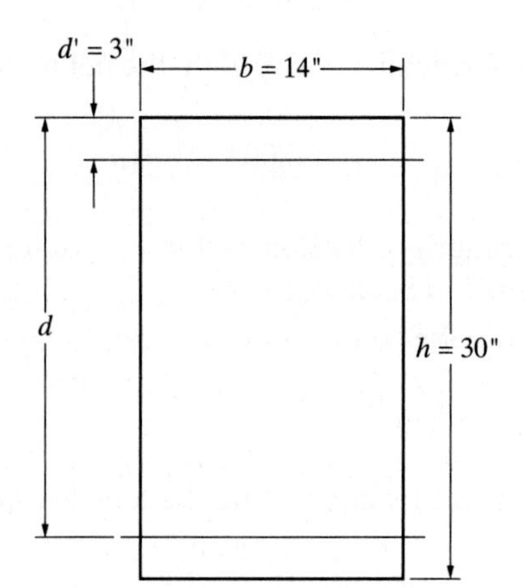

FIGURE 3-20 Sketch for Example 3-8.

Then

$$\rho_{\max} = \frac{d_t}{d}(\rho_{max} \text{ from Table A-8})$$
$$= \frac{27.6}{26}(0.01355) = 0.01438$$

The associated $\bar{k}$ (Table A-8) is 0.7177 ksi. Assuming a tension-controlled section ($\epsilon_t \geq 0.005$), ϕ will be 0.90. This assumption will be checked later.

$$\text{maximum } \phi M_n = \phi b d^2 \bar{k}$$
$$= \frac{0.90(14)(26^2)(0.7177)}{12} = 509 \text{ ft.-kip}$$

4. 509 ft.-kip < 697 ft.kip. Therefore, a doubly reinforced beam is required.
5. Provide a concrete–steel couple having ϕMn_1 of 509 ft.-kip. Therefore, $\rho = 0.01438$ and $\bar{k} = 0.7177$ ksi:

$$\text{required } A_{s1} = \rho b d = 0.01438(14)(26) = 5.23 \text{ in.}^2$$

6. The steel–steel couple must be proportioned to have moment strength equal to the remainder of the design moment:

$$\text{required } \phi M_{n2} = M_u - \phi M_{n1} = 697 - 509 = 188 \text{ ft.-kip}$$

7. Considering the steel–steel couple, we have

$$\phi M_{n2} = \phi N_{C2}(d - d')$$
$$N_{C2} = \frac{\phi M_{n2}}{\phi(d - d')} = \frac{188(12)}{0.90(263)} = 109.0 \text{ kip}$$

8. Because $N_{C2} = A'_s f'_s$, compute f'_s using the neutral axis location of the concrete–steel couple and check the strain ϵ'_s in the compression steel (see Figure 3-21).

$$a = \frac{A_{s1} f_y}{0.85 f'_c b} = \frac{5.23(60)}{0.85(3)(14)} = 8.79 \text{ in.}$$
$$c = \frac{a}{\beta_1} = \frac{8.79}{0.85} = 10.34 \text{ in.}$$
$$\epsilon'_s = \frac{(c - d')}{c}(0.003) = \frac{(10.34 - 3.00)}{10.34}(0.003)$$
$$= 0.00213$$

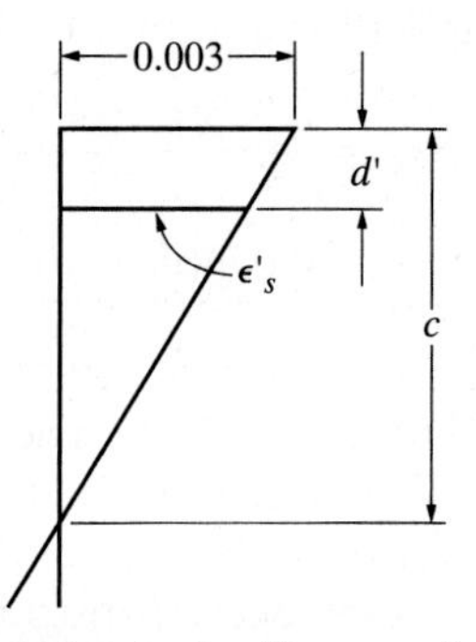

FIGURE 3-21 Concrete strain diagram for Example 3-8.

$\epsilon_y = 0.00207$ from Table A-1. Because $\epsilon'_s > \epsilon_y$, the compression steel will yield before the concrete strain reaches 0.003, and $f'_s = f_y$.

9. Determine the required compression steel:

$$\text{required } A'_s = \frac{N_{C2}}{f'_s} = \frac{N_{C2}}{f_y} = \frac{109.0}{60} = 1.82 \text{ in.}^2$$

10. Because $f'_s = f_y$, required A_{s2} = required $A'_s = 1.82$ in.^{2}
11. The total required tension steel is

$$A_s = A_{s1} + A_{s2} = 5.23 + 1.82 = 7.05 \text{ in.}^2$$

12. Select the compression steel. Two No. 9 bars will provide $A'_s = 2.00$ in.^{2}
13. Select the tension steel. Six No. 10 bars will provide $A_s = 7.62$ in.^{2} Place the tension steel in two layers of three bars each with 1 in. clear between layers. Minimum beam width for three #10 bars is 10.5 in., from Table A-3.
14. Assume No. 3 stirrups and determine the actual depth to the centroid of the bar group by considering the total depth (30 in.), required cover ($1\frac{1}{2}$ in.), stirrup size (No. 3), tension bar size (No. 10), and required 1-in. minimum clear space between layers:

$$\text{actual } d = 30 - 1.5 - 0.38 - 1.27 - 0.5 = 26.35 \text{ in.}$$

The assumed d was 26 in.

$$26.35 \text{ in.} > 26 \text{ in.}$$

15. Check d_t.

$$d_t = 30 - 1.5 - 0.38 - \frac{1.27}{2} = 27.5 \text{ in.} \approx 27.6 \text{ in.}$$

(Say O.K.)

Recalculating the design is an option. The differences will be very small.

16. Check the assumption of step 3 ($\phi = 0.90$). We will use the cross section designed (see the design sketch: Figure 3-22).

$$A_s = 7.62 \text{ in.}^2, A'_s = 2.00 \text{ in.}^2$$
$$d = 26.4 \text{ in. (step 14)}, d_t = 27.5 \text{ in. (step 15)}$$
$$d' = 1.5 + 0.38 + 1.13/2 = 2.45 \text{ in.}$$

Assume $f_s = f'_s = f_y$; $A_{s2} = A'_s$

$$a = \frac{(A_s - A'_s)f_y}{0.85f'_c b} = \frac{(7.62 - 2.00)(60)}{0.85(3)(14)} = 9.45 \text{ in.}$$

$$c = \frac{a}{\beta_1} = \frac{9.45}{0.85} = 11.12 \text{ in.}$$

$$\epsilon'_s = \frac{0.003(c - d')}{c} = \frac{0.003(11.12 - 2.45)}{11.12}$$
$$= 0.00234 > 0.00207 \quad \text{(O.K.)}$$

$$\epsilon_t = \frac{0.003(d_t - c)}{c} = \frac{0.003(27.5 - 11.12)}{11.12}$$
$$= 0.00442 > 0.004$$

Therefore, this is in the transition zone and ϕ must be reduced below 0.90:

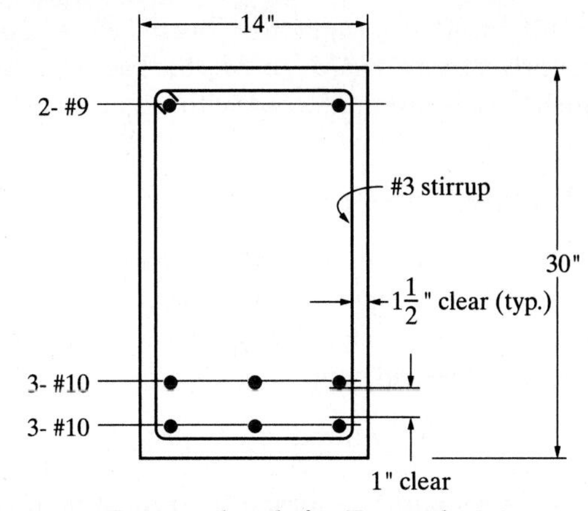

FIGURE 3-22 Design sketch for Example 3-8.

$$\phi = 0.65 + (\epsilon_t - 0.002)\left(\frac{250}{3}\right)$$
$$= 0.65 + (0.00442 - 0.002)\left(\frac{250}{3}\right) = 0.852$$

$$M_{n1} = A_{s1}f_y\left(d - \frac{a}{2}\right)$$
$$= \frac{(7.62 - 2.00)(60)\left(26.4 - \frac{9.45}{2}\right)}{12} = 609 \text{ ft.-kip}$$

$$M_{n2} = A'_s f_y(d - d') = \frac{2.00(60)(26.4 - 2.45)}{12}$$
$$= 240 \text{ ft.-kip}$$

$$\phi M_n = \phi(M_{n1} + M_{n2}) = 0.852(609 + 240)$$
$$= 723 \text{ ft.-kip}$$
$$723 \text{ ft.-kip} > 697 \text{ ft.-kip} \quad \text{(O.K.)}$$

17. Figure 3-22 is a sketch of the design.

3-12 SUMMARY OF PROCEDURE FOR DESIGN OF DOUBLY REINFORCED BEAMS (FOR MOMENT)

The size of the beam cross section is fixed.

1. Assume that $d = h - 4$ in. Estimate d_t.
2. Establish the total design moment M_u.
3. Check to see if a doubly reinforced beam is necessary. Compute maximum ϕM_n for a singly reinforced beam. Assume a tension-controlled section (net tensile strain $\epsilon_t \geq 0.005$) and use the corresponding strength-reduction factor, $\phi = 0.90$; use maximum $\bar{k}$ from the tables in Appendix A and the corresponding maximum steel ratio ρ_{max} for $\epsilon_t = 0.005$. If $d_t > d$, adjust ρ_{max} for $\epsilon_t = 0.005$ and select maximum $\bar{k}$ accordingly.

$$\text{maximum } \phi M_n = \phi b d^2 \bar{k}$$

4. If $\phi M_n < M_u$, design the beam as a doubly reinforced beam. If $\phi M_n \geq M_u$, the beam can be designed as a beam reinforced with tension steel only.

For a Doubly Reinforced Beam

5. Provide a concrete–steel couple having the maximum ϕM_n from step 3. This is ϕMn_1.

   Using ρ from step 3, find the steel required for the concrete–steel couple:

$$\text{required } A_{s1} = \rho bd$$

6. Find the remaining moment that must be resisted by the steel–steel couple:

$$\text{required } \phi M_{n2} = M_u - \phi M_{n1}$$

7. Considering the steel–steel couple, find the required compressive force in the steel (assume that $d' = 3$ in.):

$$N_{C2} = \frac{\phi M_{n2}}{\phi(d - d')}$$

8. Because $N_{C2} = A_s' f_s'$, compute f_s' so that A_s' may eventually be determined. This can be accomplished by using the neutral-axis location of the concrete–steel couple and checking the strain ϵ_s' in the compression steel with ϵ_y from Table A-1. Thus,

$$a = \frac{A_{s1} f_y}{(0.85 f_c')b}$$

$$c = \frac{a}{\beta_1}$$

$$\epsilon_s' = \frac{0.003(c - d')}{c}$$

   If $\epsilon_s' \geq \epsilon_y$, the compressive steel has yielded at the nominal moment and $f_s' = f_y$. If $\epsilon_s' < \epsilon_y$, then calculate $f_s' = \epsilon_s' E_s$ and use this stress in the following steps.

9. Because $N_{C2} = A_s' f_s'$,

$$\text{required } A_s' = \frac{N_{C2}}{f_s'}$$

10. Determine the required As_2:

$$A_{s2} = \frac{f_s' A_s'}{f_y}$$

11. Find the total tension steel required:

$$A_s = A_{s1} + A_{s2}$$

12. Select the compressive steel (f_s').
13. Select the tensile steel (A_s). Check the required beam width. Preferably, place the bars in one layer.
14. Check the actual d and compare it with the assumed d. If the actual d is slightly in excess of the assumed d, the design will be slightly conservative (on the safe side). If the actual d is less than the assumed d, the design may be on the unconservative side and an analysis and possible revision should be considered.
15. Check d_t and compare with the assumed d_t.
16. Check the ϕ value assumption.
17. Sketch the design.

3-13 ADDITIONAL CODE REQUIREMENTS FOR DOUBLY REINFORCED BEAMS

The compression steel in beams, whether it is in place to increase flexural strength or to control deflections, will act similarly to all typical compression members in that it will tend to buckle, as shown in Figure 3-23. Should this buckling occur, it will naturally be accompanied by spalling of the concrete cover. To help guard against this type of failure, the ACI Code requires that the compression bars be tied within the beam in a manner similar to that used for reinforced concrete columns (discussed in Chapter 9). Compression reinforcement in beams or girders must be enclosed by ties or stirrups. The size of the ties or stirrups is to be at least No. 3 for No. 10 longitudinal bars or smaller and No. 4 for No. 11 longitudinal bars or larger. The center-to-center spacing of the ties or stirrups is not to exceed the smaller of 16 longitudinal bar diameters, 48 tie (or stirrup) bar diameters, or the least dimension of the beam. Alternatively, welded wire fabric of equivalent area may be used. The ties or stirrups are to be used throughout the area where compression reinforcement is required (see ACI Code, Section 9.7.6.4).

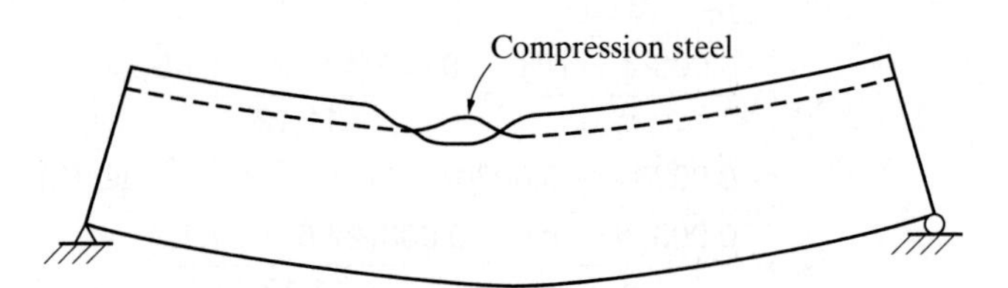

FIGURE 3-23 Possible failure mode for compression steel.

Problems

The following note applies to all the problems in this chapter. *Unless otherwise noted, assume No. 3 stirrups, $1\frac{1}{2}$-in. cover for beams, and 1-in. clear space between layers of bars. In all problems, check the net tensile strain ϵ_t in the extreme tension steel to ensure that it is within the allowable limits. Unless otherwise noted or shown, $d_t = d$.*

3-1. Find ϕM_n for the following T-beam: $b = 36$ in., $b_w = 12$ in., $h_f = 4$ in., $d = 22$ in., $f'_c = 4000$ psi, and $f_y = 60{,}000$ psi. The reinforcement is four No. 8 bars.

3-2. Rework Problem 3-1 with $b = 48$ in.

3-3. Rework Problem 3-1 with reinforcement of three No. 11 bars.

3-4. Rework Problem 3-1 with $f'_c = 5000$ psi.

3-5. The simple-span T-beam shown is part of a floor system of span length 20 ft-0 in. and beam spacing 45 in. o.c. Use $b = 45$ in., $f'_c = 3000$ psi and $f_y = 60{,}000$ psi. The bars are placed with 1-in. clear space between layers.

a. Find the practical moment strength ϕM_n.

b. How much steel would be required in this beam cross section to make the compressive stress block just completely cover the flange?

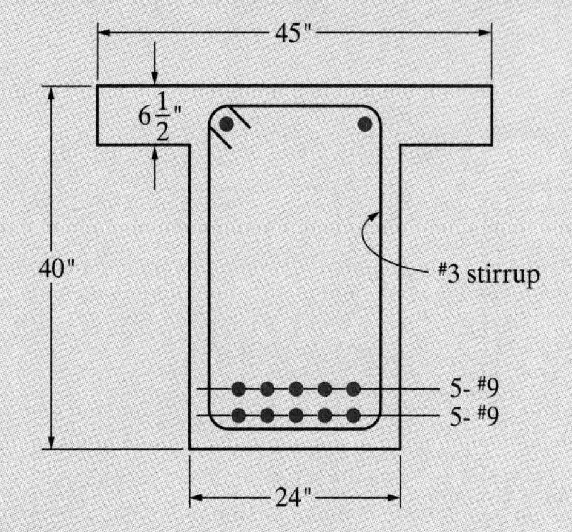

PROBLEM 3-5

3-6. Find ϕM_n for a typical T-beam in the floor system shown. The beams span 24 ft and are spaced 6′-0″ on center. Use $f'_c = 3000$ psi and $f_y = 60{,}000$ psi.

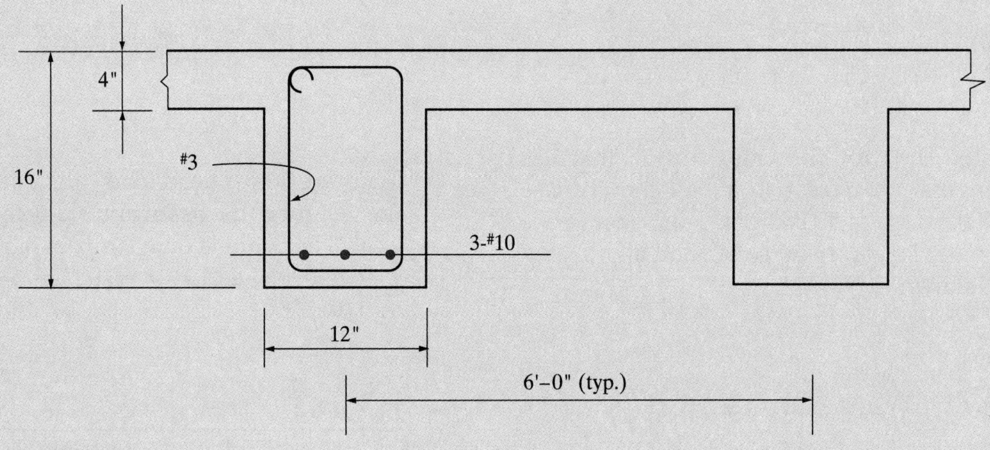

PROBLEM 3-6

3-7. Find the practical moment strength ϕM_n for the T-beam in the floor system shown. The beam span is 31 ft-6 in. Use $f'_c = 4000$ psi and $f_y = 60{,}000$ psi.

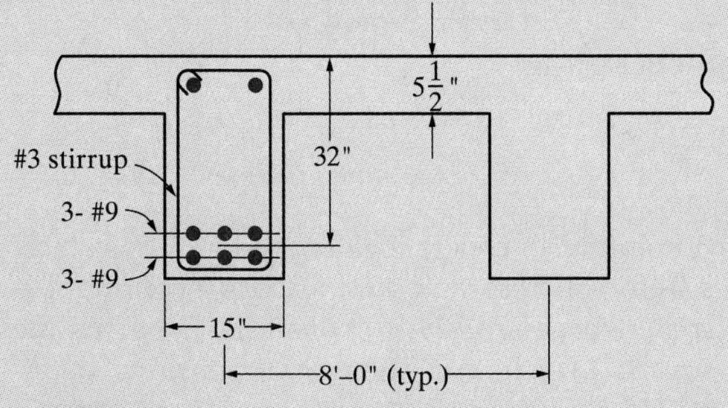

PROBLEM 3-7

3-8. The T-beam shown is on a simple span of 30 ft. Use $f'_c = 3000$ psi and $f_y = 60{,}000$ psi. No dead load exists other than the weight of the floor system.

Assume that the slab design is adequate.

a. Find the practical moment strength ϕM_n.

b. Compute the permissible service live load that can be placed on the floor (psf).

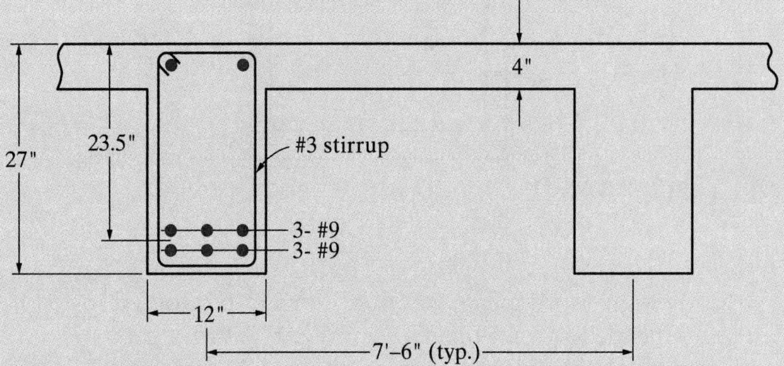

PROBLEM 3-8

3-9. The simple-span T-beam shown is part of a floor system of span length 20 ft-0 in. and beam spacing of 8 ft-0 in. o.c. Find the practical moment strength ϕM_n. Use $f_c' = 3000$ psi and $f_y = 60{,}000$ psi. Assume $d_t = 27$ in.

3-10. Find ϕM_n for the beams of cross section shown. Assume that the physical dimensions are acceptable.

a. Spandrel beam with a flange on one side only. Use $f_c' = 4000$ psi and $f_y = 60{,}000$ psi.

b. Box beam: $f_c' = 3000$ psi and $f_y = 60{,}000$ psi.

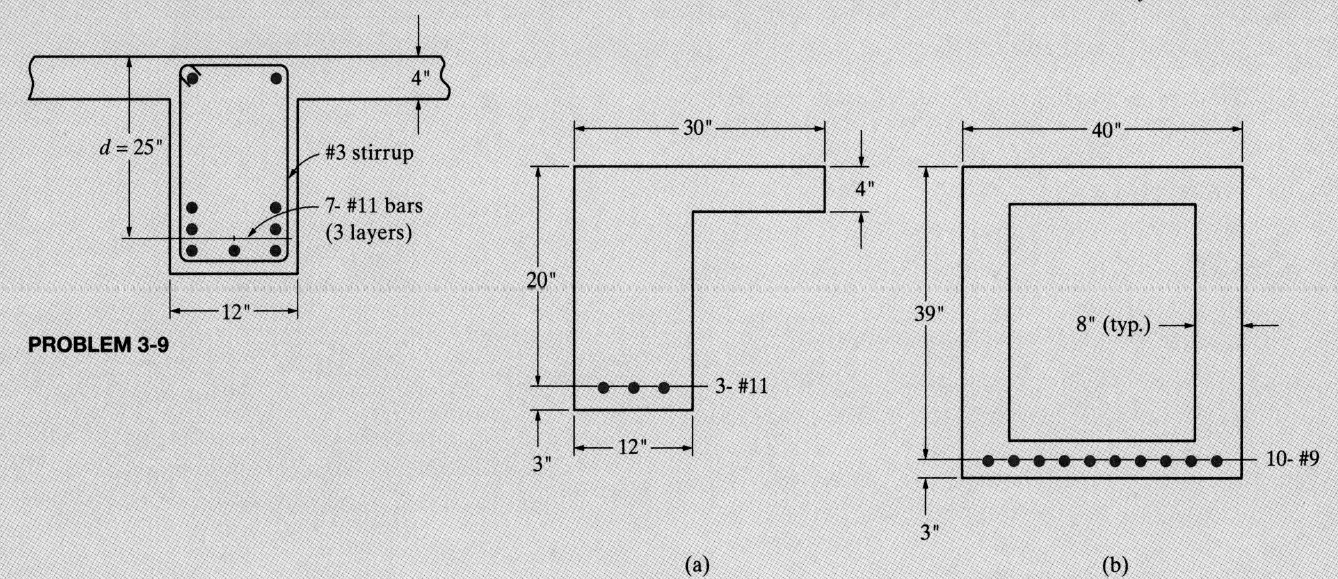

PROBLEM 3-9

(a) (b)

PROBLEM 3-10

3-11. Determine ϕM_n for the cross section that has a rectangular duct cast in it, as shown. Use $f_c' = 3000$ psi, $f_y = 60{,}000$ psi, 1-in. clear space between bar layers, $1\frac{1}{2}$-in. cover, and eight No. 7 bars, as shown.

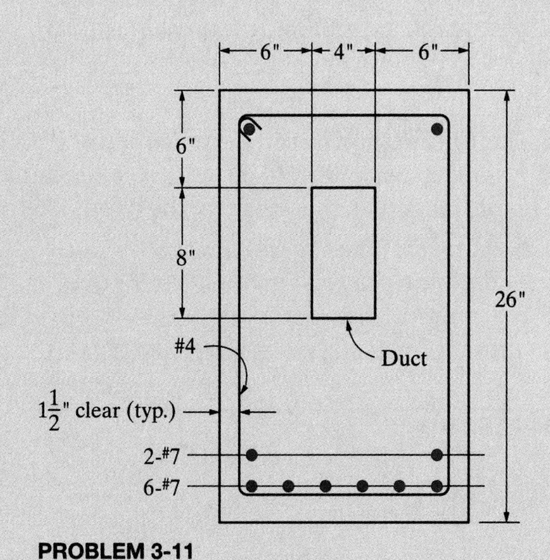

PROBLEM 3-11

3-12. Design a typical interior tension-reinforced T-beam to resist positive moment. A cross section of the floor system is shown. The service loads are 50 psf dead load (this does not include the weight of the beam and slab) and 325 psf live load. The beam is on a simple span of 18 ft. Use $f_c' = 4000$ psi and $f_y = 60{,}000$ *psi*.

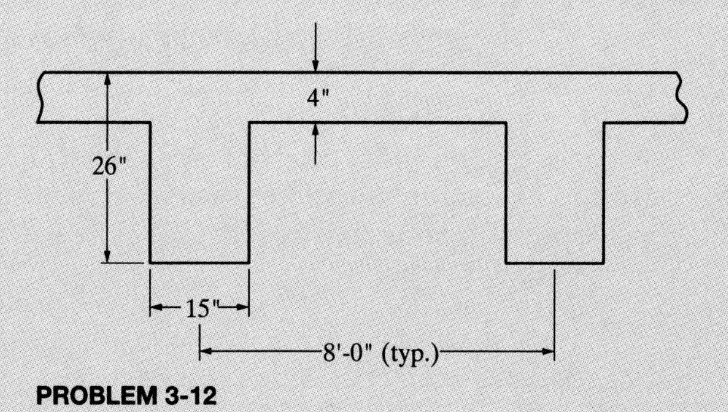

PROBLEM 3-12

3-13. A reinforced concrete floor system consists of a 3-in. concrete slab supported by continuous-span T-beams of 24-ft spans. The T-beams are spaced 4 ft-8 in. o.c. The web dimensions, determined by negative moment and shear requirements at the supports, are shown. Select the steel required at midspan to resist a total positive design moment M_u of 575 ft.-kip (this includes the weight of the floor system). Use $f_c' = 3000$ psi and $f_y = 60{,}000$ psi.

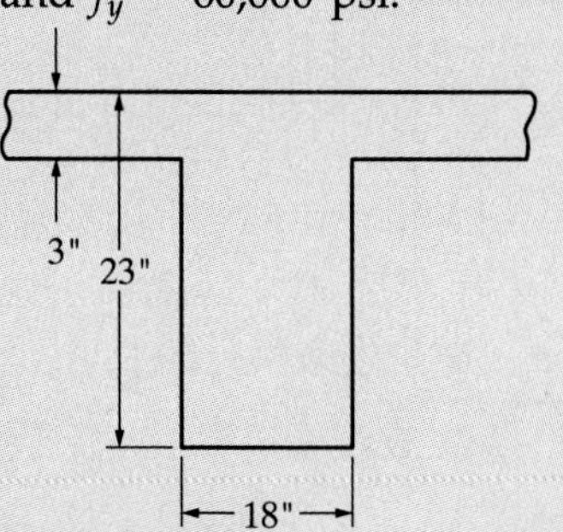

PROBLEM 3-13

3-14. A reinforced concrete floor system is to have a 4-in.-thick slab supported on 16-in.-wide beams, as shown. At one location, penetrations through the floor slab limit the effective flange width for the supporting beam to 20 in. (Use $b = 20$ in.) The positive factored moment M_u at this section is 320 ft.-kip. Design the T-beam using $f_c' = 3000$ psi and $f_y = 60{,}000$ psi.

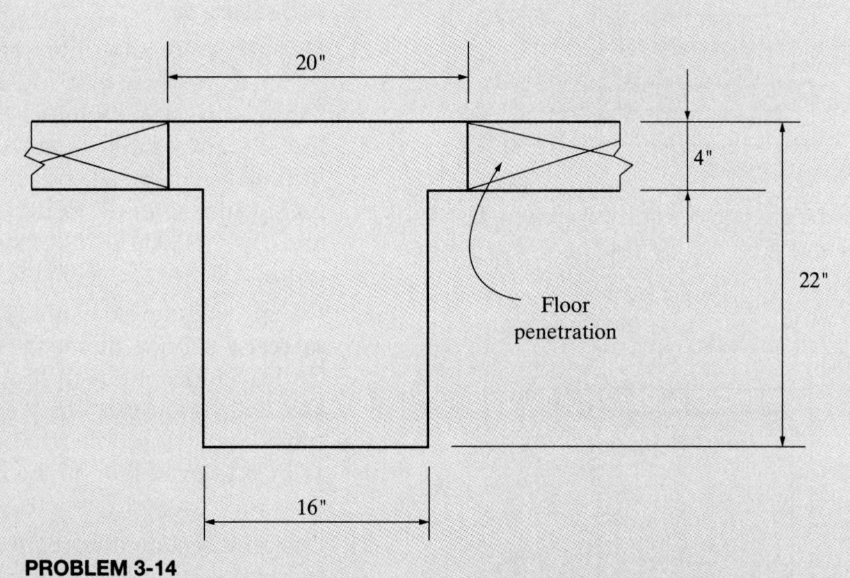

PROBLEM 3-14

3-15. Select steel for the beams of cross section shown. The positive service moments are 160 ft.-kip live load and 100 ft.-kip dead load (this includes beam weight). Use $f_c' = 3000$ psi and $f_y = 60{,}000$ psi.

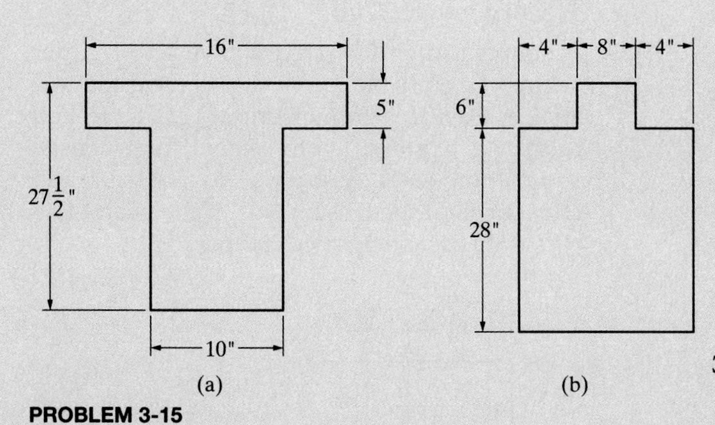

PROBLEM 3-15

3-16. Find ϕM_n for the beam of cross section shown. Use $f_c' = 4000$ psi and $f_y = 60{,}000$ psi.

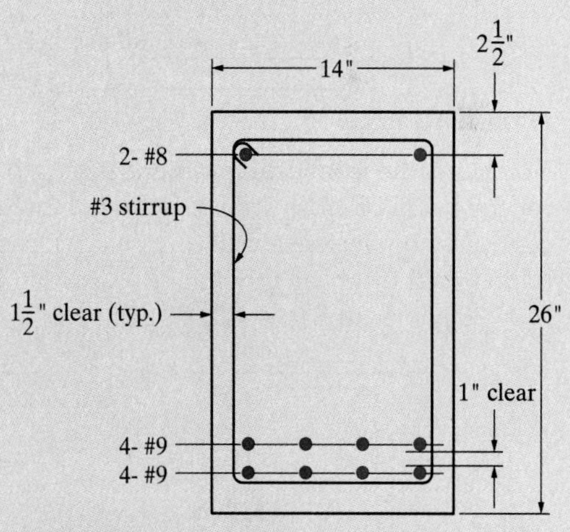

PROBLEM 3-16

3-17. The beam of cross section shown is to span 28 ft on simple supports. The uniform load on the beam (in addition to its own weight) will be composed of equal service dead load and live load. Use $f_c' = 4000$ psi and $f_y = 60{,}000$ psi.

a. Find the service loads that the beam can carry (in addition to its own weight).

b. Compare the practical moment strength ϕM_n of the beam as shown with ϕM_n of a beam of similar size reinforced with the area of steel associated with (1) $\epsilon_t = 0.005$ and (2) $\epsilon_t = 0.004$.

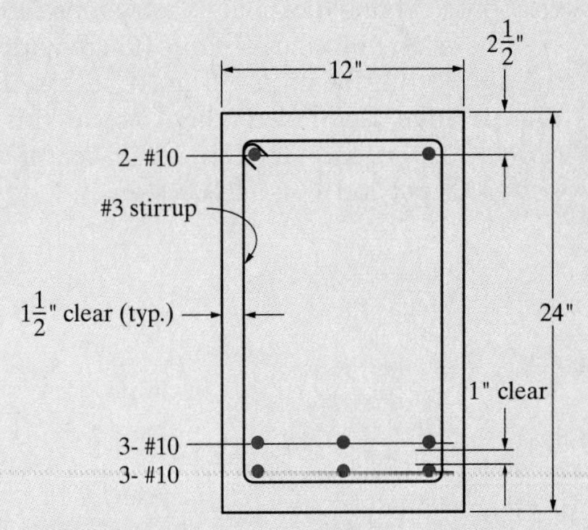

PROBLEM 3-17

3-18. Find ϕM_n for the beam cross section shown. Use $f_c' = 4000$ psi and $f_y = 60{,}000$ psi.

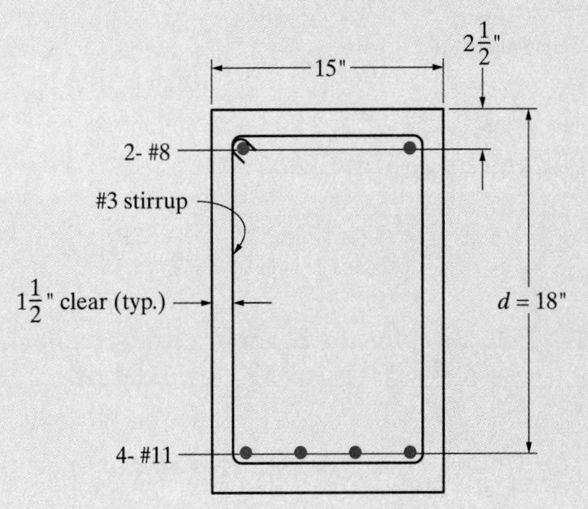

PROBLEM 3-18

3-19. Compute the practical moment strength ϕM_n for the beam of cross section shown. How much can ϕM_n be increased if four No. 8 bars are added to the top of the beam? Use $d' = 2\frac{1}{2}$ in., $f_c' = 3000$ psi, and $f_y = 60{,}000$ psi.

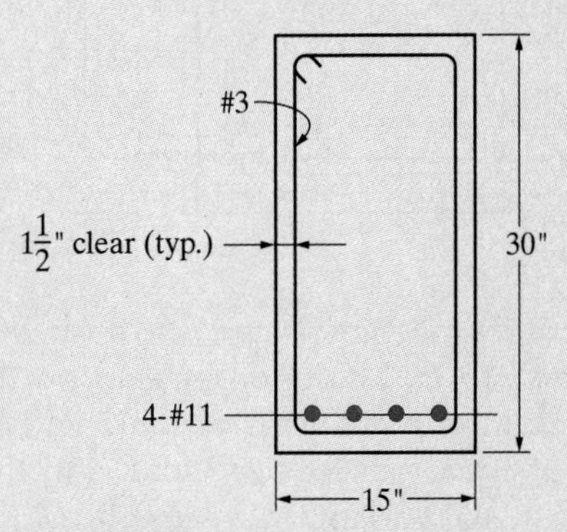

PROBLEM 3-19

3-20. Compute the practical moment strength ϕM_n for the simply supported precast inverted T-girder shown. Use $f_c' = 3000$ psi and $f_y = 60{,}000$ psi.

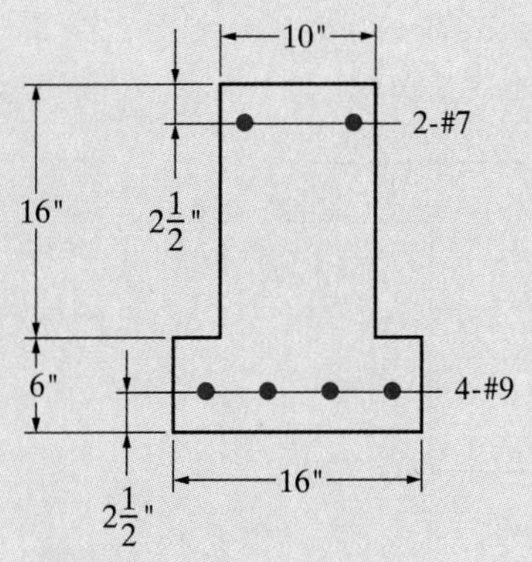

PROBLEM 3-20

3-21. Design a rectangular reinforced concrete beam to resist a total design moment M_u of 765 ft.-kip (this includes the moment due to the weight of the beam). The beam size is limited to 15 in. maximum width and 30 in. maximum overall depth. Use $f_c' = 3000$ psi and $f_y = 60{,}000$ psi. If compression steel is required, make $d' = 2\frac{1}{2}$ in.

3-22. Design a rectangular reinforced concrete beam to resist service moments of 150 ft.-kip dead load (includes moment due to weight of beam) and 160 ft.-kip due to live load. Architectural considerations require that width be limited to 11 in. and overall depth be limited to 23 in. Use $f_c' = 3000$ psi and $f_y = 60{,}000$ psi.

3-23. Design a rectangular reinforced concrete beam to carry service loads of 1.25 kip/ft dead load (includes beam weight) and 2.60 kip/ft live load. The beam is a simple span and has a span length of 18 ft. The overall dimensions are limited to width of 10 in. and overall depth of 20 in. Use $f_c' = 3000$ psi and $f_y = 60{,}000$ psi.

3-24. Redesign the beam of Problem 3-23 for tension reinforcing only and increased width. Keep an overall depth of 20 in.

3-25. A simply supported precast inverted T-girder having the cross section shown is subjected to a total positive design moment M_u of 280 ft.-kip. Select the required reinforcement (both tensile and compressive). Assume a No. 3 stirrup and $1\frac{1}{2}$ in. cover on the tension side of the beam. Use $f_c' = 4000$ psi and $f_y = 60{,}000$ psi.

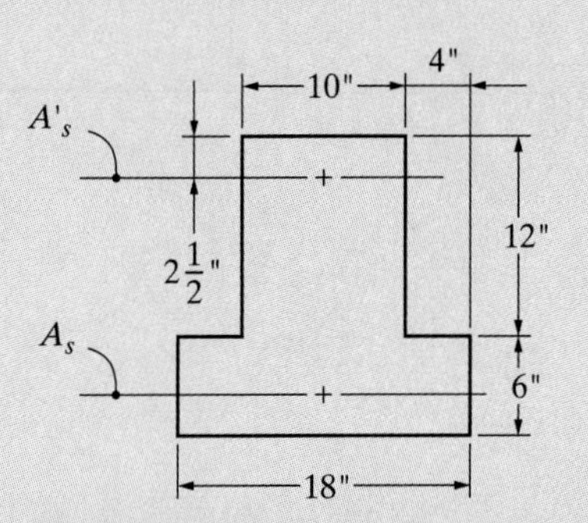

PROBLEM 3-25

SHEAR AND TORSION

4-1 Introduction
4-2 Shear Reinforcement Design Requirements
4-3 Shear Analysis Procedure
4-4 Stirrup Design Procedure
4-5 Torsion of Reinforced Concrete Members
4-6 Corbels and Brackets

4-1 INTRODUCTION

In prior chapters we have been concerned primarily with the bending strength of reinforced concrete beams and slabs. The shear forces create additional tensile stresses that must be considered. In these members steel reinforcing must be added specifically to provide additional shear strength if the shear is in excess of the shear strength of the concrete itself.

The concepts of bending stresses and shearing stresses in homogeneous elastic beams are generally discussed at great length in most strength-of-materials texts. The accepted expressions are

$$f = \frac{Mc}{I} \quad \text{and} \quad v = \frac{VQ}{Ib} \tag{4-1}$$

where f, M, c, and I are as defined in Chapter 1; v is the shear stress; V is the external shear; Q is the statical moment of area about the neutral axis; and b is the width of the cross section.

All points in the length of the beam, where the shear and bending moment are not equal to zero, and at locations other than the extreme fiber or neutral axis, are subject to both shearing stresses and bending stresses. The combination of these stresses is of such a nature that maximum normal and shearing stresses at a point in a beam exist on planes that are inclined with respect to the axis of the beam. It can be shown that maximum and minimum normal stresses exist on two perpendicular planes. These planes are commonly called the *principal planes*, and the stresses that act on them are called *principal stresses*. The principal stresses in a beam subjected to shear and bending may be calculated using the following formula:

$$f_{pr} = \frac{f}{2} \pm \sqrt{\frac{f^2}{4} + v^2} \tag{4-2}$$

where f_{pr} is the principal stress and f and v are the bending and shear stresses, respectively, calculated from Equation (4-1).

The *orientation* of the principal planes may be calculated using the following formula:

$$\tan 2\alpha = \frac{2v}{f} \tag{4-3}$$

where α is the angle measured from the horizontal.

The magnitudes of the shearing stresses and bending stresses vary along the length of the beam and with distance from the neutral axis. It then follows that the inclination of the principal planes as well as the magnitude of the principal stresses will also vary. At the neutral axis the principal stresses will occur at a 45° angle. This may be verified by Equation (4-3), substituting $f = 0$, from which $\tan 2\alpha = \infty$ and $\alpha = 45°$.

In Figures 4-1a and 4-1b, we isolate a small, square unit element from the neutral axis of a beam (where $f = 0$). The vertical shear stresses are equal and opposite on the two vertical faces by reason of equilibrium. If these were the only two stresses present, the element would rotate.

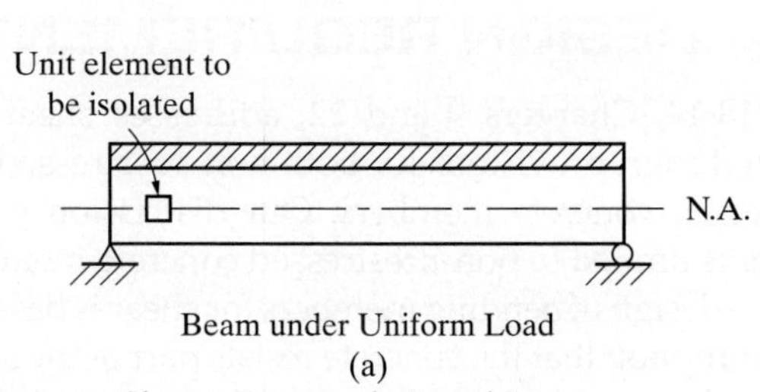

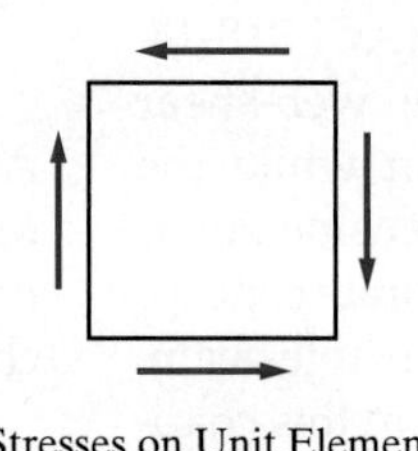

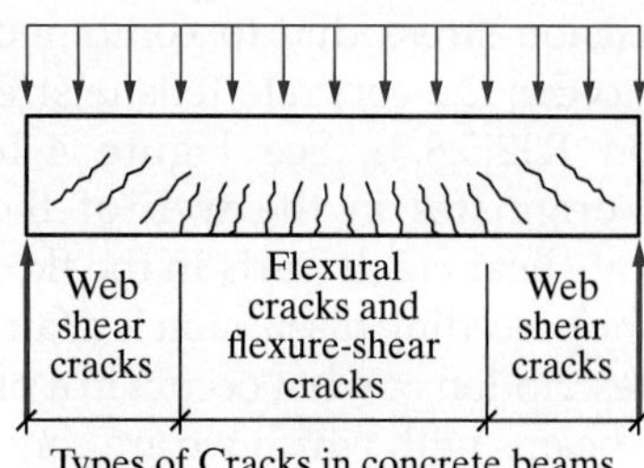

FIGURE 4-1 Shear stress relationship.

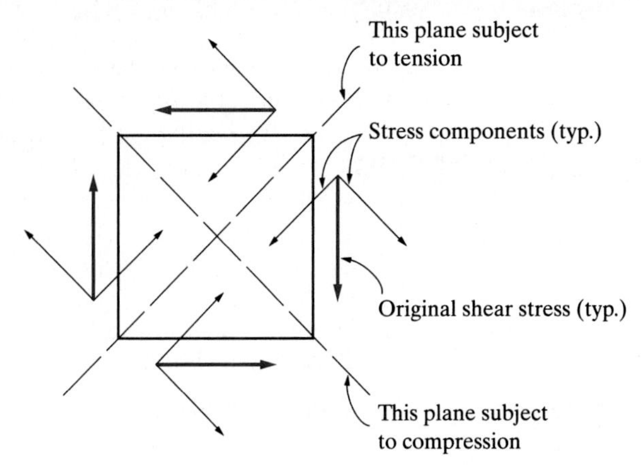

FIGURE 4-2 Effect of shear stresses on inclined planes.

Therefore, there must exist equal and opposite horizontal shear stresses on the horizontal faces and of the same magnitude as the vertical shear stresses. (The concept of horizontal shear stresses equal in magnitude to the vertical shear stresses at any point in a beam can also be found in almost any strength-of-materials text.)

If we consider a set of orthogonal planes that are inclined at 45° with respect to the original element and resolve the shear stresses into components that are parallel and perpendicular to these planes, the effect will be as shown in Figure 4-2. Note that the components combine so that one of the inclined planes is in compression while the other is in tension. Concrete is strong in compression but weak in tension, and there is a tendency for the concrete to crack on the plane subject to tension should the stress become large enough. The tensile force resulting from the tensile stress acting on a diagonal plane has historically been designated as *diagonal tension*. When it becomes large enough, it will necessitate that shear reinforcing be provided.

As stated previously, tensile stresses of various inclinations and magnitudes, resulting from either shear alone or the combined action of shear and bending, exist in all parts of a beam and must be taken into consideration in both analysis and design. There are two types of inclined cracks in a concrete flexural member: diagonal web-shear cracks which occur near the zone of maximum shear when the principal tensile stress exceeds the concrete tensile strength, and flexural-shear cracks which start out as flexural cracks and then becomes an inclined crack: they occur when the tension stress due to combined shear and bending exceeds the concrete tensile strength (ACI 318-14, Section R22.5.8.3). See Figure 4-1c. The web-shear crack originates in the web of the beam while the flexural-shear crack starts in the flexural tension zone.

The preceding discussion is a fairly accurate conceptual description of what occurs in a plain concrete beam. In the beams with which we are concerned in this chapter, where the length over which a shear failure could occur (the *shear span*) is in excess of approximately three times the effective depth, the diagonal tension failure

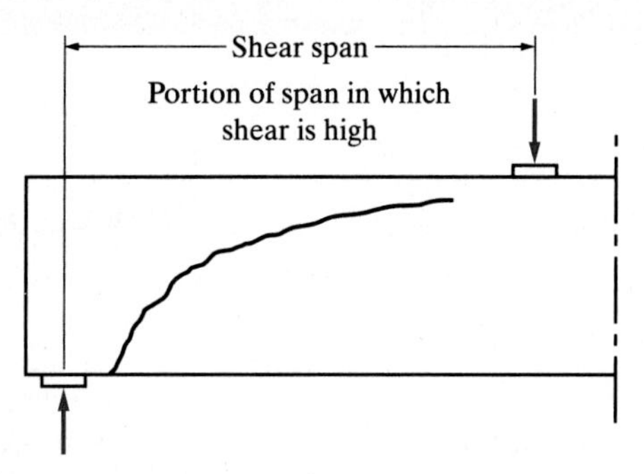

FIGURE 4-3 Typical diagonal tension failure.

would be the mode of failure in shear. Such a failure is shown in Figure 4-3. For shorter spans, the failure mode would actually be some combination of shear, crushing, and splitting. The design of deep members such as pile caps and deep beams which have relatively short shear spans (a_v/d less than or equal to 2.0) is covered in Chapter 10. For the longer shear spans in plain concrete beams, cracks due to flexural tensile stresses would occur long before cracks due to the diagonal tension. The earlier flexural cracks would initiate the failure, and shear would be of little consequence. In a concrete beam reinforced for flexure (moment) where tensile strength is furnished by steel, however, tensile stresses due to flexure and shear will continue to increase with increasing load. The steel placed in the beam to reinforce for moment is not located where the large diagonal tension stresses (due to shear) occur. The problem then becomes one of furnishing additional reinforcing steel to resist the diagonal tension stresses.

Considerable research over the years has attempted to establish the exact distribution of the shear stresses over the depth of the beam cross section. Despite extensive studies and ongoing research, the precise shear-failure mechanism is still not fully understood.

As with several previous codes, ACI 318-14 furnishes design guidelines for shear reinforcement based on the vertical shear force V_u that develops at any given cross section of a member. Although it is really the diagonal tension for which shear reinforcing must be provided, diagonal tensile forces (or stresses) are not calculated. Historically, vertical shear force (and in older codes, vertical shear stress) has been taken to be a good indicator of the diagonal tension present.

4-2 SHEAR REINFORCEMENT DESIGN REQUIREMENTS

ACI 318-14, Chapters 9 and 22, addresses shear and torsion design provisions for both non-prestressed and prestressed concrete members. Our discussion in this chapter is limited to non-prestressed concrete members.

The design of bending members for shear is based on the assumption that the concrete resists part of the shear, and any excess over and above what the concrete is capable of resisting has to be resisted by shear reinforcement.

The basic rationale for the design of the shear reinforcement, or *web reinforcement* as it is usually called in beams, is to provide steel to cross the diagonal tension cracks and subsequently keep them from opening. Visualizing this basic rationale with reference to Figure 4-3, it is seen that the web reinforcement may take several forms.

The code allows vertical stirrups and welded wire reinforcement with wires located perpendicular to the axis of the member as well as spirals, circular ties, or hoops.

Additionally, for non-prestressed members, the code allows shear reinforcement to be composed of inclined or diagonal stirrups and main reinforcement bent to act as inclined stirrups.

The most common form of web reinforcement used is the vertical stirrup. The web reinforcement contributes very little to the shear resistance *prior* to the formation of the inclined cracks but appreciably increases the *ultimate* shear strength of a bending member. After the concrete cracks, the shear strength provided by the concrete alone is due to the friction from the interlocking aggregates within the concrete, the dowel action of the longitudinal reinforcement crossing the shear crack, and the shear resistance of the uncracked concrete in the compression zone.

For members of normal-weight concrete that are subject to shear and flexure only, the amount of shear force that the concrete alone, unreinforced for shear, can resist is V_c:

$$V_c = 2\lambda\sqrt{f'_c}b_w d \qquad \textbf{[ACI Eq. (22.5.5.1)]}$$

In the expression for V_c, the terms are as previously defined with units for f'_c in psi and units for b_w and d in inches. Lambda (λ) is described in Chapter 1 and for normal-weight concrete, $\lambda = 1.0$. In this chapter we will consider only normal-weight concrete and therefore λ will be omitted. V_c will be in units of pounds. For rectangular beams, b_w is equivalent to b. The nominal shear strength of the concrete will be reduced to a dependable shear strength by applying a strength-reduction factor ϕ of 0.75 (ACI Code, Section 21.2.2). Should members be subject to other effects of axial tension or compression, other expressions for V_c can be found in Section 22.5 of the code. Also, it is permitted to calculate V_c using a more detailed calculation (ACI 318-14, Section 22.5.6.1). For most designs, it is convenient and conservative to use ACI Equation (22.5.5.1).

The design shear force is denoted V_u and results from the application of factored loads. Values of V_u are most conveniently determined using a typical shear force diagram. Theoretically, no web reinforcement should be required if $V_u \le \phi V_c$. The code, however, requires that a minimum area of shear reinforcement be provided in all reinforced concrete flexural members where V_u exceeds $\frac{1}{2}\phi V_c$ except as follows:

1. In slabs and footings
2. In concrete joist construction as defined by the ACI Code, Section 9.8
3. In beams with a total depth of not greater than 10 in., $2\frac{1}{2}$ times the flange thickness, or one-half the width of the web, whichever is greater. With regards to item no. 3, research and tests of large wide beams [1] suggests that size effect, which is not modeled by ACI Eq. (22.5.5.1), and the size of the shear cracks plays a major role in the determining the shear capacity of these beams. The researchers found that "beams without stirrups fail in shear at lower values of shear stress as the members become deeper and as the maximum aggregate becomes smaller." Therefore, they recommend that beams with an overall depth of 30 inches or more "contain at least minimum shear reinforcement" throughout the length of the beam [1].
4. In beams built integral with the slab and with total depth not exceeding 24 in.

This provision of the code is primarily to guard against those cases where an unforeseen overload would cause failure of the member due to shear. Tests have shown the shear failure of a flexural member to be sudden and without warning. In cases where it is not practical to provide shear reinforcement (e.g. spread footings and one-way slabs) and sufficient thickness is provided to resist V_u, the minimum area of web reinforcement is not required. In cases where shear reinforcement is required for strength or because $V_u > \frac{1}{2}\phi V_c$, the minimum area of shear reinforcement shall be calculated from ACI Code Table 9.6.3.3 as

$$A_v = 0.75\sqrt{f'_c}\frac{b_w s}{f_{yt}} \ge \frac{50 b_w s}{f_{yt}}$$

In the preceding equation and with reference to Figure 4-4,

A_v = total cross-sectional area of web reinforcement within a distance s; for single-loop or 2-leg stirrups, $A_v = 2A_s$, where A_s is the cross-sectional area of one leg of the stirrup bar (in.^{2}). For double loop or 4-leg stirrups, $A_v = 4A_s$. For triple loop or 6-leg stirrups, $A_v = 6A_s$. (see Figure 4-4b). Note that closed loop stirrups are used for spandrel beams and beams subject to torsion.

b_w = web width = b for rectangular sections (in.)

s = center-to-center spacing of shear reinforcement in a direction parallel to the longitudinal reinforcement (in.)

f_{yt} = yield strength of web reinforcement steel (psi)

Note that for $f'_c \le 4444$ psi, the minimum area of shear reinforcement will be controlled by

$$\frac{50 b_w s}{f_{yt}}$$

When determining the shear strength V_c of reinforced or prestressed concrete beams and concrete joist construction, the value of $\sqrt{f'_c}$ is limited to 100 psi unless minimum web reinforcement is provided.

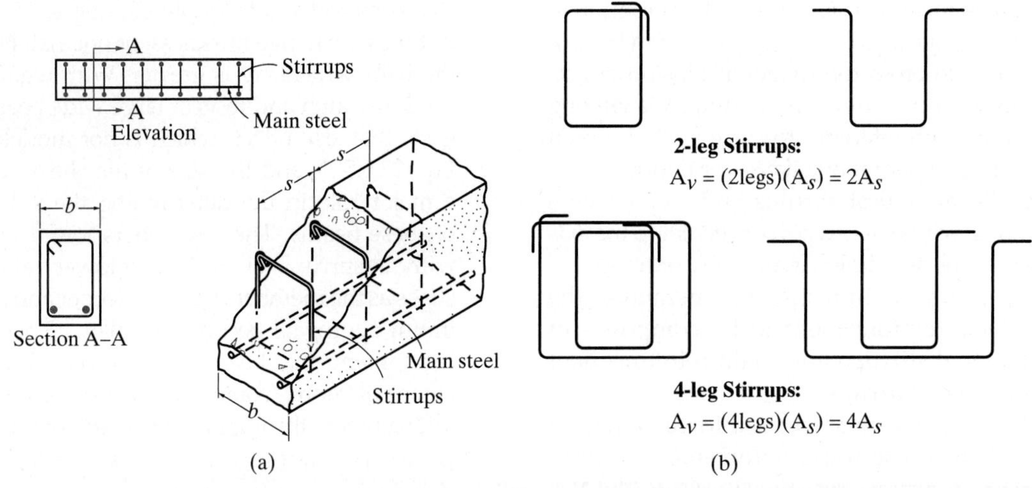

FIGURE 4-4 **(a)** Isometric section showing stirrups partially exposed. **(b)** Types of Stirrups.

In any span, that portion in which web reinforcement is theoretically necessary can be determined by using the shear (V_u) diagram. When the applied shear V_u exceeds the capacity of the concrete web ϕV_c, web reinforcement is required. In addition, according to the code, web reinforcement at least equal to the minimum required must be provided elsewhere in the span, where the applied shear is greater than one-half of ϕV_c. The ACI Code, Section 9.5.1.1(b), states that the basis for shear design at the ultimate limit state must be

$$\phi V_n \geq V_u$$

where

$$V_n = V_c + V_s \qquad \textbf{[ACI Eq. (22.5.1.1)]}$$

from which

$$\phi V_c + \phi V_s \geq V_u$$

where V_u, ϕ, and V_c are as previously defined; V_n is the total nominal shear strength; and V_s is the nominal shear strength provided by shear reinforcement. In the design process, the design of the stirrups usually follows the selection of the beam size. Therefore, V_c can be determined, as can the complete shear (V_u) diagram. The stirrups to be designed will provide the shear strength, V_s. Therefore, it is convenient to write the preceding expression as

$$\text{required } \phi\, V_s = V_u - \phi V_c$$

For vertical stirrups, V_s may be calculated from

$$V_s = \frac{A_v f_{yt} d}{s} \qquad \textbf{[ACI Eq. (22.5.10.5.3)]}$$

where all terms are as previously defined. Also, for inclined stirrups at 45°, V_s may be calculated using

$$V_s = \frac{1.414 A_v f_{yt} d}{s}$$

from ACI Equation (22.5.10.5.4) where s is the horizontal center-to-center distance of stirrups parallel to the main longitudinal steel. Note that for the inclined stirrups, the orientation of the stirrups must be such that the stirrups interrupt or cross the diagonal tension cracks.

It will be more practical if ACI Equations (22.5.10.5.3) and (22.5.10.5.4) are rearranged as expressions for spacing, because the stirrup bar size, strength, and beam-effective depth are usually predetermined. The design is then for stirrup spacing. For vertical stirrups,

$$\text{required } s = \frac{A_v f_{yt} d}{\text{required } V_s}$$

Because it is the required ϕV_s that will be conveniently determined, the preceding expression is rewritten

$$\text{required } s = \frac{\phi A_v f_{yt} d}{\text{required } \phi V_s}$$

or

$$\text{required } s = \frac{\phi A_v f_{yt} d}{V_u - \phi V_c}$$

Similarly, for 45° stirrups,

$$\text{required } s = \frac{1.414 \phi A_v f_{yt} d}{V_u - \phi V_c}$$

Note that these equations give the *maximum* spacing of stirrups based on *required strength*.

4-3 SHEAR ANALYSIS PROCEDURE

The shear analysis procedure involves checking the shear strength in an existing member and verifying that the various code requirements have been satisfied. The member may be reinforced or plain.

Example 4-1

A reinforced concrete beam of rectangular cross section shown in Figure 4-5 is reinforced for moment only (no shear reinforcement). Beam width $b = 18$ in., $d = 10.25$ in., and the reinforcing is five No. 4 bars. Calculate the maximum

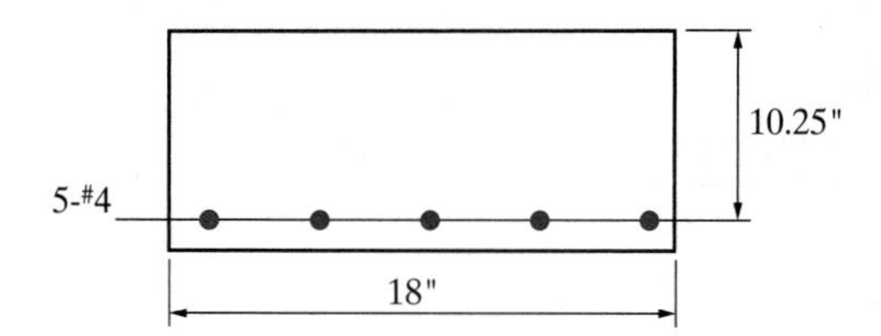

FIGURE 4-5 Cross section, Example 4-1.

factored shear force V_u permitted on the member by the ACI Code. Use $f'_c = 4000$ psi and $f_y = 60{,}000$ psi.

Solution:

Because no shear reinforcement is provided, the ACI Code, Section 9.6.3.1, requires that V_u not exceed $0.5\phi V_c$:

$$\begin{aligned}\text{maximum } V_u &= 0.5\ \phi V_c \\ &= 0.5\ \phi(2\sqrt{f'_c}b_w d) \\ &= 0.5(0.75)(2)(\sqrt{4000})(18)(10.25) \\ &= 8750 \text{ lb}\end{aligned}$$

Example 4-2

A reinforced concrete beam of rectangular cross section shown in Figure 4-6 is reinforced with seven No. 6 bars in a single layer. Beam width $b = 18$ in., $d = 33$ *in.*, single-loop No. 3 stirrups are placed 12 in. on center, and typical cover is $1\frac{1}{2}$ in. Find V_c, V_s, and the maximum factored shear force V_u permitted on this member. Use $f'_c = 4000$ psi and $f_y = 60{,}000$ psi.

Solution:

V_c and V_s will be expressed in units of kip.

$$V_c = 2\sqrt{f'_c}b_w d = \frac{2\sqrt{4000}(18)(33)}{1000} = 75.1 \text{ kip}$$

$$V_s = \frac{A_v f_{yt} d}{s} = \frac{2(0.11)(60)(33)}{12} = 36.3 \text{ kip}$$

$$\text{maximum } V_u = \phi V_c + \phi V_s = 0.75(75.1 + 36.3) = 83.6 \text{ kip}$$

In the general case of shear analysis, one must ensure that at all locations in the member, $\phi V_c + \phi V_s \geq V_u$. In addition, all other details of the reinforcement pattern must be checked to ensure that they comply with code provisions. (Refer to "Notes on Stirrup Design" in Section 4-4.)

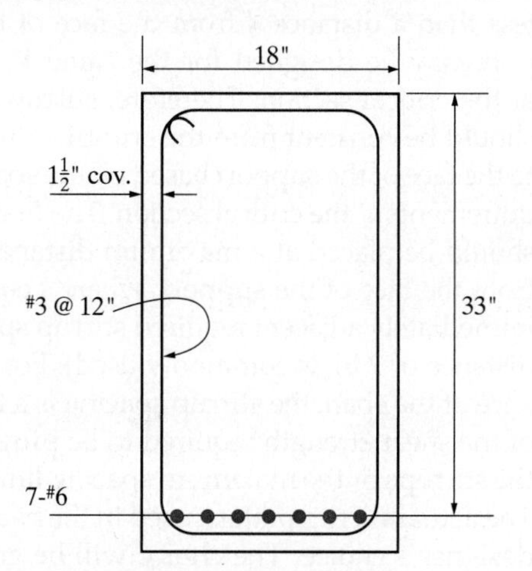

FIGURE 4-6 Cross section, Example 4-2.

4-4 STIRRUP DESIGN PROCEDURE

In the design of stirrups for shear reinforcement, the end result is a determination of stirrup size and spacing pattern. A *general procedure* may be adopted as follows:

1. Determine the shear values based on clear span and draw the factored shear diagram and determine the factored shear (V_u) at the critical section. The location of the critical section is dependent on whether confining action is present at the beam supports and is determined according to ACI 318-14, Section 9.4.3.2 (see Figure 4-7). If the uniform load on the beam in Figure 4-7a is applied at or near the bottom of the beam rather than at the top of the beam, the location of the critical section will occur at the face of the support.
2. Determine if stirrups are required.
3. Determine the length of span over which stirrups are required (assuming that stirrups *are* required).
4. On the V_u diagram, determine the area representing "required ϕV_s." This will display the required strength of the stirrups to be provided.
5. Select the size of the stirrup. See item 2a in "Notes on Stirrup Design." Find the spacing required at the critical section (a distance d from the face of the support). See "Notes on Stirrup Design" item 3b.
6. Establish the ACI Code maximum spacing requirements.
7. Determine the spacing requirements based on shear strength to be furnished by web reinforcing.
8. Establish the spacing pattern and show sketches.

Notes on Stirrup Design

1. Materials and maximum stresses
   a. To reduce excessive crack widths in beam webs subject to diagonal tension, the ACI Code, Section 20.2.2.4, limits the design yield strength of shear reinforcement to 60,000 psi. This increases to 80,000 psi for deformed welded wire reinforcing.
   b. The value of V_s must not exceed $8\sqrt{f'_c}b_w d$ irrespective of the amount of web reinforcement (ACI Code, Section 22.5.1.2).
2. Bar sizes for stirrups
   a. The most common stirrup size used is a No. 3 bar. Under span and loading conditions where the shear values are relatively large, it may be necessary to use a No. 4 bar. Rarely is anything larger than a No. 4 bar stirrup ever required, however. In large beams, multiple stirrup sets are sometimes provided in which a diagonal crack would be crossed by four or more vertical bars at one location of a beam. Single-loop stirrups, as shown in Figure 4-4, are generally satisfactory

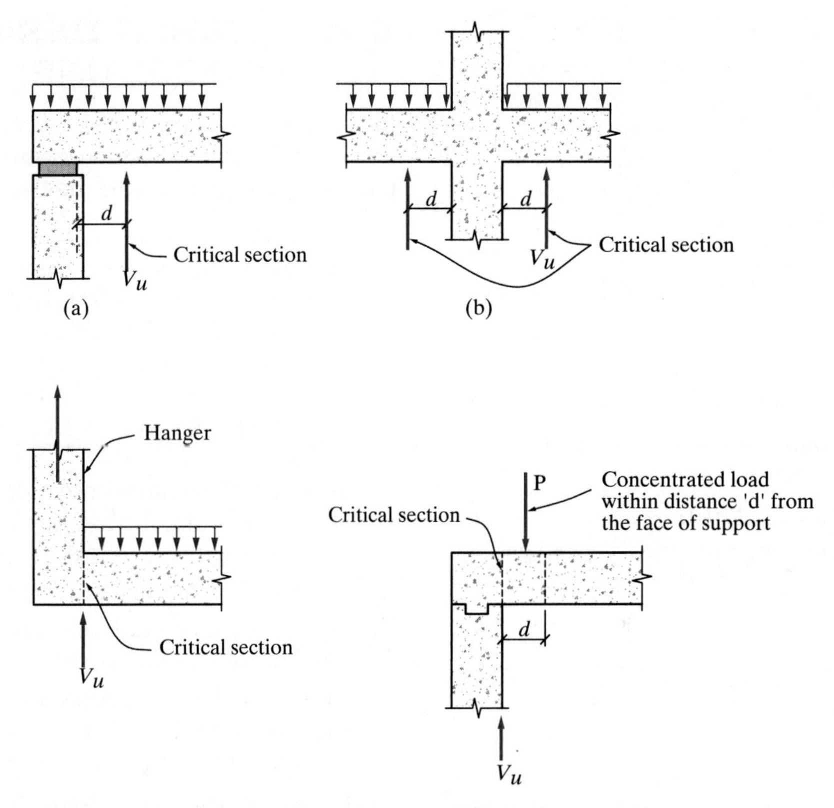

FIGURE 4-7 Location of critical section for shear.

for $b \le 24$ in.; double-loop stirrups are satisfactory for 24 in. $< b \le 48$ in.; and triple-loop stirrups are satisfactory for $b > 48$ in.

b. When conventional single-loop stirrups are used, the web area A_v provided by each stirrup is twice the cross-sectional area of the bar (No. 3 bars, $A_v = 0.22$ in.^{2}; No. 4 bars, $A_v = 0.40$ in.^{2}) because each stirrup crosses a diagonal crack twice.

c. If possible, do not vary the stirrup bar sizes; use the same bar sizes unless all other alternatives are not reasonable. Spacing should generally be varied and size held constant.

3. Stirrup spacings

a. When stirrups are required, the maximum spacing for vertical stirrups must not exceed $d/2$ or 24 in., whichever is smaller (ACI Code, Section 9.7.6.2.2). If V_s exceeds $4\sqrt{f'_c}b_w d$, the maximum spacing must not exceed $d/4$ or 12 in., whichever is smaller (ACI Code, Section 9.7.6.2.2). The maximum spacing may also be governed by the relationship for the minimum area of shear reinforcement given in ACI Code Table 9.6.3.3, which gives

$$s_{\max} = \frac{A_v f_{yt}}{0.75\sqrt{f'_c}b_w} \le \frac{A_v f_{yt}}{50 b_w}$$

b. It is usually undesirable to space vertical stirrups closer than 4 in.

c. It is generally economical and practical to compute the spacing required at several sections and to place stirrups accordingly in groups of varying spacing. Spacing values should be made to not less than 1-in. increments.

d. The code (Section 9.4.3.2) permits the maximum shear to be considered at the section a distance d from the face of the support (except for brackets, short cantilevers, and special isolated conditions), when the support reaction introduces a vertical compression into the end region of a member, and no concentrated load occurs between the face of the support and distance d from the face of the support, and the beam is loaded at or near the top. For stirrup design, the section located a distance d from the face of the support will be called the *critical section*. Sections located less than a distance d from the face of the support may be designed for the same V_u as that at the critical section. Therefore, stirrup spacing should be constant from the critical section back to the face of the support based on the spacing requirements *at* the critical section. The first stirrup should be placed at a maximum distance of $s/2$ from the face of the support, where s equals the immediately adjacent required stirrup spacing (a distance of 2 in. is commonly used). For the balance of the span, the stirrup spacing is a function of the shear strength required to be provided by the stirrups or the maximum spacing limitations.

e. The actual stirrup pattern used in the beam is the designer's choice. The choice will be governed by strength requirements and economy. Many patterns will satisfy the strength requirements.

In most cases, the shear decreases from the support to the center of the span, indicating that the stirrup spacing could be continually increased from the critical section up to the maximum spacing allowed by the code. This would create tedious design, detailing, and bar placing operations, but would nevertheless result in the least steel used. This is not warranted economically or within the framework of the philosophy outlined in Appendix B. In the usual uniformly loaded beams, no more than two or three different spacings should be used within a pattern. Longer spans or concentrated loads may warrant more detailed spacing patterns.

Example 4-3

A simply supported rectangular concrete beam shown in Figure 4-8 is 16 in. wide and has an effective depth of 25 in. The beam supports a total factored load (w_u) of 11.5 kip/ft on a clear span of 20 ft. The given load includes the weight of the beam. Design the web reinforcement. Use $f'_c = 4000$ *psi* and $f_y = 60{,}000$ psi.

Solution:

1. Draw the shear force (V_u) diagram:

$$\text{maximum } V_u = \frac{w_u\ell}{2} = \frac{11.5(20)}{2} = 115.0 \text{ kip}$$

The quantities at the critical section (25 in. or 2.08 ft from the face of the support) are designated with an asterisk. Therefore,

$$V_u^* = 115.0 - 2.08(11.5) = 91.1 \text{ kip}$$

2. Determine if stirrups are required. The ACI Code, Section 9.6.3.1, requires that stirrups be supplied if $V_u > 0.5\phi V_c$. Thus,

$$\phi V_c = \phi(2\sqrt{f'_c}b_w d) = 0.75\left[\frac{2\sqrt{4000}(16)(25)}{1000}\right] = 37.9 \text{ kip}$$

$$0.5\phi V_c = 0.5(37.9) = 18.95 \text{ kip}$$

Therefore, because 91.1 kip > 18.95 kip, stirrups are required.

3. Find the length of span over which stirrups are required. As stirrups must be provided to the point where $V_u = 0.5\phi V_c = 18.95$ kip, find where this shear exists on the V_u diagram of Figure 4-8c. From the face of the support,

$$\frac{115.0 - 18.95}{11.5} = 8.35 \text{ ft}$$

Note this location on the V_u diagram as well as the location at 6.70 ft from the face of the support, where $V_u = \phi V_c = 37.9$ kip.

4. Designate as "required ϕV_s" the area enclosed by the ϕV_c line, the V_u^* line, and the sloping V_u line. This shows

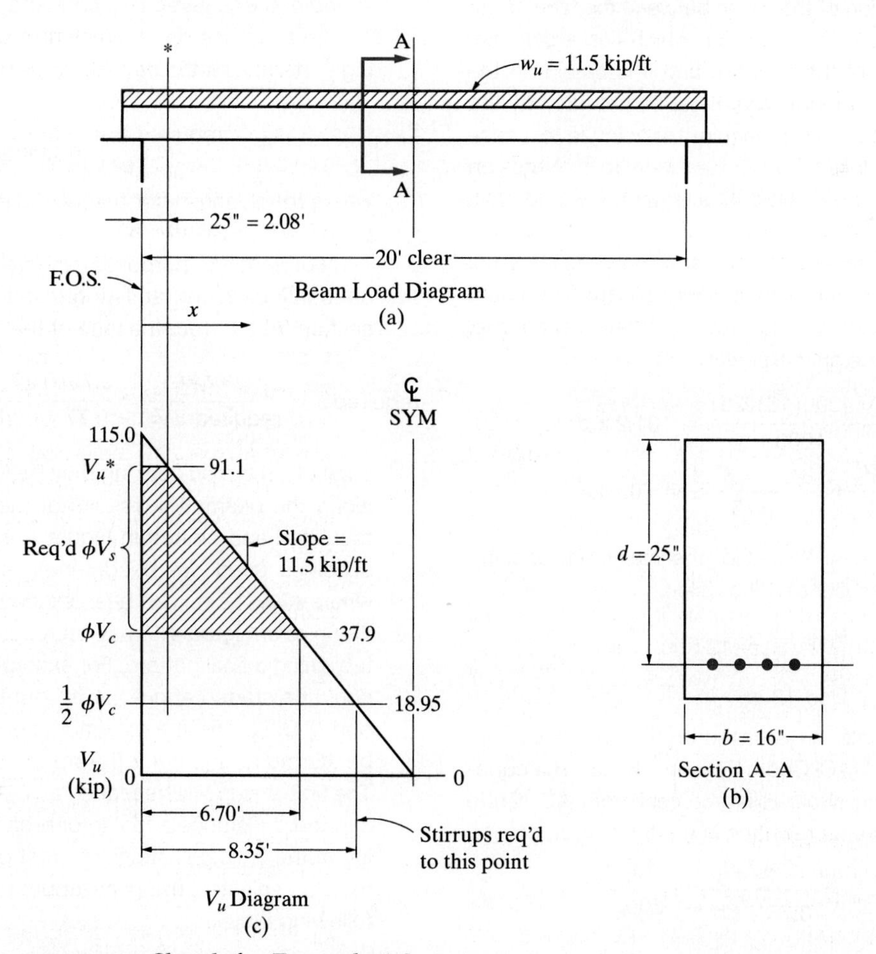

FIGURE 4-8 Sketch for Example 4-3.

the required strength of the shear reinforcing at any point along the span and graphically depicts the relationship

$$\phi V_c + \phi V_s \geq V_u$$

At any location, the required ϕV_s can be determined from the diagram as the distance between the V_u^* (or V_u) line and the ϕV_c line (the height of the crosshatched area). For this particular V_u diagram, designating the slope (kip/ft) as m, taking x (ft) from the face of the support, and considering the range 2.08 ft $\leq x \leq$ 6.70 ft, we have

$$\begin{aligned}\text{required } \phi V_s &= \text{maximum } V_u - \phi V_c - mx \\ &= 115.0 - 37.9 - 11.5x \\ &= 77.1 - 11.5x\end{aligned}$$

5. Assume a No. 3 vertical stirrup (A_v = 0.22 in.^{2}) and establish the spacing requirement at the critical section based on the required ϕV_s. At this location the stirrups will be most closely spaced. Rearranging ACI Equation (22.5.10.3) and multiplying both the numerator and denominator by the strength-reduction factor, ϕ, we obtain the required stirrup spacing as

$$\text{required } s^* = \frac{\phi A_v f_{yt} d}{\text{required } \phi V_s^*} = \frac{0.75(0.22)(60)(25)}{91.1 - 37.9} = 4.65\text{in.}$$

(Note that the denominator in the preceding expression, required ϕV_s^*, is equal to $V_u^* - \phi V_c$.)

We will use a 4-in. spacing. This is the spacing used in the portion of the beam between the face of the support and the critical section, which lies a distance d from the face of the support, and it is based on the amount of shear strength that must be provided by the shear reinforcing. Had the required spacing in this case turned out to be less than 4 in. (see item 3b in "Notes on Stirrup Design" in Section 4-4), a larger bar would have been selected for the stirrups.

6. Establish ACI Code maximum spacing requirements. Recall that if V_s is less than $4\sqrt{f'_c}b_w d$, the maximum spacing is d/2 or 24 in., whichever is smaller. Therefore, compare V_s^* at the critical section with $4\sqrt{f'_c}b_w d$:

$$4\sqrt{f'_c}b_w d = \frac{4\sqrt{4000}(16)(25)}{1000} = 101.2 \text{ kip}$$

$$V_s^* = \frac{\phi V_s^*}{\phi} = \frac{91.1 - 37.9}{0.75} = 70.9 \text{ kip}$$

Because 70.9 kip < 101.2 kip, the maximum spacing should be the smaller of d/2 or 24 in.:

$$\frac{d}{2} = \frac{25}{2} = 12.5 \text{ in.}$$

Use 12 in.

A second criterion is based on the code minimum area requirement (ACI Code, Section 9.6.3.3). The equation for minimum shear reinforcement from ACI Code Table 9.6.3.3 may be rewritten in the form

$$s_{\max} = \frac{A_v f_{yt}}{0.75\sqrt{f'_c}b_w} \leq \frac{A_v f_{yt}}{50 b_w}$$

where the units of f_{yt} should be carefully noted as psi. Evaluate these two expressions:

$$\frac{A_v f_{yt}}{0.75\sqrt{f'_c}b_w} = \frac{0.22(60{,}000)}{0.75\sqrt{4000}(16)} = 17.39 \text{ in.}$$

$$\frac{A_v f_{yt}}{50 b_w} = \frac{0.22(60{,}000)}{50(16)} = 16.50 \text{ in.}$$

Therefore, of these two, 16.50 in. controls.

Of the foregoing two maximum spacing criteria, the smaller value will control. Therefore, the 12-in. maximum spacing controls throughout the beam wherever stirrups are required.

7. Determine the spacing requirements based on shear strength to be furnished. At this point we know that the spacing required at the critical section is 4.65 in., that the maximum spacing allowed in this beam is 12 in. where stirrups are required, and that stirrups are required to 8.35 ft from the face of the support (F.O.S).

To establish a spacing pattern for the rest of the beam, the spacing required should be established at various distances from the face of support. This will permit the placing of stirrups in groups, with each group having a different spacing. The number of locations at which the required spacing should be determined is based on judgment and should be a function of the shape of the required ϕV_s portion of the V_u diagram.

To aid in the determination of an acceptable spacing pattern, a plot is developed using the formula

$$\text{required } s = \frac{\phi A_v f_{yt} d}{\text{required } \phi V_s}$$

where the denominator can be determined from the expression given in step 4.

For plotting purposes, the required spacing will arbitrarily be found at 1-ft intervals beyond the critical section. At 3 ft from the face of the support,

$$\text{required } s = \frac{\phi A_v f_{yt} d}{\text{required } \phi V_s} = \frac{0.75(0.22)(60)(25)}{77.1 - 11.5(3)} = 5.81 \text{ in.}$$

Similarly, required spacing may be found at other points along the beam. The results of these calculations are tabulated and plotted in Figure 4-9. Note that required spacings need not be determined beyond the point where $s_{\max}$ (12 in.) has been exceeded.

8. The plot of Figure 4-9 may be readily used to aid in establishing a final pattern. For example, the 5-in. spacing could be started about 2.4 ft from the face of the support (F.O.S.), and the maximum spacing of 12 in. could be started at about 4.9 ft from the face of the support. The first stirrup will be placed away from the face of the support a distance equal to one-half the required spacing at the critical section. The rest of the stirrup pattern may be selected using an approach as shown in the following table.

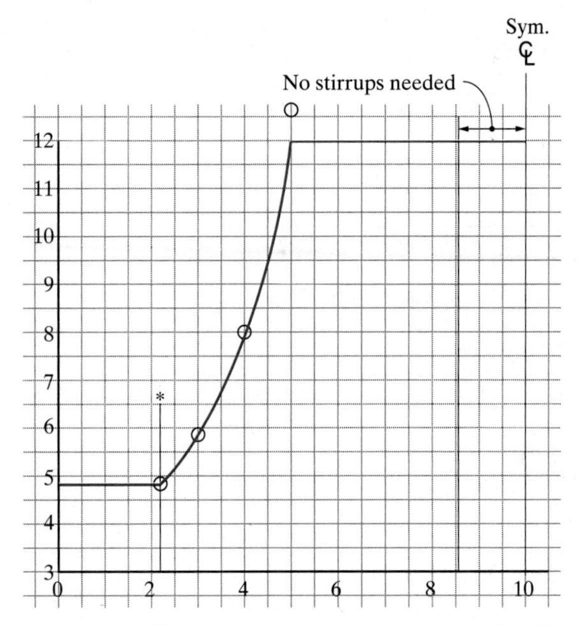

x (ft)	Req'd s (in.)
3	5.81
4	7.96
5	12.63

FIGURE 4-9 Stirrup spacing requirements for Example 4-3.

Spacing (in.)	Theoretical stopping point (from F.O.S.)	Length required to cover (in.)	Number of spaces to use	Actual length covered (in.)	Actual stopping point (inches from F.O.S.)
2	—	—	1	2	2
4	2.4′ = 29″	27	7	28	30
5	4.9′ = 59″	29	6	30	60
12	120″	60	5	60	120

The spacings and theoretical stopping points for those spacings are determined from Figure 4-9. The 4-in. spacing must run from the end of the 2-in. spacing (2 in. from the face of the support) to the theoretical stopping for the 4-in. spacing (29 in.). Therefore, the 4-in. spacings must cover 27 in. For the 4-in. spacing, the required number of spaces is then calculated from 27/4 = 6.75 spaces; use 7 spaces. This places the last stirrup of the 4-in. spacing group at 30 in. from the face of the support, where the 5-in. spacing will begin. The final pattern for No. 3 single-loop stirrups is: one at 2 in., seven at 4 in., six at 5 in., and five at 12 in. This places the last stirrup in the 12-in. spacing group on the beam centerline—thus, remembering symmetry about the centerline, providing stirrups across the full length of the beam. This is common practice and is conservative. The final stirrup pattern is shown in the design sketch of Figure 4-10.

An alternative approach to stirrup spacing is to select a desired spacing (in this case between 4 in. and 12 in.) and to compute the distance from the face of the support at which that spacing may begin. This can be accomplished by rewriting the expression

$$\text{required } s = \frac{\phi A_v f_{yt} d}{\text{required } \phi V_s}$$

$$= \frac{\phi A_v f_{yt} d}{(\max V_u - \phi V_c - mx)}$$

in the form

$$x = \frac{\max V_u - \phi V_c - \dfrac{\phi A_v f_{yt} d}{s}}{m}$$

This expression will furnish the distance x at which any desired stirrup spacing s may commence. Returning to Example 4-3, we can compute where a 5-in. spacing may begin relative to the face of the support,

$$x = \frac{115.0 - 37.9 - \dfrac{0.75(0.22)(60)(25)}{5}}{11.5} = 2.40 \text{ ft}$$

and a 12-in. spacing may begin at

$$x = \frac{115.0 - 37.9 - \dfrac{0.75(0.22)(60)(25)}{12}}{11.5} = 4.91 \text{ ft}$$

Note that Figure 4-9 may be generated equally well with these data.

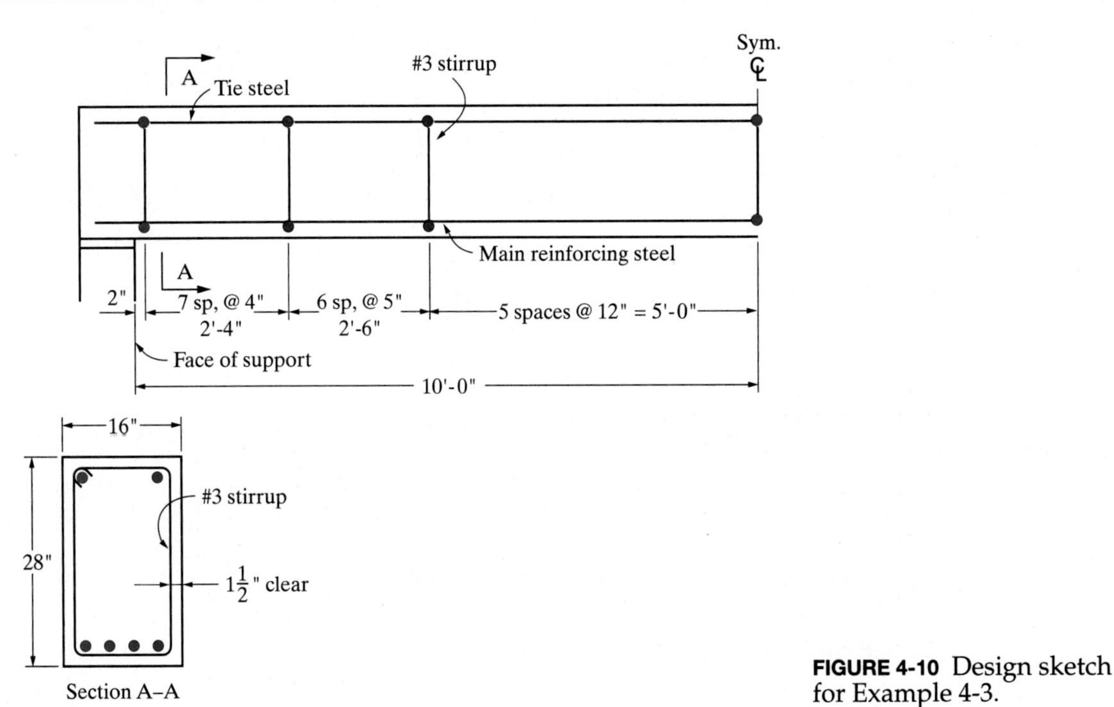

FIGURE 4-10 Design sketch for Example 4-3.

Example 4-4

A continuous reinforced concrete beam shown in Figure 4-11 is 15-in. wide and has an effective depth of 31 in. The factored loads are shown. (The factored uniform load includes the weight of the beam). Design the web reinforcement using the V_u diagram shown in Figure 4-12. Use $f'_c = 4000$ psi and $f_y = 60{,}000$ psi.

Solution:

1. Note that, by reason of symmetry, it is necessary to show only half of the V_u diagram.
2. Determine if stirrups are required:

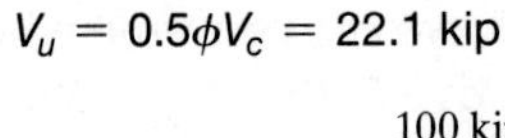

$$\phi V_c = \phi 2\sqrt{f'_c} b_w d = \frac{0.75(2\sqrt{4000})(15)(31)}{1000} = 44.1 \text{ kip}$$

$$0.5\phi V_c = 0.5(44.1) = 22.1 \text{ kip}$$

Therefore, because 104.9 kip > 22.1 kip, stirrups are required.

3. Find the length of span over which stirrups are required. Stirrups are required to the point where

$$V_u = 0.5\phi V_c = 22.1 \text{ kip}$$

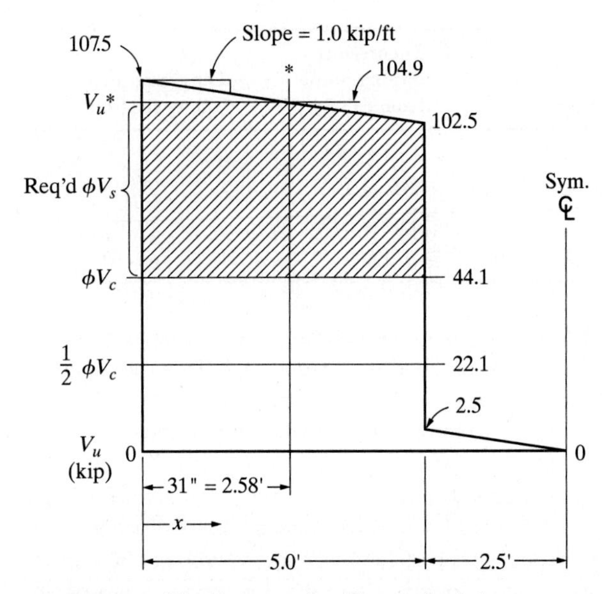

FIGURE 4-12 V_u diagram for Example 4-4.

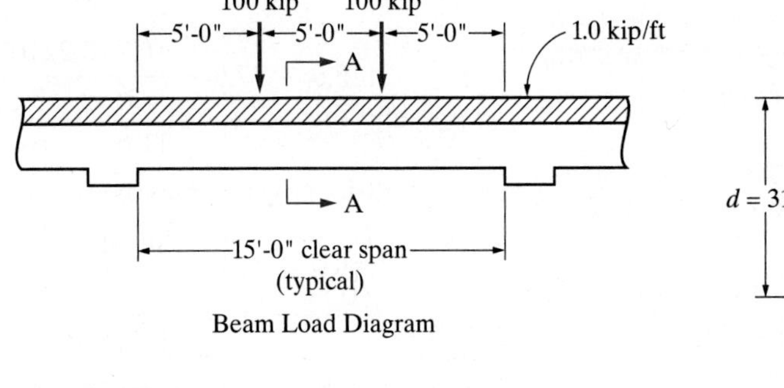

FIGURE 4-11 Sketch for Example 4-4.

From Figure 4-12, the point where V_u is equal to 22.1 kip may be determined by inspection to be at the concentrated load, 5 ft-0 in. from the face of the support. No stirrups are required between the two concentrated loads.

4. Designate "required ϕV_s" on the V_u diagram:

$$\text{required } \phi V_s = \text{maximum } V_u - \phi V_c - mx$$
$$= 107.5 - 44.1 - 1.0x$$
$$= 63.4 - 1.0x \quad \text{(applies for the range } 2.58 \text{ ft} \leq x \leq 5.0 \text{ ft)}$$

5. Assume a No. 3 vertical stirrup ($A_v = 0.22$ in.^{2}):

$$\text{required } s^* = \frac{\phi A_v f_{yt} d}{\text{required } \phi V_s^*} = \frac{0.75(0.22)(60)(31)}{104.9 - 44.1}$$
$$= 5.05 \text{ in.}$$

Use 5 in. spacing

6. Establish ACI Code maximum spacing requirements:

$$4\sqrt{f'_c} b_w d = \frac{4\sqrt{4000}(15)(31)}{1000} = 117.6 \text{ kip}$$

$$V_s^* = \frac{\phi V_s^*}{\phi} = \frac{104.9 - 44.1}{0.75} = 81.1 \text{ kip}$$

Because 81.1 kip < 117.6 kip, the maximum spacing should be the smaller of $d/2$ or 24 in.:

$$\frac{d}{2} = \frac{31}{2} = 15.5 \text{ in.}$$

Also check s_{max}. Because $f'_c < 4444$ psi,

$$s_{max} = \frac{A_v f_{yt}}{50 b_w} = \frac{0.22(60{,}000)}{50(15)} = 17.6 \text{ in.}$$

Therefore, use a maximum spacing of 15 in.

7. Determine the spacing requirements between the critical section and the concentrated load based on shear strength to be furnished. The denominator of the following formula for required spacing uses the expression for required ϕV_s from step 4:

$$\text{required } s = \frac{\phi A_v f_{yt} d}{\text{required } \phi V_s} = \frac{0.75(0.22)(60)(31)}{63.4 - 1.0x}$$

The results of these calculations are tabulated and plotted in Figure 4-13.

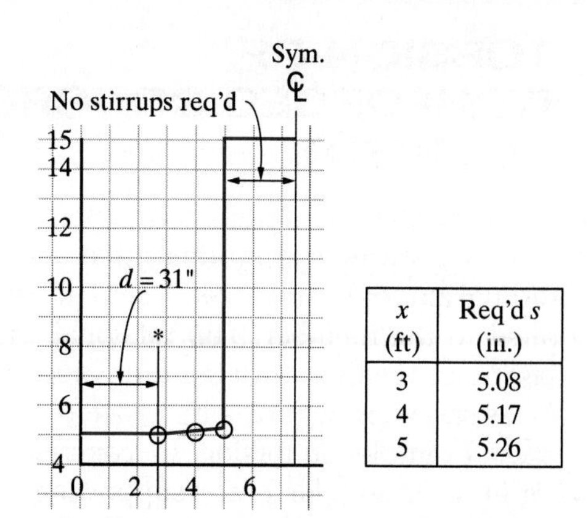

x (ft)	Req'd s (in.)
3	5.08
4	5.17
5	5.26

FIGURE 4-13 Stirrup spacing requirements for Example 4-4.

8. With reference to Figure 4-13, no stirrups are required in the portion of the beam between the point load and center of the beam. A spacing of 5 in. will be used between the face of support and the point load. In the center portion of the beam, between the point loads, stirrups will be placed at a spacing slightly less than the maximum spacing as a conservative measure and to create a convenient spacing pattern. The design sketches are shown in Figure 4-14. Note the symmetry about the span centerline.

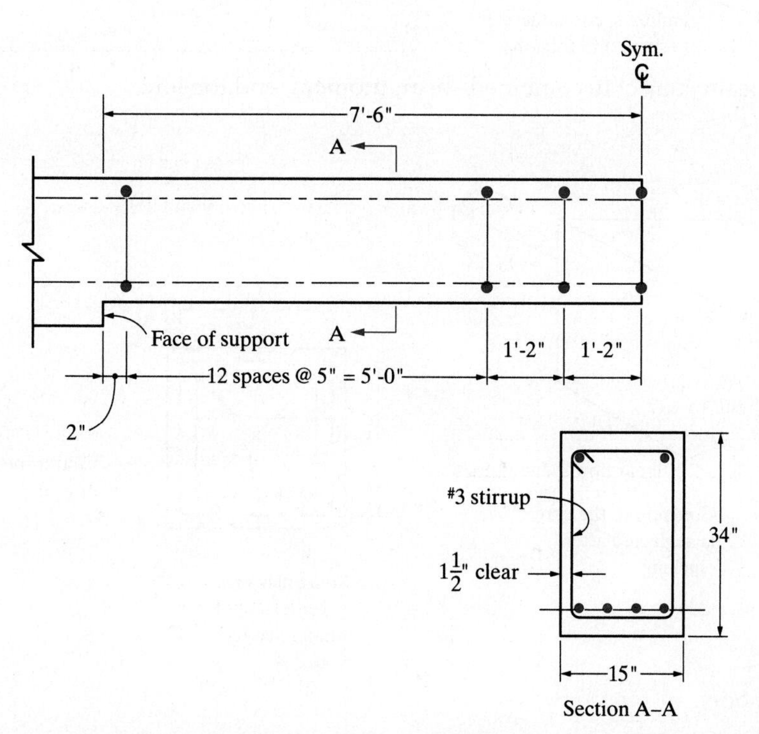

FIGURE 4-14 Design sketches for Example 4-4.

4-5 TORSION OF REINFORCED CONCRETE MEMBERS

The torsion or twisting of reinforced concrete members is caused by a torsional moment that acts about the longitudinal axis of the member due to unbalanced loads applied to the member. The torsional moment usually acts in combination with bending moment and shear force, as shown in Figure 4-15.

A typical example of torsion in concrete members occurs in a rectangular beam supporting precast hollow-core slabs (or planks). The torsion may be due to unequal live loads on adjacent spans of the hollow-core planks or due to unequal adjacent spans of the hollow-core planks supported on the beam. Torsion in such beams could also be due to the construction sequence that has the hollow-core planks fully installed on one side of the beam before being installed on the adjacent side of the beam. Rectangular and L-beams are more susceptible to torsion than T-beams. In the ACI Code, the design for torsion in solid and hollow concrete beams is based on a thin-walled tube space truss model (see Figures 4-16 and 4-17). In the thin-walled space truss model, the outer concrete cross section that is centered on the stirrups is assumed to resist the torsion while the concrete in the core is neglected, because after cracking, this core is relatively ineffective in resisting torsion. Torsional moments cause additional shear stresses that result in diagonal tension stresses in the concrete member. These diagonal tension stresses cause spiral inclined cracks to form around the surface of the concrete member, as shown in Figure 4-17. After cracking, the torsional resistance of a concrete member is provided by the outermost closed stirrups and the longitudinal reinforcement located near the surface of the beam, and this is modeled by the space truss shown in Figure 4-17, where the longitudinal reinforcement acts as the truss tension members, the stirrups act as the tension web members, and the inclined concrete struts between the diagonal cracks act as the compression web members of the space truss.

In the thin-walled tube model, the shear flow q, which is assumed to be constant around the perimeter of the beam, is equal to the product of the shear stress τ and the wall thickness, t. Using the thin-walled tube

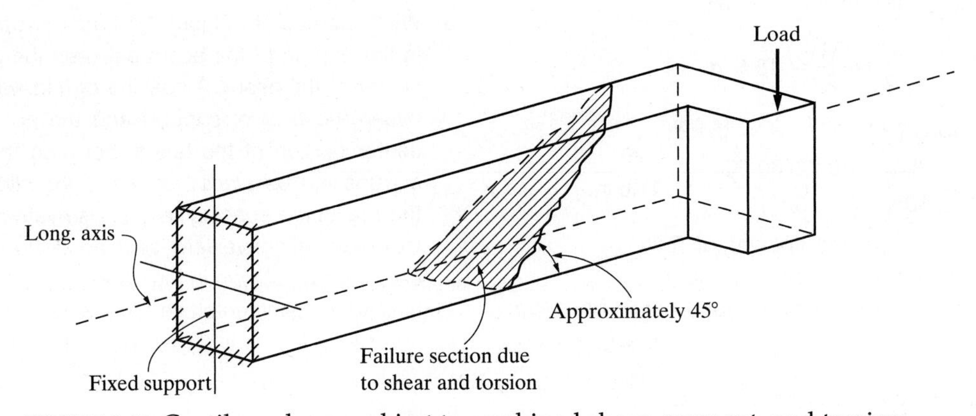

FIGURE 4-15 Cantilever beam subject to combined shear, moment, and torsion.

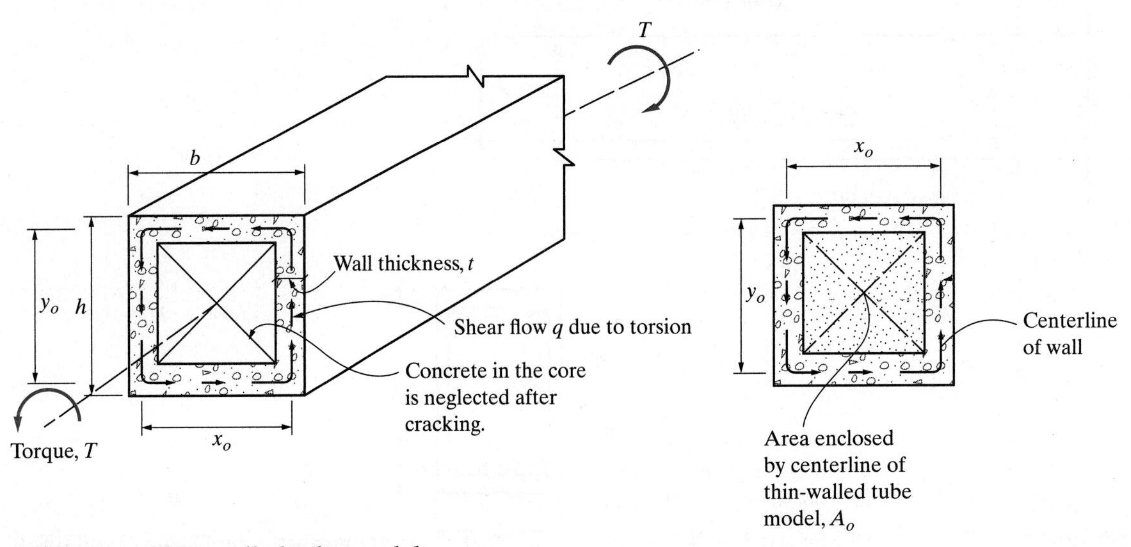

FIGURE 4-16 Thin-walled tube model.

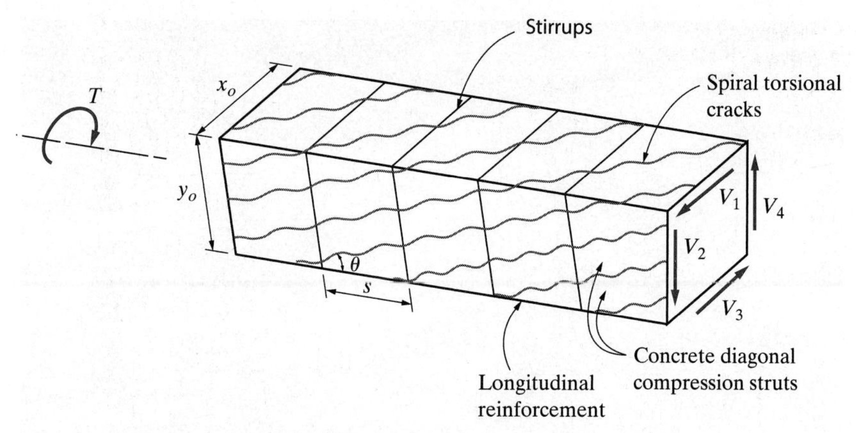

FIGURE 4-17 Space truss model for torsion.

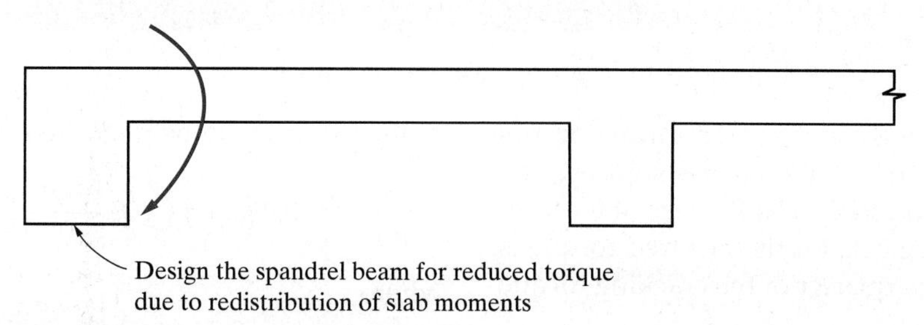

FIGURE 4-18 Compatibility torsion in spandrel beams.

model, and summing the torques, the equilibrium of torsional moments yields

$$T - (q\,x_o)y_o - (q\,y_o)x_o = 0$$

or

$$T = 2q\,x_o y_o = 2q\,A_o$$

Therefore,

$$q = T/2A_o \quad \textbf{(4-4)}$$

where

A_o = area enclosed by centerline of the shear flow path = $x_o y_o$

q = shear flow (i.e., force per unit length)

x_o and y_o are the width and height of the space truss model measured between the centerlines of the tube walls—that is, the centerlines of the longitudinal corner bars.

There are two conditions that may occur in the design of reinforced concrete members for torsion: primary or equilibrium torsion and secondary or compatibility torsion.

Compatibility Torsion

Compatibility torsion occurs in statically indeterminate structures, and the design torque, which cannot be obtained from statics alone, may be reduced due to redistribution of internal forces to maintain compatibility of deformations. Members subjected to compatibility torsion may be designed for the cracking torque multiplied by the resistance factor (i.e., ϕT_{cr}), but the redistribution of internal forces due to the reduction of the torque to ϕT_{cr} must be taken into account in the design of all the adjoining structural members. One example of compatibility torsion occurs in spandrel beams (see Figure 4-18), where the rotation of the slab is restrained by the spandrel beam. For compatibility, the restraining moment at the exterior end of the slab is equal to the uniform torsional moment per unit length on the spandrel beam. As the slab rotates and cracks, and the slab moments are redistributed, the torsional moment on the spandrel beam is reduced until it reaches the cracking torque of the spandrel beam, at which point a hinge is formed at the exterior end of the slab.

Equilibrium Torsion

For statically determinate structures, the design torque, which can be obtained from statics considerations alone, cannot be reduced because redistribution of internal moments and forces is not possible in such structures, and in order to maintain equilibrium, the full design torque has to be resisted by the beam. This is called *equilibrium torsion.* Examples of concrete members in equilibrium torsion are shown in Figures 4-19 and 4-20.

Torsion Design of Reinforced Concrete Members

The ACI Code design approach for torsion follows a similar approach to the design for shear (ACI Code, Sections 9.2.4.4 and 22.7). Like shear, the critical section for torsion is located at a distance d from the face

FIGURE 4-19 Equilibrium torsion. (George Limbrunner)

of a support, but where a concentrated torque occurs within a distance of d from the face of a support, the critical section for torsion shall be at the face of the support. Torsion can be neglected if the factored torque is less than or equal to one-quarter of the cracking torque of the beam section. That is, if

$$T_u \leq 0.25\phi T_{cr} \tag{4-5}$$

where the cracking torque for non-prestressed members not subject to axial tension or compression force is

$$T_{cr} = 4\lambda\sqrt{f'_c}\frac{(A_{cp})^2}{p_{cp}} \tag{4-6}$$

Thus, torsion may be neglected when

$$T_u \leq 0.25\phi\left(4\lambda\sqrt{f'_c}\frac{(A_{cp})^2}{p_{cp}}\right) = \lambda\phi\sqrt{f'_c}\frac{(A_{cp})^2}{p_{cp}} \tag{4-7}$$

where

ϕ = 0.75 (ACI Code, Section 21.2.1)

A_{cp} = area of outside perimeter of the cross section = bh (for rectangular beams not cast monolithic with a slab)

p_{cp} = outside perimeter of the cross section = $2(b + h)$

b and h = cross section width and depth (see Figure 4-16)

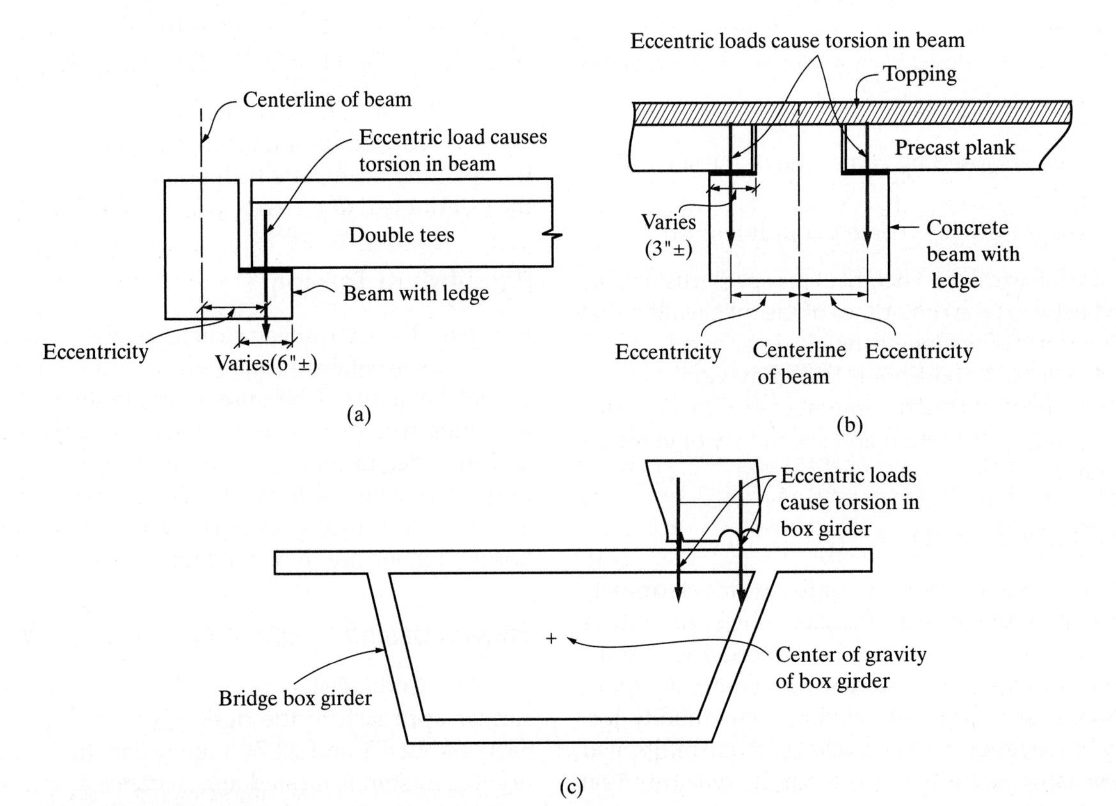

FIGURE 4-20 Equilibrium torsion.

For isolated beams cast monolithic with a slab, the area, A_{cp}, can be determined from ACI Section 9.2.4.4.

λ is the lightweight aggregate factor discussed in Chapter 1, and for normal weight concrete, $\lambda = 1.0$. When the torque is small enough such that Equation (4-5) or (4-7) is satisfied, closed stirrups are not required.

Torsional reinforcement is required to resist the full applied torsional moment as specified in ACI Section 22.7.3.1, when

$$T_u > \lambda\phi\sqrt{f'_c}\,\frac{(A_{cp})^2}{p_{cp}} \tag{4-8}$$

To reduce unsightly cracks on the surface of the beam and to prevent crushing of the surface concrete from stresses in the inclined concrete compression struts (see Figure 4-17) due to combined shear and torsion, the ACI Code requires that *solid concrete beam* cross-sectional dimensions be such that ACI Code Equations (22.7.7.1(a)) and (22.7.7.1(b)) are satisfied. The size of the beam should be increased if these relationships are not satisfied.

$$\sqrt{\left(\frac{V_u}{b_w d}\right)^2 + \left(\frac{T_u p_h}{1.7(A_{oh})^2}\right)^2} \le \phi\left(\frac{V_c}{b_w d} + 8\sqrt{f'_c}\right)$$

[ACI Eq. (22.7.7.1(a))]

Similarly, the cross-sectional dimensions of *hollow concrete beams* are limited as follows:

$$\frac{V_u}{b_w d} + \frac{T_u p_h}{1.7(A_{oh})^2} \le \phi\left(\frac{V_c}{b_w d} + 8\sqrt{f'_c}\right)$$

[ACI Eq. (22.7.7.1(b))]

where

$V_c = 2\lambda\sqrt{f'_c}\,b_w d$

A_{oh} = area enclosed by the centerline of the outermost closed stirrup $= x_1 y_1$

$p_h = 2(x_1 + y_1)$

x_1 and y_1 are the width and height of the space truss model measured to the centerline of the outermost closed stirrup, as shown in Figure 4-21. Note that for hollow sections, the shear stress due to direct shear force is directly additive to the shear stress due to torsion as indicated in ACI Eq. 22.7.7.1(b) because for hollow sections the shear stresses due to direct shear and torsion are concentrated in the two webs of the hollow section (see Figures 4-16 and 4-17). In contrast, for solid sections, the shear stress due to direct shear force is distributed across the full width of the cross-section while the shear stress due to torsion is concentrated on the outermost fibers of the cross-section, with very low torsional shear stress near the center of the solid cross-section. Consequently, the shear stress components from direct shear force and torsion are not directly additive for solid sections, hence for combined torsion and shear, ACI Eq. 22.7.7.1(a) uses the square root of the sum of the squares expression (ACI 318-14, Section R22.7.7.1). It should be noted that the nominal concrete shear strength, V_c, is assumed to be unaffected by torsion.

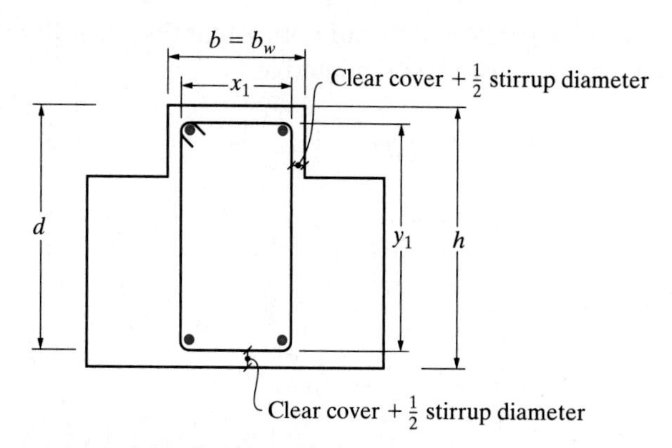

FIGURE 4-21 Definition of x_1 and y_1.

For hollow sections with varying wall thickness, ACI Equation (22.7.7.1(b)) shall be evaluated at a location where the left-hand side of the equation is at its maximum value. If the wall thickness of the hollow concrete beam, t, is less than A_{oh}/p_h at the location where the torsional stresses are being determined, the second term in ACI Equation (22.7.7.1(b)) shall be taken as $T_u/(1.7A_{oh}\,t)$.

Torsion Reinforcement

The reinforcement required for torsion shall be added to that required for other load effects that act in combination with the torsional moment, and the most restrictive spacing requirements must be satisfied (ACI Code, Section 9.5.4.3). The torsional reinforcement is determined using the space truss model shown in Figure 4-17. The yield strength of the torsional reinforcement shall not be greater than 60,000 psi per ACI Code Section 20.2.2.4. The torsional reinforcement consists of closed stirrups and longitudinal reinforcement at the corners of the beam.

Vertical Equilibrium of Forces

Considering a free body diagram of the vertical forces acting on the front wall of the space truss model (Figure 4-17), as shown in Figure 4-22, the equilibrium of the vertical forces yields

$$V_2 = A_t f_{yt}\left(\frac{y_o \cot\theta}{s}\right) \tag{4-9}$$

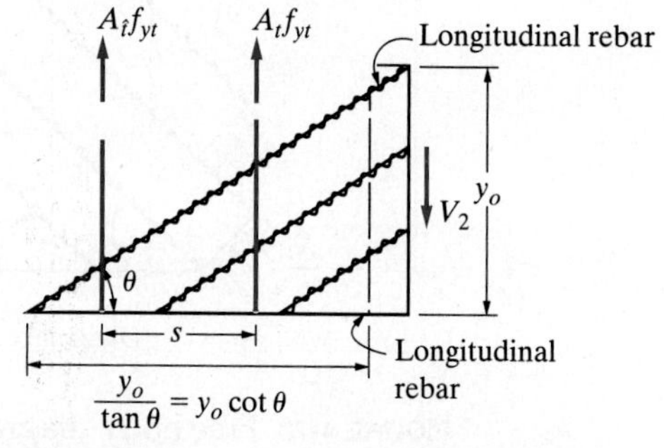

FIGURE 4-22 Free body diagram for vertical equilibrium of forces in the space truss model.

From Equation (4-4) and noting that the shear flow is constant, V_2 can be calculated as

$$V_2 = qy_o = \left(\frac{T}{2A_o}\right)y_o \qquad \textbf{(4-10)}$$

Substituting Equation (4-10) into (4-9) and rearranging, the ratio of the area of the torsional stirrup to the spacing of the stirrup is

$$\frac{A_t}{s} = \frac{T}{2A_o f_{yt}\cot\theta} \qquad \textbf{(4-11)}$$

Horizontal Equilibrium of Forces

Considering a free body diagram of the horizontal forces acting on the front wall of the space truss model (Figure 4-17), as shown in Figure 4-23, the equilibrium of the horizontal forces yields

$$A_\ell f_{y\ell} = \sum V_i\cot\theta = \sum(q\,y_i)\cot\theta = q\cot\theta\sum y_i \qquad \textbf{(4-12)}$$

Substituting Equation (4-4) into Equation (4-12) yields

$$A_\ell f_{y\ell} = \left(\frac{T}{2A_o}\right)\cot\theta\sum y_i$$

$$= \left(\frac{T}{2A_o}\right)\cot\theta\,(x_o + x_o + y_o + y_o)$$

Therefore,

$$A_\ell f_{y\ell} = \left(\frac{T}{2A_o}\right)\cot\theta\,[2(x_o + y_o)] \qquad \textbf{(4-13)}$$

Substituting for the torque, T, in Equation (4-13) using Equation (4-11) gives

$$A_\ell f_{y\ell} = \left[\left(\frac{A_t}{s}\right)f_{yt}\cot\theta\right]\cot\theta\,[2(x_o + y_o)]$$

or

$$A_\ell = \left(\frac{A_t}{s}\right)\frac{f_{yt}}{f_{y\ell}}\cot^2\theta\,[2(x_o + y_o)] \qquad \textbf{(4-14)}$$

It should be recalled that x_o and y_o are the width and height measured to the centerlines of the tube wall in the thin-walled tube model, while x_1 and y_1 are the width and height measured to the centerlines of the outermost closed stirrup. After cracking, the shear flow path is defined more by the center-to-center dimensions between the outer most closed stirrup. Therefore, in subsequent equations, assuming that $x_o = x_1$ and $y_o = y_1$ yields the equation for the longitudinal torsion reinforcement as

$$A_\ell = \left(\frac{A_t}{s}\right)\frac{f_{yt}}{f_{y\ell}}\cot^2\theta\,[2(x_1 + y_1)] \qquad \textbf{(4-15)}$$

That is,

$$A_\ell = \left(\frac{A_t}{s}\right)\frac{f_{yt}}{f_{y\ell}}\cot^2\theta\,(p_h)$$

Transverse Reinforcement Required for Torsion (Stirrups)

Using Equation (4-11) and the limit states design requirement that the design torsional strength ϕT_n be greater than or equal to the factored torsional moment, T_u yields the required torsional stirrup area,

$$\frac{A_t}{s} = \frac{T_u}{2\phi\,A_o f_{yt}\cot\theta} \qquad \textbf{(4-16)}$$

where

A_t = area of *one* leg of the torsional or outermost closed stirrups, in.^{2}
s = stirrup spacing
θ = 30° to 60°; *use* θ = 45° for non-prestressed members (ACI Code, Sections 22.7.6.1)
f_{yt} = yield strength of stirrups = 60,000 psi
T_u = factored torque at the critical section

The *critical section* for torsion is permitted to be at a distance *d* from the face of the beam support provided

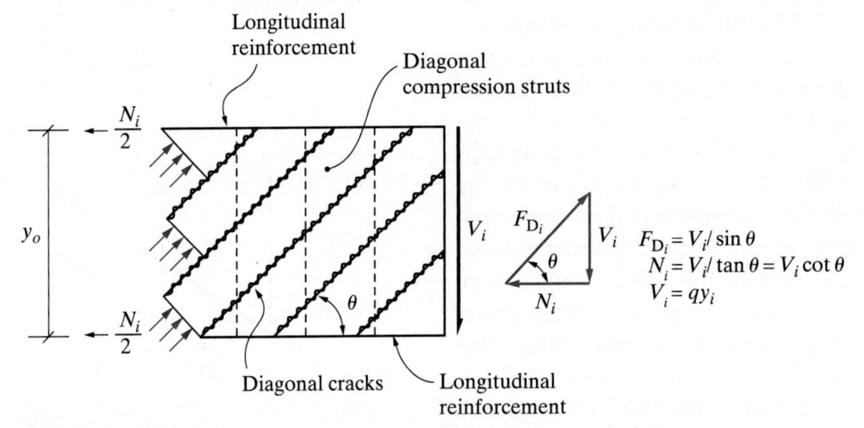

FIGURE 4-23 Free body diagram for horizontal equilibrium of forces in the space truss model.

no concentrated torque occurs within a distance of d from the face of the support. If a concentrated torque occurs within a distance d from the face of support, the critical section will be at the face of the support (ACI Code, Sections 9.4.4.2 and 9.4.4.3).

d = effective depth of the beam

$A_o \approx 0.85\ A_{oh} = 0.85\ x_1 y_1$

A_{oh} = area enclosed by the centerline of outermost closed stirrup

The nominal torsional strength, T_n, is obtained from Equation (4-11) as

$$T_n = \frac{2A_oA_tf_{yt}}{s}\cot\theta \quad \text{[ACI Eq. (22.7.6.1(a))]}$$

$$T_n \text{ should be} \leq \frac{2A_oA_\ell f_{yt}}{p_h}\tan\theta \quad \text{[ACI Eq. (22.7.6.1(b))]}$$

The *total* equivalent transverse reinforcement or stirrups required for *combined shear plus torsion* is obtained from the ACI Code, Section 9.6.4.2, as

$$\frac{A_{vt}}{s} = \frac{A_v}{s} + \frac{2A_t}{s} \geq \frac{50b_w}{f_{yt}}$$

$$\geq 0.75\sqrt{f'_c}\frac{b_w}{f_{yt}} \quad \text{(4-17)}$$

where

A_{vt} = area of *two* legs of closed stirrups required for combined shear plus torsion

A_t = area of *one* leg of closed stirrups required for torsion

A_v = area of *two* legs of closed stirrups required for shear

f_{yt} = yield strength of stirrups = 60,000 psi

b_w = width of beam stem

s = spacing of stirrups $\leq p_h/8$ and 12 in. (ACI Code, Section 9.7.6.3.3)

Note that A_v/s is a slightly modified form for the stirrup requirement for shear from Section 4-2.

Additional Longitudinal Reinforcement Required for Torsion

The additional longitudinal reinforcement required to resist torsion, and to be added to the reinforcement required for bending, is obtained from Equation 4-11 and ACI Equations (9.6.4.3(a)) and (9.6.4.3(b)) as

$$A_\ell = \left(\frac{A_t}{s}\right)p_h\frac{f_{yt}}{f_y}\cot^2\theta \quad \text{(4-18)}$$

A_ℓ should have a minimum value of the lesser of ACI Equations (9.6.4.3(a)) and (9.6.4.3(b)) reproduced below.

$$5\sqrt{f'_c}\frac{A_{cp}}{f_y} - \left(\frac{A_t}{s}\right)p_h\frac{f_{yt}}{f_y} \quad \text{[ACI Eq. (9.6.4.3(a))]}$$

$$5\sqrt{f'_c}\frac{A_{cp}}{f_y} - \left(\frac{25b_w}{f_{yt}}\right)p_h\frac{f_{yt}}{f_y} \quad \text{[ACI Eq. (9.6.4.3(b))]}$$

where

$\frac{A_t}{s}$ = torsion stirrup area from Equation (4-11)

$\geq 25b_w/f_{yt}$

$p_h = 2(x_1 + y_1)$

$A_{cp} = bh$

b_w = width of beam stem

f_{yt} = stirrup yield strength, psi

$f_{y\ell}$ = longitudinal steel yield strength, psi

The following should be noted with regard to the required torsional reinforcement:

- The additional longitudinal reinforcement must be distributed around the surface of the beam with a maximum spacing of 12 in., and there should be at least one longitudinal bar in each corner of the closed stirrups to help transfer the compression strut forces into the stirrups.
- The additional longitudinal reinforcement diameter should be at least 0.042 times the stirrup spacing (i.e., 0.042s), but not less than a $\frac{3}{8}$ in. diameter bar (ACI Code, Sections 9.7.5.1, 9.7.5.2, and 25.7.1.2) to prevent buckling of the longitudinal reinforcement due to the horizontal component of the diagonal compression strut force.
- The additional longitudinal rebar area should be added to the longitudinal rebar area required for bending.
- The closed torsional stirrups should be enclosed with 135° hooks (ACI Code, Sections 9.7.6.3.1, 25.7.1.6, and 25.7.2.5) and there should be at least one longitudinal bar enclosed by and at each corner of the stirrup. Note that 90° hooks are ineffective after the corners of the beam spall off due to torsion failure.

Torsion Design Procedure

The design procedure for torsion is as follows:

1. Determine the maximum factored concentrated or uniformly distributed torsional load and the corresponding factored gravity load that occurs simultaneously.
   Note that pattern or checkered live loading may need to be considered to maximize the torsional load and moment.
2. Determine the factored torsional moment, T_u, the factored shear, V_u, and the factored bending moment, M_u.

3. Determine the reinforcement required to resist the factored bending moment M_u.
4. Calculate the concrete shear strength, ϕV_c.
5. Determine the cracking torque, T_{cr} and if torsion can be neglected (check if $T_u \leq 0.25\, T_{cr}$).
6. Determine if the torsion in the member is caused by compatibility torsion or by equilibrium torsion:
   a. For compatibility torsion, redistribution of internal forces is possible because the torsional moment is not required to maintain equilibrium; therefore, design the member for a reduced torque of ϕT_{cr}.
   b. For equilibrium torsion, redistribution of internal forces is not possible because the torsional moment is required to maintain equilibrium; therefore, the member must be designed for the full torsional moment, T_u, calculated in step 1.
7. Check the limits of the member cross section using ACI Code Equation (22.7.7.1(a)) to prevent crushing of the diagonal concrete compression struts.
8. Determine the required torsional stirrup area, A_t/s, the stirrup area required for shear, A_v/s, and the total stirrup area required for combined shear and torsion, A_{vt}/s. Check that maximum stirrup spacing is not exceeded, and check minimum stirrup area. Using the torsional moment diagram and shear force diagram, the required stirrup spacing can be laid out to match the variation in shear and torsional moment.
9. Determine the additional longitudinal reinforcement required for torsion.
10. Draw the detail of the torsional reinforcement.

Example 4-5

Design of Beams for Torsion

The floor framing in the operating rooms in a hospital building consists of reinforced concrete beams 18″ × 24″ deep that support precast concrete planks, as shown in Figure 4-24. The clear span of the beam is 27 ft between columns. The planks are 10 in. deep with 2-in. topping and supports stud wall partitions that weigh 10 psf and mechanical/electrical equipment that weigh 5 psf. The weight of the precast planks is 70 psf. The centerline to centerline span of the planks is 30 ft on the left-hand side of the beam and 24 ft on the right-hand side of the beam. Design the beam for torsion and shear assuming normal weight concrete (i.e., $\lambda = 1.0$) and $f'_c = 4000$ psi. Assume the beam has already been designed for bending. The live load for hospital operating rooms is 60 psf.

Solution:

1. Determine the maximum factored concentrated or uniformly distributed torsional load and the corresponding factored gravity load that occurs simultaneously.

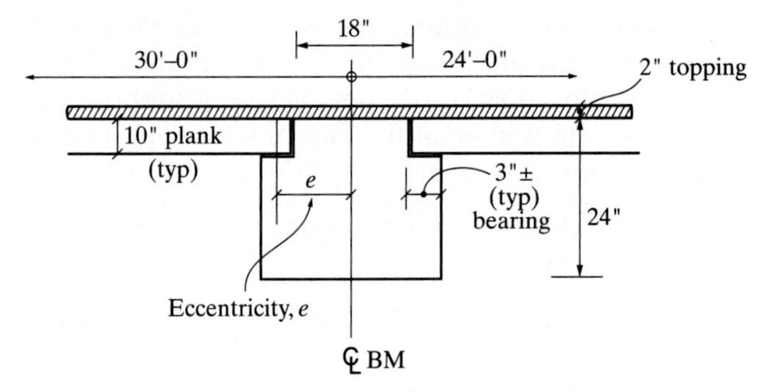

FIGURE 4-24 Beam section for Example 4-5.

Dead Load

$$10 \text{ in. plank} + 2 \text{ in. topping} = 95 \text{ psf}$$
$$\text{Mechanical and electrical} = 5 \text{ psf}$$
$$\text{Stud wall partitions} = 10 \text{ psf}$$
$$\text{Total dead load, } D = 110 \text{ psf}$$
$$\text{Floor live load (operating room), } L = 60 \text{ psf}$$

Tributary Widths (TW) of Beam

$$TW \text{ (due to the 30-ft-span hollow-core plank)} = \frac{30 \text{ ft}}{2} = 15 \text{ ft}$$

$$TW \text{ (due to the 24-ft-span hollow-core plank)} = \frac{24 \text{ ft}}{2} = 12 \text{ ft}$$

A review of Figure 4-24 shows that the torsion in this beam is equilibrium torsion caused by the eccentricity of the plank loads.

Eccentricity of the Hollow-Core Plank Load

$$\text{Eccentricity, } e = \frac{18 \text{ in.}}{2} + \frac{3 \text{ in.}}{2} = 10.5 \text{ in.} = 0.88 \text{ ft}$$

The maximum uniform torsional loading will occur due to checkerboard or partial loading on the hollow-core slabs in which the full design live load is assumed on the 30-ft-span hollow-core slab and *one-half of the design live load is assumed* on the 24-ft-span hollow-core slab. This is common practice among some designers and will generally result in a slightly more conservative design. This partial loading is similar to what is prescribed in Section 7.5 of the ASCE 7 Load Standard. The maximum torsion will be considered together with the corresponding maximum uniform vertical load that occurs at the same time.

The maximum factored uniform torsional load is

$$\begin{aligned} w_{tu} = &\{[1.2(110 \text{ psf}) + 1.6(60 \text{ psf})](15 \text{ ft}) \\ &- [1.2(110 \text{ psf}) + 1.6(\tfrac{1}{2})(60 \text{ psf})](12 \text{ ft})\} \\ &\times 0.88 \text{ ft} = 1109 \text{ ft-lb/ft} = 1.11 \text{ ft.-kips/ft} \end{aligned}$$

The corresponding maximum factored uniform vertical load is

$$\begin{aligned} w_u &= [1.2(110 \text{ psf}) + 1.6(60 \text{ psf})](15 \text{ ft}) + [1.2(110 \text{ psf}) \\ &\quad + 1.6(\tfrac{1}{2})(60 \text{ psf})](12 \text{ ft}) \\ &= 5580 \text{ lb/ft} = 5.58 \text{ kip/ft} \end{aligned}$$

2. Determine the factored torsional moment, T_u, the factored shear, V_u, and the factored bending moment, M_u:

   Assuming 2 layers of reinforcement, $d = 24$ in. $-$ 3.5 in. $= 20.5$ in. For torsion and shear, use a reduced span commencing at d from the face of the beam supports. This reduced span is

$$\ell = 27 \text{ ft} - \frac{2(20.5 \text{ in.})}{12} = 23.6 \text{ ft}$$

   Maximum design torsional moment,

$$T_u = \frac{w_{tu}\ell}{2} = \frac{1.11 \text{ ft.-kip/ft } (23.6 \text{ ft})}{2} = 13.1 \text{ ft.-kip}$$

   Maximum design shear that occurs at the same time as the maximum torsion is

$$V_u = \frac{w_u\ell}{2} = \frac{5.58 \text{ kip/ft } (23.6 \text{ ft})}{2} = 65.8 \text{ kip}$$

3. The reinforcement required to resist the bending moment is assumed to have previously been designed and is not calculated here.
4. The concrete shear strength is

$$\phi V_c = 0.75(2)\sqrt{4000}(18 \text{ in.})(20.5 \text{ in.}) = 35{,}000 \text{ lb}$$
$$= 35 \text{ kip}$$

5. Torsion can be neglected if the factored torsional moment is less than or equal to the concrete torsional strength, that is, if $T_u \le 0.25\phi T_{cr}$, where the concrete torsional strength is

$$0.25\phi T_{cr} = \lambda\phi\sqrt{f'_c}\,\frac{(A_{cp})^2}{p_{cp}}$$

$$= (1.0)(0.75)(\sqrt{4000})\,\frac{(18 \text{ in.} \times 24 \text{ in.})^2}{2(18 \text{ in.} + 24 \text{ in.})}$$

$$= 105.4 \text{ in.-kips} = 8.8 \text{ ft.-kip}$$

   Because $T_u = 13.1$ ft.-kip > 8.8 ft.-kip, this beam therefore, must be designed for torsion.
6. This is equilibrium torsion as redistribution of internal forces is *not* possible because the torsional moment is required to maintain equilibrium. The member thus must be designed for the full torsional moment, T_u, calculated in step 1.
7. Check the limits of the member cross section using ACI Code Equation (22.7.7.1(a)) to prevent crushing of the diagonal concrete compression struts:

$$b = 18 \text{ in.}$$
$$h = 24 \text{ in}$$

   Effective depth, $d = 24 - 3.5$ in. (assuming 2 layers of rebar) $= 20.5$ in.

$$\theta = 45^\circ$$
$$\phi = 0.75$$

$$x_1 = 18 \text{ in.} - (2 \text{ sides})\left(1.5\text{-in. cover} + \frac{0.5\text{-in. stirrup}}{2}\right)$$
$$= 14.5 \text{ in.}$$

$$y_1 = 24 \text{ in.} - (2 \text{ sides})\left(1.5\text{-in. cover} + \frac{0.5\text{-in. stirrup}}{2}\right)$$
$$= 20.5 \text{ in.}$$

$$f_{yt} = f_{yv} = 60{,}000 \text{ psi}$$
$$A_{oh} = x_1y_1 = (14.5 \text{ in.})(20.5 \text{ in.}) = 297.3 \text{ in.}^2$$
$$A_o = 0.85A_{oh} = 0.85x_1y_1 = 0.85(14.5 \text{ in.})(20.5 \text{ in.})$$
$$= 252.7 \text{ in.}^2$$
$$p_h = 2(x_1 + y_1) = 2(14.5 + 20.5) = 70 \text{ in.}$$
$$A_{cp} = bh = (18 \text{ in.})(24 \text{ in.}) = 432 \text{ in.}^2$$

   The limits on the beam cross-sectional dimensions will now be checked using ACI Equation (22.7.7.1(a)):

$$\sqrt{\left(\frac{V_u}{b_wd}\right)^2 + \left(\frac{T_up_h}{1.7(A_{oh})^2}\right)^2} \le \phi\left(\frac{V_c}{b_wd} + 8\sqrt{f'_c}\right)$$

   That is,

$$\sqrt{\left(\frac{65.8 \text{ kip}(1000)}{(18 \text{ in.})(20.5 \text{ in.})}\right)^2 + \left(\frac{(13.1 \text{ ft.-kip})(12{,}000)(70 \text{ in.})}{1.7(297.3 \text{ in.}^2)^2}\right)^2}$$
$$\le \frac{35 \text{ kip}(1000)}{(18 \text{ in.})(20.5 \text{ in.})} + 8(0.75)\sqrt{4000}$$

$$192.8 \text{ psi} < 474.3 \text{ psi} \qquad \text{O.K.}$$

   Thus, the diagonal concrete compression struts are not crushed and the size of the beam is adequate to resist the torsional moments.
8. Determine the required torsional stirrup area, A_t/s, the stirrup area required for shear, A_v/s, and the total stirrup area required for combined shear and torsion, A_{vt}/s. Check that maximum stirrup spacing is not exceeded, and check minimum stirrup area.

   From Equation (4-16), the torsional stirrup required is

$$\frac{A_t}{s} = \frac{T_u}{2\phi A_o f_{yt}\cot\theta} = \frac{13.1 \text{ ft.-kip}(12{,}000)}{2(0.75)(252.7)(60{,}000) \cot 45^\circ}$$
$$= 0.0069$$

   The stirrup area required to resist the maximum factored shear acting with the maximum torsion is

$$\frac{A_v}{s} = \frac{V_u - \phi V_c}{\phi f_{yv}d} = \frac{(65.8 \text{ kip} - 35 \text{ kip})(1000)}{(0.75)(60{,}000)(20.5 \text{ in.})} = 0.033$$

   The total stirrup area required (2-leg stirrups) is calculated from

$$\frac{A_{vt}}{s} = \frac{A_v}{s} + \frac{2A_t}{s} = 0.033 + 2(0.0069) = 0.047$$
$$\ge \frac{50b_w}{f_{yt}} = \frac{50(18 \text{ in.})}{60{,}000} = 0.015$$
$$\ge 0.75\sqrt{f'_c}\frac{b_w}{f_{yt}} = 0.75\sqrt{4000}\,\frac{18 \text{ in.}}{60{,}000} = 0.014 \quad \text{(O.K.)}$$

   Using No. 4 stirrups, A_{vt}(2 legs) $= 2(0.2 \text{ in.}^2) = 0.4 \text{ in.}^2$, the spacing of the stirrups required to resist the maximum combined shear and torsion is calculated as

$$s = \frac{0.4 \text{ in.}^2}{0.047} = 8.5 \text{ in. (controls)}$$

$$\leq p_h/8 = 70 \text{ in.}/8 = 8.75 \text{ in.} \qquad \text{(O.K.)}$$

$$\leq 12 \text{ in.} \qquad \text{(O.K.)}$$

Therefore, use No. 4 closed stirrups at 8-in. on center.

The shear and torsion are at their maximum values at the face of the beam support and decrease linearly to zero at the midspan of the beam; the stirrup spacing thus can be varied accordingly, as done previously in the shear design examples.

9. Additional Longitudinal Reinforcement

b_w = width of beam stem = 18 in.

f_{yt} = stirrup yield strength = 60,000 psi

f_y = longitudinal steel yield strength = 60,000 psi

$$\frac{A_t}{s} = 0.0069 \text{ (as previously calculated)} \geq \frac{25b_w}{f_{yt}}$$

$$= \frac{25(18 \text{ in.})}{60{,}000} = 0.0075$$

Therefore, use 0.0075.

From Equation (4-18) the additional longitudinal reinforcement is calculated as

$$A_\ell = \left(\frac{A_t}{s}\right)p_h\frac{f_{yt}}{f_y}\cot^2\theta = (0.0075 \text{ in.})(70 \text{ in.})\frac{60{,}000}{60{,}000}\cot^2 45°$$

$$= 0.53 \text{ in.}^2$$

$$\geq \textit{the lesser of } 5\sqrt{f'_c}\frac{A_{cp}}{f_y} - \left(\frac{A_t}{s}\right)p_h\frac{f_{yt}}{f_y} \quad \textit{and}$$

$$5\sqrt{f'_c}\frac{A_{cp}}{f_y} - \left(\frac{25b_w}{f_{yt}}\right)p_h\frac{f_{yt}}{f_y}$$

- $\geq$ *the lesser of*

$$5\sqrt{4000}\frac{432 \text{ in.}^2}{60{,}000} - (0.0075 \text{ in.})(70 \text{ in.})\frac{60{,}000}{60{,}000} \textit{ and}$$

- $$5\sqrt{4000}\frac{432 \text{ in.}^2}{60{,}000} - \left(\frac{25(18 \text{ in.})}{60{,}000}\right)(70 \text{ in.})\frac{60{,}000}{60{,}000}$$

$\geq$ *the lesser of* 1.75 in.^{2} *and* 1.75 in.^{2}

Therefore, the required additional longitudinal steel is $A_\ell = 1.75$ in.^{2}

This additional longitudinal reinforcement should be distributed at the corners of the beam but the spacing between these bars should be no greater than 12 in. Where the spacing exceeds 12 in., provide additional longitudinal bars at the midwidth or middepth of the beam as required. This longitudinal reinforcement is in addition to the reinforcement required to resist the bending moments on the beam.

If the additional reinforcement is concentrated on the top and bottom layers, therefore, the total areas of the top and bottom longitudinal reinforcement in the beam are calculated as

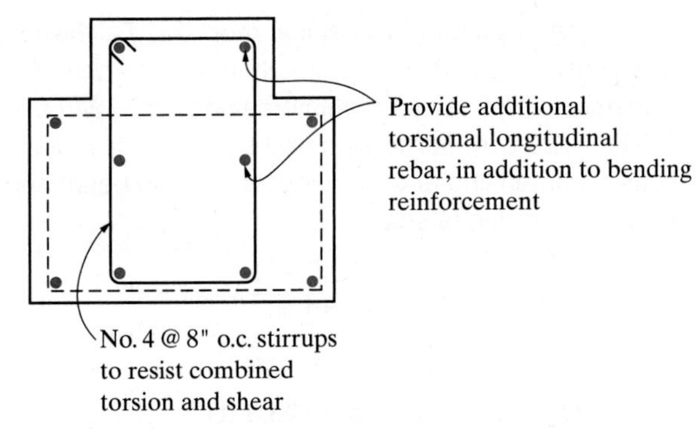

FIGURE 4-25 Beam torsional reinforcement detail.

$$A_{s,top} = A_{s,top}(\text{due to bending}) + 0.5(1.75 \text{ in.}^2)$$
$$= A_{s,top}(\text{due to bending}) + 0.88 \text{ in.}^2$$
$$A_{s,bottom} = A_{s,bottom}(\text{due to bending}) + 0.5(1.75 \text{ in.}^2)$$
$$= A_{s,bottom}(\text{due to bending}) + 0.88 \text{ in.}^2$$

However, for the beam in this example, the spacing of the longitudinal reinforcement will exceed the maximum 12 in. because the center-to-center distance between the top and bottom rebars is approximately 18 in. The additional longitudinal reinforcement should thus be distributed as follows:

$$A_{s,top} = A_{s,top}(\text{due to bending}) + \left(\frac{1}{3}\right)(1.75 \text{ in.}^2)$$
$$= A_{s,top}(\text{due to bending}) + 0.58 \text{ in.}^2$$
$$A_{s,midheight} = \left(\frac{1}{3}\right)(1.75 \text{ in.}^2) = 0.58 \text{ in.}^2$$
$$A_{s,bottom} = A_{s,bottom}(\text{due to bending}) + \left(\frac{1}{3}\right)(1.75 \text{ in.}^2)$$
$$= A_{s,bottom}(\text{due to bending}) + 0.58 \text{ in.}^2$$

The *minimum* diameter of the longitudinal reinforcement is the largest of the following:

$$0.042\, s = 0.042(8 \text{ in.}) = 0.34 \text{ in.}$$

or $\frac{3}{8}$ in. (controls)

10. Torsional reinforcement detail is shown in Figure 4-25.

4-6 CORBELS AND BRACKETS

Corbels and brackets are short cantilever deep beams that are supported off the face of columns where the distance from the face of the supporting column to the line of action of the concentrated load that is being supported (i.e., shear span, a_v) is no greater than the effective depth, d, of the corbel. Corbels and brackets are used to support loads at expansion joints, in precast construction to support precast beams, at slightly offset columns, and in retrofit situations where the corbel is cast against an existing column to support a new beam [2] (see Figures 4-26, 4-27 and 4-28). Corbels are also used where structural steel floor framing members connect to concrete core walls.

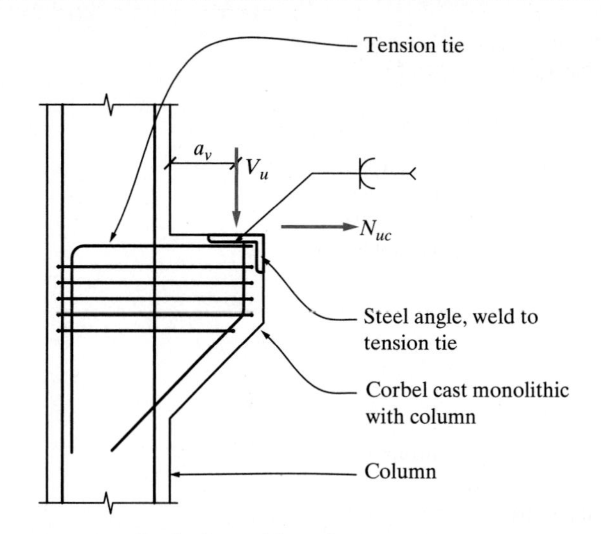

FIGURE 4-26 Corbels and brackets at expansion joints or in precast construction.

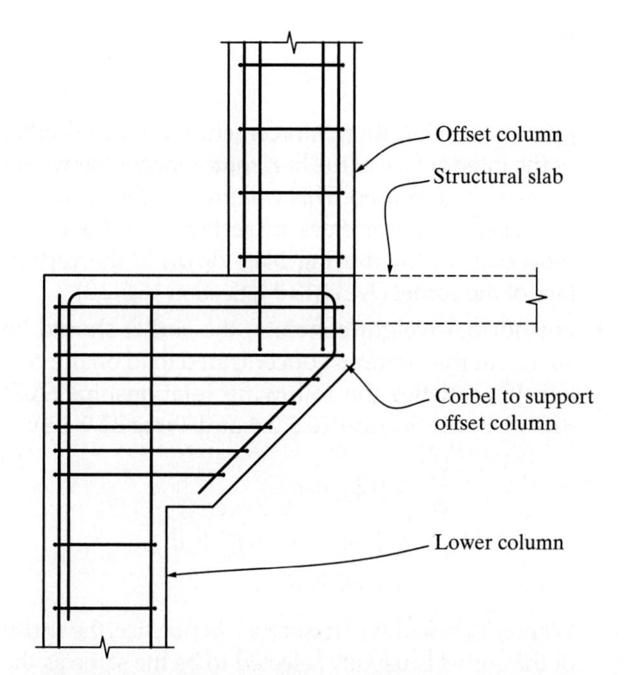

FIGURE 4-27 Corbels at offset columns.

The following possible failure modes must be considered in the design of corbels [ACI 318-14, Section R16.5.1]:

- Shearing along the vertical interface between the face of the column and the corbel
- Tension failure of the primary tension rebar
- Crushing or splitting of the compression strut
- Local failure of the concrete under the bearing plate as a result of excessive bearing stresses or shear stresses

Two methods are presented in the ACI Code for the design of corbels: the shear friction method presented in ACI 318-14 Code, Section 22.9, and the strut-and-tie method presented in ACI 318-14 Code, Chapter 23. The dimensional limits for corbels are presented in ACI 318-14 Code, Section 16.5.

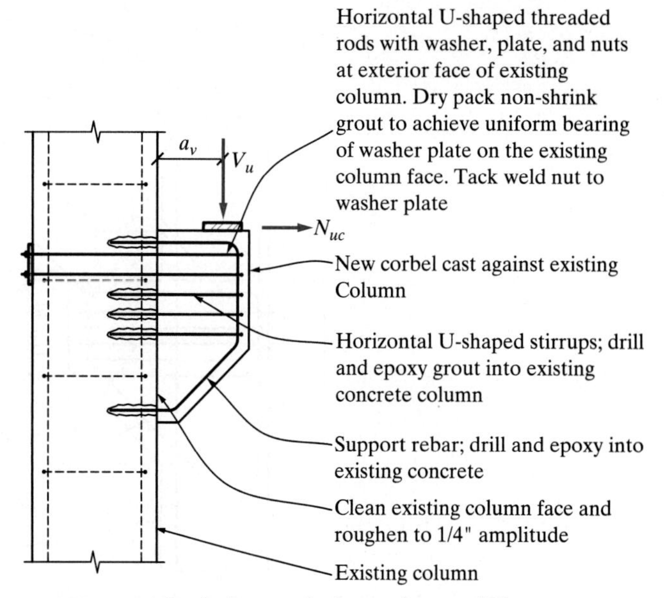

FIGURE 4-28 Corbels at existing columns [2].

In the shear friction method, the relative vertical displacement at the interface between the corbel and the column is resisted by the dowel action of the main horizontal reinforcement crossing the corbel-column interface and by the friction between the faces of the column and the corbel at the interface. The shear friction method can be used for corbels with $a_v/d \leq 1.0$ and $N_{uc} \leq V_u$ because the method has only been validated experimentally for this range of a_v/d and N_{uc} (ACI 318-14 Code, Section R16.5.1). The strut-and-tie method, introduced in Chapter 10, can be used for all corbels, but in this chapter, only the shear friction method will be covered. To prevent cracking that starts below the bearing plate and propagates vertically to the sloped surface of the corbel, the ACI Code requires the depth of the vertical face of the corbel to be at least 50% of the effective depth of the corbel at the face of the column (see Figure 4-29). The primary

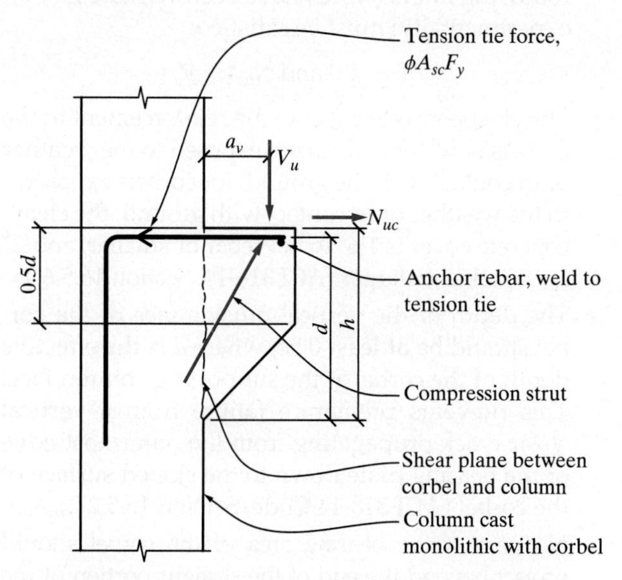

FIGURE 4-29 Corbels and brackets with anchor bar.

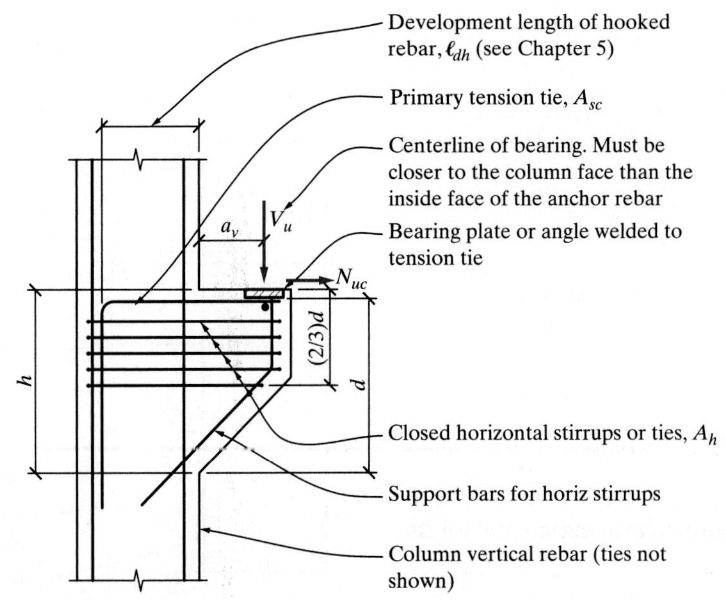

FIGURE 4-30 Corbels and brackets with steel angle anchor and bearing plate

tension tie reinforcement must be fully developed on both sides of the shear plane at the column/corbel interface. This usually requires a 90° hook in the column; in order to develop the tension tie reinforcement within the corbel, the rebar can be welded to a horizontal anchor bar near the outside face of the corbel. Alternatively, the tension tie reinforcement can be welded to an embed steel angle located at the top outside edge of the corbel; this steel angle also serves as a bearing plate for the supported load, V_u (see Figures 4-26, 4-29, and 4-30).

In accordance with ACI 318-14 Code, Section 16.5, the requirements and procedure for the design of corbels and brackets using the shear friction method are as follows:

- In order to use the shear friction method, the following limits (which have been validated experimentally) must be satisfied:

$$a_v/d \leq 1.0 \text{ and } N_{uc} \leq V_u$$

- The clear concrete cover to the reinforcement in the corbels is ¾″ for corbels not exposed to the weather or in contact with the ground; for corbels exposed to the weather or in contact with ground, the clear concrete cover is 1½″ for #5 rebar or smaller, and 2″ for #6 rebar or larger (ACI 318-14, Section 16.5.6.1).
- The depth of the vertical outside face of the corbel should be at least $0.5d$, where d is the effective depth of the corbel at the supporting column face. This prevents premature failure from a vertical shear crack propagating from the outermost edge of the bearing plate down to the sloped surface of the corbel (ACI 318-14 Code, Section 16.5.2.2).
- No part of the bearing area of the corbel should project beyond the end of the straight portion of the primary tension tie reinforcement within the corbel or the interior face of the horizontal anchor bar where an anchor bar is used. This will prevent the formation of a diagonal shear crack projecting from the outermost edge of the bearing plate down to the vertical face of the corbel (ACI 318-14, Section 16.5.2.3).
- For normal-weight concrete, the corbel should be such that the factored concentrated load on the corbel, V_u, satisfies the following relationships (ACI 318-14 Code, Section 16.5.2.4 and Table 22.9.4.4):

$$\frac{V_u}{\phi} \leq 0.2 f'_c b_w d$$

$$\leq (480 + 0.08 f'_c)\, b_w d$$

$$\leq 1600\, b_w d$$

Where, b_w = width of the corbel. In practice, the width of the corbel is usually selected to be the same as the column width. For the corresponding equations for light-weight concrete, see ACI Code Section 16.5.2.5.

- $N_{uc} \geq 0.2V_u$ (ACI 318-14 Code, Section 16.5.3.5)

This factored horizontal tension force could occur from the effects of temperature, creep, and shrinkage. The code mandates a minimum horizontal tensile force of $0.2V_u$ unless tension forces are prevented from being applied to the corbel or bracket.

- Since shear is the predominant mode of failure for corbels, a single strength reduction factor of 0.75 is used for all potential failure modes. Strength-reduction factor for corbels, ϕ = 0.75 (ACI 318-14 Code, Section R21.2.1(f))
- At the face of the supporting column (i.e., at the interface of the column and the corbel), the corbel will be subjected to the following forces (ACI 318-14 Code, Section 16.5.3.1):

- Factored vertical shear force, V_u
- Factored horizontal tension force, N_{uc} (At a minimum, $N_{uc} = 0.2\ V_u$)
- Factored moment, $M_u = V_u a_v + N_{uc}(h - d)$

- The limits for the total area of reinforcement required to resist flexure and direct tension at the top of the corbel, A_{sc}, are given as (ACI 318-14 Code, Section 16.5.5.1):

$$A_{sc} \geq A_f + A_n$$
$$\geq \left(\frac{2}{3}\right)A_{vf} + A_n$$
$$\geq 0.04\left(\frac{f'_c}{f_y}\right)(b_w d)$$

The third equation above represents the minimum amount of reinforcement that is required to cross the corbel-column interface in order to prevent sudden failure of the corbel should the corbel crack vertically at the corbel-column interface under the flexural and tensile forces (ACI 318-14, Section R16.5.5.1). The area of the direct tension tie reinforcement, A_n, is given as $A_n = \dfrac{N_{uc}}{\phi f_y}$ (ACI 318-14 Code, Section 16.5.4.3) and the area of the flexural reinforcement, A_f, required to resist the factored moment, $M_u = V_u a_v + N_{uc}(h - d)$ is determined from the flexure theory previously presented in Chapters 2 and 3 as follows:

$$A_f = \frac{M_u}{\phi f_y(d - 0.5a)}$$

$$a = \frac{A_f f_y}{0.85 f'_c\, b_w}$$

Some iteration is required to calculate A_f since the equation for A_f is a function of the depth of the stress block, a, which in turn is a function of A_f. To use the above equations, first assume a value (e.g., initially assume $a \approx 0.1$d) for the depth of the compression stress block, a, and then calculate A_f. Recalculate the stress block depth using the A_f obtained, and repeat the iterative process until the calculated depth of the stress block is close enough to the initially assumed depth of the stress block.

- Using ACI 318-14 Section 22.9.4.2, calculate the area of the shear friction reinforcement, A_{vf}, as follows using ACI Code Equation (22.9.4.2).

$$A_{vf} = \frac{V_u}{\phi\ \mu f_y}$$

where μ is the coefficient of friction with values obtained from ACI 318-14 Code, Table 22.9.4.2. For normal-weight concrete, $\mu = 1.4$ for concrete cast monolithically, which applies to concrete corbels. For light-weight concrete, $\mu = (1.4)(0.75) = 1.05$. See ACI Code Table 22.9.4.2 for values of μ for other conditions. The strength reduction factor, ϕ, for brackets and corbels is 0.75 (ACI 318-14, Table 21.2.1).

- To prevent diagonal tension failure of the corbel, closed horizontal stirrups or ties are required and these horizontal stirrups must be located with a depth of $\frac{2}{3}d$ from the primary tension tie reinforcement at the top of the corbel (ACI 318-14 Code, Section 16.5.6.6). The minimum area of the closed horizontal stirrups or ties, A_h, is given in ACI 318-14 Code Sections R16.5.5.1 and 16.5.5.2 as follows:

$$A_h = \frac{A_{vf}}{3}$$
$$\geq 0.5(A_{sc} - A_n)$$

- For the reinforcement supporting the horizontal stirrups, select nominal sizes since these are framing rebars that only provide support to the horizontal stirrups, A_h.
- The area of the bearing plate can be determined from ACI 318-14 Code Section 22.8 as follows:

$$A_{bearing\ plate} = \frac{V_u}{\phi_{bearing} 0.85 f'_c}$$

where $\phi_{bearing}$ is 0.65 (see ACI Code, Section 21.2.1(d)).

Select the bearing plate dimensions assuming a width of b_w – 4 in. (i.e., assuming the bearing plate is set in 2 inches on both sides of the corbel). The bearing plate dimension in the orthogonal direction perpendicular to the column/corbel interface $= \dfrac{A_{bearing\ plate}}{(b_w - 4\text{ in.})}$

where b_w = width of corbel.

- Check the required development length, $\ell_{dh,\ required}$, of the main tension tie reinforcement with a 90° standard hook in the column. Compare this to the actual development length provided, $\ell_{dh,\ provided}$ = column width – concrete cover – column tie diameter – diameter of column vertical rebar. Note that the hooked bars should be within the column vertical rebar cage.
- The main tension tie reinforcement must also be developed in the corbel either using an anchor bar or a steel angle to which the tension tie is welded.

Example 4-6

Corbels and Brackets

The column corbel shown in Figure 4-31 supports a service dead load of 35 kip and a service live load of 75 kip. The concrete strength, f'_c, is 4500 psi and the steel yield strength, f_y, is 60,000 psi. Design the corbel, selecting the dimensions of the corbel and all the required reinforcement per the ACI Code using the shear friction method.

Solution:

- For the corbel, the distance from the face of the supporting column to the line of action of the concentrated load that is being supported, a_v = 9 in. – 1/2 (length of leg of bearing angle) = 9 in. – (1/2)(6 in.) = 6 in.

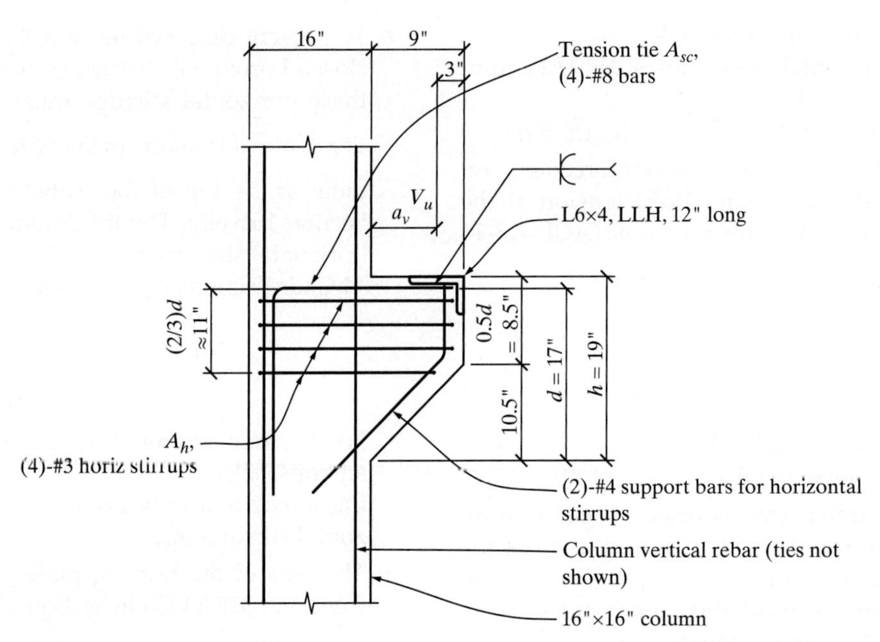

FIGURE 4-31 Corbel for Example 4-6.

- The factored vertical load, $V_u = 1.2\ (35\text{ kip}) + 1.6\ (75\text{ kip}) = 162\text{ kip}$
- The factored tension force, $N_{uc} = 0.2V_u = (0.2)(162\text{ kip}) = 32.4\text{ kip}$
- Strength-reduction factor for corbels, $\phi = 0.75$ (ACI 318-14 Code, Section 21.2.1)
- To use the shear friction method, the following limits must be satisfied:

$$a_v/d \leq 1.0 \text{ and } N_{uc} \leq V_u.$$

- The depth of the vertical outside face of the corbel should be at least $0.5d$, where d is the effective depth of the corbel at the supporting column face (ACI 318-14 Code, Section 16.5.2.2).
- No part of the bearing area of the corbel should project beyond the end of the straight portion of the primary tension tie reinforcement within the corbel or the interior face of the horizontal anchor bar where an anchor bar is used.
- The width of the column is 16 inches; Assuming the width of the corbel is equal to the column width, therefore, $b_w = 16$ inches
- For normal-weight concrete, the corbel should be such that the factored concentrated load on the corbel, V_u satisfies the following relationships (ACI 318-14 Code, Section 16.5.2.4 and Table 22.9.4.4):

$$\frac{V_u}{\phi} = \frac{162\text{ kip}}{0.75} \leq 0.2f'_c b_w d = \frac{[0.2\,(4500\text{ psi})(16\text{ in.})d]}{1000}$$

$$\leq (480 + 0.08\,f'_c)\,b_w d = \frac{[480 + 0.08\,(4500\text{ psi})](16\text{ in.})d}{1000}$$

$$\leq 1600\,b_w d = \frac{1600(16\text{ in.})d}{1000}$$

Solving the above equations yield the following values for d: 15 in.; 16.1 in.; 8.4 in.

The largest value governs, therefore, assume $d = 17$ inches.

The total depth of the bracket, $h = d +$ concrete cover to primary tie rebar $+ \frac{1}{2}$ (diameter of tie rebar)

Assuming #8 tension tie rebar and 1.5 in. cover, $h = 17 + 1.5 + \frac{1}{2}(1'') = 19$ in.

$a_v/d = 6\text{ in.}/17\text{ in.} = 0.35 < 1$, and $N_{uc} = 32.4\text{ kip} \leq V_u = 162\text{ kip}$

Therefore, the shear friction method can be used.

- At the face of the supporting column (i.e., at the interface of the column and the corbel), the corbel will be subjected to the following forces (ACI 318-14 Code, Section 16.5.3.1):
  - Factored vertical shear force, $V_u = 162$ kip
  - Factored horizontal tension force, $N_{uc} = 0.2\,V_u = 32.4$ kip
  - Factored moment, $M_u = V_u a_v + N_{uc}(h - d) = (162\text{ kip})(6\text{ in.}) + (32.4\text{ kip})\,(19\text{ in.} - 17\text{ in.}) = 1037$ kip-in.
  - $A_n = \dfrac{N_{uc}}{\phi f_y} = \dfrac{32.4\text{ kip}}{(0.75)(60\text{ ksi})} = 0.72\text{ in.}^2$ (ACI 318-14 Code, Section 16.5.4.3)

and the flexural reinforcement, A_f, required to resist the factored moment, is calculated as follows:

- $M_u = 1037$ kip-inch

Assume initial depth of the stress block, $a = 1.5$ in.

$$A_f = \frac{M_u}{\phi f_y(d - 0.5a)} = \frac{1037\text{ kip-in.}}{(0.75)(60\text{ ksi})[17'' - 0.5(1.5'')]}$$

$$= 1.42\text{ in}^2$$

$$a = \frac{A_f f_y}{0.85 f'_c b_w} = \frac{(1.42\text{ in}^2)(60\text{ ksi})}{0.85(4.5\text{ ksi})(16\text{ in.})} = 1.4\text{ in}$$

$$\approx 1.5\text{ in. assumed.} \qquad \text{O.K.}$$

Use $A_f = 1.42\text{ in.}^2$

- The coefficient of friction for corbel cast monolithic with the column and using normal-weight concrete, $\mu = 1.4$.

Using ACI 318-14, Section 22.9.4.2, calculate the area of the shear friction reinforcement, A_{vf}, as follows:

$$A_{vf} = \frac{V_u}{\phi \mu f_y} = \frac{162 \text{ kip}}{(0.75)(1.4)(60 \text{ ksi})} = 2.57 \text{ in.}^2$$

- The limits for the total area of reinforcement required to resist flexure and direct tension at the top of the corbel, A_{sc}, are given as (ACI 318-14 Code, Section 16.5.5.1):

$$A_{sc} \geq A_f + A_n = 1.42 \text{ in.}^2 + 0.72 \text{ in.}^2 = 2.14 \text{ in.}^2$$

$$\geq \left(\frac{2}{3}\right)A_{vf} + A_n = \left(\frac{2}{3}\right)(2.57 \text{ in.}^2) + 0.72 \text{ in.}^2 = 2.43 \text{ in.}^2$$

$$\geq 0.04\left(\frac{f'_c}{f_y}\right)(b_w d) = 0.04\left(\frac{4.5 \text{ ksi}}{60 \text{ ksi}}\right)(16 \text{ in.})(17 \text{ in.})$$

$$= 0.82 \text{ in.}^2$$

- Therefore, $A_{sc} = 2.43 \text{ in}^2$. Use 4 #8 bars ($A_{sc}$ provided $= 3.14 \text{ in}^2$).
- To prevent diagonal tension failure of the corbel, closed horizontal stirrups or ties are required and these horizontal stirrups must be located with a depth of $\frac{2}{3}d$ from the primary tension tie reinforcement at the top of the corbel (ACI 318-14 Code, Section 16.5.6.6). The minimum area of the closed horizontal stirrups or ties, A_h, is given in ACI 318-14 Code, Sections R16.5.5.1 and 16.5.5.2, as follows:

$$A_h = \frac{A_{vf}}{3} = \frac{2.57 \text{ in.}^2}{3} = 0.86 \text{ in.}^2$$

$$\geq 0.5(A_{sc} - A_n) = 0.5(2.43 \text{ in.}^2 - 0.72 \text{ in.}^2) = 0.86 \text{ in.}^2$$

Therefore,

$$A_h = 0.86 \text{ in.}^2$$

Use 4 #3 closed horizontal stirrups placed within $(2/3)d$ or (2/3)(17 in.) or 11 inches below the main tension tie rebar.

$$A_h \text{ provided} = (4)(2 \text{ legs})(0.11 \text{ in.}^2 \text{ per leg})$$
$$= 0.88 \text{ in.}^2 > A_h \text{ required.} \quad \text{O.K.}$$

- For the framing rebars supporting the horizontal stirrups, use 2 #4 bars.
- The area of the bearing plate can be determined from

$$A_{bearing\ plate} = \frac{V_u}{\phi_{bearing} 0.85 f'_c} = \frac{162 \text{ kips}}{(0.65)(0.85)(4.5 \text{ ksi})} = 65 \text{ in.}^2$$

$\phi_{bearing} = 0.65$. Given that the width of the corbel $b_w = 16$ inches, and assuming a 12-in.-wide bearing plate or angle which provides at least a 2 in. edge distance on each side of the plate, therefore, the length of the bearing plate $= 65 \text{ in.}^2/12 \text{ in.} = 5.5 \text{ in.} < 6$ inch angle leg provided, O.K.

- Check the required development length, $\ell_{dh,\ required}$, of the main tension tie reinforcement with a 90° standard hook in the column. Compare this to the actual development length provided, $\ell_{dh,\ provided} = 16 \text{ in.} - 1.5 \text{ in.}$ concrete cover – 3/8″ ties – 1″ (assumed column rebar diameter) = 13.1 in. Note that the hooked bar will be within the column vertical rebar cage. Reviewing what was discussed in Chapter 5, the reader can verify that the development length provided is greater than the required development length, $\ell_{dh,\ required}$, for the 4#8 tension tie rebar.
- The main tension tie reinforcement must also be developed in the corbel by welding the tension ties to the underside of the embedded L 6 × 4 angle.
- The final reinforcement detail for the corbel is shown in Figure 4-31.

References

[1] Adam Lubell, Ted Sherwood, Evan Bentz, and Michael P. Collins. "Safe Shear Design of Large, Wide Beams," Concrete International, January 2004.

[2] Terry McDonald and Homa Ghaemi. "From Office to Condominium: The Park Monroe – Adaptive Reuse of the 55/65 E. Monroe near Chicago's Millenium Park," Structure Magazine, April 2011

Problems

4-1. A reinforced concrete beam of rectangular cross section is reinforced for moment only and subjected to a shear V_u of 9000 lb. Beam width $b = 12$ in., $d = 7.25$ in., $f'_c = 3000$ psi, and $f_y = 60{,}000$ psi. Is the beam satisfactory for shear?

4-2. An 8-in.-thick one-way slab is reinforced for positive moment with No. 6 bars at 6 in. on center. Cover is 1 in. Determine the maximum shear V_u permitted. Use $f'_c = 4000$ psi and $f_y = 60{,}000$ psi.

4-3. Assume that the beam of Problem 4-1 has an effective depth of 18 in. and is reinforced with No. 3 single-loop stirrups spaced at 10 in. on center. Determine the maximum shear V_u permissible.

4-4. The simply supported beam shown is on a clear span of 30 ft. The beam carries uniformly distributed service loads of 1.9 kip/ft live load and 0.7 kip/ft dead load (excluding the weight of the beam). Additionally, the beam carries two concentrated service loads of 8 kip dead load each, one load being placed 5 ft in. from the face of each support. The beam is reinforced with No. 3 single-loop stirrups placed in the following pattern, starting at the face of the support (symmetry about midspan): one space at 2 in., five spaces at 8 in., four spaces at 10 in., and eight spaces at 12 in. Beam width $b = 15$ in., $h = 28$ in., and $d = 25.4$ in. Use $f'_c = 4000$ psi and $f_y = 60{,}000$ psi.

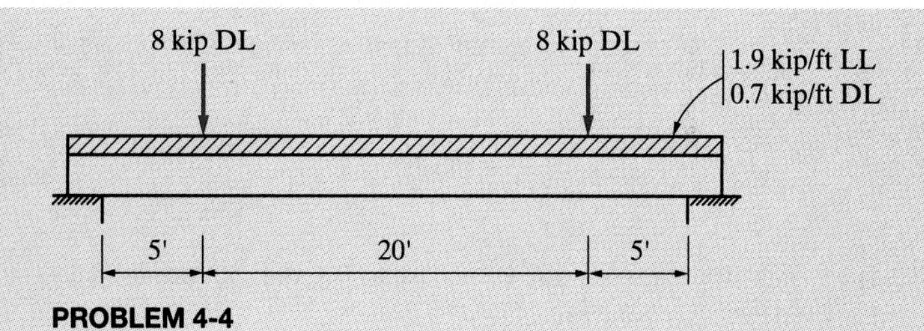

PROBLEM 4-4

a. Draw an elevation view of the beam showing the stirrup layout.
b. Draw the V_u diagram.
c. Calculate $\phi(V_c + V_s)$ for each spacing group (omit the 2-in. spacing).
d. Superimpose the results from part (c) on the V_u diagram and comment on the results.

4-5. A uniformly loaded beam is subjected to a shear V_u of 60 kip at the face of the support. Clear span is 32 ft. Beam width $b = 12$ in., $d = 22$ in., $f'_c = 4000$ psi, and $f_y = 60{,}000$ psi. Determine the maximum spacing allowed for No. 3 single-loop stirrups at the critical section.

4-6. A simply supported, rectangular, reinforced concrete beam having $d = 24$ in. and $b = 15$ in. supports a uniformly distributed load w_u of 7.50 kip/ft as shown. The given load includes the beam weight. Assume No. 3 single-loop stirrups. The concrete strength is 4000 psi and the steel is grade 60.

a. Select the stirrup spacing to use at the critical section (d distance from the face of support).
b. Determine the maximum stirrup spacing allowed for this beam.
c. Using the two spacings determined in parts (a) and (b), devise an appropriate stirrup spacing layout for this beam.

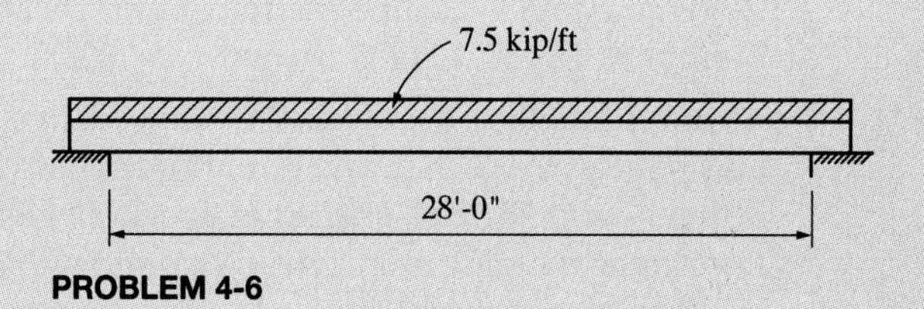

PROBLEM 4-6

4-7. A simply supported beam carries a total factored load w_u of 5.0 kip/ft. The span length is 28 ft center-to-center of supports, and the supports are 12-in. wide. Beam width $b = 14$ in., $d = 20$ in., $f'_c = 4000$ psi, and $f_y = 60{,}000$ psi. Determine spacings required for No. 3 stirrups and show the pattern with a sketch. (Recall that clear span is used for determining shears.)

4-8. A simply supported rectangular reinforced concrete beam, 13-in. wide and having an effective depth of 20 in., supports a total factored load (w_u) of 4.5 kip/ft on a 30-ft clear span. (The given load includes the weight of the beam.) Design the web reinforcement if $f'_c = 3000$ psi and $f_y = 40{,}000$ psi.

4-9. A rectangular reinforced concrete beam supports a total factored load (w_u) of 10 kip/ft on a simple clear span of 40 ft. (The given load includes the weight of the beam.) The effective depth $d = 40$ in., $b = 24$ in., $f'_c = 3000$ psi, and $f_y = 60{,}000$ psi. Design double-loop No. 3 stirrups.

4-10. Design stirrups for the beam shown. The supports are 12-in. wide, and the loads shown are service loads. The dead load includes the weight of the beam. Beam width $b = 16$ in., $d = 20$ in., $f'_c = 4000$ psi, and $f_y = 60{,}000$ psi. Sketch the stirrup pattern.

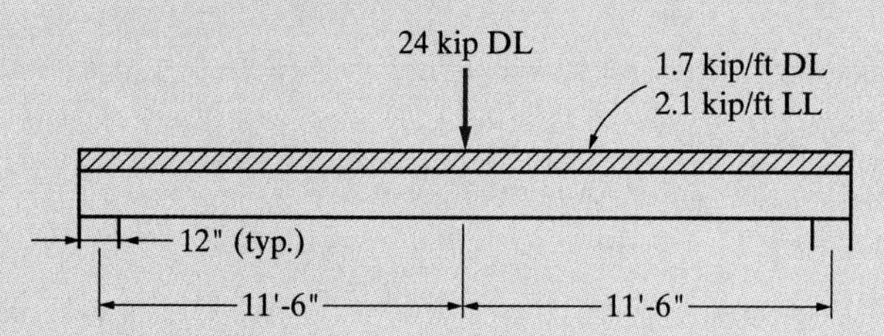

PROBLEM 4-10

4-11. For the beam shown, check moment and design single-loop stirrups. The loads shown are factored loads. Assume the supports to be 12-in. wide. Use $f'_c = 3000$ psi and $f_y = 60{,}000$ psi. The uniformly distributed load includes the beam weight.

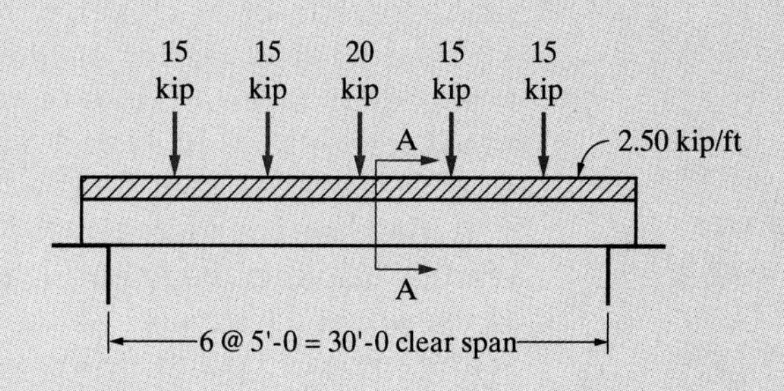

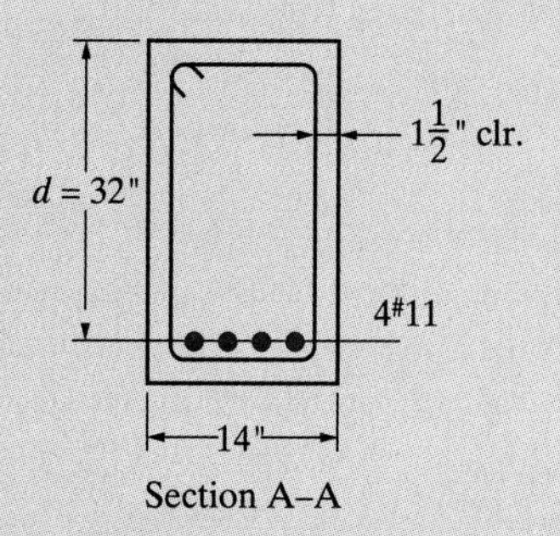

PROBLEM 4-11

4-12. Design stirrups for the beam shown. Service loads are 1.5 kip/ft dead load (includes beam weight) and 1.9 kip/ft live load. The supports are 12-in. wide. Beam width $b = 13$ in. and $d = 24$ in. for both top and bottom steel. Use $f'_c = 3000$ psi and $f_y = 60{,}000$ psi. Sketch the stirrup arrangement.

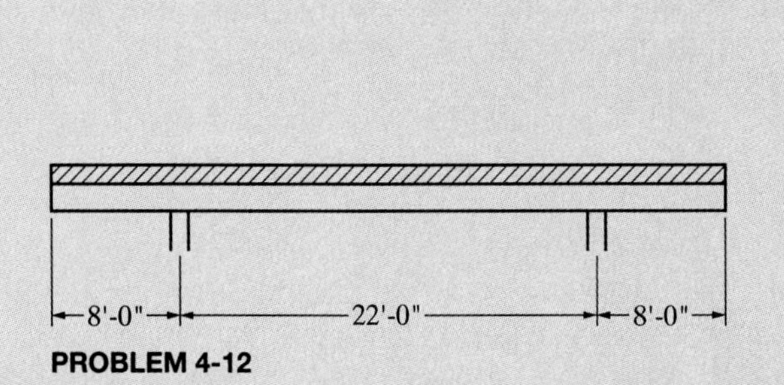

PROBLEM 4-12

4-13. Design stirrups for the beam shown. The supports are 12-in. wide, and the loads shown are factored design loads. The dead load includes the weight of the beam. Beam width $b = 14$ in., $d = 24$ in., $f'_c = 3000$ psi, and $f_y = 60{,}000$ psi. Sketch the stirrup pattern.

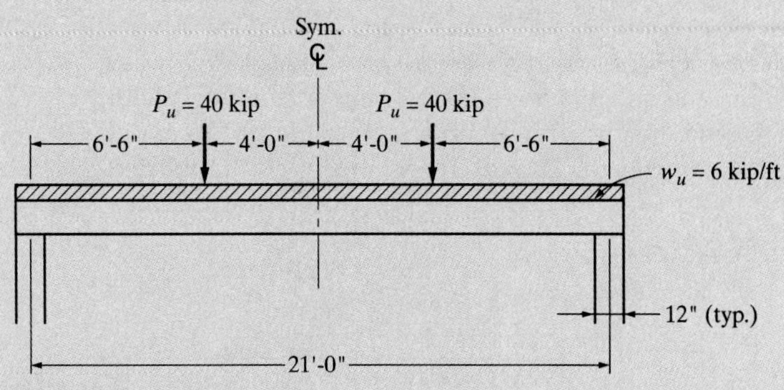

PROBLEM 4-13

4-14. The beam shown is supported on pedestals 20-in. wide. The loads are service loads, and the distributed load includes the beam weight. The clear span is 24′-4″. Use $f'_c = 4000$ psi; the reinforcing steel is A615 Grade 60. Design shear reinforcing for the beam:

a. between the point load and the left reaction

b. between the point load and the right reaction

Show design sketches, including the stirrup pattern.

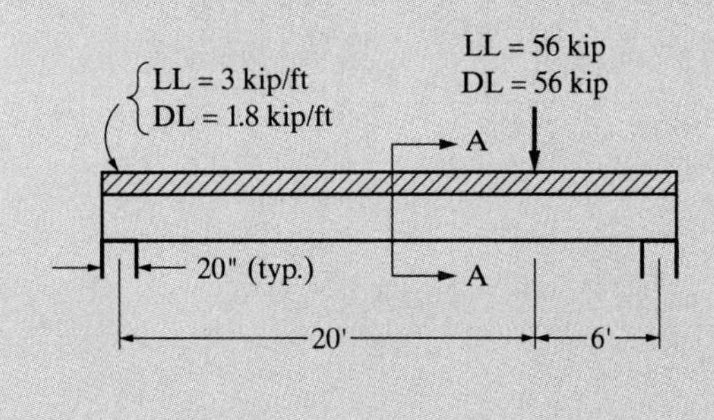

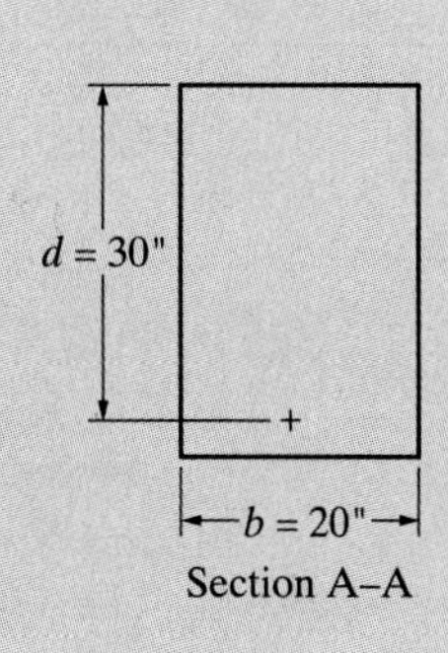

PROBLEM 4-14

4-15. Design the rectangular reinforced concrete beam for *moment and shear.* Use only tension steel for flexure. The loads shown are service loads. The uniform load is composed of 1 kip/ft dead load (does not include beam weight) and 1 kip/ft live load. The concentrated loads are dead load. Assume the supports to be 12-in. wide. Use $f'_c = 4000$ psi and $f_y = 60{,}000$ psi. Show design sketches, including the stirrup pattern.

4-16. Explain, with examples, the difference between equilibrium and compatibility torsion.

4-17. A rectangular reinforced concrete beam 14 in. × 20 in. deep is subject to a maximum factored torque of 24 ft-k. Calculate the cracking torque, T_{cr}, and determine if torsion can be neglected in the design of this beam. Assume normal-weight concrete (i.e., $\lambda = 1.0$) and $f'_c = 4000$ psi.

4-18. The floor framing in an office building consists of reinforced concrete beams 18 in. × 24 in. deep that support precast concrete planks on a 3-in. ledge similar to that shown in Figure 4-23. The clear span of the beam is 27 ft between columns. The planks are 10-in. deep with 2-in. topping and supports stud wall partitions that weigh 10 psf and mechanical/electrical equipment that weigh 5 psf. The weight of the precast planks is 70 psf. The centerline to centerline span of the planks is 30 ft on the left-hand side of the beam and 10 ft on the right-hand side of the beam. The live load on the 30-ft span is 50 psf while the live load on the 10-ft corridor span is 100 psf. Design the beam for torsion and shear assuming normal-weight concrete (i.e., $\lambda = 1.0$), $f'_c = 4000$ psi, and $f_y = 60{,}000$ psi. Assume the beam has already been designed for bending.

4-19. A reinforced concrete beam was designed with vertical stirrups (see the top figure below), but the stirrups were inadvertently fabricated with its height larger than the beam depth. To avoid having to discard all the incorrectly fabricated stirrups, the contractor's engineer has suggested using the fabricated stirrups in an inclined orientation as shown in the bottom figure below. You have been asked to review this suggested stirrup layout.

a. Is the beam stirrup arrangement suggested by the contractor's engineer adequate?

b. If the stirrup arrangement is adequate, why is it adequate? If it's not adequate, why is it not adequate? And, if it's not adequate, what is the solution?

(Note: No new stirrups will be fabricated, and no calculation is needed.)

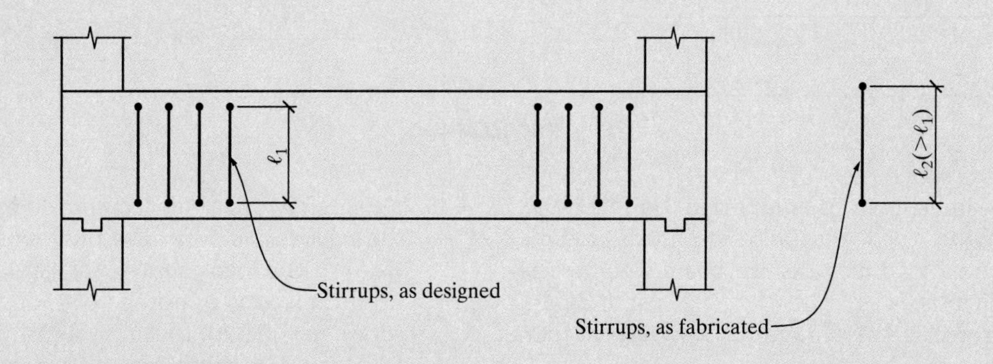

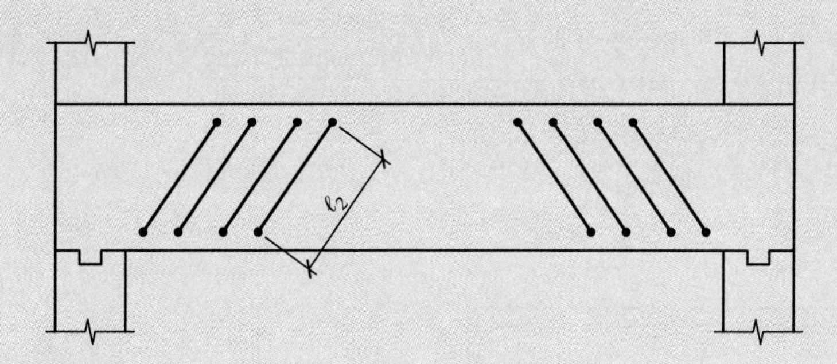

Contractor's Suggestion

PROBLEM 4-19

4-20. Refer to the given detail and shear diagram for a simple span beam. Use $f'_c = 6000$ psi, $f_y = 60{,}000$ psi

a. Determine the required spacing of the shear reinforcement in zones A and B (select uniform spacing in each zone).

b. Sketch the loading that goes along with this shear diagram.

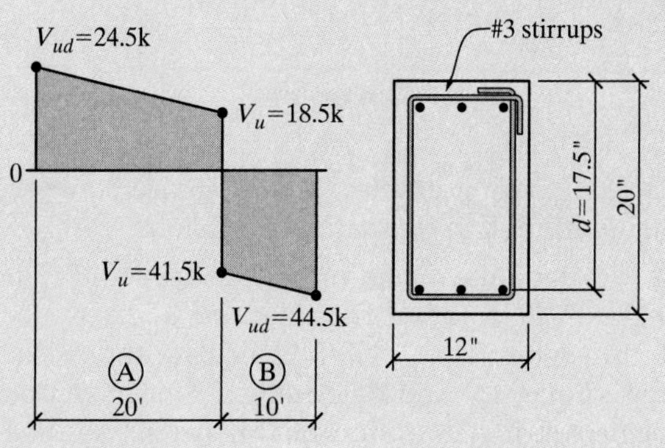

PROBLEM 4-20

4-21. Refer to the given detail and shear diagram. Use $f'_c = 3000$ psi, $f_y = 60{,}000$ psi. Determine the required spacing of the shear reinforcement along the beam.

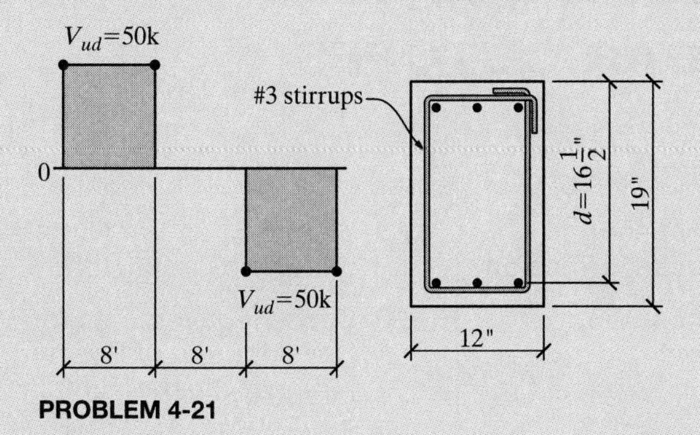

PROBLEM 4-21

4-22. Assume the clear span of a beam is 36′-0″ and has dimensions of 12″ × 24″ beam, with d = 21.5″. There are (3) concentrated loads of 23 kip (factored) that are applied at 1/4 points. Add the self-weight as a uniform load (density = 150 pcf). Draw the loading diagram and shear diagram, and determine the shear reinforcement for the beam. Sketch the complete design.

4-23. Design the corbel for a 20″ × 20″ column that supports a service dead load of 50 kip and a service live load of 110 kip. The concrete strength, f'_c, is 4500 psi and the steel yield strength, fy, is 60,000 psi. Select the dimensions of the corbel and all the required reinforcement per the ACI Code using the shear friction method.

4-24. Refer to the given detail and shear diagram for a simple span beam. Using $f'_c = 3000$ psi and $f_y = 60{,}000$ psi:

a. Determine the required spacing of the shear reinforcement in zones A, B, & C (select uniform spacing in each zone).

b. Sketch the factored loading that goes along with this shear diagram.

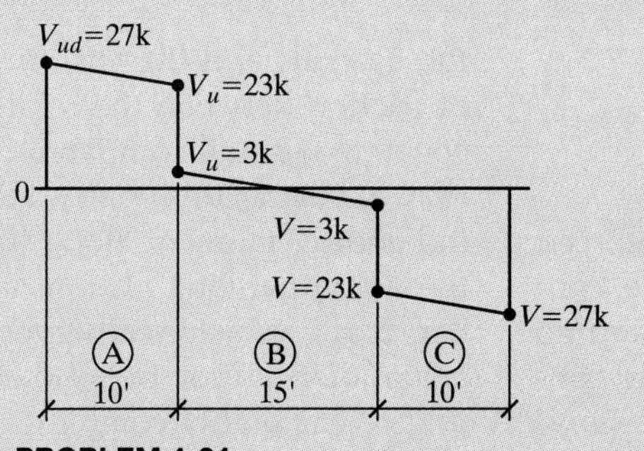

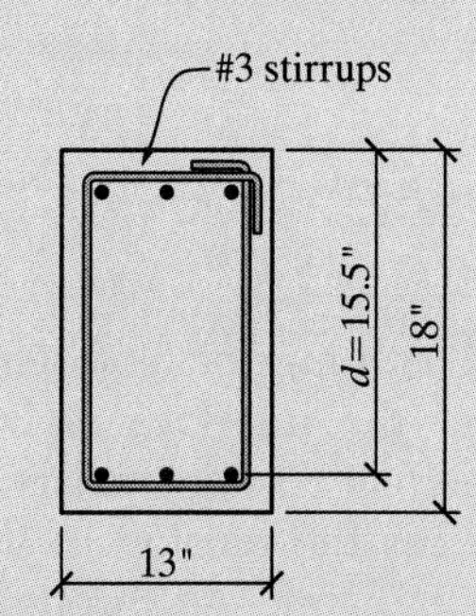

PROBLEM 4-24

CHAPTER **FIVE**

DEVELOPMENT, SPLICES, AND SIMPLE-SPAN BAR CUTOFFS

5-1 Bond Stress and Development Length: Introduction
5-2 Development Length: Tension Bars
5-3 Development Length: Compression Bars
5-4 Development Length: Standard Hooks in Tension
5-5 Development of Web Reinforcement
5-6 Splices
5-7 Tension Splices
5-8 Compression Splices
5-9 Simple-Span Bar Cutoffs and Bends
5-10 Code Requirements for Development of Positive Moment Steel at Simple Supports
5-11 Structural Integrity Reinforcement—Beams

5-1 BOND STRESS AND DEVELOPMENT LENGTH: INTRODUCTION

One of the fundamental assumptions of reinforced concrete design is that at the interface of the concrete and the steel bars, perfect bonding exists and no slippage occurs. That means the tension in the concrete is transferred from the concrete through bond to the steel reinforcement. Based on this assumption, it follows that some form of bond stress exists at the contact surface between the concrete and the steel bars. The bond action between the concrete and the deformed bar is primarily due to the wedge action of the protruding ribs of the reinforcement against the surrounding concrete (see Figure 5-1), creating equal and opposite radial forces that act on the bar ribs and the surrounding concrete. The effect of friction between the reinforcement surface and the concrete is small compared to the wedge action of the ribs. In beams, this bond stress is caused by the change in bending moment along the length of the beam and the accompanying change in the tensile stress in the bars and has historically been termed *flexural bond*. The actual distribution of bond stresses along the reinforcing steel is highly complex, due primarily to the presence of concrete cracks. Research has indicated that large local variations in bond stress are caused by flexural and diagonal cracks, and very high bond stresses have been measured adjacent to these cracks [1, 2]. These high bond stresses may result in small local slips adjacent to the cracks, with resultant widening of cracks as well as increased deflections. Generally, this will be harmless as long as failure does not propagate

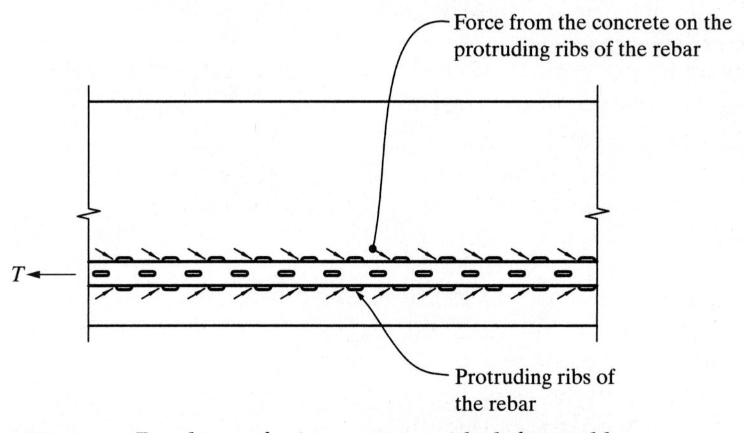

FIGURE 5-1 Bond transfer in concrete with deformed bars.

FIGURE 5-2 90° bends for development of No. 18 bars. Seabrook station, New Hampshire. (George Limbrunner)

all along the bar with resultant complete loss of bond. It is possible, if *end anchorage* is reliable, that the bond can be severed along the entire length of bar, excluding the anchorage, without endangering the carrying capacity of the beam. The resulting behavior is similar to that of a tied arch with the beam tension reinforcement acting as a tension tie anchored at the end supports.

End anchorage may be considered reliable if the bar is embedded into concrete a prescribed distance known as the *development length*, ℓ_d, of the straight bar or the reinforcement is hooked (see Figure 5-2), in which case the required development length of the hooked bar would be ℓ_{dh}. If in the beam the actual extended length of a bar is equal to or greater than this required development length, no premature bond failure will occur—that is, the predicted strength of the beam will not be controlled by bond but rather by some other factor.

Hence, the main requirement for safety against bond failure is that the length of the bar from any point of given steel stress f_s (or, at a maximum, f_y) to its nearby free end must be at least equal to its development length. If this requirement is satisfied, the magnitude of the flexural bond stress along the beam is of only secondary importance, because the integrity of the member is assured even in the face of possible minor local failures. If the actual available length is inadequate for full development of a straight bar, however, special anchorages, such as hooks, must be provided to ensure adequate strength.

The average bond stress, f_{bd}, between two cracks in a beam is the tension force in the steel reinforcement divided by the product of the sum of the circumference of the reinforcement and the length of the beam between the cracks, and can be derived as follows using:

- Equilibrium of forces between the tension force in the reinforcement at the crack and the bond force that exists between the concrete and the steel between the two cracks,
- The principles of flexural behavior from Chapter 2, and
- The relationship between shear and moment ($V = dM/dx$)

$$\text{Average bond stress, } f_{bd} = \frac{T}{\left(\sum \pi d_b\right)dx} = \frac{\dfrac{dM}{jd}}{\left(\sum \pi d_b\right)dx}$$

Since $V = dM/dx$,
The average bond stress can be rewritten as

$$f_{bd} = \frac{\dfrac{dM}{jd}}{\left(\sum \pi d_b\right)dx} = \frac{V}{\left(\sum \pi d_b\right)jd}$$

Where,

jd = the flexural lever arm
dM = the change in moment in the beam between the two cracks
dx = length of the beam between the two cracks
$\sum \pi d_b$ = sum of the circumference of all the tension bars in the beam between the two cracks
V = shear at the section

Note that f_{bd} is an average bond stress, but the actual bond stress varies considerably between the cracks [1, 2]. The smaller the diameter of the reinforcement, the higher the bond stress (strength) as is evident from the preceding equation, and in regions of zero shear, the bond stress is zero. The strain in the concrete is zero at the cracks and at a maximum near the mid-distance between the cracks. The corresponding strain in the steel reinforcement is at a maximum at the crack locations and at a minimum midway between the cracks, where the stress is fully transferred from the steel to the concrete. The stress distribution in the steel and concrete between the cracks follow a similar pattern [1].

Current design methods based on the ACI 318-14 Code do not require calculation of the bond stress and disregard high localized bond stress even though it may result in localized slip between steel and concrete adjacent to the cracks. Instead, attention is directed toward providing adequate length of embedment, past the location at which the bar is fully stressed, which will ensure development of the full strength of the bar (see Figure 5-3). The development length is the shortest length of reinforcement for which the rebar stress can increase from zero to the yield strength, f_y, without pulling out of the concrete. If the actual length of the straight rebar within the concrete cantilever beam in Figure 5-3 is less than ℓ_d, the rebar will not develop its maximum strength ($A_s f_y$). Similarly, if the length of the straight portion of the hooked bar within the column in Figure 5-3 is less than ℓ_{dh}, the hooked bar will not develop its maximum strength ($A_s f_y$). Consequently, the rebar will pull out of the concrete before its yield strength is reached.

5-2 DEVELOPMENT LENGTH: TENSION BARS

Referring to Figure 5-3, the maximum tension force, T, in the steel reinforcement is

$$T = A_s f_y = \pi \frac{(d_b)^2}{4} f_y$$

The concrete transfers this tension force into the steel reinforcement through bond stress. The force transferred through bond action (i.e., the bond force) is the product of the average bond stress, f_{bd}, and the surface area of the rebar in contact with the concrete. This surface area is the product of the sum of the circumference of the steel reinforcement and the development length, ℓ_d.

Thus, the force transferred by bond $= f_{bd}\ell_d(\pi d_b)$. Equilibrium requires that the tension force in the reinforcement be equal to the total bond force. Thus,

$$f_{bd}\ell_d(\pi d_b) = \pi \frac{(d_b)^2}{4} f_y$$

Therefore, the expression for development length, ℓ_d is given as

$$\ell_d = \left(\frac{f_y}{4 f_{bd}}\right) d_b$$

$$= K\left(\frac{f_y}{\sqrt{f'_c}}\right) d_b$$

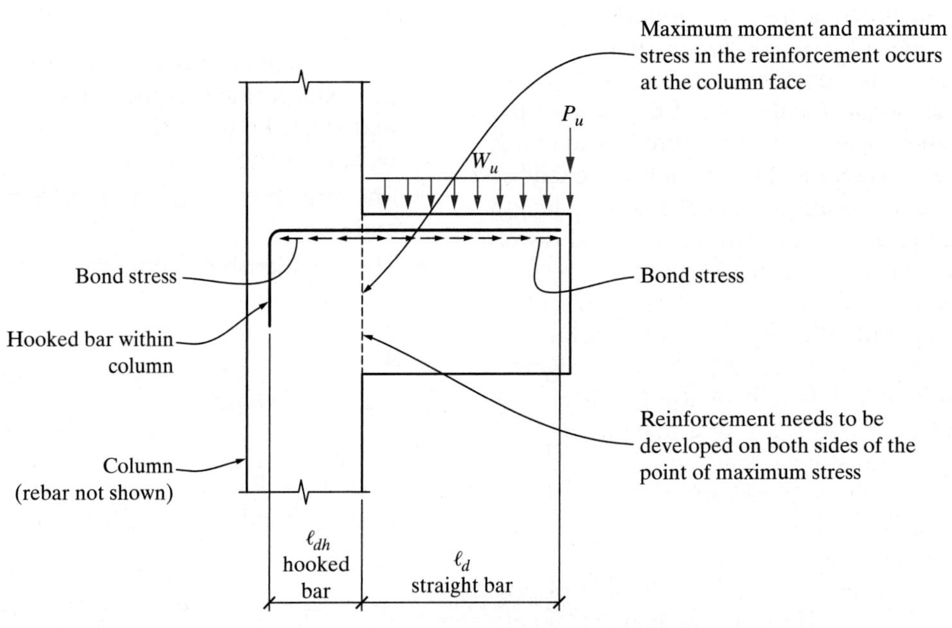

FIGURE 5-3 Development of tension reinforcement in a cantilever beam.

Where, K is an empirical factor determined from analysis of test results (ACI 318-14, Sections 25.4.2.2 and 25.4.2.3). The equations given in the ACI Code for calculating rebar development length take the same form as the preceding equation. The factors that affect the bond strength, and hence the tension development length, of a concrete flexural member are summarized below [1]:

1. Concrete compressive strength, f'_c. The higher the concrete strength, the higher the tensile strength of the concrete, and therefore, the higher the bond strength, and the smaller the required development length.
2. Steel reinforcement yield strength, f_y. The higher the steel yield strength, the higher the required development length.
3. Steel reinforcement diameter. The smaller the rebar diameter, the smaller the required development length. With regard to bond strength, it is advantageous to use more smaller diameter bars than fewer larger diameter bars for the same total area of steel.
4. Concrete cover. The smaller the side and bottom cover to the rebar, the greater the likelihood of concrete spalling and cracking due to the lack of adequate confinement to contain the radial wedging forces. Inadequate bottom covers lead to the development of vertical splitting cracks propagating from the rebar to the bottom surface. This leads to a lower bond strength and therefore a higher required development length.
5. Rebar spacing. The wider the spacing between the steel reinforcement, the better the bond between each rebar and the surrounding concrete, leading to a higher bond strength and a smaller development length. Inadequate center-to-center spacing between the tension bars leads to the development of horizontal splitting cracks between the bars that extends to the vertical surfaces of the beam. This leads to a reduced bond strength and a higher required development length.
6. Type of stress (tension or compression) in the reinforcement. The development length required for tension reinforcement is higher than that required for compression reinforcement because of the presence of flexural tension cracking.
7. Vertical location of the reinforcement within the depth of a beam or girder. When steel reinforcement is located such that there is more than 12 inches of concrete depth below the reinforcement, there is a tendency for entrapped air to gravitate to the underside of the steel rebar and for moisture to accumulate below the reinforcement due to settlement of the concrete during the concrete vibration process. This leads to reduced bond between the rebar and the surrounding concrete and thus a higher required development length. The higher the concrete slump, the greater the tendency for the concrete to settle, and therefore, the lower the bond strength.
8. Transverse reinforcement (i.e., stirrups). Confinement of the concrete around the tension rebar by transverse reinforcement helps to prevent the occurrence of horizontal and vertical splitting cracks at the tension rebar locations. This leads to a higher bond strength.
9. Lightweight concrete. The tensile strength of concrete is proportional to $\sqrt{f'_c}$ and from the equation developed earlier for development length, it can be seen that the greater the tensile strength, the lower the required development length, thus, the higher the bond strength. Conversely, the lower the tensile strength, as is the case with lightweight concrete compared to normal weight concrete, the lower the bond strength and the higher the required development length.
10. Epoxy coated rebar. The coating on epoxy coated reinforcement protects it from corrosion, but studies of the anchorage of epoxy-coated bars show that bond strength is reduced because the coating prevents adhesion and friction between the bar and the concrete. This leads to a higher required development length.

The ACI Code, Section 25.4.2.1, specifies that the development length ℓ_d for deformed bars and deformed wires in tension shall be determined using either the tabular criteria of Section 25.4.2.2 or the general equation of Section 25.4.2.3, but in either case, ℓ_d shall not be less than 12 in.

The general equation of Section 25.4.2.3 (ACI Equation [25.4.2.3a]) offers a simple approach that allows the user to see the effect of all variables controlling the development length. The tabular criteria of Section 25.4.2.2 also offer a simple, conservative approach that recognizes commonly used practical construction techniques. Based on a sampling of numerous cases, the authors have found that significantly shorter development lengths are computed using ACI Equation (25.4.2.3a) of Section 25.4.2.3. Therefore, the development length ℓ_d computation used in this text will be based on ACI Equation (25.4.2.3a):

$$\ell_d = \frac{3}{40}\left(\frac{f_y}{\lambda\sqrt{f'_c}}\right)\left[\frac{\psi_t\psi_e\psi_s}{\left(\dfrac{c_b + K_{tr}}{d_b}\right)}\right]d_b$$

in which the term $(c_b + K_{tr})/d_b$ shall not be taken greater than 2.5 and where

ℓ_d = development length (in.)

f_y = specified yield strength of non-prestressed reinforcement (psi)

f'_c = specified compressive strength of concrete (psi); the value of $\sqrt{f'_c}$ shall not exceed 100 psi (ACI Code, Section 25.4.1.4)

d_b = nominal diameter of bar or wire (in.).

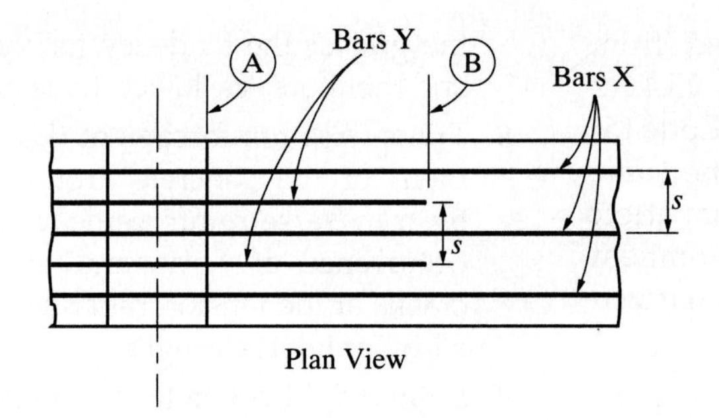

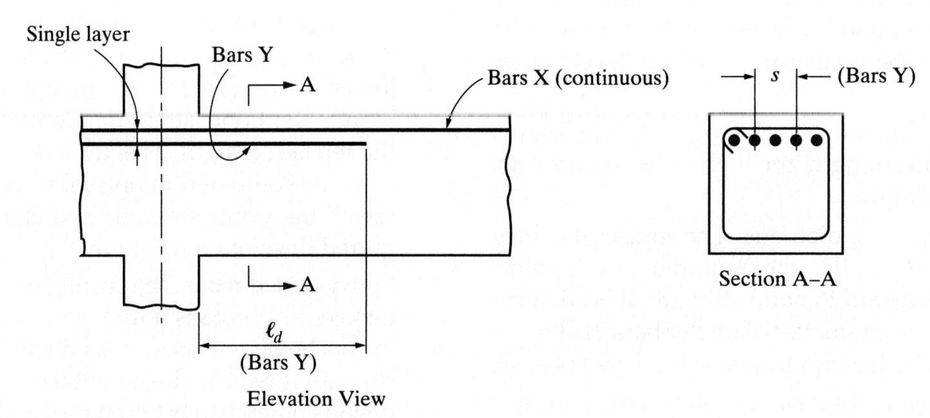

FIGURE 5-4 Spacing criteria for bars being developed.

The other factors used in ACI Equation (25.4.2.3a) are defined as follows:

1. ψ_t is a reinforcement location factor that accounts for the position of the reinforcement in freshly placed concrete.

   Where horizontal reinforcement is so placed that more than 12 in. of fresh concrete is cast in the member below the development length or splice, use $\psi_t = 1.3$ (ACI Code, Table 25.4.2.4).

   For other reinforcement, use $\psi_t = 1.0$.

2. ψ_e is a coating factor reflecting the effects of epoxy coating.

   For epoxy-coated reinforcement having cover less than $3d_b$ or clear spacing between bars less than $6d_b$, use $\psi_e = 1.5$.

   For all other conditions, use $\psi_e = 1.2$.

   For uncoated and galvanized reinforcement, use $\psi_e = 1.0$.

   The product of ψ_t and ψ_e need not be taken greater than 1.7 (ACI Code, Table 25.4.2.4).

3. ψ_s is a reinforcement size factor.

   Where No. 6 and smaller bars are used, use $\psi_s = 0.8$.

   Where No. 7 and larger bars are used, use $\psi_s = 1.0$.

4. Lambda (λ) is the lightweight-aggregate concrete factor and has been discussed in Section 1-7 of Chapter 1. For purposes of development length calculation where lightweight concrete is used, λ shall not exceed 0.75 unless average f_{ct} is specified, in which case it can be calculated (see Section 1-7).

5. The factor c_b represents a spacing or cover dimension (in.).

   The value of c_b will be the smaller of either the distance from the center of the bar to the nearest concrete surface or one-half the center-to-center spacing of the bars being developed.

   The bar spacing will be the actual center-to-center spacing between the bars if adjacent bars are all being developed at the same location. If, however, an adjacent bar has been developed at another location, the spacing to be used will be greater than the actual spacing to the adjacent bar. Note in Figure 5-4 that the spacing for bars Y may be taken the same as for bars X, because bars Y are developed in length AB, whereas bars X are developed at a location other than AB. When the bottom cover to the rebar is smaller than one-half the spacing between the longitudinal bars, vertical splitting cracks propagating from the rebar to the bottom surface occurs. When the bottom cover to the rebar is greater than one-half the horizontal spacing between longitudinal bars, horizontal splitting cracks occur between the bars and the cracks extend to the exterior vertical surfaces of the beam [1].

6. The transverse reinforcement index K_{tr} is computed from

$$\frac{40A_{tr}}{sn}$$

where

A_{tr} = total cross-sectional area of all transverse reinforcement that is within the spacing s and that crosses the potential plane of

splitting through the reinforcement being developed (in.^{2})
s = maximum center-to-center spacing of transverse reinforcement within ℓ_d (in.)
n = number of bars or wires being developed along the plane of splitting

To simplify the design, the ACI Code permits the use of $K_{tr} = 0$. This is conservative and may be used even if transverse reinforcement is present. To further simplify development length computations, we designate a portion of ACI Equation (25.4.2.3a) as K_D, where

$$K_D = \frac{3}{40}\frac{f_y}{\sqrt{f'_c}}$$

Values of K_D as a function of various combinations of f_y and f'_c are tabulated in Table 5-1.

A reduction in the required straight bar development length ℓ_d is permitted where reinforcement is in excess of that required by analysis (except where anchorage or development for f_y is specifically required or where the design includes provisions for seismic considerations). We designate this reduction factor as K_{ER}. Excess reinforcement factor K_{ER} does not apply for development of positive moment reinforcement at supports (ACI Code, Section 25.4.10.2) or for development of shrinkage and temperature reinforcement (ACI Code, Section 24.4.3.4). Although K_{ER} is not reflected in ACI Equation (25.4.2.4), it may be calculated from

$$K_{ER} = \frac{A_s \text{ required}}{A_s \text{ provided}}$$

and subsequently applied to the ℓ_d computed from ACI Equation (25.4.2.3a).

When bundled bars are used, the ACI Code, Section 25.6, stipulates that calculated development lengths are to be made for individual bars within a bundle (either in tension or in compression) and then increased by 20% for three-bar bundles and by 33% for four-bar bundles. Bundled bars consist of a group of not more than four parallel reinforcing bars in contact with each other and assumed to act as a unit. For determining the appropriate factors as discussed previously, a unit of bundled bars shall be treated as a single bar of a diameter derived from the equivalent total area.

Summary of Procedures for Calculation of ℓ_d (Using ACI Equation (25.4.2.3a))

1. Determine K_D from Table 5-1.
2. Determine applicable factors (use 1.0 unless otherwise determined).
   a. Use $\psi_t = 1.3$ for top reinforcement, when applicable.
   b. Coating factor ψ_e applies to epoxy-coated bars. Use $\psi_e = 1.5$ if cover $< 3d_b$ or clear space $< 6d_b$. Use $\psi = 1.2$ otherwise.
   c. Use $\psi_s = 0.8$ for No. 6 bars and smaller.
   d. Use $\lambda = 0.75$ for lightweight concrete with f_{ct} not specified and $\lambda = 1.0$ for normal-weight concrete. Use

   $$\lambda = f_{ct}/(6.7\sqrt{f'_c}) \le 1.0$$

   for lightweight concrete with f_{ct} specified.
3. Check $\psi_t\psi_e \le 1.7$.
4. Determine c_b, the smaller of the concrete cover or one-half the center-to-center spacing between the bars.
5. Calculate $K_{tr} = 40A_{tr}/(sn)$, or use $K_{tr} = 0$ (conservative).
6. Check $(c_b + K_{tr})/d_b \le 2.5$.
7. Calculate K_{ER} if applicable:

   $$K_{ER} = \frac{A_s \text{ required}}{A_s \text{ provided}}$$

8. Calculate the required straight bar development length, ℓ_d:

   $$\ell_d = \frac{K_D}{\lambda}\left[\frac{\psi_t\psi_e\psi_s}{\left(\frac{c_b + K_{tr}}{d_b}\right)}\right]K_{ER}\, d_b \ge 12 \text{ in.}$$

Example 5-1

Calculate the tensile development length ℓ_d required for No. 8 top bars (more than 12 in. of fresh concrete to be cast below the bars) in a lightweight-aggregate concrete beam as shown in Figure 5-5. The clear cover is 2 in., and the clear space between bars is 3 in. Use $f_y = 60{,}000$ psi and $f'_c = 4000$ psi. Stirrups are No. 4 bars. All bars are uncoated. f_{ct} is not specified.

TABLE 5-1 Coefficient K_D for ACI Code Equation (25.4.2.3a)

$$K_D = \frac{3}{40}\frac{f_y}{\sqrt{f'_c}}$$

f'_c (psi)	f_y = 40,000 psi	f_y = 50,000 psi	f_y = 60,000 psi	f_y = 75,000 psi
3000	54.8	68.5	82.2	102.7
4000	47.4	59.3	71.2	88.9
5000	42.4	53.0	63.6	79.5
6000	38.7	48.4	58.1	72.6

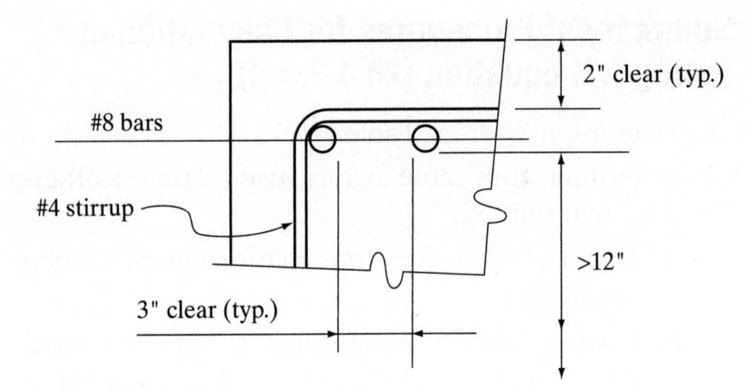

FIGURE 5-5 Partial cross section.

Solution:

Use ACI Equation (25.4.2.3a) and follow the procedural outline that precedes this example.

1. From Table 5-1, $K_D = 71.2$.
2. Establish values for the factors ψ_t, ψ_e, ψ_s, and λ.
   a. $\psi_t = 1.3$ (the bars are top bars).
   b. The bars are uncoated; $\psi_e = 1.0$.
   c. The bars are No. 8; $\psi_s = 1.0$.
   d. Lightweight-aggregate concrete is used; $\lambda = 0.75$.
3. The product $\psi_t \times \psi_e = 1.3 < 1.7$. (O.K.)
4. Determine c_b. Based on cover (center of bar to nearest concrete surface), consider the clear cover, the No. 4 stirrup diameter, and one-half the diameter of the No. 8 bar:

$$c_b = 2 + 0.5 + 0.5 = 3.0 \text{ in.}$$

   Based on bar spacing (one-half the center-to-center distance),

$$c_b = 0.5[3 + 2(0.5)] = 2.0 \text{ in.}$$

   Therefore, use $c_b = 2.0$ in.
5. In the absence of data needed for a calculation, K_{tr} may be conservatively taken as zero.
6. Check $(c_b + K_{tr})/d_b \leq 2.5$:

$$\frac{c_b + K_{tr}}{d_b} = \frac{2.0 + 0}{1.0} = 2.0 < 2.5 \qquad \text{(O.K.)}$$

7. The excess reinforcement factor is assumed not applicable and is omitted.
8. Calculate ℓ_d:

$$\ell_d = \frac{K_D}{\lambda}\left[\frac{\psi_t\psi_e\psi_s}{\left(\dfrac{c_b + K_{tr}}{d_b}\right)}\right]d_b$$

$$= \frac{71.2}{0.75}\left[\frac{1.3(1.0)(1.0)}{2.0}\right](1.0) = 61.7 \text{ in.} > 12 \text{ in.} \quad \text{(O.K.)}$$

Example 5-2

Calculate the development length required for the No. 9 bars in the top of a 15-in.-thick reinforced concrete slab (see Figure 5-6). Note that these bars are the tension reinforcement for negative moment in the slab at the supporting beam. As this is a slab, no stirrups are used. Use $f_y = 60{,}000$ psi and $f'_c = 4000$ psi (normal-weight concrete). The bars are epoxy-coated.

Solution:

1. From Table 5-1, $K_D = 71.2$.
2. Establish values for the factors ψ_t, ψ_e, ψ_s, and λ.
   a. $\psi_t = 1.3$ (the bars are top bars).
   b. The bars are epoxy-coated. Compare cover (3/4 in.) with $3d_b$. If necessary, calculate clear space and compare with $6d_b$.

$$3d_b = 3(1.13) = 3.39 \text{ in.}$$

$$0.75 \text{ in.} < 3.39 \text{ in}$$

   Therefore, use $\psi_e = 1.5$.
   c. The bars are No. 9. Use $\psi_s = 1.0$.
   d. Normal-weight concrete is used. $\lambda = 1.0$.
3. Check the product of ψ_t and ψ_e:

$$\psi_t \times \psi_e = 1.3(1.5) = 1.95 > 1.7$$

   Therefore, use $\psi_t \times \psi_e = 1.7$.

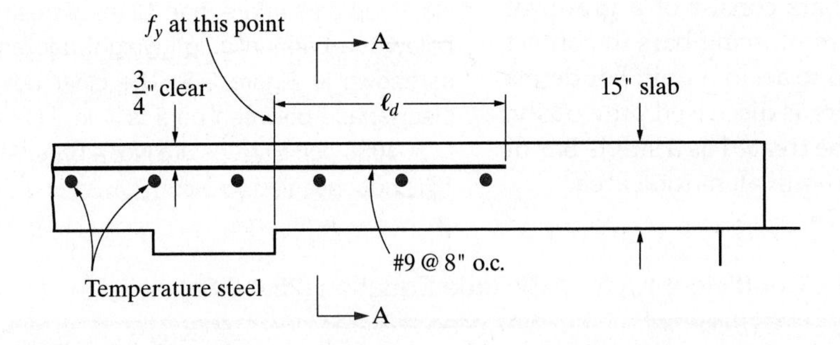

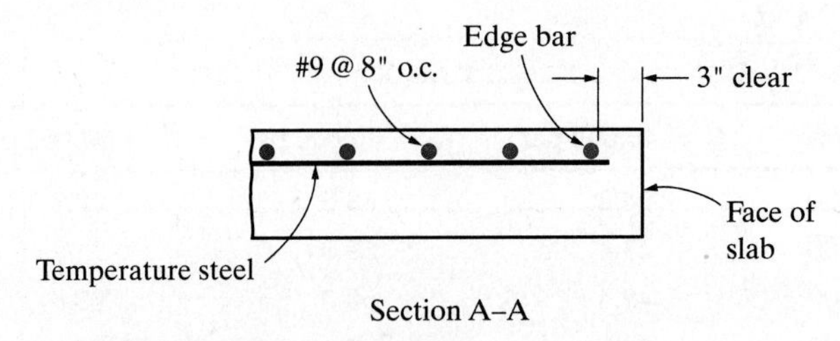

FIGURE 5-6 Sketch for Example 5-2.

4. Determine c_b. Based on cover (center of bar to nearest concrete surface), consider the clear cover and one-half the diameter of the No. 9 bar:

$$c_b = 0.75 + \frac{1.128}{2} = 1.314 \text{ in.}$$

Based on bar spacing (one-half the center-to-center distance),

$$c_b = 0.5(8.0) = 4.0 \text{ in.}$$

Therefore, use $c_b = 1.314$ in.

5. K_{tr} is taken as zero. (There is no transverse reinforcement crossing the plane of splitting.)
6. Check $(c_b + K_{tr})/d_b \le 2.5$:

$$\frac{c_b + K_{tr}}{d_b} = \frac{1.314 + 0}{1.128} = 1.165 < 2.5 \qquad \text{(O.K.)}$$

7. The excess reinforcement factor is assumed not applicable and is omitted.
8. Calculate ℓ_d (recall that $\psi_t \times \psi_e$ will be taken as 1.7 from step 3):

$$\ell_d = \frac{K_D}{\lambda}\left[\frac{\psi_t\psi_e\psi_s}{\left(\dfrac{c + K_{tr}}{d_b}\right)}\right]d_b$$

$$= \frac{71.2}{1.0}\left[\frac{1.7(1.0)}{1.165}\right](1.128) = 117.2 \text{ in.} > 12 \text{ in.} \quad \text{(O.K.)}$$

Example 5-3

Calculate the development length required for the interior two No. 7 bars in the beam shown in Figure 5-7. The two No. 7 outside bars are continuous for the full length of the beam. Use $f_y = 60{,}000$ psi and $f'_c = 4000$ psi (normal-weight concrete). The bars are uncoated. Assume that, from the design of this member, the required tension steel area was 2.28 in.^{2}.

Solution:

1. From Table 5-1, $K_D = 71.2$.
2. Establish values for the factors ψ_t, ψ_e, ψ_s, and λ.
   a. $\psi_t = 1.3$ (the bars are top bars).
   b. The bars are uncoated; $\psi_e = 1.0$.
   c. The bars are No. 7; $\psi_s = 1.0$.
   d. Normal-weight concrete is used; $\lambda = 1.0$.
3. The product $\psi_t \times \psi_e = 1.3 < 1.7$. (O.K.)
4. Determine c_b. Based on cover (center of bar to nearest concrete surface), consider the clear cover, the No. 4 stirrup diameter, and one-half the diameter of the No. 7 bar:

$$c_b = 1.5 + 0.5 + \frac{0.875}{2} = 2.44 \text{ in.}$$

Based on bar spacing (one-half the center-to-center distance),

$$c_b = \frac{13 - 2(1.5) - 2(0.5) - 2\left(\dfrac{0.875}{2}\right)}{3(2)}$$

$$= 1.354 \text{ in.}$$

Use the smaller of the two c_b values:
Therefore, use $c_b = 1.354$ in.

5. Using data on stirrups from Figure 5-7:

$$K_{tr} = \frac{40A_{tr}}{(sn)} = \frac{40(0.40)}{(6)(2)} = 1.333$$

6. Check $(c_b + K_{tr})/d_b \le 2.5$:

$$\frac{c_b + K_{tr}}{d_b} = \frac{1.354 + 1.333}{0.875} = 3.07 > 2.5$$

Therefore, use 2.5.

7. Calculate the excess reinforcement factor:

$$K_{ER} = \frac{A_s \text{ required}}{A_s \text{ provided}} = \frac{2.28 \text{ in.}^2}{2.40 \text{ in.}^2} = 0.95$$

8. Calculate ℓ_d:

$$\ell_d = \frac{K_D}{\lambda}\left[\frac{\psi_t\psi_e\psi_s}{\left(\dfrac{c_b + K_{tr}}{d_b}\right)}\right]K_{ER}d_b$$

$$= \frac{71.2}{1.0}\left[\frac{1.3(1.0)(1.0)}{2.5}\right](0.95)(0.875) = 30.8 \text{ in.} > 12 \text{ in.}$$

(O. K.)

5-3 DEVELOPMENT LENGTH: COMPRESSION BARS

Whereas tension bars produce flexural tension cracking in the concrete, this effect is not present with compression bars, and thus shorter development

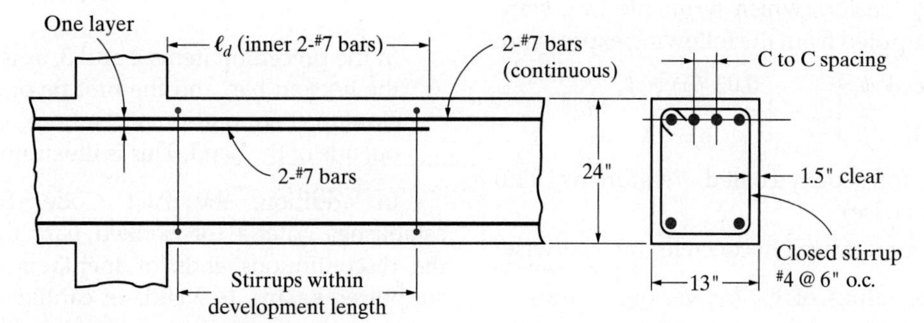

FIGURE 5-7 Sketch for Example 5-3.

lengths are allowed. For deformed bars in compression, the development length ℓ_{dc} is calculated from

$$\ell_{dc} = \left(\frac{0.02 f_y}{\lambda\sqrt{f'_c}}\right) d_b \geq 0.0003 f_y d_b$$

in inches, where all quantities are as described previously for tension development length. This value of ℓ_{dc} is tabulated in Table A-12 for $\lambda = 1.0$, grade 60 bars, and various values of f'_c. The required compression development length may be further reduced by multiplying ℓ_{dc} by the following modification factors:

1. Reinforcement in excess of that required:

$$\frac{A_s \text{ required}}{A_s \text{ provided}}$$

2. Bars enclosed within a spiral that is not less than $\frac{1}{4}$ in. in diameter and not more than 4 in. in pitch or within No. 4 ties in conformance with ACI, Section 25.7.2, and spaced at not more than 4 in. on center: Use 0.75. (For all other cases, use 1.0.)

The value of ℓ_{dc} shall not be less than 8 in.

5-4 DEVELOPMENT LENGTH: STANDARD HOOKS IN TENSION

In the event that the desired development length in tension cannot be furnished with a straight bar, it is necessary to provide mechanical anchorage at the end of the bars. Although the ACI Code (Section 25.4) allows any mechanical device to serve as anchorage if its adequacy is verified by testing, anchorage for reinforcement is usually accomplished by means of a 90° or 180° hook. The dimensions and bend radii for these hooks have been standardized by the ACI Code. Additionally, 90° and 135° hooks have been standardized for stirrups and tie reinforcement. Hooks in compression bars are ineffective and cannot be used as anchorage. Standard reinforcement hooks are shown in Figure 5-9. The bend diameters are measured on the inside of the bar.

The ACI Code, Section 25.4.3, specifies that the required development length ℓ_{dh} (see Figure 5-9) for deformed bars in tension, which terminate in a standard hook, be computed from the following expression:

$$\ell_{dh} = \left[\frac{f_y \psi_e \psi_c \psi_r}{50\lambda\sqrt{f'_c}}\right] d_b = \left[\frac{0.02\, f_y \psi_e \psi_c \psi_r}{\lambda\sqrt{f'_c}}\right] d_b$$

where ψ_e = 1.2 for epoxy-coated reinforcing (1.0 otherwise)

λ = 0.75 for lightweight concrete (1.0 otherwise)

Table A-13 gives values of ℓ_{dh} for various values of f'_c, f_y = 60,000 psi, $\psi_e = 1.0$, and $\lambda = 1.0$. The length ℓ_{dh} may be further reduced by multiplication by the following modification factors. The final ℓ_{dh} shall not be less than $8d_b$ nor less than 6 in.

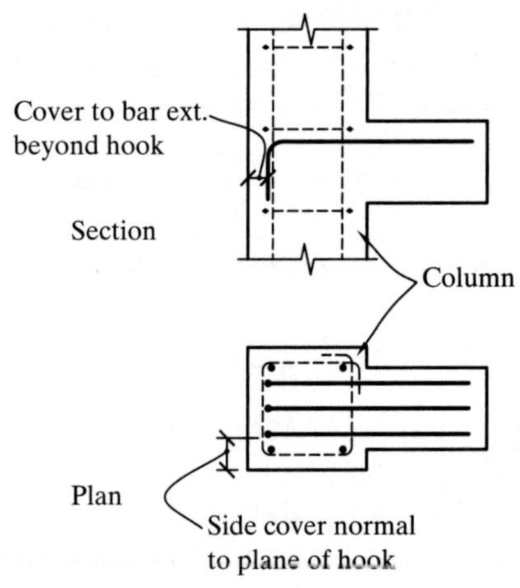

FIGURE 5-8 Covers to hooked bars.

1. Cover modification factor, ψ_c: For bars No. 11 and smaller with side cover (normal to the plane of the hook) not less than $2\frac{1}{2}$ in., and for a 90° hook with cover on the extension beyond the hook (see Figure 5-8) not less than 2 in.: 0.7. (For all other cases, use 1.0.)
2. Confining reinforcement modification factor, ψ_r:
   a. Enclosure for 90° hooks for No. 11 and smaller bars within ties or stirrups perpendicular to the bar being developed, spaced not more than $3d_b$ along ℓ_{dh}; or enclosed within ties or stirrups parallel to the bar being developed, spaced not more than $3d_b$ along the tail extension plus the bend: 0.8. (For all other cases, use 1.0.)
   b. Enclosure for 180° hooks for No. 11 and smaller bars within ties or stirrups perpendicular to the bar being developed, with spacing not greater than $3d_b$ along ℓ_{dh} of the hook (see Figure 5-10): 0.8. (For all other cases, use 1.0.)
3. Excess rebar modification factor: Where anchorage or development of f_y is not specifically required, reinforcement in excess of that required by analysis:

$$\frac{A_s \text{ (required)}}{A_s \text{ (provided)}}$$

In the preceding items 2 and 3, d_b is the diameter of the hooked bar, and the first tie or stirrup shall enclose the bent portion of the hook, within $2d_b$ of the outside of the bend. This is illustrated in Figure 5-10.

In addition, the ACI Code, Section 25.4.3.3, establishes criteria for hooked bars that terminate at the discontinuous ends of members such as simply supported beams, free ends of cantilevers, and ends of members that frame into a joint where the member does not extend beyond the joint. If the full strength (f_y) of the

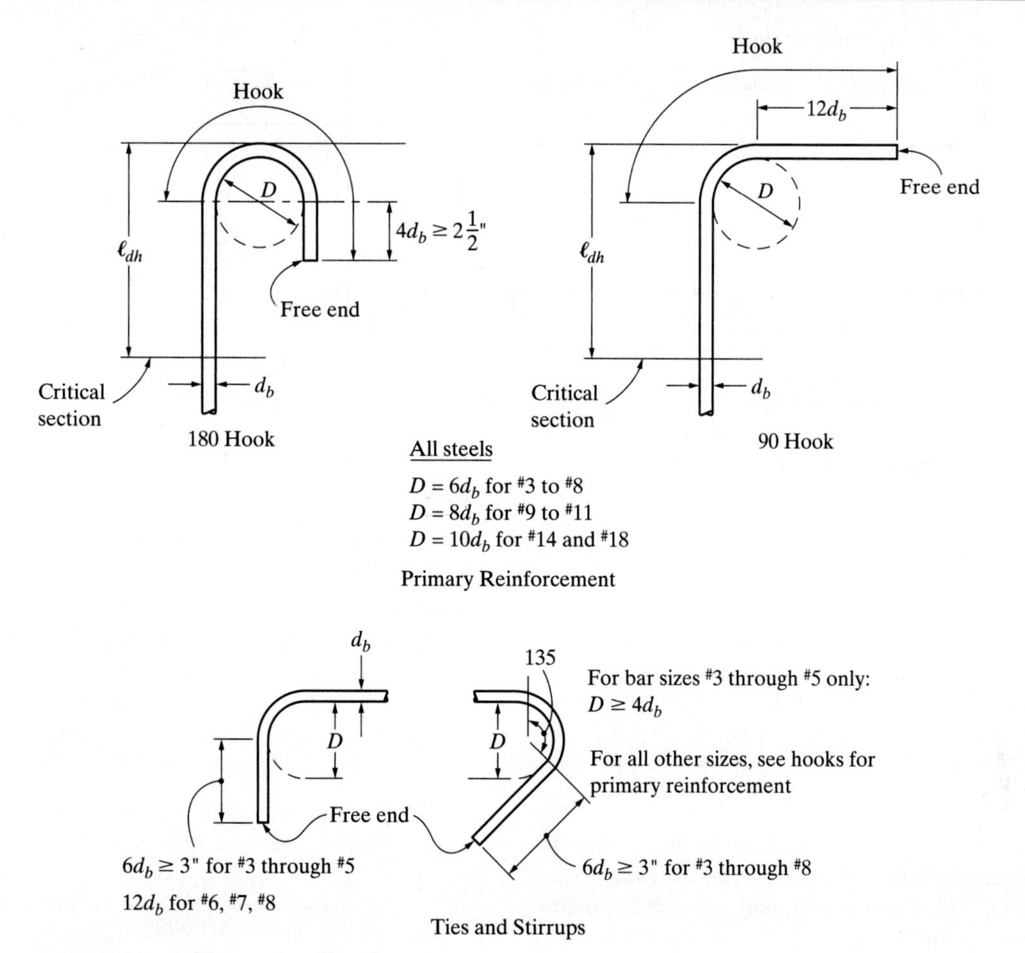

FIGURE 5-9 ACI standard hooks.

hooked bar must be developed and if both the side cover and the top (or bottom) cover over the hook are less than $2\frac{1}{2}$ in., closed ties or stirrups, perpendicular to the bar being developed, spaced at $3d_b$ maximum are required over the development length ℓ_{dh}. The first tie or stirrup shall enclose the bent portion of the hook, within $2d_b$ of the bend (see Figure 5-10). This does not apply to the discontinuous ends of slabs with concrete confinement provided by the slab continuous on both sides perpendicular to the plane of the hook. Also, the 0.8 modification factor of the preceding item 2 does not apply.

One condition in which a beam top reinforcement may not be fully developed within a column

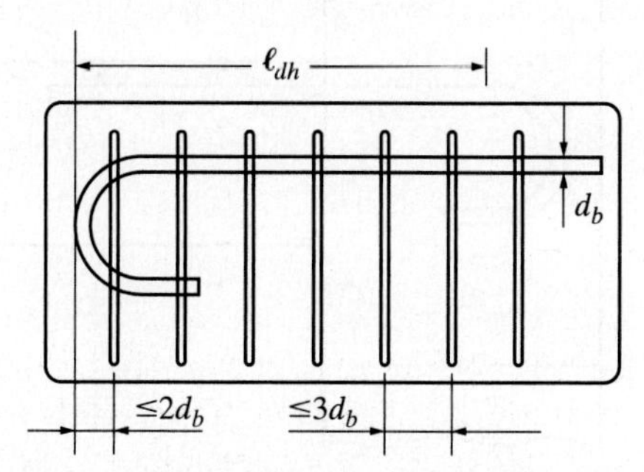

FIGURE 5-10 Enclosure for 180° hook; modification factor = 0.8.

occurs when a wide beam or girder connects to a column that has a smaller width such that some of the beam top reinforcement fall outside the width of the column. If the analysis assumes that a fully rigid connection exists between the wide beam and the column, then enough top rebar in the beam must be placed within the column width (i.e. within the column vertical reinforcement cage) to ensure adequate development of the rebar in the column and a realization of the moment capacity of the beam-column connection that was assumed in the structural analysis. The dimension of the column parallel to the beam top rebar must also be at least equal to the required development length of the hooked top reinforcement in the beam plus the concrete cover to the vertical reinforcement in the column. (See Figures 5-13 and 5-14.)

Example 5-4

Determine the anchorage or development length required for the tension (top) bars for the conditions shown in Figure 5-11. Use $f'_c = 3000$ psi (normal-weight concrete) and $f_y = 60{,}000$ psi. The No. 8 bars may be categorized as top bars. Assume a side cover on the main bars of $2\frac{1}{2}$ in. minimum. Bars are uncoated.

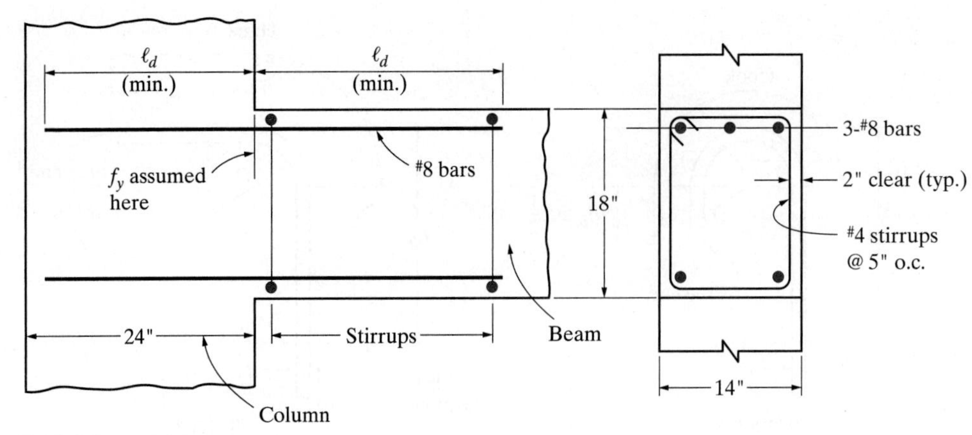

FIGURE 5-11 Sketch for Example 5-4.

Solution:

a. Anchorage of top bars into the exterior column
   1. From Table 5-1, $K_D = 82.2$.
   2. Establish values for the factors ψ_t, ψ_e, ψ_s, and λ.
      a. $\psi_t = 1.3$ (the bars are top bars).
      b. The bars are uncoated; $\psi_e = 1.0$.
      c. The bars are No. 8; $\psi_s = 1.0$.
      d. Normal-weight concrete is used; $\lambda = 1.0$.
   3. The product $\psi_t \times \psi_e = 1.3 < 1.7$. (O.K.).
   4. Determine c_b. Based on cover (center of bar to nearest concrete surface), consider the clear cover, the No. 4 stirrup diameter, and one-half the diameter of the No. 8 bar:

$$c_b = 2.0 + 0.5 + \frac{1.0}{2} = 3.0 \text{ in.}$$

   Based on bar spacing (one-half the center-to-center distance),

$$c_b = \frac{14 - 2(2.0) - 2(0.5) - 2(0.5)}{2(2)} = 2.0 \text{ in.}$$

   Use the smaller of the two c_b values: Therefore, use $c_b = 2.0$ in.
   5. Figure 5-11 shows stirrups in the beam. However, there are no stirrups in the column and K_{tr} is taken as 0.
   6. Check $(c_b + K_{tr})/d_b \leq 2.5$:

$$\frac{c_b + K_{tr}}{d_b} = \frac{2.0 + 0}{1.0} = 2.0 < 2.5 \qquad \text{(O.K.)}$$

   7. The excess reinforcement factor is assumed not applicable and is omitted.
   8. Calculate ℓ_d:

$$\ell_d = \frac{K_D}{\lambda}\left[\frac{\psi_t\psi_e\psi_s}{\left(\frac{c_b + K_{tr}}{d_b}\right)}\right]d_b$$

$$= \frac{82.2}{1.0}\left[\frac{1.3(1.0)(1.0)}{2.0}\right](1.0) = 53.4 \text{ in.} > 12 \text{ in.} \qquad \text{(O.K.)}$$

   Because 53.4 in. > 24 in. column width, use a standard hook, either a 90° hook or a 180° hook.

b. Anchorage using a standard 180° hook
   1. The development length ℓ_{dh} for the hook shown in Figure 5-12 is calculated from

$$\ell_{dh} = \frac{0.02\psi_e f_y}{\lambda\sqrt{f'_c}}d_b$$

   The bars are uncoated and the concrete is normal weight. Therefore, both ψ_e and λ are 1.0:

$$\ell_{dh} = \frac{0.02(60{,}000)}{\sqrt{3000}}(1.00) = 21.9 \text{ in.}$$

   (Check this with Table A-13.)
   2. The only applicable modification factor is based on side cover (normal to the plane of the hook) of $2\frac{1}{2}$ in. Use a modification factor of 0.7 (ACI Code, Section 12.5.3a).
   3. The required development length is then calculated from

$$\ell_{dh} = 21.9(0.7) = 15.33 \text{ in.}$$

   Check minimum:

$$\text{minimum } \ell_{dh} = 8d_b \geq 6 \text{ in.}$$

$$8d_b = 8 \text{ in.} < 15.33 \text{ in.} \qquad \text{(O.K.)}$$

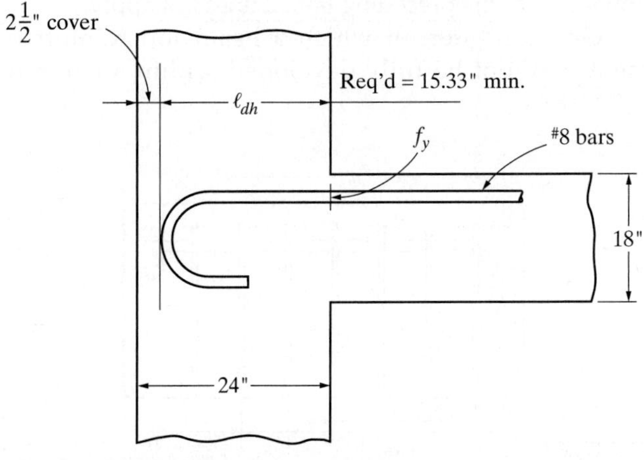

FIGURE 5-12 Sketch for Example 5-4.

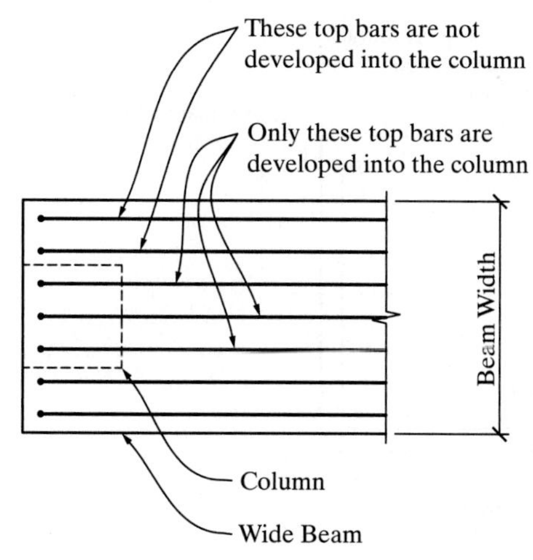

FIGURE 5-13 Development of a wide beam top rebar in a column (plan view).

The minimum width of column required is

$$15.33 + 2.5 = 17.83 \text{ in.} < 24 \text{ in.} \qquad \text{(O.K.)}$$

The hook therefore will fit into the column and the detail is satisfactory.

Anchorage into beam: The development length required if bars are straight is (conservatively) 53.4 in., as determined previously. Therefore, the bars must extend at least this distance into the span. The ACI Code has additional requirements for the extension of tension bars in areas of negative moments. These requirements are covered in the discussion on continuous construction in Chapter 6.

Adequate Development of Hooked Beam Rebar in Columns

It is very important that reinforcement be adequately developed into the supporting structural elements (e.g. columns) especially for statically determinate structures such as cantilevers (e.g. balcony slabs) which lack redundancy, with no alternate load paths. The collapse of a garage at an Atlantic City Casino Resort in 2003 [3, 4] was believed to have been caused by the lack of adequate anchorage of the top reinforcement of a cantilever slab/beam. For a cantilever beam framing into a column, the column size must be at least equal to the hooked rebar development length for the beam top reinforcement plus the concrete cover in the column. Without adequate development, the beam top reinforcement will not be fully developed, and hence the moment capacity of the beam at the beam-column joint will not be fully realized (see Figure 5-14). For cantilevered balcony slabs, improper placement of the top reinforcement - resulting in a reduced effective depth - is the leading cause of failures for these non-redundant structures. It also leads to excessive deflections of the balcony slab which could induce unanticipated forces in the balcony railing system [5].

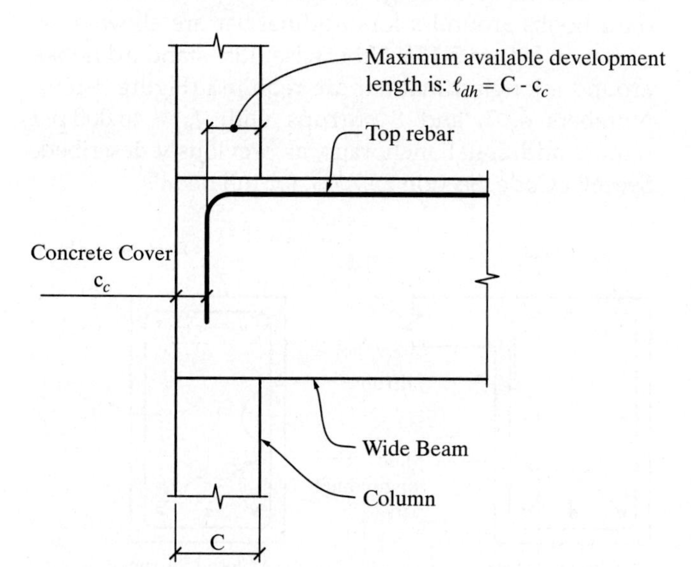

FIGURE 5-14 Development of a beam top rebar in a column.

5-5 DEVELOPMENT OF WEB REINFORCEMENT

Anchorage of web reinforcement must be furnished in accordance with the ACI Code, Section 25.7. Stirrups must be carried as close to the compression and tension surfaces as possible. Close proximity to the compression face is necessary as flexural tension cracks penetrate deeply as ultimate load is approached.

The ACI Code stipulates that ends of single-leg, simple U-, or multiple U-stirrups shall be anchored by one of the following means (see Figure 5-15):

1. For a No. 5 bar or smaller, and for Nos. 6, 7, and 8 bars of $f_y = 40{,}000$ psi or less, anchorage is provided by a standard stirrup hook, bent around a longitudinal bar (ACI Code, Section 25.7.1.3).
2. For Nos. 6, 7, and 8 stirrups with $f_y > 40{,}000$ psi, anchorage is provided by a standard stirrup hook bent around a longitudinal bar *plus* an embedment between midheight of the member and the outside end of the hook equal to or greater than

$$0.014 d_b \frac{f_y}{(\lambda \sqrt{f'_c})}$$

A 135° or a 180° hook is preferred, but a 90° hook is acceptable provided that the free end of the hook is extended the full $12d_b$ as required in ACI Code, Section 7.1.3 (ACI Code, Section 25.7.1.3).

It should be noted that the ACI standard hooks for ties and stirrups, as shown in Figure 5-15, include 90° and 135° hooks only. This does not imply that the 180° hook is not acceptable. The 135° hook is easier to fabricate. Many stirrup bending machines now in use are not designed to fabricate a 180° hook. The anchorage strength of either a 135° or 180° hook is approximately the same.

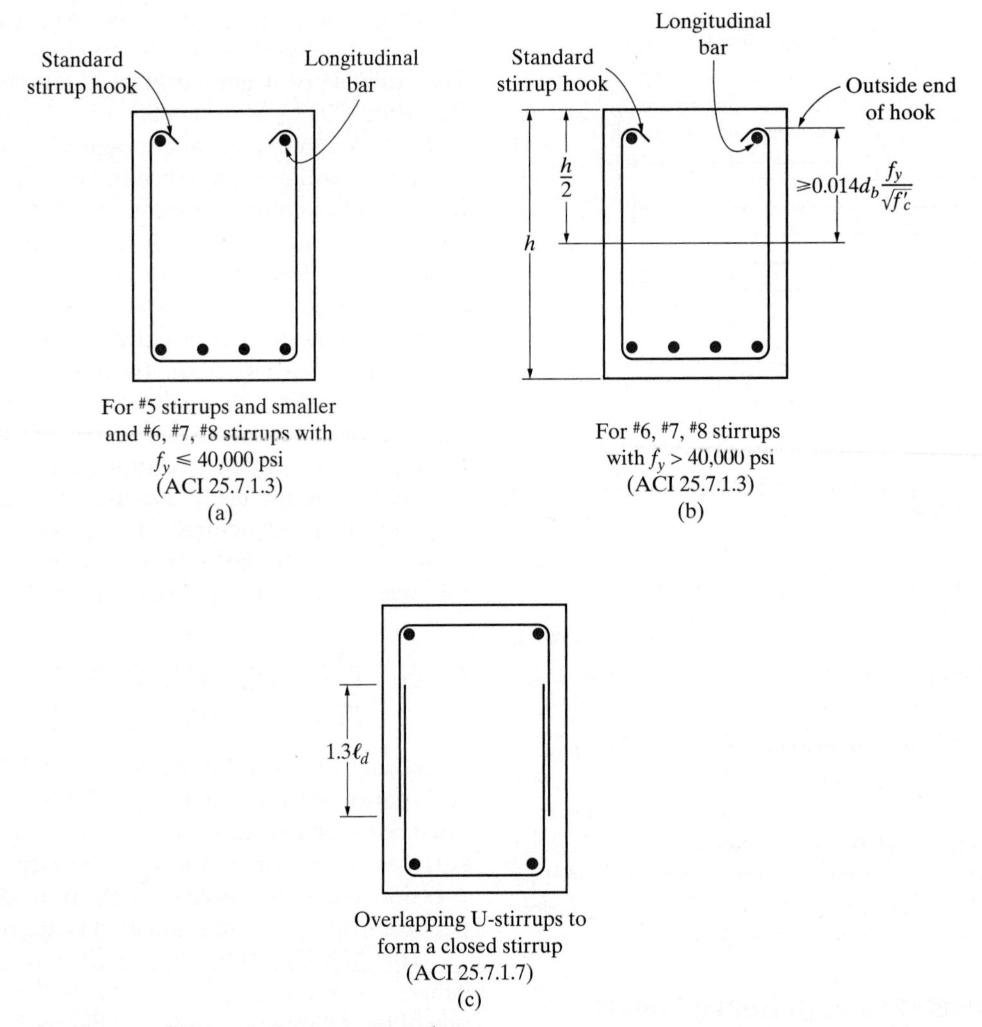

FIGURE 5-15 Web reinforcement anchorage.

In addition, the ACI Code, Section 25.7.1.7, establishes criteria with respect to lapping of double U-stirrups or ties (without hooks) to form a closed stirrup. Legs shall be considered properly spliced when lengths of lap are 1.3 ℓ_d, as depicted in Figure 5-15c. Each bend of each simple U-stirrup must enclose a longitudinal bar. If the lap of 1.3 ℓ_d cannot fit within the depth of a shallow member, provided that the depth of the member is at least 18 in., double U-stirrups may be used if each U-stirrup extends the full available depth of the member and the force in each leg does not exceed 9000 lb (i.e., $A_b fy \leq$ 9000 lb).

Where torsional reinforcing is required or desired, a commonly used alternative stirrup is the one-piece closed stirrup. Use of the one-piece closed stirrup is disadvantageous, however, in that the entire beam reinforcing (longitudinal steel and stirrups) may have to be prefabricated as a cage and then placed as a unit. This may not be practical if the longitudinal bars have to be passed between column bars. Alternatively, and at a greater cost, the longitudinal bars could be threaded through the closed stirrups and column bars. Two commonly used types of one-piece closed stirrups are shown in Figure 5-16. If spalling of the member at the transverse torsional reinforcement anchorage (hooks) is restrained by a flange or similar member, 90° standard hooks around a longitudinal bar are allowed, as shown in Figure 5-16a. Otherwise, 135° standard hooks around a longitudinal bar are required (Figure 5-16b). Numbers 6, 7, and 8 stirrups with $f_y >$ 40,000 psi require additional anchorage, as previously described. See ACI Code, Section 25.7.1.3, for full details.

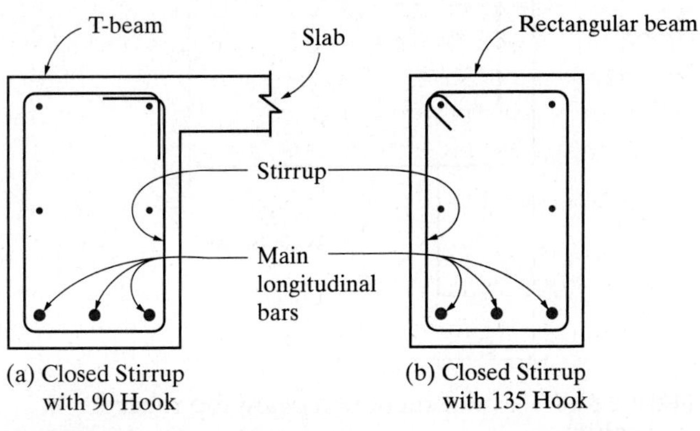

FIGURE 5-16 Closed stirrups.

5-6 SPLICES

The need to splice reinforcing steel is a reality due to the limited lengths of steel available. All bars are readily available in lengths up to 60 ft; No. 3 and No. 4 bars will tend to bend in handling when longer than 40 ft, however. Typical stock straight lengths are as follows:

No. 3 bar: 20, 40, 60 ft
No. 4 bar: 30, 40, 60 ft
Nos. 5 to 18 bars: 60 ft

Splicing may be accomplished by welding, by mechanical means, or, most commonly for No. 11 bars and smaller, by lapping bars, as shown in Figure 5-17. Lap splices may not be used for bars larger than No. 11 except for compression splices at footings, as provided in the ACI Code, Section 16.3.5.4, and for compression splices of bars of different sizes, as provided in the ACI Code, Section 25.5.5. The splice composed of lapped bars is usually more economical than the other types. The lapped bars are commonly tied in contact with each other. They may, however, be spaced apart up to one-fifth of the lap length, with an upper limit of 6 in. Splices in regions of maximum moment preferably should be avoided.

The ACI Code (Sections 1.8.1 and 26.6.1.1) requires that the design drawings show the location and length of lap splices and the type and location of mechanical and welded splices of reinforcement.

5-7 TENSION SPLICES

The required length of lap is based on the class in which the splice is categorized. The required lap length increases with increased stress and increased amount of steel spliced in close proximity. Lap length is expressed in terms of the tensile development length ℓ_d for the particular bar, as shown in Section 5-2. The 12-in. minimum for the ℓ_d calculation is not considered, nor is the excess reinforcement factor considered for tension splices, as the splice classification already reflects any excess reinforcement at the splice location.

Two classes of tension lap splices are prescribed in the ACI Code: Class A and Class B lap splices (ACI 318-14, Section 25.5.2.1). The length of the lap splice is a function of the percentage of bars spliced at the same location and the level of stress in the bars.

- Class A tension lap splice: lap length $= 1.0\ell_d$.

  Class A lap splices are permitted if the following two conditions are satisfied:

  1. Area of reinforcement provided is twice that required for the entire length of the splice, and
  2. Not more than 50% of the total reinforcement is spliced within the required lap length.

  If the two conditions above are not satisfied, then a Class B lap splice is required.

- Class B tension lap splice: lap length $= 1.3\ell_d \geq$ 12 in.

The minimum length of lap for tension lap splices is 12 in. (ACI Code, Table 25.5.2.1). Requirements for mechanical and welded splices for tensile reinforcement are contained in the ACI Code, Section 25.5.7. Mechanical or welded splices must have a yield strength in tension or compression of at least 125% of the yield strength of the spliced reinforcing bars (ACI Code, Section 25.5.7.1), and mechanical or welded splices need not be staggered except in tension tie members (ACI Code Section 25.7.3). In addition, splices in tension tie members must be made with a welded splice or mechanical splice, and their locations must be staggered a distance of at least 30 in., in accordance with the ACI Code, Section 25.5.7.4. Staggering of all non-mechanical and non-welded tension splices is encouraged. Use tension lap splices for #11 and smaller bars, and mechanical or welded splices for #14 and #18 bars.

5-8 COMPRESSION SPLICES

The ACI Code, Section 25.5.5, contains requirements for lap splices for compression bars. For $f'_c = 3000$ psi or more, the following lap lengths, in multiples of bar diameters d_b, are required:

$$f_y = 40{,}000 \text{ psi: } 20d_b$$
$$f_y = 60{,}000 \text{ psi: } 30d_b$$
$$f_y = 75{,}000 \text{ psi: } 44d_b$$

but not less than 12 in. For $f'_c < 3000$ psi, the length of lap should be increased by one-third. Within ties of specific makeup or spirals (ACI Code, Section 10.7.5.2.1), these laps may be reduced to 0.83 or 0.75, respectively, of the foregoing values, but must not be less than 12 in.

Compression splices may also be of the end-bearing type, where bars are cut square, then butted together and held in concentric contact by a suitable device. End-bearing splices must not be used except in members containing closed ties, closed stirrups, or spirals. Welded splices and mechanical connections are also acceptable and are subject to the requirements of the ACI Code, Section 25.5.5. Special splice requirements for columns are furnished by the ACI Code, Section 10.7.5. Compression lap splices shall not be used for bars larger than #11 except that compression lap splices of #14 or #18 bars to #11 or smaller bars is permitted. Where bars of different sizes are lap-spliced in compression, the compression lap splice length ℓ_{sc} shall be the greater of the Ldc of the larger bar and ℓ_{sc} of the smaller bar (ACI 318-14, Section 25.5.5.4).

5-9 SIMPLE-SPAN BAR CUTOFFS AND BENDS

The maximum required A_s for a beam is needed only where the moment is maximum. This maximum steel may be reduced at points along a bending member where the bending moment is smaller. This is usually accomplished by either stopping or bending the bars in a manner consistent with the theoretical requirements for the strength of the member, as well as the requirements of the ACI Code.

Bars can theoretically be stopped or bent in flexural members whenever they are no longer needed to resist moment. The ACI Code, Sections 7.7.3.3 and 9.7.3.3, however, requires that each bar be extended beyond the point at which it is no longer required for flexure for a distance equal to the effective depth of the member or $12d_b$, whichever is greater, except at supports of simple spans and at free ends of cantilevers. In effect, this prohibits the cutting off of a bar at the theoretical cutoff point but can be interpreted as permitting bars to be bent at the theoretical cutoff point. If bars are to be bent, a general practice that has evolved is to commence the bend at a distance equal to one-half the effective depth beyond the theoretical cutoff point. The bent bar should be anchored or made continuous with reinforcement on the opposite face of the member. It should be pointed out, however, that the bending of reinforcing bars in slabs and beams to create *truss bars* (see types 3 through 7, Figure 13-5) has fallen into disfavor over the years because of placement problems and the labor involved. It is more common to use straight bars and place them in accordance with the strength requirements.

With simple-span flexural members of constant dimensions, we can assume that the required A_s varies directly with the bending moment and that the shape of a *required A_s curve* is identical with that of the moment diagram. The moment diagram (or curve) may then be used as the required A_s curve by merely changing the vertical scale. Steel areas may be used as ordinates, but it is more convenient to use the required number of bars (assuming that all bars are of the same diameter). The necessary correlation is established by using the maximum ordinate of the curve as the maximum required A_s in terms of the required number of bars. Because of possible variations in the shape of the moment diagram, either a graphical or a mathematical approach may be more appropriate. A graphical approach requires that the moment diagram be plotted to scale. Example 5-5 lends itself to a mathematical solution.

In determining bar cutoffs, it should be remembered that the stopping of bars should be accomplished by using a symmetrical pattern so that the remaining bars will also be in a symmetrical pattern. In addition, the ACI Code, Sections 7.7.3.8 and 9.7.3.8, requires that for simple spans, at least one-third of the positive moment steel extend into the support a distance of 6 in. In practice, this requirement is generally exceeded. Normally, recommended bar details for single-span solid concrete slabs indicate that all bottom bars should extend into the support. See Figure 5-18 for recommended bar details.

Economy sometimes dictates that reinforcing steel should be cut off in a simple span. Example 5-5 shows one approach to this problem.

Example 5-5

A simple-span, uniformly loaded beam, shown in Figure 5-19a and 5-19b, requires six No. 7 bars for tensile reinforcement. If the effective depth d is 18 in., determine where the bars may be stopped. Use $f'_c = 4000$ psi and $f_y = 60{,}000$ psi. Assume that there is no excess steel (required A_s = furnished $A_s = 3.60$ in.^{2}). Assume normal-weight concrete and that the stirrups extend to the end of the beam at the support. The bars are uncoated.

Solution:

We will try to establish a bar cutoff scheme whereby the two center bars are cut first, followed by the cutting of the other two inside bars. The two corner bars are to run the full length

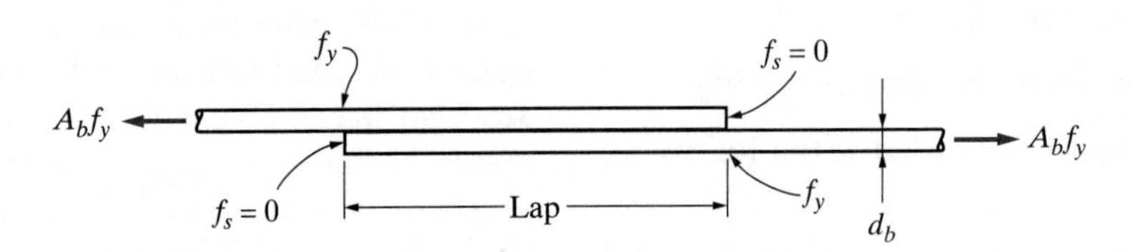

FIGURE 5-17 Stress transfer in tension lap splice.

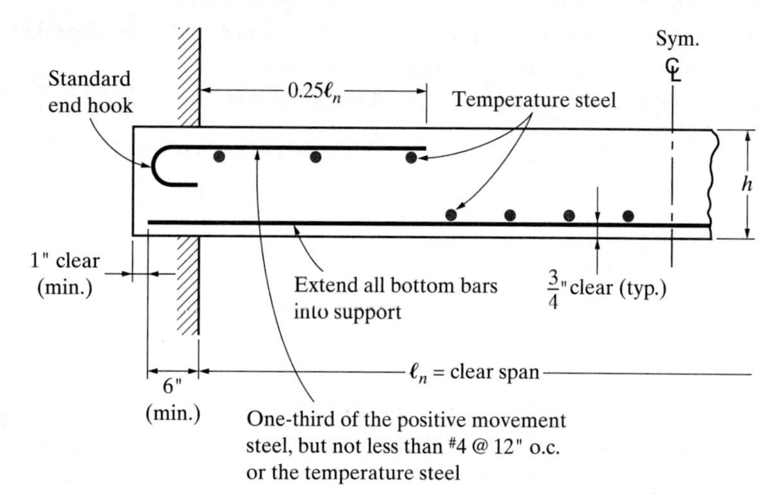

FIGURE 5-18 Recommended bar details: one-way simple-span slabs; tensile-reinforced simple-span beams similar.

of the beam. First, we check the minimum steel requirement. From Table A-5:

$$A_{s,\,min} = 0.0033\,bd = 0.0033(16.5)(18) = 0.98 \text{ in.}^2$$

The two corner bars provide a steel area of 1.20 in.^{2} Therefore, the minimum steel area requirement is met.

1. Determine where the first two bars may be cut off. The first two bars may be stopped where only four are required. This distance from the centerline of the span is designated x_1 in Figure 5-19d.

As the moment diagram is a second-degree curve (a parabola), offsets to the line tangent to the curve at the point of maximum moment vary as the squares of the distances from the centerline of the span. The solution for the distance x_1 may be formulated as follows:

$$\frac{(x_1)^2}{\left(\frac{\ell}{2}\right)^2} = \frac{y_1}{Y}$$

$$\frac{(x_1)^2}{12^2} = \frac{2 \text{ bars}}{6 \text{ bars}}$$

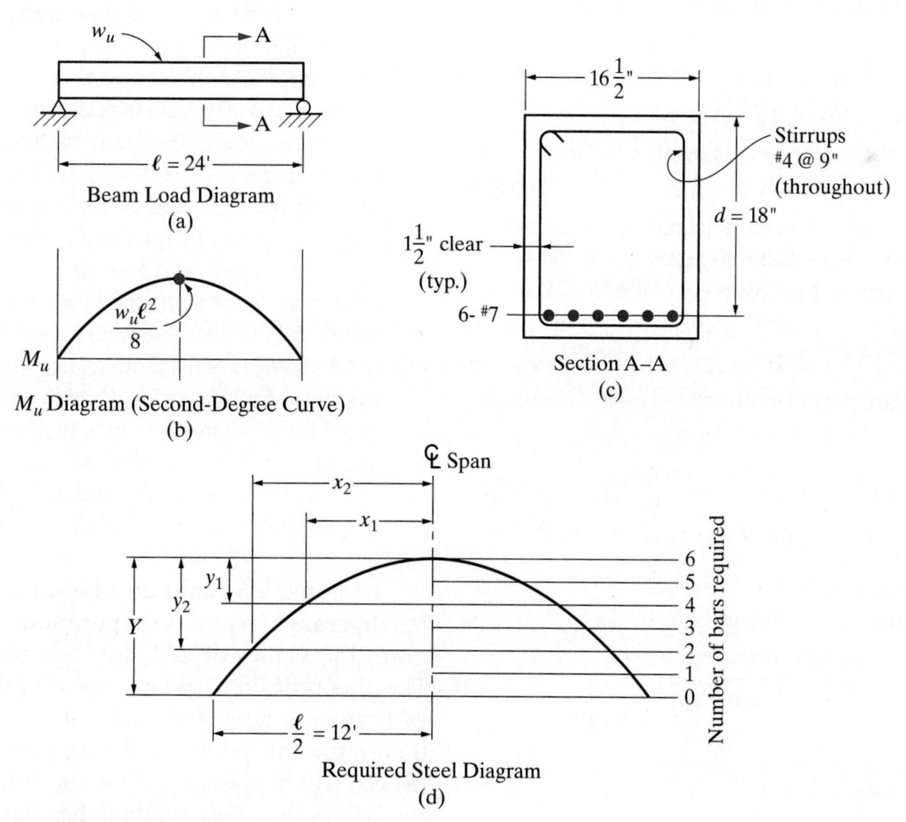

FIGURE 5-19 Sketches for Example 5-5.

from which $x_1 = 6.93$ ft. This locates the *theoretical* point where two bars may be terminated. Additionally, bars must be extended past this point a distance d or $12d_b$, whichever is larger. Thus,

$$\begin{aligned} 12 \text{ bar diameters} &= 12(0.875) \\ &= 10.5 \text{ in.} \end{aligned}$$

Because $d = 18$ in., the bars should be extended 18 in. (1.50 ft). Then the minimum distance from the centerline of the span to the cutoff of the first two bars is

$$6.93 + 1.50 = 8.43 \text{ ft}$$

2. Determine where the next two bars may be cut off. The two remaining bars, which are the corner bars, will continue into the support. With reference to Figure 5-19d:

$$\frac{(x_2)^2}{12^2} = \frac{4}{6}$$

$$x_2 = 9.80 \text{ ft}$$

Therefore, the minimum distance from the centerline of the span to the cutoff of the next two bars is

$$9.80 + 1.50 = 11.30 \text{ ft}$$

3. A check must be made of the required development length ℓ_d for the first two bars. Because the bar cutoff occurs 8.43 ft from the centerline of the span, ℓ_d must be less than 8.43 ft for the stress f_y to be developed at the centerline.
   a. From Table 5-1, $K_D = 71.2$.
   b. Establish values for the factors ψ_t, ψ_e, ψ_s, and λ.
      1. $\psi_t = 1.0$ (the bars are not top bars).
      2. The bars are uncoated; $\psi_e = 1.0$.
      3. The bars are No. 7; $\psi_s = 1.0$.
      4. Normal-weight concrete is used; $\lambda = 1.0$.
   c. The product $\psi_t \times \psi_e = 1.0 < 1.7$. (O.K.)
   d. Determine c_b. Based on cover (center of bar to nearest concrete surface), consider the clear cover, the No. 4 stirrup, and one-half the diameter of the No. 7 bar:

$$c_b = 1.5 + 0.5 + \frac{0.875}{2} = 2.44 \text{ in.}$$

   Based on bar spacing (one-half the center-to-center distance),

$$c_b = \frac{16.5 - 2(1.5) - 2(0.5) - 2\left(\frac{0.875}{2}\right)}{5(2)} = 1.163 \text{ in.}$$

   Use the smaller of the two c_b values: Therefore, use $c_b = 1.163$ in.
   e. Using data on stirrups from Figure 5-19:

$$K_{tr} = \frac{40A_{tr}}{sn} = \frac{40(0.40)}{(9)(2)} = 0.889$$

   f. Check $(c_b + K_{tr})/d_b \leq 2.5$:

$$\frac{c_b + K_{tr}}{d_b} = \frac{1.163 + 0.889}{0.875} = 2.35 < 2.5 \quad \text{(O.K.)}$$

   g. The excess reinforcement factor is neglected (conservative).
   h. Calculate ℓ_d:

$$\begin{aligned} \ell_d &= \frac{K_D}{\lambda}\left[\frac{\psi_t\psi_e\psi_s}{\left(\frac{c_b + K_{tr}}{d_b}\right)}\right]d_b \\ &= \frac{71.2}{1.0}\left[\frac{1.0(1.0)(1.0)}{2.35}\right](0.875) \\ &= 26.5 \text{ in.} > 12 \text{ in.} \quad \text{(O.K.)} \\ 26.5 \text{ in.} &= 2.21 \text{ ft} \end{aligned}$$

   As the required tensile development length of 2.21 ft is much less than the distance from the centerline to the actual cutoff point (8.43 ft), the cutoff for the first two bars is satisfactory. Similarly, if we consider the second two bars, for which ℓ_d can conservatively be taken as 2.21 ft (conservative, as the c_b distance, based on bar spacing, is larger for these two bars than for the first two cut bars), the point at which they are stressed to maximum (the theoretical cutoff point for the first two bars), measured from the centerline, is 6.93 ft. The second two bars must extend at least ℓ_d past this point. Again measured from the centerline, the bars must extend at least

$$6.93 + 2.21 = 9.14 \text{ ft}$$

   Because this is less than the actual distance to the cutoff point (11.30 ft), the cutoff for the second pair is satisfactory (see Figure 5-20).

4. The remaining pair of corner bars, representing one-third of the positive moment steel, must continue into the support. As the corner bars are stressed to f_y at a distance of 9.80 ft from the centerline of the span (where the second pair of bars may be theoretically cut), however, the development length of these bars must be satisfied from the end of the bar back to the point of maximum stress f_y. The development length required is again conservatively taken as 2.21 ft. The straight development length furnished (or available) is 2.58 ft (see Figure 5-20). Because 2.58 ft > 2.21 ft, there is adequate development length available. If the development length available were insufficient, either of two solutions could be used: (a) provide a standard 180° hook at the end of each of the two corner bars or (b) extend all four of the outside bars to the end of the beam (thereby cutting only the two center bars).

Example 5-5 could also be solved by plotting the M_u diagram to scale and superimposing on the M_u diagram the values of ϕM_n for four bars and two bars. The theoretical cut points are established where the ϕM_n lines intersect the M_u curve. For instance, the theoretical cut point for the two center bars is established where the ϕM_n line for four bars intersects the M_u curve. This method has several advantages: (1) Any excess flexural steel in the design will result in the two center bars being cut closer to the midpoint of the beam, thus saving steel; (2) the calculation of ϕM_n better reflects the available moment strength than does

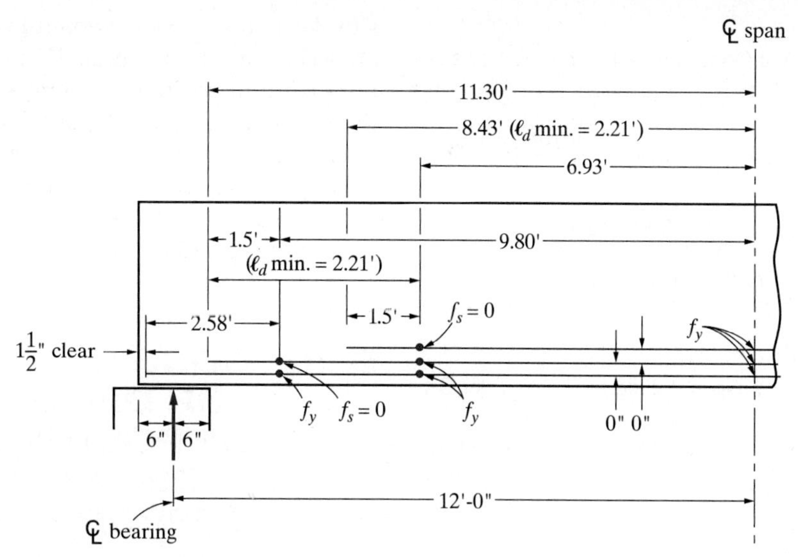

FIGURE 5-20 Sketch for Example 5-5.

the division of the M_u diagram ordinate into six equal parts (in Example 5-5); and (3) it can be used for any shape M_u diagram.

The bars in Example 5-5 are terminating in a tension zone. When this occurs, the member undergoes a reduction in shear capacity. Therefore, special shear and stirrup calculations must be performed in accordance with the ACI Code, Section 9.7.3.5, which stipulates that no flexural reinforcement shall be terminated in a tension zone unless *one* of the following conditions is satisfied:

(9.7.3.5(a)) Factored shear at the cutoff point does not exceed two-thirds of the design shear strength ϕV_n. This may be written as

$$V_u \leq \tfrac{2}{3}\phi(V_c + V_s)$$

(9.7.3.5(b)) For No. 11 bars and smaller, the continuing reinforcement provides double the area required for flexure at the cutoff point and factored shear does not exceed three-quarters of the design shear strength ϕV_n (i.e. $V_u \leq \left(\frac{3}{4}\right)\phi V_n$).

(9.7.3.5(c)) Stirrup area in excess of that required is provided along each bar terminated over a distance equal to three-fourths the effective depth of the member ($0.75d$) from the point of bar cutoff. The excess stirrup area shall not be less than $60b_w s/f_y$. The spacing s shall not exceed $d/(8\beta_b)$, where β_b is the ratio of reinforcement cutoff to total area of tension reinforcement at the section.

With respect to the preceding conditions, the first is interpreted as a check only. Although the V_s term could be varied, the code is not specific concerning the length of beam over which this V_s must exist. The second condition involves moving the cutoff point to another location. The third condition is a design method whereby additional stirrups may be introduced if the shear strength is inadequate as determined by the ACI Code, Section 9.7.3.5(a). This condition will be developed further.

It is convenient to determine the *number* of additional stirrups to be added in the length $0.75d$ along the end of the bar from the cutoff point. The required excess stirrup area is

$$A_v = \frac{60b_w s}{f_{yt}}$$

Assuming that the size of the stirrup (and therefore A_v) is known, the maximum spacing is

$$s = \frac{A_v f_{yt}}{60b_w}$$

Therefore, the number of stirrups N_s to be added in the length $0.75d$ is

$$N_s = \frac{0.75d}{s} + 1 = \frac{0.75d}{\left(\frac{A_v f_{yt}}{60b_w}\right)} + 1 = \left(\frac{45b_w d}{A_v f_{yt}}\right) + 1$$

Also, because maximum $s = d/(8\beta_b)$,

$$N_{s(\text{min.})} = \frac{0.75d}{\left(\frac{d}{8\beta_b}\right)} + 1 = 6\beta_b + 1$$

The larger resulting N_s controls.

Example 5-6

In the beam shown in Figure 5-21, the location at which the top layer of two No. 9 bars is to be terminated is in a tension zone. At the cutoff point, $V_u = 52$ kip and the No. 3 stirrups are spaced at 11 in. on center. Check shear in accordance with the ACI Code, Section 9.7.3.5, and redesign the stirrup spacing if necessary. Assume that $f'_c = 3000$ psi and $f_y = 60{,}000$ psi.

Solution:

In accordance with the ACI Code, Section 9.7.3.5(a), check $V_u \le \frac{2}{3}\phi(V_c + V_s)$:

$$\tfrac{2}{3}\phi(V_c + V_s) = \tfrac{2}{3}(0.75)\left(2\sqrt{f'_c}bd + \frac{A_v f_{yt} d}{s}\right)$$

$$= \tfrac{2}{3}(0.75)\left[2\sqrt{3000}(12)(26) + \frac{(0.22)(60{,}000)(26)}{11}\right]$$

$$= 32{,}700 \text{ lb}$$

$$= 32.7 \text{ kip} < 52 \text{ kip} \qquad \text{(N.G.)}$$

Therefore, add excess stirrups over a length of 0.75*d* along the terminated bars from the cut end in accordance with ACI 9.7.3.5(c):

$$N_s = \left(\frac{45 b_w d}{A_v f_{yt}} + 1\right) \text{ or } (6\beta_b + 1)$$

$$\frac{45(12)(26)}{0.22(60{,}000)} + 1 = 2.06$$

$$6(\tfrac{1}{3}) + 1 = 3$$

Add three stirrups over a length of $0.75d = 19.5$ in. Find the new spacing required in the 19.5-in. length:

$$\frac{19.5}{11} = 1.77 \text{ stirrups at 11-in. spacing}$$

$$\underline{+\ 3.00} \text{ additional stirrups}$$

$$4.77 \text{ stirrups in 19.5-in. length}$$

The new stirrup spacing is

$$\frac{19.5}{4.77} = 4.09 \text{ in.}$$

Use 4 in.

The stirrup pattern originally designed should now be altered to include the 4-in. spacing along the last 19.5 in. of the cut bars.

The problems associated with terminating bars in a tension zone may be avoided by extending the bars in accordance with the ACI Code, Section 9.7.3.5(c), or by extending them into the support.

In summary, a general representation of the bar cutoff requirements for positive moment steel in a simple span may be observed in Figure 5-22. If bars A are to be cut off, they must (1) project ℓ_d past the point of maximum positive moment and (2) project beyond their theoretical cutoff point a distance equal to the effective depth of the member or 12 bar diameters, whichever is greater. The remaining positive moment bars B must extend ℓ_d past the theoretical cutoff point of bars A and extend at least 6 in. into the support.

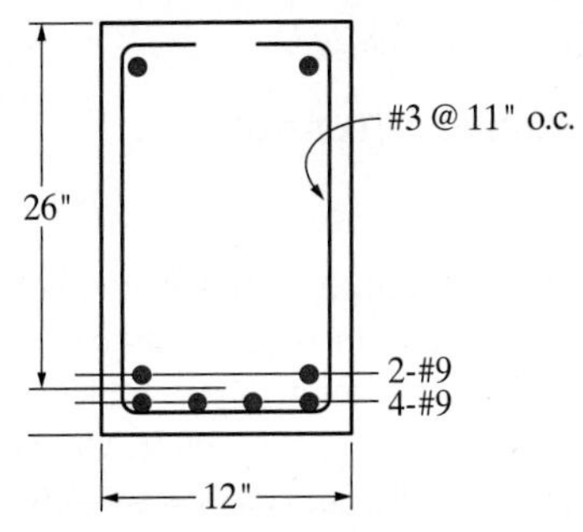

FIGURE 5-21 Sketch for Example 5-6.

5-10 CODE REQUIREMENTS FOR DEVELOPMENT OF POSITIVE MOMENT STEEL AT SIMPLE SUPPORTS

The ACI Code, Sections 7.7.3.8 and 9.7.3.8, contains requirements concerning the development of straight, positive moment bars *at simple supports* and at points of inflection. The intent of this code section is the same as the check on the development of the two bars extending into the support in Example 5-5. The method of Example 5-5 may be used if we are working with the actual moment diagram. The code approach does not require the use of a moment diagram.

The ACI Code requirement (ACI Code, Sections 7.7.3.8.3(b) and 9.7.3.8.3(b)) places a restriction on the size of the bar that may be used such that the tension development length

$$\ell_d \le \frac{M_n}{V_u} + \ell_a$$

where

M_n = nominal moment strength $\left[A_s f_y\left(d - \frac{a}{2}\right)\right]$ assuming all reinforcement at the section to be stressed to f_y

V_u = total applied design shear force at the section

ℓ_a = (at a support) the embedment length beyond the center of the support

ℓ_a = (at a point of inflection) the effective depth of the member or $12d_b$, whichever is greater

This requirement need not be satisfied for reinforcement terminating beyond the centerline of simple supports by a standard hook or a mechanical anchorage at least equivalent to a standard hook.

The effect of this code restriction is to require that bars be small enough so that they can become fully developed before the applied moment has increased to the magnitude where they *must* be capable of carrying f_y. The M_n/V_u term approximates the distance from the section in question to the location where applied moment M_u exists that is equal to ϕM_n (and where f_y must exist in the bars).

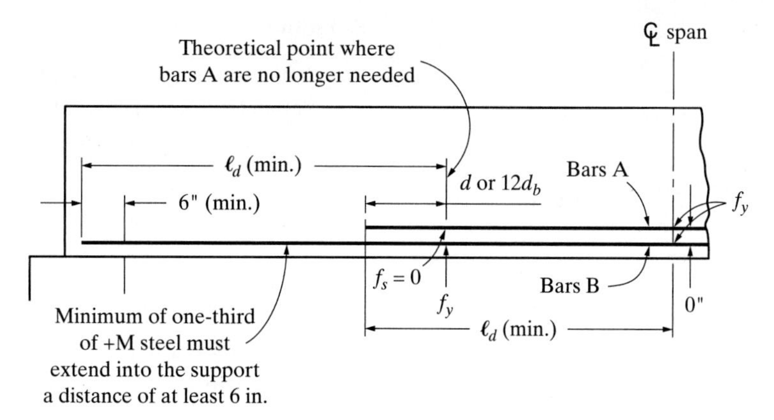

FIGURE 5-22 Bar cutoff requirements for simple spans (positive moment steel).

Therefore, the distance from the end of the bar to the point where the bar must be fully developed is $(M_n/V_u) + \ell_a$, and bars must be chosen so that their ℓ_d is less than this distance. The code allows M_n/V_u to be increased by 30% (ACI Code, Sections 7.7.3.8.3(a) and 9.7.3.8.3(a)) when the ends of the reinforcement are confined by a compressive reaction such as is found in a simply supported beam (a beam supported by a wall).

Example 5-7

At the support of a simply supported beam, a cross section exists as shown in Figure 5-23. Check the bar diameters in accordance with the ACI Code, Section 9.7.3.8.3. Assume a support width of 12 in., $1\frac{1}{2}$-in. cover, $f'_c = 4000$ psi, and $f_y = 60{,}000$ psi. Assume that V_u at the support is 80 kip. Normal-weight concrete is used. The stirrups begin at 3 in. from the face of the support. The bars are uncoated.

Solution:

Because this beam has its ends confined by a compressive reaction, we will check ACI Section 9.7.3.8.3(a):

$$\ell_d \le 1.3\left(\frac{M_n}{V_u}\right) + \ell_a$$

We next calculate the required tensile development length ℓ_d:

1. From Table 5-1, $K_D = 71.2$.
2. Establish values for the factors ψ_t, ψ_e, ψ_s, and λ.
   a. $\psi_t = 1.0$ (the bars are not top bars).
   b. The bars are uncoated; $\psi_e = 1.0$.
   c. The bars are No. 10; $\psi_s = 1.0$.
   d. Normal-weight concrete is used; $\lambda = 1.0$.
3. The product $\psi_t \times \psi_e = 1.0 < 1.7$. (O.K.)
4. Determine c. Based on cover (center of bar to nearest concrete surface), consider the clear cover, the No. 4 stirrup, and one-half the diameter of the No. 10 bar:

$$c_b = 1.5 + 0.5 + \frac{1.27}{2} = 2.64 \text{ in.}$$

   Based on bar spacing (one-half the center-to-center distance),

$$c_b = \frac{15 - 2(1.5) - 2(0.5) - 2\left(\frac{1.27}{2}\right)}{2} = 4.87 \text{ in.}$$

   Use the smaller of the two c_b values:
   Therefore, use $c_b = 2.64$ in.

5. K_{tr} is taken as zero, as stirrups do not extend to the ends of the bars.
6. Check $(c_b + K_{tr})/d_b \le 2.5$:

$$\frac{c_b + K_{tr}}{d_b} - \frac{2.64 + 0}{1.27} = 2.08 < 2.5. \qquad \text{(O.K.)}$$

7. The excess reinforcement factor is not applicable and is omitted.

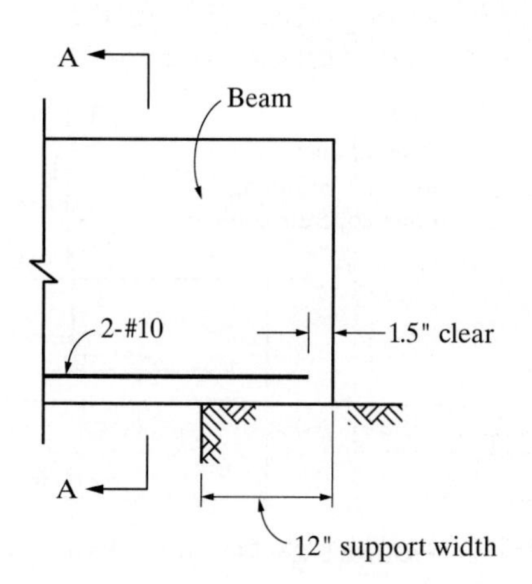

FIGURE 5-23 Sketch for Example 5-7.

8. Calculate ℓ_d:

$$\ell_d = \frac{K_D}{\lambda}\left[\frac{\psi_t\psi_e\psi_s}{\left(\frac{c_b + K_{tr}}{d_b}\right)}\right]d_b$$

$$= \frac{71.2}{1.0}\left[\frac{1.0(1.0)(1.0)}{2.08}\right](1.27)$$

$$= 43.5 \text{ in.} > 12 \text{ in.} \qquad \text{(O.K.)}$$

Now check ACI 318-14, Section 9.7.3.8.3(a):

ℓ_a = the embedment length beyond the center of the support = 12 − 1.5 − (12/2) = 4.5 in.

$$\rho = \frac{A_s}{bd} = \frac{2.54}{15(26)} = 0.0065$$

$$\overline{k} = 0.3676 \text{ ksi}$$

$$M_n = bd^2\overline{k} = 15(26)^2(0.3676) = 3728 \text{ in.-kip}$$

The maximum permissible required ℓ_d is calculated using ACI 318-14, Section 9.7.3.8.3(a) as

$$1.3\frac{M_n}{V_u} + \ell_a = 1.3\left(\frac{3728}{80}\right) + 4.5 = 65.1 \text{ in.}$$

$$65.1 \text{ in.} > 43.5 \text{ in.} \qquad \text{(O.K.)}$$

Therefore, the bar diameter is adequately small and the bar can be developed as required. (If the required ℓ_d were in excess of 65.1 in., the use of a standard hook (or a mechanical anchorage equivalent to a standard hook) beyond the centerline of support would satisfy the development requirement and this code section would not apply. Also, the use of smaller bars would result in a smaller required ℓ_d.)

Example 5-8

Tension and compression development lengths in a column footing

For the footing shown in Figure 5-24:

a. Determine the tension development length, ℓ_d for the No. 6 bottom rebars in the footing.
b. Determine the compression development length for the No. 7 column rebars
c. Determine the minimum footing thickness from bond and anchorage considerations
d. Determine the compression lap splice for the No. 7 column rebars.

Use f'_c is 4000 psi and f_y is 60,000 psi.

Solution:

a. Tension development length

Actual development length available,

$$\ell_{d,\text{ provided}} = \frac{(6\text{ ft})(12\text{ in./ft}) - 12\text{ in.}}{2} - 3\text{ in.} = 27\text{ in.}$$

#6 bars in the footing ($d_b = 0.75''$)

$$f'_c = 4000 \text{ psi}$$
$$f_y = 60000 \text{ psi}$$
$$\psi_t = \psi_e = 1.0$$
$$\lambda = 1.0$$
$$\psi_s = 0.8$$
$$K_{ER} = 1.0 \text{ (assumed)}$$
$$K_{tr} = 0$$

Determine c_b, which is the smaller of:

$$c_b = \text{cover} + \frac{d_b}{2} = 3'' + \frac{0.75''}{2} = 3.38''$$

and,

$$c_b = 0.5\left(\frac{W - (2)(\text{cover}) - d_b}{N_{\text{bars}} - 1}\right)$$

$$= 0.5\left(\frac{72'' - (2)(3'') - 0.75''}{6 - 1}\right) = 6.5''$$

Use $c_b = 3.38''$

$$\frac{c_b + K_{tr}}{d_b} < 2.5 = \frac{3.38'' + 0}{0.75''} = 4.5, \text{ therefore use } 2.5$$

for this term

$$\ell_{d,\text{ reqd}} = \frac{3}{40}\left(\frac{f_y}{\lambda\sqrt{f'_c}}\right)\left[\frac{\psi_t\psi_e\psi_s}{\frac{c_b + K_{tr}}{d_b}}\right]K_{ER}d_b$$

$$= \frac{3}{40}\left(\frac{60{,}000}{(1.0)\sqrt{4000}}\right)\left[\frac{(1.0)(1.0)(0.8)}{(2.5)}\right](1.0)(0.75)$$

$$= 17.1''$$

17.1″ < 27″, OK for tension development length

b. Compression development length of column dowels:

$$\ell_{dc} = \left(\frac{0.02f_y}{\lambda\sqrt{f'_c}}\right)d_b > 0.0003f_yd_b$$

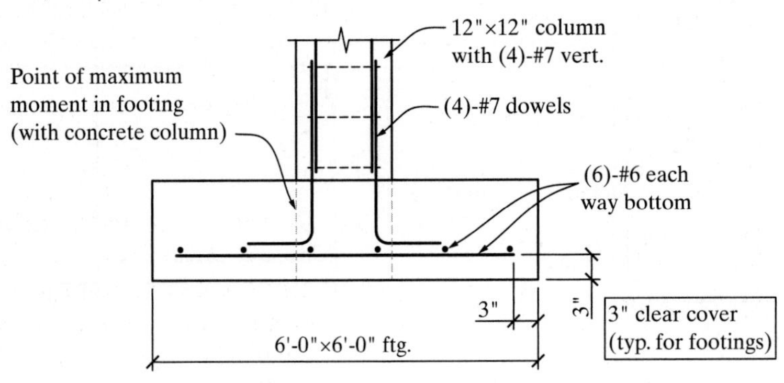

FIGURE 5-24 Detail for Example 5-8.

also

$\ell_{dc} > 8"$

$$= \left(\frac{(0.02)(60{,}000)}{(1.0)\sqrt{4000}}\right)(0.875) > (0.0003)(60{,}000)(0.875)$$

$16.6" > 15.8"$ and $16.6" > 8"$

$K_{ER} = 1.0$ and the confinement factor is 1.0

$\ell_{dc} = (16.6")(1.0)(1.0) = 16.6"$, say 17"

c. Minimum footing thickness

$$H_{min} = \ell_{dc} + \text{cover} = 17" + 3" = 20"$$

d. Compression Lap Splice

$$\ell_{sc} = 30d_b \text{ for } f_y = 60 \text{ ksi}$$

$$\ell_{sc} = (30)(0.875) = 26.3", \text{ say } 27"$$

Example 5-9

Hooked Bar development length in a beam

a. Determine the tension development length of the hooked 4 No. 9 top rebar for the 12" × 22" deep beam shown in Figure 5-25. Assume grade 60 rebar and $f'_c = 4$ksi.
b. Determine the minimum width of column to accommodate the hooked rebars. Assume #3 ties are used.
c. Check the straight embedment length of the rebar within the cantilever beam

Solution:

a. Hooked bar development

#9 bars, $d_b = 1.128"$

$\psi_e = 1.0$

$\lambda = 1.0$

$K_{ER} = 1.0$ (assumed)

$$\ell_{dh} = \left[\frac{0.02\psi_e f_y}{\lambda\sqrt{f'_c}}\right]K_{er}d_b > 8d_b \text{ and } > 6"$$

$$\ell_{dh} = \left[\frac{(0.02)(1.0)(60{,}000)}{(1.0)\sqrt{4000}}\right](1.128) = 21.4"$$

$8d_b = (8)(1.128) = 9.1" > 6"$

Use $\ell_{dh} = 21.4"$

b. Minimum width of column

$= \ell_{dh} + \text{clear concrete cover} + \text{ties} = 21.4" + 2" + 0.375" = 23.8"$, say 24"

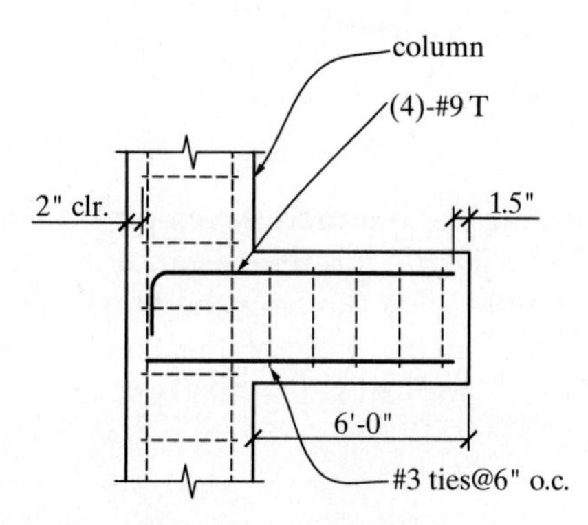

FIGURE 5-25 Detail for Example 5-9.

c. Development Length of the straight bar

Available straight bar development length,

$\ell_{d,\,available} = 72" - 1.5" = 70.5"$

$f'_c = 4000$ psi

$f_y = 60000$ psi

$\psi_t = 1.0$

$\lambda = 1.0$

$\psi_s = 1.0$

$K_{ER} = 1.0$ (assumed)

$K_{tr} - 0$

Find c_b:

$$c_b = \text{cover} + \frac{d_b}{2} = 1.5" + \frac{1.128"}{2} = 2.0"$$

$$c_b = 0.5\left(\frac{b_w - (2)(\text{cover}) - d_b}{N_{bars} - 1}\right)$$

$$= 0.5\left(\frac{12" - (2)(1.5") - 1.128"}{4 - 1}\right) = 1.31"$$

Use the smaller of the two c_b values:
Therefore, use $c_b = 1.31"$

$$K_{tr} = \frac{40A_{tr}}{sn} = \frac{(40)(2)(0.11)}{(6)(4)} = 0.366$$

$$\frac{c_b + K_{tr}}{d_b} < 2.5 = \frac{1.31" + 0.366}{1.128} = 1.48$$

$$\ell_{d,\,reqd} = \frac{3}{40}\left(\frac{f_y}{\lambda\sqrt{f'_c}}\right)\left[\frac{\psi_t\psi_e\psi_s}{\frac{c_b + K_{tr}}{d_b}}\right]K_{er}d_b = \frac{3}{40}\left(\frac{60{,}000}{(1.0)\sqrt{4000}}\right)\left[\frac{(1.0)(1.3)(1.0)}{(1.48)}\right](1.0)(1.128) = 70.1"$$

$< 70.5"$ (O.K.)

5-11 STRUCTURAL INTEGRITY REINFORCEMENT—BEAMS

The ACI Code, Section 9.7.7.1 prescribes structural integrity reinforcement for beams and girders in order to improve the redundancy and ductility of the structure, and to prevent progressive collapse such as what happened in the Alfred P. Murrah Federal Building Collapse in Oklahoma City in April 1995 [6]. Structural integrity reinforcement are also required for two-way slabs (see Chapter 6). The ACI Code prescribes the following reinforcement details for beams and girders:

Positive (+ve) Moment Regions:

a. Perimeter or spandrel beams must have closed stirrups with not less than 135 degree hooks around a continuous longitudinal rebar where spalling can occur, or a standard hook around a longitudinal rebar where spalling is restrained.
b. At least 25% of the bottom rebar (at least 2 bars) must be continuous at the supports.

c. Top and Bottom bars should be terminated with ACI standard hook at discontinuous supports and should be capable of developing f_y at the face of the support.
d. Interior beams must satisfy criteria (b) and (c) unless closed stirrups are provided.

Negative (−ve) Moment Regions:

At least 16.7% of top rebar at supports (at least 2 bars) shall be continuous at midspan.

The above requirements are usually covered in the typical details that are a part of the structural drawings. Where splices are needed to provide rebar continuity, splice top reinforcement at or near the midspan of the member, and splice bottom reinforcement at or near the support. Splices shall be Class B tension splices or mechanical or welded splices satisfying ACI Section 25.5.7.1. All longitudinal structural integrity reinforcement shall pass within the area in the column bounded by the column vertical reinforcement.

References

[1] James K. Wight and James G. MacGregor, "Reinforced Concrete—Mechanics and Design," 5th Edition, 1112 pp, Pearson, 2009.

[2] Arthur H. Nilson, David Darwin, and Charles Dolan, "Design of Concrete Structures," 13th Edition, 779 pp, McGraw Hill, 2004.

[3] Mohammad Ayub, "Investigation of the October 30, 2003, Fatal Parking Garage Collapse at Tropicana Casino Resort, Atlantic City, NJ," Report No. 2004R03, *Directorate of Construction, US Department of Labor, Occupational Health and Safety*, April, 2004, 56 pp.

[4] Eric Lipton, "Design Changes Preceded Collapse of Casino Garage," *New York Times*, April 24, 2004, p. 25.

[5] Dan Eschenasy. "Balcony Issues in High-Rise Buildings," Structure Magazine, November 2017

[6] HyunJin Kim. "Progressive Collapse Behavior of Reinforced Concrete Structures with Deficient Details," Ph.D. Dissertation, University of Texas, Austin, TX, 2006.

Problems

For the following problems, unless otherwise noted, concrete is normal weight and steel is uncoated grade 60 (f_y = 60,000 psi).

5-1. Determine the tension development length required for the No. 8 bars in the T-beam shown in Figure 3-11 of Example 3-4. Use $f_c' = 3000$ psi. Assume that the No. 3 stirrups are spaced at 8 in. throughout. Concrete is normal weight. Neglect the compression steel.

5-2. A 12-in.-thick concrete wall is supported on a continuous footing as shown. Use $f_c' = 3000$ psi. Determine if the development length is adequate if the steel is No. 6 bars at 8 in. o.c. Assume that the critical section for moment (f_y is developed) is at the face of the wall.

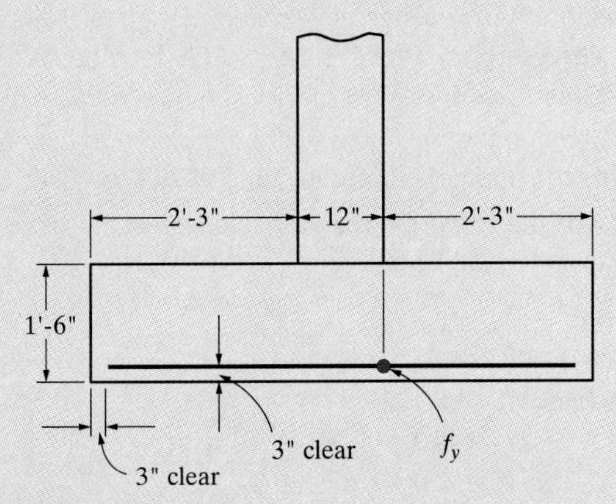

PROBLEM 5-2

5-3. The exterior balcony/canopy shown is to be constructed of lightweight concrete (f_{ct} not specified). The bars are epoxy-coated. The design-required tension steel A_s was 0.67 in.^{2}. Determine the required development length ℓ_d from the point of maximum stress in the bars. Specify the minimum required side cover. Shrinkage and temperature steel is not shown. $f_c' = 4000$ psi, $f_y = 60{,}000$ psi.

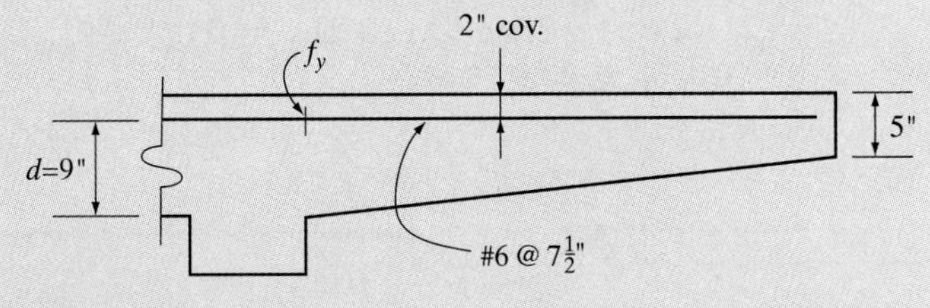

PROBLEM 5-3

5-4. Calculate the required development length (ℓ_d) into the beam for the negative moment steel shown so as to develop the tensile strength of the steel at the face of the column. Required $A_s = 2.75$ in.^{2}, and $f_c' = 4000$ psi.

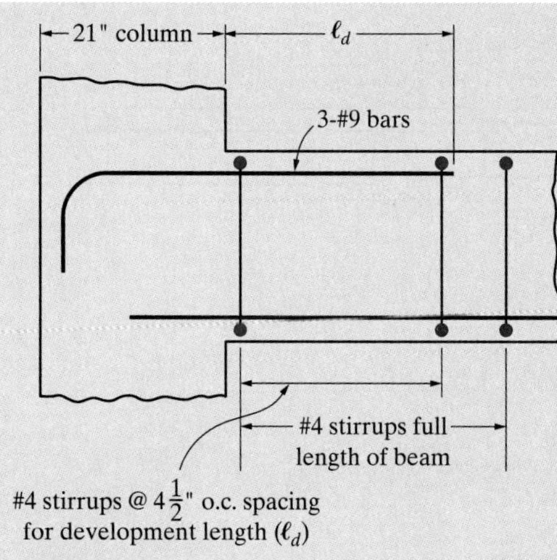

PROBLEM 5-4

5-5. Considering the anchorage of the beam bars into the column, determine the largest bar that can be used without a hook. Use $f'_c = 4000$ psi. Clear space between bars is 3 in. (minimum). Side cover is 2 in.

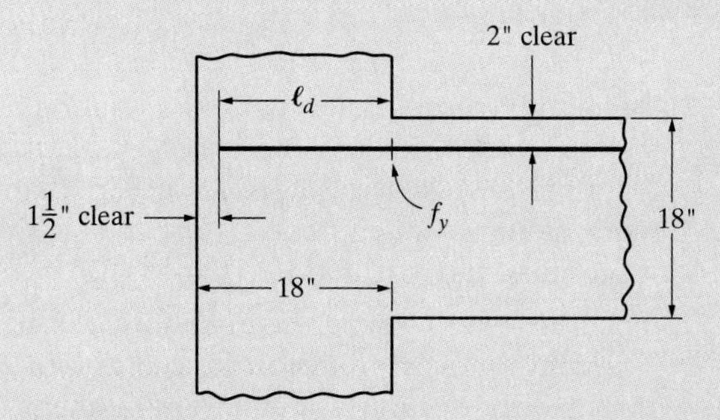

PROBLEM 5-5

5-6. Seven No. 11 vertical compression bars extend from a column into the supporting footing. Use $f'_c = 5000$ psi. The A_s required was 10.2 in.^{2} There is no lateral reinforcing enclosing the vertical bars in the footing. Determine the required compression development length.

5-7. Determine the development length required in the column for the bars shown. If the available development length is not sufficient to develop the tensile strength of the steel (f_y), design an anchorage using a 180° hook and check its adequacy. Use $f'_c = 4000$ psi. The clear space between the No. 10 bars is 3 in. with a side cover of $2\frac{1}{2}$ in.

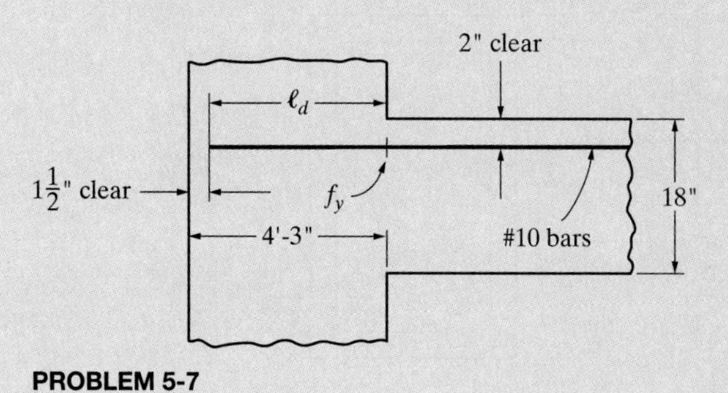

PROBLEM 5-7

5-8. The tension bars in the stem of the reinforced concrete retaining wall are No. 9 at 7 in. o.c. and are to be lap spliced to similar dowels extending up from the footing. Required $A_s = 1.50\,\text{in.}^2/\text{ft.}$ Use $f'_c = 3000$ psi. Cover on the bars from the rear face of the wall is 2 in., and cover from the end of the wall to the edge bar, measured in the plane of the bars, is 3 in.

a. Calculate the required length of splice.

b. Find the hook development length ℓ_{dh} required for a 180° hook in the footing.

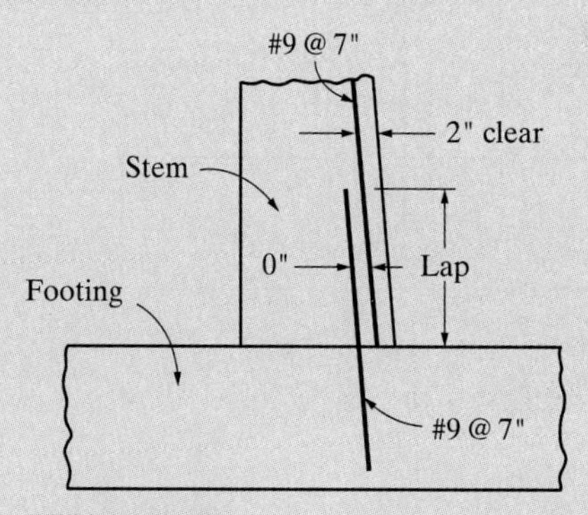

PROBLEM 5-8

5-9. Design the tension splices at points A and B in the beam shown, assuming, for purposes of this problem, that the splices must be located as shown. Positive M_u at A is 120 ft.-kip; negative M_u at B is 340 ft.-kip. Assume that 50% of the steel is spliced at the designated locations. Use $f'_c = 4000$ psi. Assume $\phi = 0.90$ for moment.

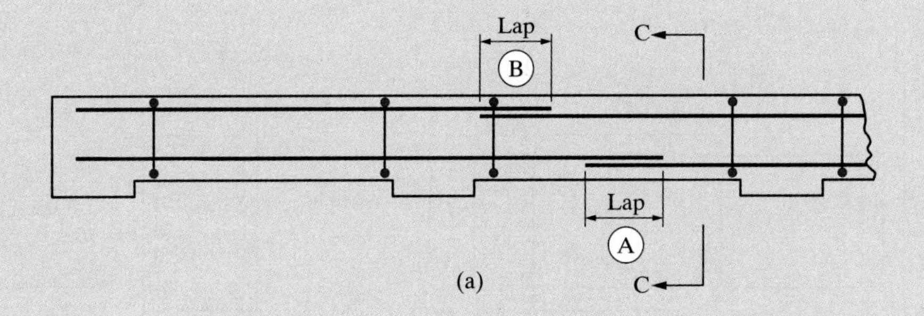

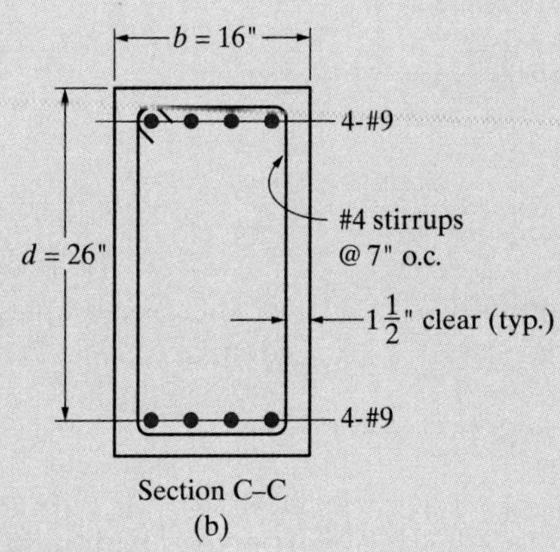

PROBLEM 5-9

5-10. A wall is reinforced with No. 7 bars at 12 in. o.c. in each face as shown. The bars are in compression and are to be spliced to dowels of the same size and spacing in the footing. Use $f_c' = 3000$ psi. Determine the required length of splice.

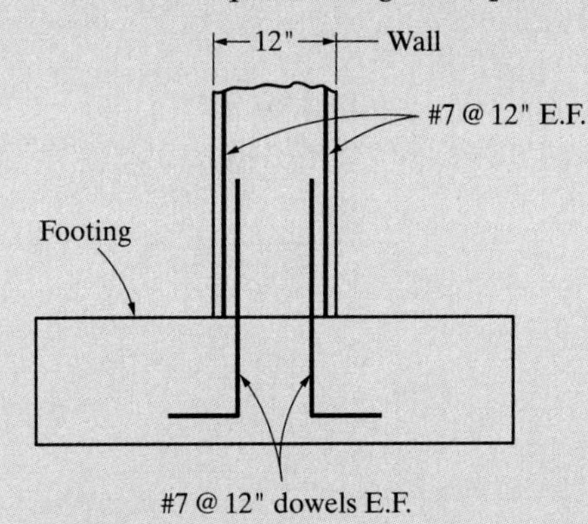

PROBLEM 5-10

5-11. Number 9 compression bars in a column are to be lap spliced. There is no excess steel, and the bars are enclosed by a spiral of $\frac{3}{8}$-in.-diameter wire having a pitch of 3 in. Use $f_c' = 4000$ psi. Determine the required length of splice.

5-12. For the overhanging beam shown, find the theoretical cutoff point for the center No. 9 bar. Assume the section of maximum moment to be at the face of support. Use $f_c' = 5000$ psi.

5-13. A simply supported, uniformly loaded beam carries a total factored load of 4.8 kip/ft (this includes the beam weight) on a clear span of 34 ft. The cross section has been designed and is shown. The required tension steel area for the design was 5.48 in.^{2} The supports are 12-in. wide. Use $f_c' = 3000$ psi.

a. Determine the cutoff location for the two center bars (indicated in the sketch).

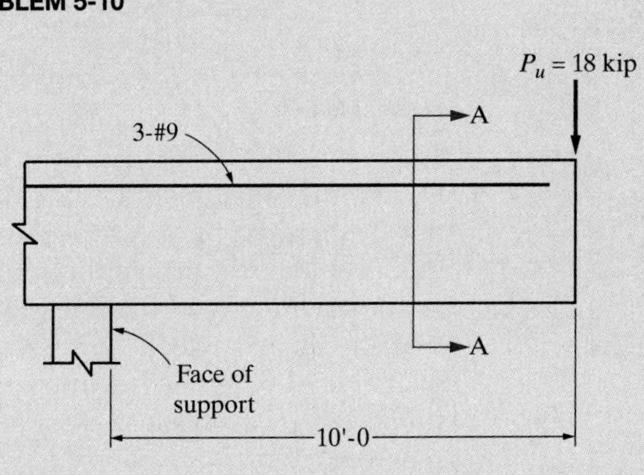

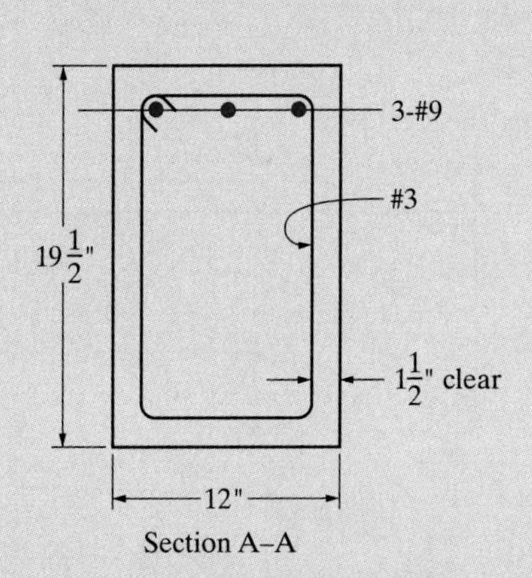

PROBLEM 5-12

b. Design the shear reinforcement and check whether code requirements are met for bars terminated in a tension zone. Redesign the stirrups if necessary.

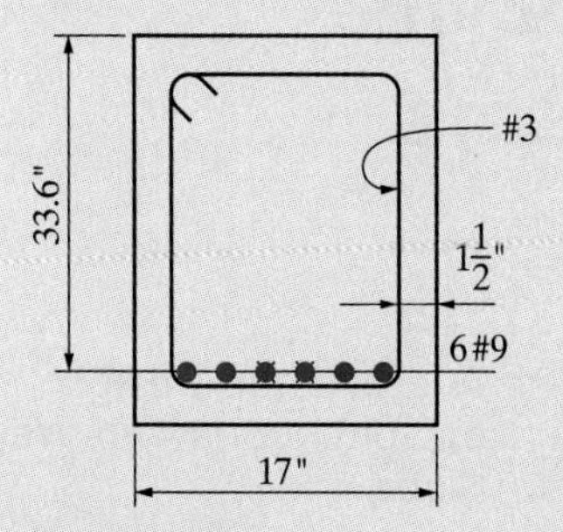

PROBLEM 5-13

5-14. The beam shown is to carry a total factored uniform load of 10 kip/ft (includes beam weight) and factored concentrated loads of 20 kip. Use $f_c' = 4000$ psi. The supports are 12-in. wide.

a. Design a rectangular beam (tension steel only).

b. Determine bar cutoffs (use the graphical approach).

c. Design the shear reinforcement and check bar cutoffs in tension zones if necessary.

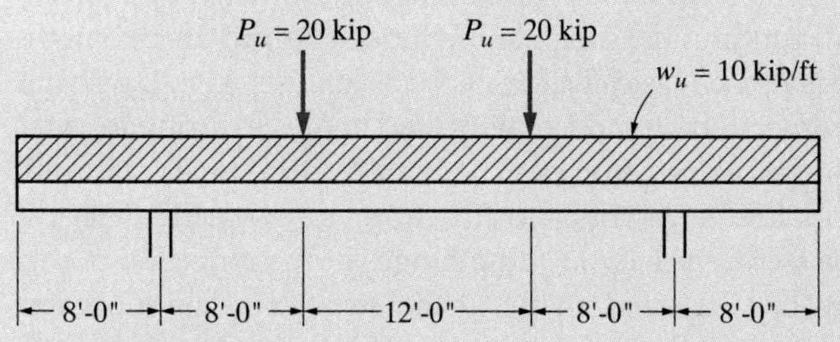

PROBLEM 5-14

5-15 For the footing detail shown below with an uplift load, assume #7 bars and use $(c + K_{tr})/d_b = 2.5$

a. Determine the required development length for both the hooked and straight bars

b. Explain why the answers are different for (a)

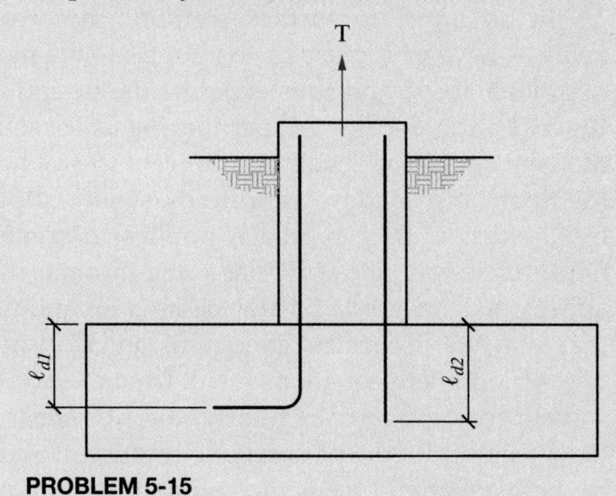

PROBLEM 5-15

5-16 For the wall and footing detail shown below, use $f_c' = 3{,}500$ psi, $f_y = 60{,}000$ psi. Also, for the #6 wall reinforcing, the design requires 0.5 in.^{2}/ft.

a. Determine the lap splice length (Note: consider the loading on the wall and determine if it is in tension or compression).

b. Determine the development length in compression and tension for the hooked bars; is the footing depth adequate?

c. Determine if the footing width "W" is adequate for the development length of the footing bars in tension.

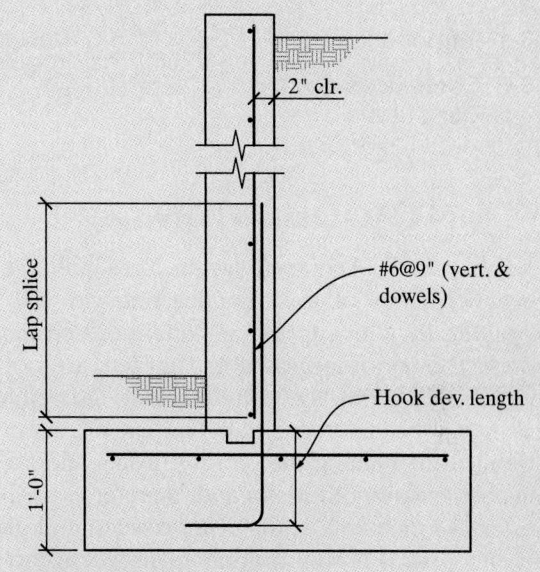

PROBLEM 5-16

5-17 For the Wall detail shown below:

a. Determine the lap splice length.

b. Determine the development length in compression and tension for the hooked bars; is the footing depth adequate?

c. Determine if the footing width "W" is adequate for the development of the footing bars in tension.

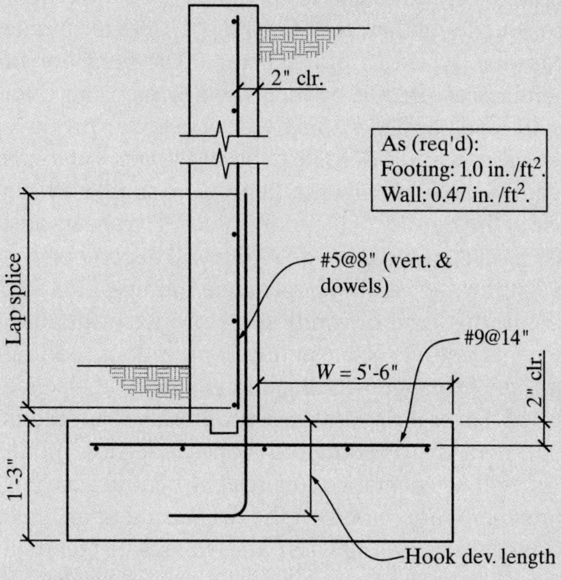

PROBLEM 5-17

CONTINUOUS ONE-WAY AND TWO-WAY FLOOR SYSTEMS

6-1 Introduction

6-2 Continuous-Span Bar Cutoffs

6-3 Design of Continuous One-Way Floor Systems

6-4 Analysis and Design of Continuous Two-Way Slabs

6-1 INTRODUCTION

A common form of concrete cast-in-place building construction consists of a continuous one-way slab cast monolithically with supporting continuous beams and girders. The concrete beams and girders are not cast monolithically with the columns, but the reinforcement in the beams and girders are extended into the columns, thus making the beams/girders and the columns behave monolithically and, therefore, capable of transferring moments at the beam-column and girder-column joints. This is in contrast to steel or wood construction where the connections between the beams and the girders, and the beams/girders and the columns are usually simply supported connections, resulting in a statically determinate structure, unless a moment connection is deliberately included in the structure. In concrete framing systems, all members contribute in carrying the floor load to the supporting columns (see Figures 3-1 and 6-19). The slab steel reinforcement runs through the beams, the beam steel runs through the girders, and the steel from both the beams and girders runs through the columns. The result is that the whole floor system is tied together, forming a highly indeterminate and complex type of rigid structure. The behavior of the members is affected by their rigid connections. Not only will loads applied directly *on* a member produce moment, shear, and a definite deflected shape, but loads applied to *adjacent* members will produce similar effects because of the rigidity of the connections. The shears and moments transmitted through a joint will depend on the relative stiffnesses of all the members framing into that joint. With this type of condition, a precise evaluation of moments and shears resulting from a floor loading is excessively time-consuming and is outside the scope of this text. Several classical structural analysis methods (e.g., moment distribution and slope deflection methods), as well as several commercial structural analysis computer software (based on the displacement or force methods of structural analysis), are available to facilitate these analysis computations. In the gravity load analysis of continuous one-way slabs or continuous beams, pattern- or skip-live loading must be considered. The aim is to vary the position of the live load (the position of the dead load is not varied) in order to obtain the most critical loading condition or scenario that produces the maximum factored shear and moment at each midspan and support location on the continuous member.

To simplify the calculation of moments and shears in continuous concrete members, approximate methods can be used for lateral load analysis (e.g., the portal frame method) and gravity load analysis during the preliminary design phase. These methods reduce a statically indeterminate concrete structure to a statically determinate structure by locating hinges at the inflection points on the deflected shape of the structure. The inflection points are the points on each member where a change in curvature occurs. In the portal method for lateral load analysis, the inflection points are assumed to occur at the mid-lengths of the beams and columns, and for gravity load analyses, the points of inflection or hinges can be assumed to occur at a distance of 10 to 20-percent of the beam length from each end of the beams.

In an effort to further simplify the structural analysis beyond the use of the approximate methods described above and thus expedite the design phase, the ACI Code, Section 6-5, permits the use of standard moment and shear equations (see Table 6-1) whenever the span and loading conditions satisfy stipulated requirements. This approach applies to continuous non-prestressed one-way slabs and beams. It is an approximate method and may be used for buildings of the usual type of construction, spans, and story heights. The ACI moment equations result from the product of a coefficient and $w_u \ell_n^2$. Similarly, the ACI shear equations result from the product of a coefficient and $w_u \ell_n$. In these equations, w_u is the factored design uniform load and ℓ_n is the *clear span* for positive moment (and shear) and the average of two adjacent clear spans for

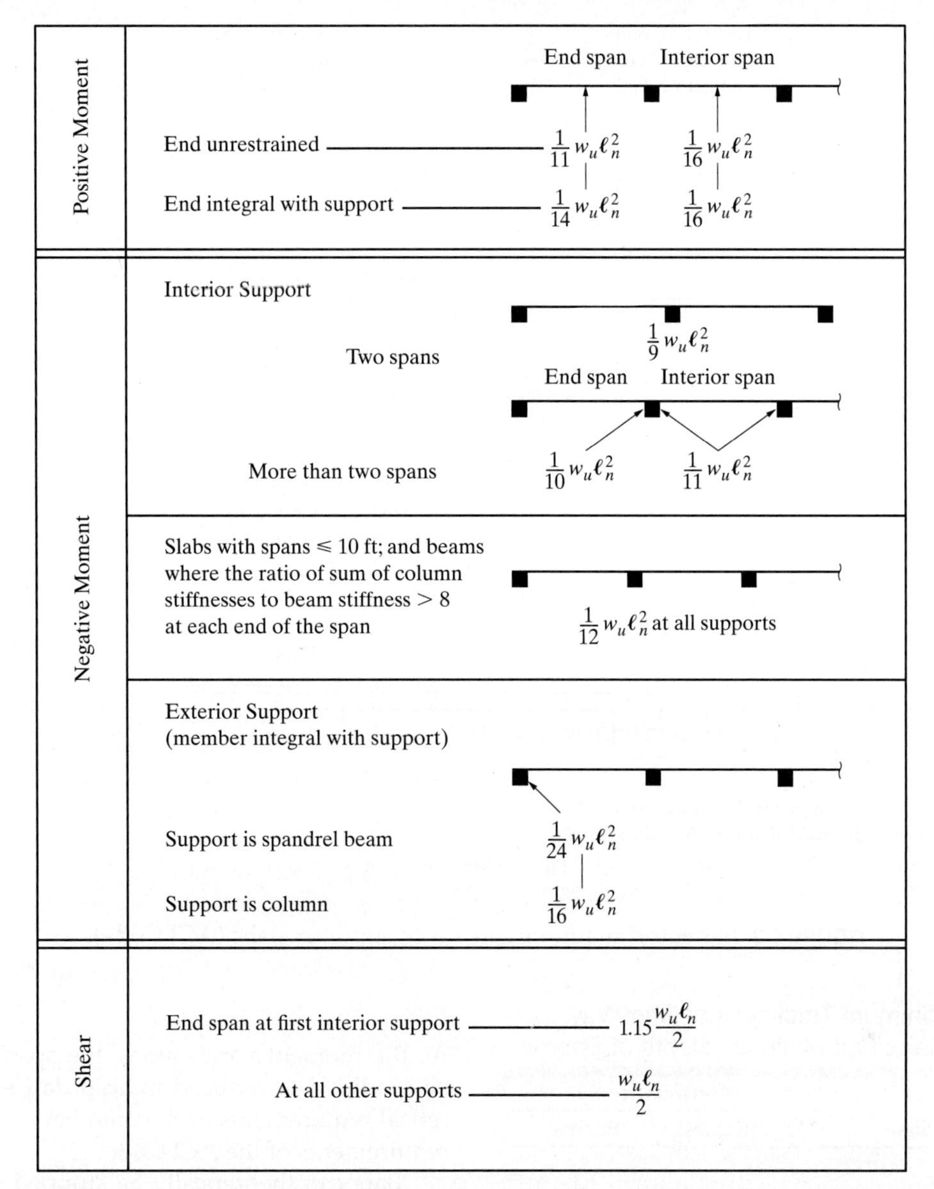

FIGURE 6-1 ACI Code coefficients and equations for shear and moment for continuous beams and one-way slabs.

negative moment. The application of the equations is limited to the following:

1. The equations can be used for two or more approximately equal spans (with the larger of two adjacent spans not exceeding the shorter by more than 20%).
2. Loads must be uniformly distributed (therefore, girders are excluded).
3. The maximum allowable ratio of live load to dead load is 3:1 (based on service loads).
4. Members must be prismatic.

The ACI moment and shear coefficients give the envelopes of the maximum moments and shears, respectively, at critical locations on the flexural member. These maximum moments and shears at the different critical locations along the flexural member do not necessarily occur under the same loading condition. These shear and moment equations generally give reasonably conservative values for the stated conditions. If more precision is required, or desired, for economy, or because the stipulated conditions are not satisfied, a more theoretical and precise analysis must be made. The moment and shear equations are depicted in Figure 6-1. Their use will be demonstrated later in this chapter.

For approximate moments and shears for girders, see Section 14-2.

For reinforced concrete beams and slabs of normal weight concrete and f_y of 60,000 psi not supporting deflection sensitive elements (i.e., elements likely to be damaged by large deflections), the minimum slab thickness and minimum beam depths from ACI Code Tables 7.3.1.1. and 9.3.1.1. are given in Table 6-1.

Another common form of cast-in-place concrete construction are column-supported two-way slabs which will be discussed in Section 6-4.

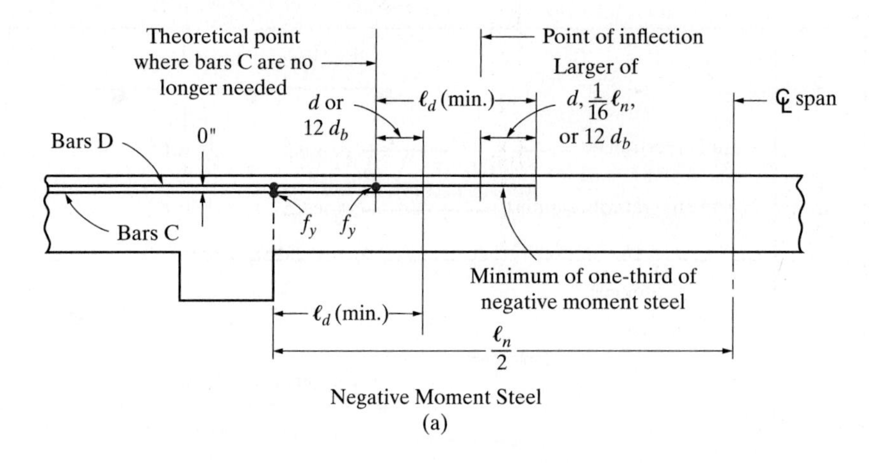

Negative Moment Steel
(a)

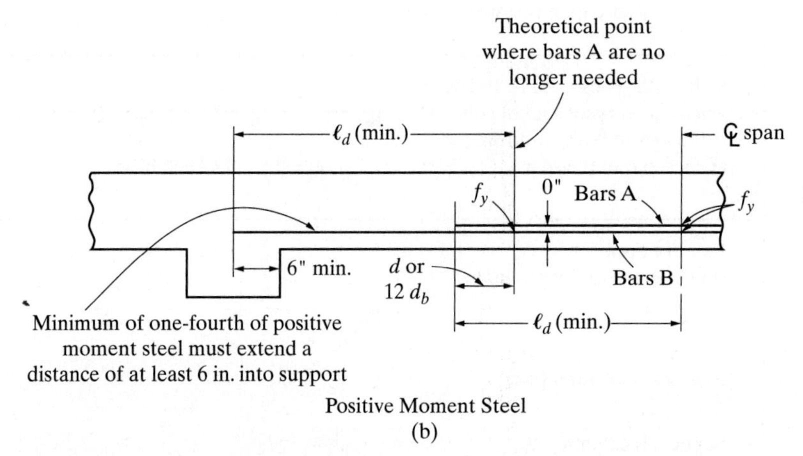

Positive Moment Steel
(b)

FIGURE 6-2 Bar cutoff requirements for continuous slabs (ACI Code).

TABLE 6-1 Minimum Thickness of One-Way Slabs and Minimum Depth of Beams

	Minimum *h*	
Support Conditions	**One-way slabs**	**Beams**
Simply Supported	$\frac{\ell}{20}$	$\frac{\ell}{16}$
One end continuous	$\frac{\ell}{24}$	$\frac{\ell}{18.5}$
Both ends continuous	$\frac{\ell}{28}$	$\frac{\ell}{21}$
Cantilever	$\frac{\ell}{10}$	$\frac{\ell}{8}$

[1]Adapted from ACI Tables 7.3.1.1. and 9.3.1.1

[2]For f_y other than 60,000 psi, multiply the minimum depth from the table by $(0.4 + f_y/100{,}000)$

[3]For lightweight concrete with density, w_c, between 90 and 115 psf, multiply the minimum depth from the table by the greater of (a) $1.65 - 0.005w_c$ and (b) 1.09.

6-2 CONTINUOUS-SPAN BAR CUTOFFS

Using a design approach similar to that for simple spans, the area of main reinforcing steel required at any given point is a function of the design moment. As the moment varies along the span, the steel may be modified or reduced in accordance with the theoretical requirements of the member's strength and the requirements of the ACI Code.

Bars can theoretically be stopped or bent in flexural members whenever they are no longer needed to resist moment. A general representation of the bar cutoff requirements for continuous spans (both positive and negative moments) is shown in Figure 6-2.

In continuous members the ACI Code, Sections 7.7.3.8 and 9.7.3.8, requires that a minimum of one-fourth of the positive moment steel be extended into the support a distance of at least 6 in. The ACI Code, Sections 7.7.3.8.4 and 9.7.3.8.4, also requires that at least one-third of the negative moment steel be extended beyond the extreme position of the point of inflection a distance not less than one-sixteenth of the clear span, the effective depth of the member *d*, or 12 bar diameters, whichever is greater. If negative moment bars C (Figure 6-2a) are to be cut off, they must extend at least a full development length ℓ_d beyond the face of the support. In addition, they must extend a distance equal to the effective depth of the member or 12 bar diameters, whichever is larger, beyond the theoretical cutoff point defined by the

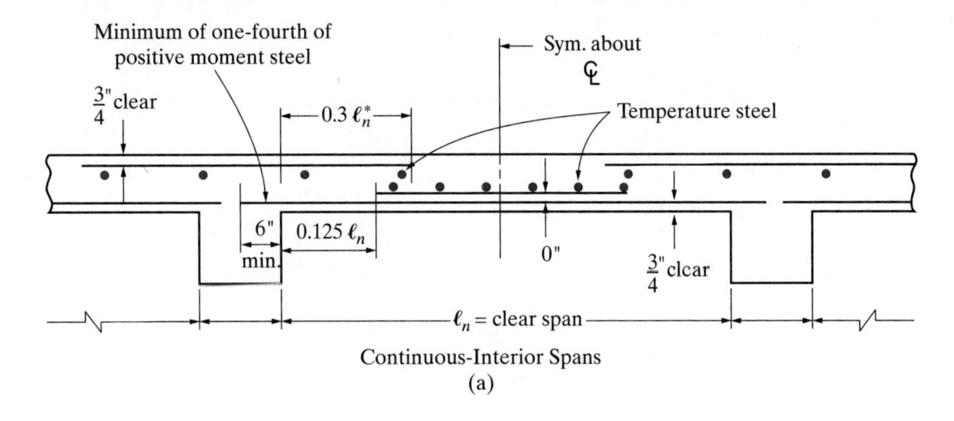

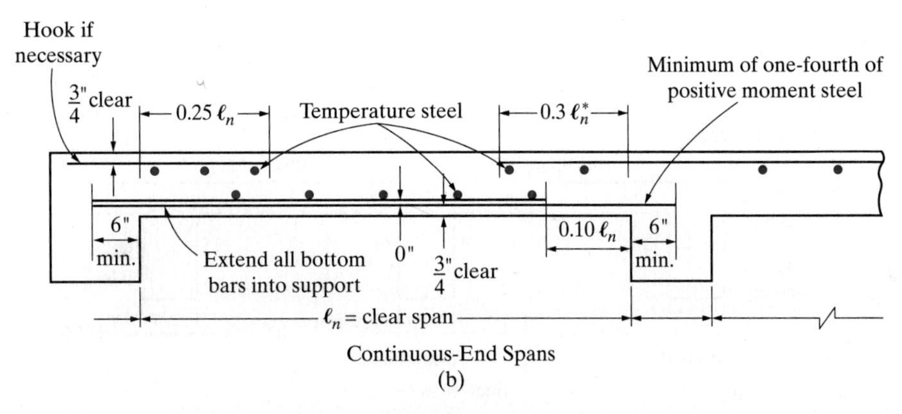

*If adjacent spans have different span lengths, use the larger of the two.

FIGURE 6-3 Recommended bar details and cutoffs, one-way slabs; tensile reinforced beams similar.

moment diagram. The remaining negative moment bars D (minimum of one-third of total negative steel) must extend at least ℓ_d beyond the theoretical point of cutoff of bars C and, in addition, must extend a distance equal to the effective depth of the member, 12 bar diameters, or one-sixteenth of the clear span, whichever is greater, past the point of inflection. Where negative moment bars are cut off before reaching the point of inflection, the situation is analogous to the simple beam cutoffs where the reinforcing bars are being terminated in a tension zone. The reader is referred to Example 5-5.

In Figure 6-2b, if positive moment bars A are to be cut off, they must project ℓ_d past the point of maximum positive moment as well as a distance equal to the effective depth of the member or 12 bar diameters, whichever is larger, beyond their theoretical cutoff point. Recall that the location of the theoretical cutoff point depends on the amount of steel to be cut and the shape of the applied moment diagram. The remaining positive moment bars B must extend ℓ_d past the theoretical cutoff point of bars A and extend at least 6 in. into the support. Comments on terminating bars in a tension zone again apply. Additionally, the size of the positive moment bars at the point of inflection must meet the requirements of the ACI Code, Sections 7.7.3.8.3 and 9.7.3.8.3.

Because the determination of cutoff and bend points constitutes a relatively time-consuming chore, it has become customary to use defined cutoff points that experience has indicated are safe. These defined points may be used where the ACI moment coefficients have application but must be applied with judgment where parameters vary. The recommended bar details and cutoffs for continuous spans are shown in Figure 6-3.

6-3 DESIGN OF CONTINUOUS ONE-WAY FLOOR SYSTEMS

One common type of floor system consists of a continuous, cast-in-place, one-way reinforced concrete slab supported by monolithic, continuous reinforced concrete beams. Assuming that the floor system parameters and loading conditions satisfy the criteria for application of the ACI Code coefficients, the design of the system may be based on these coefficients. Example 6-1 furnishes a complete design of a typical continuous one-way slab and beam floor system. Note that the ACI moment and shear coefficients cannot be used for girders since girders are subjected primarily to concentrated loads from the beam reactions.

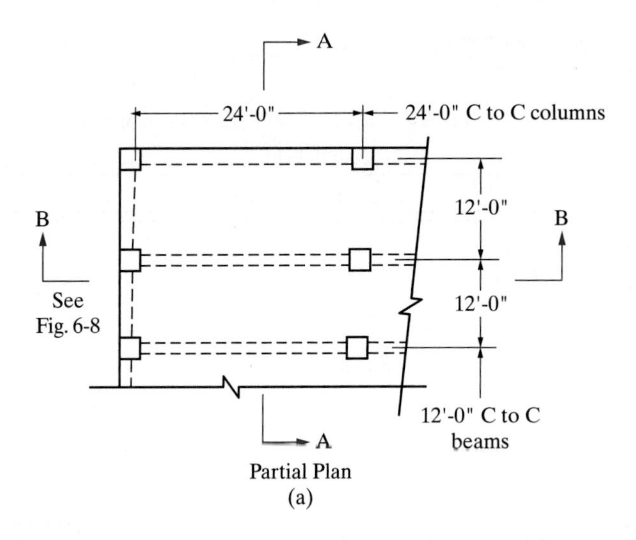

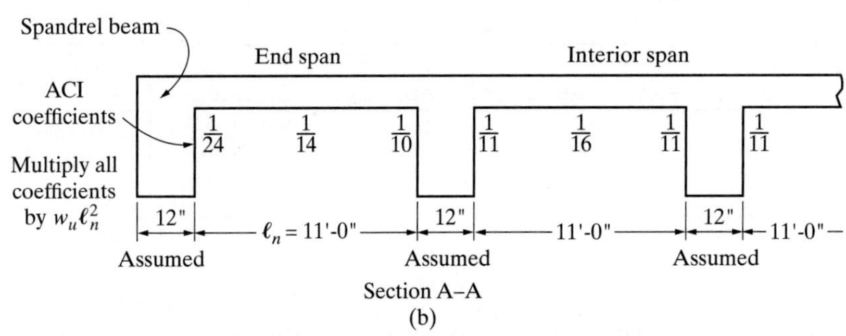

FIGURE 6-4 Sketches for Example 6-1.

Example 6-1

The floor system shown in Figure 6-4 consists of a continuous one-way slab supported by continuous beams. The service loads on the floor are 25 psf dead load (does not include weight of slab) and 250 psf live load. Use $f'_c = 3000$ psi (normal-weight concrete) and $f_y = 60{,}000$ psi. The bars are uncoated.

a. Design the continuous one-way floor slab.
b. Design the continuous supporting beam.

Solution:

The primary difference in this design from previous flexural designs is that, because of continuity, the ACI coefficients and equations will be used to determine design shears and moments. The reader should verify that the three conditions (see Section 6-1) required for the ACI coefficients to be used are indeed satisfied for this problem.

a. **Continuous one-way floor slab (see Section A-A) in Figure 6-4:**

1. Determine the slab thickness: The slab will be designed to satisfy the ACI minimum thickness requirements from Table 7.3.1.1 of the Code and this thickness will be used to estimate slab weight.

   With both ends continuous,

$$\text{minimum } h = \frac{1}{28}\ell_n = \frac{1}{28}(11)(12) = 4.71 \text{ in.}$$

   With one end continuous,

$$\text{minimum } h = \frac{1}{24}\ell_n = \frac{1}{24}(11)(12) = 5.5 \text{ in.}$$

   Try a $5\frac{1}{2}$-in.-thick slab. Design a 12-in.-wide segment ($b = 12$ in.).

2. Determine the load:

$$\text{slab dead load} = \frac{5.5}{12}(150) = 68.8 \text{ psf}$$

$$\text{total dead load} = 25.0 + 68.8 = 93.8 \text{ psf}$$

$$\begin{aligned} w_u &= 1.2w_{DL} + 1.6w_{LL} \\ &= 1.2(93.8) + 1.6(250) \\ &= 112.6 + 400 \\ &= 512.6 \text{ psf} \quad \text{(design load)} \end{aligned}$$

   Because we are designing a slab segment that is 12-in. wide, the foregoing loading is the same as 512.6 lb/ft or 0.513 kip/ft.

3. Determine the moments and shears: Moments are determined using the ACI moment equations. Refer to Figures 6-1 and 6-4. Thus

$$+M_u = \frac{1}{14}w_u\ell_n^2 = \frac{1}{14}(0.513)(11)^2 = 4.43 \text{ ft.-kip}$$

$$+M_u = \frac{1}{16}w_u\ell_n^2 = \frac{1}{16}(0.513)(11)^2 = 3.88 \text{ ft.-kip}$$

$$-M_u = \frac{1}{10}w_u\ell_n^2 = \frac{1}{10}(0.513)(11)^2 = 6.21 \text{ ft.-kip}$$

$$-M_u = \frac{1}{11}w_u\ell_n^2 = \frac{1}{11}(0.513)(11)^2 = 5.64 \text{ ft.-kip}$$

$$-M_u = \frac{1}{24}w_u\ell_n^2 = \frac{1}{24}(0.513)(11)^2 = 2.59 \text{ ft.-kip}$$

Similarly, the shears are determined using the ACI shear equations. In the end span at the face of the first interior support,

$$V_u = 1.15\frac{w_u\ell_n}{2} = 1.15(0.513)\left(\frac{11}{2}\right) = 3.24 \text{ kip}$$

whereas at all other supports,

$$V_u = \frac{w_u\ell_n}{2} = 0.513\left(\frac{11}{2}\right) = 2.82 \text{ kip}$$

4. Design the slab: Using the assumed slab thickness of $5\frac{1}{2}$ in., find the approximate d. Assume No. 5 bars for main steel and $\frac{3}{4}$-in. cover for the bars in the slab. Thus

$$d = 5.5 - 0.75 - 0.31 = 4.44 \text{ in.}$$

5. Design the steel reinforcing: Assume a tension-controlled section ($\varepsilon_t \geq 0.005$) and $\phi = 0.90$. Select the point of maximum moment. This is a negative moment and occurs in the end span at the first interior support, and

$$M_u = \frac{w_u\ell_n^2}{10} = 6.21 \text{ ft.-kip}$$

$$\phi M_n = \phi bd^2\bar{k}$$

Because for design purposes $M_u = \phi M_n$ as a limit, then

$$\text{required } \bar{k} = \frac{M_u}{\phi bd^2} = \frac{6.21(12)}{0.90(12)(4.44)^2}$$

$$= 0.3500 \text{ ksi}$$

From Table A-8,

$$\rho = 0.0063 < \rho_{max} = 0.01355 \qquad \text{(O.K.)}$$

$$\text{required } A_s = \rho bd = 0.0063(12)(4.44)$$

$$A_s = 0.34 \text{ in.}^2$$

As the steel area required at all other points will be less, the preceding process will be repeated for the other points. The expression

$$\text{required } \bar{k} = \frac{M_u}{\phi bd^2}$$

can be simplified because all values are constant except M_u:

$$\text{required } \bar{k} = \frac{M_u(12)}{0.9(12)(4.44)^2} = \frac{M_u}{17.74}$$

where M_u must be in ft-kip. In the usual manner, the required steel ratio ρ and the required steel area A_s may then be determined. The results of these calculations are listed in Table 6-2.

Minimum reinforcement for slabs of constant thickness is the same as that required for shrinkage and temperature reinforcement:

$$\text{minimum required } A_s = 0.0018bh$$

$$= 0.0018(12)(5.5) = 0.12 \text{ in.}^2$$

Also, maximum $\rho = 0.01355$, corresponding to a net tensile strain ε_t of 0.005 (see Table A-8). Therefore, the slab steel requirements for flexure as shown in Table 6-1 are within acceptable limits. The shrinkage and temperature steel may be selected based on the preceding calculation, therefore,

$$\text{use No. 3 bars at 11 in. o.c. } (A_s = 0.12 \text{ in.}^2)$$

Recall that the maximum spacing allowed for the main reinforcement in the slab is 16.5 in. (the smaller of $3h$ or 18 in.) and for temperature and shrinkage reinforcement is the smaller of $5h$ or 18 in. Because $5h = 5(5.5) = 27.5$ in., the 18 in. would control for temperature and shrinkage steel and the spacing is acceptable.

6. Check the shear strength: From step 3, maximum $V_u = 3.24$ kip at the face of the support. A check of shear at the face of the support, rather than at the critical section that is at a distance equal to the effective depth of the member from the face of the

TABLE 6-2 Slab Steel Area Requirements

Location	Moment equation	$\bar{k}$ (ksi)	Required ρ	As (in.^{2}/ft)
End span				
At spandrel	$-\frac{1}{24}w_u\ell_n^2$	0.1460	0.0025	0.13
Midspan	$+\frac{1}{14}w_u\ell_n^2$	0.2497	0.0044	0.24
Interior spans				
Interior support	$-\frac{1}{11}w_u\ell_n^2$	0.3179	0.0057	0.30
Midspan	$+\frac{1}{16}w_u\ell_n^2$	0.2187	0.0038	0.20

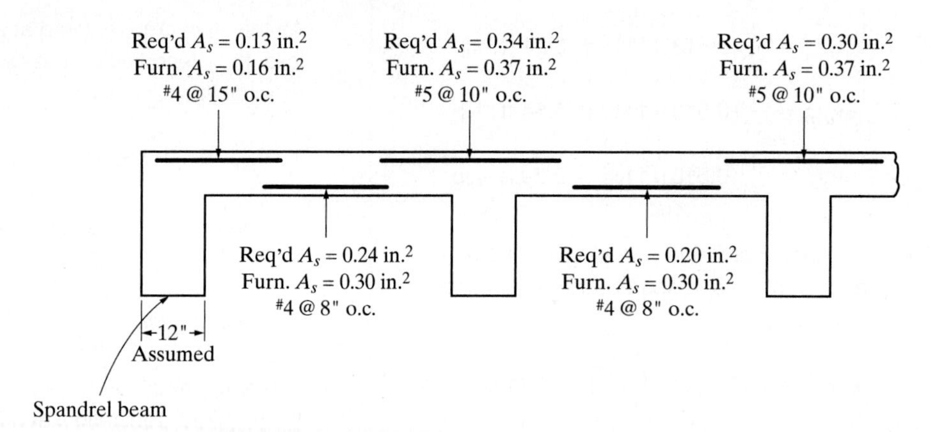

FIGURE 6-5 Work sketch for Example 6-1.

support, is conservative. Slabs are not normally reinforced for shear; therefore,

$$\phi V_n = \phi V_c = \phi 2\sqrt{f'_c}\, b_w d$$

$$= \frac{0.75(2\sqrt{3000})(12)(4.44)}{1000} = 4.38 \text{ kip}$$

$$\phi V_n > V_u$$

Therefore, the thickness is O.K.

7. Select the main steel: The maximum spacing for the main steel in the one-way slab is 16.5 in. (the smaller of $3h$ or 18 in.).

   Using Table A-4, establish a pattern in which the number of bar sizes and the number of different spacings are kept to a minimum. A work sketch (see Figure 6-5) is recommended to establish steel pattern and cutoff points. With regard to the steel selection, note that in the positive moment areas of both the end span and interior span, No. 4 bars at 9 in. could be used. If alternate bars were terminated, however, the spacing of the bars remaining would exceed the maximum spacing of 16.5 in. The use of No. 4 bars at 8 in. avoids this problem. The steel selected is conservative.

8. Check anchorage into the spandrel beam: The steel is No. 4 bars at 15 in. o.c. Refer to the procedure for development length calculation in Section 5-2.

   a. From Table 5-1, $K_D = 82.2$.

   b. Establish values for the factors ψ_t, ψ_e, ψ_s, and λ.

      1. $\psi_t = 1.3$ (the bars are top bars).
      2. The bars are uncoated; $\psi_e = 1.0$.
      3. The bars are No. 4; $\psi_s = 0.8$.
      4. Normal-weight concrete is used; $\lambda = 1.0$.

   c. The product $\psi_t \times \psi_e = 1.3 < 1.7$. (O.K.)

   d. Determine c_b. Based on cover (center of bar to nearest concrete surface),

$$c_b = \frac{3}{4} + \frac{0.5}{2} = 1 \text{ in.}$$

   Based on bar spacing (one-half the center-to-center distance),

$$c_b = \frac{1}{2}(15) = 7.5 \text{ in.}$$

   Therefore, use $c_b = 1.0$ in.

   e. K_{tr} is taken as zero. There is no transverse steel that crosses the potential plane of splitting.

   f. Check $(c_b + K_{tr})/d_b \le 2.5$:

$$\frac{c_b + K_{tr}}{d_b} = \frac{1.0 + 0}{0.5} = 2.0 < 2.5 \qquad \text{(O.K.)}$$

   g. Calculate the excess reinforcement factor:

$$K_{ER} = \frac{A_s \text{ required}}{A_s \text{ provided}} = \frac{0.130}{0.160} = 0.813$$

   h. Calculate ℓ_d:

$$\ell_d = \frac{K_D}{\lambda}\left[\frac{\psi_t\psi_e\psi_s}{\left(\dfrac{c_b + K_{tr}}{d_b}\right)}\right]K_{ER}d_b$$

$$= \frac{82.2}{1.0}\left[\frac{1.3(1.0)(0.8)}{2.0}\right](0.813)(0.5)$$

$$= 17.4 \text{ in.} > 12 \text{ in.} \qquad \text{(O. K.)}$$

   Use $\ell_d = 18$ in. (minimum).

9. Because the 18-in. length cannot be furnished, a hook will be provided. Determine if a 180° standard hook will be adequate.

   a. Calculate ℓ_{dh}.

$$\ell_{dh} = \left(\frac{0.02\psi_e f_y}{\lambda\sqrt{f'_c}}\right)d$$

   b. The bars are uncoated and the concrete is normal weight. Therefore, ψ_e and λ are both 1.0.

   c. Modification factors are as follows:

      1. Assume the concrete side cover is $2\frac{1}{2}$ in. normal to the plane of the hook; use 0.7.
      2. For excess steel, use

$$\frac{A_s \text{ required}}{A_s \text{ provided}} = \frac{0.13}{0.16} = 0.813$$

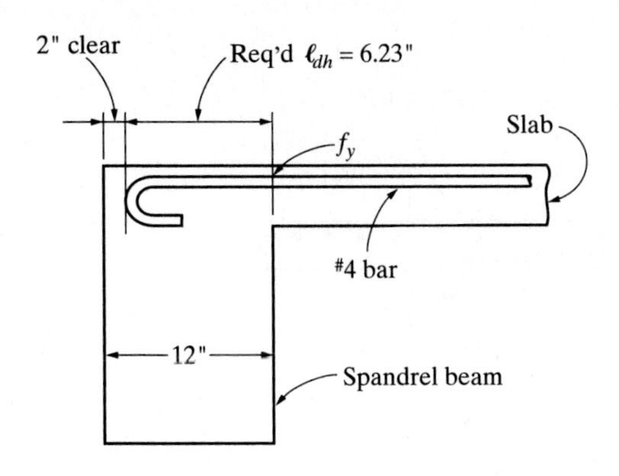

FIGURE 6-6 Hook detail for Example 6-1.

d. Therefore, the required development length is

$$\ell_{dh} = \left(\frac{0.02(1.0)(60{,}000)}{(1.0)\sqrt{3000}}\right)(0.50)(0.7)(0.813)$$
$$= 6.23 \text{ in.}$$

Minimum ℓ_{dh} is 6 in. or $8d_b$, whichever is greater:

$$8d_b = 8\left(\frac{1}{2}\right) = 4 \text{ in.}$$

Therefore, the minimum is 6 in.:

$$6.23 \text{ in.} > 6 \text{ in.} \qquad \text{(O.K.)}$$

e. Check the total width of beam required at the discontinuous end (see Figure 6-6):

$$6.23 + 2 = 8.23 \text{ in.} < 12 \text{ in.} \qquad \text{(O.K.)}$$

10. Determine the bar cutoff points: For the normal type of construction for which the typical bar cutoff points shown in Figure 6-3 are used, the cutoff points are located so that all bars terminate in compression zones. Thus, the requirements of the ACI Code, Section 7.7.3.5, need not be checked, and the recommended bar cutoff points, as shown in Figure 6-3, are used.
11. Prepare the design sketches: The final design sketch for the slab is shown in Figure 6-7. For clarity, the interior and end spans are shown separately.

b. **Continuous beam supporting the one-way slab (see Section B-B in Figures 6-4a and 6-8):** The second part of Example 6-1 involves the design of the continuous supporting beam. From Figure 6-4a, it is seen that these beams span between columns. The ACI coefficients to be used for moment determination are shown in Figure 6-8.

1. Determine the loading:

$$\text{service live load} = 250 \text{ psf} \times 12 = 3000 \text{ lb/ft}$$
$$\text{service dead load} = 25 \text{ psf} \times 12 = 300 \text{ lb/ft}$$
$$\text{weight of slab} = \left(\frac{5.5}{12}\right)(150)(12) = 825 \text{ lb/ft}$$

Assuming a beam width of 12 in. and an overall depth of 30 in. for purposes of member weight estimate (see Figure 6-9),

$$\text{weight of beam} = \frac{12(30 - 5.5)}{144}(150)$$
$$= 306.3 \text{ lb/ft}$$
$$\text{total service live load} = 3000 \text{ lb/ft}$$
$$\text{total service dead load} = 1431.3 \text{ lb/ft} \quad \text{say, } 1431 \text{ lb/ft}$$

2. Calculate the design load:

$$w_u = 1.2w_{DL} + 1.6w_{LL}$$
$$= 1.2(1431) + 1.6(3000)$$
$$= 1717 + 4800$$
$$= 6517 \text{ lb/ft} \quad \text{say, } 6.5 \text{ kip/ft}$$

The loaded beam is depicted in Figure 6-10.

3. Calculate the design moments and shears. The design moments and shears are calculated by using the ACI equations:

$$+M_u = \frac{1}{14}w_u\ell_n^2 = \frac{1}{14}(6.5)(22.67)^2 = 238.6 \text{ ft.-kip}$$
$$+M_u = \frac{1}{16}w_u\ell_n^2 = \frac{1}{16}(6.5)(22.67)^2 = 208.8 \text{ ft.-kip}$$
$$-M_u = \frac{1}{10}w_u\ell_n^2 = \frac{1}{10}(6.5)(22.67)^2 = 334.1 \text{ ft.-kip}$$
$$-M_u = \frac{1}{11}w_u\ell_n^2 = \frac{1}{11}(6.5)(22.67)^2 = 303.7 \text{ ft.-kip}$$
$$V_u = \frac{w_u\ell_n}{2} = \frac{6.5(22.67)}{2} = 73.7 \text{ kip}$$
$$V_u = 1.15\frac{w_u\ell_n}{2} = 1.15(6.5)\left(\frac{22.67}{2}\right) = 84.7 \text{ kip}$$

4. Design the continuous beam: Establish concrete dimensions based on the maximum bending moment. This occurs in the end span at the first interior support where the negative moment $M_u = w_u\ell_n^2/10$. Since the top of the beam is in tension, the design will be that of a rectangular beam.
   a. Maximum moment (negative) = 334.1 ft.-kip.
   b. From Table A-5, assume that $\rho = 0.0090$ (which is less than ρ_{max} of 0.01355 from Table A-8). A check of minimum steel required will be made shortly.
   c. From Table A-8, $\bar{k} = 0.4828$ ksi.
   d. Assume that $b = 12$ in.:

$$\text{required } d = \sqrt{\frac{M_u}{\phi b\bar{k}}} = \sqrt{\frac{334.1(12)}{0.9(12)(0.4828)}}$$
$$= 28.0 \text{ in.}$$
$$d/b \text{ ratio} = \frac{28.0}{12} = 2.34, \quad \text{which is within the acceptable range}$$

   e. Check the estimated beam weight assuming one layer of No. 11 bars and No. 3 stirrups:

$$\text{required } h = 28.0 + \frac{1.41}{2} + 0.38 + 1.5$$
$$= 30.6 \text{ in.}$$

Use $h = 31$ in. with an assumed d of 28 in. Also, check the minimum h from the ACI Code, Table 9.3.1.1:

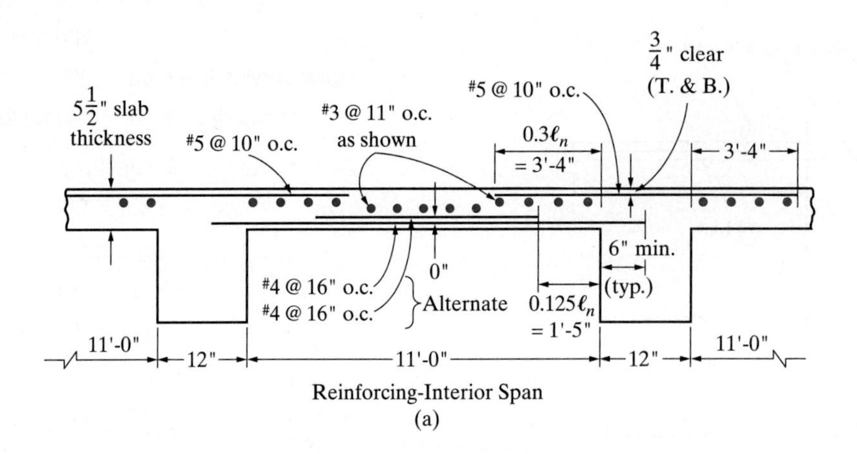

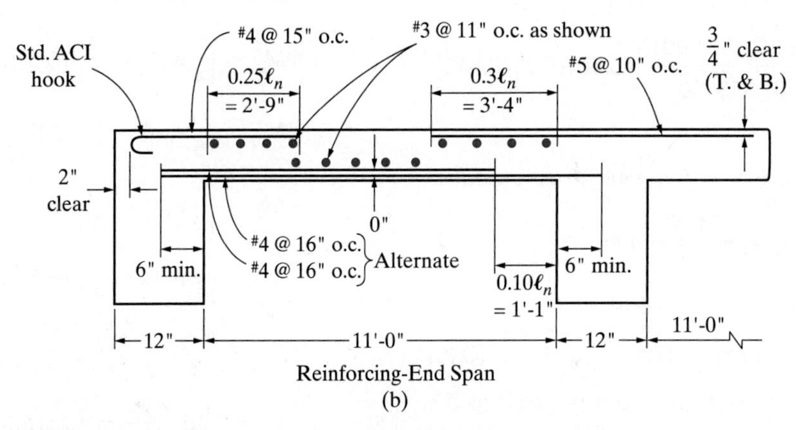

FIGURE 6-7 Design sketches for Example 6-1.

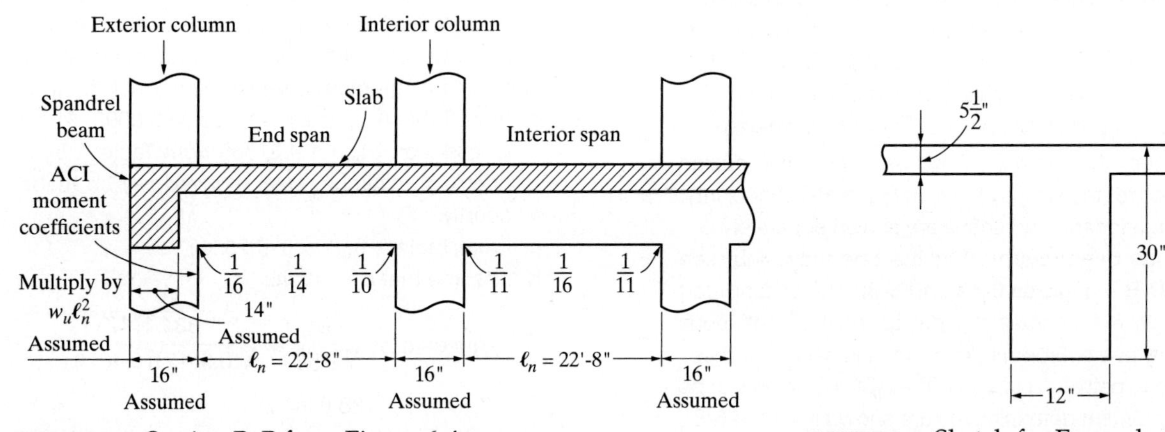

FIGURE 6-8 Section B–B from Figure 6-4a.

FIGURE 6-9 Sketch for Example 6-1.

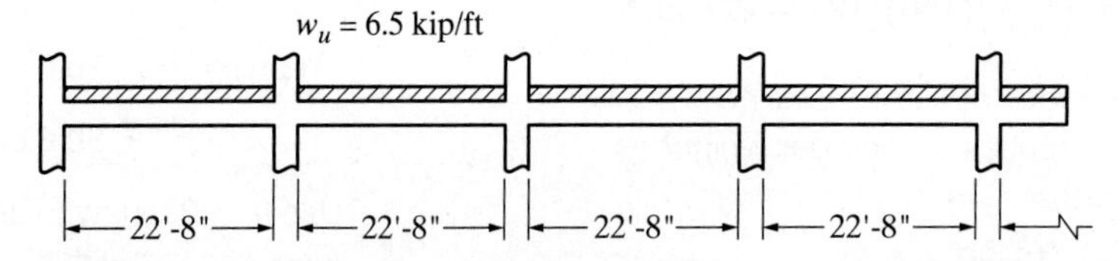

FIGURE 6-10 Beam design load and spans.

$$\text{minimum } h = \frac{1}{18.5}(22.67)(12)$$
$$= 14.7 \text{ in.} < 31 \text{ in.} \quad \text{(O.K.)}$$

Note that the estimated beam weight based on $b = 12$ in. and $h = 30$ in. is slightly on the low side but may be considered acceptable.

f. Design the steel reinforcing for points of negative moment as follows:

At the negative moment locations, the compression zone is at the bottom of the beam, and therefore, the width of the compression zone in these regions is equal to the width of the beam web, $b_w = 12$ in.

1. At the first interior support, based on an assumed ρ of 0.0090 (see the preceding step 4),

$$\text{required } A_s = \rho bd$$
$$= 0.0090(12)(28) = 3.0 \text{ in.}^2$$

From Table A-5,

$$A_{s,\min} = 0.0033(12)(28) = 1.11 \text{ in.}^2$$
$$3.0 \text{ in.}^2 > 1.11 \text{ in.}^2 \quad \text{(O.K.)}$$

2. At the other interior supports, $-M_u =$ 303.7 ft.-kip:

$$\text{required } \bar{k} = \frac{M_u}{\phi bd^2} = \frac{303.7(12)}{0.9(12)(28)^2}$$
$$= 0.4304 \text{ ksi}$$

From Table A-8, $\rho = 0.0079$. Check ρ with $\rho_{\max}$:

$$0.0079 < 0.01355 \quad \text{(O.K.)}$$

Therefore,

$$\text{required } A_s = \rho bd$$
$$= 0.0079(12)(28) = 2.65 \text{ in.}^2$$

Check the required A_s with $A_{s,\min}$. From Table A-5,

$$A_{s,\min} = 0.0033(12)(28) = 1.11 \text{ in.}^2$$
$$2.65 \text{ in.}^2 > 1.11 \text{ in.}^2 \quad \text{(O.K.)}$$

3. At the exterior support (exterior column), $-M_u = 208.8$ ft.-kip:

At the support, the width of the compression zone, $b = b_w = 12''$

$$\text{required } \bar{k} = \frac{M_u}{\phi bd^2} = \frac{208.8(12)}{0.9(12)(28)^2}$$
$$= 0.2959 \text{ ksi}$$

From Table A-8, $\rho = 0.0053$. Check ρ with $\rho_{\max}$:

$$0.0053 < 0.01355 \quad \text{(O.K.)}$$

Therefore,

$$\text{required } A_s = \rho bd$$
$$= 0.0053(12)(28) = 1.78 \text{ in.}^2$$

Check the required A_s with $A_{s,\min}$. From Table A-5,

$$A_{s,\min} = 0.0033(12)(28) = 1.11 \text{ in.}^2$$
$$1.78 \text{ in.}^2 > 1.11 \text{ in.}^2 \quad \text{(O.K.)}$$

g. Design steel reinforcing for points of positive moment as follows:

At points of positive moment, the top of the beam is in compression; therefore, the design will be that of a T-beam and the width of the compression zone is equal to the effective width of the T-beam.

1. End-span positive moment:
   a. Design moment $= 238.6$ ft.-kip $= M_u$.
   b. Effective depth $d = 28$ in. (see negative moment design).
   c. Effective flange width:

$$b_w + \frac{1}{4}\text{span length} = 12 + 0.25(22.67)(12) = 80 \text{ in.}$$
$$b_w + 16h_f = 12 + 16(5.5) = 100 \text{ in.}$$
$$\text{beam spacing} = 144 \text{ in.}$$

Use an effective flange width $b = 80$ in. (see Figure 6-11).

   d. Assuming total flange in compression, $\varepsilon_t = 0.005$, and $\phi = 0.90$:

$$\phi M_{nf} = \phi(0.85f'_c)bh_f\left(d - \frac{h_f}{2}\right)$$
$$= 0.9(0.85)(3)(80)(5.5)\left(\frac{28 - 5.5/2}{12}\right)$$
$$= 2125 \text{ ft.-kip}$$

   e. Because $2125 > 238.6$, the member behaves as a wide rectangular T-beam with $b = 80$ in. and $d = 28$ in.
   f. $\text{required } \bar{k} = \frac{M_u}{\phi bd^2}$

$$= \frac{238.6(12)}{0.9(80)(28)^2} = 0.0507 \text{ ksi}$$

   g. From Table A-8

$$\text{required } \rho = 0.0010$$

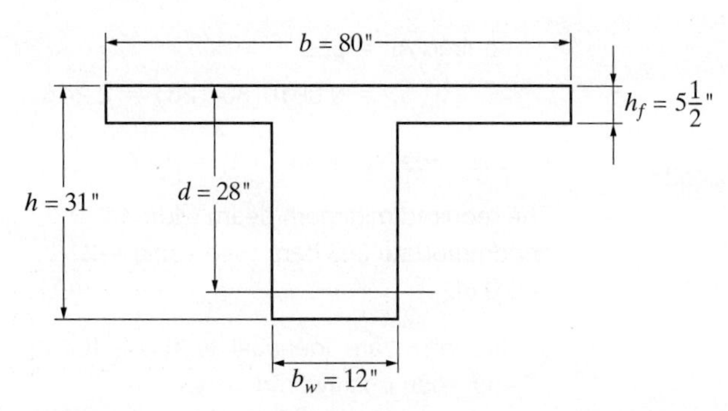

FIGURE 6-11 Beam cross section.

h. The required steel area is

$$\text{required } A_s = \rho bd$$
$$= 0.0010(80)(28) = 2.24 \text{ in.}^2$$

i. Use three No. 8 bars ($A_s = 2.37$ in.^{2}):

The required minimum beam width to accommodate 3#8 bars (see Table A-3) = 9.0 in. (O.K.)

j. Check d. With a No. 3 stirrup and $1\frac{1}{2}$-in. cover,

$$d = 31 - 1.5 - 0.38 - 1.00/2$$
$$= 28.6 \text{ in.} > 28 \text{ in.} \quad \text{(O.K.)}$$

k. Check the required A_s with $A_{s,min}$. From Table A-5,

$$A_{s,min} = 0.0033(12)(28) = 1.11 \text{ in.}^2$$
$$2.37 \text{ in.}^2 > 1.11 \text{ in.}^2 \quad \text{(O.K.)}$$

i. Check ε_t and ϕ. From Table A-8, with $\rho = 0.001$, $\varepsilon_t \geq 0.005$. Therefore, the assumed ϕ of 0.90 is O.K.

2. Interior span positive moment:
   a. Design moment = 208.8 ft.-kip = M_u.
   b. through (e) See the end-span positive moment computations. At the positive moment location, the width of the compression zone is equal to the effective width of the T-beam. Therefore, use an effective flange width $b = 80$ in. and an effective depth $d = 28$ in. Also, for total flange in compression, $\phi M_n = 2125$ ft.-kip $> M_u$. Therefore, this member also behaves as a rectangular T-beam.
   c.

$$\text{required } \bar{k} = \frac{M_u}{\phi bd^2}$$
$$= \frac{208.8(12)}{0.9(80)(28)^2} = 0.0444 \text{ ksi}$$

   d. From Table A-8

$$\text{required } \rho = 0.0010$$

   e. The required steel area is

$$\text{required } A_s = \rho bd$$
$$= 0.0010(80)(28) = 2.24 \text{ in.}^2$$

   f. Use three No. 8 bars ($A_s = 2.37$ in.^{2}):

The required minimum beam width to accommodate 3#8 bars (see Table A-3) = 9.0 in. (O.K.)

   g. through i. are identical to those for the end-span positive moment.

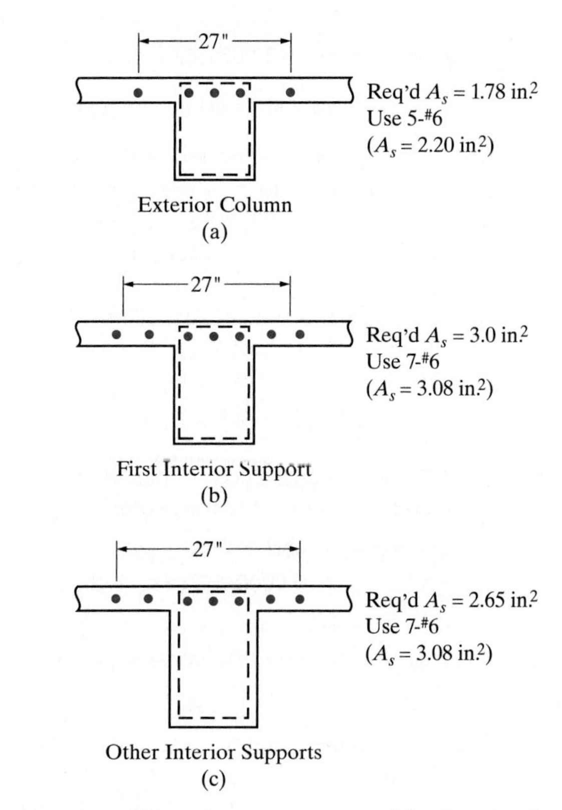

FIGURE 6-12 Negative moment steel for beam of Example 6-1.

5. Check the distribution of negative moment steel: The ACI Code, Section 24.3.4, requires that where flanges are in tension, a part of the main tension reinforcement be distributed over the effective flange width or a width equal to one-tenth of the span, whichever is smaller. The use of smaller bars spread out into part of the flange will also be advantageous where a beam is supported by a spandrel girder or exterior column and the embedment length for the negative moment steel is limited. Thus

$$\frac{\text{span}}{10} = \frac{22.67(12)}{10} = 27 \text{ in.}$$
$$\text{effective flange width} = b = 68 \text{ in.}$$

Therefore, distribute the negative moment bars over a width of 27 in. Figure 6-12 shows suitable bars and patterns to use to satisfy the foregoing code requirement and furnish the cross-sectional area of steel required for flexure.

The ACI Code also stipulates that if the effective flange width exceeds one-tenth of the span, some longitudinal reinforcement shall be provided in the outer portions of the flange. In this design no additional steel will be furnished. In the authors' opinion, this requirement is satisfied by the slab temperature and shrinkage steel (see Figures 6-7 and 6-18).

6. Prepare the work sketch. A work sketch is developed in Figure 6-13, which includes the bars previously chosen.

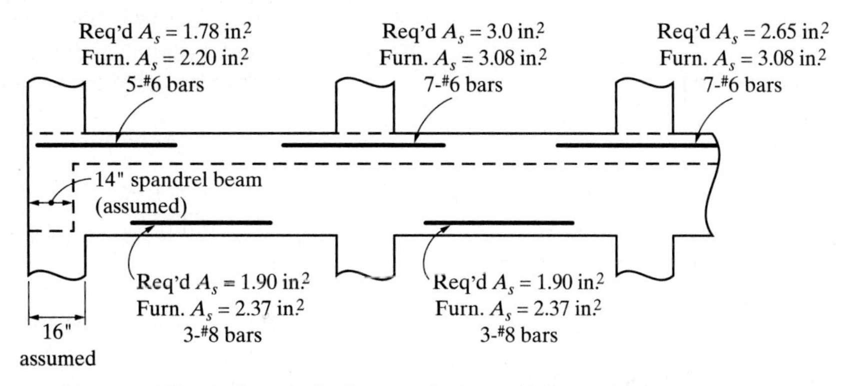

FIGURE 6-13 Work sketch for beam design of Example 6-1.

7. Check the anchorage or development of the beam top reinforcement into the exterior column:
   a. From Table 5-1, $K_D = 82.2$.
   b. Establish values for the factors ψ_t, ψ_e, ψ_s, and λ.
      1. $\psi_t = 1.3$ (the bars are top bars).
      2. The bars are uncoated; $\psi_e = 1.0$.
      3. The bars are No. 6; $\psi_s = 0.8$.
      4. Normal-weight concrete is used; $\lambda = 1.0$.
   c. The product $\psi_t \times \psi_e = 1.3 < 1.7$. (O.K.)
   d. Determine c_b. Based on cover (center of bar to nearest concrete surface), assume $1\frac{1}{2}$-in. clear cover and a No. 3 stirrup:

$$c_b = 1.5 + 0.375 + \frac{0.75}{2} = 2.25 \text{ in.}$$

   Based on bar spacing (one-half the center-to-center distance), refer to Figures 6-12. Consider the three No. 6 top bars in the beam that are located within the No. 3 closed loop stirrup in Figure 6-12a. Here

$$c_b = \left[\frac{12 - 2(1.5) - 2(0.375) - 0.75}{2}\right]\left(\frac{1}{2}\right)$$
$$= 1.875 \text{ in.}$$

   Therefore, use $c_b = 1.875$ in.
   e. K_{tr} may be conservatively taken as zero.
   f. Check $(c_b + K_{tr})/d_b \leq 2.5$:

$$\frac{c_b + K_{tr}}{d_b} = \frac{1.875 + 0}{0.75} = 2.5 (\leq 2.5) \quad \text{(O.K.)}$$

   g. Determine the excess reinforcement factor:

$$K_{ER} = \frac{A_s \text{ required}}{A_s \text{ provided}} = \frac{1.78}{2.20} = 0.809$$

   h. Calculate ℓ_d:

$$\ell_d = \frac{K_D}{\lambda}\left[\frac{\psi_t\psi_e\psi_s}{\left(\frac{c_b + K_{tr}}{d_b}\right)}\right] d_b K_{ER}$$
$$= \frac{82.2}{1.0}\left[\frac{1.3(1.0)(0.8)}{2.5}\right](0.75)(0.809)$$
$$= 20.7 \text{ in.} > 12 \text{ in.} \quad \text{(O.K.)}$$

   Use $\ell_d = 21$ in. (minimum). This is the required straight bar development length.

   With 2.0 in. clear at the end of the bar, the embedment length available in the column is $16.0 - 2.0 = 14.0$ in.

8. Because 21 in. > 14.0 in., a hook is required. Determine if a 90° standard hook will be adequate.
   a. From Table A-13, $\ell_{dh} = 16.4$ in.
   b. Modification factors (MF) to be used are
      1. Assume concrete cover $\geq 2\frac{1}{2}$ in. and cover on the bar extension beyond the hook = 2 in.; use 0.7.
      2. For excess steel, use

$$\frac{\text{required } A_s}{\text{provided } A_s} = \frac{1.78}{2.20} = 0.809$$

   c. The required development length for the hook is

$$\ell_{dh} = 16.4(0.7)(0.809) = 9.29 \text{ in.}$$

   Minimum ℓ_{dh} is 6 in. or $8d_b$, whichever is greater:

$$8d_b = 8\left(\frac{3}{4}\right) = 6 \text{ in.}$$

$$9.29 \text{ in.} > 6 \text{ in.} \quad \text{(O.K.)}$$

   d. Check the total width of column required (see Figure 6-14):

$$9.29 + 2 = 11.3 \text{ in.} < 16 \text{ in.} \quad \text{(O.K.)}$$

For other points along the continuous beam, use bar cutoff points recommended in Figure 6-3 and as shown in Figure 6-18.

9. Prepare the stirrup design for the beam: Established values are $b_w = 12$ in., effective depth $d = 28$ in., $f'_c = 3000$ psi, and $f_y = 60{,}000$ psi.
   a. The shear force V_u diagram may be observed in Figure 6-15. Note that the shear diagram is unsymmetrical with respect to the centerline of the span, and the point of zero shear does not occur at the midspan of the beam. Furthermore, the point of zero shear for the right-hand section of the end span of the beam occurs at 13.03 ft. (that is, 84.7 kip/w_u = 84.7/6.5) from the face

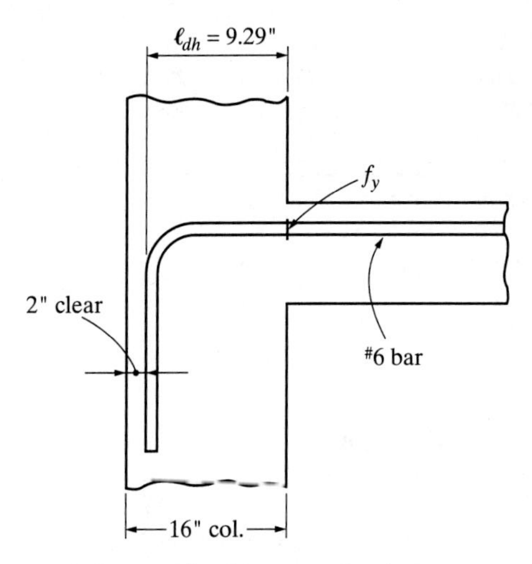

FIGURE 6-14 Anchorage at column.

of the first interior support. It should be noted that the point of zero shear for the left-hand section of the beam occurs at 11.34 ft. (that is, 73.7 kip/6.5) from the interior face of the end support. The points of zero shears for both sections of the beam occur at different locations because of the approximate nature of the ACI shear coefficients as pointed out earlier. The stirrup design will be based on shear in the interior portion of the end span where the maximum values occur. The resulting stirrup pattern will be used throughout the continuous beam. Only the applicable portion of the V_u diagram is shown.

b. Determine if stirrups are required:

$$\phi V_c = \phi 2\sqrt{f'_c}\, b_w d = \frac{0.75(2\sqrt{3000})(12)(28)}{1000}$$
$$= 27.6 \text{ kip}$$
$$\frac{1}{2}\phi V_c = \frac{1}{2}(27.6) = 13.8 \text{ kip}$$

At the critical section d distance (28 in.) from the face of the support,

$$V_u^* = 84.7 - \frac{28}{12}(6.5) = 69.5 \text{ kip}$$

(Quantities at the critical section are designated with an asterisk.) Stirrups are required because

$$V_u^* > \frac{1}{2}\phi V_c \quad (\textit{i.e.}\ 69.5 \text{ kip} > 13.8 \text{ kip})$$

c. Find the length of span over which stirrups are required. Stirrups are required to the point where

$$V_u = \frac{1}{2}\phi V_c = 13.8 \text{ kip}$$

From Figure 6-15 and referencing from the face of the support, V_u = 13.8 kip at

$$\frac{84.7 - 13.8}{6.5} = 10.91 \text{ ft}$$

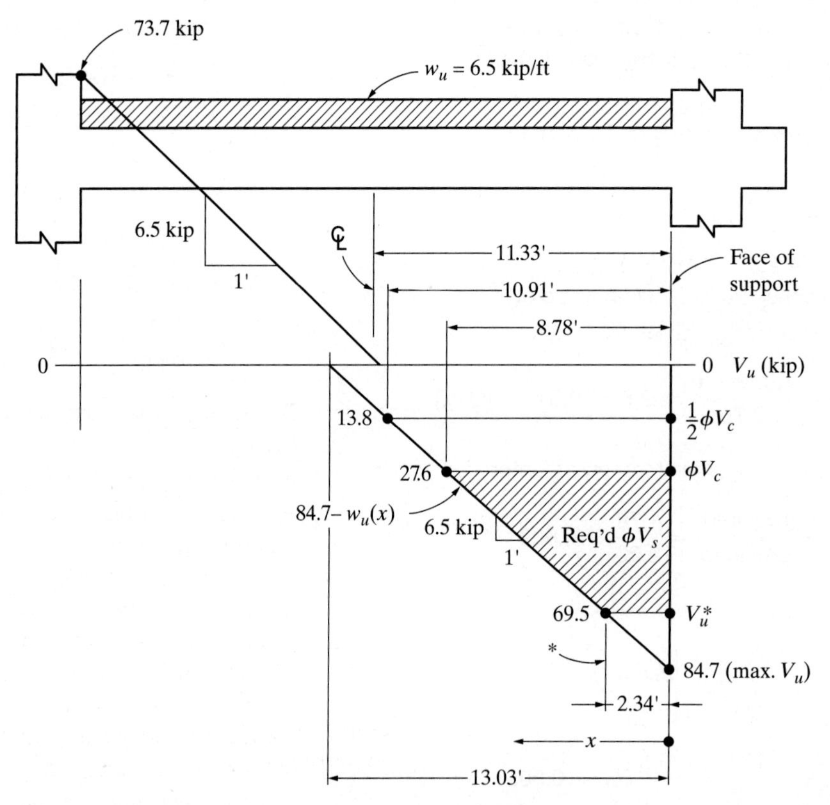

FIGURE 6-15 V_u diagram for stirrup design (Example 6-1).

The distance from the face of the support to where $V_u = \phi V_c = 27.6$ kip is

$$\frac{84.7 - 27.6}{6.5} = 8.78 \text{ ft}$$

d. On the V_u diagram, designate the area between the ϕV_c line, the V_u^* line, and the sloping V_u line as "Req'd ϕV_s." At locations between 2.34 ft and 8.78 ft from the face of the support, the required ϕV_s varies. Designating the slope of the V_u diagram as m (kip/ft) and taking x (ft) from the face of the support ($2.34 \leq x \leq 8.78$) yields

$$\begin{aligned}\text{required } \phi V_s &= \text{maximum } V_u - \phi V_c - mx\\ &= 84.7 - 27.6 - 6.5x\\ &= 57.1 - 6.5x\end{aligned}$$

e. Assume a No. 3 vertical stirrup ($A_v = 0.22$ in.^{2}):

$$\begin{aligned}\text{required } s^* &= \frac{A_v f_{yt} d}{V_s}\\ &= \frac{\phi A_v f_{yt} d}{\text{required } \phi V_s^*}\\ &= \frac{\phi A_v f_{yt} d}{V_u^* - \phi V_c}\\ &= \frac{0.75(0.22)(60)(28)}{69.527.6} = 6.62 \text{ in.}\end{aligned}$$

Use $6\frac{1}{2}$-in. spacing between the critical section and the face of the support.

f. Establish ACI Code maximum spacing requirements:

$$4\sqrt{f'_c}\, b_w d = \frac{4\sqrt{3000}\,(12)(28)}{1000} = 73.6 \text{ kip}$$

Calculating V_s^* at the critical section yields

$$\begin{aligned}\phi V_s^* &= V_u^* - \phi V_c\\ &= 69.5 - 27.6 = 41.9 \text{ kip}\end{aligned}$$

$$V_s^* = \frac{\phi V_s^*}{\phi} = \frac{41.9}{0.75} = 55.9 \text{ kip}$$

Because 55.9 kip < 73.6 kip, the maximum spacing should be the smaller of $d/2$ or 24 in.:

$$\frac{d}{2} = \frac{28}{2} = 14 \text{ in.}$$

Also check:

$$s_{max} = \frac{A_v f_{yt}}{0.75\sqrt{f'_c}\, b_w} \leq \frac{A_v f_{yt}}{50 b_w}$$

$$\frac{A_v f_{yt}}{0.75\sqrt{f'_c}\, b_w} = \frac{0.22(60{,}000)}{0.75\sqrt{3000}(12)} = 26.7 \text{ in.}$$

and

$$\frac{A_v f_{yt}}{50 b_w} = \frac{0.22(60{,}000)}{50(12)} = 22.0 \text{ in.}$$

Therefore, use a maximum spacing of 14 in.

g. Determine the spacing requirements based on shear strength to be furnished: The denominator of the following formula for required spacing uses the expression for required ϕV_s from step d:

$$\begin{aligned}\text{required } s &= \frac{\phi A_v f_{yt} d}{\text{required } \phi V_s}\\ &= \frac{0.75(0.22)(60)(28)}{57.1 - 6.5x} = \frac{277.2}{57.1 - 6.5x}\end{aligned}$$

The results for several arbitrary values of x are shown tabulated and plotted in Figure 6-16.

h. Using Figure 6-16, the stirrup pattern shown in Figure 6-17 is developed. Despite the lack of symmetry in the shear diagram, the stirrup pattern is symmetrical with respect to the centerline of the span. This is conservative and will be used for all spans.

The design sketches are shown in Figure 6-18. As with the slab design, the typical bar cutoff points of Figure 6-3 are used for this beam. All bars, therefore, terminate in compression zones, and the requirements of the ACI Code, Section 9.7.3.5, need not be checked. Although the beam reinforcement obtained from the preceding design has been presented in the form of sketches (see Figure 6-18), in practice, the beam reinforcement is also presented in a tabular format as a beam schedule. An example of a beam schedule is shown in Figures 13-2 and 13-3.

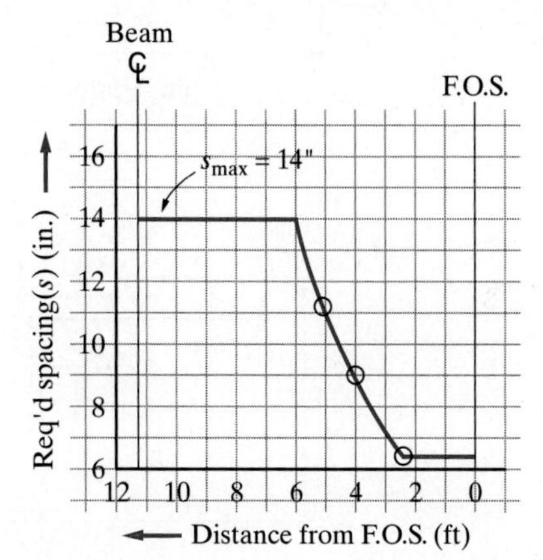

x (ft)	Req'd s (in.)
2.34	6.62
4	8.91
5	11.27
6	15.3

FIGURE 6-16 Stirrup spacing requirements for Example 6-1.

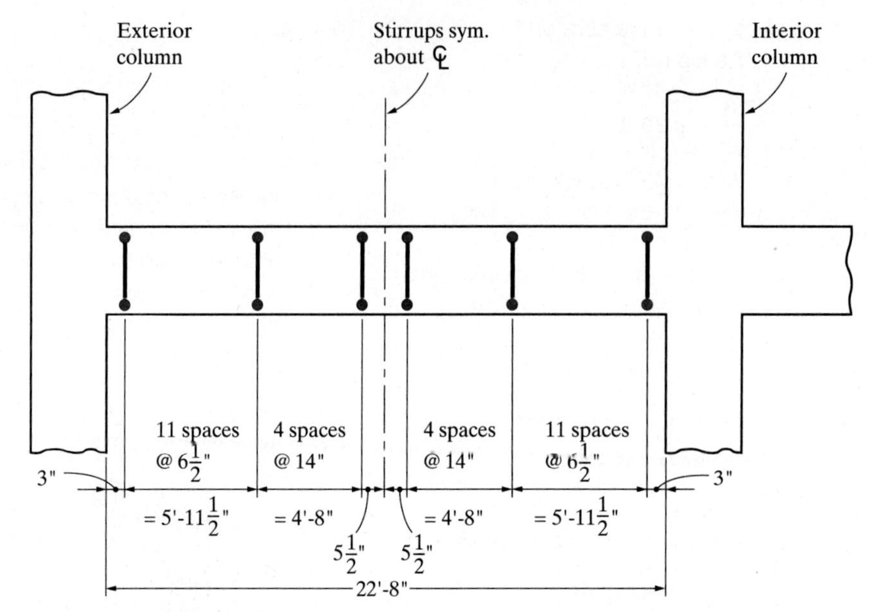

FIGURE 6-17 Stirrup spacing for Example 6-1 end span (interior spans similar).

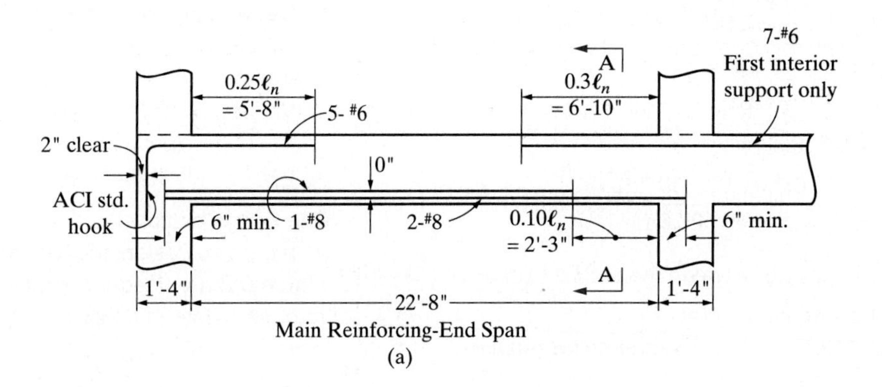

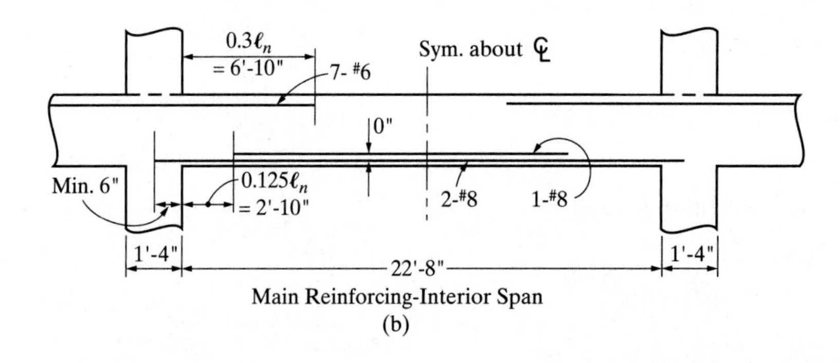

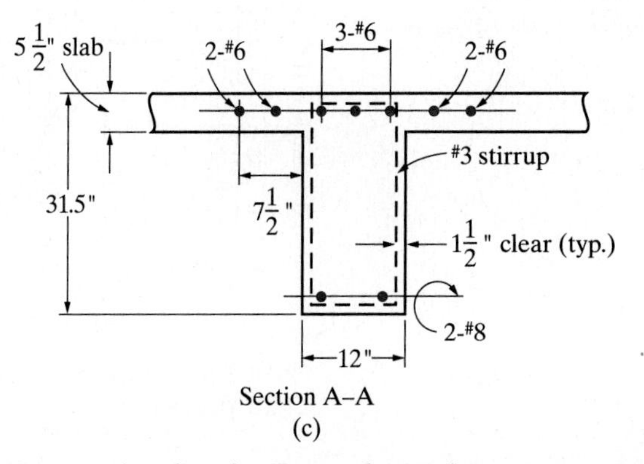

FIGURE 6-18 Design sketches for Example 6-1 (stirrup spacings not shown).

6-4 ANALYSIS AND DESIGN OF CONTINUOUS TWO-WAY SLABS

As discussed in Chapter 2, two-way slabs are horizontal or planer structural elements that support gravity loads by bending in two orthogonal directions, and for which the ratio of the long side to the short side (ℓ_1/ℓ_2) of the rectangular panel is less than 2. These slabs can be supported on all four sides of a panel by walls or beams, or directly by columns at the corners of the panel. In edge-supported two-way slabs (see Figure 1-12), the uniform load is is shared between the two orthogonal directions of the slab panel, and the shorter direction of the panel supports a higher proportion of the load (see Section 14-10). In contrast, for column-supported two-way slabs, 100% of the uniform load on the slab is supported in each orthogonal direction. The more common type of two-way slabs are those supported by columns. There are a variety of column-supported two-way slab systems, and the basic types that are commonly used are illustrated in Figure 6-19. A flat plate is the simplest type and consists of a slab that is supported directly on columns with or without column capitals. A flat slab is a flat plate with drop panels with or without column capitals, and is stronger than a flat plate because of the drop panel or slab thickening at the columns that increases the shear capacity of the slab. Drop panel thickness varies from 2¼-in. to 8-in. based on the size of plyform and sawn lumber that are used for forming the slab and drop panel. A flat slab with beams is a two-way slab with beams between the columns in the two orthogonal directions. A waffle slab is another variation of a two-way slab with a waffle-type soffit; they are relatively stiff and are recommended for floors that are subject to floor vibrations, but they have higher installation costs due to the complicated forming required. Flat plates and flat slabs are the more widely used two-way slabs in high-rise building construction. Square bay sizes are more economical than rectangular bay sizes. A comparison of of the practical range of sizes of flat plates and flat slabs is shown in Table 6-3 [3].

There are several advantages of a two-way slab system and one of them is the consistently flat soffit that allows for a cleaner installation of utility lines and other items hung from the underside of the slab. They are also widely used in high-rise building construction because of the reduction in headroom that is achievable with flat plates and

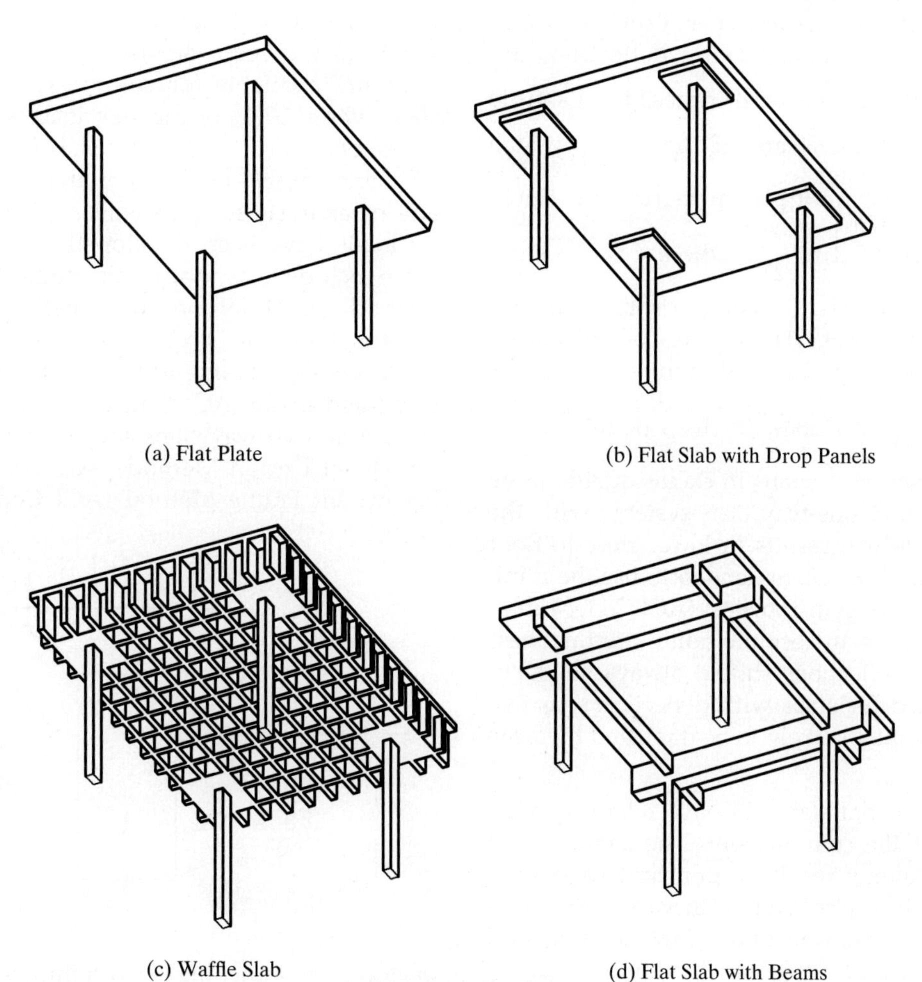

FIGURE 6-19 Types of two-way slabs.

TABLE 6-3 Practical Range of Sizes for Flat Plates and Flat Slabs [3]

Flat plate		Flat slab	
SDL = 20 psf L = 50 psf	SDL = 20 psf L = 100 psf	SDL = 20 psf L = 50 psf	SDL = 20 psf L = 100 psf
15 ft × 15 ft bay size 6 in. thick slab- 14 in. square column	15 ft × 15 ft bay size 6½ in. thick slab- 14 in. square column	20 ft × 20 ft bay size 7 in. thick slab 2¼ in. deep drop panel 20 in. square column	20 ft × 20 ft bay size 7 in. thick slab 2¼ in. deep drop panel 22 in. square column
30 ft × 30 ft bay size 12 in. thick slab- 32 in. square column	30 ft × 30 ft bay size 14 in. thick slab- 34 in. square column	35 ft × 35 ft bay size 12 in. thick slab 4¼ in. deep drop panel 38 in. square column	35 ft × 35 ft bay size 12 in. thick slab 6¼ in. deep drop panel 40 in. square column

[1]Assumes f'_c of 4000 psi and f_y of 60,000 psi

flat slabs which leads to a reduction in the overall height of the building, and hence a reduction in the overall cost of the project. Two-way slabs are often used in hospitals where a fair amount of medical equipment are hung from the underside of the structure. One of the more significant advantages of a two-way slab system is that a higher span-to-depth ratio can be achieved compared to one-way slabs since the load distribution occurs in two orthogonal directions. Per Table 6-1 (ACI Code Table 7.3.1.1) and assuming deflections are not calculated, the minimum thickness for a simple span one-way slab is $\frac{\ell}{20}$ and if the one-way slab is continuous over multiple supports, the thickness can be decreased to $\frac{\ell}{28}$, where ℓ is the clear span between supports. By comparison, Table 6-4 (ACI Code Table 8.3.1.1) requires a minimum thickness for two-way slabs without drop panels of $\frac{\ell}{30}$, and $\frac{\ell}{36}$ for a slab with drop panels. Thus, two-way slabs can have spans in excess of 30% more than the equivalent one-way slab systems with the same thickness, which results in lower floor-to-floor heights. Note that it is recommended to use the minimum thicknesses given in Table 6-1 (ACI Code, Tables 7.3.1.1 and 8.3.1.1) as an absolute minimum thickness of the slab, and deflections should always be calculated and checked for slabs with deflection sensitive elements (i.e., elements likely to be damaged by large deflections).

A flat plate or flat slab will have relatively high shear stresses at the columns since the concentrated reaction at the column results in punching (two-way) shear stress which results in a tendency for the column to punch through the slab under high loading (see Figure 6-20). The use of a drop panel, column capital, or a shear cap helps to mitigate the punching shear issue.

The elastic behavior of two-way slabs under gravity loads can be described by a fourth order differential equation [1]. The solution of the differential equation that defines plate behavior under gravity loads is complex and not trivial. This is further complicated when the two-way slabs are supported on columns rather than on stiff beams or walls on all four sides of the slab panel. In contrast to beams or one-way slabs where closed form solutions can be found that describe the distribution of moments and shears in the beam or one-way slab, no such solutions are available for column-supported two-way slabs where 100% of the slab load is carried in each orthogonal direction of the slab [1]. In lieu of any available "exact" methods of analysis, several methods of analysis for two-way slabs subject to gravity loads have been developed over the years, and these include: elastic analysis methods, the yield line method, the Hillerborg strip method, the finite element method, the Direct Design Method (DDM), and the Equivalent Frame Method [1]. The two methods presented in the ACI Code for analysis of column-supported two-way slabs subject to gravity loads are the Direct Design Method (ACI Code 8.10) and the Equivalent Frame Method (ACI Code 8.11). In this

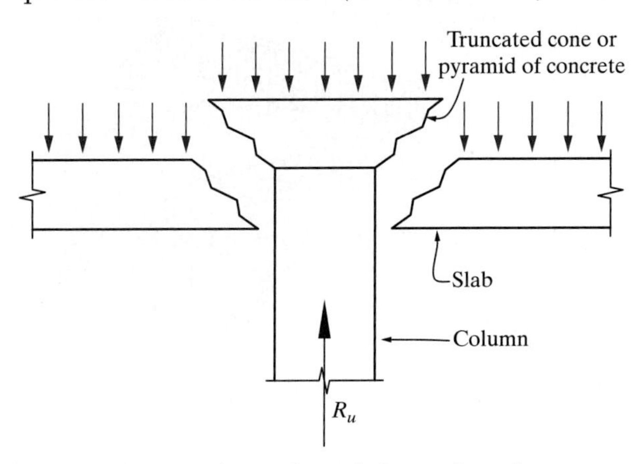

FIGURE 6-20 Punching shear failure of a column-supported two-way slab.

text, the focus will be on the direct design method since it is a much simpler analysis method for two-way slabs compared to the equivalent frame method, though the step-by-step procedure for the equivalent frame method of analysis will be covered later in this Chapter. In order to use the Direct Design Method, several conditions must be satisfied (ACI Section 8.10.2) and these are summarized below:

- The slab has at least three continuous spans in each direction.
- The slab panels are generally rectangular with a maximum aspect ratio (long side to short side) of 2.0.
- Column offsets are limited to 10% of the span in the offset direction from either axis between successive column centerlines. That is, the offset of a column from the regular rectangular grid must not exceed 10% of the regular column-to-column dimension in the direction of the offset within the slab panels where the column offset occurs (see Figure 6-40).
- The lengths of successive spans are such that the shorter span is at least two-thirds of the longer span.
- Only uniformly distributed gravity loads are applied.
- The ratio of unfactored live load to the unfactored dead load is no greater than 2.0. This obviates the need to check for the effects of pattern live loading. Where pattern live loading analysis needs to be considered (i.e., $L/D > 2$), the equivalent frame method is used (ACI Code, Section 8.11).
- When the slab is supported on beams on all four sides of the panel, the relative stiffness of the beams to the stiffness of the floor slab should be between 0.2 and 5.0.

When analyzing two-way slabs, the slab is divided into design strips in both orthogonal directions, and the design strips are centered on the column grid lines; the design strip is the slab strip bounded by one-half the distance to the columns to the right and left of the column line under consideration. The design strip consists of a column strip and two half middle strips. The column strip is centered on the column line and has a width of the smaller of 0.25 ℓ_1 or 0.25 ℓ_2, where ℓ_2 is the width of the design strip and ℓ_1 is the span of the slab in the longitudinal direction of the slab (see Figure 6-21). The moments in the middle strip are generally smaller than those in the column strip.

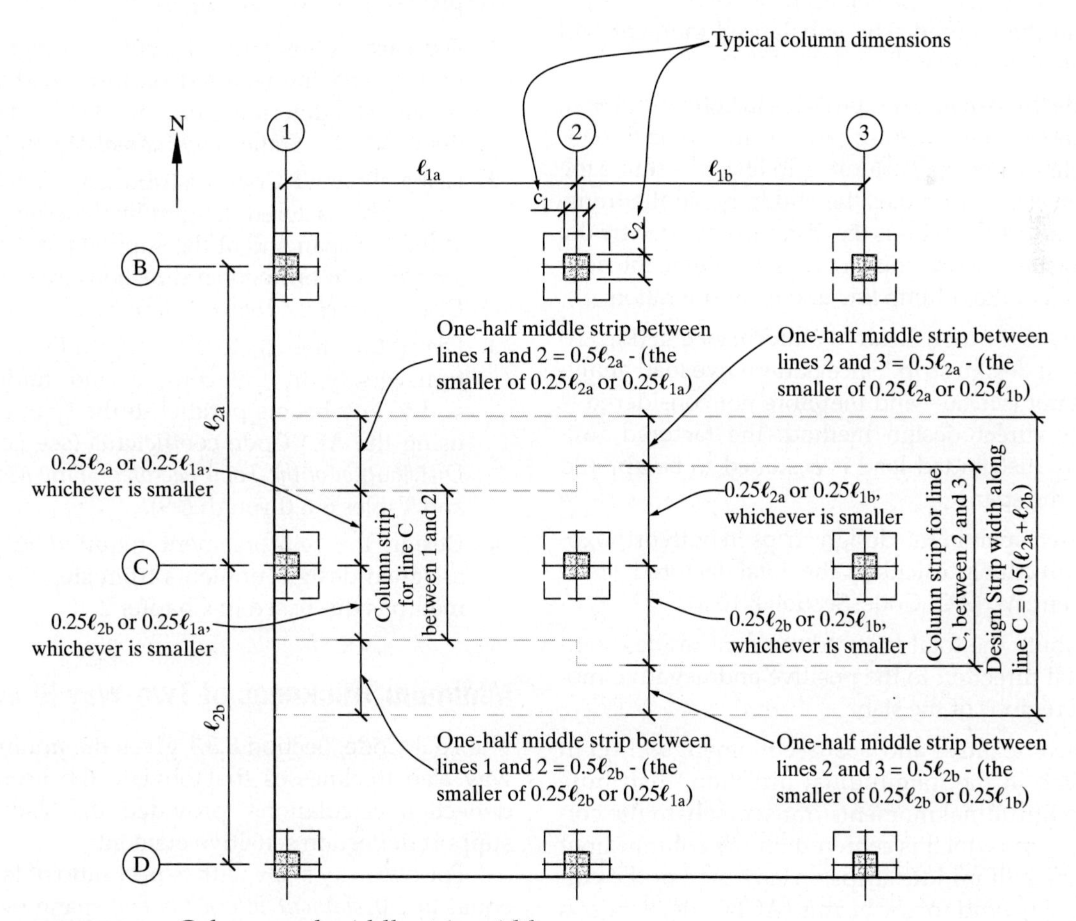

FIGURE 6-21 Column and middle strip widths.

The direct design method allows for the quick calculation of approximate moments and shears due to gravity loads at critical locations in the column and middle strips without the need for complex computer models or the calculations that are needed for the moment distribution analysis in the equivalent frame method. In general, the critical locations are the points of maximum stress for regular loading and framing conditions using the previously stated limitations. Those critical locations are at the faces of supports and at the midspan.

The steps involved in the direct design method (DDM) are as follows (ACI Section R8.10):

1. Determine the minimum slab thickness, h, to avoid deflection calculations (see *Minimum Thickness of Two-Way Slabs*). Where deflection sensitive elements are supported by the slab, the slab deflections should be calculated and checked against the Code specified limits.
2. Determine the total factored uniform load on the slab in psf.
3. Check one-way shear and two-way (punching) shear and ensure that the slab has adequate thickness to resist direct shear, and make allowance in the slab thickness to resist additional shear stresses due to transfer of unbalanced moments from the slab to the column. The unbalanced moment will be calculated later.
4. Divide the slab into design strips in both orthogonal directions. The design strips are centered on the column lines. For each column grid line, determine the design strip width (i.e., the width, ℓ_2 , in the transverse direction). This is the distance bounded by the mid distance to the adjacent column lines to the right and left of the column line under consideration.
5. Calculate the total factored uniform load in (kip/ft) for each design strip. Since pattern live load analysis is not critical—and therefore not considered—in the direct design method, the factored uniformly distributed load is assumed to be applied to all the spans.
6. For each span in the design strips in both orthogonal directions, calculate the total factored static moment, M_0 (ACI Code, Section 8.10.3).
7. Distribute the total factored moment in the longitudinal direction to the positive and negative moment regions of the slab.
8. At each negative and positive moment section in the slab in the longitudinal direction, distribute the longitudinal moments transversely to the column strip (which is centered on the column line) and the half middle strips on each side of the column strip, and to any beams (ACI Code, Sections 8.10.5 and 8.10.6). This process is carried out for each design strip in both orthogonal directions.
9. Using the moments from step (8) and the width of each strip, determine the required area of tension steel in the column and middle strips using the methods discussed in Chapter 2
10. Show the slab reinforcement provided in both orthogonal directions on a plan of the two-way slab similar to that shown in Figure 6-39.

Column and Middle Strips (ACI Code, Sections 8.4.1.5 and 8.4.1.6)

The widths of the column strips are determined as shown in Figure 6-21. The column strip width on each side of the column line is the smaller of 0.25 ℓ_1 or 0.25 ℓ_2, whichever is smaller. The determination of the column strip width for the E-W design strips is illustrated in Figure 6-21. It can be seen that the width of the column strip may sometimes vary from span to span if the span of the slab varies, but in practice and to ease the difficulty of placing the slab reinforcement on site, some engineers prefer to use a uniform column strip width throughout the entire length of the building.

The ACI Code prescribes the longitudinal moment coefficients for the direct design method as well as the transverse distribution coefficients for these longitudinal moments between the column and middle strips. The procedure for calculating the moments is as follows:

1. For each column line in both orthogonal directions, using the factored uniform load calculated previously, determine the total factored static moment, M_o (see *Static Longitudinal Moment, M_o*).
2. Using the ACI Code distribution coefficients, calculate the factored longitudinal design moments at the midspan and at the supports for each design strip in both orthogonal directions (see *Longitudinal Distribution of the Total Factored Static Moment, M_o*).
3. Distribute the factored longitudinal moments transversely to the column and middle strips and to any beams parallel to the span of the slab using the ACI Code coefficients (see *Longitudinal Distribution of the Total Factored Static Moment, M_o*, and Tables 6-6 through 6-9).
4. Design the reinforcement required to resist the assigned design moments from step (3) using the methods discussed in Chapter 2.

Minimum Thickness of Two-Way Slabs

The ACI Code, Section 8.3.1 gives the minimum two-way slab thicknesses that obviate the need for slab deflection calculations, provided the slab does not support deflection sensitive elements.

For two-way slabs with aspect ratio of less than or equal to 2.0, *without interior beams* spanning between supports, the minimum slab thickness shall meet the requirements of Table 6-4 and the requirements in

TABLE 6-4 Minimum Thickness of Reinforced Concrete Two-Way Slabs *Without* Interior Beams[1]

	Slabs without drop panels			Slabs with drop panels[3]		
	Exterior slab panel			Exterior slab panels		
Yield strength of slab reinforcement[2], f_y (psi)	Without edge beams	With edge beams[4] ($\alpha_f \geq 0.80$)	Interior slab panel	Without edge beams	With edge beams[4] ($\alpha_f \geq 0.80$)	Interior slab panels
40,000	$\frac{\ell_n}{33}$	$\frac{\ell_n}{36}$	$\frac{\ell_n}{36}$	$\frac{\ell_n}{36}$	$\frac{\ell_n}{40}$	$\frac{\ell_n}{40}$
60,000	$\frac{\ell_n}{30}$	$\frac{\ell_n}{33}$	$\frac{\ell_n}{33}$	$\frac{\ell_n}{33}$	$\frac{\ell_n}{36}$	$\frac{\ell_n}{36}$
75,000	$\frac{\ell_n}{28}$	$\frac{\ell_n}{31}$	$\frac{\ell_n}{31}$	$\frac{\ell_n}{31}$	$\frac{\ell_n}{34}$	$\frac{\ell_n}{34}$

Adapted from ACI Code, Table 8.3.1.1.

[1] ℓ_n is the clear span in the long direction of the slab panel measured face to face of the supports (in.).

[2] For f_y values between the tabulated values, linear interpolation is allowed.

[3] Drop panels must satisfy the thickness and plan length requirements in ACI Code 8.2.4.

[4] If the value of the relative stiffness (α_f) of the edge beam to the corresponding slab, calculated in accordance with ACI Code, Section 8.10.2.7, is less than 0.8, the edge beam shall be neglected and the slab treated as a panel without edge beam.

[5] For slabs ***without edge beams or with edge beams having α_f less than 0.8***, increase the minimum thickness obtained from Table 6-4 by 10% (ACI Code, Section 8.3.1.2.1).

(a) or (b) below unless the calculated deflection satisfies the limits in ACI Code, Section 8.3.2:

a. Slabs with drop panels that do *not meet* the length and depth requirements of the ACI Code 8.2.4 (see Column Capitals, Drop Panels, and Shear Caps): $h \geq 5$ in.
b. Slabs with drop panels that *meet* the length and depth requirements of the ACI Code: $h \geq 4$ in.

The minimum thickness obtained from Tables 6-4 and 6-5 should be rounded to the nearest ½ inch. Note that the above minimum slab thicknesses do not apply to slabs with heavy sustained live loads such as storage loads or mechanical room loads. For such slabs, the two-way slab deflections should be calculated in accordance with ACI Code, Section 8.3.2. For two-way slabs with beams between supports, the minimum slab thickness shall conform to Table 6-5.

TABLE 6-5 Minimum Thickness of Reinforced Concrete Two-Way Slabs *with* Beams Spanning Between Supports on All Four Sides of the Panel

α_{fm}	Minimum thickness, h	Expressions
$\alpha_{fm} \leq 0.2$	Use Table 6-4, but ≥ 5 in. for slab without drop panels ≥ 4 in. for slab with drop panels	(a)
$0.2 \leq \alpha_{fm} \leq 2.0$	$\dfrac{\ell_n\left(0.8 + \dfrac{f_y}{200{,}000}\right)}{36 + 5\beta(\alpha_{fm} - 0.2)} \geq 5$ in.	(b)[4] (c)
$\alpha_{fm} > 2.0$	$\dfrac{\ell_n\left(0.8 + \dfrac{f_y}{200{,}000}\right)}{36 + 9\beta} \geq 3.5$ in.	(d)[4] (e)

Adapted from ACI Code, Table 8.3.1.2.

[1] ℓ_n is the clear span in the *long direction* of the slab panel measured face to face of the supports (in.).

[2] β is the ratio of the clear spans in the long direction to the short direction of the slab panel.

[3] α_{fm} is the average value of the relative stiffness of beam to slab (i.e., α_f) for all beams on the four edges of a slab panel calculated per ACI Code, Section 8.10.2.7.

[4] For slab panels with the long-to-short span ratio greater than 2.0, the minimum thickness in Expressions (b) and (d) in Table 6-5 may give unreasonably large values because of the use of the long span in the equations. For such cases, the minimum thickness expressions for one-way slabs will be more appropriate to use (ACI Code, Section R8.3.1.2). See Table 6-1.

[5] For slab panels with beams between supports on all four sides of the panel, provide edge beams with a relative stiffness, α_f, greater than or equal to 0.80; otherwise, the minimum thickness from Expressions (b) and (d) in Table 6-5 shall be increased by 10% in the exterior slab panel (ACI Code, Section 8.3.1.2.1).

Static Longitudinal Moment, M_o

In general, the total factored static moment is calculated from ACI Code, Eq. 8.10.3.2 as follows:

In the ℓ_1 direction, the factored static moment,

$$M_o = \frac{w_u \ell_2 (\ell_{n1})^2}{8}$$

In the ℓ_2 direction, the factored static moment,

$$M_o = \frac{w_u \ell_1 (\ell_{n2})^2}{8}$$

where,

ℓ_{n1} is the clear span (face-to-face of the columns) in the ℓ_1 direction

ℓ_{n2} is the clear span (face-to-face of the columns) in the ℓ_2 direction

For the two-way slab in Figure 6-21, the calculation of the total static longitudinal moment, M_o, for the East-West design strips and for the North-South design strips will now be illustrated. For example, the static moment along grid line C in the E-W direction are given as follows (see Figure 6-21):

Between grid lines 1 and 2:

$$M_o = w_u \frac{1}{2}(\ell_{2a} + \ell_{2b}) \frac{\left(\ell_{1a} - \frac{c_1}{2} - \frac{c_1}{2}\right)^2}{8}$$

where, $\left(\ell_{1a} - \frac{c_1}{2} - \frac{c_1}{2}\right) \geq 0.65\ell_{1a}$

Between grid lines 2 and 3:

$$M_o = w_u \frac{1}{2}(\ell_{2a} + \ell_{2b}) \frac{\left(\ell_{1b} - \frac{c_1}{2} - \frac{c_1}{2}\right)^2}{8}$$

where, $\left(\ell_{1b} - \frac{c_1}{2} - \frac{c_1}{2}\right) \geq 0.65\ell_{1b}$

In the orthogonal (N-S) direction, the static moments along grid line 2 are given as follows (see Figure 6-21):

Between grid lines B and C:

$$M_o = w_u \frac{1}{2}(\ell_{1a} + \ell_{1b}) \frac{\left(\ell_{2a} - \frac{c_2}{2} - \frac{c_2}{2}\right)^2}{8}$$

where, $\left(\ell_{2a} - \frac{c_2}{2} - \frac{c_2}{2}\right) \geq 0.65\ell_{2a}$

Between grid lines C and D:

$$M_o = w_u \frac{1}{2}(\ell_{1a} + \ell_{1b}) \frac{\left(\ell_{2b} - \frac{c_2}{2} - \frac{c_2}{2}\right)^2}{8}$$

where, $\left(\ell_{2b} - \frac{c_2}{2} - \frac{c_2}{2}\right) \geq 0.65\ell_{2b}$

Similar static moments can be calculated for all the other E-W and N-S design strips.

For each span, the static equilibrium relationship between the static longitudinal moment, $M_{u,o}$, and the longitudinal moments at the critical sections (i.e., at the supports and midspan) of the slab in the ℓ_1 direction is given as follows:

$$M_{u,o} = \frac{w_u \ell_2 \ell_{n1}^2}{8} = \frac{1}{2}[M_{u,-ve(\text{left end support})} + M_{u,-ve(\text{right end support})}] + M_{u,+ve\,(\text{midspan})} \leq \phi M_n$$

Similarly, for the ℓ_2 direction, the static equilibrium relationship is given as follows:

$$M_{u,o} = \frac{w_u \ell_1 \ell_{n2}^2}{8} = \frac{1}{2}[M_{u,-ve(\text{left end support})} + M_{u,-ve(\text{right end support})}] + M_{u,+ve\,(\text{midspan})} \leq \phi M_n$$

In the static equilibrium equations above, use only the magnitude of the moments.

Column Capitals, Drop Panels, and Shear Caps

Column Capital: These are conical shaped concrete cast with the slab on top of the column with the sole aim of increasing the slab punching shear capacity (ACI Code 8.2.5). The column capital increases the perimeter of the critical section for punching shear at the slab-column joint (see Figure 6-22).

Drop Panels: This is an additional slab thickening below the slab at the column locations (i.e., in the negative moment regions of the slab). The ACI Code specifies the minimum depth and minimum lengths in both orthogonal directions for drop panels. Drop panels are used for the following purposes:

- To increase the perimeter of the critical section for punching shear at the column and thus reduce the shear stress or increase the shear capacity.
- To increase the effective depth of the slab top reinforcement at the columns, and thus lead to a reduction in the required area of negative reinforcement over the columns.

To qualify as a drop panel, the following conditions must be satisfied (see Figures 6-23(a) and 6-23(b)):

- The thickness of the drop panel below the slab must be at least one-fourth the slab thickness.
- The plan dimension of the drop panel in each orthogonal direction shall be at least one-sixth the span length of the slab in each direction. Where the drop panels do not meet the minimum dimension limits imposed by ACI Code, Section 8.2.4, the drop

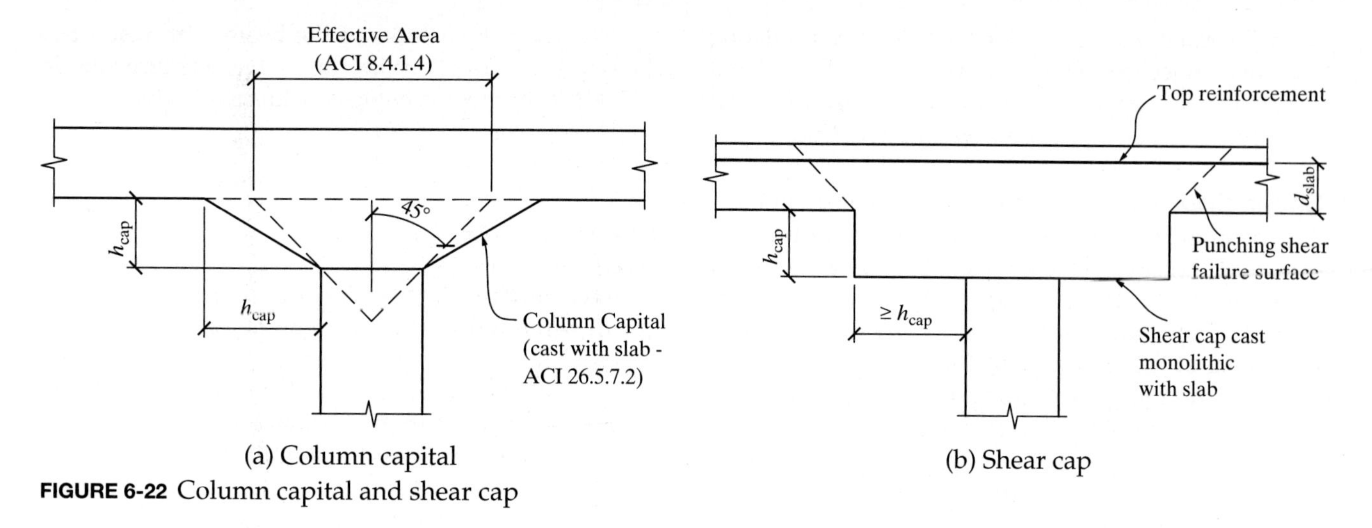

FIGURE 6-22 Column capital and shear cap

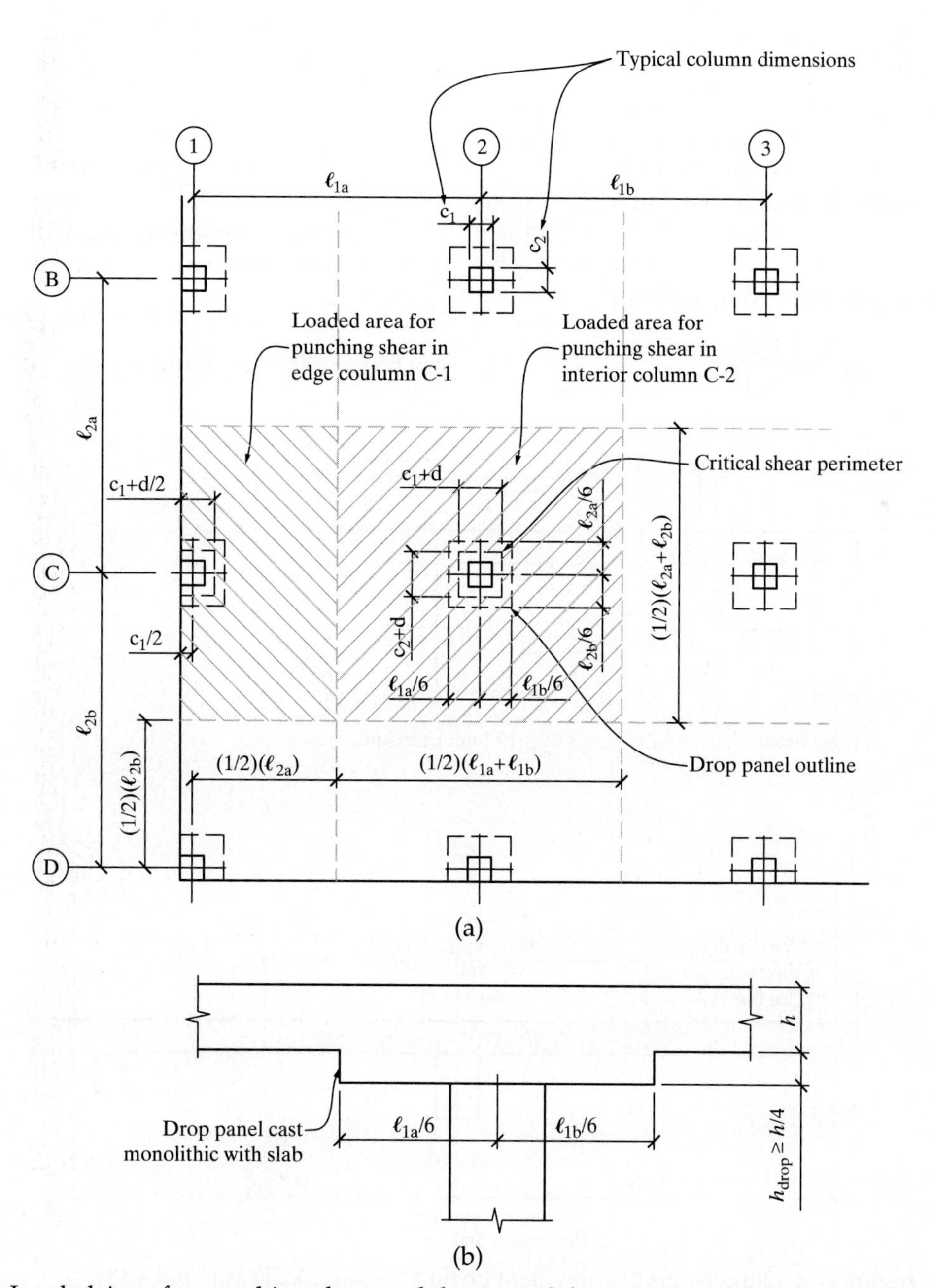

FIGURE 6-23 (a) Loaded Area for punching shear and drop panel dimensions. (b) Drop panel thickness.

panel cannot be used to increase the effective depth, but a projection of the drop panel can be considered as a shear cap which is used to increase the shear strength (ACI Code, Sections 8.2.4 and 8.2.5).

Shear Cap: These are drop panels that do not satisfy the minimum plan dimensions in both orthogonal directions specified in ACI Code, Section 8.2.4. Shear caps can only be used to increase the c_1 and c_2 dimensions, and thus increase the critical perimeter for two-way shear at the slab-column connection which leads to a reduction in the punching shear stress; they cannot be used to increase the effective depth for calculating the amount of negative reinforcement required in the slab at the column locations. They also cannot be used in the deflection calculations.

Relative Stiffness of Beam and Slab

To account for the effect of beam stiffness on the deflection of two-way slabs supported by beams, the stiffnesses of the beams relative to the slab are calculated as follows (ACI Code, Section 8.10.2.7(b)):

The relative stiffness of the beam with respect to the slab in the ℓ_1 direction (i.e., the direction in which the moment is being considered) is given as

$$\alpha_{f1} = \frac{E_{cb} I_{b1}}{E_{cs} I_{s1}}$$

The relative stiffness of the beam with respect to the slab in ℓ_2 direction (i.e., the transverse direction in which the moment is being considered) is given as

$$\alpha_{f2} = \frac{E_{cb} I_{b2}}{E_{cs} I_{s2}}$$

where

E_{cb} = modulus of elasticity of the beam concrete.
E_{cs} = modulus of elasticity of the slab concrete.
I_{b1} = moment of inertia about the centroidal axis of the gross section of the equivalent beam that spans in the ℓ_1 direction.
I_{b2} = moment of inertia about the centroidal axis of the gross section of the equivalent beam that spans in the ℓ_2 direction.
I_{s1} = moment of inertia about the centroidal axis of the slab in the ℓ_1 direction.
I_{s2} = moment of inertia about the centroidal axis of the slab in the ℓ_2 direction.

The code definitions of the equivalent beam and slab are shown in Figure 6-24.

In practice, since the slab and the beam in reinforced concrete structures are cast monolithically, $E_{cb} = E_{cs}$, therefore $\frac{E_{cb}}{E_{cs}} = 1.0$ and hence the previous equations can be simplified as follows:

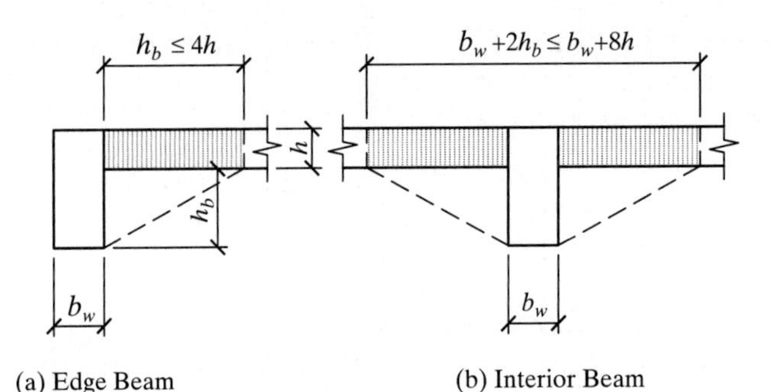

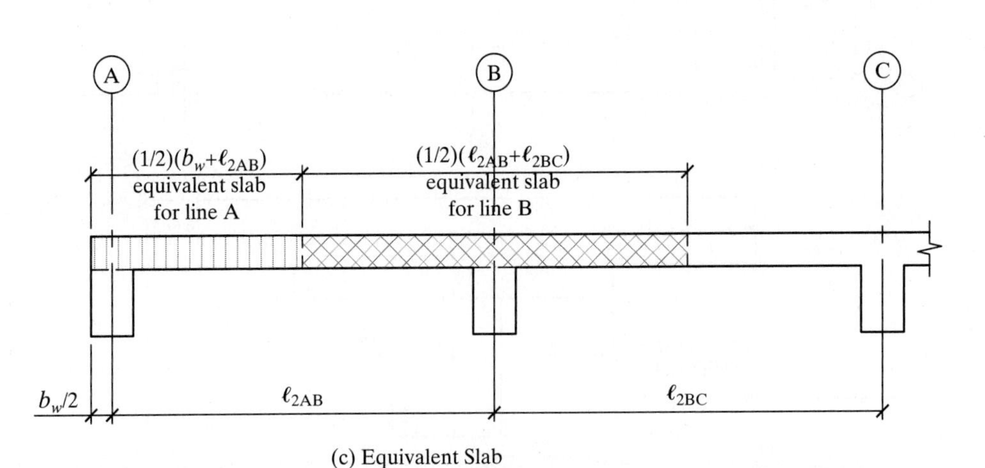

(c) Equivalent Slab

FIGURE 6-24 Dimensions of equivalent beam and slab (ACI Code R.8.4.18).

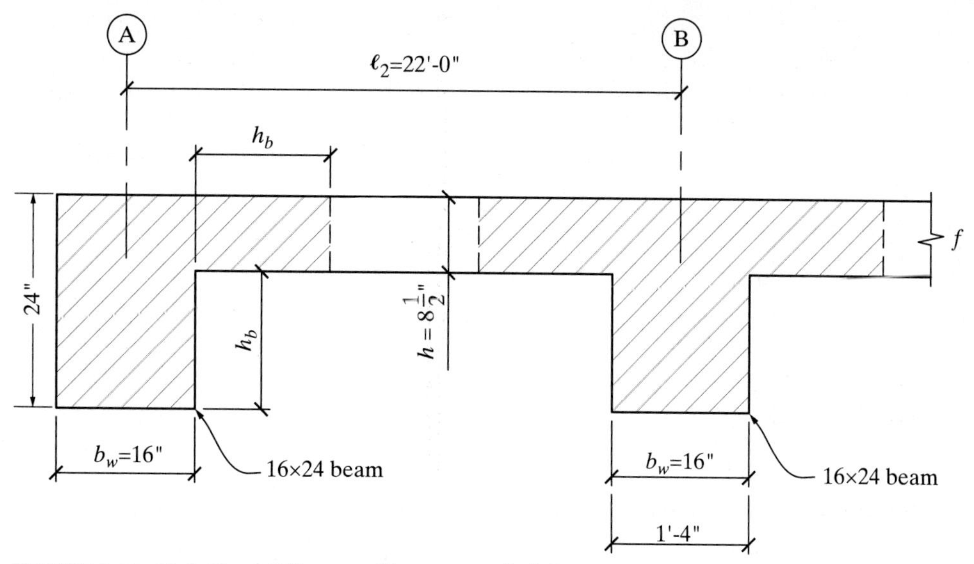

FIGURE 6-25 Relative stiffness of beams and slabs.

$$\alpha_{f1} = \frac{I_{b1}}{I_{s1}}$$

$$\alpha_{f2} = \frac{I_{b2}}{I_{s2}}$$

For flat plates, where there are no longitudinal beams, $\alpha_{f1} = 0$; and where there are no transverse or edge beams, $\alpha_{f2} = 0$. For a slab panel with beams between supports on all the four edges of the panel, the beams in the two orthogonal directions shall satisfy the following equation to ensure that the moment distribution assumed in the direct design method does not deviate significantly from the elastic distribution of moments (ACI Code, Eq. 8.10.2.7a):

$$0.2 \leq \frac{\alpha_{f1}\ell_2^2}{\alpha_{f2}\ell_1^2} \leq 5.0$$

where

ℓ_1 = length of the span in the direction that moments are being determined, measured center-to-center of supports (inches).

ℓ_2 = length of the span in the direction perpendicular to ℓ_1, measured center-to-center of supports (inches) (i.e., the transverse width of the design strip).

Example 6-2

Calculate the relative stiffness of the edge beam and the interior beam for the slab-beam cross-section in Figure 6-25. The slab span being designed is ℓ_1 and is perpendicular to the ℓ_2 direction.

Solution:

Longitudinal *edge* beam along grid line A:

The distance of the centroid of the edge beam section shown in Figure 6-26 from the top of the beam is calculated as

$$\frac{(16)(24)\left(\frac{24}{2}\right) + (15.5)(8.5)\left(\frac{8.5}{2}\right)}{(16)(24) + (15.5)(8.5)} = 10.02 \text{ in.}$$

from the top of the beam

The gross moment of inertia about the centroidal axis of the gross section of the equivalent beam that spans in the ℓ_1 direction is:

$$I_{b1} = (16)\frac{(24)^3}{12} + (16)(24)\left(\frac{24}{2} - 10.02\right)^2 + (15.5)\frac{(8.5)^3}{12} + (15.5)(8.5)\left(10.02 - \frac{8.5}{2}\right)^2$$

$$= 25{,}087 \text{ in.}^4$$

The corresponding equivalent slab section for the edge beam is shown in Figure 6-27 and the gross moment of inertia of the slab section is calculated as

$$I_{s1} = (140)\frac{(8.5)^3}{12} = 7{,}165 \text{ in.}^4$$

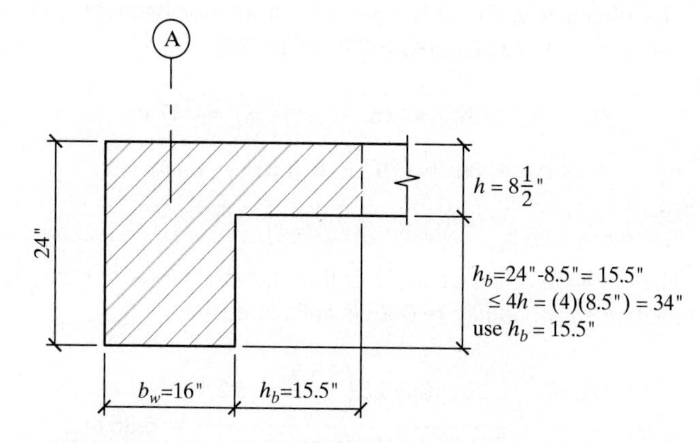

FIGURE 6-26 Equivalent longitudinal edge beam along grid line A.

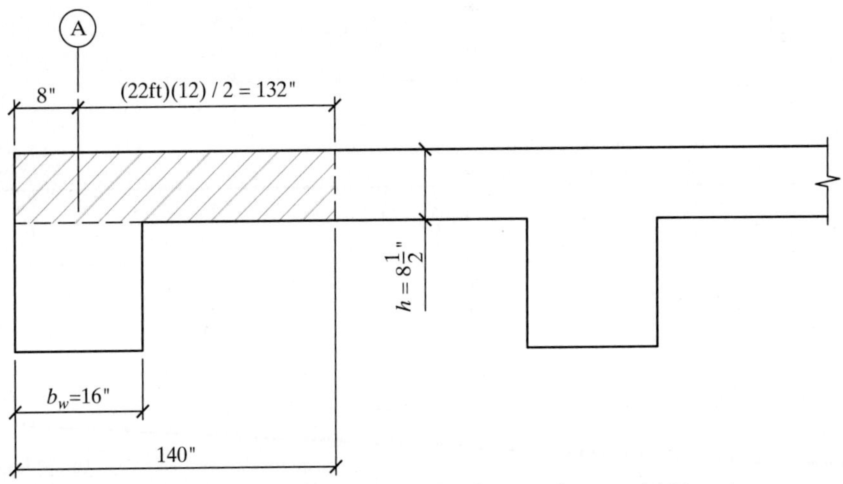

FIGURE 6-27 Equivalent slab for the edge beam along grid line A.

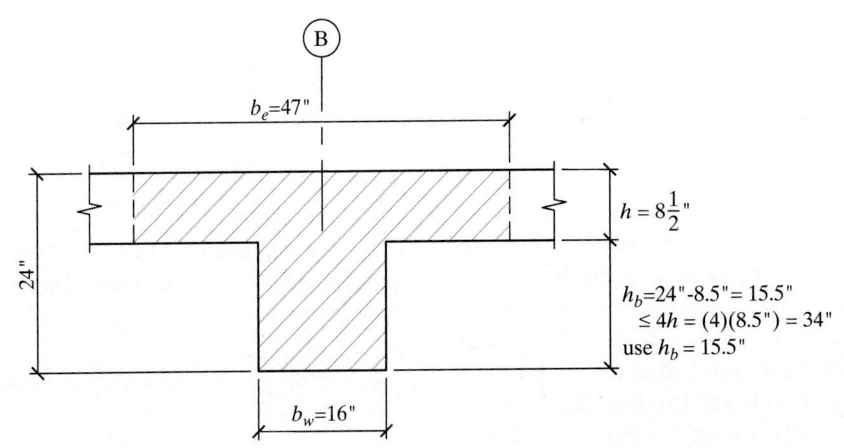

FIGURE 6-28 Equivalent longitudinal interior beam along grid line B.

The relative stiffness of the edge beam along grid line A is

$$\alpha_{f1} = \frac{I_{b1}}{I_{s1}} = \frac{25{,}087}{7{,}165} = 3.5$$

Longitudinal *interior* beam along grid line B:
The effective width of the equivalent interior beam, b_e, is calculated as (see Figures 6-24b and 6-28)

$$b_e = b_w + 2h_b = 16 + 2(15.5) = 47 \text{ in.}$$
$$\leq b_w + 8h = 16 + 8(8.5) = 84 \text{ in.}$$

Therefore, use $b_e = 47$ in.

The distance of the centroid of the equivalent interior beam section shown in Figure 6-28 is calculated as

$$\frac{(47)(8.5)\left(\frac{8.5}{2}\right) + (16)(15.5)\left(\frac{15.5}{2} + 8.5\right)}{(47)(8.5) + (16)(15.5)} = 8.85 \text{ in.}$$

from the top of the beam

The gross moment of inertia about the centroidal axis of the equivalent beam that spans in the ℓ_1 direction is:

$$I_{b1} = (47)\frac{(8.5)^3}{12} + (47)(8.5)\left(8.85 - \frac{8.5}{2}\right)^2 + (16)\frac{(15.5)^3}{12} + (16)(15.5)\left(\frac{15.5}{2} + 8.5 - 8.85\right)^2$$
$$= 29{,}404 \text{ in.}^4$$

The corresponding equivalent slab section for the interior beam along grid line B is shown in Figure 6-29 and the gross moment of inertia of the slab section is calculated as

$$I_{s1} = (264)\frac{(8.5)^3}{12} = 13{,}511 \text{ in.}^4$$

The relative stiffness of the interior beam is

$$\alpha_{f1} = \frac{I_{b1}}{I_{s1}} = \frac{29{,}404}{13{,}511} = 2.18$$

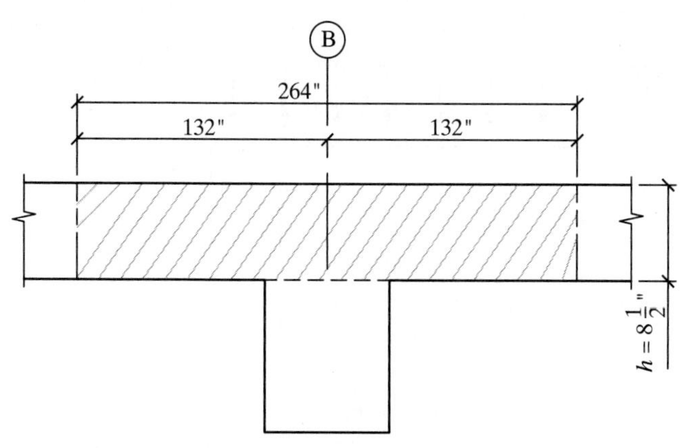

FIGURE 6-29 Equivalent slab section for interior beam along grid line B.

Longitudinal Distribution of the Total Factored Static Moment, M_o

The total factored static moment, M_o, within each span are distributed in the direction of the span of the slab (i.e., in the longitudinal direction) between the positive and negative moment regions of the slab as shown in Figures 6-30a, 6-30b, and 6-30c (ACI Code, Section 8.10.4). For interior spans, the critical negative moment used in design is the higher of the factored negative moment on each side of the slab support (ACI Code, Section 8.10.4.5). The edge beams and edge of slabs shall be designed to resist their share of the exterior negative moment (ACI Code, Section 8.10.4.6).

Transverse Distribution of the Longitudinal Moments Between Column Strip, Middle Strip, and Longitudinal Beams

The longitudinally distributed moments in Figure 6-30a through 6-30c are further distributed transversely between the column strip, the middle strip, and longitudinal beams, if any. The coefficients for the transverse distributions of the interior negative moment, the exterior negative moment, and the positive moment to the column strip are given in Tables 6-6 through 6-9. For two-way slabs with longitudinal beams between supports, the portion of the column strip moment not resisted by the beams must be resisted by the slab within the column strip (ACI Code, Section 8.10.5.6).

The remaining moments that are not distributed to the column strip (at the interior support, the exterior support, and the positive moment regions) are assigned to the half middle strips on both sides of the column strip.

One-Way Shear in Two-Way Slabs

Similar to one-way slabs, two-way slabs need to be designed for one-way shear due to gravity loads. The critical section for one-way shear in column-supported two-way reinforced concrete slabs can be assumed to occur at a distance d from the face of the column provided the following conditions are satisfied (ACI Code, Sections 8.4.3.1 and 8.4.3.2):

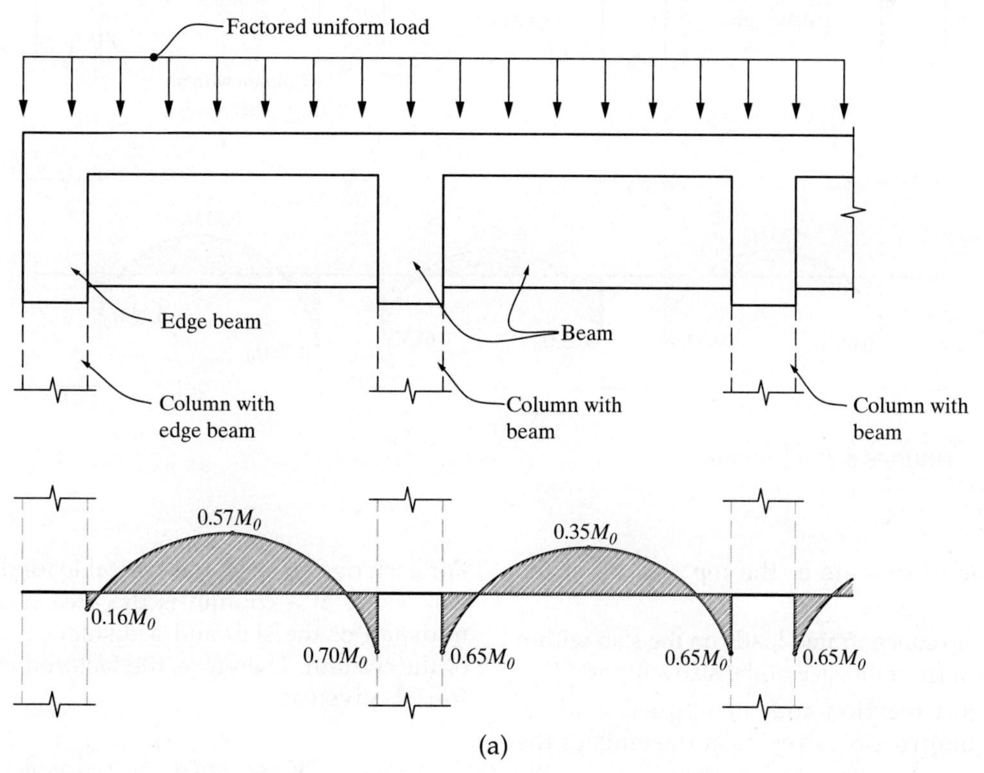

FIGURES 6-30 Longitudinal distribution of static moment, M_o.

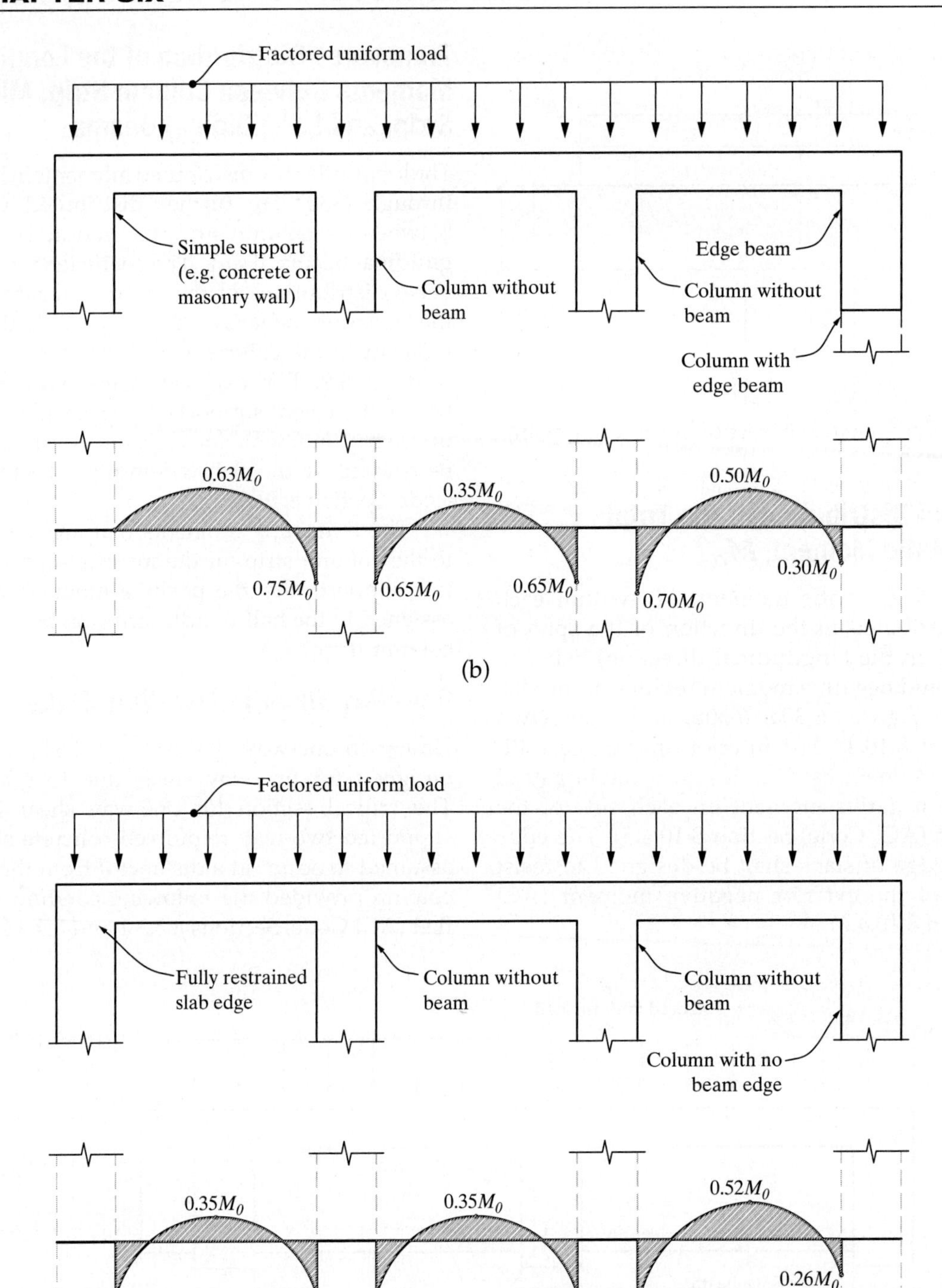

FIGURES 6-30 *Continued*

- The applied loads are at the top surface of the slab.
- There are no concentrated loads on the slab within a distance d from the face of the slab support.
- The support reaction and the applied load results in compression stresses at the ends of the slab.

For each design strip, the critical loaded area for one-way shear at a column is the area bounded by the midspan of the slab and a distance d from the face of the column. Therefore, the factored one-way shear force is given as

$$V_u = w_u \ell_2\left(\frac{\ell_1}{2} - \frac{c_1}{2} - d\right)$$

The limit state design equation requires that the factored shear be less than or equal to the one-way shear capacity of the slab. Therefore,

$$V_u \leq \phi V_c = 2\phi\lambda\sqrt{f'_c}b_w d$$

where,

The strength reduction factor, $\phi = 0.75$ (ACI Code, Table 21.2.1)

b_w = width of the design strip = ℓ_2

One-way shear rarely governs the design of two-way slabs. In any case, the thickness of the slab should be selected so that the one-way shear capacity, ϕV_c, is greater than the one-way shear demand, V_u, since it is not usual nor practical to reinforce a slab for one-way shear.

Punching Shear in Two-Way Slabs

In addition to one-way shear, column-supported slabs are also subject to punching or two-way shear which is a failure mode where the column and a truncated cone or pyramid of concrete punches through the slab as a result of the concentrated reaction and load transfer at the column (see Figure 6-20). Shear stresses occur in the slab near the columns due to direct shear from the gravity loads and also due to the unbalanced moment at the column that is transferred from the slab to the column through shear. The critical section for punching shear occurs at a distance of $d/2$ from the face of the column on all sides of the column (ACI Code, Section 22.6.4.1). For a flat plate or flat slab without beams, the factored direct two-way shear, V_u, is equal to the factored uniform load on the slab multiplied by the difference between the column tributary area and the area of the critical section (see the shaded areas in Figure 6-23a).

Critical Section for Punching Shear The critical section for punching shear in flat plates is at a distance of $d/2$ from the face of the column all around the column (see Figures 6-31 and 6-32) or from the edge of the column capital or shear cap. The critical perimeter for an interior column in direct shear is,

$$b_o = 2(c_1 + d) + 2(c_2 + d)$$

TABLE 6-6 The Portion of the Interior Negative Moment Distributed Transversely to the Column Strip (ACI Code, Table 8.10.5.1)

$\alpha_{f1}\frac{\ell_2}{\ell_1}$	$\frac{\ell_2}{\ell_1}$ = 0.5	1.0	2.0
0	0.75	0.75	0.75
≥1.0	0.90	0.75	0.45

[1] α_{f1} is the relative stiffness of longitudinal interior beam, if any, compared to the slab
[2] Linear interpolation is permitted between values in this table
[3] $\alpha_{f1} = \frac{I_{b1}}{I_{s1}}$ (see Relative Stiffness of Beam and Slab)

TABLE 6-7 The Portion of the Exterior Negative Moment Distributed Transversely to the Column Strip (ACI Code, Table 8.10.5.2)

$\alpha_{f1}\frac{\ell_2}{\ell_1}$	β_t	$\frac{\ell_2}{\ell_1}$ = 0.5	1.0	2.0
0	0	1.0	1.0	1.0
	≥2.5	0.75	0.75	0.75
≥1.0	0	1.0	1.0	1.0
	≥2.5	0.90	0.75	0.45

[1] Linear interpolation is permitted between the values shown
[2] $\beta_t = \frac{E_{cb}C}{2E_{cs}I_s}$ [ACI Eq. 8.10.5.2a]

$C = \sum\left(1 - 0.63\frac{x}{y}\right)\frac{x^3y}{3}$ [ACI Eq. 8.10.5.2b]

x = smaller dimension of a rectangular area; y = larger dimension of a rectangular area

TABLE 6-8 The Portion of the Positive Moment Distributed Transversely to the Column Strip (ACI Code, Table 8.10.5.5)

$\alpha_{f1}\frac{\ell_2}{\ell_1}$	$\frac{\ell_2}{\ell_1}$ = 0.5	1.0	2.0
0	0.60	0.60	0.60
≥1.0	0.90	0.75	0.45

[1] Linear interpolation is permitted between the values shown

TABLE 6-9 The Portion of the Column Strip Moment Distributed to the Longitudinal Beam (ACI Code, Table 8.10.5.7.1)

$\alpha_{f1}\frac{\ell_2}{\ell_1}$	Portion of column strip moment distributed to longitudinal beam
0	0
≥1.0	0.85

[1] Linear interpolation is permitted between the values shown
[2] Beam to resist assigned moment plus moment due to the directly applied load on the beam including the weight of the beam stem (ACI Code, Section 8.10.5.7.2).

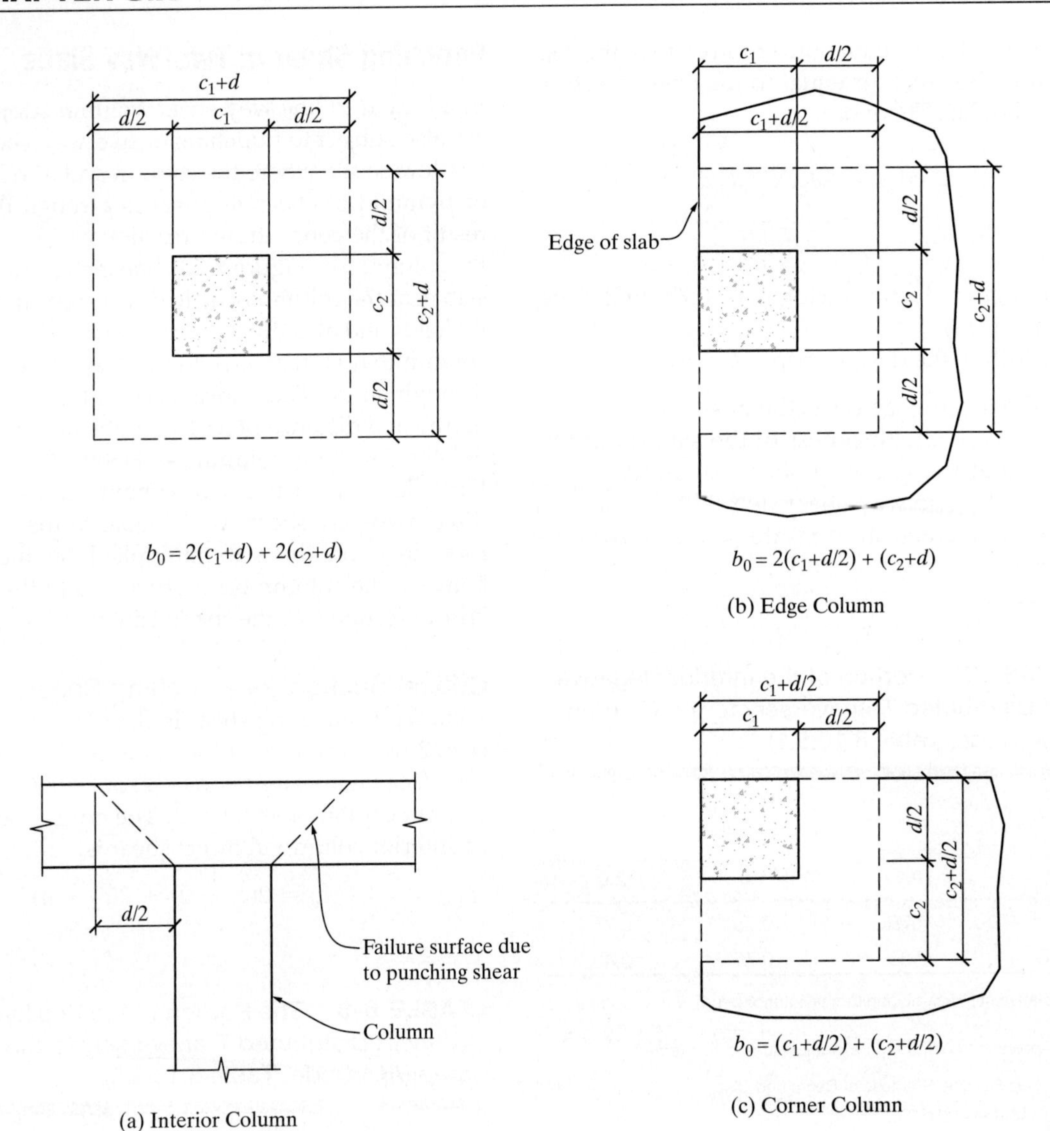

FIGURE 6-31 Critical sections for punching or two-way shear.

$$= 2c_1 + 2c_2 + 4d$$

Similar equations for the critical perimeter in direct shear can be derived for edge and corner columns as shown in Figures 6-31b and 6-31c. To allow for some reserve shear capacity due to additional shear stresses from unbalanced moments (that will be calculated later), a common rule of thumb is to increase the slab depth required for direct shear by 10% at interior columns, 40% at edge columns, and 70% at corner columns [2].

Critical Section at Drop Panels (ACI Code, Section 22.6.4.1) When drop panels are used, two critical sections need to be checked: The first critical section is around the column, and the second critical section is around the drop panel as shown in Figure 6-32.

The nominal concrete shear strength in two-way shear due to gravity loads (without unbalanced moments) is the smallest of the following equations derived from ACI Code, Table 22.6.5.2:

$$V_c = 4\lambda\sqrt{f'_c}b_o d$$

$$V_c = \left(2 + \frac{4}{\beta}\right)\lambda\sqrt{f'_c}b_o d$$

$$V_c = \left(2 + \frac{\alpha_s d}{b_o}\right)\lambda\sqrt{f'_c}b_o d$$

where

β = ratio of the long cross-sectional dimension to the short cross-sectional dimension of the column, concentrated load or reaction

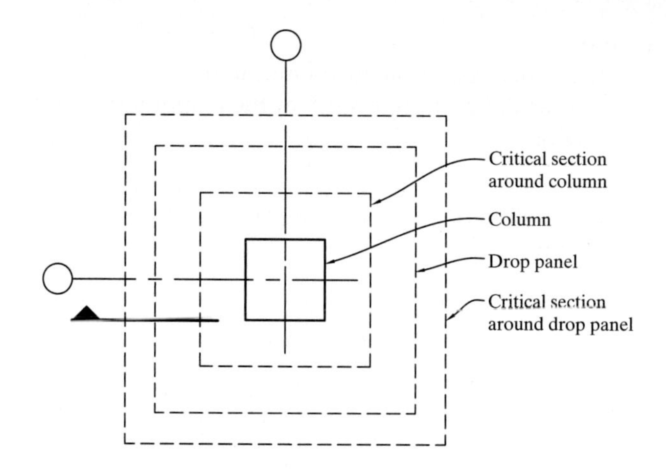

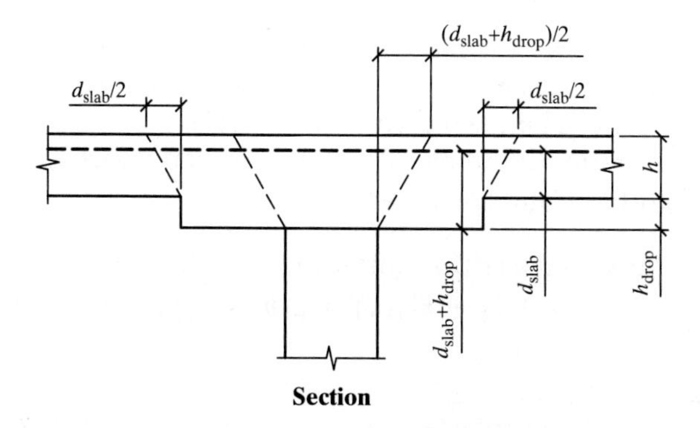

FIGURE 6-32 Critical sections around a column and drop panel.

b_o = perimeter of the critical section for punching or two-way shear

α_s = 40 for interior columns
30 for edge columns
20 for corner columns

λ = modification factor for lightweight concrete
= 1.0 for normal weight concrete

The shear capacity is ϕV_c and the shear demand or factored shear force is V_u. The limit state design equation requires that the demand be less than or equal to the capacity. Therefore,

$$V_u \leq \phi V_c$$

The strength reduction factor for shear, $\phi = 0.75$ (ACI Code, Table 21.2.1)

As can be observed from the punching shear strength equation, the punching shear strength can be increased by:

- Increasing the concrete strength, f'_c.
- Increasing the effective depth of the slab at the columns. This can be achieved by using a thicker slab for flat plates or adding a drop panel for flat slabs.
- Increasing the perimeter of the critical section, b_o since $b_o = 2c_1 + 2c_2 + 4d$, the perimeter can be increased by increasing the size of the column (i.e., c_1 and c_2), and/or by increasing the effective depth of the slab at the columns.

Note that the shear demand, V_u, above only considers direct shear, but additional shear force (and shear stress) will be present where unbalanced moments are transferred from the slab to the column. The preceding equation is used to determine the minimum thickness of the two-way slab that will satisfy the punching shear requirements without the need for shear reinforcement. For two-way slabs that do not satisfy the punching shear strength requirements above, shear reinforcement in the form of stirrups or headed studs have to be provided. In this text, we assume that the slab thickness will be selected to avoid the need for shear reinforcement. The design of shear reinforcement for two-way slabs is outside the scope of this text; the author should refer to ACI Code, Sections 22.6.6 through 22.6.9 for the design of shear reinforcement in two-way slabs.

Combined Shear and Unbalanced Moment Transfer in Slab-Column Connections

The transfer of unbalanced moments from the slab to the column at the slab-column connections causes non-uniform shear stress distribution, with shear stresses that are larger on one side of the column and lower on the opposite side. The unbalanced moments are usually more critical at the corner and edge columns than at an interior column, while the direct shear is typically highest at the interior columns. In the analyses of slab-column connections for combined shear and unbalanced moments, a linear distribution of shear stress around the critical section is assumed in the ACI Code [1]. A fraction, γ_f, of the factored unbalanced moment, M_u, is transferred by flexure from the slab to the column; the effective width of the slab for transferring the moment, $\gamma_f M_u$, is equal to the width of the column plus $3h$, where h is the slab thickness (see Figure 6-33). The flexural reinforcement required should be symmetrically placed within the effective width; closer spacing of the reinforcement or additional reinforcement can be used to meet this requirement. The remaining unbalanced moment, $\gamma_v M_u$, is transferred by eccentricity of shear. The values of the fraction of the unbalanced moment transferred by flexure and eccentricity of shear are defined as

$$\gamma_f = \frac{1}{1 + \left(\frac{2}{3}\right)\sqrt{\frac{b_1}{b_2}}} \qquad \text{[ACI Code, Eq. 8.4.2.3.2]}$$

$$\gamma_v = 1 - \gamma_f \qquad \text{[ACI Code, Eq. 8.4.4.2.2]}$$

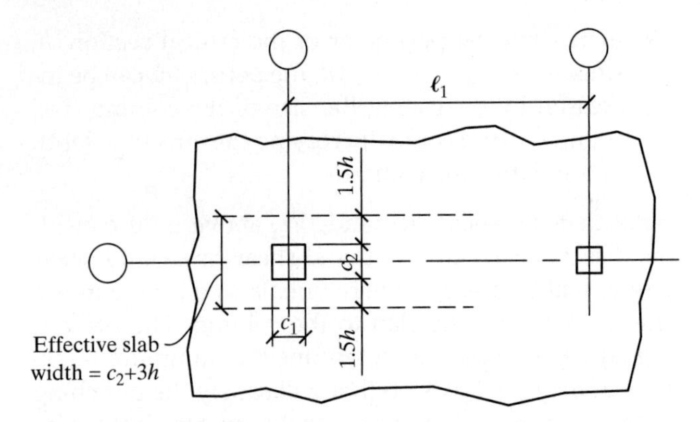

FIGURE 6-33 Effective width of slab for transfer of unbalanced moment.

Where b_1 and b_2 are the dimensions of the punching shear critical perimeter parallel and perpendicular, respectively, to the span of the slab (see Figures 6-34 and 6-35).

Interior columns: $b_1 = c_1 + d;\ \ b_2 = c_2 + d$

Corner columns: $b_1 = c_1 + 0.5d;\ b_2 = c_2 + 0.5d$

Edge columns: $b_1 = c_1 + 0.5d;\ b_2 = c_2 + d$

Where, c_1 is the column cross-sectional dimension parallel to the span of the slab, and c_2 is the column cross-sectional dimension in the transverse direction. When shear caps are used, the values of c_1 and c_2 will reflect the increased critical shear perimeter, and for an interior column, the revised c_1 will be $c_{1\text{column}} + 2h_{\text{shear-cap}}$. Similarly, the revised c_2 will be $c_{2\text{column}} + 2h_{\text{shear-cap}}$.

For columns where $b_1 = b_2$, γ_f will be 0.6 and γ_v is 0.4. For this case, 60% of the unbalanced moment is transferred from the slab to the column by flexure, and the remaining 40% is transferred by eccentricity of the shear about the centroid of the critical section. A linear distribution of the shear stress about the centroid of the critical section is assumed for the factored shear stress caused by unbalanced moment, $\gamma_v M_u$ (ACI Code, Section 8.4.4.2.3) as shown in Figure 6-34 for interior and edge columns, where M_u is the factored unbalanced moment transferred from the slab to the column.

For edge columns, the unbalanced gravity load moments to be transferred from the slab to the column must be at least $0.3M_o$. That is, $M_u \geq 0.30M_o$ (ACI Code, Section 8.10.7.3).

The maximum factored shear stresses under combined shear and unbalanced moment are given as:

$$v_{u,AB} = \frac{V_u}{A_c} + \frac{\gamma_v M_u c_{AB}}{J_c} \leq v_{\text{allowable}}$$

$$v_{u,CD} = \frac{V_u}{A_c} - \frac{\gamma_v M_u c_{CD}}{J_c} \leq v_{\text{allowable}}$$

where

M_u = factored unbalanced moment

γ_v is as defined previously at the beginning of this section.

The allowable concrete shear stress in two-way shear is the smallest of the following equations from ACI Code, Table 22.6.5.2:

$$v_c = 4\phi\lambda\sqrt{f'_c}$$

$$v_c = \phi\left(2 + \frac{4}{\beta}\right)\lambda\sqrt{f'_c}$$

$$v_c = \phi\left(2 + \frac{\alpha_s d}{b_o}\right)\lambda\sqrt{f'_c}$$

$$\phi = 0.75 \text{ for shear}$$

The parameter, J_c, which is analogous to the polar moment of inertia of the critical section about the centroidal axis can be calculated as follows [1]:

For **Interior Columns**—ACI Code, Section R8.4.4.2.3 (see Figures 6-34(a) and 6-35):

$$\begin{aligned} A_c &= \text{area of the critical surface for two-way shear} \\ &= (b_1 + b_2 + b_1 + b_2)(d) = 2(b_1 + b_2)d \\ &= 2(c_1 + d + c_2 + d) \\ &= 2d(c_1 + c_2 + 2d) \\ c_{AB} &= c_{CD} = (c_1 + d)/2 \end{aligned}$$

$$J_c = J_c = \frac{d(c_1 + d)^3}{6} + \frac{(c_1 + d)d^3}{6} + \frac{d(c_2 + d)(c_1 + d)^2}{2}$$

For **Edge Columns with moment perpendicular to the edge** (see Figure 6-34(b)):

A_c = area of the critical surface for two-way shear

$$A_c = \left(c_1 + \frac{d}{2}\right)d + \left(c_1 + \frac{d}{2}\right)d + (c_2 + d)d$$

$$= (2c_1 + c_2 + 2d)d$$

$$c_{AB} = \frac{(c_2 + d)(d)(0) + \left(c_1 + \frac{d}{2}\right)(d)\left(\frac{c_1 + \frac{d}{2}}{2}\right)(2\ \textit{sides})}{(c_2 + d)(d) + \left(c_1 + \frac{d}{2}\right)(d)(2\ \textit{sides})}$$

$$= \frac{\left(c_1 + \frac{d}{2}\right)^2(d)}{d(2c_1 + c_2 + 2d)}$$

$$c_{CD} = c_1 + \frac{d}{2} - c_{AB}$$

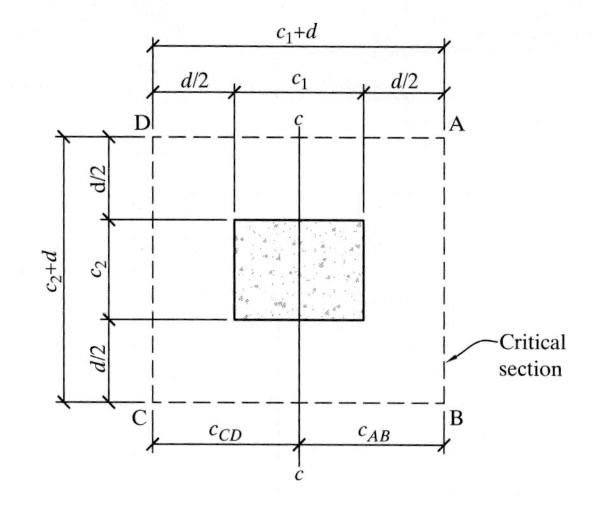

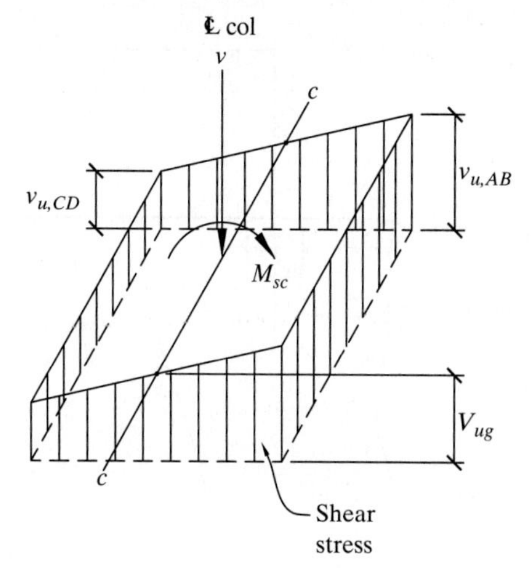

(a) Interior Column

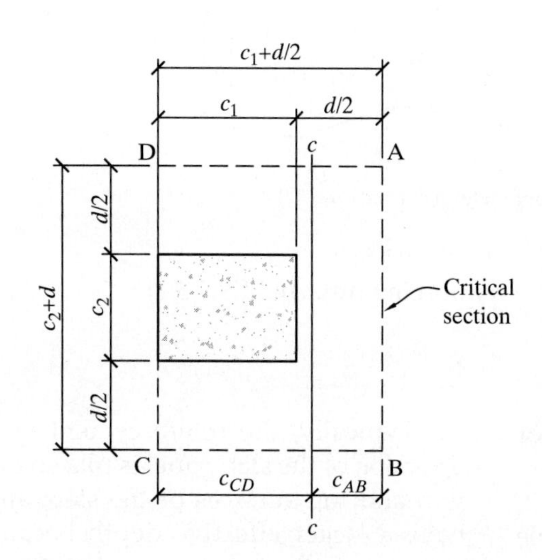

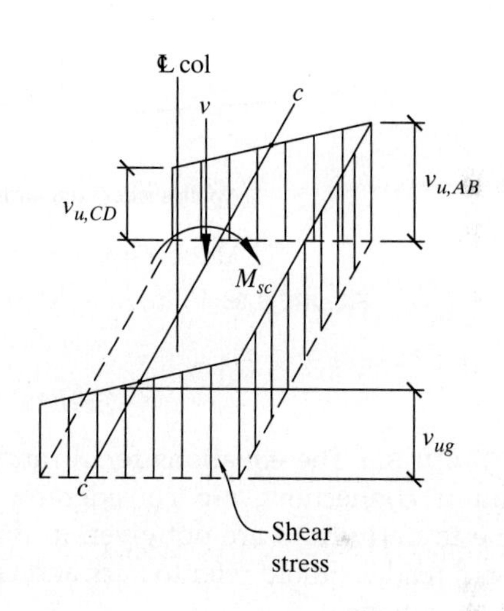

(b) Edge Column

FIGURES 6-34 Assumed distribution of shear stress for combined shear and moment (ACI Figure R8.4.4.2.3).

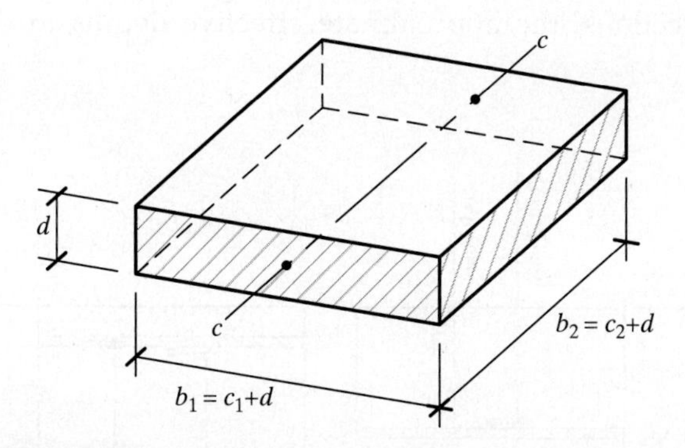

FIGURES 6-35 Critical section for punching shear at interior column.

$$J_c = \frac{d\left(c_1 + \frac{d}{2}\right)^3 (2\ sides)}{12} + \left(c_1 + \frac{d}{2}\right)(d)\left(\frac{c_1 + \frac{d}{2}}{2} - c_{AB}\right)^2$$

$$(2\ sides) + \frac{\left(c_1 + \frac{d}{2}\right)(d)^3 (2\ sides)}{12} + (c_2 + d)(d)(c_{AB})^2$$

$$= \frac{d\left(c_1 + \frac{d}{2}\right)^3}{6} + 2d\left(c_1 + \frac{d}{2}\right)\left(\frac{c_1 + \frac{d}{2}}{2} - c_{AB}\right)^2$$

$$+ \frac{\left(c_1 + \frac{d}{2}\right)(d)^3}{6} + (c_2 + d)(d)(c_{AB})^2$$

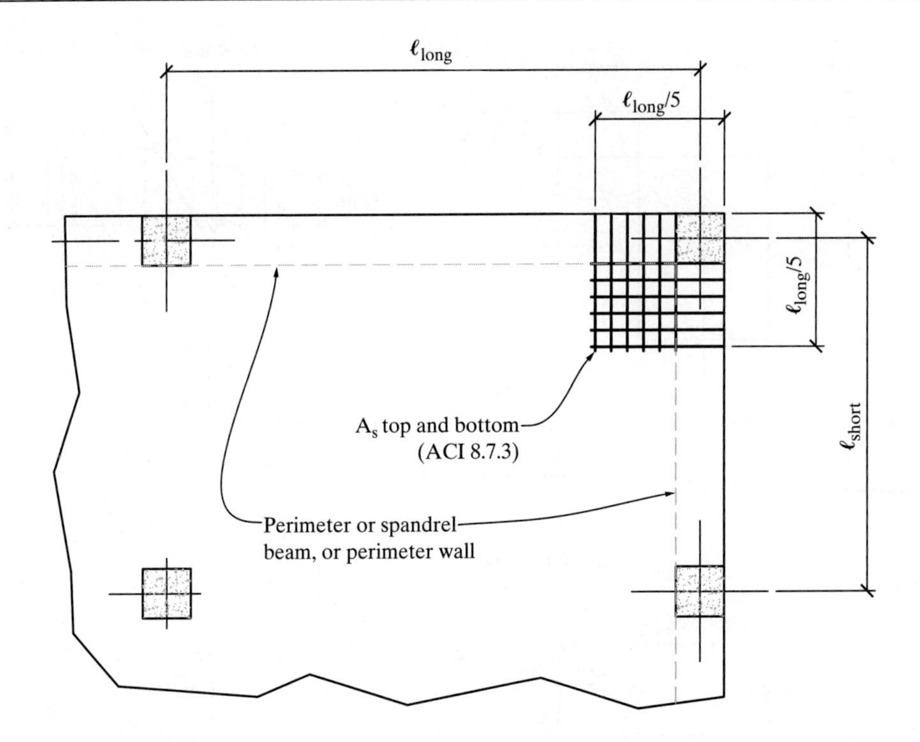

FIGURE 6-36 Plan view, slab corner reinforcing (ref. ACI R8.7.3.1).

Corner Columns: The equations for A_c and J_c for corner column connections are considerably more cumbersome to derive and are not given in this text. The interested reader should refer to Park and Gamble [1] for these equations.

Corner Reinforcement in Beam or Wall-Supported Two-Way Slabs (ACI 8.7.3)

When a two-way slab is supported on perimeter walls or beams with α_f greater than 1.0, reinforcement shall be provided at the top and bottom of the slab at the exterior corner of the slab to resist a factored moment equal to the maximum factored positive moment per unit width in the slab panel. The reinforcement shall be extended in both orthogonal directions from the edge of the slab a distance of 20% of the longer end span of the slab panel as shown in Figure 6-36 (ACI Code, Section R8.7.3.1).

Effective Depth of Two-Way Slabs

Since the top and bottom reinforcement in two-way slabs are placed in two orthogonal directions, the effective depth in the two directions will be different (see Figure 6-37).

Typically, the reinforcement spanning the long direction of the slab panel is placed nearest to the bottom and top surfaces of the slab, and therefore will have a larger effective depth because of the larger moments in the longer span direction. Since the effective depth is a variable that is needed to design two-way slabs, an effective depth needs to be assumed for both directions. Assuming interior clear cover to the reinforcement of ¾″ and assuming #6 rebar in both directions, the approximate effective depths in the

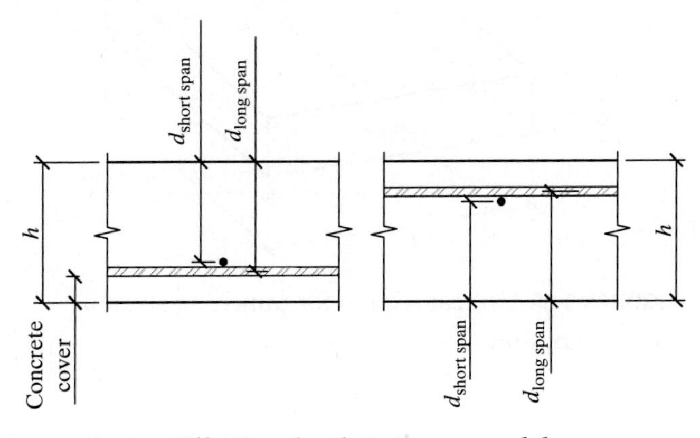

FIGURE 6-37 Effective depth in two-way slab.

short and long directions of the slab panel are (see Figure 6-37):

$$d_{long} = h - \frac{3''}{4} - \frac{0.75''}{2} = h - 1.13''$$

$$d_{short} = h - \frac{3''}{4} - 0.75 - \frac{0.75''}{2} = h - 1.9''$$

$$d_{average} = h - 1.5''$$

For the calculation of the shear strength, ϕV_c, the average effective depth should be used (ACI Code, Section 22.6.2.1).

Structural Integrity Reinforcement in Two-Way Slabs (ACI 8.7.4.2)

To prevent progressive collapse of column-supported two-way slabs, all the bottom reinforcement in the column strip in each direction shall be continuous. Continuity can be achieved by using class B tension lap splice or spliced with a fully mechanical or fully welded tension splice located in the regions shown in Figure 6-38 (ACI Code, Section 8.7.4.2.1). At least two bottom bars in the column strip must pass within the area bounded by the column vertical reinforcement, and these structural integrity reinforcement must be adequately

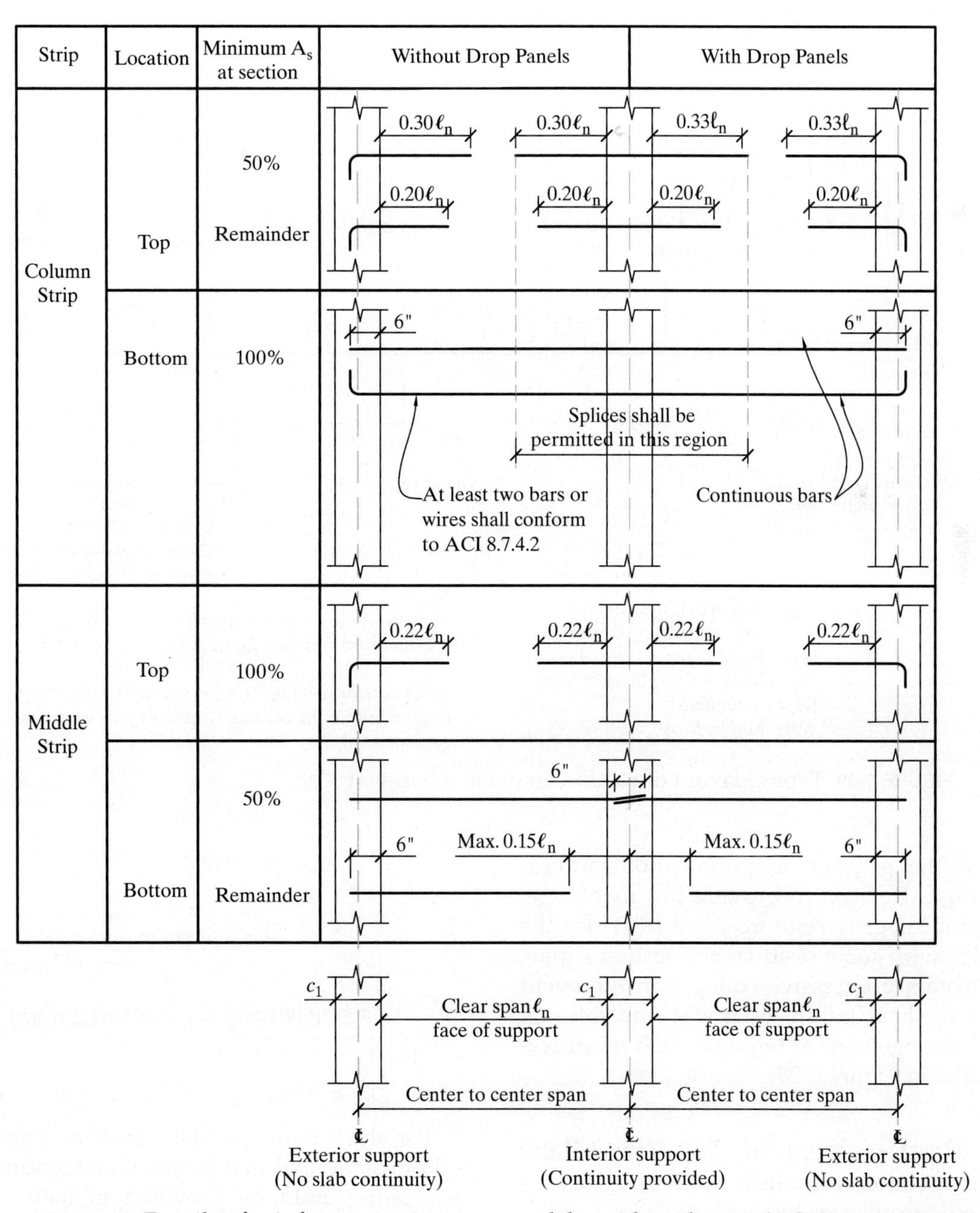

FIGURE 6-38 Details of reinforcement in two-way slabs without beams (ACI Figure 8.7.4.1.3a).

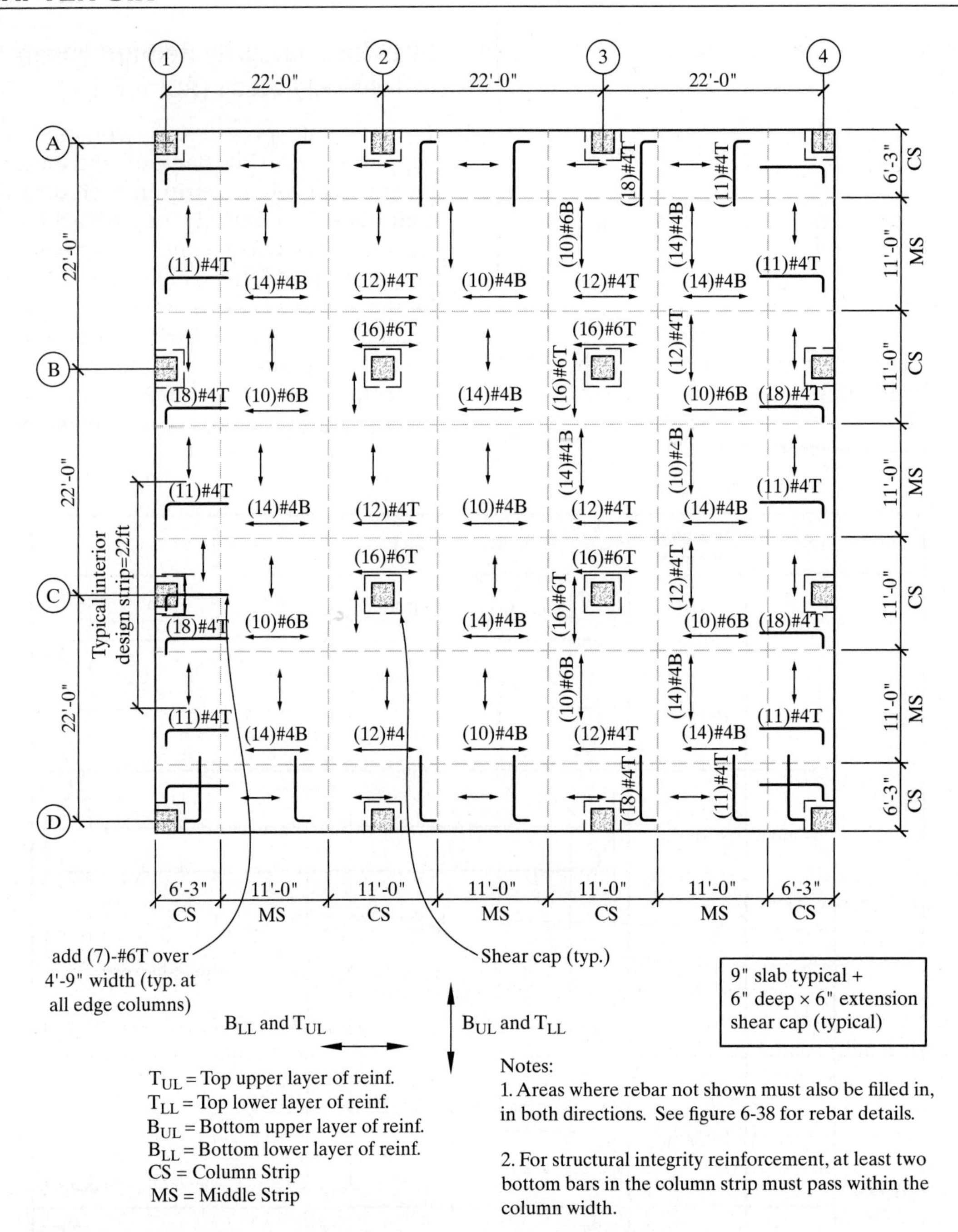

FIGURE 6-39 Typical layout of reinforcement in a two-way slab.

anchored at the exterior supports, and standard hooks are typically used to provide the anchorage. The structural integrity reinforcement provides the two-way slab with some residual strength or capacity that prevents progressive collapse in the event that punching shear failure occurs at one column. The typical arrangement of two-way slab reinforcement is shown in Figure 6-39.

Minimum Reinforcement in Two-Way Slabs

The minimum reinforcement helps to control cracking in the slab and are given as follows:

For $f_y \geq 60{,}000$ psi:

$$A_{s\,\min} \geq \frac{0.0018(60{,}000)}{f_y} A_g$$

$$= \frac{0.0018(60{,}000)}{f_y}(12\text{ in.})(h) \quad \text{in.}^2/\text{ft}$$

$$\geq 0.0014A_g = 0.0014(12\text{ in.})(h) \quad \text{in.}^2/\text{ft}$$

For $f_y < 60{,}000$ psi:

$$A_{s\,\min} \geq 0.002\,A_g = 0.002(12\text{ in.})(h) \quad \text{in.}^2/\text{ft}$$

For thick two-way slabs such as transfer slabs, podium slabs, and mat foundations, continuous reinforcement should be provided in both orthogonal directions at the top and bottom faces of the thick slab (ACI Code, Section R8.6.1.1). This ensures that any

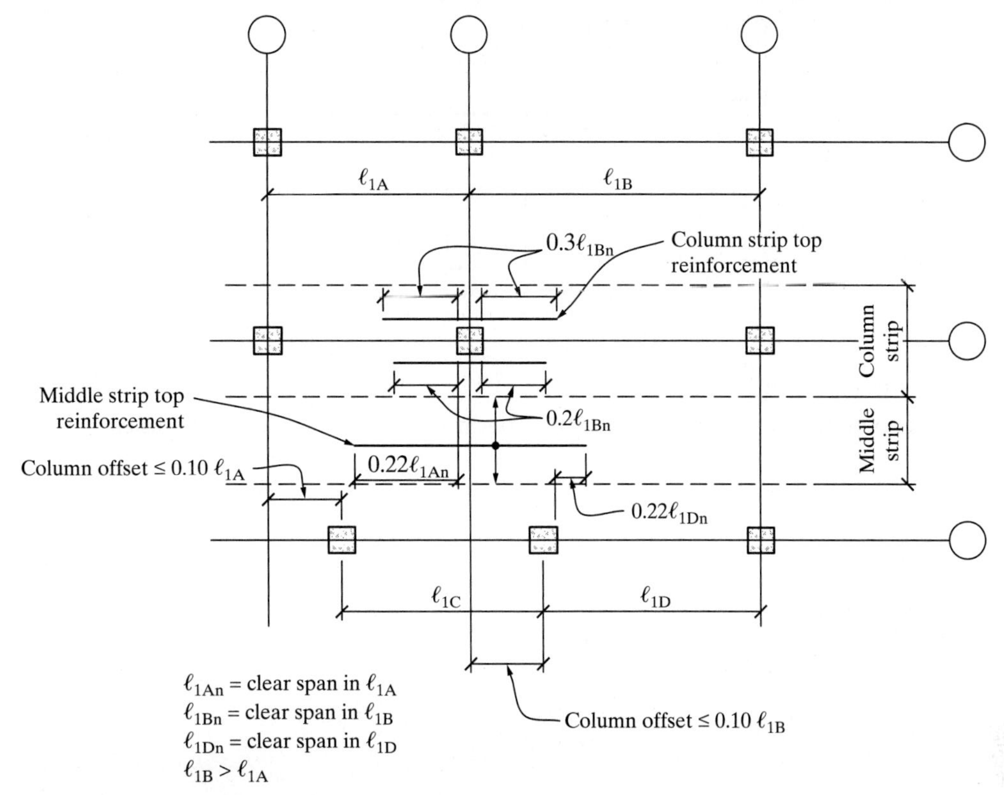

FIGURE 6-40 Reinforcement layout in two-way slabs with column offset, and maximum column offset.

potential punching shear cracks is intercepted by the tension reinforcement.

Maximum Reinforcement Spacing in Two-Way Slabs (ACI 8.7.2.2) The maximum spacing of the reinforcement in two-way slabs at critical sections is the smaller of $2h$ and 18 inches, where h is the thickness of the slab. Limiting the spacing of the reinforcement helps to control cracking and also provides resistance against heavy concentrated loads (e.g., forklift wheel loads) acting over small areas (ACI Code, Section R8.7.2.2) (Figure 6.39).

The layout of the column strip and middle strip top reinforcement in two-way slabs with column offset, and the maximum permitted column offset for the direct design method are shown in Figure 6.40.

Openings in Two-Way Slabs

The reduction effects of openings and free edges on the critical perimeter for punching shear in a two-way slab are illustrated in Figure 6-41. The effect of openings in two-way slabs located within a column strip or located within a distance of 10h (h is the slab depth) from the column is obtained by projecting two radial lines originating from the center of the closest column to the outermost edges of the opening; and the length of the critical perimeter within the radial lines is considered ineffective and is neglected in the shear strength calculations. Slabs reinforced with shear heads and with openings located within a column strip or located within $10h$ of the concentrated load or column must meet the requirements of ACI Code, Section 22.6.9.9 (ACI Code, Section 8.5.4.2d).

The allowable sizes of openings in two-way slabs are shown in Figure 6-42. Typical reinforcement at slab openings are shown in Figure 6-43.

Effect of Transverse Edge Beams

Where there is no transverse edge beam, all of the exterior support negative moment in the ℓ_1 direction is assigned to the column strip (see Table 6-7). When there is an edge beam, the value of the torsional stiffness parameter, β_t, of the edge beam determines how much of the exterior negative moment is assigned to the middle strip. The torsional stiffness parameter, β_t, accounts for the torsional stiffness of the edge beam or wall; for the moment in the ℓ_1 direction in Figure 6-44, this would be the beam or wall along grid line 1. For a slab supported along the edge on a masonry wall or on a concrete wall that is monolithic with the slab, the values of β_t are given in Figure 6-44 (ACI Code, Section R8.10.5.2). If the width of the edge column or wall is at least $0.75\,\ell_2$, the exterior negative moment shall be distributed uniformly across the transverse length, ℓ_2 (ACI Code, Section 8.10.5.4). The torsional stiffness parameter for edge beams is defined as

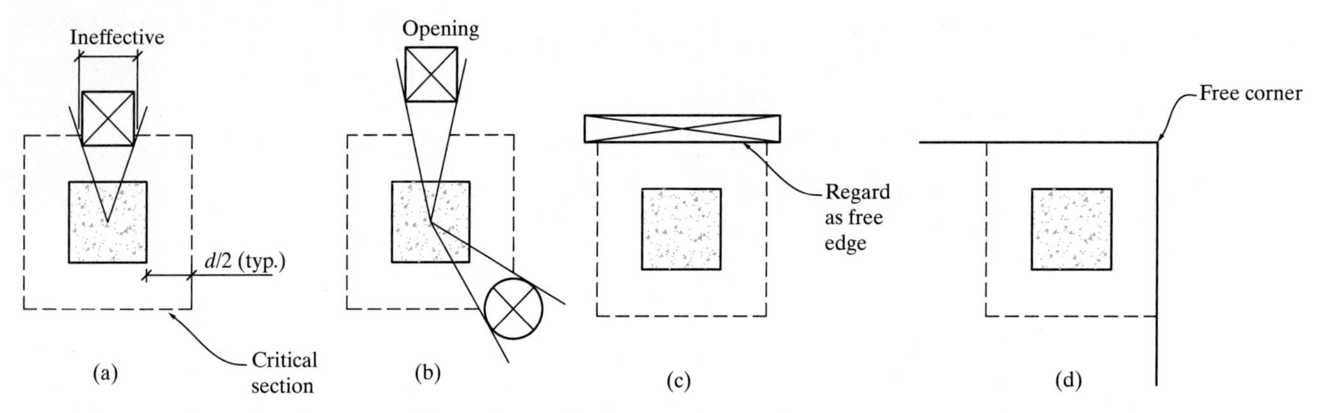

FIGURE 6-41 Effect of openings and free edges-effective perimeter shown with dashed lines (ACI Fig. R22.6.4.3).

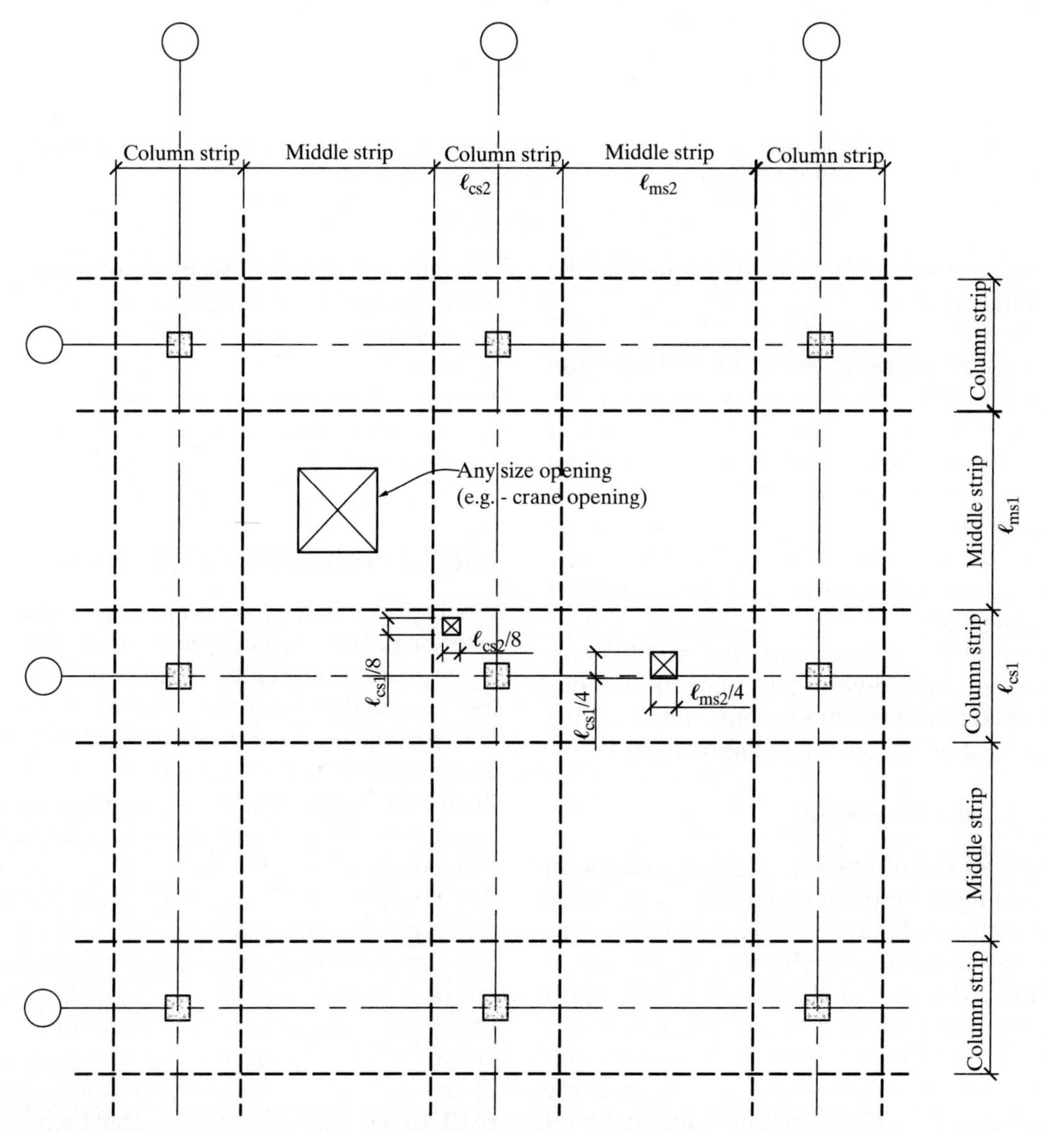

FIGURE 6-42 Allowable sizes of openings in two-way slabs.

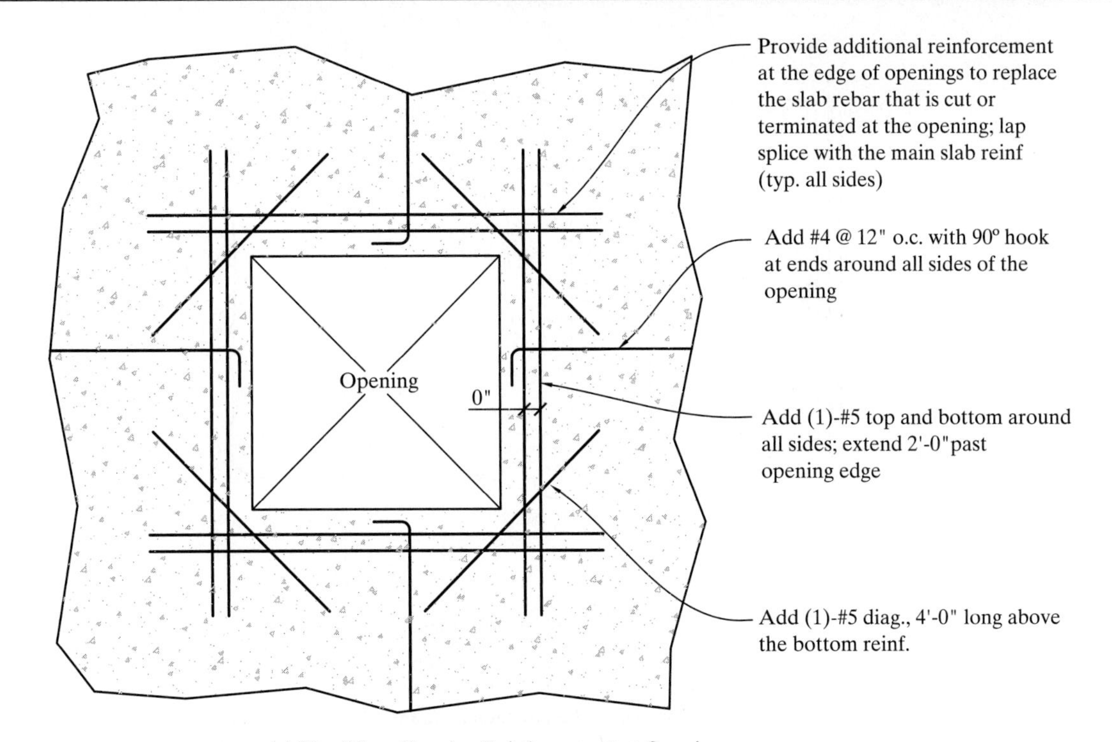

(a) Plan View Showing Reinforcement at Openings

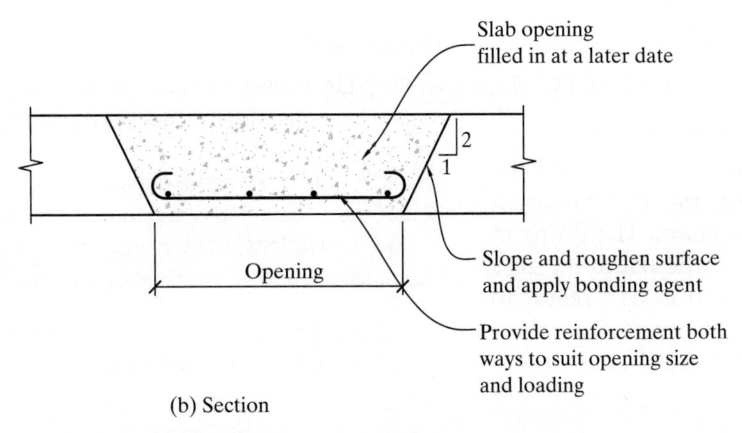

(b) Section

FIGURE 6-43 Typical reinforcement layout at slab openings.

$$\beta_t = \frac{E_{cb}C}{2E_{cs}I_s} \qquad \text{[ACI Eq. 8.10.5.2a]}$$

where

$$C = \sum\left(1 - 0.63\frac{x}{y}\right)\frac{x^3y}{3} \qquad \text{[ACI Eq. 8.10.5.2b]}$$

where

x = smaller dimension of a rectangular area
y = larger dimension of a rectangular area
I_s = moment of inertia of the slab

The other terms in the equations have been defined previously in this chapter.

Slab Systems with Beams Along All Four Edges of a Panel

For slabs with beams along all the four edges of each rectangular slab panel, the distribution of the uniform gravity loads to the supporting beams is as shown in Figure 6-45. In addition, each beam shall resist the shears caused by the factored loads directly supported by the beams (e.g., wall loads) including the weight of the beam stem (ACI Code, Section 8.10.8.3).

When a beam has a stiffness ratio $\alpha_{f1}\frac{\ell_2}{\ell_1}$ greater than or equal to 1.0, the beam resists 100% of the shear force from the tributary are in Figure 6-45.

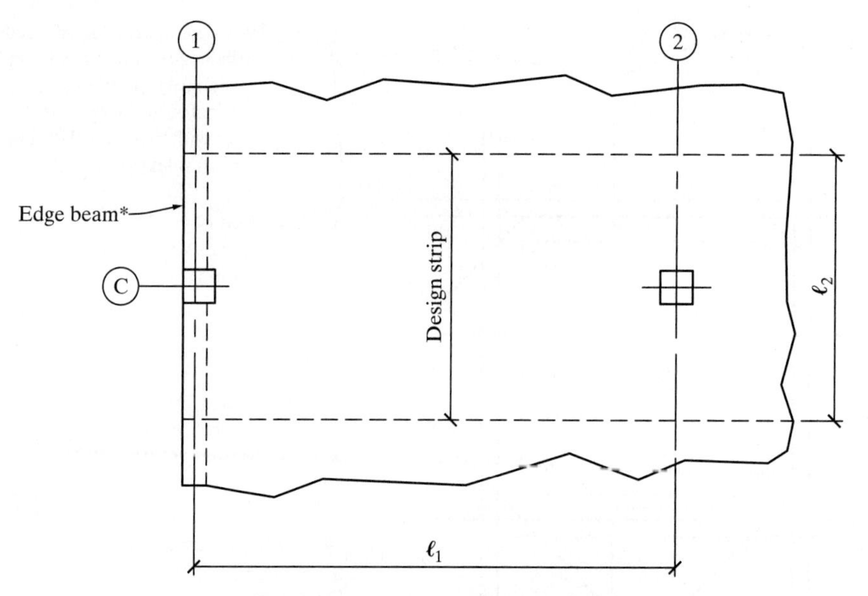

*For slab supported on wall at the edge (Line C)

$\beta_t = 0$ for masonry wall (since wall has no torsional resistance)
$= 2.5$ for concrete wall

ℓ_1 = length of span in the direction moment is being determined

ℓ_2 = transverse width of the design strip, perpendicular to the ℓ_1 direction

FIGURE 6-44 Effect of transverse edge beam on moment distribution.

Therefore, the portion of shear resisted by a beam in beam-supported slabs is $\alpha_{f1}\frac{\ell_2}{\ell_1} \leq 1.0$ (see ACI Code, Table 8.10.8.1). When the beam stiffness ratio, $\alpha_{f1}\frac{\ell_2}{\ell_1}$, is less than 1.0, only a portion of the shear from the tributary area in Figure 6-45 plus the direct load on the beam will be resisted by the beam. The uniform load on the remaining portion of the tributary area will be resisted by the slab, and this will result in shear stresses at the slab-column connection similar to what obtains for flat plates and flat slabs.

Example 6-3

A flat plate shown in Figure 6-46 is supported on 18 in. square columns with 6 in. deep shear caps on a 22 ft × 22 ft column grid and the slab is continuous over at least three spans in both orthogonal directions. f'_c is 4000 psi and f_y is 60,000 psi. Assuming a superimposed dead load of 30 psf (includes partitions, mechanical and electrical, ceiling, and floor finishes) and an occupancy live load of 100 psf,

a. Check if the direct design method can be used for the analysis of this two-way slab.

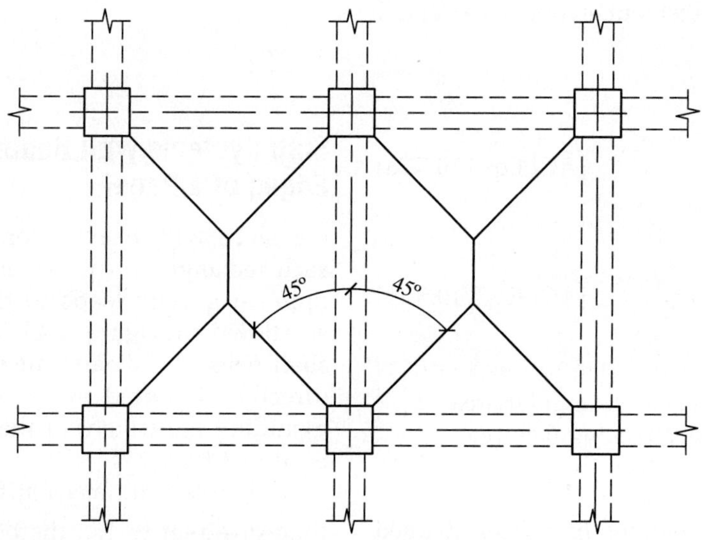

FIGURE 6-45 Tributary area for shear on interior beams (ACI Code Figure 8.10.8.1).

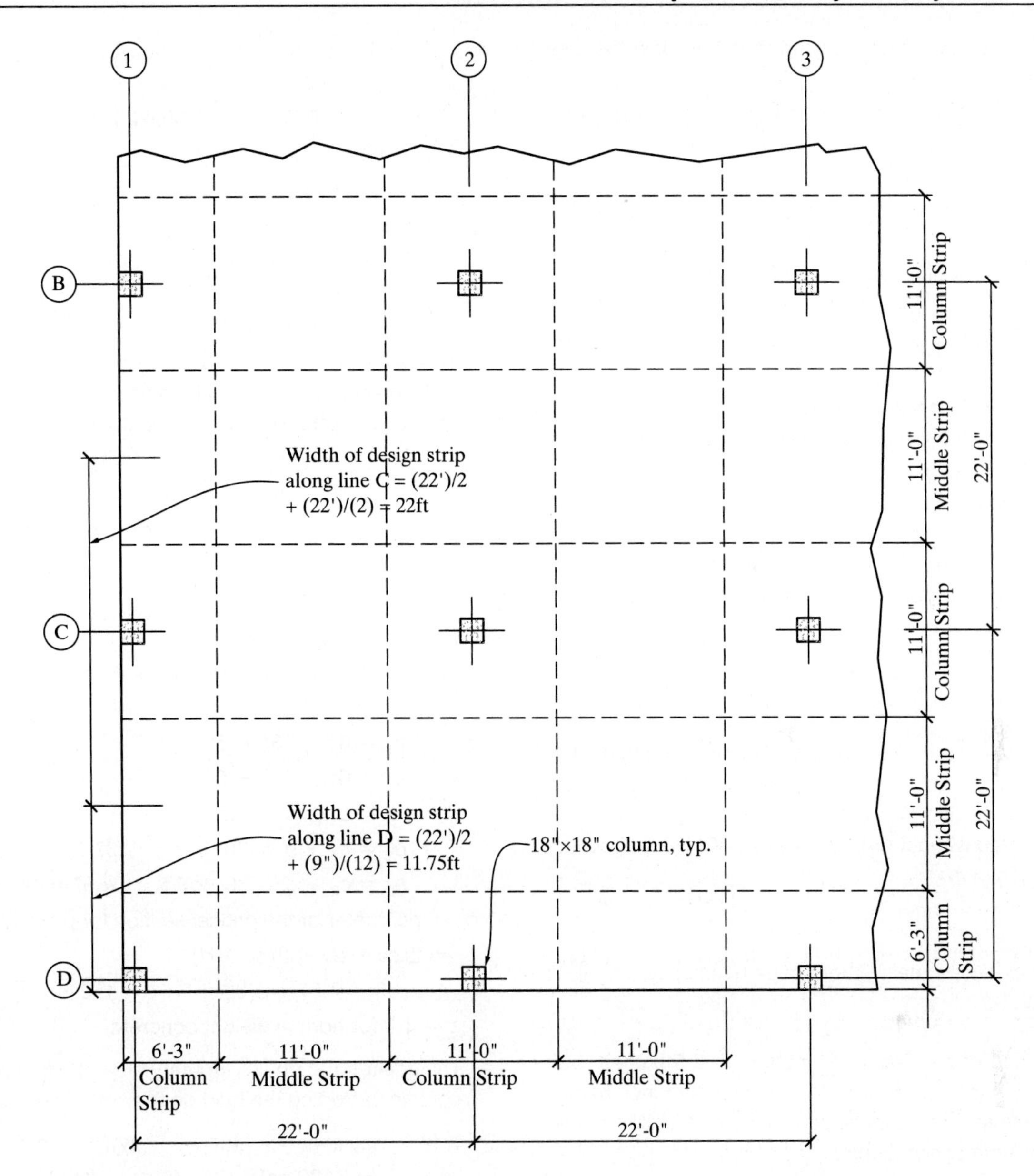

FIGURE 6-46

b. Select the slab thickness.
c. Calculate the total factored load on the slab in psf.
d. Divide the slab into column and middle strips and determine the widths of the column and middle strips.
e. Calculate the factored static moment for the East-West design strip along grid line C.
f. Calculate the effective depths of the slab reinforcement and the average effective depth.
g. Calculate the direct shear demand, V_u, at a typical interior column and check against the shear strength, ϕV_c.
h. Calculate the factored longitudinal design moments for each negative and positive regions of the slab for the East-West design strip along grid line C.
i. Using the moments from part (h), calculate the factored design moments distributed transversely to the column and middle strips, and to the longitudinal beams, if any.
j. Select the reinforcement for the column and middle strip for the design strip along line C.
k. Check the shear plus unbalanced moment transfer at the typical interior column C-2.
l. Check the shear plus unbalanced moment transfer at the typical edge column C-1. Assume an exterior cladding load of 25 psf and a floor-to-floor height of 10 ft.

Solution:

a. Check if the direct design method can be used (see *Analysis and Design of Continuous Two-Way Slabs*)
   1. The ratio of the unfactored live load to unfactored dead load (see load calculations below) is w_L/w_D, $= 100 \text{ psf}/143 \text{ psf} = 0.70 < 2.0$. OK, therefore, pattern live load is not critical and does not need to be checked.
   2. There are at least 3 continuous spans in each direction. OK

3. There is no column offset, therefore offset is less than 10%. OK
4. Ratio of long side to short side of rectangular panel measured center to center of supports $= 22\text{ ft}/22\text{ ft} = 1.0 < 2.0$. OK
5. The lengths of successive spans are equal, therefore. OK
6. The loads are uniformly distributed.
7. The slab is not supported on beams on all four sides of the panels, therefore, the relative stiffness criteria does not apply.
8. Therefore, per ACI Code, Section 8.10.2, the direct design method can be used for the analysis of this two-way slab.

b. Select slab minimum thickness

$$f_y = 60{,}000 \text{ psi}$$

Minimum thickness for two-way slabs without beams: The clear span in the long direction which is the span face-to-face of columns is,

$$\ell_n = 22' - \left(\frac{18/2}{12}\right) - \left(\frac{18/2}{12}\right) = 20.5'$$

Since there are no beams spanning between supports on all sides of a slab panel, use Table 6-4 (see *Minimum Thickness of Two-Way Slabs*)

h_{min} for slab without drop panels per ACI Code, Section 8.2.4: 5 in.

For an interior panel, $h_{min} = \frac{\ell_n}{33} = \frac{20.5(12)}{33} = 7.5$ in

For an exterior panel (without edge beams), $h_{min} = \frac{\ell_n}{30} = \frac{20.5(12)}{30} = 8.2\,in$

Rounding to the nearest ½ inch, $h = 8.5$ inches, but **use $h = 9$ inches** to allow for the expected increase in shear stress from unbalanced moment transfer at the slab-column connections.

c. Calculate the factored uniform load on the slab in psf

Slab self-weight $= (9"/12)(150 \text{ pcf}) = 113$ psf

Superimposed dead load $= 30$ psf

Total dead load $= 143$ psf

Live load $= 100$ psf

The factored uniform load, $w_u = 1.2(143 \text{ psf}) + 1.6(100\, psf) =$ **332 psf**

d. For column and middle strips, see Figure 6-45. The width of the typical interior column strip on each side of the column line (ACI Code, Section 8.4.1.5)

$=$ smaller of $0.25\ell_1$ and $0.25\ell_2 = 0.25(22\text{ ft}) = 5.5$ ft

Therefore, the total width of a typical interior column strip $= 5.5\text{ ft} + 5.5\text{ ft} = 11$ ft

The width of each interior design strip $= (22\text{ ft}/2) + (22\text{ ft}/2) = 22$ ft Therefore, the total width of the middle strip for each design strip $= 22\text{ ft} - 11\text{ ft} = 11$ ft

That is, for each 22 ft wide design strip, there is a 5.5. ft (i.e., 11 ft/2) wide middle strip on each side of the column strip. See Figure 6-46 for the layout of the column and middle strips.

e. Factored static design moment, M_o:

$$\ell_{n1} = 20.5'$$
$$\geq 0.65\,\ell_1 = (0.65)(22') = 14.4'$$

Therefore, use $\ell_{n1} = 20.5'$

From *Static Longitudinal Moment, M_o*, the factored static moment is calculated as

$$M_o = \frac{w_u \ell_2 \ell_n^2}{8} = \frac{(332\text{ psf})(22')(20.5')^2}{8}$$
$$= 384 \text{ ft-kip}$$

This is the factored total static moment for a typical 22 ft wide interior design strip along grid line C.

f. Effective depths of reinforcement (see *Effective Depth of Two-Way Slabs*)

$$d_{E-W} = h - 1.1" = 9 - 1.1 = 7.9"$$
$$d_{N-S} = h - 1.9" = 9 - 1.9 = 7.1"$$
$$d_{average} = h - 1.5" = 9 - 1.5 = 7.5"$$

g. Direct shear at typical interior column (see Figure 6-47) Without the shear cap, $c_1 = c_2 = 18"$ (i.e., the size of the column), but because of the shear cap that is 6" wide (all around the column), the revised

$$c_1 = 6" + 18" + 6" = 30"$$

Also, $c_2 = 6" + 18" + 6" = 30"$

$$d = d_{average} = 7.5"$$
$$b_1 = c_1 + d = 30 + 7.5 = 37.5" = 3.13 \text{ ft}$$
$$b_2 = c_2 + d = 30 + 7.5 = 37.5" = 3.13 \text{ ft}$$

$b_o =$ perimeter of the critical section for punching shear

$= 2(c_1 + d) + 2(c_2 + d)$

$= 2(30 + 7.5) + 2(30 + 7.5) = 150$ in.

$\lambda = 1.0$ for normal weight concrete

The shear force on the tributary area of a typical interior column excluding the load on the critical perimeter is

$$V_u(2 - \text{way shear}) = w_u \ell_1 \ell_2 - w_u(c_1 + d)(c_2 + d)$$
$$= (332 \text{ psf})[(22')(22') - (3.13')(3.13')]$$
$$= 157{,}435 \text{ lb} = 157 \text{ kip}$$

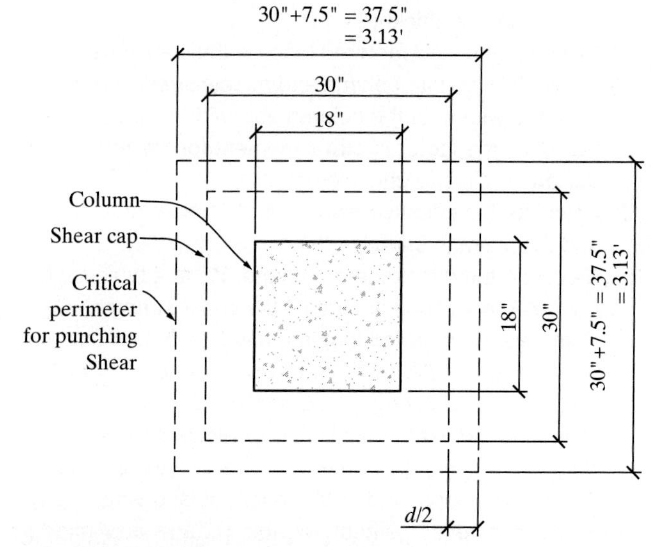

FIGURE 6-47 Critical section for shear.

β = ratio of the long cross-sectional dimension to the short cross-sectional dimension of the column, concentrated load or reaction $= 18''/18'' = 1.0$

$\alpha_s = 40$, for interior columns

The concrete shear strength in two-way shear is the smallest of the following equations derived from ACI Code, Table 22.6.5.2:

$$\phi V_c = 4\phi\lambda\sqrt{f'_c}b_o d = (4)(0.75)(1)(\sqrt{4000\,\text{psi}})(150'')(7.5'') = 214\,\text{kip}$$

$$\phi V_c = \phi\left(2 + \frac{4}{\beta}\right)\lambda\sqrt{f'_c}b_o d = (0.75)\left(2 + \frac{4}{1}\right)(1)(\sqrt{4000\text{ psi}})(150'')(7.5'') = 320\text{ kip}$$

$$\phi V_c = \phi\left(2 + \frac{\alpha_s d}{b_o}\right)\lambda\sqrt{f'_c}b_o d = (0.75)\left(2 + \frac{40(7.5)}{150}\right)(1)(\sqrt{4000\,\text{psi}})(150'')(7.5'') = 214\,\text{kip}$$

Therefore, use $\phi V_c = 214$ kip $> V_u = 157$ kip OK.

The slab is adequate for direct shear at the interior column.

h. Longitudinal distribution of the static Moment, M_o, parallel to the span of the slab:

$M_o = 384$ ft-kip (see part (e)).

$0.26M_o = 100$ ft-kip (see Figure 6-48)

$0.52M_o = 200$ ft-kip

$0.70M_o = 269$ ft-kip

$0.65M_o = 250$ ft-kip

$0.35M_o = 134$ ft-kip

The longitudinal moments are summarized in Table 6-10

i. Transverse distribution of the longitudinal moments from part (h) to the column strip, longitudinal beams, if any, and to the middle strips (see *Longitudinal Distribution of the Total Factored Static Moment, M_o*):

There are no interior longitudinal beams in this slab, therefore, $\alpha_f = 0$

Since there are no edge beams in this slab, $\beta_t = 0$

$$\frac{\ell_2}{\ell_1} = \frac{22'}{22'} = 1.0$$

Portion of **interior** negative moment, M_u, in column strip $= 0.75$ (see Table 6-6)

Therefore, the portion of **interior** negative moment in the middle strip (i.e., the two half middle strips) $= 1 - 0.75 = 0.25$

Portion of **exterior** negative moment, M_u, in column strip $= 1.0$ (see Table 6-7)

Therefore, the portion of **exterior** negative moment in the middle strip (i.e., the two half middle strips) $= 1 - 1 = 0$

Portion of **positive** moment, M_u, in column strip $= 0.6$ (see Table 6-8)

Therefore, the portion of **positive** moment in the middle strip (i.e., the two half middle strips) $= 1 - 0.6 = 0.4$

There are no longitudinal beams, therefore, no moments are transferred to the beams. The transverse distribution of moments for the design strip along grid line C is summarized in Table 6-11.

j. For the column strip and middle strip reinforcement, see Table 6-11.

k. Shear plus unbalanced moment transfer at the **typical interior column C-2:**

The unbalanced moment transferred at a typical interior column C-2 is, $M_u = 269$ ft-kip $- 250$ ft-kip $= 19$ ft-kip (see Table 6-11 and Figure 6-49)

From part (g), $V_u = 157$ kip

Without the shear cap, $c_1 = c_2 = 18"$ (i.e., the size of the column), but because of the shear cap that is 6" wide (all around the column), the revised

$$c_1 = 6'' + 18'' + 6'' = 30''$$

$$c_2 = 6'' + 18'' + 6'' = 30''$$

$$c_1/c_2 = 30/30 = 1.0$$

$$d = d_{\text{average}} = 7.5''$$

$$b_1 = c_1 + d = 30 + 7.5 = 37.5'' = 3.13\text{ ft}$$

$$b_2 = c_2 + d = 30 + 7.5 = 37.5'' = 3.13\text{ ft}$$

$$b_1/b_2 = 1.0$$

$b_o =$ perimeter of the critical section for punching shear $= 2(c_1 + d) + 2(c_2 + d) = 2(30 + 7.5) + 2(30 + 7.5) = 150$ in

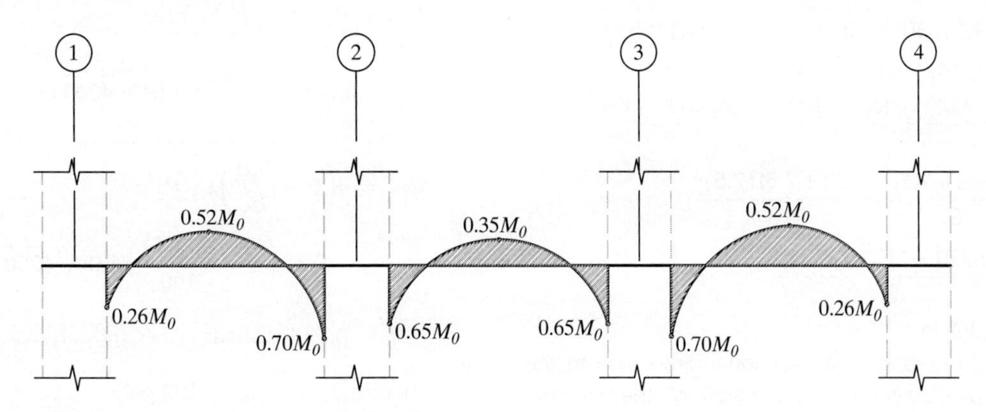

FIGURE 6-48 Longitudinal distribution of static moment, M_o.

TABLE 6-10 Longitudinal Distribution of the Static Moment, M_o, Parallel to the Span of the Slab (i.e., M_u)

	Grid line 1	Midspan	Grid line 2 (left)	Grid line 2 (right)	Midspan	Grid line 3 (left)	Grid line 3 (right)	Midspan	Grid line 4
Negative moment (ft-kip)	100		269	250		250	269		100
Positive moment (ft-kip)		200			134			200	

TABLE 6-11 Transverse Distribution of Moment, M_u, Perpendicular to the Span of the Slab

	Grid line 1	Midspan	Grid line 2 (left)	Grid line 2 (right)	Midspan	Grid line 3 (left)	Grid line 3 (right)	Midspan	Grid line 4
Transverse distribution factors for column strip	1.0	0.6	0.75	0.75	0.6	0.75	0.75	0.6	1.0
Column strip moments in ft-kip. (width = 11 ft = 132 in.)	100	120	202	202	81	202	202	120	100
Transverse distribution factors for middle strip	0	0.4	0.25	0.25	0.4	0.25	0.25	0.4	0
Middle strip moment in ft-kip. (width = 11 ft = 132 in.)	0	80	68	68	54	68	68	80	0
Area of reinforcement, A_s, required [1] (Column strip)	18#4T	10#6B	16#6T	16#6T	14#4B	16#6T	16#6T	10#6B	18#4T
Area of reinforcement, A_s, required [1] (Middle strip)	11#4T	14#4B	12#4T	16#6T	10#4B	12#4T	12#4T	14#4B	11#4T

[1]Calculate the area of reinforcement using the method presented in Chapter 2. See the reinforcement layout in Figure 6-39.

$$\gamma_f = \frac{1}{1+\left(\frac{2}{3}\right)\sqrt{\frac{b_1}{b_2}}} = \frac{1}{1+\left(\frac{2}{3}\right)\sqrt{1.0}} = 0.60$$

[ACI Code, Eq. 8.4.2.3.2]

$$\gamma_v = 1 - 0.60 = 0.40$$

[ACI Code, Eq. 8.4.4.2.2]

A_c = area of the critical surface for two-way shear

$$= 2d(c_1 + c_2 + 2d)$$

$$= (2)(7.5)[30 + 30 + 2(7.5)] = 1125 \text{ in.}^2$$

$$c_{AB} = c_{CD} = (c_1 + d)/2 = (30 + 7.5)/2 = 18.75 \text{ in.}$$

$$J_c = \frac{d(c_1+d)^3}{6} + \frac{(c_1+d)d^3}{6} + \frac{d(c_2+d)(c_1+d)^2}{2}$$

$$J_c = \frac{(7.5)(30+7.5)^3}{6} + \frac{(30+7.5)(7.5)^3}{6} + \frac{(7.5)(30+7.5)(30+7.5)^2}{2}$$

$$= 266{,}309 \text{ in.}^4$$

β = ratio of the long cross-sectional dimension to the short cross-sectional dimension of the column, concentrated load or reaction = 18"/18" = 1.0 = 18"/18" = 1.0

The allowable concrete shear stress in two-way shear is the smallest of the following equations from ACI Code, Table 22.6.5.2:

$$\phi v_c = 4\phi\lambda\sqrt{f'_c} = 4(0.75)(1.0)\sqrt{4000 \text{ psi}} = 190 \text{ psi}$$

$$\phi v_c = \phi\left(2 + \frac{4}{\beta}\right)\lambda\sqrt{f'_c} =$$

$$(0.75)\left(2 + \frac{4}{1.0}\right)(1.0)\sqrt{4000 \text{ psi}} = 285 \text{ psi}$$

$$\phi v_c = \phi\left(2 + \frac{\alpha_s d}{b_o}\right)\lambda\sqrt{f'_c} =$$

$$(0.75)\left(2 + \frac{40(7.5)}{150}\right)(1.0)\sqrt{4000 \text{ psi}} = 190 \text{ psi}$$

$$\phi = 0.75 \text{ for shear}$$

Therefore, $\nu_{\text{allowable}}$ = 190 psi

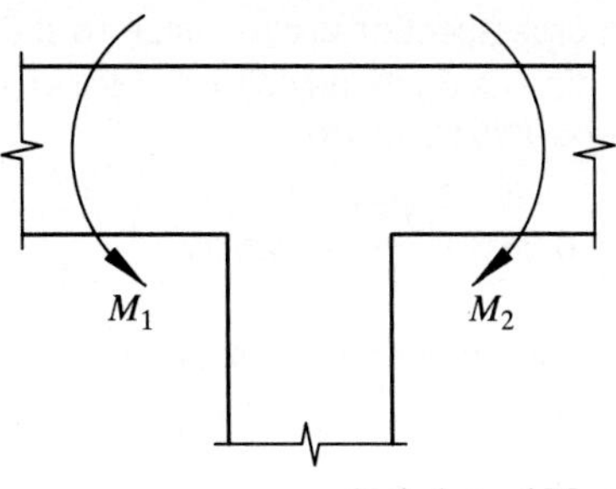

(a) Interior Column

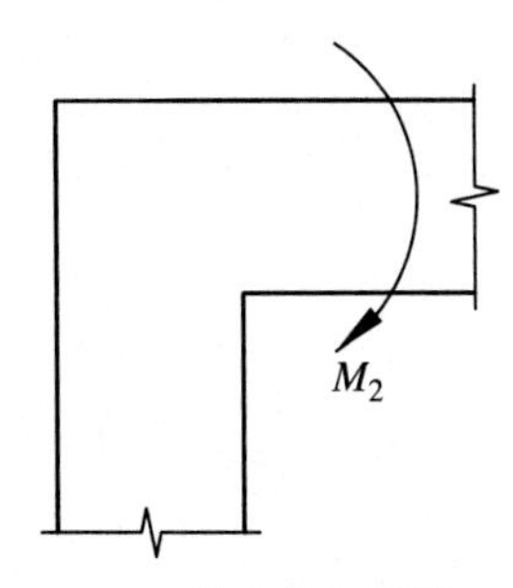

(b) Exterior Column

FIGURE 6-49 Unbalanced moments at columns.

The maximum factored shear stresses under combined shear and unbalanced moment are:

$$\nu_{u,AB} = \frac{V_u}{A_c} + \frac{\gamma_v M_u c_{AB}}{J_c}$$

$$= \frac{(157)(1000)}{1125} + \frac{(0.40)(19)(12{,}000)(18.75)}{266{,}309}$$

$$= 146 \text{ psi} \leq \nu_{\text{allowable}} = 190 \text{ psi} \quad \text{O.K.}$$

$$\nu_{u,CD} = \frac{V_u}{A_c} - \frac{\gamma_v M_u c_{CD}}{J_c}$$

$$= \frac{(157)(1000)}{1125} - \frac{(0.40)(19)(12{,}000)(18.75)}{266{,}309}$$

$$= 133 \text{ psi} \leq \nu_{\text{allowable}} = 190 \text{ psi O.K.}$$

Therefore, the shear stress at the interior column due to combined direct shear and unbalanced moment is adequate.

Calculate the moment transferred by flexure at slab-column connection C-2:

Since γ_f = 0.60, the unbalanced moment transferred from the slab to the column through flexure is

$$M_f = \gamma_f M_u = (0.6)(19 \text{ ft-kip}) = 11.4 \text{ ft-kip}$$

This is the moment transferred by flexure at the slab-column connection at column C-2, and the effective transfer width for transfer of this moment by flexure is

$$= c_2 + 3h = 30'' + (3)(9'') = 57'' = 4.75 \text{ ft (see ACI Code, Section 8.4.2.3.3)}$$

Therefore, M_f = 11.4 ft-kip/(4.75 ft) = 2.4 ft-kip/ft width

By inspection, this moment is very small and does not require any additional reinforcement.

I. Shear plus unbalanced moment transfer at the **typical edge column C-1**

The cladding load, w_{cladding} = 25 psf(10 ft.) = 250 lb/ft

Without the shear cap, $c_{1,\text{ column}} = c_{2,\text{ column}}$ = 18″ (i.e., the size of the column), but because of the shear cap that is 6″ wide (all around the edge column), the revised

$$c_1 = 6'' + 18'' = 24'' = 2.0 \text{ ft}$$

$$c_2 = 6'' + 18'' + 6'' = 30'' = 2.5 \text{ ft}$$

$$c_1/c_2 = 24/30 = 0.8$$

$$d = d_{\text{average}} = 7.5''$$

$$b_1 = c_1 + d/2 = 24'' + 7.5''/2 = 27.75'' = 2.31 \text{ ft}$$

$$b_2 = c_2 + d = 30'' + 7.5'' = 37.5'' = 3.13 \text{ ft}$$

$$b_1/b_2 = 27.75''/37.5'' = 0.74$$

b_o = perimeter of the critical section for punching shear at the edge column C-1 is

$$= 2(c_1 + d/2) + (c_2 + d) = 2(24 + 7.5/2). + (30 + 7.5) = 93 \text{ in.}$$

$$\gamma_f = \frac{1}{1 + \left(\frac{2}{3}\right)\sqrt{\frac{b_1}{b_2}}} = \frac{1}{1 + \left(\frac{2}{3}\right)\sqrt{0.74}} = 0.64$$

[ACI Code, Eq. 8.4.2.3.2]

$$\gamma_v = 1 - 0.64 = 0.36$$ [ACI Code, Eq. 8.4.4.2.2]

The factored shear at the **exterior edge column, C-1**, including the slab uniform load and the exterior cladding load is given as

$$V_u = w_u \ell_2 \left(\frac{\ell_1}{2} + \frac{c_{1,\text{column}}}{2}\right) - w_u\left(c_1 + \frac{d}{2}\right)(c_2 + d) + 1.2(w_{\text{cladding}})(\ell_2 - c_{2,\text{column}})$$

$$= (332 \text{ psf})(22 \text{ ft})\left(\frac{22 \text{ ft}}{2} + \frac{1.5 \text{ ft}}{2}\right)$$

$$- (332 \text{ psf})\left(2.5 \text{ ft} + \frac{(7.5''/12)}{2}\right)(2.5 \text{ ft}) + (7.5''/12)) + 1.2(250 \text{lb/ft})(22 \text{ ft} - 1.5 \text{ ft})$$

$$= 91038 \text{ lb} = 91 \text{ kip}$$

Note that this shear force acts around the critical perimeter, but for simplification, assume that V_u will act along the face of the column (i.e., face AB in Figure 6-34(b))

Unbalanced moment transferred at a typical exterior edge column C-1 $= M_u + V_u e \geq 0.30 M_o$ (ACI Code, Section 8.10.7.3)

The negative moment, M_u, at the edge column on Grid line 1 is obtained from Table 6-10 as 100 ft-kip (see Figure 6-50)

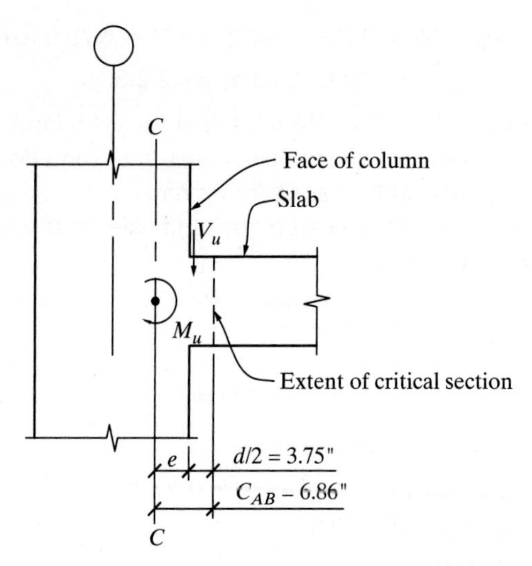

FIGURE 6-50 Elevation of slab-column connection showing moment and shear force.

The geometric properties of the critical section are calculated as follows:

For **Edge Columns with moment perpendicular to the edge** (see Figure 6-34(b)):

A_c = area of the critical surface for two-way shear

$$A_c = (2c_1 + c_2 + 2d)d$$

$$= [2(24") + 30" + 2(7.5")](7.5") = 698 \text{ in.}^2$$

$$c_{AB} = \frac{\left(c_1 + \frac{d}{2}\right)^2 (d)}{d(2c_1 + c_2 + 2d)}$$

$$= \frac{\left(24" + \frac{7.5"}{2}\right)^2 (7.5")}{(7.5")(2(24") + 30" + 2(7.5"))}$$

$$= 8.28 \text{ in.}$$

$$c_{CD} = c_1 + \frac{d}{2} - c_{AB} = 24" + \frac{7.5"}{2} - 8.28" = 19.47"$$

$$J_c = \frac{d\left(c_1 + \frac{d}{2}\right)^3}{6} + 2d\left(c_1 + \frac{d}{2}\right)\left(\frac{c_1 + \frac{d}{2}}{2} - c_{AB}\right)^2$$

$$+ \frac{\left(c_1 + \frac{d}{2}\right)(d)^3}{6} + (c_2 + d)(d)(c_{AB})^2$$

$$= \frac{(7.5")\left(24" + \frac{7.5"}{2}\right)^3}{6} + 2(7.5")\left(24" + \frac{7.5"}{2}\right)$$

$$\left(\frac{24" + \frac{7.5"}{2}}{2} - 8.28"\right)^2 + \frac{\left(24" + \frac{7.5"}{2}\right)(7.5")^3}{6}$$

$$+ (30" + 7.5")(7.5")(8.28")^2$$

$$= 60{,}975 \text{ in.}^4$$

β = ratio of the long cross-sectional dimension to the short cross-sectional dimension of the column, concentrated load or reaction = 18"/18" = 1.0

$$e = c_{AB} - \frac{d}{2} = 8.28" - \frac{7.5"}{2} = 4.53"$$

The factored static moment (previously calculated in part (e)), M_o = 384 ft-kip

The total unbalanced moment
$= M_u + V_u e$ = 100 ft-kip + (91 kip) (4.53″/12)
= **134 ft-kip**
$\geq 0.30M_o$ = 0.30(384 ft-kip) = 115 ft-kip

Therefore, the total unbalanced moment = 134 ft-kip

The allowable concrete shear stress in two-way shear is the smallest of the following equations from *Combined Shear and Unbalanced Moment Transfer in Slab-Column Connections*:

$$\phi v_c = 4\phi\lambda\sqrt{f'_c} = 4(0.75)(1.0)\sqrt{4000 \text{ psi}} = 190 \text{ psi}$$

$$\phi v_c = \phi\left(2 + \frac{4}{\beta}\right)\lambda\sqrt{f'_c} = (0.75)\left(2 + \frac{4}{1.0}\right)(1.0)\sqrt{4000 \text{ psi}}$$

$$= 285 \text{ psi}$$

$$\phi v_c = \phi\left(2 + \frac{\alpha_s d}{b_o}\right)\lambda\sqrt{f'_c} =$$

$$(0.75)\left(2 + \frac{40(7.5)}{150}\right)(1.0)\ \sqrt{4000 \text{ psi}} = 190 \text{ psi}$$

ϕ = 0.75 for shear

Therefore, $\nu_{allowable}$ = 190 psi

The maximum factored shear stresses under combined shear (V_u = 91 kip), and unbalanced moment (M_u = 134 ft-kip) are:

$$\nu_{u,AB} = \frac{V_u}{A_c} + \frac{\gamma_v M_u c_{AB}}{J_c} = \frac{(91)(1000)}{698} + \frac{(0.36)(134)(12{,}000)(8.28")}{60{,}975}$$

$$= 209 \text{ psi} > \nu_{allowable} = 190 \text{ psi Not O.K.}$$

$$\nu_{u,CD} = \frac{V_u}{A_c} - \frac{\gamma_v M_u c_{CD}}{J_c} = \frac{(91)(1000)}{698} - \frac{(0.36)(134)(12{,}000)(8.28")}{60{,}975}$$

$$= 52 \text{ psi} \leq \nu_{allowable} = 190 \text{ psi O.K.}$$

There is a 10% overstress in shear at the edge column C-1 due to combined shear and unbalanced moment. Several measures could be taken to rectify this:

1. Increase the depth and therefore, the extent of the shear cap in order to increase the dimensions of the critical section (i.e., c_1 and c_2).
2. Increase the allowable shear stress by increasing the slab concrete strength.
3. The column size could be increased to 20” square column.

Calculate the moment transferred by flexure at the slab-column connection at edge column C-1:

Since γ_f = 0.64, the unbalanced moment transferred from the slab to the column through flexure is

$$M_f = \gamma_f M_u = (0.64)(134 \text{ ft-kip}) = 86 \text{ ft-kip}$$

This is the moment transferred by flexure at the slab-column connection at column C-1, and the effective transfer width for transfer of this moment by flexure is $= c_2 + 3h = 30'' + (3)(9'') = 57'' = 4.75$ ft (see *Combined Shear and Unbalanced Moment Transfer in Slab-Column Connections*)

Therefore, $M_f = 86$ ft-kip$/(4.75$ ft$) = 18.1$ ft-kip/ft width

Using the methods from Chapter 2,

$$M_u \le \phi M_n = \phi\, bd^2\bar{k}$$

ϕ for bending $= 0.9$

$b = 12''$ since the moment is calculated above in ft-kip per ft width

$d = 7.5''$

$$M_u \le \phi M_n = \phi bd^2\bar{k}$$

$$\text{Therefore, } \phi bd^2\bar{k} \ge M_u = (18.1 \text{ ft-kip/ft})(12)$$

$$\bar{k} = \frac{18.1(12)}{(0.9)(12'')(7.5'')^2} = 0.358$$

From Table A10 in Appendix A, ρ = 0.0063

$$A_{s,reqd} = \rho bd = 0.0063(12 \text{ in.})(7.5'') = 0.567 \text{ in.}^2/\text{ft}$$

Total area of steel required over the transfer width 4.75 ft $= (0.567 \text{ in.}^2/\text{ft})(4.75 \text{ ft}) = 2.7 \text{ in.}^2$

Provide additional 7#6T over the 4.75 ft width centered on the column at the edge columns

See Figure 6-39 for the layout of the two-way slab reinforcement.

Example 6-4

1. If drop panels are to be used in the two-way slab in Example 6-3, determine the minimum size of the drop panel for a typical interior column that meets the requirements of the ACI Code.
2. Discuss the implications of a drop panel in an existing building not meeting the Code-specified length and depth requirements.

Solution:

1. The drop panel minimum size that satisfies the ACI Code is illustrated in Figure 6-51.

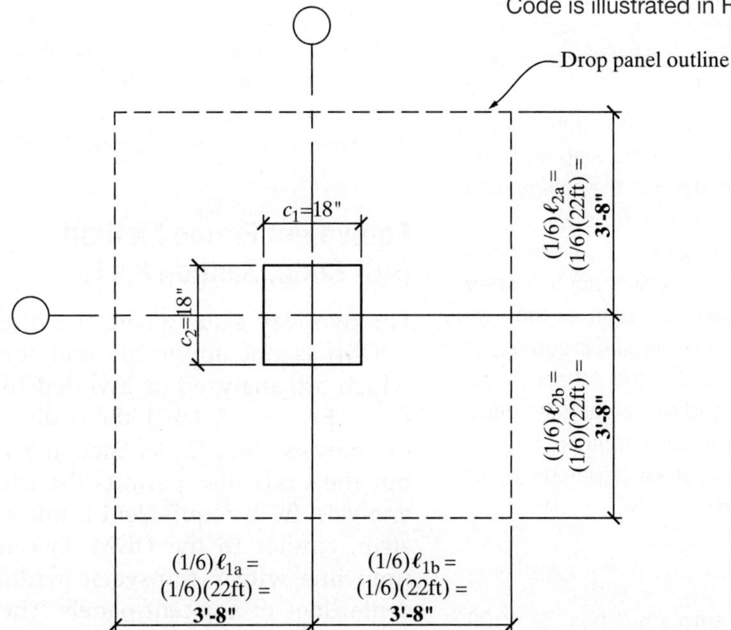

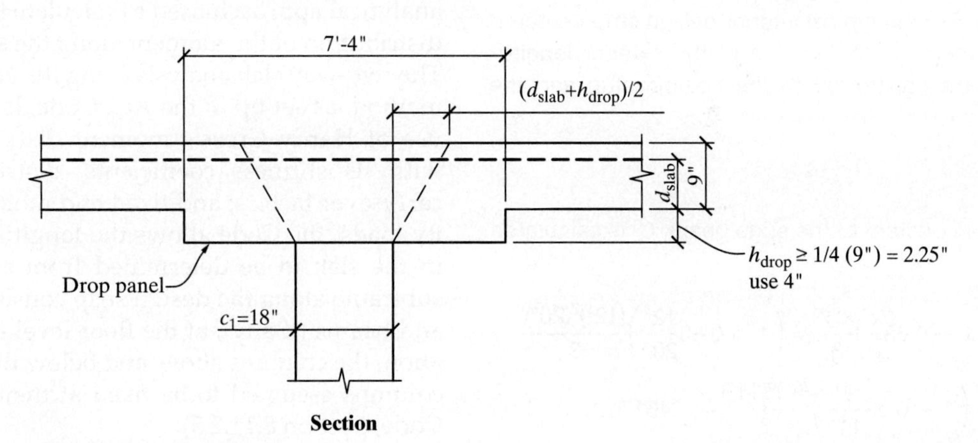

FIGURE 6-51 Drop panel dimensions.

Note that in contrast to the shear cap that was used in Example 6-3 which did not meet the drop panel requirements and thus could only be used to increase the critical shear perimeter, the drop panel in this example on the other hand meets the length and depth requirements specified in the Code; therefore, the drop panel can be used to:

a. Increase the effective depth in the negative moment region and thus reduce the amount of negative reinforcement required, in addition to increasing the shear perimeter and helping to reduce the shear stress. With the 4″ deep drop panel (see Figure 6-51), the average effective depth in the negative moment region, $d = 9'' + 4''\text{ drop} - 1.5'' = 11.5''$ (Compare to $d = 7.5''$ in Example 6-3)
b. Due to the increased effective depth and the code specified length of the drop panel, the critical perimeter for shear around the column, b_o, is increased because of the drop panel (see Figure 6-51). $b_o = 2(c_1 + c_2 + 2d) = 2(18 + 18 + 2(11.5)) = 118$ in.
c. The drop panel can be included in the slab deflection calculations, and the increased depth reduces the slab deflections.

2. If the code specified length and depth of the drop panel is not satisfied, then:
   - The depth of the drop panel cannot be counted upon to increase the effective depth of the slab and thus reduce the negative reinforcement at the column. The effective depth will be that of the slab alone.
   - The drop panel that do not meet the requirements of ACI Code, Section 8.2.4 will essentially act as a shear cap, provided the shear cap extension beyond the column face is at least equal to the shear cap depth. Shear caps cannot be included in the deflection calculations nor can they be used to increase the effective depth of the slab in the negative moment regions. Shear caps will increase the shear perimeter by increasing the value of c_1 and c_2.

Example 6-5

Assume the two-way slab in Example 6-3 has 12″ × 20″ deep spandrel beams along grid lines 1, 4, A, and D (see Figure 6-52). For the typical interior design strip centered on line 2, determine the portion of the exterior longitudinal moments transferred to the column strip and the middle strip.

Solution:

The torsional constant of the edge beam, C, is calculated as

$$C = \sum\left(1 - 0.63\frac{x}{y}\right)\frac{x^3y}{3} = \left(1 - 0.63\frac{12''}{20''}\right)\frac{(12'')^3(20'')}{3} + \left(1 - 0.63\frac{9''}{11''}\right)\frac{(9'')^3(11'')}{3} = 8461 \text{ in.}^4$$

The corresponding slab moment of inertia, I_s, is

$$I_s = (141'')\frac{(9'')^3}{12} = 8566 \text{ in.}^4$$

The torsional stiffness parameter for the edge beam is,

$$\beta_t = \frac{E_{cb}C}{2E_{cs}I_s} = \frac{8461}{2(8566)} = 0.49 \quad \text{[ACI Eq. 8.10.5.2a]}$$

Since there is no interior beam (i.e., beam along line C),

$$\alpha_{f1} = 0$$

$$\text{therefore, } \alpha_{f1}\frac{\ell_2}{\ell_1} = 0$$

Using Table 6-7, and entering the table with $\frac{\ell_2}{\ell_1} = 1.0$, $\alpha_{f1}\frac{\ell_2}{\ell_1} = 0$, and $\beta_t = 0.49$, and interpolating yields the portion of the exterior negative moment, M_u, assigned to the column strip to be: $-(0.1)(0.49) + 1.0 = \mathbf{0.95}$

Therefore only 5% of the exterior negative moment is transferred to the middle strips.

Note: When an edge beam has a high torsional stiffness, a higher percentage of the exterior negative moment is transferred to the middle strip. According to Table 6-7, for a square slab panel, depending on the torsional stiffness of the edge beam, up to 25% of the exterior negative moment can be transferred to the middle strip.

Equivalent Frame Method (ACI Code, Section 8.11)

For two-way slabs where the direct design method (DDM) is not applicable and for mat foundations which are analyzed as inverted two-way slabs (ACI Code, Section 13.3.4.2), the equivalent frame method of analysis (ACI Code, Section 8.11) has to be used, but the Code also permits the use of finite element analysis. In the equivalent frame method, the design strip, similar to the DDM, is centered on the column line with a transverse width bounded by the centerlines of adjacent panels. The equivalent frame method differs from the direct design method in the analytical approach used to calculate the longitudinal distribution of the moment along the span of the slab. The two-way slab analysis using the equivalent frame method, as set up in the ACI Code, lends itself to the use of Hardy Cross's moment distribution method with its stiffness coefficients, distribution factors, carry-over factors, and fixed end moments. For gravity loads, the Code allows the longitudinal moments in the slab to be determined from an analysis of a subframe along the design strip consisting of the slab and beams (if any), at the floor level under consideration, the columns above and below the slab, with the columns assumed to be fixed at their far ends (ACI Code, Section 8.11.2.5).

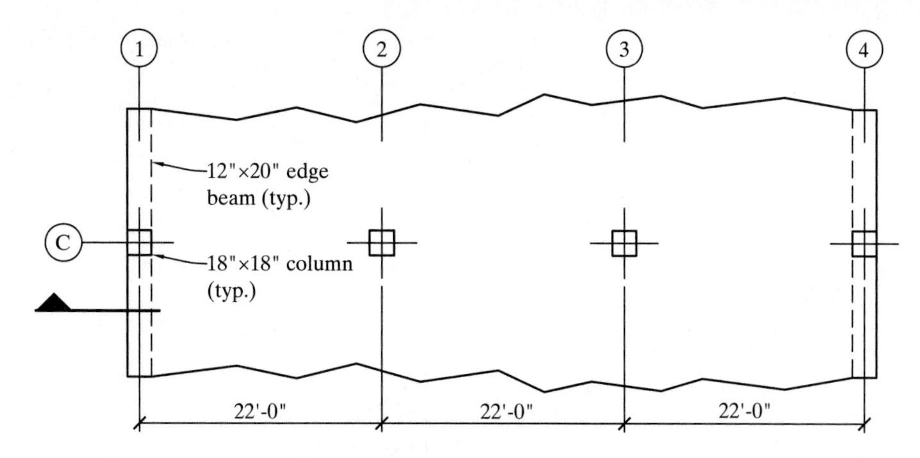

(a) Slab part-plan showing edge beam

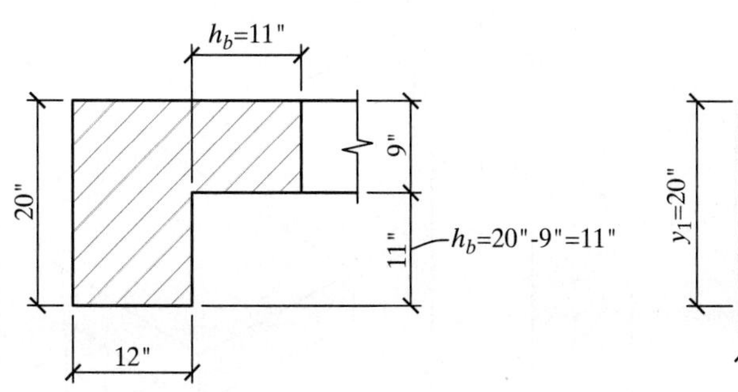

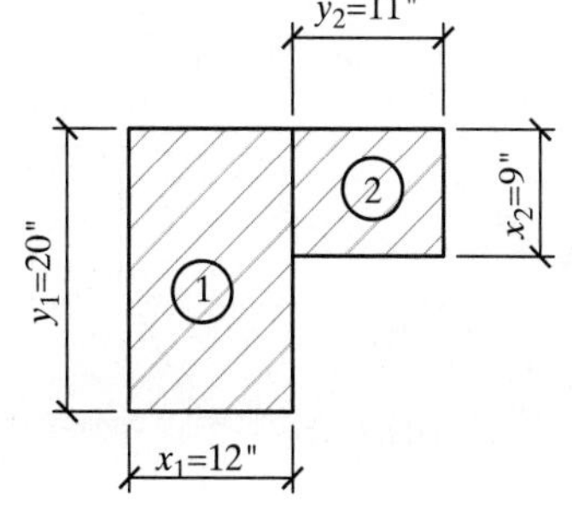

(b) Equivalent edge beam

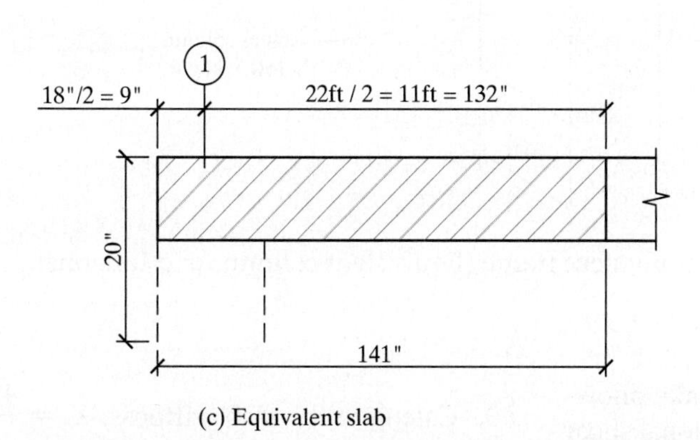

(c) Equivalent slab

FIGURE 6-52 Edge beam and slab for Example 6-5.

The equivalent frame consists of the following component parts (see Figure 6-53):

- The horizontal slab strip including any longitudinal beams.
- The columns or walls extending a story height above and below the slab (for roof slabs the column or wall only extends below the slab). The columns are usually considered fixed at their ends. Concrete walls whose longitudinal direction is perpendicular to the longitudinal direction of the slab do not provide any restraint to the slab, but instead they act like knife edge supports. Concrete walls whose longitudinal direction is parallel to the longitudinal direction of the slab provides moment restraint to the slab, similar to columns since the walls will be bending about their strong axis.
- Transverse elements at the columns that help to transfer the moment from the slab to the column through the torsional resistance of the transverse beams and the slab.

The procedure for the equivalent frame method of analysis for two-way slabs is as follows:

1. Determine the minimum slab thickness, h, to avoid deflection calculations (see *Minimum Thickness of Two-Way Slabs*).
2. Check one-way shear and two-way (punching) shear and ensure that the slab has adequate

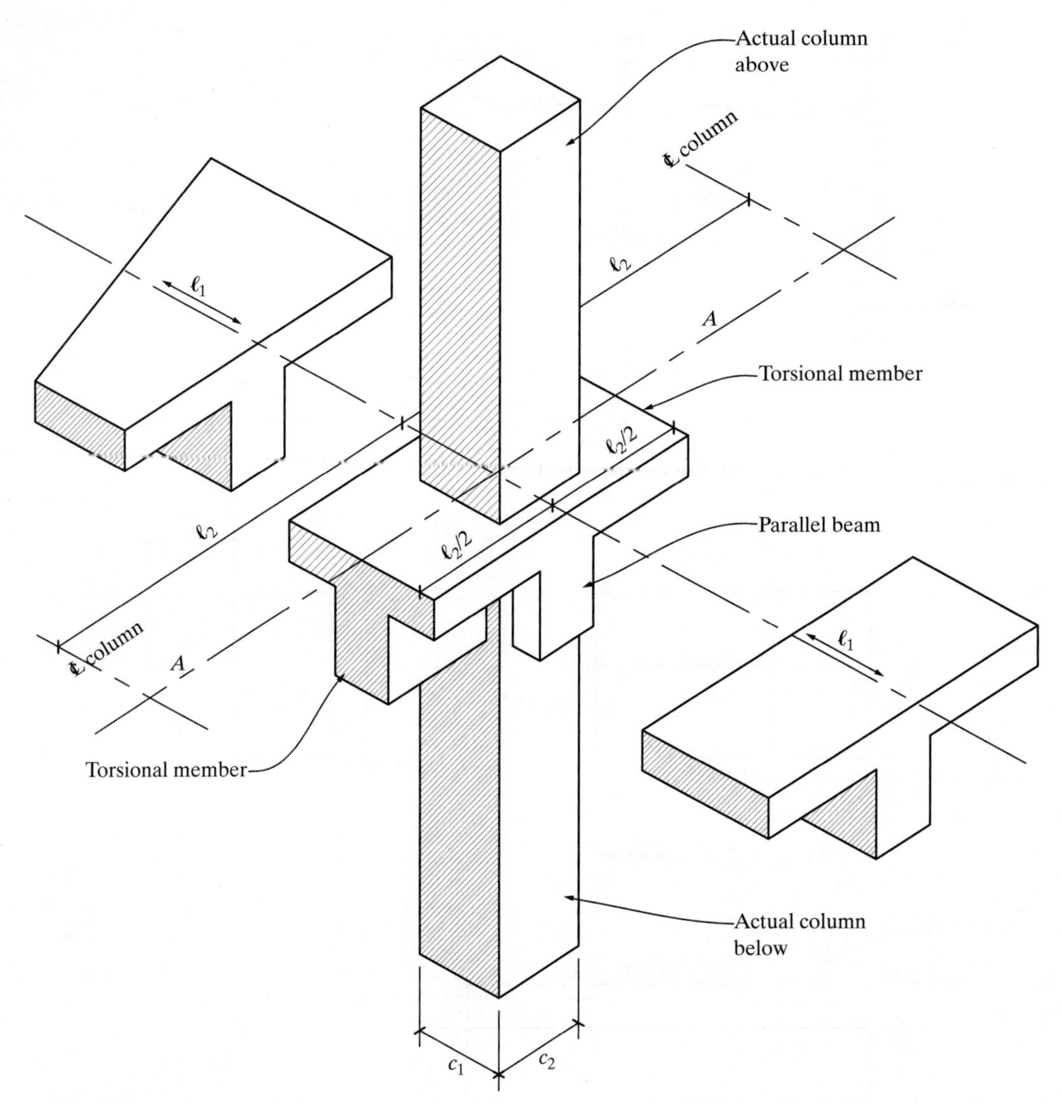

FIGURE 6-53 Components of the equivalent frame (Equivalent columns and torsional members, ref. ACI Fig. R8.11.4).

thickness to resist direct shear, and make allowance in the slab thickness to resist additional shear stresses due to transfer of unbalanced moments from the slab to the column. To allow for some reserve shear capacity to resist the additional shear stresses caused by unbalanced moments, ref. [2] recommends to increase the required depth for perimeter shear by 10% at interior columns, 40% at edge columns and 70% at corner columns.

3. Divide the slab into equivalent frames (design strips) in both orthogonal directions and for each design strip and at each slab-column connection, the stiffness parameters are calculated.
4. Different live load patterns are considered in addition to a uniform dead load over the entire length of the design strip, and the fixed end moments are calculated that will be used in the moment distribution analyses.
5. Calculate the slab stiffness, $K_s = \frac{4E_cI_s}{\ell_1}$, where E_c is the modulus of elasticity of the slab concrete, ℓ_1 is the span of the slab, and the slab moment of inertia, $I_s = \frac{(\ell_2)h^3}{12}$. Note that the gross sectional properties are used to calculate the moment of inertia for all members of the equivalent frame (ACI Code, Sections 8.11.3.3 and 8.11.4.3).
6. Calculate the stiffness of the column above and below the floor under consideration, $K_{column} = \frac{4E_{col}I_{column}}{\ell_{column}}$, where ℓ_{column} is the floor to floor height of the column, E_{col} is the modulus of elasticity of the column concrete, and the moment of inertia of the column, $I_{column} = \frac{(c_2)(c_1)^3}{12}$.

7. Calculate the torsional constant, C, for the transverse torsional member, where $C = \sum\left(1 - 0.63\frac{x}{y}\right)\frac{x^3y}{3}$.
   a. If there is no transverse beam, the transverse torsional member would consist of the slab with a width equal to the dimension of the column, c_1, parallel to the span of the slab, and a depth equal to the slab depth, h.
   b. Where there is a transverse beam framing into the column, the equivalent beam dimensions are determined in accordance with ACI Code, Section 8.4.1.8, similar to the direct design method (See Figure 6-24).
   c. For a typical interior column, there will be two torsional members, one on the near side and the other on the far side of the column.
8. Calculate the torsional stiffness, K_t , of the torsional members where $K_t = \sum\frac{9E_cC}{\ell_2\left(1 - \frac{c_2}{\ell_2}\right)^3}$ (ACI Code, Section R8.11.5).
9. Calculate the stiffness of the equivalent column, K_{ec}. The equivalent column consists of the columns above and below the level under consideration and the torsional members on the near side and far sides of the column. $\frac{1}{K_{ec}} = \frac{1}{\sum K_{\text{column}}} + \frac{1}{\sum K_t}$

   and thus, $K_{ec} = \frac{\sum K_{\text{column}}}{1 + \sum\frac{K_{\text{column}}}{K_t}}$

   Note:
   a. For a roof slab, there will be one column below and no column above; for a typical floor slab, there will be two columns, one below the floor and a second above the floor.
   b. For an exterior design strip (e.g., line D in Figure 6-46), there will be just one torsional member (on the far side), whereas for a typical interior design strip (e.g., line C in Figure 6-46), there will be two torsional members, one on the near face of the column and the second on the far face of the column.
10. Knowing the modulus of elasticity $E_c = 57{,}000\sqrt{f_c}$, calculate the equivalent column stiffness, K_{ec}, and the slab stiffness, K_s, in units of kip-ft/unit rotation. A typical exterior joint will consists of the equivalent column, K_{ec} and a slab stiffness Ks. A typical interior column will consists of the equivalent column stiffness, and two slab stiffnesses (i.e., the slab to the left and the slab to the right of the column). These stiffnesses are used to determine the distribution factors, DF, at each joint that will be used in the moment distribution analysis.
11. These stiffness factors can then be used in the moment distribution process after all the fixed end moments are calculated. A carry-over factor of 0.5 is assumed, ignoring the non-prismatic column section, and the final end moments and the midspan positive moments in the design strip are obtained from the moment distribution analysis.
12. The ACI Code allows the sum of the absolute value of the average negative moments at the ends of a span and the midspan positive moment to be no greater than the maximum static moment within the span, $M_o = \frac{w_u\ell_2\ell_{n1}^2}{8}$ (ACI Code, Section 8.11.6.5).
13. The longitudinal moments obtained from the equivalent frame analyses can be distributed transversely to the column strips, longitudinal beams, and the middle strips using the same coefficients that are used for the direct design method (see Tables 6-6 through 6-9) provided the following condition is satisfied: $0.2 \leq \frac{\alpha_{f1}\ell_2^2}{\alpha_{f2}\ell_1^2} \leq 5.0$ (ACI Code Eq. 8.10.2.7a and Section 8.11.6.6), where $\alpha_{f1} = \frac{I_{b1}}{I_{s1}}$ and $\alpha_{f2} = \frac{I_{b2}}{I_{s2}}$ (see *Relative Stiffness of Beam and Slab*).

The analysis of two-way slabs by the equivalent frame method is more complex and time consuming than the direct design method, therefore, in design practice, two-way slab analyses using the equivalent frame method are carried out using computer analysis and design software.

References

[1] R. Park and W.L. Gamble "Reinforced Concrete Slabs," John Wiley & Sons, New York, 1980.
[2] ACI Committee 340, "Slab Design in Accordance with ACI 318-77," ACI Publication SP-17(73)(S), American Concrete Institute, Detroit, 1978.
[3] David A. Fanella "Concrete Floor Systems: Guide to Estimating and Economizing," Portland Cement Association, Skokie, Illinois, 2000.

Problems

For the following problems, all concrete is normal weight and all steel is grade 60 (f_y = 60,000 psi), unless otherwise noted.

6-1. For the one-way slab shown, determine all moments and shears for service loads of 100 psf dead load (includes the slab weight) and 300 psf live load.

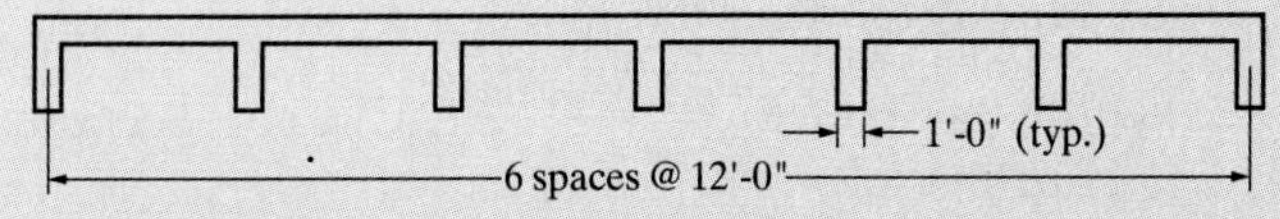

PROBLEM 6-1

6-2. For the continuous beam shown, determine all moments and shears for service loads of 2.00 kip/ft dead load (includes weight of beam and slab) and 3.00 kip/ft live load.

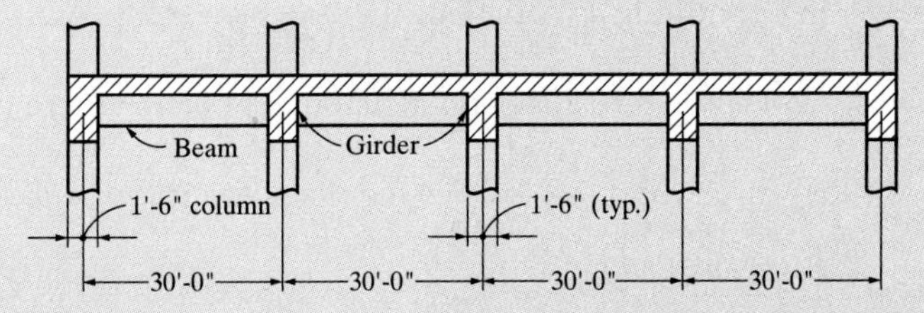

PROBLEM 6-2

6-3. Using the ACI coefficients, design a continuous reinforced concrete beam that will span four supports. The end spans are 24-ft long; the center span is 28-ft long (spans are measured center to center of supports). The exterior support (a spandrel beam) and the interior supporting girders have widths b of 18 in. The service loads are 0.90 kip/ft dead load (not including the weight of the beam) and 1.10 kip/ft live load. Use f'_c = 3000 psi. Use only tension reinforcing. Design for moment only. Use the recommended bar cutoffs shown in Figure 6-3.

6-4. A floor system is to consist of beams, girders, and a slab; a partial floor plan is shown. Service loads are to be 45 psf dead load (does *not* include the weight of the floor system) and 160 psf live load. Use f'_c = 4000 psi.

a. Design the continuous one-way slab.

b. Design the continuous beam along column line 2.

Be sure to include complete design sketches.

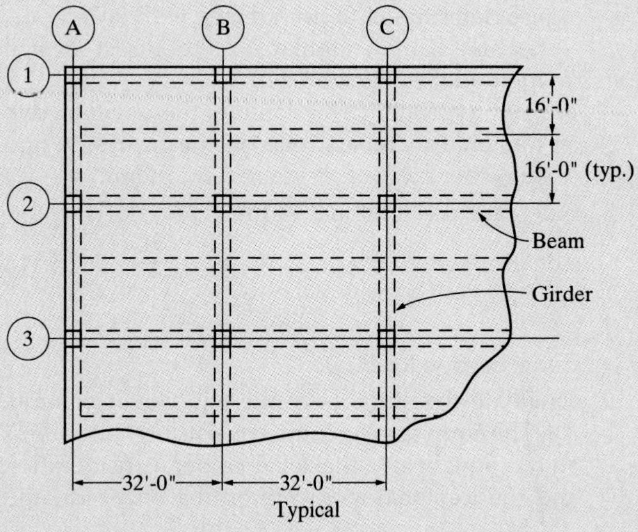

PROBLEM 6-4

6-5. Assume an 8″ one-way slab continuous over 4 total supports. The **Clear span between supports is ℓ_n = 16′−0″. Use f'_c = 3000 psi and f_y = 60,000 psi. Assume d = 6″ to the main reinforcing**

a. Determine the moment and shear values using the ACI coefficients. Draw each diagram assuming a knife-edge support (i.e., ignore the width of the supporting beams).

b. Determine the required flexural reinforcing (bar size and spacing) in the slab at the first interior support as shown below.

6-6. Assume a 4″ one-way slab continuous over three or more supports with a clear span between supports of ℓ_n = 7 ft. The total factored uniform load supported by the slab is 520 psf. Assume the main reinforcing is centered in the slab section and use f'_c = 3000 psi.

a. Determine the required flexural reinforcing in the slab at the first interior support.

b. Sketch your design showing the main reinforcing and the slab span.

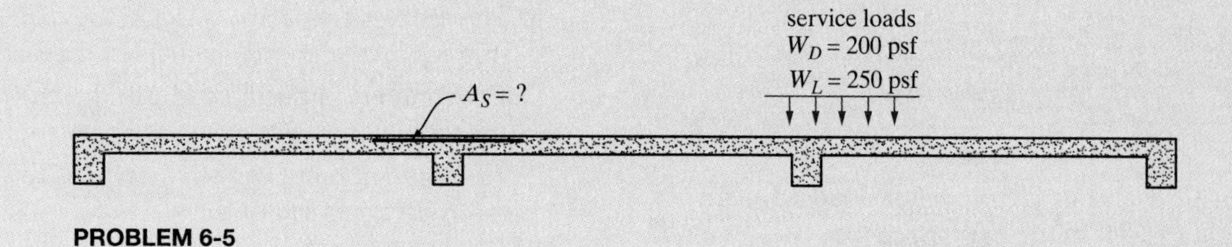

PROBLEM 6-5

6-7. The floor plan for an office building calls for a two-way flat slab supported on 18-in. square columns on a 30 ft × 30 ft column grid. Design the two-way floor slab: select the slab thickness, the dimensions of the drop panel and column capital, if any, and the slab reinforcement. Lay out the slab reinforcement in the column and middle strips in both orthogonal directions. Assume f'_c is 4000 psi and f_y is 60,000 psi.

6-8. A 10-in. thick two-way flat plate is supported on 20-in. square columns on 22 ft × 22 ft column grid, and the edge of the slab coincides with the exterior face of the columns. Assume f'_c is 4000 psi and f_y is 60,000 psi. Determine the punching shear capacity, ϕV_c and the factored direct shear, V_u, for a typical interior column, a typical edge column, and a typical corner column.

6-9. The floor plan for an office building calls for a two-way flat plate supported on 24-in. square columns on a 24 ft × 24 ft column grid. Design the two-way floor slab: select the slab thickness and the slab reinforcement. Lay out the slab reinforcement in the column and middle strips in both orthogonal directions. Assume f'_c is 4000 psi and f_y is 60,000 psi.

6-10. An existing 12 in. thick flat slab in an industrial building is supported on 48 in. (E-W). × 24 in. (N-S) columns on a 33 ft (E-W) × 27 ft (N-S) grid with 4" deep drop panels. The plan dimensions of the drop panels in the two orthogonal directions satisfy the Code-defined minimum limits. The superimposed dead load on the slab including concrete topping and mechanical and electrical equipment is 60 psf, and the live load is unknown and is to be determined. The concrete strength, f'_c, is 3000 psi and the steel yield strength is 40,000 psi. The existing slab reinforcement in the exterior bay are given as follows:

E-W Direction (i.e., center to center span = 33 ft)

Width of column strip = 13.5 ft

Width of two-half middle strips = 13.5 ft

Exterior discontinuous end of slab: Column strip: 10#7; Middle strip: 10#6

Midspan of end span of slab: Column strip: 16#8; Middle strip: 14#8

Interior continuous end of slab: Column strip: 11#8; Middle strip: 14#7

N-S Direction (i.e., center to center span = 27 ft)

Width of column strip = 16.5 ft

Width of two-half middle strips = 16.5 ft

Exterior discontinuous end of slab: Column strip: 12#7; Middle strip: 12#6

Midspan of end span of slab: Column strip: 16#8; Middle strip: 14#8

Interior continuous end of slab: Column strip: 12#8; Middle strip: 14#7

1. Calculate the total unfactored dead load on the slab in psf.
2. Does the depth of the drop panels satisfy the ACI Code requirements?
3. Considering only direct shear at a typical interior column, determine the live load capacity of the slab in psf.
4. For the **E-W direction**, determine the following:
   - **a.** The moment capacity, ϕM_n, in the *column strip* at the *exterior discontinuous end* of the slab
   - **b.** The moment capacity, ϕM_n, in the *middle strip* at the *exterior discontinuous end* of the slab
   - **c.** The total moment capacity, ϕM_n, at the *exterior discontinuous end* of the slab
   - **d.** The moment capacity, ϕM_n, in the *column strip* at the *midspan of the end span* of the slab
   - **e.** The moment capacity, ϕM_n, in the *middle strip* at the *midspan of the end span* of the slab
   - **f.** The total moment capacity, ϕM_n, at the *midspan of the end span* of the slab
   - **g.** The moment capacity, ϕM_n, in the *column strip* at the *first interior continuous end* of the slab
   - **h.** The moment capacity, ϕM_n, in the *middle strip* at the *first interior continuous end* of the slab
   - **i.** The total moment capacity, ϕM_n, at the *first interior continuous end* of the slab
   - **j.** Using the results from above and the following static equilibrium relationship between the static moment, M_o, and the moments (or moment capacities) at the critical sections of the slab:

$$M_{u,o} = \frac{w_u \ell_n^2}{8} \le \phi M_{n,o} = \frac{1}{2}\left[\phi M_{n,-ve(\text{left end sup port})} + \phi M_{n,-ve(\text{right end sup port})}\right] + \phi M_{n,+ve\ (\text{midspan})},$$

determine the unfactored live load capacity of the slab in psf in the E-W direction.

5. For the **N-S direction**, determine the following:
   a. The moment capacity, ϕM_n, in the *column strip* at the *exterior discontinuous end* of the slab
   b. The moment capacity, ϕM_n, in the *middle strip* at the *exterior discontinuous end* of the slab
   c. The total moment capacity, ϕM_n, at the *exterior discontinuous end* of the slab
   d. The moment capacity, ϕM_n, in the *column strip* at the *midspan of the end span* of the slab
   e. The moment capacity, ϕM_n, in the *middle strip* at the *midspan of the end span* of the slab
   f. The total moment capacity, ϕM_n, at the *midspan of the end span* of the slab
   g. The moment capacity, ϕM_n, in the *column strip* at the *first interior continuous end* of the slab
   h. The moment capacity, ϕM_n, in the *middle strip* at the *first interior continuous end* of the slab
   i. The total moment capacity, ϕM_n, at the *first interior continuous end* of the slab
   j. Using the results from above and the following static equilibrium relationship between the static moment, M_o, and the moments (or moment capacities) at the critical sections of the slab:

$$M_{u,o} = \frac{w_u \ell_n^2}{8} \le \phi M_{n,o} = \frac{1}{2}\left[\phi M_{n,-ve(\text{left end sup port})} + \phi M_{n,-ve(\text{right end sup port})}\right] + \phi M_{n,+ve\ (\text{midspan})},$$

   determine the unfactored live load capacity of the slab in psf in the N-S direction.

6. Based on your analyses for both the E-W and N-S directions, what is your recommended live load capacity (in psf) for this two-way slab?

CHAPTER **SEVEN**

SERVICEABILITY

7-1 Introduction
7-2 Deflections
7-3 Calculation of I_{cr}
7-4 Immediate Deflection
7-5 Long-Term Deflection
7-6 Procedure for Calculating the Deflection of Simply Supported and Continuous Beams and Slabs
7-7 Procedure for Calculating the Deflection of Continuous Girders
7-8 Deflection Control Measures in Reinforced Concrete Structures
7-9 Crack Control
7-10 Floor Vibrations
7-11 Gross and Cracked Section Properties of Concrete Sections

7-1 INTRODUCTION

The ACI Code requires that bending members have structural strength adequate to support the anticipated factored design loads (ACI Code Sections 4.4.4 and 4.6) and that they have adequate performance at service load levels (ACI Code Section 4.7). Adequate performance, or *serviceability*, relates to deflections and cracking in reinforced concrete beams and slabs. It is important to realize that serviceability is to be assured at *service load levels*, not at ultimate strength. At service loads, deflections should be held to specified limits because of many considerations, among which are aesthetics, effects on nonstructural elements such as windows and partitions, undesirable vibrations, and proper functioning of roof drainage systems. Any cracking should be limited to hairline cracks for reasons of appearance and to ensure protection of reinforcement against corrosion.

7-2 DEFLECTIONS

The deflection of concrete flexural members is affected by the following factors: the magnitude and distribution of the load on the member; the span of the member and support conditions; the percentage of reinforcement; member section properties; properties of the concrete material; and the degree and extent of flexural cracking in the member. Guidelines for the control of deflections are found in the ACI Code, Section 24.2. In addition, Table 24.2.2 of the Code indicates the maximum permissible deflections. Note that the permissible live load deflections for spandrel or edge beams supporting exterior cladding may need to be more stringent than those given in ACI Code Table 24.2.2, depending on the span of the beam and the performance requirements of the exterior cladding. A specific live load deflection limit of say 3/8-inch for spandrel beams regardless of the span is recommended to ensure adequate performance of the cladding [4]. For the purpose of following the Code guidelines, either of two methods may be used: (1) using the minimum thickness (or depth of member) criteria as established in Tables 7.3.1.1 and 9.3.1.1 of the Code, which will result in sections that are sufficiently deep and stiff so that deflections will not be excessive; and (2) calculating expected deflections using standard deflection formulas in combination with the Code provisions for moment of inertia and the effects of the load/time history of the member.

Minimum thickness (depth) guidelines are simple and direct and should be used whenever possible. Note that the tabulated minimum thicknesses apply to non-prestressed, one-way members that do not support and are not attached to partitions or other construction likely to be damaged by large deflections. For members not within these guidelines, deflections *must* be calculated.

For the second method, in which deflections are calculated, the ACI Code stipulates that the members should have their deflections checked at *service load levels*. Therefore, the *properties* at service load levels must be used. Under service loads, concrete flexural members still exhibit generally elastic-type behavior (see Figure 1-1) but will have been subjected to cracking in tension zones at any point where the applied moment is large enough to produce tensile stress in excess of concrete tensile strength. The cross section

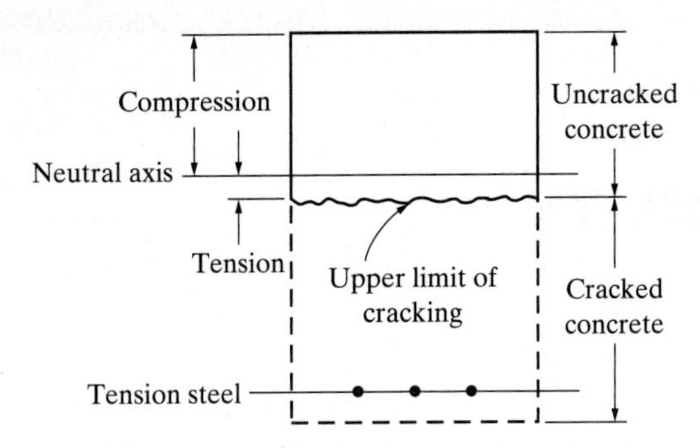

FIGURE 7-1 Typical cracked cross section.

for moment-of-inertia determination then has the shape shown in Figure 7-1.

The moment of inertia of the cracked section of Figure 7-1 is designated I_{cr}. It is determined based on the assumption that the concrete is cracked to the neutral axis. In other words, the concrete is assumed to have no tensile strength, and the small tension zone below the neutral axis and above the upper limit of cracking is neglected.

The moment of inertia of the cracked section described represents one end of a range of values that may be used for deflection calculations. At the other end of the range, as a result of a small bending moment and a maximum flexural stress less than the modulus of rupture, the full uncracked section may be considered in determining the moment of inertia to resist deflection. This is termed the *moment of inertia of the gross cross section* and is designated I_g.

In reality, both I_{cr} and I_g occur in a bending member in which the maximum moment is in excess of the cracking moment. The I_{cr} occurs at or near the cracks, whereas the I_g occurs between the cracks. Research has indicated that due to the presence of the two extreme conditions, a more realistic value of a moment of inertia lies somewhere between these values. The ACI Code recommends that deflections be calculated using an *effective moment of inertia, I_e,* where

$$I_g > I_e > I_{cr}$$

Once the effective moment of inertia is determined, the member deflection may be calculated by using standard deflection expressions. The effective moment of inertia will depend on the values of the moment of inertia of the gross section and the moment of inertia of the cracked section. The gross (or uncracked) moment of inertia for a rectangular shape may be easily calculated from

$$I_g = \frac{bh^3}{12}$$

This expression neglects the presence of any reinforcing steel.

7-3 CALCULATION OF I_{cr}

The moment of inertia of the cracked cross section can also be calculated in the normal way once the problem of the differing materials (steel and concrete) is overcome. To accomplish this, the steel area will be replaced by an equivalent area of concrete A_{eq}. This is a fictitious concrete that *can* resist tension. The determination of the magnitude of A_{eq} is based on the theory (from strength of materials) that when two differing elastic materials are subjected to equal strains, the stresses in the materials will be in proportion to their moduli of elasticity.

In Figure 7-2, ϵ_1 is the compressive strain in the concrete at the top of the beam and ϵ_2 is a tensile strain at the level of the steel. Using the notation

f_s = tensile steel stress

$f_{c(\text{tens})}$ = theoretical tensile concrete stress at the level of the steel

E_s = steel modulus of elasticity (29,000,000 psi)

E_c = concrete modulus of elasticity

the following relationships can be established:

$$\epsilon_2 = \frac{f_s}{E_s} \quad \text{and} \quad \epsilon_2 = \frac{f_{c(\text{tens})}}{E_c}$$

Equating these two expressions and solving for f_s yields

$$\frac{f_s}{E_s} = \frac{f_{c(\text{tens})}}{E_c}$$

$$f_s = \frac{E_s}{E_c}[f_{c(\text{tens})}]$$

The ratio E_s/E_c is normally called the *modular ratio* and is denoted n. Therefore,

$$f_s = nf_{c(\text{tens})}$$

Values of n may be taken as the nearest whole number (but not less than 6). Values of n for normal-weight concrete are tabulated in Table A-6.

As we are replacing the steel (theoretically) with an equivalent concrete area, the equivalent concrete

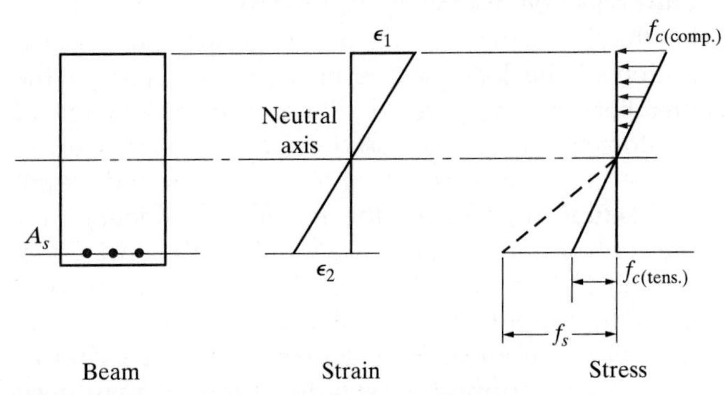

FIGURE 7-2 Beam bending: elastic theory.

area A_{eq} must provide the same tensile resistance as that provided by the steel.

Therefore,

$$A_{eq} f_{c(\text{tens})} = f_s A_s$$

Substituting, we obtain

$$A_{eq} f_{c(\text{tens})} = n f_{c(\text{tens})} A_s$$

from which

$$A_{eq} = nA_s$$

This defines the equivalent area of concrete with which we are replacing the steel. Another way of visualizing this is to consider the steel to be transformed into an equivalent concrete area of nA_s. The resulting *transformed concrete cross section* is composed of a single (although hypothetical) material and may be dealt with in the normal fashion for neutral-axis and moment-of-inertia determinations. Figure 7-3 depicts the transformed section. Because the steel is normally assumed to be concentrated at its centroid, which is a distance d from the compression face, the replacing equivalent concrete area must also be assumed to act at the same location. Therefore, the representation of this area is shown as a thin rectangle extending out past the beam sides. Recalling that the cross section is assumed cracked up to the neutral axis, the resulting effective area is as shown in Figure 7-4 and the neutral axis will be located a distance $\bar{y}$ down from a reference axis at the top of the section.

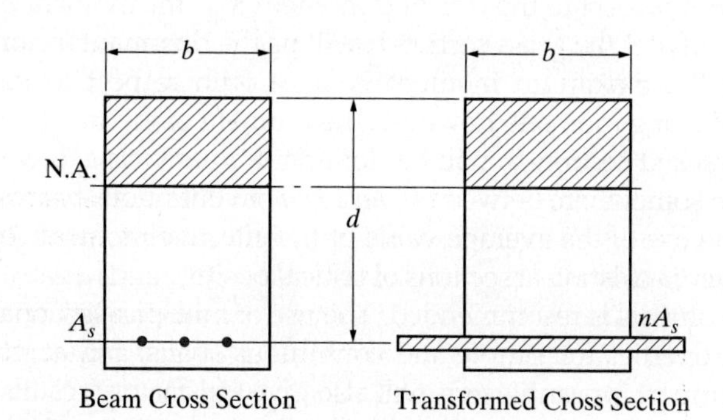

FIGURE 7-3 Method of transformed section.

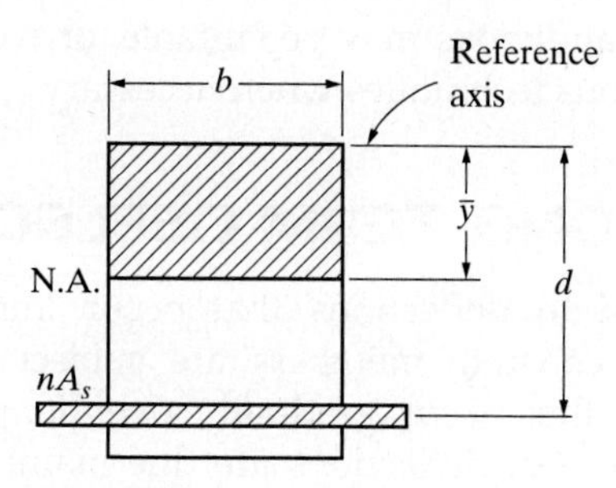

FIGURE 7-4 Effective area.

The neutral axis may be determined by taking a summation of moments of the effective areas about the reference axis:

$$\bar{y} = \frac{\Sigma(Ay)}{\Sigma A} = \frac{(b\bar{y})\frac{\bar{y}}{2} + nA_s d}{b\bar{y} + nA_s}$$

$$b(\bar{y}^2) + nA_s\bar{y} = \frac{b(\bar{y}^2)}{2} + nA_s d$$

$$\frac{b(\bar{y}^2)}{2} + nA_s\bar{y} - nA_s d = 0$$

This is a quadratic equation of the form $ax^2 + bx + c = 0$, and it may be solved either by completion of the square, as was done in Example 3-7, or by using the formula for roots of a quadratic equation, which will result in the following useful expression:

$$\bar{y} = \frac{nA_s\left[\sqrt{1 + 2\frac{bd}{nA_s}} - 1\right]}{b}$$

Once the neutral axis is located, the moment of inertia (I_{cr}) may be found using the familiar transfer formula from engineering mechanics. The equations for the gross and cracked section properties for rectangular and flanged sections are summarized in Section 7-11.

Example 7-1

Find the cracked moment of inertia for the cross section shown in Figure 7-5a. Use $A_s = 2.00$ in.^{2}, $n = 9$ (Table A-6), and $f'_c = 3000$ psi.

Solution:

With reference to the transformed section, shown in Figure 7-5b, the neutral axis is located as follows:

$$\bar{y} = \frac{nA_s\left[\sqrt{1 + 2\frac{bd}{nA_s}} - 1\right]}{b}$$

$$= \frac{9(2.0)\sqrt{1 + 2\frac{(8)(17)}{9(2.0)}} - 1}{8}$$

$$= 6.78 \text{ in.}$$

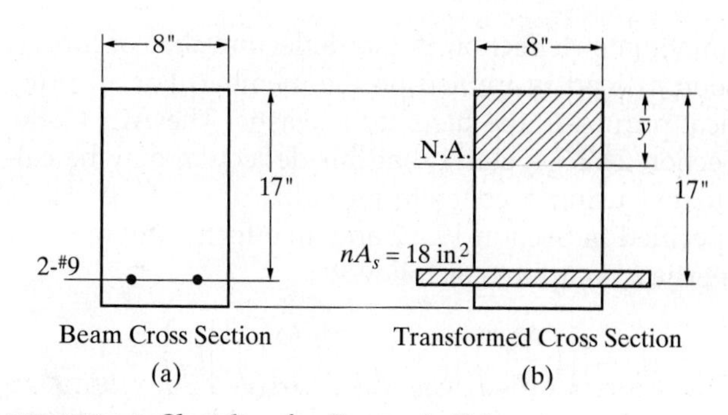

FIGURE 7-5 Sketches for Example 7-1.

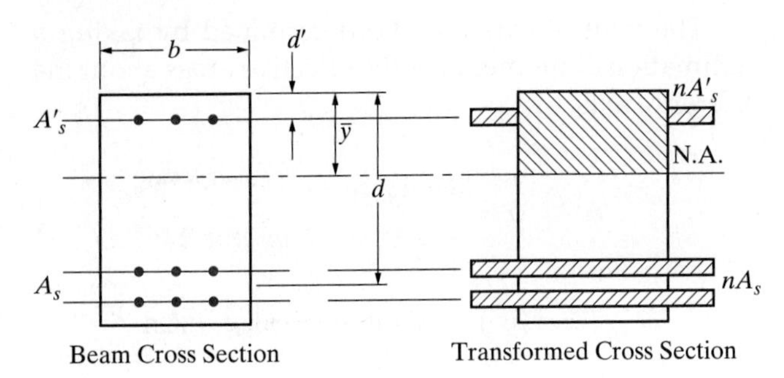

FIGURE 7-6 Doubly reinforced beam.

The moment of inertia of the cracked section may now be found (all units are inches):

$$I_{cr} = \frac{b\bar{y}^3}{3} + nA_s(d - \bar{y})^2$$

$$= \frac{8(6.78)^3}{3} + 18(17 - 6.78)^2$$

$$= 831.1 + 1880$$

$$= 2711 \text{ in.}^4$$

If the beam cross section contains compression steel, this steel may also be transformed and the neutral-axis location and cracked moment-of-inertia calculations can be carried out as before. Because the compression steel displaces concrete that is in compression, it should theoretically be transformed using $(n - 1)A'_s$ rather than nA'_s. For deflection calculations, which are only approximate, however, the use of nA'_s will not detract from the accuracy expected. The resulting transformed section will appear as shown in Figure 7-6.

The neutral axis location $\bar{y}$ may be determined from the solution of

$$\frac{b}{2}\bar{y}^2 + nA'_s\bar{y} - nA'_s d' - nA_s d + nA_s\bar{y} = 0$$

and the moment of inertia with respect to the neutral axis from

$$I_{cr} = \frac{b\bar{y}^3}{3} + nA_s(d - \bar{y})^2 + nA'_s(\bar{y} - d')^2$$

7-4 IMMEDIATE DEFLECTION

Immediate deflection is the deflection that occurs as soon as load is applied on the member. For all practical purposes, the member is elastic. The ACI Code, Section 24.2.3.4, states that this deflection may be calculated using a concrete modulus of elasticity E_c as specified in Section 19.2.2 and an effective moment of inertia I_e computed as follows:

$$I_e = \left\{\left(\frac{M_{cr}}{M_a}\right)^3 I_g + \left[1 - \left(\frac{M_{cr}}{M_a}\right)^3\right] I_{cr}\right\} \le I_g$$

[ACI Eq. (24.2.3.5a)]

where

I_e = effective moment of inertia
I_{cr} = moment of inertia of the cracked section transformed to concrete
I_g = moment of inertia of the gross (uncracked) concrete cross section about the centroidal axis, neglecting all steel reinforcement
M_a = maximum moment in the member at the stage for which the deflection is being computed
M_{cr} = moment that would initially crack the cross section computed from

$$M_{cr} = \frac{f_r I_g}{y_t} \qquad \text{[ACI Eq. (24.2.3.5b)]}$$

where

f_r = modulus of rupture for the concrete = $7.5\lambda\sqrt{f'_c}$ [ACI Code Eq. (19.2.3.1)]
λ = 1.0 for normal-weight concrete
= 0.85 for sand-lightweight concrete
= 0.75 for all-lightweight concrete

Values for the modulus of rupture for normal-weight concrete are tabulated in Table A-6.

y_t = distance from the neutral axis of the uncracked cross section (neglecting steel) to the extreme tension fiber

Inspection of the formula for the effective moment of inertia will show that if the maximum moment is low with respect to the cracking moment M_{cr}, the moment of inertia of the gross section I_g will be the dominant factor. If the maximum moment is large with respect to the cracking moment, however, the moment of inertia of the cracked section I_{cr} will be dominant. In any case, I_e will lie somewhere between I_{cr} and I_g. For continuous beams, the use of the average value of the effective moments of inertia existing at sections of critical positive and negative moments is recommended. The use of midspan sectional properties for simple and continuous spans, and at the support for cantilevers, will also give satisfactory results.

The actual calculation of deflections will be made using the standard deflection methods for elastic members. Deflection formulas of the type found in standard handbooks may be suitable, or we may use more rigorous techniques when necessary.

7-5 LONG-TERM DEFLECTION

In addition to deflections that occur immediately, reinforced concrete members are subject to added deflections that occur gradually over long periods. These additional deflections are due mainly to creep and shrinkage and may eventually become excessive.

The additional (or long-term) deflections are computed based on two items: (1) the amount of sustained dead and live load and (2) the amount of compression reinforcement in the beam. The additional long-term deflections may be estimated as follows:

$$\Delta_{LT} = \lambda_\Delta \Delta_i = \left(\frac{\xi}{1 + 50\rho'}\right)\Delta_i$$

where

Δ_{LT} = additional long-term deflection

Δ_i = immediate or instantaneous deflection due to sustained loads

λ_Δ = a multiplier for additional long-term deflections (λ_Δ accounts for the effect of creep of concrete) $= \left(\frac{\xi}{1 + 50\rho'}\right)$ [ACI Eq. (24.2.4.1.1)]

ρ' = non-prestressed compression reinforcement ratio (A'_s/bd)

ξ = time-dependent factor for sustained loads, that is, the creep factor measured relative to time t = 0:

5 years or more	2.0
12 months	1.4
6 months	1.2
3 months	1.0

Note: At 5 years, the incremental long-term deflection of the member is calculated using a creep or time-dependent factor, ξ, of 2.0. "Deflection sensitive (non-structural) elements" (DSEs) that are supported by the member which are installed during construction will undergo this long-term deflection. If, however, the installation of the DSEs is delayed till, for example, after 12 months from when the concrete member was initially cast, the incremental deflections that the DSEs will undergo would be the difference between the long-term deflection calculated with a creep factor, ξ, of 2.0 (i.e., at 5 years) and the long-term deflection calculated with a creep factor, ξ, of 1.4 (i.e., at 12 months). Thus, it is advantageous to install DSEs after much of the creep deflection in a concrete member has occurred. Examples of DSEs include sensitive equipments such as medical equipments, plastered ceiling, unreinforced masonry partition walls, glazing, expensive flooring such as terrazo or marble, and operable partition walls.

Δ_i = immediate or instantaneous deflection due to sustained loads $= \Delta_{DL} + \alpha\Delta_{LL}$

α = percentage of live load sustained.

The total long-term deflection (*which excludes the instantaneous dead load deflection*) is the sum of the instantaneous live load deflection plus the long-term deflection due to the dead load and the sustained portion of the live load, and is given as

$$\Delta_{TT} = \Delta_{LL} + \Delta_{LT}$$
$$= \Delta_{LL} + \left(\frac{\xi}{1 + 50\rho'}\right)\Delta_i$$
$$= \Delta_{LL} + \lambda_\Delta(\Delta_{DL} + \alpha\Delta_{LL})$$

Some judgment will be required in determining just what portion of the live loads should be considered *sustained*. In a residential application, 20% sustained live load (i.e. $\alpha = 0.2$) might be a logical estimate, whereas in storage facilities, 100% sustained live load (i.e. $\alpha = 1.0$) would be reasonable.

The calculated deflections must not exceed the maximum permissible deflections that are found in the ACI Code, Table 24.2.2. This table sets permissible deflections in terms of fractions of span length. These limitations guard against damage to the various parts of the system (both structural and nonstructural parts) as a result of excessive deflection. In the case of attached nonstructural elements, only the deflection that takes place after such attachment needs to be considered.

Example 7-2

Calculate the maximum deflection of the simply supported non-prestressed reinforced concrete beam having a cross-section shown in Figure 7-7 and determine which of the ACI Code deflection criteria will be satisfied. The beam is subjected to maximum unfactored moments of $M_{DL} = 20$ ft.-kip and $M_{LL} = 15$ ft.-kip. Assume a 50% sustained live load and a sustained load time period of more than 5 years. Assume normal-weight concrete. Use $f'_c = 3000$ psi and $f_y = 60{,}000$ psi. The beam is on a simple span of 30 ft.

Solution:

1. The maximum *service* moments at midspan are

$$M_{DL} = 20 \text{ ft.-kip} \quad \text{and} \quad M_{LL} = 15 \text{ ft.-kip}$$
$$M_{a,m} = M_{DL} + M_{LL} = 35 \text{ ft-kip.}$$

2. Check the beam depth based on the ACI Code Table 9.3.1.1:

$$\text{minimum } h = \frac{\ell}{16} = \frac{30(12)}{16} = 22.5 \text{ in.}$$

Because 22.5 in. > 16.5 in., deflection must be calculated.

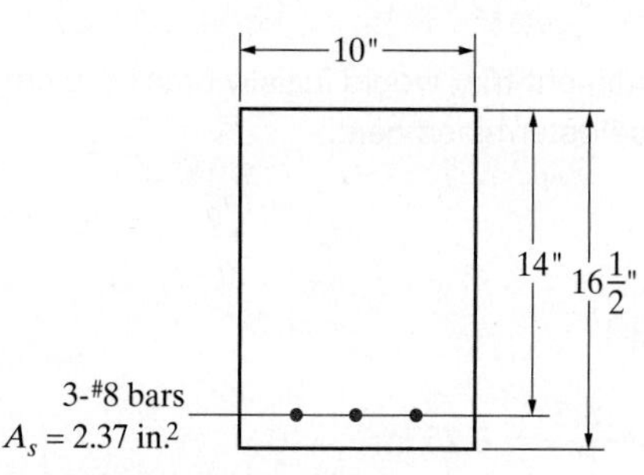

FIGURE 7-7 Sketch for Example 7-2.

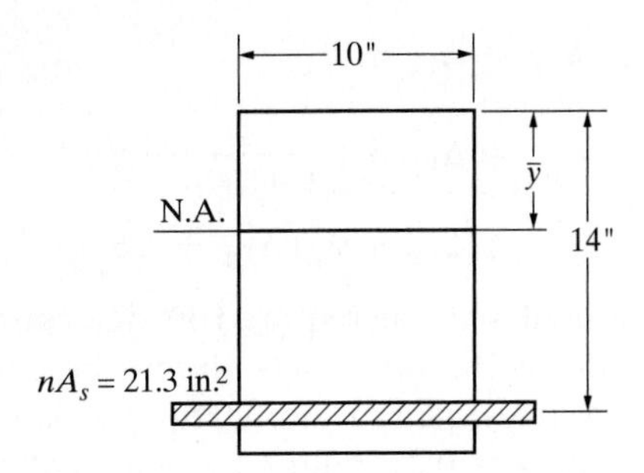

FIGURE 7-8 Transformed section for Example 7-2.

3. The effective moment of inertia will be calculated using

$$I_e = \left(\frac{M_{cr}}{M_a}\right)^3 I_g + \left[1 - \left(\frac{M_{cr}}{M_a}\right)^3\right] I_{cr}$$

Therefore, we first must compute the various terms within the expression. The moment of inertia of the cracked transformed section will be determined with reference to Figure 7-8.

The steel area of 2.37 in.^{2} and a modular ratio n of 9 (see Table A-6) will result in a transformed area of

$$nA_s = 9(2.37) = 21.3 \text{ in.}^2$$

The neutral-axis location is determined as follows:

$$\bar{y} = \frac{nA_s\left[\sqrt{1 + 2\frac{bd}{nA_s}} - 1\right]}{b}$$

$$= \frac{21.3\left[\sqrt{1 + 2\frac{(10)(14)}{21.3}} - 1\right]}{10}$$

$$= 5.88 \text{ in.}$$

The moment of inertia of the cracked transformed section is then determined:

$$I_{cr} = \frac{10(5.88)^3}{3} + 21.3(14 - 5.88)^2$$

$$= 2082 \text{ in.}^4$$

Determination of the moment of inertia of the gross section results in

$$I_g = \frac{bh^3}{12} = \frac{1}{12}(10)(16.5)^3 = 3743 \text{ in.}^4$$

The moment that would initially crack the cross section may be determined next:

$$M_{cr} = \frac{f_r I_g}{y_t}$$

where

$$y_t = \frac{16.5}{2} = 8.25 \text{ in.}$$

$$f_r = 7.5\lambda\sqrt{f'_c} = 0.411 \text{ ksi (from Table A-6 for } \lambda = 1)$$

Therefore,

$$M_{cr} = \frac{0.411(3743)}{8.25(12)} = 15.5 \text{ ft.-kip}$$

We will assume that the maximum total unfactored moment in the beam at the midspan, $M_{a,m} = M_a = 35$ ft.-kip, will establish the crack pattern in the beam. This is not in strict accordance with the ACI Code, which indicates that M_a should be the maximum moment occurring at the stage the deflection is computed. It is logical that, as the cracking pattern is irreversible, the use of the effective moment of inertia based on the full maximum moment is more realistic. This approach is conservative and will furnish a lower I_e, which will subsequently result in a larger computed deflection.

4. Determine the effective moment of inertia:

$$I_e = \left(\frac{M_{cr}}{M_a}\right)^3 I_g + \left[1 - \left(\frac{M_{cr}}{M_a}\right)^3\right] I_{cr}$$

$$= \left(\frac{15.5}{35}\right)^3 3743 + \left[1 - \left(\frac{15.5}{35}\right)^3\right](2082) = 2226 \text{ in.}^4$$

5. Compute the immediate or instantaneous dead load deflection ($M_{DL} = 20$ ft.-kip):

From structural analysis, the maximum deflection of a simply supported beam is,

$$\Delta_{DL} = \frac{5w_{DL}\ell^4}{384E_cI_e} = \frac{5M_{DL}\ell^2}{48E_cI_e}$$

where M is the moment due to a uniform load and E_c may be found in Table A-6. Thus

$$\Delta_{DL} = \frac{5(20 \text{ ft.-kip})(30 \text{ ft})^2(1728 \text{ in.}^3/\text{ft}^3)}{48(3120 \text{ kip/in.}^2)(2226 \text{ in.}^4)} = 0.467 \text{ in.}$$

6. Compute the immediate or instantaneous live load deflection ($M_{LL} = 15$ ft.-kip). By proportion,

$$\Delta_{LL} = \frac{M_{LL}}{M_{DL}}\Delta_{DL} = \frac{15}{20}(0.467) = 0.350 \text{ in.}$$

7. The total immediate or instantaneous dead plus live load (DL + LL) deflection is

$$0.467 + 0.350 = 0.817 \text{ in.}$$

8. The long-term (LT) deflection (DL + sustained LL) multiplier is

$$\lambda_\Delta = \frac{\xi}{1 + 50\rho'} = \frac{2.0}{1 + 0} = 2.0$$

Because $M_{DL} = 20$ ft.-kip and 50% $M_{LL} = 7.5$ ft.-kip, the sustained moment for long-term deflection = 27.5 ft.-kip. Then

$$\Delta_{LT} = \frac{27.5}{20}(0.467)(2.0) = 1.28 \text{ in.}$$

9. A comparison of actual deflections to maximum permissible deflections may now be made. In the comparison, made in Table 7-1, the maximum permissible deflections

TABLE 7-1 Permissible Versus Actual Deflection (Example 7-2)

Case	Maximum permissible deflection*	Actual computed deflection
1	$\frac{\ell}{180} = \frac{30(12)}{180} = 2$ in.	$\Delta_{LL} = 0.35$ in. (immediate LL)
2	$\frac{\ell}{360} = \frac{30(12)}{360} = 1$ in.	$\Delta_{LL} = 0.35$ in. (immediate LL)
3	$\frac{\ell}{480} = \frac{30(12)}{480} = 0.75$ in.	$\Delta_{LL} + \Delta_{LT} = 0.35 + 1.28 = 1.63$ in.
4	$\frac{\ell}{240} = \frac{30(12)}{240} = 1.5$ in.	$\Delta_{LL} + \Delta_{LT} = 0.35 + 1.28 = 1.63$ in.

*From ACI 318-14, Table 24.2.2.

are from the ACI Code, Table 24.2.2. Case 1 in Table 7-1 applies to flat roofs not supporting or attached to nonstructural elements likely to be damaged by large deflections. Case 2 applies to floors not supporting or attached to nonstructural elements likely to be damaged by large deflections. As the permissible deflection is not exceeded in case 1 or case 2, the beam of Example 7-2 is limited to usage as defined by those two cases.

7-6 PROCEDURE FOR CALCULATING THE DEFLECTION OF SIMPLY SUPPORTED AND CONTINUOUS BEAMS AND SLABS

The step-by-step procedure for calculating the long-term deflections of simply supported and continuous beams and slabs subject to uniformly distributed loads is as follows [4]:

1. Calculate the service dead and live load moments at the midspan of the most critical span or the span of interest. (i.e., calculate $M_{DL,m}$ and $M_{LL,m}$). For continuous beams and slabs, also calculate the unfactored moments at the supports. For continuous members, these service load moments are usually determined using the ACI moment coefficients (see Chapter 6).
   a. Assume that a portion of the live load is permanently sustained. The percentage of live load sustained is denoted as α (see Section 7-5).
2. Calculate the cracked moment of inertia, I_{cr}, and the gross moment of inertia, I_g, at the midspan (and at the supports for continuous members).
3. Calculate the effective moment of inertia, I_e, using ACI Equation (24.2.3.5a).
   a. Calculate the cracking moment, M_{cr}, for the beam at the midspan using ACI Equation (24.2.3.5b) and the effective moment of inertia, $I_{e,m\ beam}$, at the midspan using M_{cr}/M_{am} and ACI Equation (24.2.3.5a).
      i. The *total* service load moment at midspan, $M_{a,m} = M_{DL,m} + M_{LL,m}$.
      ii. If $M_{a,m} \le M_{cr}$, the member is uncracked and therefore $I_e = I_g$
   b. For continuous members, repeat Step 3a at the member supports and obtain the effective moment of inertia at the supports, $I_{e,s}$.
   c. The effective moment of inertia at the midspan and supports, respectively (i.e., $I_{e,m}$ and $I_{e,s}$), are combined as follows to obtain a weighted average moment of inertia for the member, I_e (average) [4]:
      – Beam or slab with one end continuous: $I_e(\text{average}) = 0.85I_{e,m} + 0.15I_{e,s}$
      – Beam or slab with both ends continuous: $I_e(\text{average}) = 0.70I_{e,m} + 0.15(I_{e,s1} + I_{e,s2})$
   d. Use the I_e (average) in the member deflection calculations. For simply supported members, there are no moments at the end supports, therefore, $I_e(\text{average}) = I_{e,m}$

where,

$I_{e,m}$ = effective moment of inertia at the midspan, calculated using the section properties and the unfactored moment at the midspan,

$I_{e,s}$ = effective moment of inertia at the one continuous end, calculated using the section properties and the unfactored moment at this support,

$I_{e,s1}$ = effective moment of inertia at the left continuous end, calculated using the section properties and the unfactored moments at the left support,

$I_{e,s2}$ = effective moment of inertia at the right continuous end, calculated using the section properties and the moments at the left support.

Note: For many practical situations, especially for prismatic continuous slabs, beams and girders, it is accurate enough to use only the midspan properties and midspan moments in calculating the effective moment of inertia because the member stiffness at midspan predominates (ACI Code, Section 24.2.3.7). In this case, $I_e(\text{average}) \approx I_{e,m}$. This simplified approach is adopted in the examples in this text.

4. Calculate the instantaneous maximum deflection:
   - For **simply supported** beams or slabs with uniformly distributed loads, the instantaneous maximum deflections, calculated using the midspan moments, are

$$\Delta_{DL} = K\left(\frac{5}{48}\right)\frac{M_{DL,m}\ell^2}{E_cI_e}$$

$$\Delta_{LL} = K\left(\frac{5}{48}\right)\frac{M_{LL,m}\ell^2}{E_cI_e} = \frac{M_{LL,m}}{M_{DL,m}}\Delta_{DL}$$

where,

K = deflection constant = 1.0 for simply supported members

$M_{DL,m}$ = maximum service dead load moment at midspan $= \dfrac{w_{DL}\ell^2}{8}$

$M_{LL,m}$ = maximum service live load moment at midspan $= \dfrac{w_{LL}\ell^2}{8}$

   - For **continuous** beams and slabs with uniformly distributed loads, the instantaneous maximum deflections at the midspan are calculated as

$$\Delta_{DL} = K\left(\frac{5}{48}\right)\frac{M_{DL,m}\ell^2}{E_cI_e}$$

$$\Delta_{LL} = K\left(\frac{5}{48}\right)\frac{M_{LL,m}\ell^2}{E_cI_e} = \frac{M_{LL,m}}{M_{DL,m}}\Delta_{DL}$$

$$\Delta_{DL+LL} = K\left(\frac{5}{48}\right)\frac{M_{a,m}\ell^2}{E_cI_e}$$

$$K = 1.2 - 0.2\frac{M_o}{M_{a,m}}$$

$$M_o = \frac{w_{DL+LL}\,\ell^2}{8}$$

where M_o is the statical moment.

$M_{a,m}$ can be obtained directly using the ACI moment coefficients.

5. The total long-term deflection (*which excludes the instantaneous dead load deflection*) is the sum of the instantaneous live load deflection plus the long-term deflection due to dead load and the sustained portion of the live load, and is given as

$$\Delta_{TT} = \Delta_{LL} + \lambda_\Delta(\Delta_{DL} + \alpha\Delta_{LL})$$

6. Check the permissible deflection limits from ACI 318-14, Table 24.2.2:
   a. For roof and floor flexural members not supporting deflection sensitive elements, $\Delta_{TT} \leq \dfrac{\ell}{240}$

   b. For roof and floor flexural members supporting deflection sensitive elements, $\Delta_{TT} \leq \dfrac{\ell}{480}$

   c. For roof and floor flexural members not supporting deflection sensitive elements, $\Delta_{LL} \leq \dfrac{\ell}{180}$

   d. For roof and floor flexural members supporting deflection sensitive elements, $\Delta_{TT} \leq \dfrac{\ell}{480}$

7-7 PROCEDURE FOR CALCULATING THE DEFLECTION OF CONTINUOUS GIRDERS

The ACI coefficients are only applicable to flexural members supporting uniformly distributed loads. For beams supporting concentrated loads or for girders—which support the concentrated reactions from the beams, the ACI moment coefficients are not applicable. Therefore, a plane frame structural analysis software is typically used to analyze the girder-column subframe with the appropriate effective moment of inertia considered. The procedure for calculating deflections for girders is as follows:

1. Perform a computer-based structural analysis of the girder-column sub-frame under service (i.e. unfactored) dead and live loads. The loads on the girder usually consist of unfactored uniform self-weight of the girder stem and the unfactored concentrated dead and live load reactions from the beams framing into the girder. The maximum deflection in a continuous girder with equal spans will typically occur in the end span. Pattern live loading must be considered.
2. From the computer analysis results, obtain the service dead load moment M_{DL} and the service live load moment M_{LL} at the midspan of the end span of the continuous girder. Similarly, obtain the moments at the supports of the girder.

Midspan moments: $M_{a,m} = M_{DL,m} + M_{LL,m}$
Support Moments: $M_{a,s} = M_{DL,s} + M_{LL,s}$

3. Calculate the effective moment of inertia, I_e.
   a. Calculate the cracking moment M_{cr} for the girder at the midspan using ACI Equation (24.2.3.5b) and the girder effective moment of inertia $I_{e,\,m\,girder}$ at the midspan using $M_{cr}/M_{a,m}$ and ACI Equation (24.2.3.5a).
      i. The *maximum* service load moment at midspan, $M_{a,m} = M_{DL,m} + M_{LL,m}$
      ii. If $M_{a,m} \leq M_{cr}$, the member is uncracked and therefore, $I_e = I_g$
   b. Repeat Step 3a at the girder supports and obtain the effective moment of inertia at the supports, $I_{e,s}$.
   c. The effective moment of inertia at the midspan and supports, respectively (i.e., $I_{e,m}$ and $I_{e,s}$), are combined as follows to obtain a weighted average moment of inertia for the girder, I_e (average) [4]:

      –Girder with one end continuous: $I_e(\text{average}) = 0.85\, I_{e,m} + 0.15\, I_{e,s}$

      –Girder with both ends continuous: $I_e(\text{average}) = 0.70\, I_{e,m} + 0.15\, (I_{e,s1} + I_{e,s2})$

      Use the I_e(average) in the girder deflection calculations.

      where,

      $I_{e,m}$ = effective moment of inertia at the midspan, calculated using the section properties and moments at the midspan

      $I_{e,s}$ = effective moment of inertia at the one continuous end, calculated using the section properties and the moments at the continuous end

      $I_{e,s1}$ = effective moment of inertia at the left continuous end, calculated using the section properties and the moments at the left end support

      $I_{e,s2}$ = effective moment of inertia at the right continuous end, calculated using the section properties and the moments at the right end support

      As discussed previously, since the girder midspan stiffness is dominant, the effective moment of inertia calculated at midspan can be used and will yield sufficiently accurate results. Thus, $I_e(\text{girder}) \approx I_{e,m}$.

4. The computer analysis in Step 1 is repeated using the effective moment of inertia $I_{e,\,girder}$ obtained in Step 3 for all the girders, and a moment of inertia for the columns of 70% of the gross moment of inertia of the columns (i.e., $I_{e\,columns} = 0.70 I_{g\,columns}$).

   From the computer analysis, the instantaneous dead load deflection, Δ_{DL} and the instantaneous live load deflection Δ_{LL} at the midspan of the end-span of the girder are obtained.

5. The total long-term deflection (*which excludes the instantaneous dead load deflection*) is the sum of the instantaneous live load deflection plus the long-term deflection due to dead load and the sustained portion of the live load, and is given as

$$\Delta_{TT} = \Delta_{LL} + \lambda_{\Delta}(\Delta_{DL} + \alpha\Delta_{LL})$$

6. Check the permissible deflection limits from ACI 318-14, Table 24.2.2.

Example 7-3

Deflection of Continuous T-beam or L-Beams

A reinforced concrete T-beam in an office building is continuous over three equal center-to-center spans of 30 ft (assume a clear span of 29 ft). The beam is 14″ × 22″ deep with a 6″ thick slab and is reinforced with 4#8 bottom bars at the midspan of the end span, and 4#8 top bars at the first interior support. The service dead load on the beam, which includes the beam self-weight is 1.40 kip/ft and the service live load is 0.5 kip/ft. Calculate the maximum deflection at the midspan of the end span of the continuous beam and compare with the ACI Code deflection limits assuming the beam is supported by girders. As an approximation, use only the midspan section properties in calculating the beam deflections, and assume the effective flange width of the beam at midspan, $b_e = 87''$. Use $f'_c = 4000$ psi and $f_y = 60{,}000$ psi.

Solution:

Assuming knife-edge supports for the continuous beam, and since all the three requirements for using the ACI coefficients are satisfied, the ACI moment coefficients can be used to calculate the service load moments (i.e., unfactored moments) at the supports and at midspan. Furthermore, as stated in the problem, the effective moment of inertia at the midspan will be used.

1. Calculate the service dead and live load moments, M_{DL} and M_{LL} at the midspan of the end span
   Clear span of the beam, $\ell_n = 29$ ft
   Using the ACI moment coefficients from Chapter 6, the unfactored moment at midspan due to dead load is,

$$M_{DL,m} = \frac{1}{11}(1.40 \text{ kip/ft})(29 \text{ ft})^2 = 107 \text{ ft.-k}$$

   Similarly, the unfactored moment at midspan due to live load is,

$$M_{LL,m} = \frac{1}{11}(0.5 \text{ kip/ft})(29 \text{ ft})^2 = 39 \text{ ft.-k}$$

2. Calculate T-section properties at midspan
   Effective flange width of the T-beam at midspan is given as, $b_e = 87''$

$$n = \frac{E_s}{E_c} = \text{modular ratio}$$

$$= \frac{(29)(10^6)}{57000\sqrt{4000}} = 8.1$$

The area of the tension reinforcement, A_S for 4#8 bars at midspan = 3.16 in.^{2}

The effective depth at midspan and at the support, $d = 22'' - 2.5'' = 19.5''$. The pertinent dimensions are:

$$b_w = 14''$$
$$h_f = 6''$$
$$h = 22''$$
$$d = 19.5''$$

Using the equations from Section 7-11, calculate the following sectional properties for the T-beam:

$$C = \frac{b_w}{nA_s} = \frac{14''}{(8.1)(3.16)} = 0.55$$

$$Z = \frac{(b_e - b_w)h_f}{nA_s} = \frac{(87 - 14)(6)}{(8.1)(3.16)} = 17.1$$

$$y_t = 22 - 0.5\left(\frac{[(87 - 14)(6)^2 + 14(22)^2]}{[(87 - 14)(6) + 14(22)]}\right) = 15.7''$$

$$\bar{y} = \frac{\left[\sqrt{0.55(2x19.5 + 6x17.1) + (1 + 17.1)^2}\right] - (1 + 17.1)}{0.55}$$
$$= 3.7''$$

$$I_g = \frac{(87 - 14)(6)^3}{12} + \frac{14(6)^3}{12} + 6(87 - 14)$$
$$(22 - 0.5[6] - 15.7)^2 + 14(22)(15.7 - 0.5[22])^2$$
$$= 25,310 \text{ in.}^4$$

$$I_{cr} = \frac{(87 - 14)(6)^3}{12} + \frac{14(3.7)^3}{3} + (87 - 14)(6)(3.7 - 0.5[6])^2$$
$$+ (8.04)(3.16)(19.5 - 3.7)^2$$
$$= 8155 \text{ in.}^4$$

3. Calculate the effective moment of inertia, I_e at the midspan using the section properties calculated at the midspan

$$I_{g,m} = 25310 \text{ in.}^4$$
$$I_{cr,m} = 8155 \text{ in.}^4 \quad \text{(see Step 2)}$$

y_t at midspan = 15.7″ (calculated previously in Step 2)

Cracking stress, $f_r = 7.5\sqrt{4000}$ psi = 474 psi

The cracking moment at midspan is,

$$M_{cr} = \frac{f_r I_g}{y_t} = \frac{(474)(25,310)}{15.7}$$
$$= 764,136 \text{ in.-lb} = 64 \text{ ft.-kip}$$

Total service load moment at midspan,

$$M_{a,m} = M_{DL,m} + M_{LL,m}$$
$$= 107 \text{ ft.-kip} + 39 \text{ ft.-kip} = 146 \text{ ft.-kip}$$

Use ACI 318-14, Equation 24.2.3.5a from Section 7-4 to calculate the effective moment of inertia, I_e:

$$\left[\frac{M_{cr}}{M_{a,m}}\right]^3 = \left[\frac{64}{146}\right]^3 = 0.084$$

Therefore, the effective moment of inertia at the midspan is,

$$I_{e,m} = \{(0.084)(25,310) + [1 - (0.084)](8155)\} \leq 25,310$$
$$= 9600 \text{ in.}^4 < 25,310 \text{ in.}^4 \qquad \text{(O.K.)}$$

This effective moment of inertia will be used to calculate the beam deflections.

4. Calculate the midspan deflections at the end span of the continuous beam

Instantaneous dead load deflection: The statical moment due to dead load is

$$M_{o,DL} = \frac{w_{DL}\ell^2}{8} = \frac{(1.40)(29)^2}{8} = 148 \text{ ft.-kip}$$

The service dead load moment at midspan due to service loads was calculated in Step 1 and is

$$M_{a,DL,m} = 107 \text{ ft.-kip}$$

For a continuous beam with a uniformly distributed load, the deflection constant, K, is

$$K = 1.2 - 0.2\frac{M_o}{M_{a,m}} = 1.2 - 0.2\left(\frac{148}{107}\right) = 0.93$$

The instantaneous or immediate dead load deflection, Δ_{DL}, including the effect of cracking is,

$$\Delta_{DL} = K\left(\frac{5}{48}\right)\frac{M_{DL,m}\ell^2}{E_c I_e} = 0.93\left(\frac{5}{48}\right)\frac{(107)(12,000)(29 \times 12)^2}{(57,000\sqrt{4000})(9600)}$$
$$= 0.43''$$

Instantaneous live load deflection: The statical moment due to live load is

$$M_{o,LL} = \frac{w_{LL}\ell^2}{8} = \frac{(0.5)(29)^2}{8} = 53 \text{ ft.-kip}$$

The service live load moment at midspan due to service loads was calculated in Step 1 using the ACI moment coefficients and is

$$M_{a,LL,m} = 39 \text{ ft.-k}$$

For a continuous beam with a uniformly distributed load, the deflection constant, K, is

$$K = 1.2 - 0.2\frac{M_o}{M_{a,m}} = 1.2 - 0.2\left(\frac{53}{39}\right) = 0.93$$

The instantaneous or immediate live load deflection, Δ_{LL}, including the effect of cracking is,

$$\Delta_{LL} = K\left(\frac{5}{48}\right)\frac{M_{LL,m}\ell^2}{E_c I_e} = 0.93\left(\frac{5}{48}\right)\frac{(39)(12,000)(29 \times 12)^2}{(57,000\sqrt{4000})(9600)}$$
$$= 0.16''$$

5. For an office building, the percentage of live load sustained can be assumed to be approximately 25%.

Therefore, $\alpha = 0.25$

For long-term deflections, use the creep factor after 5 years, therefore, $\xi = 2.0$. Since there is no compression steel, $\rho' = 0$, the long-term deflection multiplier is

$$\lambda_\Delta = \left(\frac{\xi}{1 + 50\rho'}\right) = \frac{2.0}{1 + 50(0)} = 2.0$$

The total long-term deflection (*which excludes the instantaneous dead load deflection*) is calculated as

$$\begin{aligned}\Delta_{TT} &= \Delta_{LL} + \Delta_{LT} \\ &= \Delta_{LL} + \lambda_\Delta(\Delta_{DL} + \alpha\Delta_{LL}) \\ &= 0.16'' + 2.0(0.43'' + 0.25[0.16'']) = 1.1''\end{aligned}$$

This long-term deflection as well as the instantaneous live load deflection will now be compared to the ACI Code deflection limits (ACI 318-14, Table 24.2.2).

a. For roof and floor flexural members not supporting deflection sensitive elements, $\Delta_{TT} = 1.1''$ *must be* $\leq \frac{\ell}{240} = \frac{29(12)}{240} = 1.45''$ (O.K.)

b. For roof and floor flexural members supporting deflection sensitive elements, $\Delta_{TT} = 1.1''$ *must be* $\leq \frac{\ell}{480} = \frac{29(12)}{480} = 0.73''$ (Not Good)

c. For roof and floor flexural members not supporting deflection sensitive elements, $\Delta_{LL} = 0.16''$ must be $\leq \frac{\ell}{180} = \frac{29(12)}{180} = 1.93''$ (O.K.)

d. For roof and floor flexural members supporting deflection sensitive elements, $\Delta_{LL} = 0.16''$ *must be* $\leq \frac{\ell}{480} = \frac{29(12)}{480} = 0.73''$ (O.K.)

The deflection in the end span of this continuous beam satisfies all the ACI Code deflection limits except the second limit. To reduce the long-term deflections, we could add compression steel to the beam at the midspan and recalculate the long-term deflection.

Assume 4#8 bars are added at the top of the beam at midspan, therefore the compression reinforcement ratio,

$$\rho' = \frac{A_s}{b_w d} = \frac{3.16\ in^2}{(14'')(19.5'')} = 0.0116$$

The revised long-term multiplier is

$$\lambda_\Delta = \left(\frac{\xi}{1 + 50\rho'}\right) = \frac{2.0}{1 + 50(0.0116)} = 1.27$$

There is a reduction of the long-term multiplier from 2.0 (without compression steel) to 1.27 with compression steel. To prevent buckling of the reinforcement, the compression steel will have to be confined with stirrups in accordance with ACI Code, Section 25.7.2.

With compression steel, the long-term deflection (excluding the instantaneous dead load deflection) is,

$$\Delta_{TT} = 0.16'' + 1.27(0.43'' + 0.25[0.16'']) = 0.76''$$

The revised long-term deflection still slightly exceeds the ACI Code limit if deflection sensitive elements are supported by the beam, but it is only about 4% over the limit, and in practice, this small excess deflection (0.03 in.) can be ignored.

7-8 DEFLECTION CONTROL MEASURES IN REINFORCED CONCRETE STRUCTURES

There are several ways for controlling deflections in reinforced concrete flexural members, and some are feasible during the design phase and others are carried out during the construction phase. The control measures include the following:

- Add compression steel to the beam, but the compression steel has to be tied with closed stirrups to prevent buckling, and the tie spacing must conform with ACI 318-14, Section 25.7.2. Note that compression steel cannot be used to control deflections in slabs since the compression bars cannot be confined. The advantageous effect of compression steel in reducing long-term deflections is reflected by the compression rebar ratio, ρ', being in the denominator of the equation for the long-term deflection multiplier, λ_Δ, in Section 7-5.
- Provide additional tension reinforcement which increases the cracked moment of inertia, I_{cr}, which in turn leads to a higher effective moment of inertia, I_e.
- Increase the slab thickness and the beam depth. Increasing the width of a beam is not as effective as increasing the beam depth in achieving an increase in the effective moment of inertia.
- To reduce the impact of incremental long-term deflections, the installation of the DSEs should be delayed to allow a significant proportion of the creep deflection of the member to occur prior to the installation.
- Deflections can also be controlled through the following construction practices:
  - Do not load the concrete structure until it reaches its design strength.
  - Provide adequate shoring to the concrete member during construction until it reaches its design strength.
  - Ensure that the reinforcement in the concrete member is not displaced or misplaced during the concrete pour.

7-9 CRACK CONTROL

With the advent of higher-strength reinforcing steels, where more strain is required to produce the higher stresses, cracking of reinforced concrete flexural members has become more troublesome. Some of the reasons why crack control is necessary include aesthetics considerations and the need to avoid the perception in the public's mind that the structure is in imminent danger of collapse due to the presence of noticeable cracks.

It seems logical that cracking would have an effect on corrosion of the reinforcing steel. However, there is no clear correlation between corrosion and surface crack widths in the usual range found in structures with reinforcement stresses at service load levels. Further, there is no clear experimental evidence available regarding the crack width beyond which a corrosion danger exists. Exposure tests indicate that concrete quality, adequate consolidation, and ample concrete cover may be more important in corrosion considerations than is crack width.

Rather than a small number of large cracks, it is more desirable to have only hairline cracks and to accept more numerous cracks, if necessary. To achieve this, the current ACI Code (Section 24.3) directs that the flexural tension reinforcement be well distributed in the maximum tension zones of a member. ACI Code Section 24.3.2 contains a provision for maximum spacing s that is intended to control surface cracks to a width that is generally acceptable in practice. The maximum spacing (see ACI Code Table 24.3.2) is limited to

$$s = 15\left(\frac{40{,}000}{f_s}\right) - 2.5c_c \leq 12\left(\frac{40{,}000}{f_s}\right)$$

where

s = center-to-center spacing of flexural tension reinforcement nearest to the tension face, in.

f_s = calculated stress, psi. This may be taken as $\frac{2}{3}$ of the specified yield strength.

c_c = clear cover from the nearest surface in tension to the surface of the flexural tension reinforcement, in.

ACI 318, Section 24.3.5, cautions that if a structure is designed to be watertight or if it is to be subjected to very aggressive exposure, the provisions of Section 24.3.2 are not sufficient and special investigations and precautions are required.

Example 7-4

Check the steel distribution for the beam shown in Figure 7-9 to establish whether reasonable control of flexural cracking is accomplished in accordance with the ACI Code, Section 24.3. Use f_y = 60,000 psi. Assume d = 30 in.

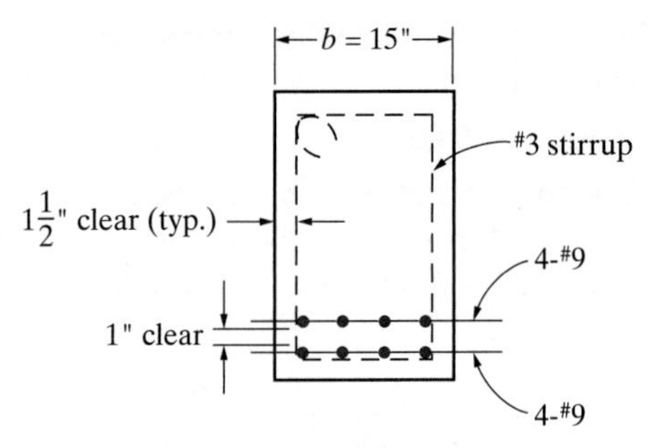

FIGURE 7-9 Sketch for Example 7-3.

Solution:

1. Calculate the center-to-center spacing between the No. 9 bars:

$$s = \frac{15 - 2(1.5) - 2(0.375) - 2\left(\frac{1.128}{2}\right)}{3} = 3.37 \text{ in.}$$

2. Assume positive moment and calculate the concrete clear cover from the bottom (tension) face of the beam to the surface of the nearest tension reinforcement:

$$c_c = 1.5 + 0.375 = 1.875 \text{ in.}$$

3. Calculate f_s using $\frac{2}{3}$ of f_y:

$$f_s = \frac{2}{3}f_y = \frac{2}{3}(60{,}000) = 40{,}000 \text{ psi}$$

4. Calculate maximum spacing allowed for deformed bars using ACI Code Table 24.3.2:

$$s = 15\left(\frac{40{,}000}{f_s}\right) - 2.5c_c = 15\left(\frac{40{,}000}{40{,}000}\right) - 2.5(1.875)$$
$$= 10.31 \text{ in.}$$

Check the upper limit of the equation for deformed bars from ACI Code Table 24.3.:

$$12\left(\frac{40{,}000}{f_s}\right) = 12\left(\frac{40{,}000}{40{,}000}\right) = 12 \text{ in.} > 10.31 \text{ in.} \quad \text{(O.K.)}$$

And lastly:

$$3.37 \text{ in.} < 10.31 \text{ in.} \quad \text{(O.K.)}$$

When beams are relatively deep, there exists the possibility for surface cracking in the tension zone areas away from the main reinforcing. ACI 318-14, Section 9.7.2.3, requires, for beams having depths h in excess of 36 in., the placing of longitudinal skin reinforcing along both side faces for a distance $h/2$ from the tension face of the beam. The spacing s between these longitudinal bars or wires shall not exceed the spacing as provided in ACI 318-14, Table 24.3.2. Bar sizes ranging from No. 3 to No. 5 (or welded wire reinforcement with a minimum area of 0.1 in.^{2} per foot of depth) are typically used.

Example 7-5

Select skin reinforcement for the cross section shown in Figure 7-10a. Flexural tension reinforcement is 5 No. 9 bars and f_y = 60,000 psi.

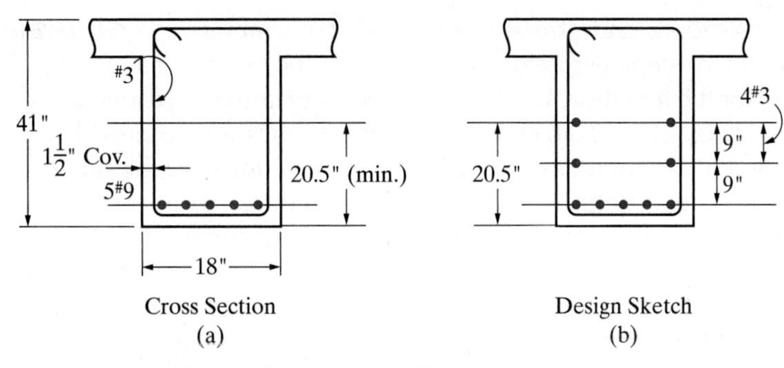

FIGURE 7-10 Sketch for Example 7-4.

Solution:

Because $h > 36$ in., skin reinforcement is required. The skin reinforcing must cover the tension surface for a minimum of $h/2$, or 20.5 in., up from the tension face of the beam. This is shown in Figure 7-10a. Assume No. 3 bars ($A_b = 0.11$ in.^{2}) for the skin reinforcing and calculate the maximum spacing s as follows:

$$c_c = 1.5 + 0.375 = 1.875 \text{ in.}$$

$$f_s = \frac{2}{3}f_y = \frac{2}{3}(60{,}000) = 40{,}000 \text{ psi}$$

$$s = 15\left(\frac{40{,}000}{f_s}\right) - 2.5c_c = 15\left(\frac{40{,}000}{40{,}000}\right) - 2.5(1.875)$$

$$= 10.31 \text{ in.}$$

Check upper limit:

$$12\left(\frac{40{,}000}{f_s}\right) = 12\left(\frac{40{,}000}{40{,}000}\right) = 12 \text{ in.} > 10.31 \text{ in. (O.K.)}$$

The distance from the tension face of the beam to the centroid of the tension reinforcement is

$$1.5 + 0.375 + \frac{1.128}{2} = 2.44 \text{ in.}$$

Therefore, the required number of spaces N is

$$N = \frac{20.5 - 2.44}{10.31} = 1.75 \text{ spaces} \quad \text{(Use 2 spaces)}$$

The actual spacing provided is

$$\frac{20.5 - 2.44}{2 \text{ spaces}} \approx 9 \text{ in.} < 10.31 \text{ in.} \qquad \text{(O.K.)}$$

The design is shown in Figure 7-10b.

7-10 FLOOR VIBRATIONS

Concrete floors are not as susceptible to issues related to floor vibrations as compared to structures framed with wood or steel, but attention needs to be given to the design of concrete floors for floor vibrations. There are several factors that need to be considered in the assessment of floor vibrations, and some of the basic terms used in the assessment are as follows: [1, 2]:

Damping: This is a measure of how much and how quickly vibrations will dissipate within a structure. Damping in a structural system is usually expressed as a percentage of critical damping, and critical damping is that which is required to bring the system to rest in one-half cycle. In general, the presence of full or partial height partitions is the critical factor in the value assumed for damping. For floors with few or no partitions, the damping ratio is 0.02. For partial height partitions, the value is 0.03 and for full height partitions, the value is 0.05. This damping ratio is directly proportional to the acceleration of the floor, so its value needs to be carefully determined.

Period: Time, in seconds, for one complete cycle of oscillation.

Frequency: Number of oscillations per second in hertz (Hz) or cycles per second, $f = 1/T$. A person walking at a pace of two steps per second is said to walk at a frequency of 2 Hz or two cycles per second.

Forcing Frequency: The frequency, in hertz, of the applied dynamic force. For rhythmic activity such as dancing or aerobics, the forcing frequency of the activity can occur between 1.5 Hz and 3.0 Hz for the first harmonic, 3.0 Hz–5.5 Hz for the second harmonic, and 6.0 Hz–8.25 Hz for the third harmonic.

Harmonic: An integer multiple of a forcing frequency. Any forcing frequency can have an infinite number of harmonics, but human activities are generally limited to a maximum of three harmonics. For example, for an applied forcing frequency of 2 Hz, the first harmonic is 2 Hz, the second harmonic is 4 Hz, and the third harmonics 6 Hz.

Natural Frequency: The frequency at which a structure vibrates when it is displaced and then suddenly released from at-rest state. This is also called free vibration since no external forces are applied. The natural frequency of a structure is proportional to its stiffness. The natural frequency of a structure should be greater than the forcing frequency causing the vibration excitation to avoid resonance.

Resonance: A phenomenon where the forcing frequency, (or a harmonic multiple of the forcing frequency), of the dynamic activity coincides with one of the natural frequencies, f_n, of the structure. This causes very large displacements, velocity, acceleration, and stresses. For a person walking at a pace of two steps per second, the floor will have to be checked for the first three harmonics of this forcing frequency (i.e., 2 Hz, 4 Hz, and 6 Hz)

There are three basic types of vibrations that are common in buildings [1, 2]: walking, rhythmic, and sensitive equipment vibrations. Most concrete structures satisfy walking vibrations requirements [3], and so this criteria is typically not a critical concern in the design process. To satisfy walking vibrations, the peak acceleration of the floor system needs to be less than a prescribed limit that has been established through research. The acceleration limit is described in terms of a percentage of the acceleration due to gravity, g. This limit, which is defined as a_0/g, is limited to 0.005 or 0.5% for basic walking vibrations. For shopping malls and indoor footbridges, this limit is increased to 0.015, and for outdoor footbridges, the limit is 0.05. For dining and dancing, the acceleration limit is 1.5%g; for rythmic activities, the acceleration limit is 5%g [1]. The most critical component of the vibration analysis is the stiffness of the floor system, which is captured in the natural frequency equation for the floor system, defined as:

$$f_n = 0.18\sqrt{\frac{g}{\Delta_c + \Delta_j + \Delta_g}}$$

where

f_n = natural frequency of the floor system
g = acceleration due to gravity, 386 in./s^2
Δ_c = column axial deformation (in.)
Δ_j = joist deflection (in.)
Δ_g = girder deflection (in.)

The girder and joist deflections can be calculated using the dynamic elastic modulus of concrete due to the transient nature of vibrations [2, 3].

The natural frequency of more complicated systems such as flat plates, two-way joists, and grillage systems can be found in reference [2].

The most common variables used to control vibrations are the span and the cross-sectional size of the members. Adjusting the modulus of elasticity, for example, would help with vibration issues, but typically it is more economical to adjust the beam and girder depths to control vibrations.

It is always advisable to locate any rhythmic activity on the ground floor of a structure where possible. For activities on elevated floors that could cause rhythmic vibrations, such as dancing, jumping exercises, aerobics, concerts, and sporting events, it is often more economical to utilize a two-way floor system [3]. To avoid floor vibration problems during rhythmic activities, the natural frequency of the floor system should be greater than a specified minimum value. For any given activity, there are several harmonics of vibration. For example, the participants in a lively concert might induce a forcing frequency of 2.0 Hz for the first harmonic and 4.0 Hz for the second harmonic. The floor would have to be checked for both harmonics. For most rhythmic activities, not more than three harmonics would need to be checked. The natural frequency limits are the same as stated before for the forcing frequencies of certain activities.

H. Bachmann [5] recommends the following minimum natural frequencies for reinforced concrete structures subject to human-induced vibrations:

- Pedestrian bridge: 4.6 Hz.
- Office buildings: 7.5 Hz
- Gymnasiums and sports halls: 7.5 Hz
- Dance halls and concert halls without fixed seating: 6.5 Hz
- Concert halls with fixed seating: 3.4 Hz

For floor structures that support sensitive equipment, the floor motion is expressed in terms of velocity, because the design criteria for equipment often corresponds to a constant velocity over the frequency range considered. As was the case with rhythmic vibrations, it is often more economical to utilize a two-way floor system to control equipment vibrations. The natural frequency of the floor system is still a critical factor in the assessment of a floor for the support of sensitive equipment. The maximum velocity of the floor system is defined as [1]:

$$V = 2\pi f_n X_{\max}$$

where

V = maximum velocity of the floor system, μ-in./s
f_n = natural frequency of the floor system
$X_{\max}$ = maximum floor displacement based on a forcing function

The forcing function is based on the pace of a person walking, which can vary between 50 and 100 steps per minute. Common ways to mitigate vibrations due to walking would be to locate the equipment away from corridors or to limit the length of corridors, which has the effect of slowing the walking speed of the occupants.

A floor would be considered acceptable to support sensitive equipment if the maximum velocity of the floor system is less than a prescribed value, which can be obtained from reference [1] or from the manufacturer of the equipment. Computer floors, for example, can have a velocity limit of 8000μ-in./s and operating rooms may have a vibrational velocity limit of 4000μ − in./s. The most sensitive equipment can be found in a Class E microelectronics manufacturing facility, where the maximum allowable vibrational velocity is about 130μ-in./s [6, 7]. This would require a floor structure that has a natural frequency close to 50 Hz [6]. Floor structures that support microelectronics manufacturing are often designed as a two-way system such as a

TABLE 7-2 Minimum Flat Plate Thickness and Maximum Span for Human-induced Floor Vibrations[1]

Minimum flat plate thickness (in.)	Cause of vibration excitation	Maximum Span (ft)
7.0	Walking vibration	22.4
	Rhythmic vibration – Dinning & dancing	23.9
	Rhythmic vibration – Lively concert; sporting event	22.4
	Rhythmic vibration – Jumping exercises, Aerobics	19.6
7.5	Walking vibration	23.9
	Rhythmic vibration – Dinning & dancing	24
	Rhythmic vibration – Lively concert; sporting event	23.5
	Rhythmic vibration – Jumping exercises, Aerobics	20.6
8.0	Walking vibration	24
	Rhythmic vibration – Dinning & dancing	24
	Rhythmic vibration – Lively concert; sporting event	24
	Rhythmic vibration – Jumping exercises, Aerobics	21.4
8.5	Walking vibration	24
	Rhythmic vibration – Dinning & dancing	24
	Rhythmic vibration – Lively concert; sporting event	24
	Rhythmic vibration – Jumping exercises, Aerobics	22.2
9.0	Walking vibration	24
	Rhythmic vibration – Dinning & dancing	24
	Rhythmic vibration – Lively concert; sporting event	24
	Rhythmic vibration – Jumping exercises, Aerobic	23
9.5	Walking vibration	24
	Rhythmic vibration – Dinning & dancing	24
	Rhythmic vibration – Lively concert; sporting event	24
	Rhythmic vibration – Jumping exercises, Aerobics	24
10.0	Walking vibration	24
	Rhythmic vibration – Dinning & dancing	24
	Rhythmic vibration – Lively concert; sporting event	24
	Rhythmic vibration – Jumping exercises, Aerobics	24

[1]Adapted from Ref. [7]

TABLE 7-3 Minimum Flat Plate Thickness and Maximum Span for Sensitive Equipment Vibrations[1]

Minimum Flat plate thickness (in.)	Sensitive Equipment Vibrational Velocity Limit (μ inch/sec)	Maximum Span (ft)
7.5	**V = 8000**	18
8.0	**V = 8000**	19
	V = 4000	16
8.5	**V = 8000**	20
	V = 4000	17
9.0	**V = 8000**	21
	V = 4000	18
9.5	**V = 8000**	22
	V = 2000	16
10.0	**V = 8000**	24

[1]Adapted from Ref. [7]

waffle slab [6–8]. For preliminary design, the minimum thicknesses and maximum spans for flat plate systems for various vibration excitations can be obtained from Tables 7-2 and 7-3 [7]. For more stringent vibrational velocity limits as would be required in microelectronics facilities where thicker waffle slabs or two-way joists have to be used, the reader should refer to Ref.[8].

7-11 GROSS AND CRACKED SECTION PROPERTIES OF CONCRETE SECTIONS

The principle of transformed sections from mechanics of materials was used to derive the cracked section properties of concrete sections in Section 7-3. The equations for gross moment of inertia and cracked moment of inertia for rectangular and flanged (i.e., T- or L-shape) concrete sections are summarized below [4].

Section properties for rectangular concrete sections with tension reinforcement only:

$$I_{cr} = \frac{b\bar{y}^3}{3} + nA_s(d - \bar{y})^2$$

$$\bar{y} = \frac{[\sqrt{(2dB + 1)} - 1]}{B}$$

$$B = \frac{b}{nA_s}$$

$$n = \frac{E_s}{E_c} = \text{modular ratio}$$

A_s = area of the tension reinforcement

$$I_g = \frac{bh^3}{12}$$

$$y_t = 0.5h$$

$\bar{y}$ is the distance from the compression face to the neutral axis.

Section properties for T or L-sections with tension reinforcement only:

$$I_{cr} = \frac{(b_e - b_w)h_f^3}{12} + \frac{b_w \bar{y}^3}{3} + (b_e - b_w)(h_f)(\bar{y} - 0.5h_f)^2 + nA_s(d - \bar{y})^2$$

$$\bar{y} = \frac{\left[\sqrt{C(2d + h_f Z) + (1 + Z)^2}\right] - (1 + Z)}{C}$$

$$C = \frac{b_w}{nA_s}$$

$$Z = \frac{(b_e - b_w)h_f}{nA_s}$$

$$y_t = h - 0.5\left(\frac{[(b_e - b_w)h_f^2 + b_w h^2]}{[(b_e - b_w)h_f + b_w h]}\right)$$

$$I_g = \frac{(b_e - b_w)h_f^3}{12} + \frac{b_w h^3}{12} + h_f(b_e - b_w)(h - 0.5h_f - y_t)^2 + b_w h(y_t - 0.5h)^2$$

$n = \dfrac{E_s}{E_c}$ = modular ratio

A_s = area of the tension reinforcement

$\bar{y}$ = distance from the *compression* face to the neutral axis of the cracked section

y_t = distance from the *tension* face to the neutral axis of un-cracked section

References

[1] Thomas M. Murray, David E. Allen, and Eric E. Ungar. American Institute of Steel Construction. *Steel Design Guide Series 11: Floor Vibrations Due to Human Activity*. Chicago, IL: AISC, 2003.

[2] Fanella, D.A. and Mota, M. *Design Guide for Vibrations of Reinforced Concrete Floor Systems*. Schaumburg, IL: Concrete Reinforcing Steel Institute, 2014.

[3] Fanella, D.A. and Mota, M. "Vibration of Reinforced Concrete Floor Systems." *Structure Magazine*, April 2015.

[4] David A. Fanella, Javeed A. Munshi, and Basile G. Rabbat. *Notes on ACI 318-99*. Portland Cement Association, Skokie, IL, 1999.

[5] H. Bachmann. "Case Studies of Structures with Man-induced Vibrations," ASCE Journal of Structural Engineering, Vol. 118, Issue 3, March 1992.

[6] Hal Amick and Paulo J. M. Monteiro. "Construction of Nanotechnology Facilities," Concrete International, pp. 68–72, March 2004.

[7] David A. Fanella and Michael Mota. "Vibration Excitations—Part 1: How to Select a Reinforced Concrete Floor System," Structure Magazine, September 2017.

[8] David A. Fanella and Michael Mota. "Vibration Excitations—Part 2: How to Select a Reinforced Concrete Floor System," Structure Magazine, September 2017.

Problems

For the following problems, unless otherwise noted, assume normal-weight concrete.

7-1. Locate the neutral axis and calculate the moment of inertia for the cracked transformed cross sections shown. Use $f'_c = 4000$ psi and $f_y = 60{,}000$ psi.

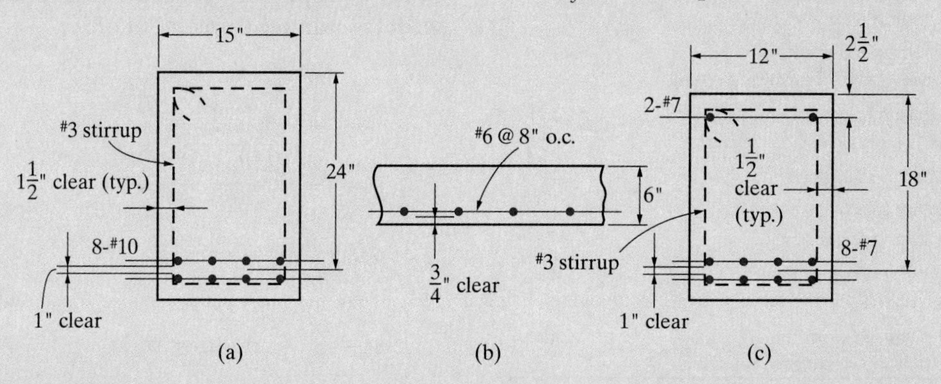

PROBLEM 7-1

7-2. Find I_g and I_{cr} for the T-beam shown. The effective flange width is 60 in., and $f'_c = 4000$ psi.

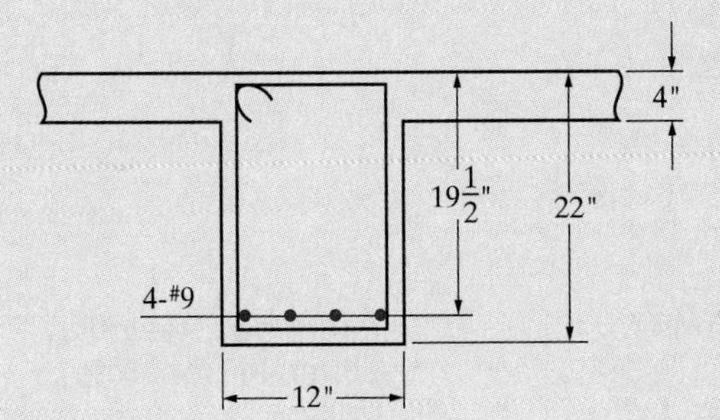

PROBLEM 7-2

7-3. The beam of cross section shown is on a simple span of 20 ft and carries service loads of 1.5 kip/ft dead load (includes beam weight) and 1.0 kip/ft live load. Use $f'_c = 3000$ psi and $f_y = 60{,}000$ psi.

a. Compute the immediate deflection due to dead load and live load.

b. Compute the long-term deflection due to the dead load. Assume the time period for sustained loads to be in excess of 5 years.

c. Develop a spreadsheet to solve this problem and to confirm your calculations.

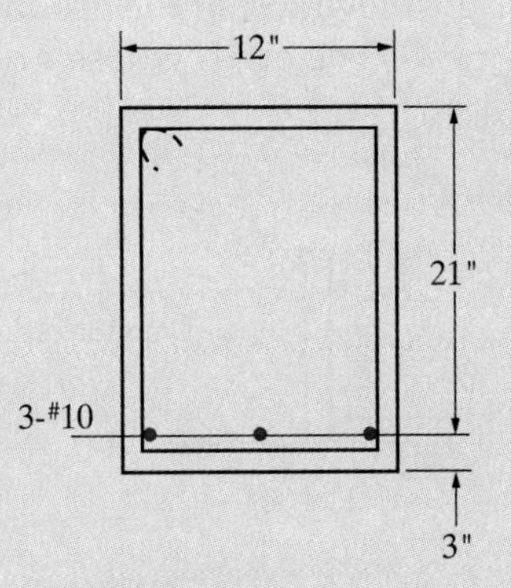

PROBLEM 7-3

7-4. Rework Problem 7-3 using a beam on a simple span of 26 ft, service load of 0.8 kip/ft dead load (includes beam weight) over the full span, and a point live load of 12 kip at midspan. Assume tensile reinforcing to be three No. 9 bars and $f'_c = 4000$ psi.

7-5. The floor beam shown is on a simple span of 16 ft. The beam supports nonstructural elements likely to be damaged by large deflections. The service loads are 0.6 kip/ft dead load (does not include the beam weight) and 1.40 kip/ft live load. Assume that the live load is 60% sustained for 6-month periods. Use $f'_c = 3000$ psi and $f_y = 60{,}000$ psi.

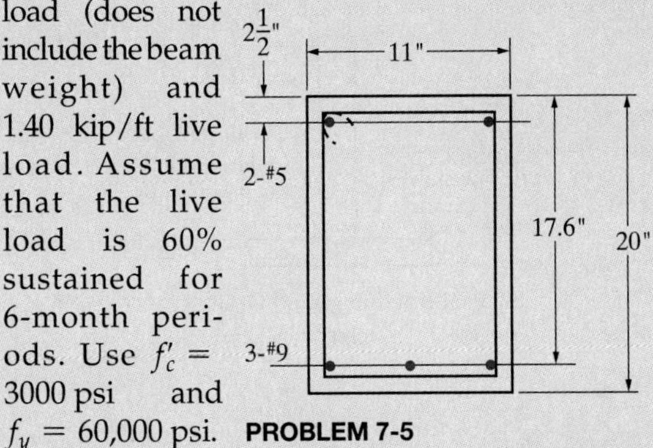

PROBLEM 7-5

a. Check the beam for deflections.

b. If the beam is unsatisfactory, redesign it so that it meets both flexural strength and deflections requirements.

7-6. Calculate the expected long-term deflection due to dead load and sustained live load for (a) the slab and (b) the beam of the floor system designed for Problem 6-4. Assume 10% sustained live load and a time period in excess of 5 years.

7-7. Check the cross sections of Problems 7-1(a) and (b) for acceptability under the ACI Code provisions for distribution of flexural reinforcement (crack control).

7-8. Check the distribution of flexural reinforcement for the members designed in Problems 2-21 and 2-25. If necessary, redesign the steel.

7-9. Refer to the beam diagram and detail shown. The loads shown are service level and the self-weight is included in the dead load. Assume 20% of the live load is permanently sustained for more than 5 years. Use $f'_c = 4000$ psi, $f_y = 60{,}000$ psi

a. Find the effective moment of inertia, I_e

b. Determine the deflections due to dead, live, and total loads; compare with L/480

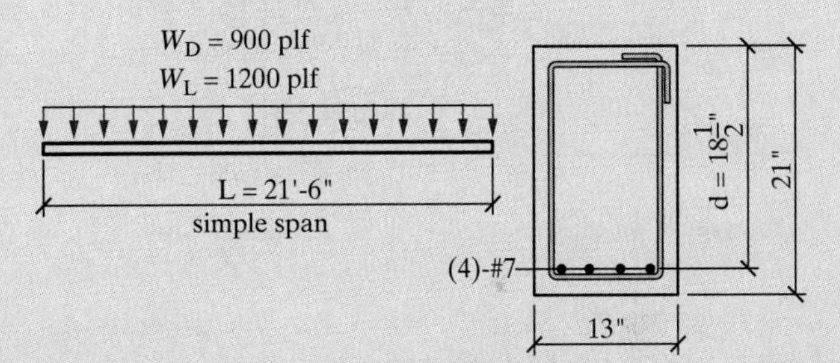

PROBLEM 7-9

7-10. Refer to the concrete beam diagram and detail shown. The loads shown are service level and the self-weight is included in the dead load. Assume 25% of the live load is permanently sustained for more than 5 years. Use $f'_c = 4000$ psi, $f_y = 60{,}000$ psi

a. Find the effective moment of inertia, I_e

b. Determine the deflections due to dead, live, and total loads; is the beam adequate for total load deflection? Compare with L/240

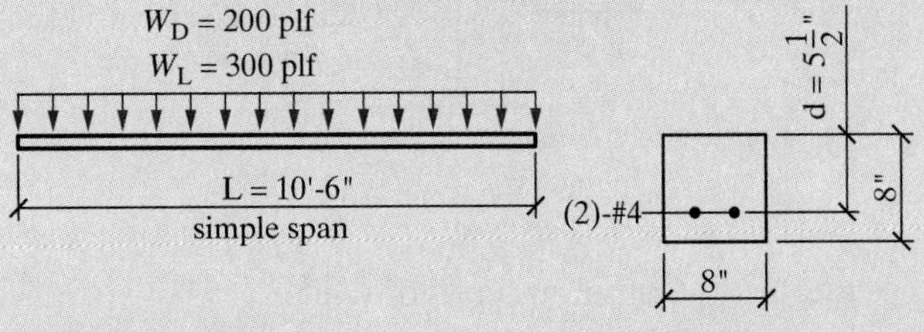

PROBLEM 7-10

WALLS

8-1 Introduction
8-2 Lateral Forces on Retaining Walls
8-3 Design of Reinforced Concrete Cantilever Retaining Walls
8-4 Design Considerations for Bearing Walls
8-5 Design Considerations for Basement Walls
8-6 Lateral Load Resisting Systems in Concrete Buildings
8-7 Concrete Moment Frames
8-8 Shear Walls

8-1 INTRODUCTION

Walls are generally used to provide lateral support for an earth fill, embankment, or some other material and to support vertical loads. Some of the more common types of walls are shown in Figure 8-1. One primary purpose for these walls is to maintain a difference in the elevation of the ground surface on each side of the wall. The earth whose ground surface is at the higher

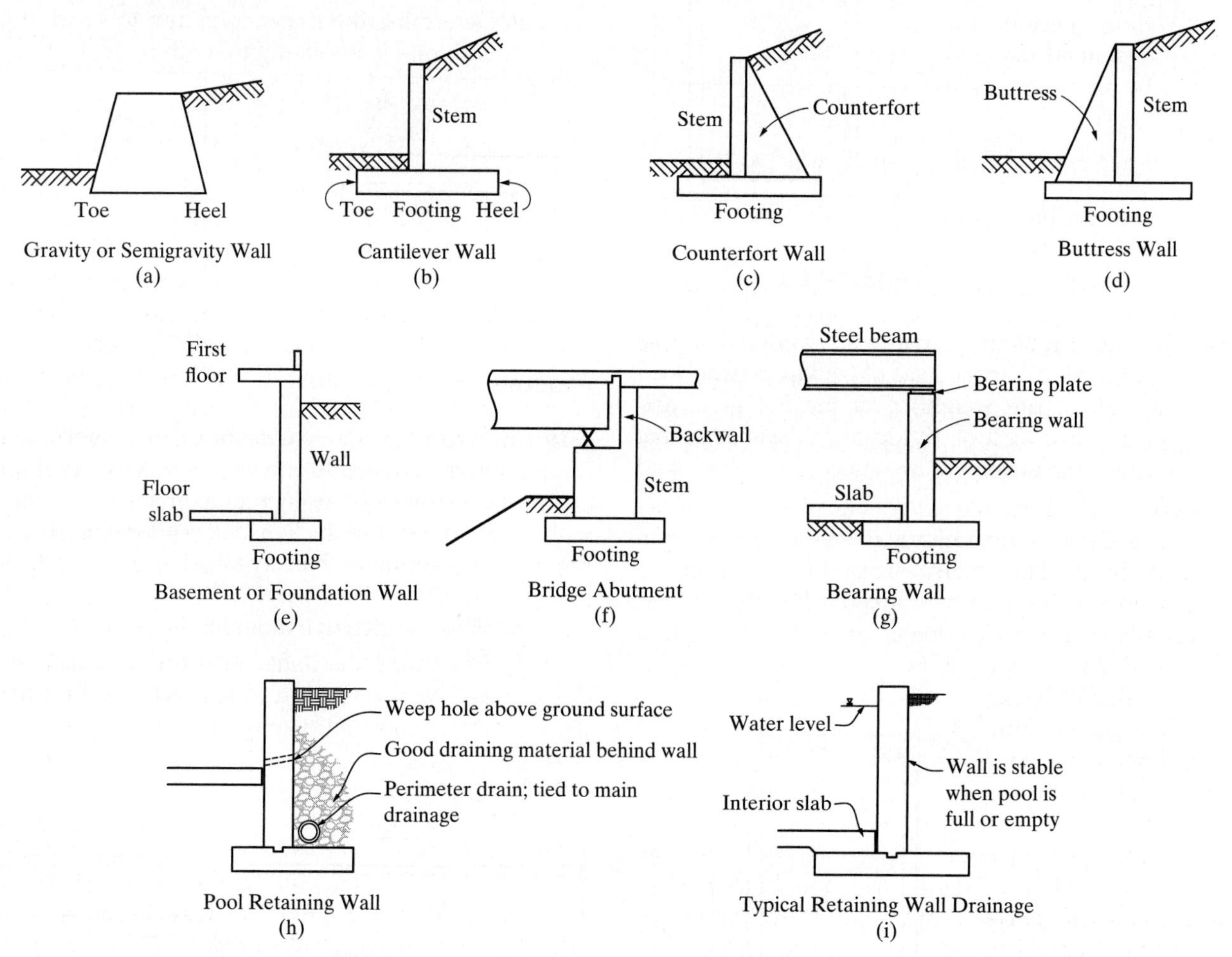

FIGURE 8-1 Common types of walls.

elevation is commonly called the *backfill*, and the wall is said to retain this backfill.

All the walls shown in Figure 8-1 have applications in either building or bridge projects. They do not necessarily behave in an identical manner under load, but still serve the same basic function of providing lateral support for a mass of earth or other material that is at a higher elevation behind the wall than the earth or other material in front of the wall. Hence, they all may be broadly termed *retaining structures* or *retaining walls*. Some retaining walls may support vertical loads in addition to the lateral loads from the retained materials.

The gravity wall (Figure 8-1a) depends mostly on its own weight for stability. It is usually made of plain concrete and is used for walls up to approximately 10 ft in height. The semigravity wall is a modification of the gravity wall in which small amounts of reinforcing steel are introduced. This, in effect, reduces the massiveness of the wall.

The cantilever wall (Figures 8-1b and 8-2) is the most common type of retaining structure and generally is used for walls in the range from 10 to 25 ft in height. It is so named because its individual parts (toe, heel, and stem) behave as, and are designed as, cantilever beams. Aside from its stability, the capacity of the wall is a function of the strength of its individual parts.

The counterfort wall (Figure 8-1c) may be economical when the wall height is in excess of 25 ft. The counterforts are spaced at intervals and act as tension members to support the stem. The stem is then designed as a continuous member spanning horizontally between the counterforts.

The buttress wall (Figure 8-1d) is similar to the counterfort wall except that the buttresses are located on the side of the stem opposite to the retained material and act as compression members to support the stem. The counterfort wall is more commonly used because it has a clean, uncluttered exposed face and allows for more efficient use of space in front of the wall.

The basement or foundation wall (Figure 8-1e) may act as a cantilever retaining wall. The first floor may provide an additional horizontal reaction similar to the basement floor slab, however, thereby making the wall act as a vertical beam. This wall would then be designed as a simply supported member spanning between the first floor and the basement floor slab.

The bridge abutment (Figure 8-1f) is similar in some respects to the basement wall. The bridge superstructure induces horizontal as well as vertical loads, thus altering the normal cantilever behavior.

The bearing wall (Figure 8-1g) may exist with or without lateral loads. A bearing wall may be defined as a wall that supports any vertical load in addition to its own weight. Depending on the magnitudes of the vertical and lateral loads, the wall may have to be designed for combined bending and axial compression. Bearing walls and basement walls are further discussed later in this chapter.

Figure 8-1h shows a swimming pool retaining wall and Figure 8-1i shows the typical drainage provided at the perimeter of retaining walls. The drainage is provided by means of weep holes that are spaced at regular intervals along the length of the wall.

8-2 LATERAL FORCES ON RETAINING WALLS

The design of a retaining wall must account for all the applied loads. The load that presents the greatest problem and is of primary concern is the lateral earth pressure induced by the retained soil. The comprehensive earth pressure theories evolving from the original Coulomb and Rankine theories can be found in almost any textbook on soil mechanics.

The magnitude and direction of the pressures as well as the pressure distribution exerted by a soil backfill upon a wall are affected by many variables. These variables include, but are not limited to, the type of backfill used, the drainage of the backfill material, the level of the water table, the slope of the backfill material, added loads applied on the backfill, the degree of soil compaction, and movement of the wall caused by the action of backfill.

Figure 8-2 shows a retaining wall under construction. An important consideration is that water must be prevented from accumulating in the backfill material. Walls are rarely designed to retain saturated material, which means that proper drainage must be provided. It is generally agreed that the best backfill material behind a retaining structure is a well-drained, cohesionless material. Hence, it is the condition that is usually specified and designed for. Materials that contain combinations of types of soil will act like the predominant material. An example of an acceptable backfill used behind basement and retaining walls is "well graded sand and gravel or crusher stone having a maximum size of 3 in. and no more than 7% of particles passing the #200 sieve." For basement walls, the perimeter drainage system could consist of a 4-in. diameter perimeter drain wrapped in geotextile material that acts as a filter and located in a bed of gravel just above the underside of the wall footing are provided around the outside perimeter of the building near the basement wall footing. Drainage for retaining walls can be provided using a sloped 3-in. diameter pipe through the thickness of the stem of the retaining wall near the bottom of the retaining wall stem that act as weep holes to drain water from the back of the retaining wall to the front of the wall. Typical horizontal spacing of the weep holes is 8 ft on centers and

FIGURE 8-2 Cantilever retaining wall ($h_w = 12'\text{-}0''$). (George Limbrunner)

located vertically within 1 ft from the top of the retaining wall footing. These drainage systems for basement and retaining walls help to prevent the buildup of hydrostatic pressures behind these walls as well as help to prevent uplift pressures on the slab-on-grade in buildings with basement walls. Where the natural water table is only a few feet below grade, the retained soil behind retaining walls and basement walls will be saturated; in this situation, the walls must be designed for the hydrostatic pressures in addition to the lateral pressures from the saturated soil.

The lateral earth pressure can exist and develop in three different categories: active state, at rest, and passive state. If a wall is absolutely rigid, earth pressure at rest will develop. If the wall should deflect or move a very small amount away from the backfill, active earth pressure will develop and in effect reduce the lateral earth pressure occurring in the at-rest state. Should the wall be forced to move toward the backfill for some reason, passive earth pressure will develop and increase the lateral earth pressure appreciably above that occurring in the at-rest state. As indicated, the magnitude of earth pressure at rest lies somewhere between active and passive earth pressures.

Under normal conditions, earth pressure at rest is of such a magnitude that the wall deflects slightly, thus relieving itself of the at-rest pressure. The active pressure results. For this reason, retaining walls are generally designed for active earth pressure due to the retained soil.

Because of the involved nature of a rigorous analysis of an earth backfill and the variability of the material and conditions, assumptions and approximations are made with respect to the nature of lateral pressures on a retaining structure. It is common practice to assume linear active and passive earth pressure distributions. The pressure intensity is assumed to increase with depth as a function of the weight of the soil in a manner similar to that which would occur in a fluid. Hence, this horizontal pressure of the earth against the wall is frequently called an *equivalent fluid pressure*. Experience has indicated that walls designed on the basis of these assumptions and those of the following discussion are safe and relatively economical.

Level Backfill

If we consider a level backfill (of well-drained cohensionless soil), the assumed pressure diagram is shown in Figure 8-3. The unit pressure intensity p_y in any plane a distance y down from the top is

$$p_y = K_a w_e y$$

Therefore, the total active earth pressure acting on a 1-ft width of wall may be calculated as the product of the average pressure on the total wall height h_w and the area on which this pressure acts:

$$H_a = \tfrac{1}{2} K_a w_e h_w^2 \tag{8-1}$$

where K_a, the *coefficient of active earth pressure*, has been established by both Rankine and Coulomb to be

$$K_a = \frac{1 - \sin\phi}{1 + \sin\phi} = \tan^2\left(45^\circ - \frac{\phi}{2}\right) \tag{8-2}$$

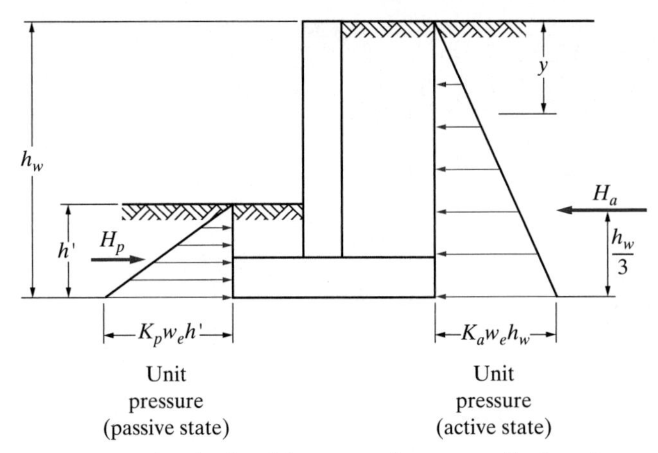

FIGURE 8-3 Analysis of forces acting on walls: level backfill.

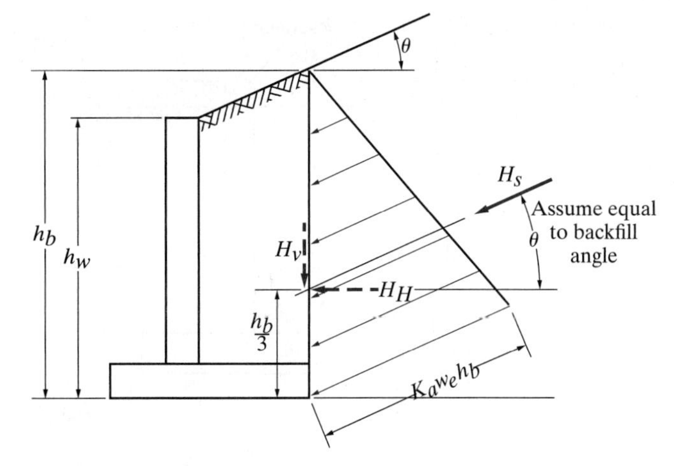

FIGURE 8-4 Analysis of forces acting on walls: sloping backfill.

and

$$w_e = \text{unit weight of earth (lb/ft}^3)$$

$$\phi = \text{angle of internal friction (soil on soil)}$$

K_a usually varies from 0.27 to 0.40. The term $K_a w_e$ in Equation (8-1) is generally called an *equivalent fluid weight*, because the resulting pressure is identical to that which would occur in a fluid of that weight (units are lb/ft^3).

In a similar manner, the total passive earth pressure force may be established as

$$H_p = \tfrac{1}{2} K_p w_e (h')^2 \qquad \textbf{(8-3)}$$

where h' is the height of earth and K_p is the *coefficient of passive earth pressure:*

$$K_p = \frac{1 + \sin\phi}{1 - \sin\phi} = \tan^2\left(45° + \frac{\phi}{2}\right) = \frac{1}{K_a}$$

Note that K_p usually varies from 2.5 to 4.0.

The total force in each case is assumed to act at one-third the height of the triangular pressure distribution, as shown in Figure 8-3.

Sloping Backfill

If we consider a sloping backfill, the assumed active earth pressure distribution is shown in Figure 8-4, where $H_s = \frac{1}{2} K_a w_e h_b^2$, h_b is the height of the backfill at the back of the footing, and K_a is the coefficient of active earth pressure. Thus

$$K_a = \cos\theta \left(\frac{\cos\theta - \sqrt{\cos^2\theta - \cos^2\phi}}{\cos\theta + \sqrt{\cos^2\theta - \cos^2\phi}} \right)$$

where θ is the slope angle of the backfill and ϕ is as previously defined. Note that H_s is shown acting parallel to the slope of the backfill.

For walls approximately 20 ft in height or less, it is recommended that the horizontal force component H_H simply be assumed equal to H_s and be assumed to act at $h_b/3$ above the bottom of the footing, as shown in Figure 8-4. The effect of the vertical force component H_V is neglected. This is a conservative approach.

Assuming a well-drained, cohensionless soil backfill that has a unit weight of 110 lb/ft^3 and an internal friction angle ϕ of 33°40′, values of equivalent fluid weight for sloping backfill may be determined as listed in Table 8-1.

Level Backfill with Surcharge

Loads are often imposed on the backfill surface behind a retaining wall. They may be either live loads or dead loads. These loads are generally termed a *surcharge* and theoretically may be transformed into an equivalent height of earth.

A uniform surcharge over the adjacent area adds the same effect as an additional (equivalent) height of earth. This equivalent height of earth h_{su} may be obtained by

$$h_{su} = \frac{w_s}{w_e}$$

where

$$w_s = \text{surcharge load (lb/ft}^2)$$

$$w_e = \text{unit weight of earth (lb/ft}^3)$$

TABLE 8-1 $K_a w_e$ Values for Sloping Backfill

θ (deg)	$K_a w_e$ (lb/ft^3)
0	32
10	33
20	38
30	54

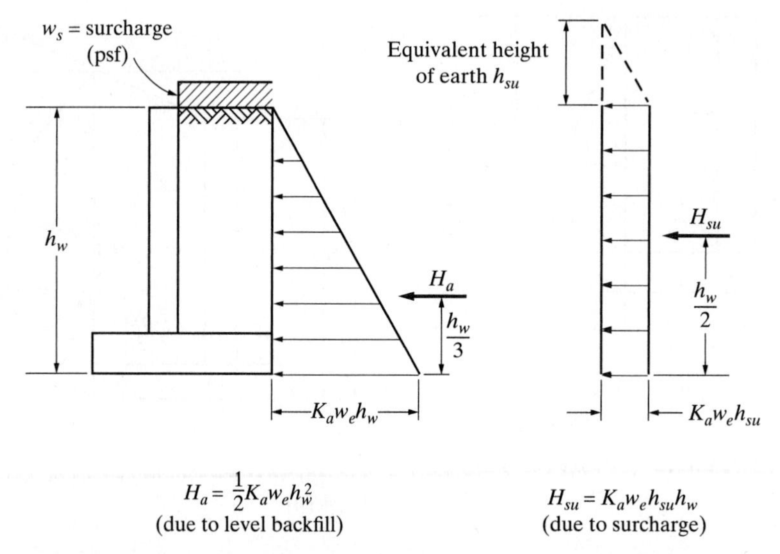

FIGURE 8-5 Forces acting on wall: level backfill and surcharge.

In effect, this adds a rectangle of pressure behind the wall with a total lateral surcharge force assumed acting at its midheight, as shown in Figure 8-5. Surcharge loads far enough removed from the wall cause no additional pressure acting on the wall.

8-3 DESIGN OF REINFORCED CONCRETE CANTILEVER RETAINING WALLS

A retaining wall must be stable as a whole, and it must have sufficient strength to resist the forces acting on it. Four possible *modes of failure* will be considered. *Overturning about the toe*, point O, as shown in Figure 8-6, could occur due to lateral loads. The stabilizing moment must be sufficiently in excess of the overturning moment so that an adequate factor of safety against overturning is provided. The factor of safety should never be less than 1.5 and should preferably be 2.0 or more. *Sliding on the base of the footing*, surface OM in Figure 8-6, could also occur due to lateral loads. The resisting force is based on an assumed coefficient of friction of concrete on earth. The factor of safety against sliding should never be less than 1.5 and should preferably be 2.0 or more. *Excessive soil pressure* under the footing will lead to undesirable settlements and possible rotation of the wall. Actual soil pressures should not be allowed to exceed specified allowable pressures, which depend on the characteristics of the underlying soil. The *structural failure of component parts of the wall* such as stem, toe, and heel, each acting as a cantilever beam, could occur. These must be designed to have sufficient strength to resist all anticipated loads.

A *general design procedure* for specific, known conditions may be summarized as follows:

1. Establish the general shape of the wall based on the desired height and function.
2. Establish the site soil conditions, loads, and other design parameters. This includes the determination of allowable soil pressure, earth-fill properties for active and passive pressure calculations, amount of surcharge, and the desired factors of safety.
3. Establish the tentative proportions of the wall.
4. Analyze the stability of the wall. Check factors of safety against overturning and sliding and compare actual soil pressure with allowable soil pressure.
5. Assuming that all previous steps are satisfactory, design the component parts of the cantilever retaining wall, stem, toe, and heel as cantilever beams.

Using a procedure similar to that used for one-way slabs, the analysis and design of cantilever retaining walls is based on a 12-in. (1-ft)-wide strip measured along the length of wall. The preliminary proportions of a cantilever retaining wall may be obtained from the following rules of thumb (see Figure 8-6):

1. Footing width L: Use $\frac{1}{2}h_w$ to $\frac{2}{3}h_w$.
2. Footing thickness h: Use $\frac{1}{10}h_w$.
3. Stem thickness G (at top of footing): Use $\frac{1}{12}h_w$.
4. Toe width A: Use $\frac{1}{4}L$ to $\frac{1}{3}L$.
5. Use a minimum wall batter of $\frac{1}{4}$ in./ft to improve the efficiency of the stem as a bending member and to decrease the quantity of concrete required.
6. The top of stem thickness D should not be less than 10 in.

The given rules of thumb will usually result in walls that can reasonably be designed. Depending on the

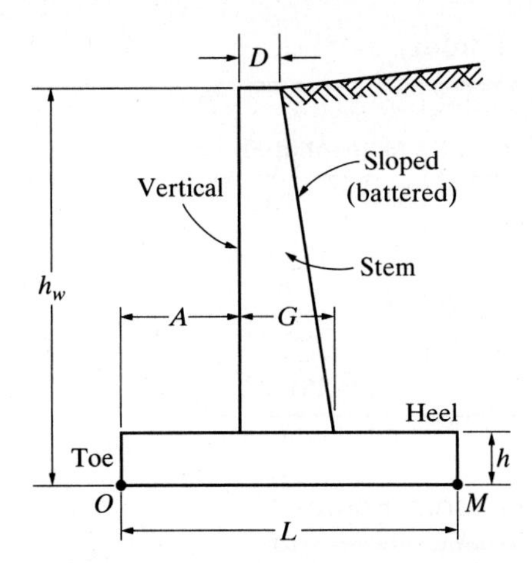

FIGURE 8-6 Cantilever retaining wall proportions.

specific conditions, however, dimensions may have to be adjusted somewhat to accommodate such design criteria as reinforcement limits, shear strength, anchorage, and development. One common alternative design approach is to assume a footing thickness and then immediately design the stem thickness for an assumed steel ratio. Once the stem thickness is established, the wall stability can be checked. Whichever procedure is used, adjustment of dimensions during the design is not uncommon.

Example 8-1

Design a retaining wall for the conditions shown in Figure 8-7. Use $f'_c = 3000$ psi and $f_y = 60{,}000$ psi. Other design data are given in step 2. Assume normal-weight concrete.

Solution:

1. The general shape of the wall, as shown, is that of a cantilever wall, because the overall height of 18 ft is within the range in which this type of wall is normally economical.
2. The design data are unit weight of earth $w_e = 100$ lb/ft^3, allowable soil pressure = 4000 psf, equivalent fluid weight $K_a w_e = 30$ lb/ft^3, and surcharge load $w_s = 400$ psf. The desired minimum factor of safety against overturning is 2.0 and against sliding is 1.5.

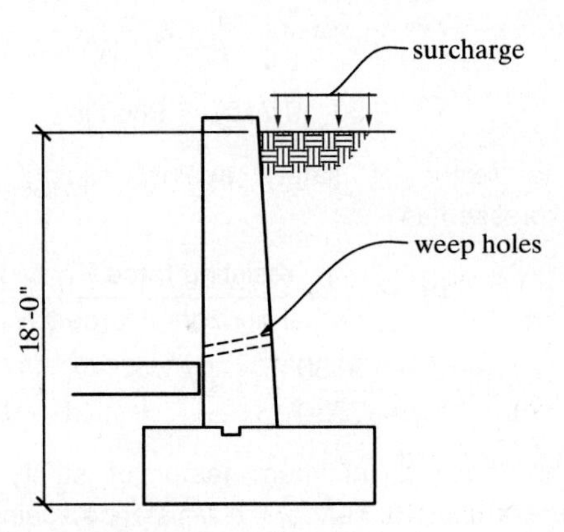

FIGURE 8-7 Sketch for Example 8-1.

3. Establish preliminary proportions for the wall.

   a. Footing width:

   $$\tfrac{1}{2} \text{ to } \tfrac{2}{3} \text{ of wall height}$$

   $$\tfrac{1}{2}(18) \text{ to } \tfrac{2}{3}(18) = 9 \text{ to } 12 \text{ ft}$$

   Use 11 ft-0 in.

   b. Footing thickness:

   $$\tfrac{1}{10}(18) = 1.8 \text{ ft}$$

   Use 1 ft-9 in.

   c. Stem thickness at top of footing:

   $$\tfrac{1}{12}(18) = 1.5 \text{ ft}$$

   Use 1 ft-6 in.

   d. Toe width:

   $$\tfrac{1}{4} \text{ to } \tfrac{1}{3} \text{ footing width}$$

   $$\tfrac{1}{4}(11) \text{ to } \tfrac{1}{3}(11) = 2.75 \text{ to } 3.67 \text{ ft}$$

   Use 3 ft-0 in.

   e. Use a batter for the rear face of wall approximately $\tfrac{1}{2}$ in./ft.

   f. Top of stem thickness, based on G and a batter of $\tfrac{1}{2}$ in./ft:

   $$D = G - (h_w - h)\tfrac{1}{2}$$
   $$= 18 \text{ in.} - (18 \text{ ft} - 1.75 \text{ ft})\tfrac{1}{2} \text{ in./ft} = 9.88 \text{ in.}$$

   Use 10 in. Therefore, the calculated batter is

   $$\frac{\text{total batter}}{\text{stem height}} = \frac{18 \text{ in.} - 10 \text{ in.}}{18 \text{ ft} - 1.75 \text{ ft}} = 0.492 \text{ in./ft}$$

   The preliminary wall proportions are shown in Figure 8-8.

4. For the stability analysis, use unfactored weights and loads in accordance with the ACI Code, Section 13.3.1.1 or 13.4.1.1.

   a. *Factor of safety against overturning:* The tendency of the wall to overturn is a result of the horizontal loads acting on the wall. An assumption is made that in

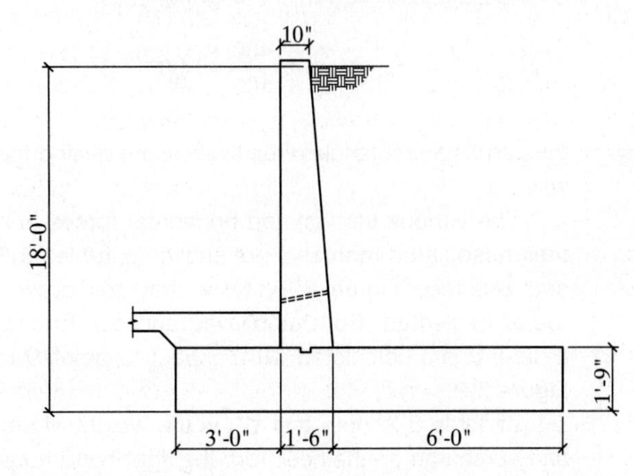

FIGURE 8-8 Preliminary wall proportions for Example 8-1.

TABLE 8-2 Stabilizing Moments (Vertical Forces)

Force	Magnitude (lb)	Lever arm (ft)	Moment (ft-lb)
W_1	$0.833(16.25)(150) =$ 2030	3.42	6940
W_2	$7.17(16.25 + 4.0)(100) =$ 14,520	7.42	107,700
W_3	$(\frac{1}{2})(16.25)(0.67)(50) =$ 272	4.05	1102
W_4	$11.0(1.75)(150) =$ 2890	5.5	15,900
	$\Sigma W =$ 19,710		$\Sigma M =$ 131,600

TABLE 8-3 Overturning Moments (Horizontal Forces)

Force	Magnitude (lb)	Lever arm (ft)	Moment (ft-lb)
H_1	$1/2(30)(18)^2 = 4860$	6.0	29,200
H_2	$4(30)(18) = 2160$	9.0	19,440
	$\Sigma H = 7020$		$\Sigma M =$ 48,600

overturning, the wall will rotate about the toe, and the horizontal loads are said to create an *overturning moment* about the toe. Any vertical loads will tend to create rotation about the toe in the opposite direction and are, therefore, said to provide a *stabilizing moment*.

The factor of safety (FS) against overturning is then expressed as

$$FS = \frac{\text{stabilizing moment}}{\text{overturning moment}}$$

The required minimum factor of safety against overturning is normally governed by the applicable building code. A minimum of 1.5 is generally considered good practice. The passive earth resistance of the soil in front of the wall is generally neglected in stability computations because of the possibility of its removal by erosion or excavation.

As discussed previously, the surcharge may be converted into an equivalent height of earth,

$$h_{su} = \frac{w_s}{w_e} = \frac{400}{100} = 4 \text{ ft}$$

thus adding a rectangle of earth pressure behind the wall.

The various vertical and horizontal forces and their associated moments are shown in Tables 8-2 and 8-3 (see Figure 8-9). Note that soil on the toe is neglected. Both stabilizing and overturning moments are calculated with respect to point *O* in Figure 8-9.

In Table 8-2, note that W_2 is the weight of soil and surcharge on the heel from the right-hand edge of the heel to the vertical dashed line in the stem. The weight difference between reinforced concrete and soil (50 lb/ft^3) in the triangular portion that is the back of the stem is represented by W_3.

The factor of safety against overturning is

$$FS = \frac{131{,}600}{48{,}600} = 2.71 > 2.0 \qquad \text{(O.K)}$$

b. *Factor of safety against sliding:* The tendency of the wall to slide is primarily a result of the horizontal forces, whereas the vertical forces cause the frictional resistance against sliding.

The total frictional force available (or resisting force) may be expressed as

$$F = f(\Sigma W)$$

where f = coefficient of friction between the concrete and soil and $\sum W$ = summation of vertical forces (see Figure 8-9). A typical value for the coefficient of friction is $f = 0.50$. Then the resisting force F can be calculated:

$$F = f(\Sigma W)$$

$$0.50(19{,}710) = 9860 \text{ lb}$$

The factor of safety against sliding may be expressed as

$$FS = \frac{\text{resisting force } F}{\text{actual horizontal force } \Sigma H}$$

$$= \frac{9860}{7020} = 1.40$$

The required minimum factor of safety for this problem is 1.5; hence, the resistance against sliding is *inadequate*.

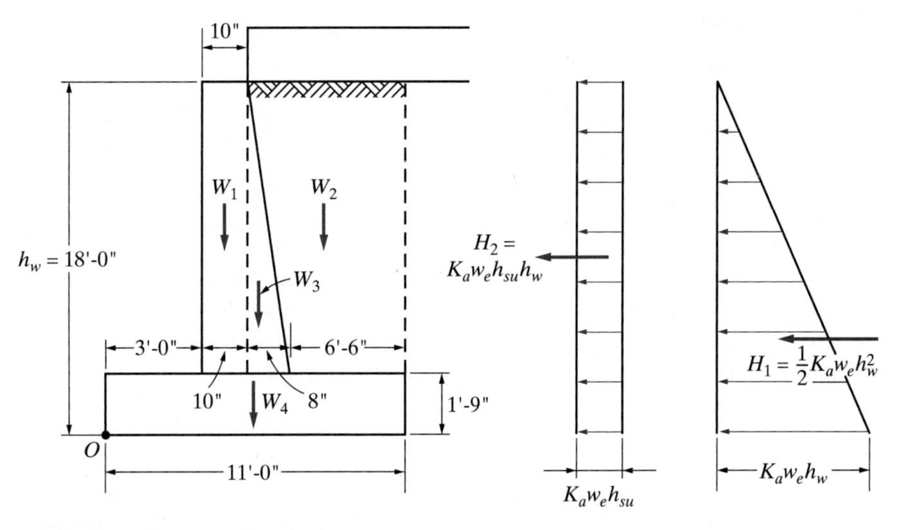

FIGURE 8-9 Stability analysis diagram.

One solution to this problem is to reproportion the wall until the requirements are met. Rather than change the wall, we will use a base shear key to mobilize the passive resistance of the soil and in effect increase the resisting force F and subsequently increase the factor of safety. The design of the base shear key will be one of the last steps in the design of the retaining wall.

c. *Soil pressures and location of resultant force:* The soil pressure under the footing of the wall is a function of the location of the resultant force, which in turn is a function of the vertical forces and horizontal forces. It is generally desirable and usually required that for walls on soil, the resultant of all the forces acting on the wall must lie within the middle third of the base. When this occurs, the resulting pressure distribution could be either triangular, rectangular, or trapezoidal in shape, with the soil in compression under the entire width of the footing. The resulting maximum foundation pressure must not exceed the safe bearing capacity of the soil.

The first step is to locate the point at which the resultant of the vertical forces $\sum W$ and the horizontal forces $\sum H$ intersects the bottom of the footing. This may be visualized with reference to Figure 8-10. Note that $\sum W$ and $\sum H$ are first combined to form the resultant force R. Because R may be moved anywhere along its line of action, it is moved to the bottom of the footing, where it is then resolved back into its components $\sum W$ and $\sum H$. The moment about point O due to the resultant force acting at the bottom of the footing must be the same as the moment effect about point O of the components $\sum W$ and $\sum H$. Therefore, using the components in the plane of the bottom of the footing, the location of the resultant that is a distance x from point O may be determined. Note that the moment due to $\sum H$ at the bottom of the footing is zero:

$$\Sigma W(x) = \Sigma Wm - \Sigma Hn$$

As may be observed, $\sum Wm$ is the stabilizing moment of the vertical forces with respect to point O and $\sum Hn$ is the overturning moment of the horizontal forces with respect to the same point. Hence, this expression may be rewritten as

$$\Sigma W(x) = \text{stabilizing } M - \text{overturning } M$$

$$x = \frac{\text{stabilizing } M - \text{overturning } M}{\Sigma W}$$

With reference to Figure 8-9,

$$x = \frac{131{,}600 - 48{,}600}{19{,}710} = 4.21 \text{ ft from toe}$$

Therefore, the eccentricity e with respect to the centerline of the footing is

$$e = 5.5 - 4.21 = 1.29 \text{ ft}$$

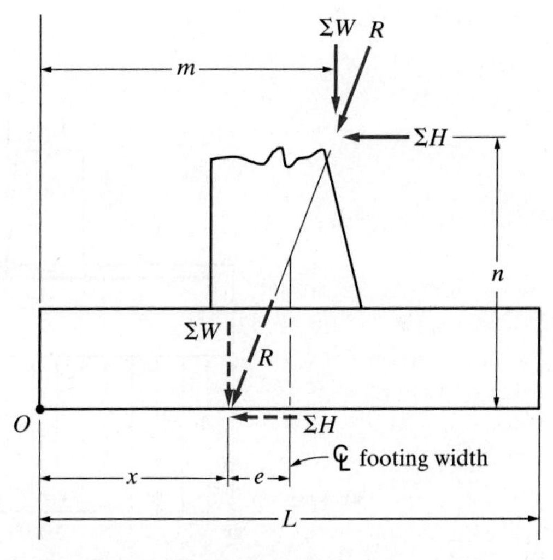

FIGURE 8-10 Resultant of forces acting on wall.

With a footing length of 11 ft, the middle third has a length of $\frac{11}{3} = 3.667$ ft. Eccentricity e is measured from the centerline of the footing and because

$$e = 1.29 \text{ ft} \leq \frac{3.667 \text{ ft}}{2} = 1.834 \text{ ft}$$

the resultant lies in the middle third of the footing, and the resulting pressure distribution is trapezoidal. If the resultant intersected the base at the edge of the middle third (i.e., $e = \frac{L}{6} = 1.833$ ft), the pressure distribution would have been triangular, and if $e = 0$, the pressure distribution would have been rectangular, indicating a uniform soil pressure distribution.

The pressures may now be calculated considering ΣW applied as an eccentric load on a rectangular section 11 ft long by 1 ft wide (this is a *typical* 1-ft-wide strip of the wall footing). The pressures are obtained by using the basic equations for bending and axial compression:

$$p = \frac{P}{A} \pm \frac{Mc}{I}$$

where

p = unit soil pressure intensity under the footing
P = total vertical load (ΣW)
A = footing cross-sectional area [$(L)(1.0)$]
M = moment due to eccentric load [$(\Sigma W)(e)$]
c = distance from centerline of footing to outside edge ($L/2$)
I = moment of inertia of footing with respect to its centerline [$(1.0)(L)^3/12$]

The expression given previously may be rewritten as

$$p = \frac{\Sigma W}{(L)(1.0)} \pm \frac{(\Sigma W)(e)(L/2)}{1.0(L)^3/12}$$

Simplifying and rearranging yields

$$p = \frac{\Sigma W}{L}\left(1 \pm \frac{6e}{L}\right)$$

Substituting, we obtain

$$p = \frac{19{,}710}{11}\left[1 \pm \frac{6(1.29)}{11}\right]$$
$$= 1792(1 \pm 0.704)$$
$$\text{maximum } p = 1792(1.704) = 3050 \text{ psf}$$
$$\text{minimum } p = 1792(0.296) = 530 \text{ psf}$$

The resulting pressures are less than 4000 psf and are, therefore, satisfactory. The resulting pressure distribution beneath the footing is shown in Figure 8-11.

5. Design the component parts.
   a. *Design of heel*: The load on the heel is primarily earth dead load and surcharge, if any, acting vertically downward. With the assumed straight-line pressure distribution under the footing, the downward load is reduced somewhat by the upward-acting pressure. For design purposes, the heel is assumed to be a cantilever beam 1 ft in width and with a span length equal to 6 ft-6 in. fixed at the rear face of the wall (point A in Figure 8-11).

   The weight of the footing is

$$1.75(150)(6.50) = 1706 \text{ lb}$$

   The earth and surcharge weight is

$$6.50(20.25)(100) = 13{,}160 \text{ lb}$$

   The slope of pressure distribution under the footing is

$$\frac{3050 - 530}{11} = 229 \text{ psf/linear foot}$$

   Thus far, all analysis has been based on service (unfactored) loads because allowable soil pressures are determined using a certain factor of safety against reaching pressures that will cause unacceptable settlements. Reinforced concrete, however, is designed on the basis of factored loads. The ACI Code,

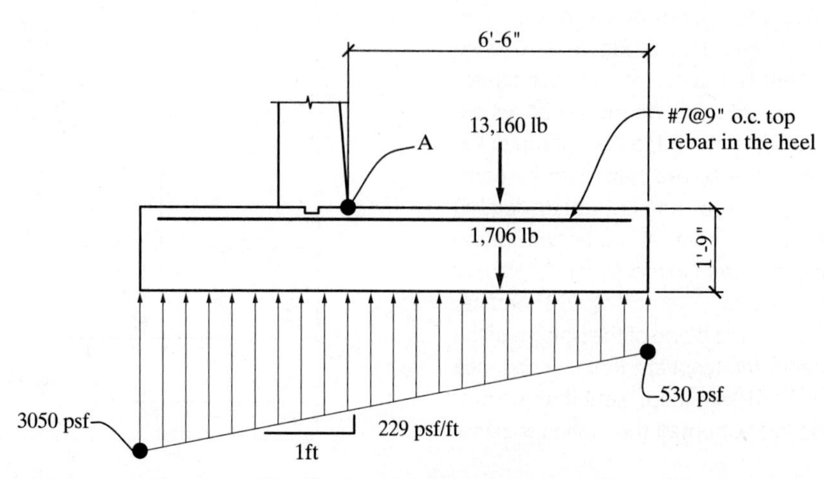

FIGURE 8-11 Pressure distribution under footing. And top rebar in the heel.

Section 5.3.8, specifies that where lateral earth pressure H must be included in design and adds to the primary load effect, the strength U shall be at least equal to $1.2D + 1.6L + 1.6H$. Where the lateral earth pressure H opposes the primary load effect, a load factor of 0.9 should be used on H when the lateral earth pressure is permanent and a load factor of zero on H for all other conditions.

A conservative approach to the design of the heel is to use factored loads, as required, and to ignore the relieving effect of the upward pressure under the heel. This is the unlikely condition that would exist if there occurred a lateral force overload (and no associated increased vertical load) causing uplift of the heel.

The *maximum moment* may be obtained by taking a summation of moments about point A (utilizing service loads):

$$M = (13{,}160 + 1706)\frac{6.50}{2} = 48{,}300 \text{ ft-lb}$$

The bending of the heel is such that tension occurs in the top of the footing.

The *maximum shear* may be obtained by taking a summation of vertical forces on the heel side of point A (using service loads):

$$V = 13{,}160 + 1706 = 14{,}870 \text{ lb}$$

The maximum moment and shear must be modified for strength design. As the loads on the heel are predominantly dead load, a load factor of 1.2 is used. This in effect considers the surcharge to be a dead load. This is acceptable, however, because of the conservative nature of the design. Thus

$$M_u = 48{,}300(1.2) = 58{,}000 \text{ ft-lb}$$

$$V_u = 14{,}870(1.2) = 17{,}840 \text{ lb}$$

The *footing size and reinforcement* (for heel) will be determined next. Because it commonly occurs that shear strength will be the controlling factor with respect to footing thickness, the shear will be checked first. The heel effective depth available, assuming 2-in. cover (ACI Code, Section 7.7.1) and No. 8 bars, is

$$d = 21 - 2 - 0.5 = 18.5 \text{ in.}$$

The shear strength ϕV_n of the heel, if no shear reinforcing is provided, is the shear strength of the concrete alone:

$$\phi V_n = \phi V_c = \phi(2\sqrt{f'_c})bd$$
$$= 0.75(2)\sqrt{3000}(12)(18.5)$$
$$\phi V_n = 18{,}240 \text{ lb}$$

Therefore, $\phi V_n > V_u$ (O.K.).

The *tensile reinforcement* requirement may now be determined in the normal way. Assuming $\phi = 0.90$:

$$\text{required } \bar{k} = \frac{M_u}{\phi bd^2} = \frac{58.0(12)}{0.9(12)(18.5)^2}$$
$$= 0.1883 \text{ ksi}$$

Therefore, from Table A-8, the required $\rho = 0.0033$, $\epsilon_t > 0.005$, and $\phi = 0.90$ (O.K.). The required steel area is then calculated from

$$\text{required } A_s = \rho bd = 0.0033(12)(18.5) = 0.73 \text{ in.}^2$$

The minimum area of steel required by the ACI Code, Section 9.6.1, may be obtained using Table A-5:

$$A_{s,\min} = 0.0033(12)(18.5) = 0.73 \text{ in.}^2$$

As discussed in Section 2-8, $A_{s,\min}$ must be provided wherever reinforcement is needed, except where such reinforcement is at least one-third greater than that required by analysis (see the ACI Code, Section 9.6.1.3).

The ACI Code, Sections 7.6.1.1 and 7.7.2.3, also permits the use of a minimum reinforcement equal to that required for shrinkage and temperature steel in structural slabs of uniform thickness as furnished in the ACI Code, Section 24.4.1. This will always be somewhat less than that required by the ACI Code, Section 9.6.1, but not necessarily less than that specified by the ACI Code, Section 9.6.1.3. For footings, the shrinkage and temperature steel requirement for slabs of uniform thickness will be used as an absolute minimum in this text.

For the heel reinforcement for this retaining wall, because the required steel area is equal to $A_{s,\min}$, we will select No. 7 bars at 9 in. o.c. ($A_s = 0.80$ in.^{2}).

b. *Design of toe*: The load on the toe is primarily a result of the soil pressure distribution on the bottom of the footing acting in an upward direction. For design purposes, the toe is assumed to be a cantilever beam 1 ft in width and with a span length equal to 3 ft-0 in. fixed at the front face of the wall (point B in Figure 8-12). The soil on the top of the toe is conservatively neglected. Forces and pressures are calculated with reference to Figure 8-12.

As the reinforcing steel will be placed in the bottom of the footing, the effective depth available, assuming a 3-in. cover and No. 8 bars, is

$$d = 21 - 3 - 0.5 = 17.5 \text{ in.}$$

The weight of footing for the toe design is

$$1.75(3.0)(150) = 788 \text{ lb}$$

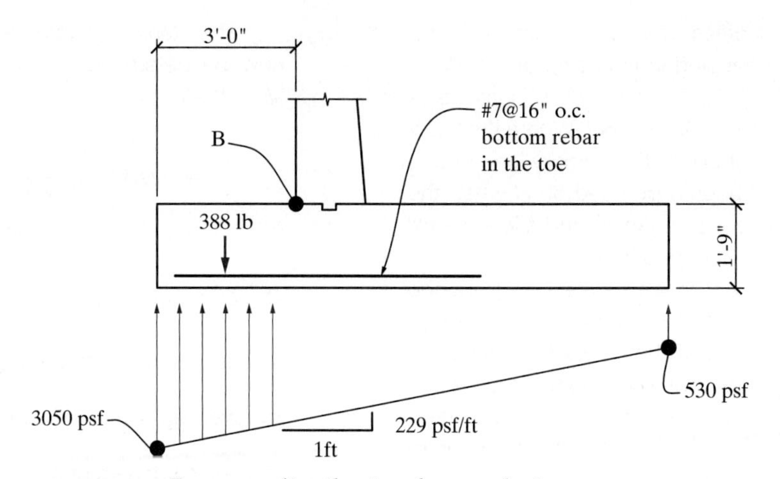

FIGURE 8-12 Pressure distribution for toe design.

The soil pressure directly under point *B*, recalling that the slope of the pressure diagram is 229 psf/ft, is

$$3050 - 3.0(229) = 2360 \text{ psf}$$

The *design moment* M_u may be obtained by a summation of moments about point *B* in Figure 8-12. Both the moment and the shear must be modified for strength design. According to the ACI Code, Section 5.3.8, a load factor of 1.6 should be used for horizontal earth pressure and for live load, whereas 1.2 should be used for dead load. Because the soil pressure under the toe is largely the result of horizontal earth pressure, however, a conservative procedure of using 1.6 is recommended. In addition, as the weight of the footing reduces the effect of the horizontal earth pressure, a factor of 0.9 is used for the footing dead load, as recommended by the ACI Code (Section 5.3.8).

The design moment M_u is then calculated as

$$M_u = 1.6\left(\tfrac{1}{2}\right)(3050)(3.0)^2\left(\tfrac{2}{3}\right) + 1.6\left(\tfrac{1}{2}\right)(2360)(3.0)^2\left(\tfrac{1}{3}\right) - 0.9(788)\left(\tfrac{3.0}{2}\right)$$

$$= 19{,}240 \text{ ft-lb}$$

The bending of the toe is such that tension occurs in the bottom of the footing.

The *design shear* V_u is obtained by a summation of vertical forces on the toe side of point *B* and applying load factors, as previously discussed:

$$V_u = 1.6\left(\tfrac{1}{2}\right)(3050)(3.0) + 1.6\left(\tfrac{1}{2}\right)(2360)(3.0) - 0.9(788)$$

$$= 12{,}270 \text{ lb}$$

Footing size and reinforcement, based on the requirements of the toe, are treated as they were for the heel. The shear strength ϕV_n of the toe is calculated from

$$\phi V_n = \phi V_c = \phi(2\sqrt{f'_c})bd$$

$$= 0.75(2)\sqrt{3000}(12)(17.5)$$

$$\phi V_n = 17{,}250 \text{ lb}$$

$$\phi V_n > V_u$$

Therefore, the thickness of the footing is satisfactory.

Based on the determined M_u, and assuming $\phi = 0.90$:

$$\text{required } \bar{k} = \frac{M_u}{\phi b d^2} = \frac{19.24(12)}{0.9(12)(17.5)^2}$$

$$= 0.0698 \text{ ksi}$$

Therefore, from Table A-8,

$$\text{required } \rho = 0.0012$$

Note that $\epsilon_t > 0.005$. Therefore, $\phi = 0.90$. The required steel area is

$$\text{required } A_s = \rho b d = 0.0012(12)(17.5) = 0.25 \text{ in.}^2$$

The minimum area of steel required may be obtained using Table A-5:

$$A_{s,\min} = 0.0033(12)(17.5) = 0.69 \text{ in.}^2$$

Because required $A_s < A_{s,\min}$, other minimum steel criteria should be checked to establish a controlling minimum value:

1. Provide one-third additional reinforcing as outlined in the ACI Code, Section 9.6.1.3:

$$A_{s,\min} = 1.33(0.25) = 0.33 \text{ in.}^2$$

2. Check the required steel area based on the absolute minimum of shrinkage and temperature steel required for structural slabs of uniform thickness (ACI Code, Section 24.4.3.1):

$$\text{required } A_s = 0.0018 \text{ bh}$$

$$= 0.0018(12)(21) = 0.45 \text{ in.}^2$$

As we consider the shrinkage and temperature steel requirement as an absolute minimum, use No. 7 bars at 16 in. o.c. ($A_s = 0.45$ in.^{2}).

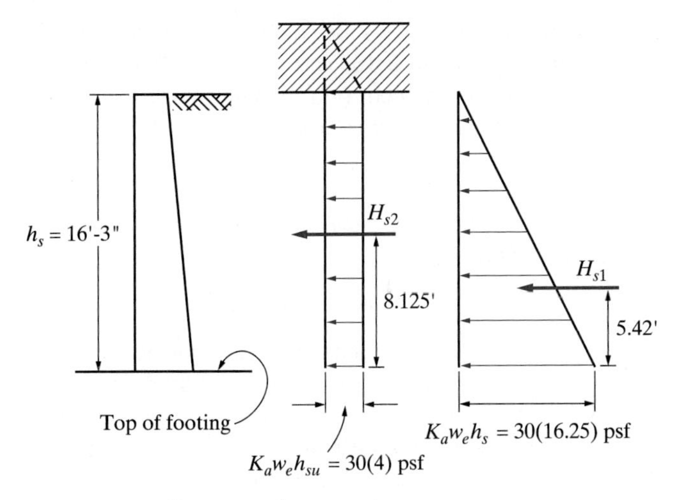

FIGURE 8-13 Forces acting on stem.

c. *Design of stem*: The load on the stem is primarily lateral earth pressure acting (in this problem) horizontally. For design purposes, the stem is assumed to be a vertical cantilever beam 1 ft in width and with a span length equal to 16 ft-3 in. fixed at the top of the footing.

The loads acting on the stem from the top of the wall to the top of the footing are depicted in Figure 8-13. The magnitudes of the horizontal forces are

$$H_{s1} = \tfrac{1}{2}(30)(16.25)^2 = 3960 \text{ lb}$$

$$H_{s2} = 4(30)(16.25) = 1950 \text{ lb}$$

The design moment M_u in the stem may be obtained by a summation of moments about the top of the footing. Both shears and moments must be modified for strength design. Because the forces are due to lateral earth pressure, a factor of 1.6 should be used. M_u may then be calculated as

$$M_u = 1.6(3960)(5.42) + 1.6(1950)(8.125)$$
$$= 59{,}700 \text{ ft-lb}$$

The design shear V_u is obtained by a summation of the horizontal forces acting on the stem above the top of the footing:

$$V_u = 1.6(3960) + 1.6(1950)$$
$$= 9460 \text{ lb}$$

The *stem size and reinforcement* requirements will be determined next. Because the stem is assumed to be a cantilever slab, V_u as a limit may equal ϕV_n, where $V_n = V_c$. It is not practical to reinforce the stem for shear; therefore, if shear strength is inadequate, stem thickness must be increased. The effective depth of the stem at the top of the footing, assuming 2-in. cover and No. 8 bars, is

$$d = 18 - 2 - 0.5 = 15.5 \text{ in.}$$

The shear strength ϕV_n of the stem (at the top of the footing), if no shear reinforcement is provided, is the shear strength of the concrete alone:

$$\phi V_n = \phi V_c = \phi 2\sqrt{f'_c}bd$$
$$= 0.75(2)\sqrt{3000}(12)(15.5)$$
$$\phi V_n = 15{,}280 \text{ lb}$$
$$\phi V_n > V_u$$

Thus, the stem thickness need not be increased. Based on M_u at the bottom of the stem and assuming $\phi = 0.90$,

$$\text{required } \bar{k} = \frac{M_u}{\phi bd^2} = \frac{59.7(12)}{0.9(12)(15.5)^2}$$
$$= 0.2761 \text{ ksi}$$

Therefore, from Table A-8, the required $\rho = 0.0049$, and $\epsilon_t > 0.005$, so $\phi = 0.90$. The required steel area is

$$\text{required } A_s = \rho bd = 0.0049(12)(15.5) = 0.91 \text{ in.}^2$$

The minimum area of steel required may be obtained using Table A-5:

$$A_{s,\min} = 0.0033(12)(15.5) = 0.61 \text{ in.}^2$$

Because required $A_s > A_{s,\min}$, the calculated required A_s controls. Use No. 7 bars at $7\frac{1}{2}$ in. o.c. ($A_s = 0.96$ in.^{2}).

The *stem reinforcement pattern* will be investigated more closely. As the moment and shear vary along the height of the wall, it is only logical that the steel requirements also vary. A pattern is usually created whereby some of the stem reinforcement is cut off where it is no longer required. The actual pattern used is the choice of the designer, based on various practical constraints.

A typical approach to the creation of a stem reinforcement pattern is first to draw the complete M_u diagram for the stem and then to determine bar cutoff points using the procedure discussed in Chapter 5. Stem moments due to the external loads will be calculated at 5 ft from the top of the wall and at 10 ft from the top of the wall. These distances are arbitrary and should be chosen so that a sufficient number of points will be available to draw the M_u diagram.

With reference to Figure 8-14, the moment at 5 ft from the top of the wall is calculated by first determining the horizontal forces:

$$H_{s1} = \tfrac{1}{2}(30)(5.0)^2 = 375 \text{ lb}$$
$$H_{s2} = 4(30)(5.0) = 600 \text{ lb}$$

Summing the moments about a plane 5 ft below the top of the wall and introducing the appropriate load factor,

$$M_u = 1.6(375)\left(\frac{5.0}{3}\right) + 1.6(600)\left(\frac{5.0}{2}\right)$$
$$= 3400 \text{ ft-lb}$$

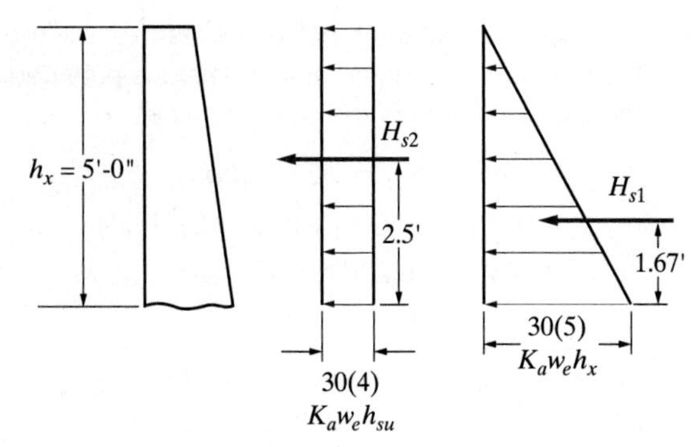

FIGURE 8-14 Stem analysis for top 5 ft. of retaining wall.

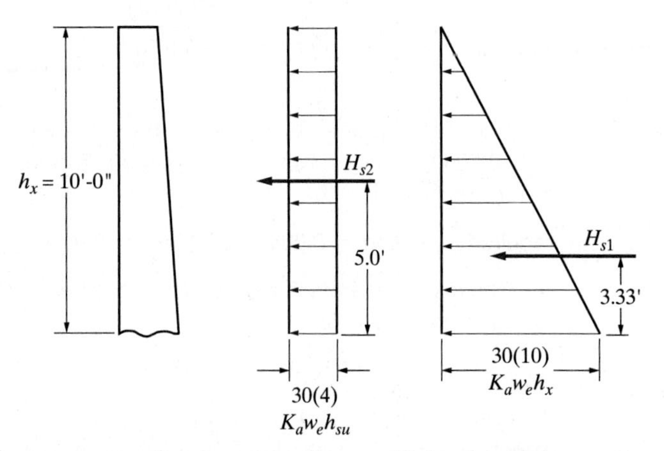

FIGURE 8-15 Stem analysis for top 10 ft. of retaining wall.

With reference to Figure 8-15, the moment at 10 ft below the top of the wall may be calculated:

$$H_{s1} = \tfrac{1}{2}(30)(10.0)^2 = 1500 \text{ lb}$$
$$H_{s2} = 4(30)(10.0) = 1200 \text{ lb}$$

and

$$M_u = 1.6(1500)\left(\frac{10.0}{3}\right) + 1.6(1200)\left(\frac{10.0}{2}\right)$$
$$= 17{,}600 \text{ ft-lb}$$

The three M_u values may now be plotted as the solid curve in Figure 8-16. This diagram will reflect both the design moment and moment strength with respect to the distance from the top of the wall and must be plotted to scale.

The moment strength ϕM_n of the stem at various locations may be computed from the expression

$$\phi M_n = \phi b d^2 \overline{k}$$

Because we are primarily attempting to establish a cutoff location for some vertical stem reinforcement, the moment strength will be based on a pattern of alternate vertical bars being cut off. With alternate bars cut off, the remaining reinforcement pattern is No. 7 bars at 15 in. o.c., which furnishes an $A_s = 0.48$ in.^{2} Caution must be exercised that the spacing of the remaining bars does not exceed three times the wall thickness or 18 in., whichever is less.

The moment strength will be computed at the top of the wall and at the top of the footing, neglecting minimum steel requirements at this time. As the rear face of the wall is battered, the effective depth varies and may be computed as follows:

$$\text{top of wall: } d = 10 - 2 - 0.5 = 7.5 \text{ in.}$$
$$\text{top of footing: } d = 18 - 2 - 0.5 = 15.5 \text{ in.}$$

At the top of the wall, with half of the stem bars cut off:

$$\rho = \frac{A_s}{bd} = \frac{0.48}{12(7.5)} = 0.0053$$

Therefore, from Table A-8, $\overline{k} = 0.2982$ ksi. Then

$$\phi M_n = \frac{0.9(12)(7.5)^2}{12}(0.2982)$$
$$= 15.10 \text{ ft.-kip}$$

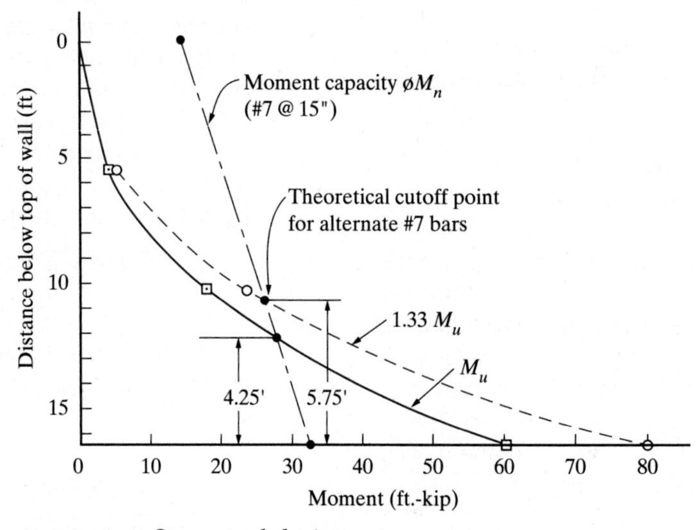

FIGURE 8-16 Stem steel design.

At the top of the footing,

$$\rho = \frac{A_s}{bd} = \frac{0.48}{12(15.5)} = 0.0026$$

Therefore, from Table A-8, $\overline{k} = 0.1512$ ksi. Then

$$\begin{aligned}\phi M_n &= \frac{0.9(12)(15.5)^2}{12}(0.1512)\\ &= 32.7 \text{ ft.-kip}\end{aligned}$$

These two values (ϕM_n) may now be plotted in Figure 8-16. A straight line will be used to approximate the variation of ϕM_n for No. 7 bars at 15 in. o.c. from the top to the bottom of the wall.

The intersection of the ϕM_n line and the M_u solid curve establishes a theoretical cutoff point for alternate vertical bars based on strength requirements. This location may be determined graphically by scaling on Figure 8-16. The scaled distance is 12.0 ft below the top of the wall or 4 ft-3 in. above the top of the footing. The No. 7 bar must be extended past this theoretical cutoff point a distance equal to the effective depth of the member or 12 bar diameters, whichever is larger.

Because the effective depth of the stem steel at the top of the footing is 15.5 in. and the wall batter is approximately 0.492 in./ft, the effective depth of the stem steel at 4.25 ft above the top of the footing may be calculated as follows:

$$\begin{aligned}d &= 15.5 - 4.25(0.492) = 13.4 \text{ in.}\\ 12d_b &= 12(0.875) = 10.5 \text{ in.}\end{aligned}$$

Therefore, the required length of bar above the top of the footing would be

$$4.25 + \frac{13.4}{12} = 5.37 \text{ ft}$$

Use 5 ft 5 in. (5.42 ft).

Check the required minimum area of steel, $A_{s,min}$, at the theoretical cutoff point to ensure that discontinuing half of the steel will not violate this code requirement. Once alternate bars have been terminated, the remaining steel will be No. 7 bars at 15 in. ($A_s = 0.48$ in.^{2}). At the theoretical cutoff point 4.25 ft above the top of the footing, the effective depth $d = 13.4$ in. From Table A-5,

$$\begin{aligned}A_{s,min} &= 0.0033bd\\ &= 0.0033(12)(13.4) = 0.53 \text{ in.}^2\end{aligned}$$

This is unsatisfactory, because the steel area provided is less than $A_{s,min}$. However, recall from previous discussions that the minimum reinforcing requirement need not be applied if the reinforcing provided is one-third greater than that required by analysis. Therefore, a simple solution is to move the cutoff point upward to where the reinforcing provided is one-third greater than required. This can be accomplished by drawing a $1.33 \times M_u$ curve and reestablishing the theoretical cutoff point for alternate No. 7 bars. At this cutoff point, one-third more steel is provided than is required; therefore, the minimum steel requirement does not apply.

Multiplying the previously determined design moments M_u by 1.33, the following moments are obtained:

At 5 ft below the top of the wall:

$$M_u = 1.33(3400) = 4520 \text{ ft-lb}$$

At 10 ft below the top of the wall:

$$M_u = 1.33(17{,}600) = 23{,}400 \text{ ft-lb}$$

At the top of the footing:

$$M_u = 1.33(59{,}700)\text{ft-lb} = 79{,}400 \text{ ft-lb}$$

Plotting this moment curve (see the dashed line on Figure 8-16) and scaling the distance above the top of the footing to theoretical cutoff point, we arrive at a value of 5.75 ft, or 5 ft-9 in. The bars must extend past this theoretical cutoff point a distance equal to the greater of the effective depth of the member or 12 bar diameters.

$$\begin{aligned}d &= 15.5 - 5.75(0.492) = 12.67 \text{ in.}\\ 12d_b &= 12(0.875) = 10.5 \text{ in.}\end{aligned}$$

Adding 12.67 in. to the previously determined theoretical cutoff point above the top of the footing, we have

$$5.75 + \frac{12.67}{12} = 6.81 \text{ ft}$$

Use 6 ft-10 in. (6.83 ft).

Therefore, terminate alternate No. 7 bars at 6 ft-10 in. above the top of the footing.

The ACI Code, Sections 7.7.3.5 and 9.7.3.5, stipulate that flexural reinforcement must not be terminated in a tension zone unless one of several conditions is satisfied. One of these conditions is that the shear at the cutoff point does not exceed two-thirds of the shear permitted. Therefore, check the shear at the actual cutoff point, 6.83 ft above the top of the footing (9.42 ft below the top of the wall; see Figure 8-17).

For the shear strength calculation, the wall thickness at a height of 6.83 ft above the top of the footing is

$$18.0 - 6.83(0.492) = 14.64 \text{ in.}$$

from which d may be calculated as

$$\begin{aligned}d &= 14.64 - 2 - 0.5 = 12.14 \text{ in.}\\ H_{s1} &= \tfrac{1}{2}(30)(9.42)^2 = 1331 \text{ lb}\end{aligned}$$

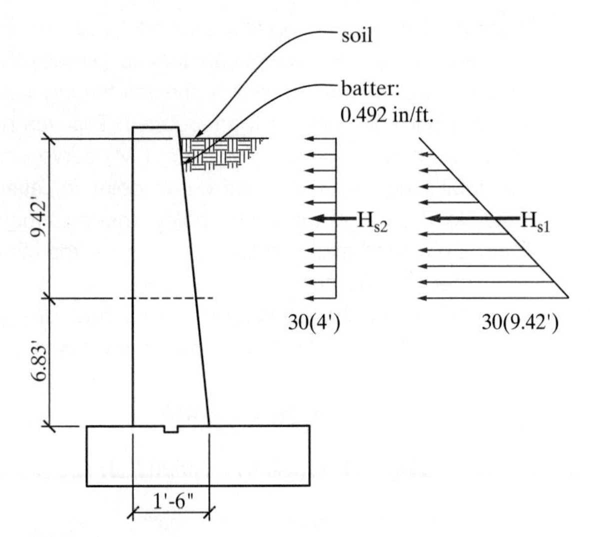

FIGURE 8-17 Shear at bar cutoff point.

$$H_{s2} = 4(30)(9.42) = 1130 \text{ lb}$$
$$\text{total} = 2460 \text{ lb}$$
$$V_u = 2460(1.6) = 3940 \text{ lb}$$
$$\tfrac{2}{3}\phi V_n = \tfrac{2}{3}\phi V_c = \tfrac{2}{3}\phi(2\sqrt{f'_c})bd$$
$$= \tfrac{2}{3}(0.75)(2)\sqrt{3000}(12)(12.14)$$
$$\tfrac{2}{3}\phi V_n = 7980 \text{ lb}$$
$$\tfrac{2}{3}\phi V_n > V_u$$

In summary, alternate vertical stem reinforcing bars may be stopped at 6 ft-10 in. above the top of the footing (see the typical wall section shown in Figure 8-21 later in this chapter).

d. *Additional design details*: Check the anchorage requirements for the stem steel (No. 7 bars at $7\frac{1}{2}$ in. o.c.). Assume all steel is uncoated. Note that anchorage requirements into the stem and into the footing may differ for these bars. Anchorage length required in the stem will be equal to or greater than that required in the footing and will also impact the splice length. Therefore, stem anchorage will be checked.

1. From Table 5-1, $K_D = 82.2$.
2. Establish values for the factors ψ_t, ψ_e, ψ_s, λ.
   a. $\psi_t = 1.0$ (the bars are not top bars).
   b. The bars are uncoated; $\psi_e = 1.0$.
   c. The bars are No. 7.; $\psi_s = 1.0$.
   d. Normal-weight concrete is used; $\lambda = 1.0$.
3. The product $\psi_t \times \psi_e = 1.0 < 1.7$. (O.K.)
4. Determine c_b. Based on cover (center of bar to nearest concrete surface), consider the clear cover and one-half the diameter of the No. 7 bar:

$$c_b = 2 + \frac{0.875}{2} = 2.44 \text{ in.}$$

Based on bar spacing (one-half the center-to-center distance):

$$c_b = 0.5(7.5) = 3.75 \text{ in.}$$

Therefore, use $c_b = 2.44$ in.

5. K_{tr} is taken as zero. There is no transverse steel crossing the potential plane of splitting.
6. Check $(c_b + K_{tr})/d_b \le 2.5$:

$$\frac{c_b + K_{tr}}{d_b} = \frac{2.44 + 0}{0.875} = 2.79 > 2.5$$

Use 2.5.

7. Calculate the excess reinforcement factor:

$$K_{ER} = \frac{A_s \text{ required}}{A_s \text{ provided}} = \frac{0.91}{0.96} = 0.95$$

8. Calculate ℓ_d.
   a. Omitting K_{ER} (this calculated value of ℓ_d will be used shortly for splice length determination):

$$\ell_d = \frac{K_D}{\lambda}\left[\frac{\psi_t\psi_e\psi_s}{\left(\frac{c_b + K_{tr}}{d_b}\right)}\right]d_b$$
$$= \frac{82.2}{1.0}\left(\frac{1.0(1.0)(1.0)}{2.5}\right)(0.875)$$
$$= 28.8 \text{ in.} > 12 \text{ in.} \qquad \text{(O.K.)}$$

   b. Including K_{ER} from step 7, the required anchorage length is

$$\ell_d = 28.8(0.95) = 27.4 \text{ in.} > 12 \text{ in.} \qquad \text{(O.K.)}$$

Use $\ell_d = 28$ in. (minimum).

With the footing thickness equal to 21 in. and a minimum of 3 in. of cover required for the steel at the bottom of the footing, the anchorage length available is $21 - 3 = 18$ in. This anchorage length is not adequate. Rather than hook the bars, however, the bars will be extended into a footing base shear key that will be used to increase the sliding resistance of the wall (see the check on factor of safety against sliding).

The *length of splice* required for main stem reinforcing steel (see the typical wall section shown in Figure 8-21 later in this chapter) may be calculated recognizing that the class B splice is applicable for this condition (see Chapter 5, Section 5-7). Therefore, the required length of the splice is calculated from

$$1.3(28.8) = 37.4 \text{ in.}$$

Use 38 in. Note that the preceding calculation omits the effect of excess reinforcement.

Stem face steel in the form of horizontal and vertical reinforcement will be provided as per the ACI Code, Section 11.6. The minimum horizontal reinforcement is specified in Section 11.6.1. Although the minimum vertical reinforcement requirement of Section 11.6.1 does not strictly apply to reinforced concrete cantilever retaining walls, it is considered good practice to provide some vertical bars in the exposed face of the wall. The minimum recommended steel for deformed bars not larger than No. 5 with specified yield strength of not less than 60,000 psi is as follows:

horizontal bars (per foot of height of wall):

$$A_s = 0.0020\,bt$$

vertical bars (per foot of wall horizontally):

$$A_s = 0.0012\,bt$$

where t = thickness of the wall.

The code also stipulates that walls more than 10-in. thick, except basement walls, must have reinforcement for each direction in each face of the wall. The exposed face must have a minimum of one-half and a maximum of two-thirds the total steel required for each direction. The maximum spacing for the steel must not exceed three times the wall thickness nor 18 in. (ACI Code, Sections 11.7.2.1 and 11.7.3.1) for both vertical bars and horizontal bars. For the front face of the wall (exposed face), use two-thirds of the total steel required and an average stem thickness = 14 in.

For horizontal steel,

$$0.0020(12)(14) = 0.34\text{ in.}^2\text{/ft of height}$$
$$= 0.34(0.67) = 0.23\text{ in.}^2$$

Use No. 4 bars at 10 in. ($A_s = 0.24$ in.^{2}).
For vertical steel,

$$0.0012(12)(14) = 0.20\text{in.}^2\text{/ft of horizontal length}$$
$$= (0.20)(0.67) = 0.14\text{ in.}^2$$

Use No. 4 bars at 17 in. ($A_s = 0.14$ in.^{2}) in the front face. No additional vertical steel is needed in the rear face.

Rear face of wall, not exposed, for horizontal steel,

$$0.34(0.33) = 0.11\text{ in.}^2$$

Use No. 4 bars at 18 in. ($A_s = 0.13$ in.^{2}).

Longitudinal reinforcement in the footing should provide a steel area equal to that required for shrinkage and temperature in slabs. This is a conservative and acceptable approach because temperature and shrinkage exposure are ordinarily less severe for footings than for slabs. These bars serve chiefly as bar supports and spacers to hold the main steel in place during construction. Thus

$$\text{required } A_s = 0.0018(11\text{ ft})(1.75\text{ ft})(144\text{ in.}^2\text{/ft})$$
$$= 4.99\text{ in.}^2$$

Use 12 No. 6 bars ($A_s = 5.28$ in.^{2}), as shown in Figure 8-20 later in this chapter.

Transverse reinforcement in the footing need not run the full width of the footing. Proper anchorage length must be provided from the point of maximum tensile stress, however.

Check anchorage for the top transverse (heel) steel in the footing (No. 7 bars at 9 in. o.c.). This calculation for ℓ_d follows the eight-step procedure presented in Chapter 5, Section 5-2, and is summarized as follows:

1. $K_D = 82.2$
2. $\psi_t = 1.3$, $\psi_e = 1.0$, $\psi_s = 1.0$, $\lambda = 1.0$
3. $\psi_t \times \psi_e = 1.3$ (O.K.)
4. $c_b = 2.44$ in.
5. $K_{tr} = 0$
6. $(c_b + K_{tr})/d_b = 2.79$ in. Use 2.5 in.
7. $K_{ER} = 0.91$
8. $\ell_d = 34$ in.

Check the anchorage of the bottom transverse (toe) steel in the footing (No. 7 bars at 16 in. o.c.). As before, this ℓ_d calculation follows the procedure presented in Chapter 5 and is summarized as follows:

1. $K_D = 82.2$
2. $\psi_t = 1.0$, $\psi_e = 1.0$, $\psi_s = 1.0$, $\lambda = 1.0$
3. $\psi_t \times \psi_e = 1.0$ (O.K.)
4. $c_b = 3.44$ in.
5. $K_{tr} = 0$
6. $(c_b + K_{tr})/d_b = 3.93$ Use 2.5
7. $K_{ER} = 1.0$
8. $\ell_d = 28.8$ in. Use 29 in.

These details are included in the typical wall section, Figure 8-21.

The stem and the footing are elements cast at different times and a *shear key* will be used between the two. This is common practice. We will use a depressed key formed by a 2×6 plank (dressed dimensions $1\frac{1}{2} \times 5\frac{1}{2}$), as shown in Figure 8-21. The need for a shear key is questionable, because considerable slip is required to develop the key for purposes of lateral force transfer. It may be considered as an added mechanical factor of safety, however.

The shear-friction design method of the ACI Code Section 22.9 should be used to design for the transfer of the horizontal force between the stem and the footing. This approach eliminates the need for the traditional shear key. The shear-friction approach assumes that all of the horizontal force will be transferred through friction that develops on the contact surface between the two elements. The magnitude of the force that can be so transmitted will depend on the characteristics of the contact surfaces and on the existence of adequate shear-friction reinforcing A_{vf} crossing those surfaces. The angle at which the shear-friction reinforcing crosses the contact surface also plays a role. For our example, we assume the dowels will be placed perpendicular to the top of the footing.

The code allows for two possible contact-surface conditions that could exist in a situation such as between stem and footing in our retaining wall, either not intentionally roughened, though clean and free of laitance, or intentionally roughened. The latter assumption requires the interface to be roughened to a full amplitude of approximately $\frac{1}{4}$ in. This may be accomplished by raking of the fresh concrete or by some other means.

Assuming normal-weight concrete and shear-friction reinforcement placed perpendicular to the interface, the nominal shear strength (or friction force that resists sliding) may be computed from

$$V_n = A_{vf} f_y \mu \qquad \text{[ACI Eq. (22.9.4.2)]}$$

where

A_{vf} = area of shear-friction reinforcement

μ = coefficient of friction in accordance with the ACI Table 22.9.4.2; it may be taken as 1.0 for normal-weight concrete placed against hardened concrete roughened as described previously

Note that A_{vf} is steel that is provided for the shear-friction development and that it is in addition to any other steel already provided. This additional steel will be provided in the form of *dowels*. (A *dowel* is defined as a short bar that connects two separately cast sections of concrete.) The dowels will be placed approximately perpendicular to the shear plane.

As a limit,

$$V_u = \phi V_n$$

where V_u is shear force applied at the cracked plane. Substituting for V_n,

$$V_u = \phi A_{vf} f_y \mu$$

Solving for A_{vf},

$$\text{required } A_{vf} = \frac{V_u}{\phi f_y \mu} = \frac{9460}{(0.75)(60{,}000)(1.0)} = 0.21 \text{ in.}^2 \text{ per ft}$$

It is desirable to distribute the shear-friction reinforcement across the width of the contact surface. Therefore, provide $0.21/2 = 0.11$ in.^{2}/ft in each face. Use No. 4 bars at 18 in. in each face. This will provide 0.13 in.^{2}/ft in each face for a total of 0.26 in.^{2}/ft. Then

$$\text{shear strength } V_n = A_{vf} f_y \mu = (0.26)(60{,}000)(1.0) = 15{,}600 \text{ lb}$$

The ACI Code, Sections 22.9.4.1 and 22.9.4.4, stipulate that the maximum V_n for concrete placed monolithically or against hardened concrete intentionally roughened as previously described, shall not exceed the smallest of

$$0.2f'_c A_c, (480 + 0.08f'_c)A_c, \quad \text{and} \quad 1600A_c$$

where A_c is the contact area resisting the shear transfer, and for all other cases, V_n shall not exceed the smaller of

$$0.2f'_c A_c \quad \text{and} \quad 800\, A_c$$

Thus, checking the upper limit for V_n:

$$0.2f'_c A_c = 0.2(3000)(12)(18) = 129{,}600 \text{ lb}$$

$$(480 + 0.08f'_c)A_c = [480 + 0.08(3000)](12)(18) = 155{,}500 \text{ lb}$$

$$1600A_c = 1600(12)(18) = 346{,}000 \text{ lb}$$

$$\text{Calculated } V_n = 15{,}600 \text{ lb} \ll 129{,}600 \text{ lb} \quad \text{(O.K.)}$$

The development length for the No. 4 dowels (18 in. o.c.) must be furnished into both the stem and the footing. These are considered to be tension bars. This calculation for ℓ_d follows the eight-step procedure presented in Chapter 5, Section 5-2, and is summarized as follows:

1. $K_D = 82.2$
2. $\psi_t = 1.0, \psi_e = 1.0, \psi_s = 0.8, \lambda = 1.0$
3. $\psi_t \times \psi_e = 1.0$ (O.K.)
4. $c_b = 2.25$ in.
5. $K_{tr} = 0$
6. $(c_b + K_{tr})/d_b = 4.50$ Use 2.5
7. $K_{ER} = 0.808$
8. $\ell_d = 10.6$ in. < 12 in. Use 12 in.

The No. 4 dowels are depicted in Figure 8-18.

6. Design the footing base shear key. The *footing base shear key* (sometimes called a *bearing lug*) is primarily used to prevent a sliding failure. The magnitude of the additional resistance to sliding offered by the key is questionable and is a function of the subsoil material. The key, cast in a narrow trench excavated below the bottom of footing elevation, becomes monolithic with the footing. The excavation for a key will generally disturb the subsoil during construction and conceivably

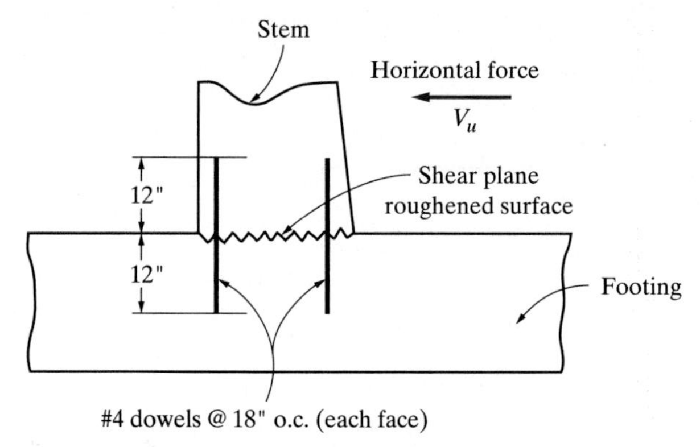

FIGURE 8-18 Shear-friction reinforcement.

will do more harm than good. Hence the use of the key for the purpose intended is controversial.

An acceptable design approach is to use the passive earth resistance H_p in front of the key (from the bottom of the footing to the bottom of the key) as the additional resistance to sliding (see Figure 8-19). This neglects any earth in front of the footing and reflects the case where excavation or scour has removed the earth to the level of the bottom of the footing.

The passive earth resistance may be expressed in terms of an equivalent fluid weight. Because $K_a w_e = 30\ \text{lb/ft}^3$ and $w_e = 100\ \text{lb/ft}^3$,

$$K_a = \frac{30}{100} = 0.3$$

The coefficient of passive earth resistance is

$$K_p = \frac{1 + \sin\phi}{1 - \sin\phi} = \frac{1}{K_a} = \frac{1}{0.3} = 3.33$$

and

$$K_p w_e = 3.33(100) = 333\ \text{lb/ft}^3$$
$$H_p = \tfrac{1}{2} K_p w_e h_k^2 = \tfrac{1}{2}(333)h_k^2$$

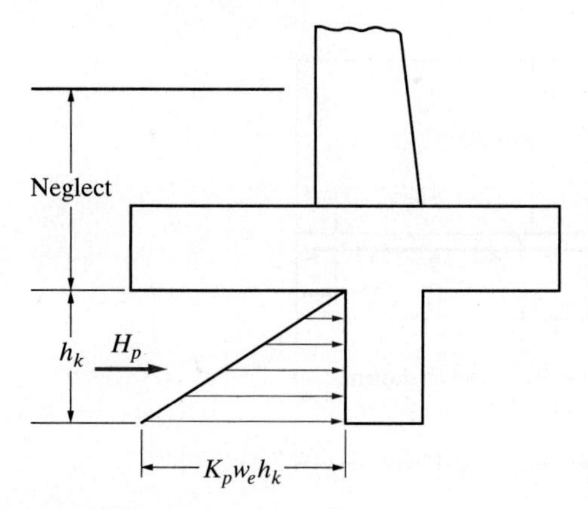

FIGURE 8-19 Base shear key force analysis.

The unfactored horizontal force $\Sigma H = 7020$ lb. Therefore, the required resistance to sliding that will furnish a factor of safety of 1.5 is

$$1.5(7020) = 10{,}530\ \text{lb}$$

The frictional resistance furnished is 9860 lb (see the discussion of stability analysis). Thus, the resistance that must be furnished by the base shear key and the passive earth resistance is

$$10{,}530 - 9860 = 670\ \text{lb}$$

With reference to Figure 8-19, the height h_k required to furnish this resistance may be obtained by establishing horizontal equilibrium ($\Sigma H = 0$).

$$\tfrac{1}{2}(333)(h_k)^2 = 670$$
$$h_k^2 = 4.02$$
$$\text{required } h_k = 2.01\ \text{ft}$$

Use 2 ft-0 in.

The key must be designed for moment and shear. The worst case would be the situation where excavation had *not* taken place and the full height of earth in front of the wall was available to develop passive pressure. This is shown in Figure 8-20. A 1-ft depth of earth has been assumed on top of the footing. Assuming the key as a vertical cantilever beam, taking a summation of moment about the plane of the bottom of footing, and applying a load factor of 1.6,

$$M_u = 1.6(916)(2.0)(1.0) + 1.6\tfrac{1}{2}(666)(2.0)(\tfrac{2}{3})(2.0)$$
$$= 4350\ \text{ft-lb}$$

Assume a 10-in. width of key and a No. 8 bar, which provides an effective depth $d = 10 - 3 - 0.5 = 6.5$ in. Assume $\phi = 0.90$.

$$\text{required } \bar{k} = \frac{M_u}{\phi b d^2} = \frac{4350}{0.9(12)(6.5)^2} = 0.1144\ \text{ksi}$$

From Table A-8, the required $\rho = 0.0020$, $\epsilon_t > 0.005$; therefore, $\phi = 0.90$. Now we can calculate the required steel area:

$$\text{required } A_s = \rho b d = 0.0020(12)(6.5) = 0.16\ \text{in.}^2$$

The minimum area of steel required by the ACI Code, Sections 9.6.1.1 and 9.6.1.2, may be obtained using Table A-5:

$$A_{s,\min} = 0.0033(12)(6.5) = 0.26\ \text{in.}^2$$

Therefore, $A_{s,\min}$ of 0.26 in.^{2} controls. This is a very small amount of steel, and it will be more practical to extend some existing stem bars into the key. All stem bars must be extended into the key for anchorage reasons, however. Therefore, all the bars will be extended to within 3 in. of the bottom of the key. This provides 42 in. of anchorage below the bottom of the stem for the No. 7 bars and exceeds the requirement of 28 in. (see typical wall section in Figure 8-21). The steel furnished in the key, therefore, is No. 7 bars at $7\tfrac{1}{2}$ in. o.c. ($A_s = 0.96$ in.^{2}).

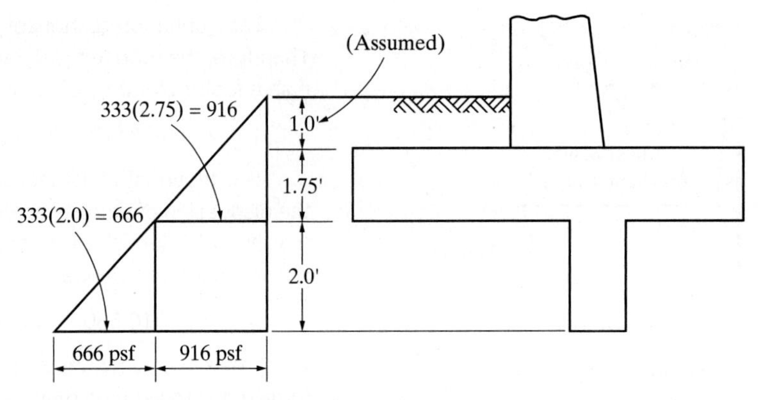

FIGURE 8-20 Base shear key force design.

If we conservatively neglect the effect of the greater cover (3.0 in.) in the key, calculations for the required anchorage length at the top of the key are identical to those for the stem, except that the excess reinforcement factor $K_{ER} = 0.26/0.96 = 0.27$, from which the final ℓ_d calculation yields

$$\ell_d = 28.8(0.27) = 7.8 \text{ in.} < 12 \text{ in.}$$

Use $\ell_d = 12$ in. The available anchorage length is $24 - 3 = 21$ in., which exceeds the requirements of 12 in.

The shear strength of the key is

$$\begin{aligned} \phi V_n &= \phi V_c \\ &= \phi(2\lambda\sqrt{f'_c})bd \\ &= 0.75(2)(1.0)(\sqrt{3000})(12)(6.5) \\ &= 6410 \text{ lb} \end{aligned}$$

The factored shear is

$$V_u = (1.6)(916)(2.0) + 1.6\left(\tfrac{1}{2}\right)(666)(2) = 4000 \text{ lb}$$

$$\phi V_n > V_u \qquad \text{(O.K.)}$$

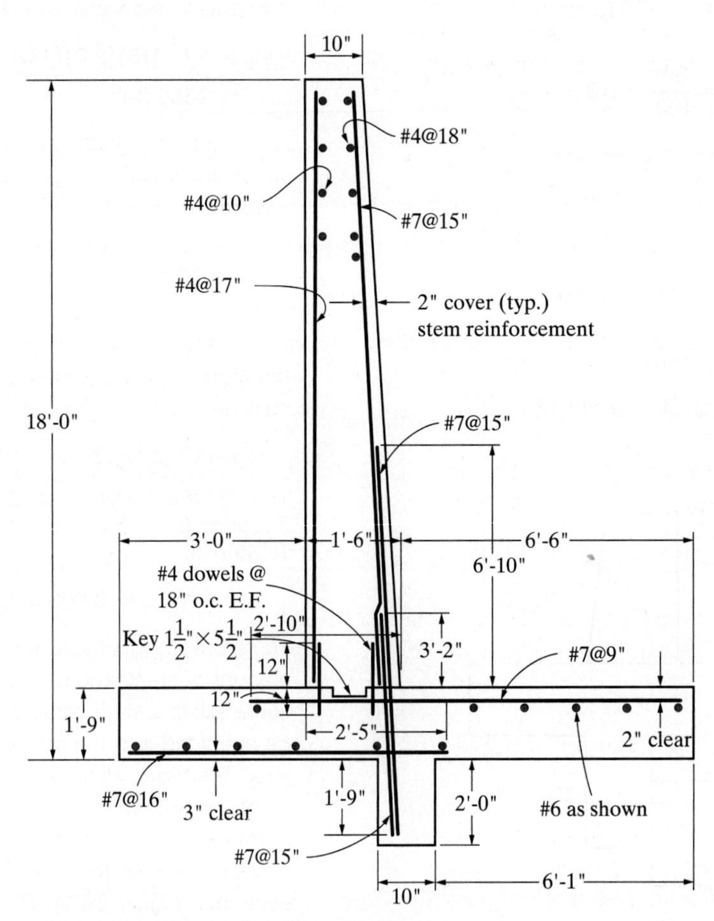

FIGURE 8-21 Typical wall section.

8-4 DESIGN CONSIDERATIONS FOR BEARING WALLS

Bearing walls (Figure 8-1g) were briefly described at the beginning of this chapter as those walls that carry vertical load in addition to their own weight. Recommendations for the empirical design of such walls are presented in Chapter 11 of the ACI Code and apply primarily to relatively short walls spanning vertically and subject to vertical loads only, such as those resulting from the reactions of floor or roof systems supported on walls. Walls, other than short walls carrying "reasonably concentric" loads, should be designed as compression members for combined axial load and flexure in accordance with ACI 318-14, Chapter 22. "Reasonably concentric" implies that the resultant factored load falls within the middle third of the cross section.

The design axial load strength or capacity of such a wall will be

$$\phi P_n = 0.55\phi f'_c A_g\left[1 - \left(\frac{k\ell_c}{32h}\right)^2\right] \quad \text{[ACI Eq. (11.5.3.1)]}$$

where

ϕ = strength-reduction factor corresponding to to compression-controlled sections in accordance with ACI Code, Section 21.2.2; 0.75 for members with spiral reinforcement and 0.65 for other reinforced members

h = thickness of wall (in.)

ℓ_c = vertical distance between supports (in.)

A_g = gross area of section (in.^{2})

k = effective length factor

The effective length factor k shall be

1. For walls braced top and bottom against lateral translation and
   a. Restrained against rotation at one or both ends (top and/or bottom) 0.8
   b. Unrestrained against rotation at both ends 1.0
2. For walls not braced against lateral translation 2.0

Where the wall is subject to concentrated loads, the effective length of wall for each concentration must not exceed the center-to-center distance between loads nor exceed the width of bearing plus four times the wall thickness.

The following requirements applicable to bearing walls are prescribed, among others, by the ACI Code, Chapter 11:

1. Reinforced concrete bearing walls must have a thickness of at least $\frac{1}{25}$ of the unsupported height or width, whichever is shorter, and not less than 4 in.
2. Thickness of nonbearing walls shall not be less than 4 in. nor less than $\frac{1}{30}$ times the least distance between members that provide lateral support.
3. The area of horizontal reinforcement must be at least 0.0025 times the area of the wall (0.0025bh per foot) and the area of vertical reinforcement not less than 0.0015 times the area of the wall (0.0015bh per foot), where b = 12 in. and h is the wall thickness. These values may be reduced to 0.002 and 0.0012, respectively, if the reinforcement is deformed bars not larger than No. 5 with f_y not less than 60,000 psi or if the reinforcement is welded wire reinforcement larger than W31 or D31.
4. Exterior basement walls and foundation walls must not be less than $7\frac{1}{2}$-in. thick.
5. Reinforced concrete walls must be anchored to intersecting elements such as floors and roofs or to columns, pilasters, buttresses, intersecting walls, and to footings.
6. Walls more than 10-in. thick, except for basement walls, must have reinforcement in each direction for each face. The exterior surface shall have a minimum of one-half and a maximum of two-thirds of the total steel required, with the interior surface having the balance of the reinforcement.
7. Vertical reinforcement must be enclosed by lateral ties if in excess of 0.01 times the gross concrete area or when it is required as compression reinforcement.

Additionally, from the ACI Code, Section 22.8.3, the design-bearing strength of concrete under a bearing plate may be taken as $\phi(0.85f'_c A_1)$, where A_1 is the loaded area. An exception occurs when the supporting surface is wider on all sides than the loaded area. In that case, the foregoing expression for bearing strength may be multiplied by $\sqrt{A_2/A_1} \le 2.0$, where A_2 is a concentric and geometrically similar support area that is the lower base of a frustum (upper base of which is A_1) of a pyramid having 1:2 sloping sides and fully contained within the support.

Example 8-2

Design a reinforced concrete bearing wall to support a series of steel wide-flange beams at 8 ft-0 in. o.c. Each beam rests on a bearing plate 6 in. × 12 in. The wall is braced top and bottom against lateral translation. Assume the bottom end fixed against rotation. The wall height is 15 ft and the design (factored) load P_u from each beam is 115 kip. Use f'_c = 3000 psi and f_y = 60,000 psi. See Figure 8-22.

Solution:

1. Assume an 8-in.-thick wall with full concentric bearing.

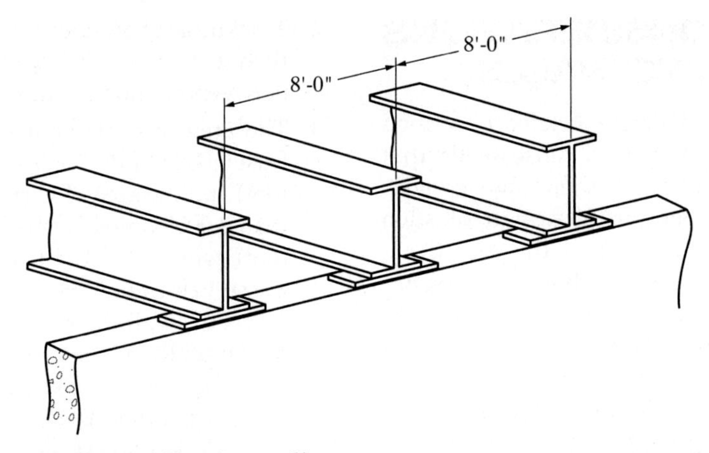

FIGURE 8-22 Bearing wall.

2. From our previous discussion, the bearing strength of the concrete under the bearing plate, neglecting the $\sqrt{A_2/A_1}$ multiplier, is

$$\phi(0.85)f'_c A_1 = 0.65(0.85)(3000)(6)(12)$$
$$= 119{,}300 \text{ lb}$$
$$\text{factored bearing load } P_u = 115{,}000 \text{ lb}$$
$$119{,}300 > 115{,}000 \qquad \text{(O.K.)}$$

3. The effective length of the wall (ACI Code, Section 11.2.3.1) must not exceed the center-to-center distance between loads nor the width of bearing plus four times the wall thickness. Beam spacing = 96 in. Thus

$$12 + 4(8) = 44 \text{ in.}$$

Therefore, use 44 in.

4. The minimum thickness required is $\frac{1}{25}$ times the shorter of the unsupported height or width. Assume, for our case, that width does not control. Then

$$h_{min} = \frac{\ell_c}{25} = \frac{15(12)}{25} = 7.2 \text{ in.}$$

Also, $h_{min} = 4$ in. Therefore, the 8-in. wall is satisfactory.

5. The capacity of wall is calculated from

$$\phi P_n = 0.55\phi f'_c A_g\left[1 - \left(\frac{k\ell_c}{32h}\right)^2\right]$$

The strength-reduction factor ϕ for compression-controlled sections is discussed in Section 2-9 of this text. In this case, because the wall will not contain spiral lateral reinforcing, ϕ is taken as 0.65.

$$\phi P_n = 0.55(0.65)(3)(44)(8)\left\{1 - \left[\frac{0.8(15)(12)}{32(8)}\right]^2\right\}$$
$$= 378(1 - 0.316)$$
$$= 258 \text{ kip}$$
$$\phi P_n > P_u \qquad \text{(O.K.)}$$

6. The reinforcing steel (ACI Code, Table 11.6.1), assuming No. 5 bars or smaller, can be found: For the vertical reinforcement per foot of wall length,

$$\text{required } A_s = 0.0012\,bh$$
$$= 0.0012(12)(8) = 0.12 \text{ in.}^2$$

For the horizontal reinforcement per foot of wall height,

$$\text{required } A_s = 0.002\,bh$$
$$= 0.002(12)(8) = 0.19 \text{ in.}^2$$

The maximum spacing of reinforcement must not exceed three times the wall thickness nor 18 in. (ACI Code, Sections 11.7.2.1 and 11.7.3.1). Thus

$$3(8) = 24 \text{ in.}$$

Use 18 in.

To select reinforcing,

vertical steel: Use No. 4 bars at 18 in. ($A_s = 0.13$ in.^{2})

horizontal steel: Use No. 4 bars at 12 in. ($A_s = 0.20$ in.^{2})

The steel may be placed in one layer, as the wall is less than 10-in. thick (see Figure 8-23).

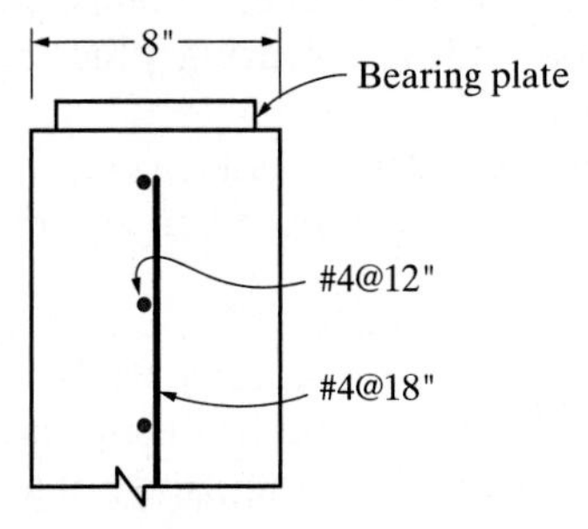

FIGURE 8-23 Section of wall.

8-5 DESIGN CONSIDERATIONS FOR BASEMENT WALLS

The active soil pressure coefficient, K_a, is used for the design of retaining walls, while the at-rest lateral pressure coefficient, K_o, is used for the design of basement walls due to the relatively smaller displacement of a basement wall compared to a retaining wall. K_a is typically smaller that K_o. As an example, for a granular native soil with angle of internal friction of 30 degrees at a particular site, the specified K_a was 0.33, the passive pressure coefficient, K_p was 2.50, and the at-rest lateral soil pressure coefficient, K_o was 0.50.

A basement wall is a type of retaining wall in which lateral support is assumed to be provided at the bottom and top of the wall by the basement floor slab and the first-floor construction, respectively. As previously mentioned, the wall would be designed as a simply supported member with a loading diagram and moment diagram as shown in Figure 8-24.

The design approach for the wall is similar to that of one way slabs that was discussed in Chapter 2. For buildings with multi-level basements, the basement wall is analyzed similar to a continuous one-way slab subject to a linearly varying load due to the lateral soil pressure. The basement floor slabs act as the lateral supports for the wall. The main vertical reinforcement in the wall needed to resist the lateral soil pressure will be located on the inside face of the basement wall between the basement floor levels, and on the outside face of the wall at the basement floor supports.

If the wall is part of a bearing wall, the vertical load will relieve some of the tension in the vertical reinforcement. This may be neglected because its effect may be small compared with the uncertainties in the assumption made regarding the loads. If the vertical load is of a permanent nature and of significant magnitude, its effect should be considered in the design. The wall should then be designed as a member subject to combined bending and axial loads.

When a part of the basement wall is above ground, the lateral bending moment may be small and may be computed as shown in Figure 8-25. This assumes that the wall is spanning in a vertical direction. Depending on the type of construction, the basement wall may also span in a horizontal direction and may behave as a slab reinforced in either one or two directions. If the wall design assumes two horizontal reactions, as shown in Figures 8-24 and 8-25, caution must be exercised that the two supports are in place prior to backfilling behind the wall.

8-6 LATERAL LOAD RESISTING SYSTEMS IN CONCRETE BUILDINGS

The lateral loads that act on structures include wind loads, seismic loads, unbalanced lateral earth pressure, hydrostatic pressure (and uplift), and flood loads, but the most common lateral loads that act on building structures are wind loads, seismic loads, and lateral earth pressure. For a discussion of the methods for calculating

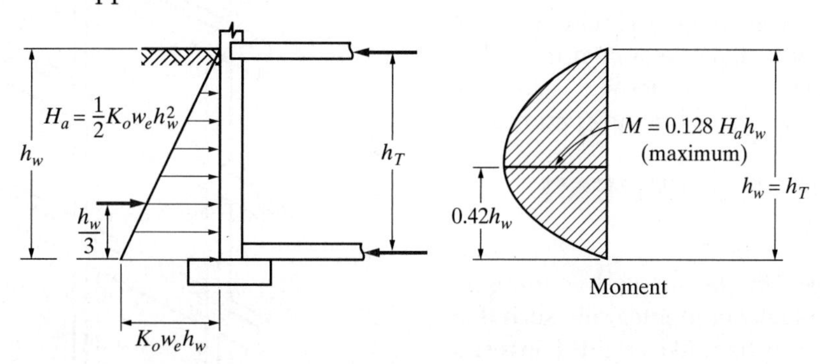

FIGURE 8-24 Basement wall with full height backfill: forces and moment diagram.

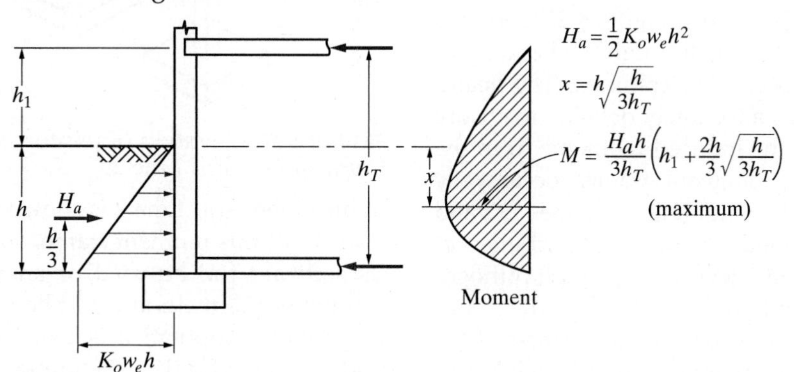

FIGURE 8-25 Basement wall with partial backfill: forces and moment diagram.

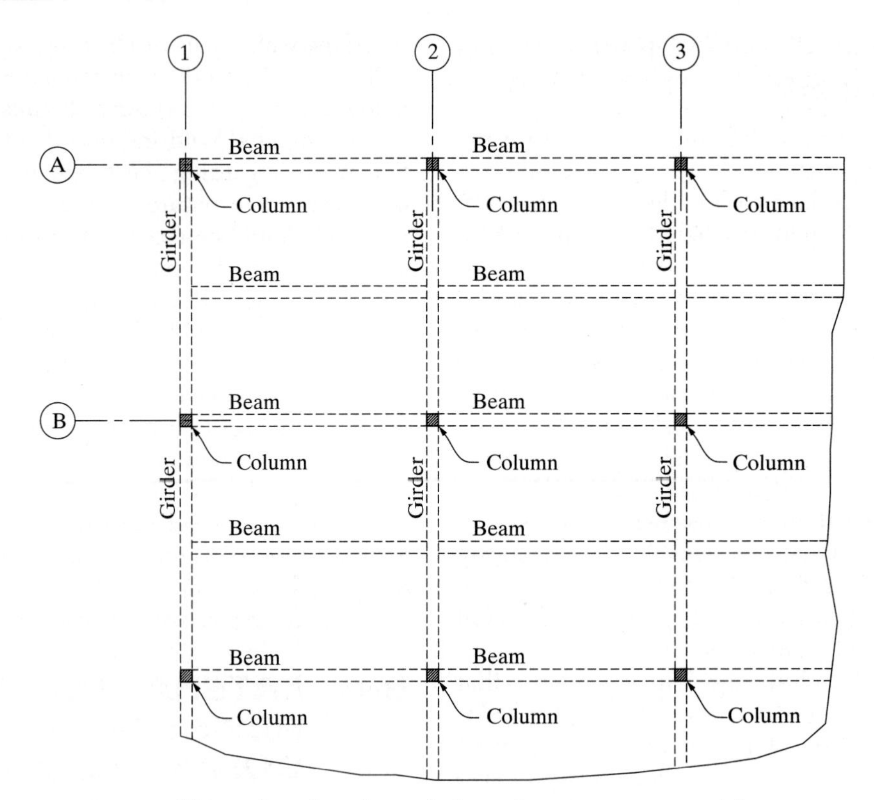

FIGURE 8-26 Floor plan showing reinforced concrete moment frames.

the lateral loads that act on buildings, the reader should refer to References [3, 4]. The two most common types of lateral load resisting systems used in concrete buildings are reinforced concrete moment frames and reinforced concrete shear walls. These lateral-load-resisting-elements may be combined in a building or one or the other element may be the sole system used. The lateral-load-resisting-elements can be located within the interior of the building or along the building perimeter.

8-7 CONCRETE MOMENT FRAMES

The connections between the beams and columns in concrete buildings are usually built integrally such that the beam-to-column joints behave like "rigid" joints and thus have the capacity to resist moments due to lateral loads. Moment frames consists of girders or beams rigidly connected to the columns (see Figure 8-27) and they resist the lateral load moments through the bending of the concrete beams/girders and columns. The beam-to-column joints must be adequately detailed to ensure their ability to transfer the induced moments from the lateral loads. Naturally, moment frames occur only along the column lines in a building (e.g., lines 1, 2 and 3 and lines A and B in Figure 8-26). The distribution of the lateral load to each moment frame in each orthogonal direction will depend on the stiffness of the floor/roof diaphragms and the relative lateral stiffness of the moment frames and any shear walls that may be present. The elevation of the reinforced concrete moment frame along Grid Line 2 is shown in Figure 8-27. In the analysis of this moment frame, any plane frame analysis software can be used or an approximate method like the portal method could be used. In modeling the foundation supports, it is usual to model these supports as pinned unless the foundations are deliberately fixed against rotation by, for example, by socketing and

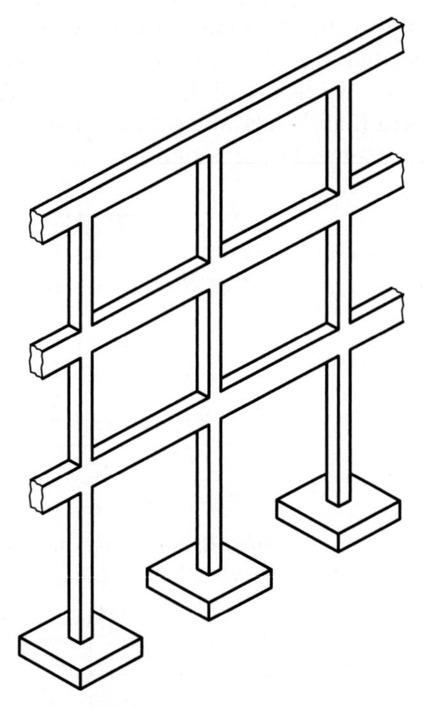

FIGURE 8-27 Elevation of reinforced concrete moment frame along Grid Line 2.

dowelling the footing into a rock foundation where the footing bears directly on bedrock, or by supporting the building on piles or caissons (see Chapter 10).

8-8 SHEAR WALLS

Concrete or masonry walls fixed at their base are used to resist lateral wind and seismic loads in building structures parallel to the plane of the shear wall, in addition to supporting gravity loads. These lateral load resisting elements may consist of single walls located internally within the building or on the exterior face of the building, or they could be the stair and elevator core walls. Shear walls are very efficient lateral load resisting elements that resist lateral loads by acting as a vertical cantilever. Sometimes, the walls may be perforated by door or window openings or corridors, and when these vertical wall segments are connected together with horizontal wall segments or deep beams spanning the door, window or corridor openings, they are called "coupled shear" walls (see Figure 8-29c). The coupling beams connecting the shear wall segments are subjected to large shear forces

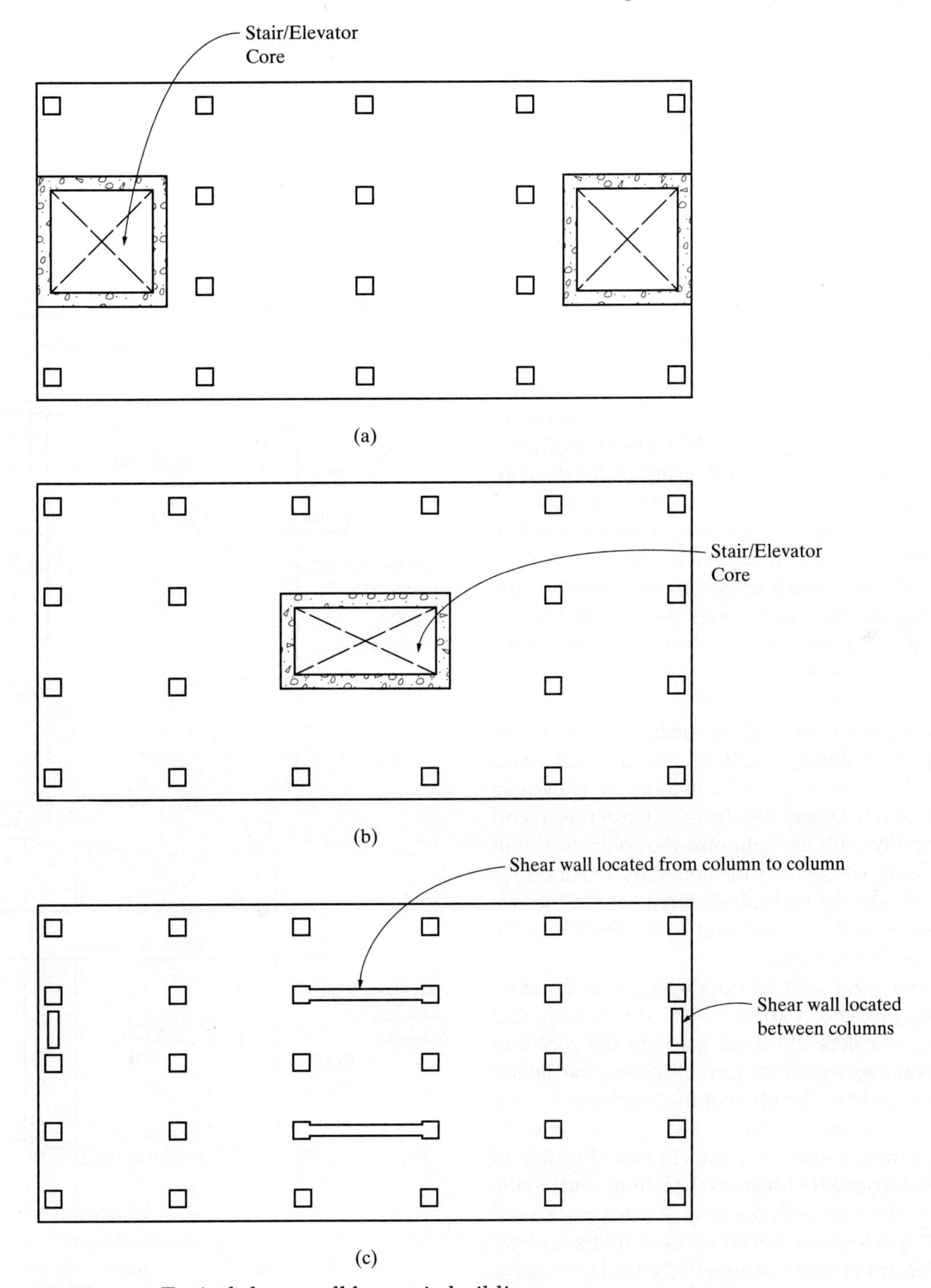

FIGURE 8-28 Typical shear wall layout in buildings.

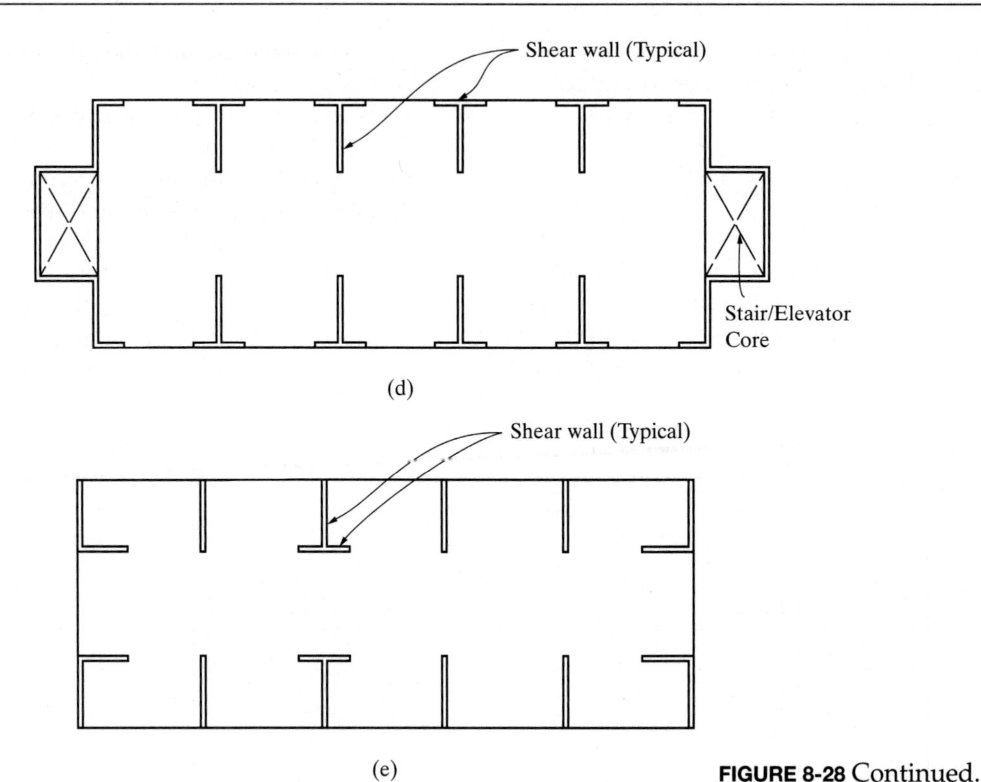

FIGURE 8-28 Continued.

and deep coupling beams with clear span-to-depth ratios less than 4 require confined diagonal reinforcement (see ACI 318-14 Code, Section 18.10.7). The strength of a coupled shear wall depends on the stiffness of the coupled walls and the stiffness of the coupling beams and lies between the strength of a moment frame and that of an unperforated shear wall. The analysis of coupled shear walls, which is much more cumbersome than the analysis of segmented cantilevered shear walls, can be accomplished using structural analysis software, or the equivalent frame method if an approximate method is desired [1]. Figures 8-28 and 8-29 show some typical shear wall layout and typical shear wall elevations, respectively, in building structures. Single shear walls could be laid out between column lines as shown in Figure 8-29a and b. Where the shear wall extends to and is built integrally with the columns, the column at both ends of the wall will serve as the boundary members for the shear wall and the vertical reinforcement in the columns will serve as the vertical end reinforcement in the shear wall (see Figure 8.29a).

The lateral wind load on buildings act on the exterior cladding perpendicular to the wind direction, and the cladding transfers the wind loads to the roof and floor diaphragms, which in turn transfers the lateral loads through in-plane bending of the diaphragm to the lateral force resisting systems - LFRS (i.e., shear walls and moment frames) that are parallel to the direction of the wind load. The LFRS transfers the lateral loads to the foundations. The load path for seismic loads starts with the ground acceleration which causes predominantly lateral movement in the structure. In the Code, the effect of the ground acceleration on the building is modeled as a set of equivalent lateral forces acting on the building

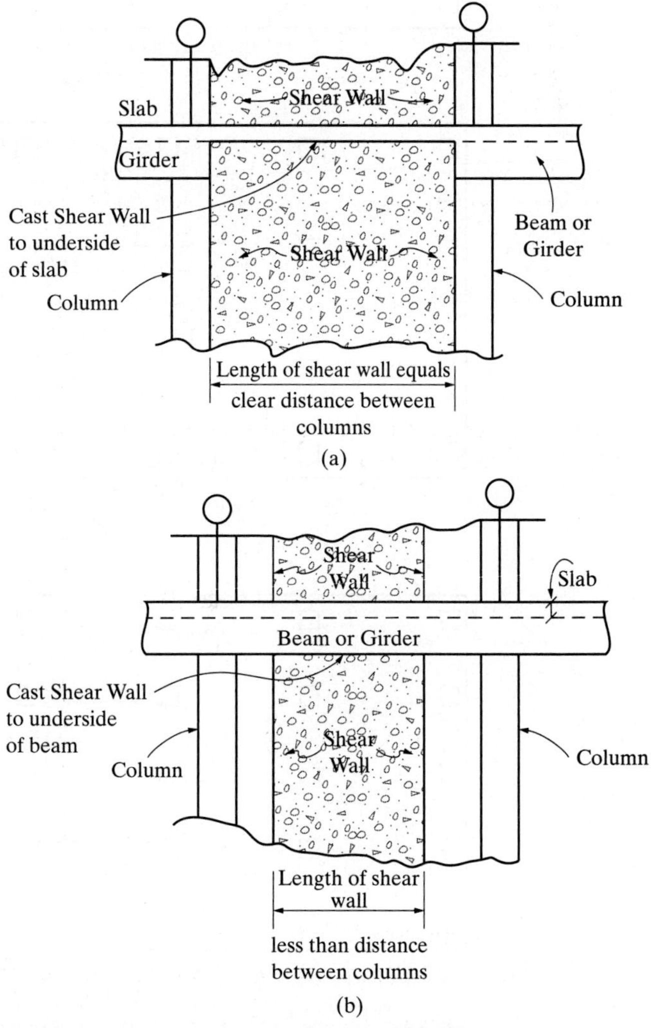

FIGURE 8-29a, b Typical shear wall elevations in concrete buildings.

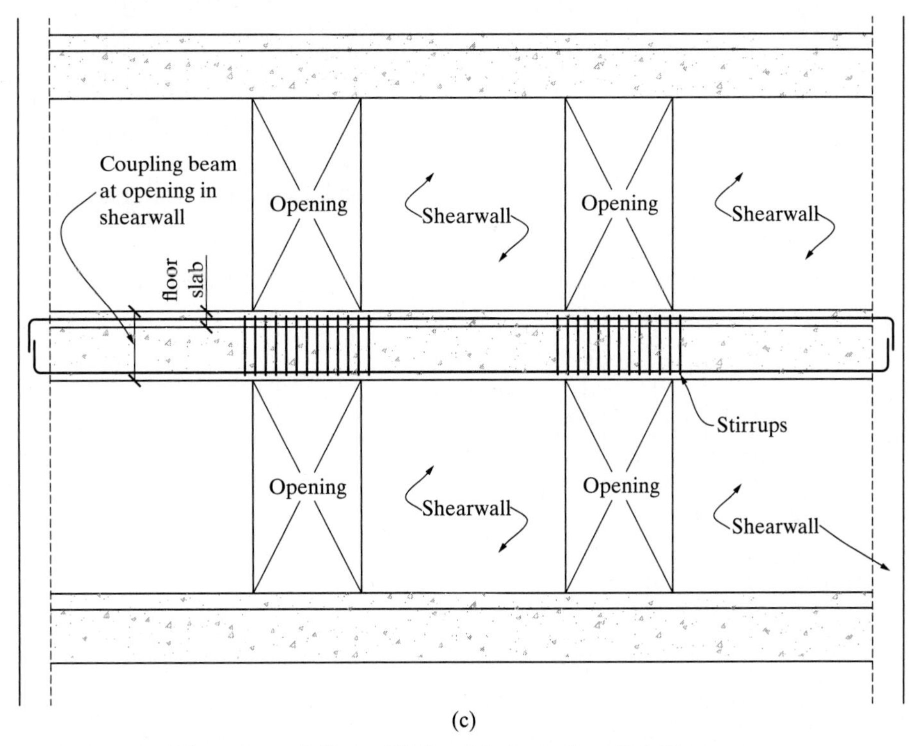

FIGURE 8-29c Elevation of Shear Wall with Coupling Beams.

at the floor and roof levels. The lateral forces in the diaphragms are transmitted to the foundation through the LFRS similar to the load path for wind loads.

Typical concrete buildings have floor and roof slabs that can be considered rigid in the horizontal plane, and thus the floors and roofs can be modeled as rigid diaphragms. See Chapter 14 for further discussion of rigid and flexible diaphragms. For buildings with rigid diaphragms, the lateral load acting on the building is distributed to the lateral force resisting elements parallel to the lateral load in proportion to the stiffness of the lateral force resisting element. For shear walls, the stiffness, K, consists of a flexural stiffness component and a shear stiffness component, and is calculated as

$$K = \frac{3EI}{h_w^3} + \frac{GA}{1.2h_w}$$

If the total lateral force on the building at level x is F_x, the lateral force at level x is distributed to each shear wall that is parallel to the lateral load as follows:

$$F_{\text{shear wall},x} = \frac{K_{\text{wall}}}{\Sigma K_{\text{wall}}} F_x$$

where

h_w = overall or total height of shear wall from top of footing to the top of the wall (see Figure 8-30a)

E = modulus of elasticity

I = moment of inertia of the wall about the strong axis $= \dfrac{h\ell_w^3}{12}$

ℓ_w = overall length of shear wall (see Figure 8-30a)

h = wall thickness (minimum practical wall thickness is 8 in.; see Figure 8-30b)

A = gross cross-sectional area of the wall $= h\ell_w$

G = shear modulus of elasticity $= \dfrac{E}{2(1+\nu)}$

ν = Poisson'sratio for concrete ≈ 0.20.

K_{wall} = stiffness of the shear wall being considered

ΣK_{wall} = sum of the stiffnesses of all the shear walls *parallel* to the lateral load

The following practical considerations should be taken into account when laying out shear walls in concrete buildings:

- Locate shear walls to minimize the effect on architectural features in the building such as doors and windows.
- Utilize stair and elevator shaft walls as shear walls. Shear walls can also be located on the outer perimeter of a building, but this may reduce the number of available windows in a building and, therefore, lead to a reduction in natural light and exterior views.
- Locate shear walls in each orthogonal direction as symmetrically as possible to minimize twisting or torsional deformations of the building from lateral loads. If a symmetrical arrangement is *not* feasible

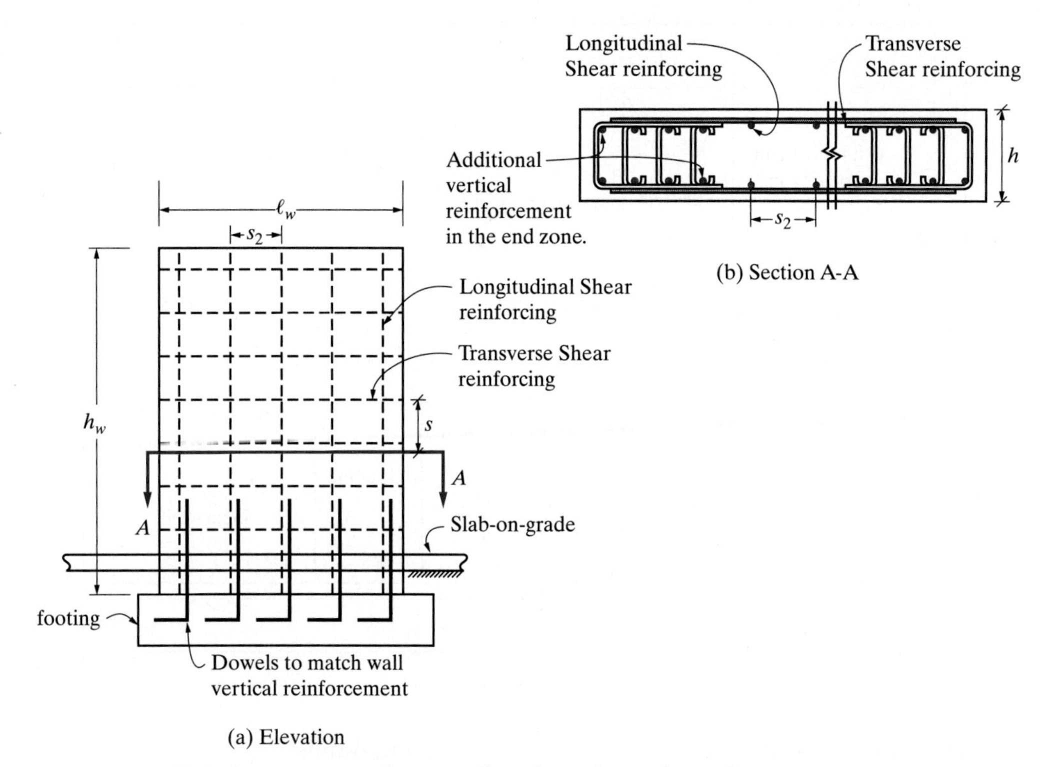

FIGURE 8-30 Reinforcement in shear wall without boundary element.

because of architectural or other constraints, the building should be analyzed for the resulting in-plane *twisting forces*, and these would lead to additional lateral forces in the shear walls.

- Shear walls and/or moment frames are required in both orthogonal directions of the building.

Shear Wall Design Considerations

The shear wall design considerations in this chapter pertain to ordinary reinforced concrete shear walls as presented in ACI Code, Section 11.5.4. The design considerations for ductile or special shear walls used in highly seismic zones are presented in ACI Code, Section 18.10. The following load effects should be considered in the design of shear walls:

- The horizontal *shear* force which is maximum at the base of the wall.
- The *bending moment* which is maximum at the base of the wall. This produces compression at the end zone at one end of the wall and tension at the end zone at the opposite end. The location of the tension and compression forces will change depending on the direction of the lateral load.
- The *gravity* or *vertical loads* (i.e. the cumulative roof and floor dead and live loads) that cause compression on the wall.

For low-rise concrete buildings, the effect of the interaction of the gravity loads with the shear or bending moment capacity of the wall will be minimal, and therefore, it is practical to consider the load effects just listed separately for low-rise buildings. However, for high-rise buildings where the gravity load on the wall could be substantial, the interactions between the axial load and bending moment and shear capacities of the wall have to be considered, and in that case, the shear wall is usually designed as a "column" with combined axial load and moment, and a column-type interaction diagram is used.

Reinforcement in Shear Walls

The reinforcement in shear walls consists of distributed horizontal reinforcement used to resist shear forces in the wall, distributed vertical reinforcement used to resist gravity loads and to control shrinkage and cracking, and concentrated vertical end reinforcement used to resist the bending moment due to lateral loads. To avoid congestion of the vertical reinforcement at the end of the shear wall, mechanical splices are used for No. 10 and larger bars, and the splices are located above every other floor level [2]. The typical shear wall reinforcement is shown in the wall elevation and section in Figure 8-30, and the typical reinforcement details at the corners and at the ends of shear walls are shown in Figures 8-31(a) and 8-31(b). Figure 8-31(c) shows shear walls with boundary elements at the ends of the wall which acts integrally with the shear wall stem to resist the lateral forces. For shear walls with flanges, the effective flange width can be determined using ACI 318-14 Code, Section 18.10.5.2. (see Figure 8-31(c)). The

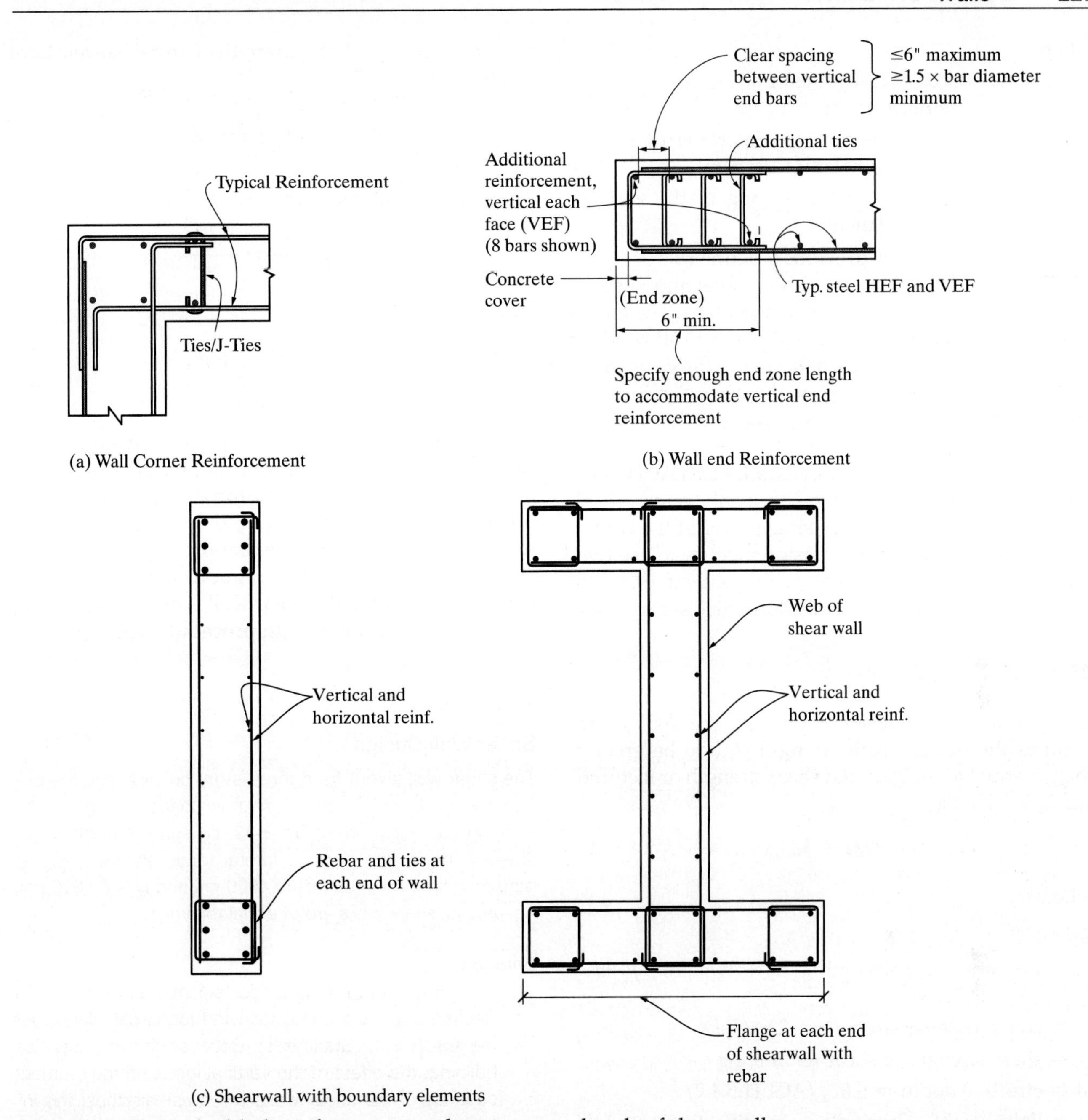

FIGURE 8-31 Typical added reinforcement at the corners and ends of shear walls.

effective flange width (for both tension and compression) beyond the face of the wall web is equal to the smaller of one-half the distance to an adjacent wall web and 25% of the overall wall height.

Minimum Reinforcement in Shear Walls

The minimum ratio of distributed *transverse* or *horizontal* reinforcement, ρ_t in the wall to the gross cross-sectional area of the wall perpendicular to the reinforcement is given in ACI Code, Section 11.6:

$$\rho_t = \frac{A_v}{sh} \geq 0.0025$$

The minimum ratio of distributed *vertical* or *longitudinal* reinforcement, ρ_ℓ in the wall to the gross cross-sectional area of the wall perpendicular to the reinforcement is given in ACI Code, Section 11.6:

$$\rho_\ell = \frac{A_v}{s_2 h} = 0.0025 + 0.5\left(2.5 - \frac{h_w}{\ell_w}\right)(\rho_t - 0.0025)$$

$$\geq 0.0025 \qquad \text{[ACI Eq. (11.6.2)]}$$

but need not be greater than ρ_t.

where

A_v = area of distributed transverse reinforcement

s = center-to-center spacing of the transverse or horizontal reinforcement

s_2 = center-to-center spacing of the vertical reinforcement

and h, ℓ_w, and h_w are as previously defined.

The size of the distributed horizontal and vertical reinforcement is usually No. 4 or larger bars. The maximum spacing of the *horizontal* reinforcement allowed by the Code (s_{maximum}) is the smallest of $\ell_w/5$, $3h$, or 18 in, and the maximum spacing of the *vertical* reinforcement allowed by the Code ($s_{2\ \text{maximum}}$) is the smallest of $\ell_w/3$, $3h$, or 18 in.

For 8-in.- or 10-in.-thick walls, although ACI Code, Section 11.7.2.3 implies that the horizontal and vertical reinforcement be placed in a single layer at the center of the wall, it is common in design practice to use two layers of reinforcement (i.e., reinforcement on both faces of the wall) for 8-in.- and 10-in.-thick shear walls.

Strength of Shear Walls

In the limit states design for shear, the ACI Code requires the design shear strength, ϕV_n to be greater than or equal to the required shear strength or factored shear, V_u. That is,

$$\phi V_n \geq V_u$$

where

$\phi = 0.75$

V_n = nominal shear strength $= V_c + V_s \leq 10\sqrt{f'_c}\,hd$ (ACI 11.5.4.3)

V_c = concrete shear strength $= 2\lambda\sqrt{f'_c}\,hd$

V_s = shear strength of shear reinforcing

d = effective depth $= 0.8\ell_w$ (ACI 11.5.4.2)

λ = lightweight concrete modification factor defined in Chapter 1 and is 1.0 for normal-weight concrete

When $V_u \leq \phi V_c$, minimum *horizontal* reinforcing should be provided in the shear wall per ACI Sections 11.6.2 and 11.7.3.

When $V_u > \phi V_c$, the required *horizontal* reinforcing is calculated as follows:

The limit states design equation for shear requires that

$$\phi V_n = \phi V_c + \phi V_s \geq V_u$$

Therefore, $\phi V_s \geq V_u - \phi V_c$, and as a limit

$$V_s = \frac{V_u - \phi V_c}{\phi}$$

From Chapter 4, the strength of any shear reinforcing is determined from

$$V_s = \frac{A_v f_y d}{s}$$

Therefore, by substitution,

$$\frac{V_u - \phi V_c}{\phi} = \frac{A_v f_y d}{s}$$

and

$$\frac{A_v}{s} = \frac{V_u - \phi V_c}{\phi f_y d}$$

After selecting the size of the horizontal reinforcement (usually No. 4 or larger bars), the cross-sectional area of the horizontal reinforcement, A_v, can be obtained as well as the required spacing of the reinforcement using the above equation.

The end zone vertical reinforcement due to the bending moment in the shear wall is determined using the rectangular beam design procedure in Chapter 2.

Example 8-3

Shear Wall Design

The shear wall layout for a three-story building and the unfactored north–south lateral seismic loads acting on the building are shown in Figure 8-32. Design the north–south shear walls for the seismic lateral loads shown. Assume normal-weight concrete; f'_c is 4000 psi and f_y is 60,000 psi. Assume all shear walls are of equal thickness.

Solution:

1. The maximum load factor for seismic loads from ACI Section 5.3.1 is 1.0 (1.0 for wind loads calculated per the ASCE 7–10 Standards). Because this is a low-rise building, the effect of the vertical loads on the moment capacity of the shear walls will be small; thus, the interaction between the gravity loads and the moment is assumed to be negligible and hence the load effects on the wall will be considered separately.
2. As the length, thickness, and total height of the north–south shear walls, W1 and W2, are equal, their stiffness, K, will also be equal. The factored seismic lateral forces for shear walls $W1$ or $W2$, including the seismic load factor, are calculated as follows:

$$\text{Roof: } F_r = 1.0 \times \frac{K}{2K}(50\text{ kip}) = 25\text{ kip}$$

$$\text{Third floor: } F_3 = 1.0 \times \frac{K}{2K}(50\text{ kip}) = 25\text{ kip}$$

$$\text{Second floor: } F_2 = 1.0 \times \frac{K}{2K}(100\text{ kip}) = 50\text{ kip}$$

For high-rise buildings, the wall reinforcement is typically specified or designed for two or three story lifts,

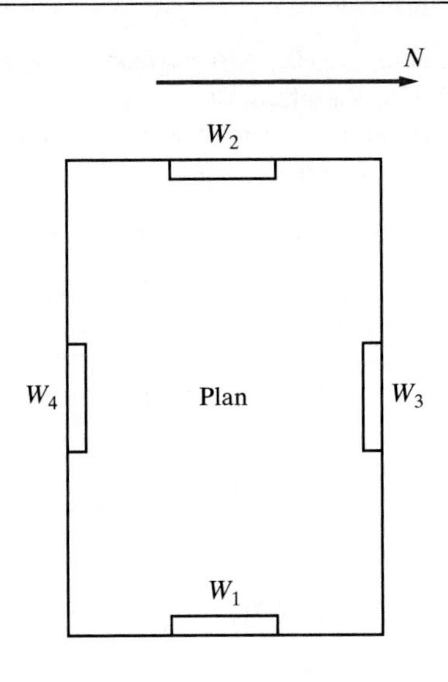

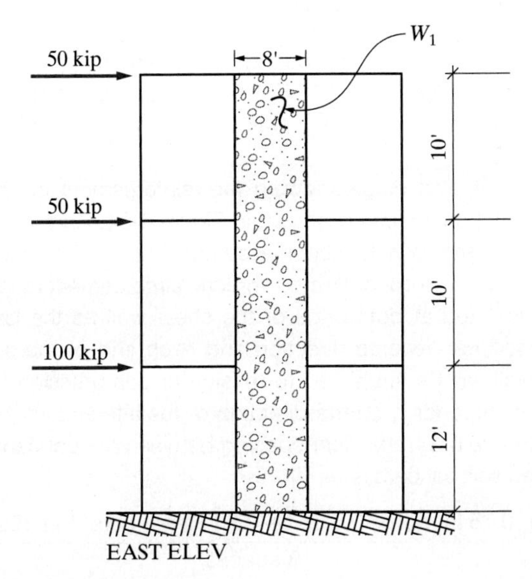

FIGURE 8-32 Shear wall layout and unfactored lateral loads for Example 8-3.

but for this low-rise building, only the reinforcement at the base of the wall will be designed and this reinforcement will be used throughout the full height of the shear wall. The factored shear wall lateral loads and load effects are shown in Figure 8-33.

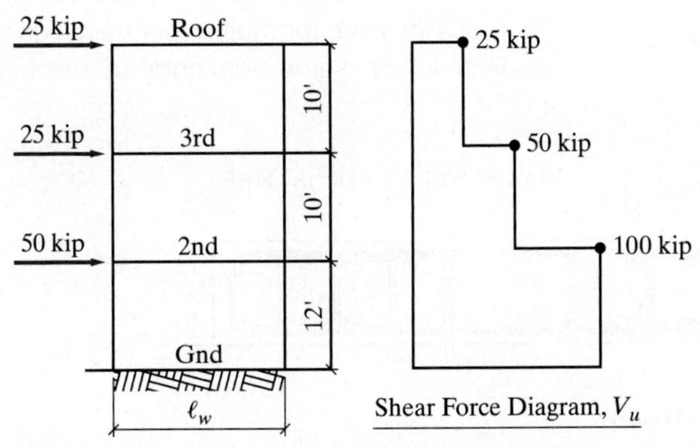

FIGURE 8-33 Shear wall lateral loads and load effects.

3. Select wall thickness: Assume $h = 8$ in. (the reinforcement will be placed on both faces of the wall).
4. Check the maximum allowed shear strength of the wall. The effective depth, $d = 0.8 \times$ length of shear wall $= (0.8)(8\text{ ft})(12\text{in./ft}) = 76.8$ in.

$$\text{Total maximum allowable shear strength} = 10\phi\sqrt{f'_c}\,hd$$

$$= 10(0.75)\sqrt{4000}(8\text{ in.})(76.8\text{ in.}) = 291.4\text{ kip}$$

Required shear strength V_u at the base of the shear wall

$$= 100\text{ kip} < 291.4\text{ kip} \qquad \text{(O.K.)}$$

5. Calculate shear strength provided by concrete alone.

$$\phi V_c = 2\phi\sqrt{f'_c}hd = 2(0.75)\sqrt{4000}(8\text{ in.})(76.8\text{ in.})$$
$$= 58.3\text{ kip}$$

Because $V_u = 100\text{ kip} > \phi V_c$, shear reinforcement, therefore, is required in the wall.

6. Determine the required *horizontal* shear reinforcement, A_v:

$$\frac{A_v}{s} = \frac{V_u - \phi V_c}{\phi f_y d} = \frac{100\text{ kip} - 58.3\text{ kip}}{(0.75)(60\text{ ksi})(76.8\text{ in.})} = 0.0121$$

Try No. 4 horizontal bars on both faces of the wall; therefore, $A_v = 2\text{ faces} \times 0.2\text{ in.}^2 = 0.4\text{ in.}^2$. The required spacing of the horizontal shear reinforcement is

$$s\text{ required} = \frac{0.4}{0.0121} = 33\text{ in.}$$

The maximum spacing of the horizontal reinforcement that is allowed by the Code (s maximum) is the smallest of the following:

- $\frac{\ell_w}{5} = \frac{96\text{ in.}}{5} = 19.2$ in.
- $3h = (3)(8\text{ in.}) = 24$ in.
- 18 in. (Controls)

Therefore, Try $s = 18$ in.

The corresponding horizontal reinforcement ratio provided is

$$\rho_t = \frac{A_v}{sh} = \frac{0.4\text{ in.}^2}{(18\text{ in.})(8\text{ in.})} = 0.0028 \geq 0.0025 \qquad \text{(O.K.)}$$

For the distributed horizontal reinforcement, provide No. 4 HEF @ 18 in. o.c. (HEF = horizontal each face of wall.)

7. Determine distributed vertical shear reinforcement. The required vertical reinforcement ratio is

$$\rho_\ell = \frac{A_v}{s_2 h} = 0.0025 + 0.5\left(2.5 - \frac{h_w}{\ell_w}\right)(\rho_t - 0.0025)$$

$$= 0.0025 + 0.5\left(2.5 - \frac{32\text{ ft}}{8\text{ ft}}\right)(0.0028 - 0.0025)$$

$$= 0.00228$$

ρ_ℓ must not be less than 0.0025 and need not be greater than ρ_t. Therefore, use $\rho_\ell = 0.0025 = A_v/s_2h$.

$$\frac{A_v}{s_2} - (0.0025)(h) = (0.0025)(8\text{ in.}) = 0.02\text{ in.}$$

Try No. 4 vertical bar both face of wall; therefore, $A_v = 2$ faces $\times$ 0.2 in.^{2} = 0.4 in.^{2}

Therefore, the required spacing of the vertical shear reinforcement is

$$s_2 \text{ required} = \frac{0.4}{0.02} = 20\text{ in.}$$

The maximum spacing vertical reinforcement allowed by the Code (s_2 maximum) is the smallest of the following:

- $\frac{\ell_w}{3} = \frac{96\text{ in.}}{3} = 32\text{ in.}$
- $3h = (3)(8\text{ in.}) = 24\text{ in.}$
- 18 in. (Controls)

Because maximum spacing = 18 in. < s_2 required = 20 in.; therefore, use s_2 = 18 in.

For the distributed vertical reinforcement, provide No. 4 VEF @ 18 in. o.c. (VEF = vertical each face of wall.)

8. Design the shear wall for flexure or bending and determine the end zone vertical reinforcement. The maximum factored bending moment at the base of the wall due to the factored seismic lateral load is

$$M_u = 1950\text{ ft.-kip}$$

The limit states design equation for flexure requires that $\phi M_n = \phi hd^2\bar{k} \geq M_u$. Initially, assume $\phi = 0.9$ (this will be checked later after ϵ_t is determined) and then calculate the required $\bar{k}$ as follows:

$$\bar{k} = \frac{M_u}{\phi hd^2} = \frac{1950(12)}{(0.9)(8\text{ in.})(76.8\text{ in.})^2} = 0.55$$

From Table A-10, we obtain $\rho = 0.0101$ and $\epsilon_t \gg 0.005$; therefore, $\phi = 0.9$, as initially assumed. The concentrated vertical reinforcement required at each end zone of the shear wall is

$$A_s\text{required} = \rho hd = (0.0101)(8\text{ in.})(76.8\text{ in.}) = 6.21\text{ in.}^2$$

The minimum area of concentrated steel required for bending at the ends of the shear wall is

$$A_{s,\min} = \frac{3\sqrt{f'_c}}{f_y}hd \geq \frac{200}{f_y}hd$$

$$= \frac{3\sqrt{4000}}{60{,}000}(8)(76.8) = 1.94\text{ in.}^2$$

$$\geq \frac{200}{60{,}000}(8)(76.8) = 2.05\text{ in.}^2$$

Because A_s required = 6.21 in.^{2} > $A_{s,\min}$ = 2.05 in.^{2}, A_s required = 6.21 in.^{2}

Try 8 No. 8 vertical reinforcement at each end of the wall (i.e., 4 No. 8 VEF) Total area of steel provided at each end of the wall = 6.32 in.^{2} > A_s required (O.K.).

A plan detail showing the reinforcement provided in the shear wall is shown in Figure 8-34. An end zone length of 16 in. has been assumed. This will be checked below. The concentrated vertical reinforcement must be provided at both ends of the shear wall as the lateral load can reverse direction and each end of the shear wall will be subjected to tension or compression forces depending on the direction of the lateral load. From Figure 8-34, the clear spacing between the concentrated vertical bars is

$$\frac{16\text{ in.} - 0.75\text{ in. cover} - 0.5\text{ in. tie} - 0.5\text{ in. tie} - 4\text{ bars}(1\text{ in. diameter})}{3\text{ spaces}}$$

$$= 3.42\text{ in.} > 1.5(1\text{ in. diameter}) = 1.5\text{ in.} \qquad \text{(O.K.)}$$

Therefore, the assumed end zone length of 16 in. is adequate.

Note that the building in the preceding example had no basement floor levels, so the overturning moment at the base of the shear wall was the sum of the moments of the lateral forces about the base of the shear wall. Similarly, the total base shear was the sum of all the lateral forces acting on the shear wall. However, for multi-story buildings with basement floor levels, some horizontal restraint will be provid-

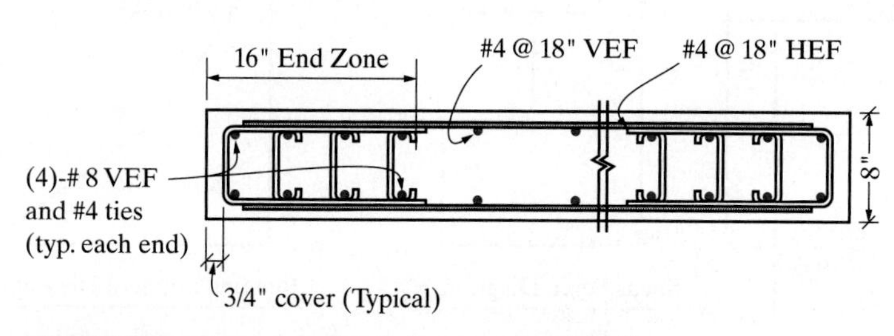

FIGURE 8-34 Shear wall reinforcement detail.

ed to the shear wall at the ground floor level by the ground floor and the basement floor slabs acting as diaphragms in transferring some of the lateral force to the basement walls, which act as shear walls in a direction parallel to its length. The basement walls are typically much longer in length compared to the core walls of the building; therefore, they are much stiffer than the shear walls and thus attract a larger portion of the lateral load at the ground floor and basement levels. The amount of reaction provided at the ground floor level and hence the corresponding moment and lateral shear at the base of the building shear wall will depend on the in-plane stiffness of the ground floor slab diaphragm and the relative in-plane stiffness of the basement walls compared to the in-plane stiffness of the building shear walls (see Figure 8-35). The use

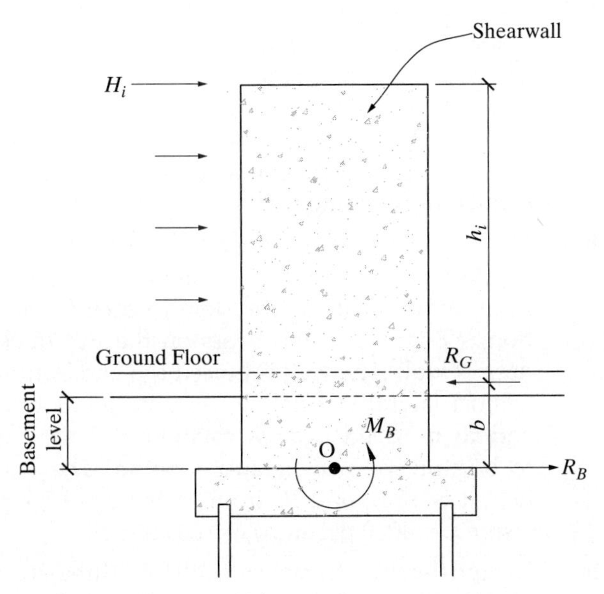

FIGURE 8-35 Shear wall forces due to restraint at ground floor level.

of the basement walls in resisting a portion of the lateral loads reduces the otherwise large moments at the base of the building in the core or shear walls.

Using equilibrium equations $\Sigma M_O = 0$ *and* $\Sigma F_x = 0$ in Figure 8-35 gives the ground floor restraint and base shear as

$$R_G = \frac{\Sigma H_i(h_i + b) - M_B}{b} \quad \text{and} \quad R_B = \frac{\Sigma H_i(h_i) - M_B}{b}$$

Note that for simplicity we have not shown the vertical forces on the wall and the vertical reaction at the foundation level. These do not impact the above analysis. Theoretically, R_G can vary from a small value up to a maximum of ΣH_i depending on the ratio of the in-plane stiffness of the basement wall to the in-plane stiffness of the building core or shear walls. The reinforcement required to transfer the reaction, R_G, from the shear wall to the ground floor slab or diaphragm is critical and must be properly detailed. The base shear in the basement walls will be resisted by a combination of friction between the basement wall footing and the soil (or by the lateral resistance of piles or caissons where the basement wall is supported on deep foundations) and by the passive soil pressure against the basement walls from the surrounding soil (see Chapter 10).

Spread or strip footings may only be adequate for shear walls in low rise buildings because of the detrimental effect of the wall rotation caused by the overturning moments, given the fact that the base of a shear wall is usually assumed to be fixed. For mid- and high rise buildings, the shear walls are usually supported on mat, pile or caisson foundations, rather than on shallow spread footings [2]. The design of foundations is discussed in Chapter 10.

References

[1] Angelo Mattacchione, "Equivalent Frame Method Applied to Concrete Shearwalls," *Concrete International*, November 1991, pp. 65–72.

[2] David G. Kittridge, "High Rise Shearwall Analysis, Design and Constructibility Guidelines," ASCE Structures Congress, 2002.

[3] Abi Aghayere and Jason Vigil, Structural Steel Design: A Practice-oriented Approach, 2nd Edition, Pearson, Hoboken, NJ, 2015, 499 pp.

[4] David A. Fanella, Structural Loads – 2012 IBC and ASCE/SEI 7, International Code Council.

Problems

8-1. Compute the active earth pressure horizontal force on the wall shown for the following conditions. Use $w_e = 100\ \text{lb/ft}^3$.

Case	ϕ	θ	w_s (psf)	h_w(ft)
(a)	25	0	400	15
(b)	28	10	0	18
(c)	30	0	200	20
(d)	33	20	0	25

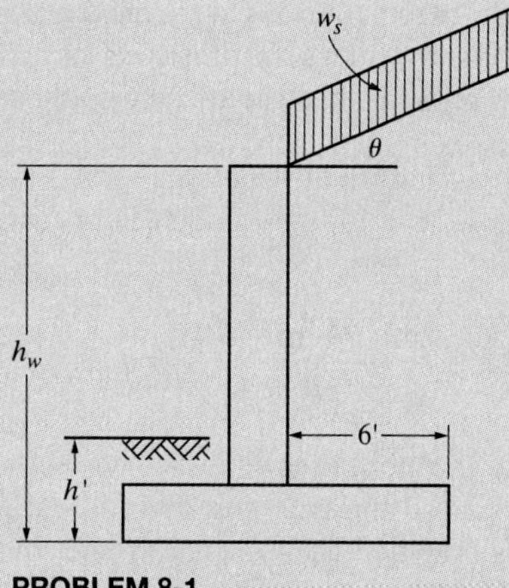

PROBLEM 8-1

8-2. Find the passive earth pressure force in front of the wall for Problem 8-1(a), if $h' = 4$ ft.

8-3. For the wall shown, determine the factors of safety against overturning and sliding and determine the soil pressures under the footing. Use $K_a = 0.3$ and $w_e = 100$ lb/ft^3. The coefficient of friction $f = 0.50$.

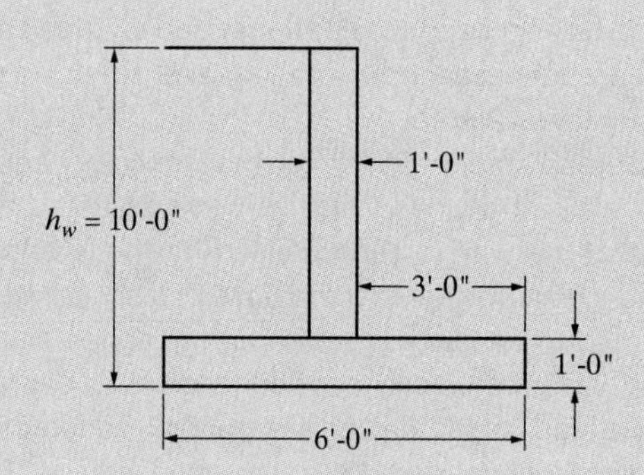

PROBLEM 8-3

8-4. Same as Problem 8-3, but the toe is 2 ft-0 in.

8-5. In Problem 8-1(c), assume a footing depth of 2 ft. Design the stem steel for M_u at the top of the footing. Check the anchorage into the footing. Disregard other stem steel details. Use $f'_c = 4000$ psi and $f_y = 60{,}000$ psi. Use thickness at top of stem = 12 in. and thickness at bottom of stem = 1 ft − 9 in.

8-6. Completely design the cantilever retaining wall shown. The height of wall h_w is 22 ft, the backfill is level, the surcharge is 600 psf, $w_e = 100$ lb/ft^3, $h' = 4$ ft, $K_a = 0.30$, and the allowable soil pressure is 4 ksf. Use a coefficient of friction of concrete on soil of 0.5, $f'_c = 3000$ psi, and $f_y = 60{,}000$ psi. The required factor of safety against overturning is 2.0 and against sliding is 1.5. The wall batter should be between $\frac{1}{4}$ and $\frac{1}{2}$ in./ft. The design is to be in accordance with the ACI Code (318-14).

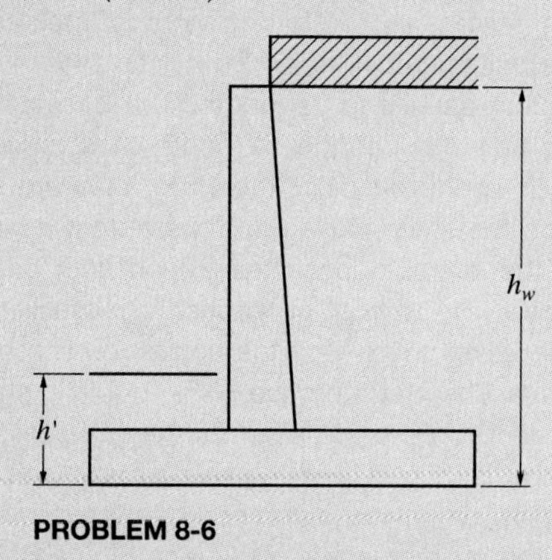

PROBLEM 8-6

8-7. Develop a spreadsheet application that will determine the location of the theoretical cut-off point for alternate bars in the stem of Example 8-1 (refer to Figure 8-16). Set up the spreadsheet so that a user could input any typical design data for a reinforced concrete cantilever retaining wall.

8-8. Design a reinforced concrete bearing wall to support a series of precast single tees spaced 7 ft-6 in. on centers. The stem of each tee section is 8-in. wide and bears on the full thickness of wall. The wall is braced top and bottom against lateral translation. Assume the bottom end to be fixed against rotation and the top to be unrestrained against rotation. The wall height is 14 ft, and P_u from each tee is 65 kip. Use $f'_c = 4000$ psi and $f_y = 60{,}000$ psi.

8-9. Design the first story shear wall for a three-story building with unfactored north–south lateral wind loads of 20 kip, 40 kip, 40 kip at the roof, third floor, and second floor, respectively. The floor-to-floor height is 12 ft and all shear walls have equal thickness and an equal length of 15 ft. Assume normal-weight concrete; f'_c is 3000 psi and f_y is 60,000 psi. Draw the detail of the shear wall reinforcement showing the required vertical end zone reinforcement and the distributed vertical and horizontal face reinforcement.

8-10. Refer to the given details for a shear wall. Use $f'_c = 5000$ psi, $f_y = 60{,}000$ psi

a. Determine the required reinforcing for bending. Sketch a plan detail of your design.

b. Determine the required horizontal and vertical reinforcing (bar size and spacing) for the wall. Sketch a wall elevation with the reinforcing.

8-11. Refer to the given plan detail and loading diagram. The Loads shown are factored.

a. Determine if the shear wall is adequate for flexural strength

b. Determine if the shear wall is adequate for shear strength

c. Determine if the wall has adequate horizontal and vertical reinforcing at each face of the wall.

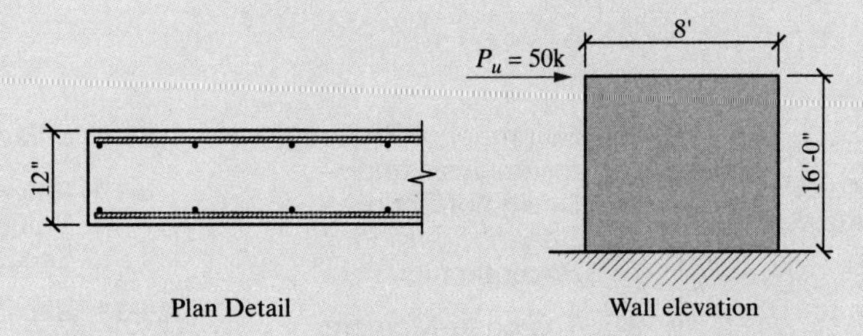

PROBLEM 8-10

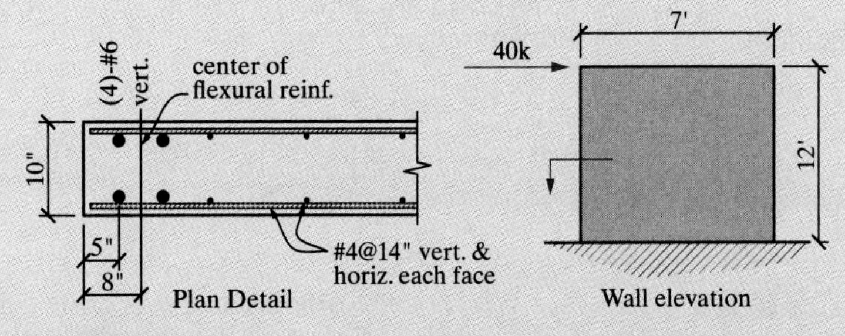

PROBLEM 8-11

CHAPTER **NINE**

COLUMNS

9-1 Introduction
9-2 Strength of Reinforced Concrete Columns: Small Eccentricity
9-3 Code Requirements Concerning Column Details
9-4 Analysis of Short Columns: Small Eccentricity
9-5 Design of Short Columns: Small Eccentricity
9-6 Summary of Procedure for Analysis and Design of Short Columns with Small Eccentricities
9-7 The Load-Moment Relationship
9-8 Columns Subjected to Axial Load at Large Eccentricity
9-9 ϕ Factor Considerations
9-10 Analysis of Short Columns: Large Eccentricity
9-11 Biaxial Bending
9-12 The Slender Column
9-13 Concrete Column Schedule

9-1 INTRODUCTION

The main vertical load-carrying members in buildings are called *columns*. The ACI Code defines a column as a member used primarily to support axial compressive loads and with a height at least three times its least lateral dimension. The code further defines a *pedestal* as an upright compression member having a ratio of unsupported height to least lateral dimension of 3 or less. The code definition for columns will be extended to include members subjected to combined axial compression and bending moment (in other words, eccentrically applied compressive loads), because, for all practical purposes, no column is truly axially loaded.

The three basic types of reinforced concrete columns are shown in Figure 9-1. *Tied columns* (Figure 9-1a and 9-2) are reinforced with longitudinal bars enclosed by horizontal, or lateral, ties placed at specified spacings. *Spiral columns* (Figure 9-1b) are reinforced with longitudinal bars enclosed by a continuous, rather closely spaced, steel spiral. The spiral is made up of either wire or bar and is formed in the shape of a helix. A third type of reinforced concrete

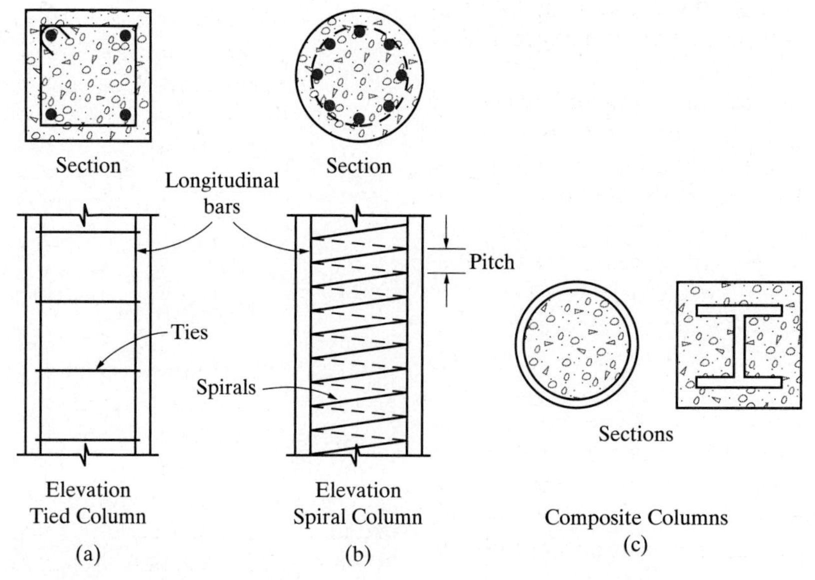

FIGURE 9-1 Column types.

FIGURE 9-2 Foundation columns, turbine generator building. Seabrook Station, New Hampshire. (Courtesy of George Limbrunner)

column, a *composite column*, is shown in Figure 9-1c. This type of column encompasses compression members reinforced longitudinally with structural steel shapes, pipes, or tubes with or without longitudinal bars. Code requirements for composite compression members are found in Sections 6.2.5.2, 6.6.4.4.5, 10.3.1.6, 10.5.2.2, 10.6.1.2, 10.7.3.2, 10.7.6.1.4, 20.4.2.2, 22.3.3, and 22.4.2.1. Our discussion is limited to the first two types: tied and spiral columns. Tied columns are generally square, rectangular, or circular, whereas spiral columns are normally circular. This is not a hard-and-fast rule, however, as square, spirally reinforced columns, and circular tied columns do exist, as do other shapes, such as octagonal and L-shaped columns.

We initially discuss the analysis and design of columns that are *short*. A column is said to be short when its length is such that lateral buckling need not be considered. The ACI Code does, however, require that the length of columns be a design consideration. It is recognized that as length increases, the usable strength of a given cross section is decreased because of the buckling problem. By their very nature, concrete columns are more massive and, therefore, stiffer than their structural steel counterparts. For this reason, slenderness is less of a problem in reinforced concrete columns. It has been estimated that more than 90% of typical reinforced concrete columns existing in braced frame buildings may be classified as short columns, and slenderness effects may be neglected.

9-2 STRENGTH OF REINFORCED CONCRETE COLUMNS: SMALL ECCENTRICITY

If a compressive load P is applied coincident with the longitudinal axis of a symmetrical column, it theoretically induces a uniform compressive stress over the cross-sectional area. If the compressive load is applied a small distance e away from the longitudinal axis, however, there is a tendency for the column to bend due to the moment $M = Pe$. This distance e is called the *eccentricity*. Unlike the zero eccentricity condition, the compressive stress is not uniformly distributed over the cross section but is greater on one side than the other is. This is analogous to our previously discussed eccentricity of applied loads with respect to retaining wall footings, where the eccentrically applied load resulted in a nonuniform soil pressure under the footing.

We consider an *axial load* to be a load that acts parallel to the longitudinal axis of a member but *need not* be applied at any particular point on the cross section, such as a centroid or a geometric center. The column that is loaded with a compressive axial load at zero eccentricity is probably nonexistent, and even the axial load/small eccentricity (axial load/small moment) combination is relatively rare. Nevertheless, we first consider the case of columns that are loaded with compressive axial loads at *small eccentricities*, further defining this situation as that in which the induced moments, although they are present, are so small that they are of little significance. Earlier codes have defined small eccentricity as follows:

For spirally reinforced columns: $e/h \leq 0.05$
For tied columns: $e/h \leq 0.10$

where h is the column dimension perpendicular to the axis of bending.

The fundamental assumptions for the calculation of column axial load strength (small eccentricities) are that at nominal strength the concrete is stressed to $0.85f'_c$ and the steel is stressed to f_y. For the cross sections shown in Figure 9-1a and b, the nominal axial load strength at small eccentricity is a straightforward sum of the forces existing in the concrete and longitudinal steel when each of the materials is stressed to its maximum. The following ACI notation will be used:

A_g = gross area of the column section (in.^{2})
A_{st} = total area of *longitudinal* reinforcement (in.^{2})
P_0 = nominal, or theoretical, axial load strength at zero eccentricity
P_n = nominal, or theoretical, axial load strength at given eccentricity
P_u = factored applied axial load at given eccentricity

For convenience, we will use the following longitudinal steel reinforcement ratio:

ρ_g = ratio of total longitudinal reinforcement area to cross-sectional area of column (A_{st}/A_g)

The nominal, or theoretical, axial load strength for the special case of zero eccentricity may be written as

$$P_0 = 0.85f'_c(A_g - A_{st}) + f_yA_{st}$$

This theoretical strength must be further reduced to a maximum usable axial load strength using two different strength reduction factors.

Extensive testing has shown that spiral columns are tougher than tied columns, as depicted in Figure 9-3. Both types behave similarly up to the column yield point, at which time the outer shell spalls off. At this point, the tied column fails through crushing and shearing of the concrete and through outward buckling of the bars between the ties. The spiral column, however, has a core area within the spiral that is effectively laterally supported and continues to withstand load. Failure occurs only when the spiral steel yields following large deformation of the column. Naturally, the size and spacing of the spiral steel will affect the final load at failure. Although both columns have exceeded their usable strengths once the outer shell spalls off, the ACI Code recognizes the greater tenacity of the spiral column in Section 21.2.1, where it directs that a strength-reduction factor, ϕ, of 0.75 be used for a spiral column, whereas the strength-reduction factor for the tied column is 0.65.

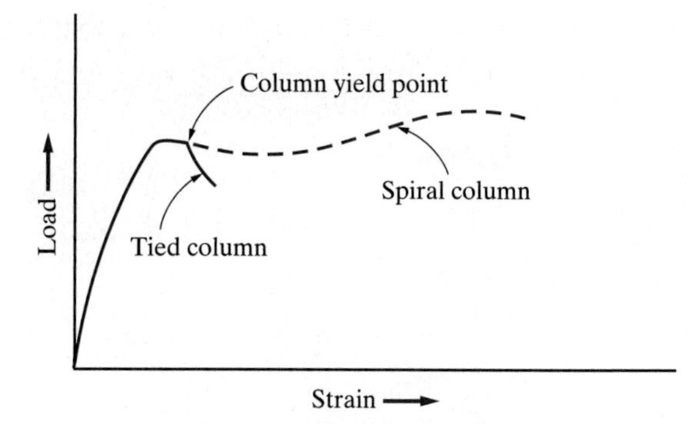

FIGURE 9-3 Load-strain relationship for columns.

The code directs that the basic load-strength relationship be

$$\phi P_n \geq P_u$$

where P_n is the nominal axial load strength at a given eccentricity and ϕP_n is designated as the *design axial load strength*. Logically, for the case of zero eccentricity, if it could exist, P_n would equal P_0. The ACI Code recognizes that no practical column can be loaded with zero eccentricity, however. Therefore, in addition to imposing the strength reduction factor ϕ, the code directs that the nominal axial load strengths be further reduced by factors of 0.80 and 0.85 for tied and spiral columns, respectively, to account for moments due to accidental or small eccentricities. This results in the following expressions for usable axial load strengths (see ACI Code, Tables 22.4.2.1).

For spiral columns,

$$\phi P_{n(\max)} = 0.85\phi[0.85f'_c(A_g - A_{st}) + f_yA_{st}]$$

[ACI Eq. (22.4.2.2)]

For tied columns,

$$\phi P_{n(\max)} = 0.80\phi[0.85f'_c(A_g - A_{st}) + f_yA_{st}]$$

[ACI Eq. (22.4.2.2)]

These expressions provide the magnitude of the maximum design axial load strength that may be realized from any column cross section. This will be the design axial load strength at small eccentricity. Should the eccentricity (and the associated moment) become larger, ϕP_n will have to be reduced, as shown in Section 9-9. It may be recognized that the code equations for $\phi P_{n(\max)}$ provide for an extra margin of axial load strength. This will, in effect, provide some reserve strength to carry small moments.

9-3 CODE REQUIREMENTS CONCERNING COLUMN DETAILS

Main (longitudinal) reinforcing should have a cross-sectional area so that ρ_g will be between 0.01 and 0.08. The minimum number of longitudinal bars is four within rectangular or circular ties, three within triangular ties, and six for bars enclosed by spirals. The foregoing requirements are stated in the ACI Code, Sections 10.6.1.1 and 10.7.3.1. Although not mentioned in the present code, the 1963 code recommended a minimum bar size of No. 5.

The clear distance between longitudinal bars must not be less than 1.5 times the nominal bar diameter nor $1\frac{1}{2}$ in. (ACI Code, Section 25.2.3). This requirement also holds true where bars are spliced. Table A-14 may be used to determine the maximum number of bars allowed in one row around the periphery of circular or square columns.

Cover shall be $1\frac{1}{2}$ in. minimum over primary reinforcement, ties, or spirals (ACI Code, Section 20.6.1.3.1).

Tie requirements are discussed in detail in the ACI Code, Section 25.7.2 and are presented as follows:

- The minimum size of column ties is No. 3 for longitudinal bars No. 10 and smaller; otherwise, minimum tie size is No. 4 (see Table A-14 for a suggested tie size). Usually, No. 5 is a maximum.
- The center-to-center spacing of ties should not exceed
    - the smaller of 16 longitudinal bar diameters,
    - 48 tie-bar diameters,
    - or the least column dimension.
- Furthermore, rectilinear ties shall be arranged so that every corner and alternate longitudinal bar will have lateral support provided by the corner of a tie having an included angle of not more than 135°, and no bar shall be farther than 6 in. clear on each side from such a laterally supported bar. See Figure 9-4 for typical rectilinear tie arrangements.

Individual circular ties are permitted per ACI Code 25.7.2.4 where longitudinal column reinforcements are located around the perimeter of a circle. The ends of the circular ties should be terminated with standard hooks around a longitudinal column bar and should overlap by at least 6 in. The overlaps around the ends of adjacent circular ties should be staggered.

Spiral requirements are discussed in the ACI Code, Section 25.7.3. The minimum spiral size is $\frac{3}{8}$ in. in diameter for cast-in-place construction ($\frac{5}{8}$ in. is usually maximum). Clear space between spirals must not exceed 3 in. or be less than 1 in. The *spiral steel ratio* ρ_s must not be less than the value given by

$$\rho_{s(\min)} = 0.45\left(\frac{A_g}{A_{ch}} - 1\right)\frac{f'_c}{f_{yt}} \quad \text{[ACI Eq. (25.7.3.3)]}$$

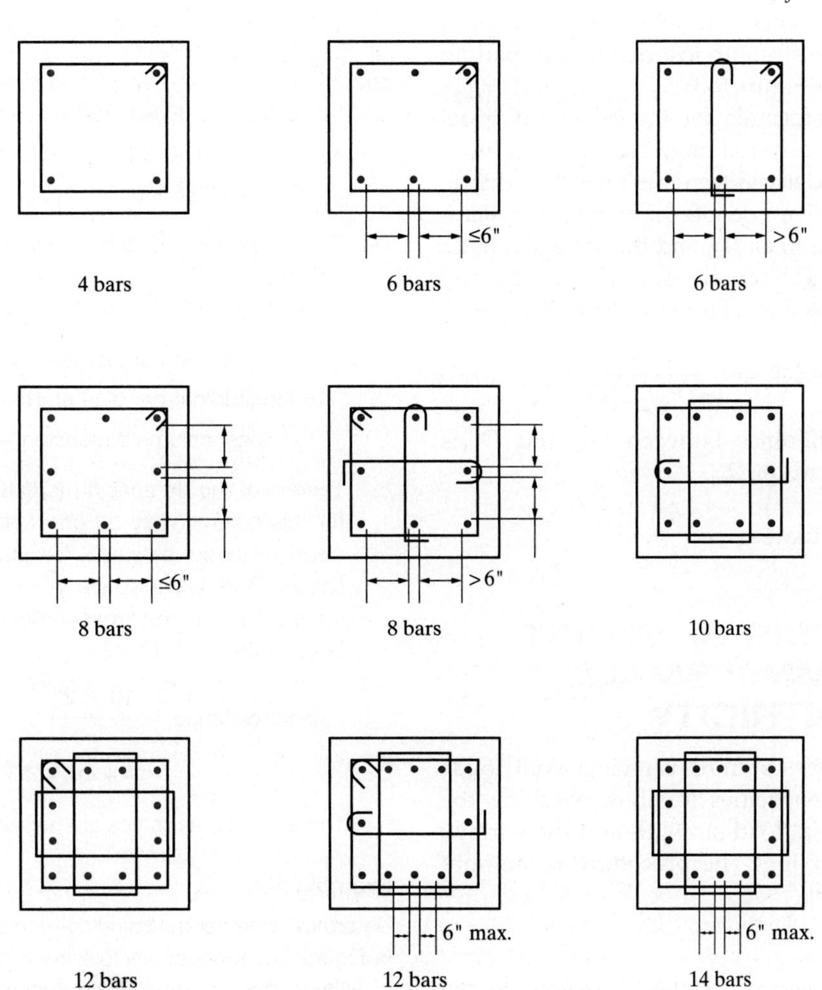

FIGURE 9-4 Typical tie arrangements.

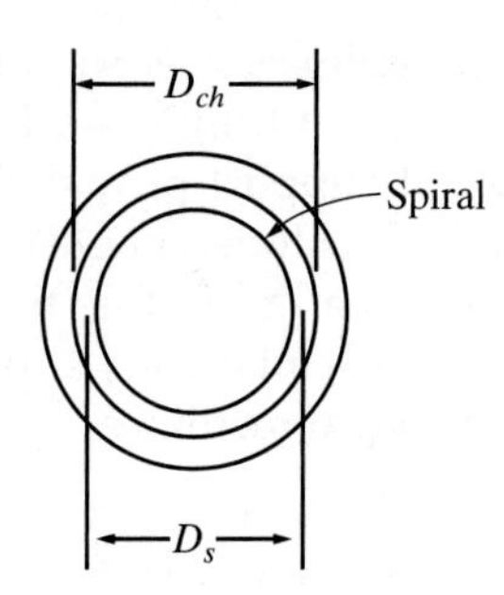

FIGURE 9-5 Definition of D_{ch} and D_s.

where

$$\rho_s = \frac{\text{volume of spiral steel in one turn}}{\text{volume of column core in height } (s)}$$

s = center-to-center spacing of spiral (in.) (sometimes called the *pitch*)

A_g = gross cross-sectional area of the column (in.^{2})

A_{ch} = cross-sectional area of the core (in.^{2}) (out-to-out of spiral)

f_{yt} = *spiral* steel yield point (psi) $\leq$ 60,000 psi

f'_c = compressive strength of concrete (psi)

This particular spiral steel ratio will result in a spiral that will make up the strength lost due to the spalling of the outer shell (see Figure 9-3).

An approximate formula for the calculated spiral steel ratio in terms of physical properties of the column cross section may be derived from the preceding definition of ρ_s. In Figure 9-5, we denote the overall core diameter (out-to-out of spiral) as D_{ch} and the spiral diameter (center to center) as D_s. The cross-sectional area of the spiral bar or wire is denoted A_{sp}. From the definition of ρ_s:

$$\text{calculated } \rho_s = \frac{A_{sp}\pi D_s}{(\pi D_{ch}^2/4)(s)}$$

If the small difference between D_{ch} and D_s is neglected, then in terms of D_{ch},

$$\text{calculated } \rho_s = \frac{4A_{sp}}{D_{ch}s}$$

9-4 ANALYSIS OF SHORT COLUMNS: SMALL ECCENTRICITY

The analysis of short columns carrying axial loads that have small eccentricities involves checking the maximum design axial load strength and the various details of the reinforcing. The procedure is summarized in Section 9-6.

Example 9-1

Find the maximum design axial load strength for the tied column of cross section shown in Figure 9-6. Check the ties. Assume a short column. Use f'_c = 4000 psi and f_y = 60,000 psi for both longitudinal steel and ties.

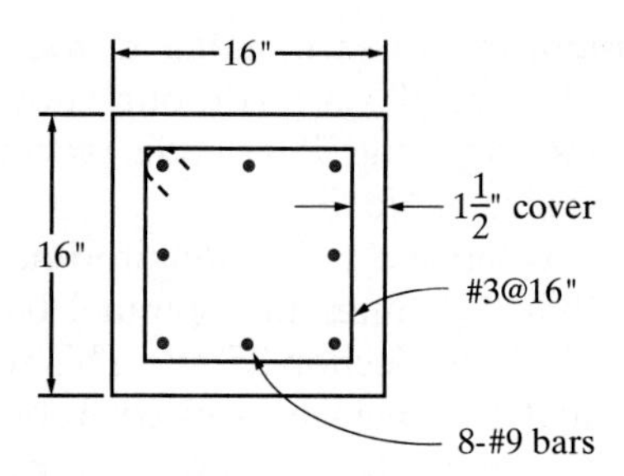

FIGURE 9-6 Sketch for Example 9-1.

Solution:

1. Check the steel ratio for the longitudinal steel:

$$\rho_g = \frac{A_{st}}{A_g} = \frac{8.00}{(16)^2} = 0.0313$$

$$0.01 < 0.0313 < 0.08 \qquad \text{(O.K.)}$$

2. From Table A-14, using a 13-in. core (column size less cover on each side), the maximum number of No. 9 bars is eight. Therefore, the number of longitudinal bars is satisfactory.
3. The maximum design axial load strength may now be calculated:

$$\begin{aligned}\phi P_{n(\max)} &= 0.80\phi[0.85f'_c(A_g - A_{st}) + f_y A_{st}]\\ &= 0.80(0.65)[0.85(4)(256 - 8) + 60(8)]\\ &= 688 \text{ kip}\end{aligned}$$

4. Check the ties. Tie size of No. 3 is acceptable for longitudinal bar size up to No. 10. The spacing of the ties must not exceed the smaller of

$$\begin{aligned}\text{48 tie-bar diameters} &= 48\left(\tfrac{3}{8}\right) = 18 \text{ in.}\\ \text{16 longitudinal-bar diameters} &= 16(1.128) = 18 \text{ in.}\\ \text{least column dimension} &= 16 \text{ in.}\end{aligned}$$

Therefore, the tie spacing is O.K. The tie arrangement for this column may be checked by ensuring that the clear distance between longitudinal bars does not exceed 6 in. Clear space in excess of 6 in. would require additional ties in accordance with the ACI Code, Section 25.7.2.3. Thus

$$\text{clear distance} = \frac{16 - 2\left(1\tfrac{1}{2}\right) - 2\left(\tfrac{3}{8}\right) - 3(1.128)}{2}$$

$$= 4.4 \text{ in.} < 6 \text{ in.}$$

Therefore, no extra ties are needed.

Example 9-2

Determine whether the spiral column of cross section shown in Figure 9-7 is adequate to carry a factored axial load (P_u) of 540 kip. Assume small eccentricity. Check the spiral. Use f'_c = 4000 psi and f_y = 60,000 psi.

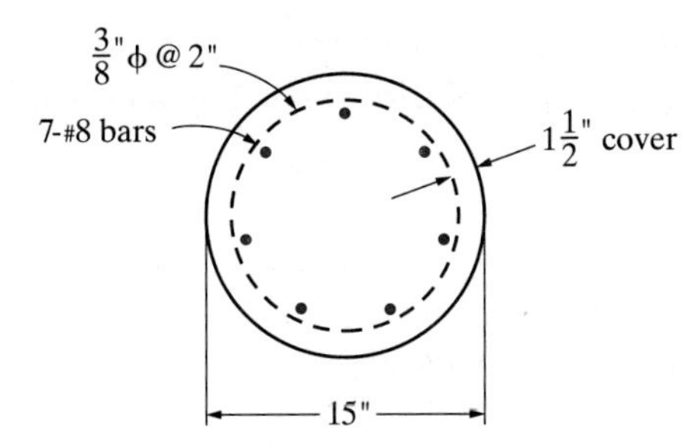

FIGURE 9-7 Sketch for Example 9-2.

Solution:

1. From Table A-2, $A_{st} = 5.53$ in.^{2}, and from Table A-14, a diameter of 15 in. results in a circular area $A_g = 176.7$ in.^{2}. Therefore,

$$\rho_g = \frac{5.53}{176.7} = 0.0313$$

$$0.01 < 0.0313 < 0.08 \qquad \text{(O.K.)}$$

2. From Table A-14, using a 12-in. core, seven No. 8 bars are satisfactory.
3. Find the maximum design axial load strength:

$$\begin{aligned} \phi P_{n(\max)} &= 0.85\phi[0.85f'_c(A_g - A_{st}) + f_y A_{st}] \\ &= 0.85(0.75)[0.85(4)(176.7 - 5.53) + 60(5.53)] \\ &= 583 \text{ kip} \end{aligned}$$

The strength is O.K. because 583 kip > 540 kip.

4. Check the spiral steel. The $\frac{3}{8}$-in. diameter is O.K. (ACI Code, Section 25.7.3.2; and Table A-14). The minimum ρ_s is calculated using Table A-14 for the value of A_{ch}:

$$\begin{aligned} \rho_{s(\min)} &= 0.45\left(\frac{A_g}{A_{ch}} - 1\right)\frac{f'_c}{f_{yt}} \\ &= 0.45\left(\frac{176.7}{113.1} - 1\right)\frac{4}{60} = 0.0169 \end{aligned}$$

$$\text{actual } \rho_s = \frac{4A_{sp}}{D_{ch}s} = \frac{4(0.11)}{12(2)} = 0.0183$$

$$0.0183 > 0.0169 \qquad \text{(O.K.)}$$

The clear distance between spirals must not be in excess of 3 in. nor less than 1 in.:

$$\text{clear distance} = 2.0 - \tfrac{3}{8} = 1\tfrac{5}{8} \text{ in.} \qquad \text{(O.K.)}$$

The column is satisfactory for the specified conditions.

9-5 DESIGN OF SHORT COLUMNS: SMALL ECCENTRICITY

The design of reinforced concrete columns involves the proportioning of the steel and concrete areas and the selection of properly sized and spaced ties or spirals. Because the ratio of steel to concrete area must fall within a given range ($0.01 \le \rho_g \le 0.08$), the strength equation given in Section 9-2 is modified to include this term. For a tied column,

$$\phi P_{n(\max)} = 0.80\phi[0.85f'_c(A_g - A_{st}) + f_y(A_{st})]$$

$$\rho_g = \frac{A_{st}}{A_g}$$

from which

$$A_{st} = \rho_g A_g$$

Therefore,

$$\begin{aligned} \phi P_{n(\max)} &= 0.80\phi[0.85f'_c(A_g - \rho_g A_g) + f_y \rho_g A_g] \\ &= 0.80\phi A_g[0.85f'_c(1 - \rho_g) + f_y \rho_g] \end{aligned}$$

As

$$P_u \le \phi P_{n(\max)}$$

an expression can be written for required A_g in terms of the material strengths, P_u and ρ_g. For tied columns,

$$\text{required } A_g = \frac{P_u}{0.80\phi[0.85f'_c(1 - \rho_g) + f_y \rho_g]}$$

Similarly for spiral columns,

$$\text{required } A_g = \frac{P_u}{0.85\phi[0.85f'_c(1 - \rho_g) + f_y \rho_g]}$$

It should be recognized that there can be many valid choices for the size of column that will provide the necessary strength to carry any load P_u. A low ρ_g will result in a larger required A_g and vice versa. Other considerations will normally affect the practical choice of column size. Among them are architectural requirements and the desirability of maintaining column size from floor to floor so that forms may be reused.

The procedure for the design of short columns for loads at small eccentricities is summarized in Section 9-6.

Example 9-3

Design a square tied column to carry axial service loads of 320 kip dead load and 190 kip live load. There is no identified applied moment. Assume that the column is short. Use ρ_g about 0.03, $f'_c = 4000$ psi, and $f_y = 60{,}000$ psi.

Solution:

1. Material strengths and approximate ρ_g are given.
2. The factored axial load is

$$P_u = 1.6(190) + 1.2(320) = 688 \text{ kip}$$

3. The required gross column area is

$$\begin{aligned} \text{required } A_g &= \frac{P_u}{0.80\phi[0.85f'_c(1 - \rho_g) + f_y \rho_g]} \\ &= \frac{688}{0.80(0.65)[0.85(4)(1 - 0.03) + 60(0.03)]} \\ &= 260 \text{ in.}^2 \end{aligned}$$

4. The required size of the square column would be

$$\sqrt{260} = 16.1 \text{ in.}$$

Use a 16-in.-square column. This choice will require that the actual ρ_g be slightly in excess of 0.03:

$$\text{actual } A_g = (16 \text{ in.})^2 = 256 \text{ in.}^2$$

5. The load on the concrete area (this is approximate since ρ_g will increase slightly) is

$$\begin{aligned}\text{load on concrete} &= 0.80\phi(0.85f'_c)A_g(1 - \rho_g)\\ &= 0.80(0.65)(0.85)(4)(256)(1 - 0.03)\\ &= 439 \text{ kip}\end{aligned}$$

Therefore, the load to be carried by the steel is

$$688 - 439 = 249 \text{ kip}$$

Because the maximum design axial load strength of the steel is ($0.80\ \phi A_{st} f_y$), the required steel area may be calculated as

$$\text{required } A_{st} = \frac{249}{0.80(0.65)(60)} = 7.98 \text{ in.}^2$$

We will distribute bars of the same size evenly around the perimeter of the column and must, therefore, select bars in multiples of four. Use eight No. 9 bars ($A_{st} = 8.0 \text{ in.}^2$). Table A-14 indicates a maximum of eight No. 9 bars for a 13-in. core (O.K.).

6. Design the ties. From Table A-14, select a No. 3 tie. The spacing must not be greater than

$$\begin{aligned}\text{48 tie-bar diameters} &= 48\left(\tfrac{3}{8}\right) = 18 \text{ in.}\\ \text{16 longitudinal-bar diameters} &= 16(1.128) = 18.0 \text{ in.}\\ \text{least column dimension} &= 16 \text{ in.}\end{aligned}$$

Use No. 3 ties spaced 16 in. o.c. Check the arrangement with reference to Figure 9-8. The clear space between adjacent bars in the same face is

$$\frac{16 - 3 - 0.75 - 3(1.13)}{2} = 4.43 \text{ in.} < 6.0 \text{ in.}$$

Therefore, no additional ties are required by the ACI Code, Section 25.7.2.3.

7. The design sketch is shown in Figure 9-8.

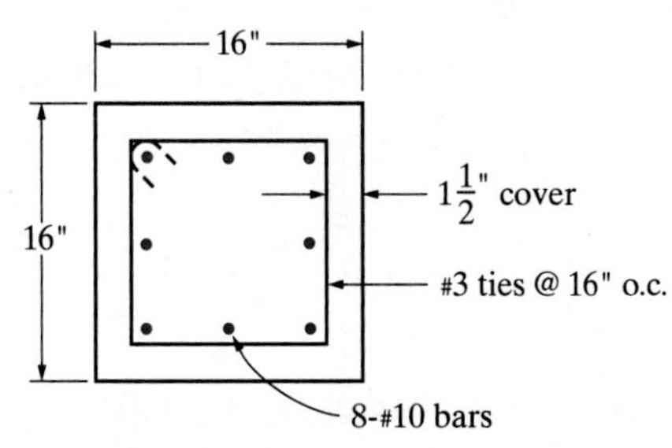

FIGURE 9-8 Design sketch for Example 9-3.

Example 9-4

Redesign the column of Example 9-3 as a circular, spirally reinforced column.

Solution:

1. Use $f'_c = 4000$ psi, $f_y = 60{,}000$ psi, and ρ_g approximately 0.03.
2. $P_u = 688$ kip, as in Example 9-3.
3.

$$\begin{aligned}\text{required } A_g &= \frac{P_u}{0.80\phi[0.85f'_c(1 - \rho_g) + f_y\rho_g]}\\ &= \frac{688}{0.80(0.75)[0.85(4)(1 - 0.03) + 60(0.03)]}\\ &= 212 \text{ in.}^2\end{aligned}$$

4. From Table A-14, use an 18-in.-diameter column. Thus

$$A_g = 254.5 \text{ in.}^2$$

5.

$$\begin{aligned}\text{Load on concrete} &= 0.85\phi(0.85)f'_cA_g(1 - \rho_g)\\ &= 0.85(0.75)(0.85)(4)(254.5)(1 - 0.03)\\ &= 535 \text{ kip}\\ \text{load on steel} &= 688 - 535 = 153 \text{ kip}\\ \text{required } A_{st} &= \frac{153}{0.85\phi f_y}\\ &= \frac{153}{0.85(0.75)60} = 4.00 \text{ in.}^2\end{aligned}$$

Use seven No. 7 bars ($A_{st} = 4.20 \text{ in.}^2$). Table A-14 indicates a maximum of 13 No. 7 bars for a circular core 15 in. in diameter (O.K.).

6. Design the spiral. From Table A-14, select a $\frac{3}{8}$-in. -diameter spiral. The spacing will be based on the required spiral steel ratio. Here A_{ch} is from Table A-14. Thus

$$\begin{aligned}\rho_{s(\min)} &= 0.45\left(\frac{A_g}{A_{ch}} - 1\right)\frac{f'_c}{f_{yt}}\\ &= 0.45\left(\frac{254.5}{176.7} - 1\right)\frac{4}{60} = 0.0132\end{aligned}$$

The maximum spiral spacing may be found by setting the calculated ρ_s equal to $\rho_{s(\min)}$:

$$\text{calculated } \rho_s = \frac{4A_{sp}}{D_{ch}s}$$

from which

$$\begin{aligned}s_{\max} &= \frac{4A_{sp}}{D_{ch}\rho_{s(\min)}}\\ &= \frac{4(0.11)}{15(0.0132)} = 2.22 \text{ in.}\end{aligned}$$

Use a spiral spacing of 2 in. The clear space between spirals must not be less than 1 in. nor more than 3 in.:

$$\text{clear space} = 2 - \tfrac{3}{8} = 1\tfrac{5}{8} \text{ in.} \qquad \text{(O.K.)}$$

7. The design sketch is shown in Figure 9-9.

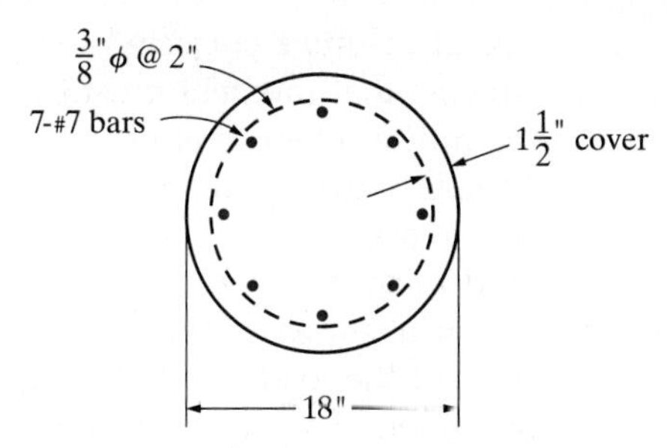

FIGURE 9-9 Design sketch for Example 9-4.

9-6 SUMMARY OF PROCEDURE FOR ANALYSIS AND DESIGN OF SHORT COLUMNS WITH SMALL ECCENTRICITIES

Analysis

1. Check ρ_g within acceptable limits:

$$0.01 \leq \rho_g \leq 0.08$$

The above maximum column reinforcing ratio of $\rho_{g(maximum)}$ of 0.08 assumes the use of mechanical splices for the column longitudinal reinforcement which ensures that there will be no congestion of the rebar within the column. However, where lap splices are used, the maximum column reinforcing ratio is cut in half to 0.04 since the area of steel at each spliced section is doubled if all the rebars are spliced at the same location.

2. Check the number of bars within acceptable limits for the clear space (see Table A-14). The minimum number is four for bars with rectangular or circular ties and six for bars enclosed by spirals.
3. Calculate the maximum design axial load strength $\phi P_{n(\max)}$. See Section 9-2.
4. Check the lateral reinforcing. For ties, check size, spacing, and arrangement. For spirals, check size, ρ_s, and clear distance.

Design

1. Establish the material strengths. Establish the desired ρ_g (if any).
2. Establish the factored axial load P_u.
3. Determine the required gross column area A_g.
4. Select the column dimensions. Use full-inch increments.
5. Find the load carried by the concrete and the load required to be carried by the longitudinal steel. Determine the required longitudinal steel area. Select the longitudinal steel.
6. Design the lateral reinforcing (ties or spiral).
7. Sketch the design.

9-7 THE LOAD-MOMENT RELATIONSHIP

The equivalency between an eccentrically applied load and an axial load-moment combination is shown in Figure 9-10. Assume that a force P_u is applied to a cross section at a distance e (eccentricity) from the centroid, as shown in Figure 9-10a and b. Add equal and opposite forces P_u at the centroid of the cross section (Figure 9-10c). The original eccentric force P_u may now be combined with the upward force P_u to form a couple, P_ue, that is a pure moment. This will leave remaining one force, P_u, acting downward at the centroid of the cross section. It can, therefore, be seen that if a force P_u is applied with an eccentricity e, the situation that results is identical to the case where an axial load of P_u at the centroid and a moment of P_ue are simultaneously applied (Figure 9-10d). If we define M_u as the factored moment to be applied on a *compression* member along with a factored axial load of P_u at the centroid, the relationship between the two is

$$e = \frac{M_u}{P_u}$$

We have previously spoken of a column's nominal axial load strength at zero eccentricity P_0. Temporarily neglecting the strength reduction factors that must be applied to P_0, let us assume that we may apply a load P_u with zero eccentricity on a column of nominal axial load strength P_0 (where $P_u = P_0$). If we now move the load P_u away from the zero eccentricity position a distance of e, the column must resist the load P_u and, in addition, a moment P_ue. Because $P_u = P_0$ (in our

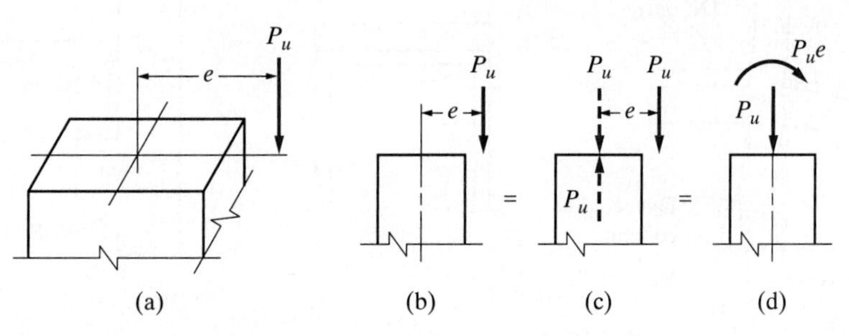

FIGURE 9-10 Load-moment-eccentricity relationship.

hypothetical case), it is clear that there is no additional strength to carry the moment and that the column is overloaded. For this particular column to not be overloaded when subjected to load at eccentricity e, we must reduce P_u to the point where the column can carry both P_u and $P_u e$. The amount of the required decrease in P_u will depend on the magnitude of the eccentricity.

The preceding discussion must be modified because the ACI Code imposes $\phi P_{n(\max)}$ as the upper limit of axial load strength for any column. Nevertheless, the strength of any column cross section is such that it will support a broad spectrum of load and moment (or load and eccentricity) combinations. In other words, we may think of a column cross section as having many different axial load strengths, each with its own related moment strength.

9-8 COLUMNS SUBJECTED TO AXIAL LOAD AT LARGE ECCENTRICITY

At one time the ACI Code stipulated that compression members be designed for an eccentricity e of not less than $0.05h$ for spirally reinforced columns or $0.10h$ for tied columns, but at least 1 in. in any case. Here h is defined as the overall dimension of the column. These specified minimum eccentricities were originally intended to serve as a means of reducing the axial load design strength of a section in pure compression. The effect of the minimum eccentricity requirement was to limit the *maximum* axial load strength of a compression member.

Under the 2014 ACI Code, as we have discussed, the maximum design axial load strength $\phi P_{n(\max)}$ is obtained from ACI Equation (22.4.2.2). These two equations apply when eccentricities are not in excess of *approximately* the $0.10h$ and $0.05h$ minimum eccentricity limits previously discussed. Therefore, *small eccentricities* may be considered as those eccentricities up to about $0.10h$ and $0.05h$ for tied and spiral columns, respectively. We will consider cases of *large eccentricities* as those in which ACI Equation (22.4.2.2) no longer apply and where ϕP_n must be reduced below $\phi P_{n(\max)}$.

The occurrence of columns subjected to eccentricities sufficiently large so that moment must be a design consideration is common. Even interior columns supporting beams of equal spans will receive unequal loads from the beams due to applied live load patterns. These unequal loads could mean that the column must carry both load and moment, as shown in Figure 9-11a, and the resulting eccentricity of the loads could be appreciably in excess of our definition of small eccentricity. Another example of a column carrying both load and moment is shown in Figure 9-11b. In both cases, the rigidity of the joint will require the column to rotate along with the end of the beam that it is supporting. The rotation will induce moment in the column. A third and very practical example can be found in precast work (Figure 9-11c), where the beam reaction can clearly be seen to be eccentrically applied on the column through the column bracket.

9-9 ϕ FACTOR CONSIDERATIONS

Columns discussed so far have had strength-reduction factors applied in a straightforward manner. That is, $\phi = 0.75$ for spiral columns, and $\phi = 0.65$ for tied columns. These ϕ factors correspond to the compression-controlled strain limit or a net tensile strain in the extreme tension reinforcement, $\epsilon_t \leq 0.002$. Eccentrically loaded columns, however, carry both axial load and moment. For values of ϵ_t larger than 0.002, the ϕ equations from ACI Code, Table 21.2.2, discussed in Chapter 2, will give higher values than indicated above. The ϕ equations are repeated here as follows:

Tied columns:

$$\phi = 0.65 + (\epsilon_t - 0.002)\left(\frac{250}{3}\right)$$

$$0.65 \leq \phi \leq 0.90$$

Spiral columns:

$$\phi = 0.75 + (\epsilon_t - 0.002)\left(\frac{200}{3}\right)$$

$$0.75 \leq \phi \leq 0.90$$

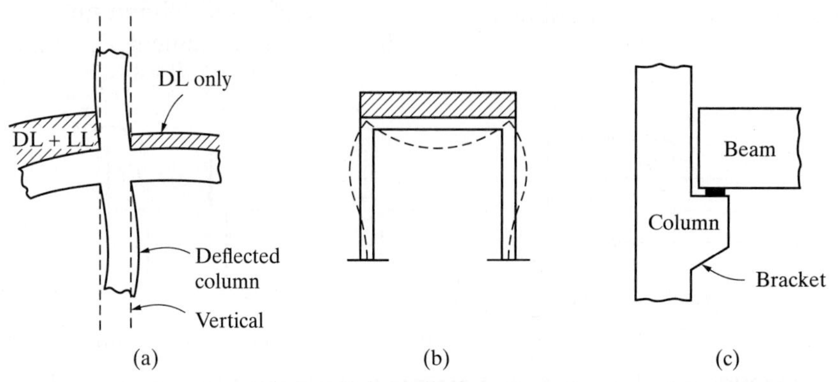

FIGURE 9-11 Eccentrically loaded columns.

9-10 ANALYSIS OF SHORT COLUMNS: LARGE ECCENTRICITY

The first step in our investigation of short columns carrying loads at large eccentricity is to determine the strength of a given column cross section that carries loads at various eccentricities. This may be thought of as an analysis process. For this development, we will find the design axial load strength ϕP_n, where P_n is defined as the nominal axial load strength at a given eccentricity.

Example 9-5

Find the design axial load strength ϕP_n for the tied column for the following conditions: (a) small eccentricity ($e = 0$ to $0.10\,h$); (b) $e = 5$ in.; (c) the balanced strain condition or compression-controlled strain limit, $\epsilon_t = 0.002$; (d) $\epsilon_t = 0.004$; (e) the tension-controlled strain limit, $\epsilon_t = 0.005$; and (f) pure moment. The column cross section is shown in Figure 9-12. Assume a short column. Bending is about the Y–Y axis. Use $f'_c = 4000$ psi and $f_y = 60{,}000$ psi.

Solution:

a. The analysis of the small eccentricity condition is similar to the analyses of Examples 9-1 and 9-2. We can calculate the design axial load strength from

$$\begin{aligned}\phi P_n &= \phi P_{n(\max)}\\ &= 0.80\phi[0.85f'_c(A_g - A_{st}) + f_y A_{st}]\\ &= 0.80(0.65)[0.85(4)(280 - 6) + 60(6)]\\ &= 672 \text{ kip}\end{aligned}$$

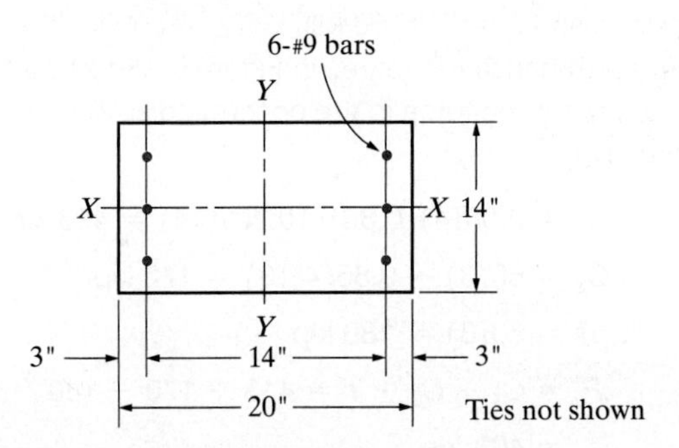

FIGURE 9-12 Column cross section for Example 9-5.

The corresponding maximum moment:

$$\phi M_n = \phi P_n e = 672(0.10)\left(\frac{20}{12}\right) = 112 \text{ ft.-kip}$$

b. The situation of $e = 5$ in. is shown in Figure 9-13. In part (a) of this example, all steel was in compression. As eccentricity increases, the steel on the side of the column away from the load is subjected to less compression. Therefore, there is some value of eccentricity at which this steel will change from compression to tension. Because this value of eccentricity is not known, the strain situation shown in Figure 9-14 is *assumed* and will be verified (or disproved) by calculation.

The assumptions at nominal strength are

1. Maximum concrete strain $= 0.003$.
2. $\epsilon'_s > \epsilon_y$. Therefore, $f'_s = f_y$.
3. ϵ_s is tensile.
4. $\epsilon_s < \epsilon_y$. Therefore, $f_s < f_y$.

The unknown quantities are P_n and c.

Using basic units of kip and inches, the tensile and compressive forces are evaluated. Force C_2 is the force in the compressive steel accounting for the concrete displaced by the steel. With reference to Figure 9-14c,

$$C_1 = 0.85f'_c ab = 0.85(4)(0.85c)(14) = 40.46c$$

$$\begin{aligned}C_2 &= f_y A'_s - 0.85f'_c A'_s = A'_s(f_y - 0.85f'_c)\\ &= 3[60 - 0.85(4)] = 169.8\end{aligned}$$

$$\begin{aligned}T &= f_s A_s = \epsilon_s E_s A_s = 87\left(\frac{d-c}{c}\right)A_s\\ &= 87\left(\frac{17-c}{c}\right)3 = 261\left(\frac{17-c}{c}\right)\end{aligned}$$

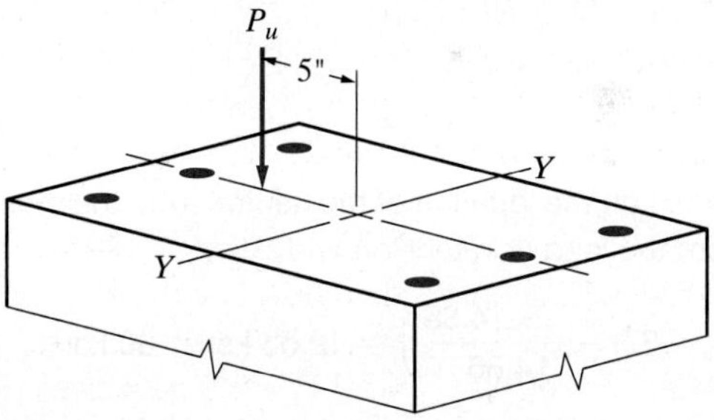

FIGURE 9-13 Example 9-5b, $e = 5$ in.

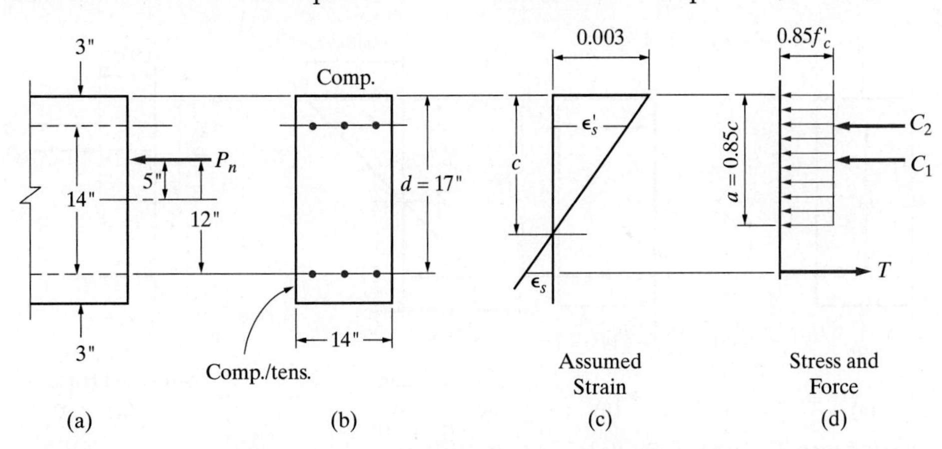

FIGURE 9-14 Stress and strain distribution for Example 9-5b, $e = 5$ in.

From Σ forces = 0 in Figure 9-14,

$$P_n = C_1 + C_2 - T$$
$$= 40.46c + 169.8 - 261\left(\frac{17 - c}{c}\right)$$

From Σ moments = 0, taking moments about T in Figure 9-14,

$$P_n(12) = C_1\left(d - \frac{a}{2}\right) + C_2(14)$$
$$P_n = \frac{1}{12}\left[40.46(c)\left(17 - \frac{0.85c}{2}\right) + 169.8(14)\right]$$

The preceding two equations for P_n may be equated and the resulting cubic equation solved for c by trial or by some other iterative method. The solution will yield $c = 14.86$ in., which will result in a value for P_n (from either equation) of 733 kip. The net tensile strain in the extreme tension reinforcement may be calculated from

$$\epsilon_t = 0.003\left(\frac{d - c}{c}\right) = 0.003\left(\frac{17 - 14.86}{14.86}\right)$$
$$= 0.00043 < 0.002$$

For $\epsilon_t \leq 0.002$, the corresponding tied column strength-reduction factor ϕ is 0.65. Therefore,

$$\phi P_n = 0.65(733)$$
$$= 476 \text{ kip}$$

Check the assumptions that were made:

$$\epsilon'_s = \left(\frac{14.86 - 3}{14.86}\right)(0.003) = 0.0024$$
$$\epsilon_y = 0.00207$$

Because $\epsilon'_s > \epsilon_y$,

$$f'_s = f_y \qquad \text{(O.K.)}$$

Based on the location of the neutral axis, the steel away from the load *is* in tension and

$$f_s = 87\left(\frac{17 - 14.86}{14.86}\right) = 12.53 \text{ ksi} < 60 \text{ ksi} \qquad \text{(O.K.)}$$

All assumptions are verified.

We may also determine the design moment strength for an eccentricity of 5 in. as follows:

$$\phi P_n e = \frac{476(5)}{12}$$
$$= 198 \text{ ft.-kip}$$

Therefore, the given column has a design load-moment combination strength of 476 kip axial load and 198 ft.-kip moment. This assumes that the moment is applied about the Y–Y axis.

c. The compression-controlled strain limit (balanced condition) exists when the concrete reaches a strain of 0.003 at the same time the extreme tension steel reaches a strain of 0.002, as shown in Figure 9-15c. Here P_b is defined as nominal axial load strength at the balanced condition, e_b is the associated eccentricity, and c_b is the distance from the compression face to the balanced neutral axis.

Using the strain diagram in Figure 9-15, we may calculate the value of c_b:

$$\frac{0.003}{c_b} = \frac{0.002}{17 - c_b}$$

from which $c_b = 0.6(17) = 10.2$ in.

For $\epsilon_t = 0.002$, the tied column strength-reduction factor ϕ is 0.65.

We then may determine ϵ_s:

$$\epsilon'_s = \frac{7.2}{10.2}(0.003) = 0.0021$$

Because 0.0021 > 0.00207, the compression steel has yielded and $f'_s = f_y = 60$ ksi.

Summarize the forces in Figure 9-15d and let C_2 account for the force in the concrete displaced by the three No. 9 bars:

$$C_1 = 0.85(4)(0.85)(10.20)(14) = 413 \text{ kip}$$
$$C_2 = 60(3) - 0.85(4)(3) = 170 \text{ kip}$$
$$T = 60(3) = 180 \text{ kip}$$
$$P_b = C_1 + C_2 - T = 413 + 170 - 180$$
$$= 403 \text{ kip}$$

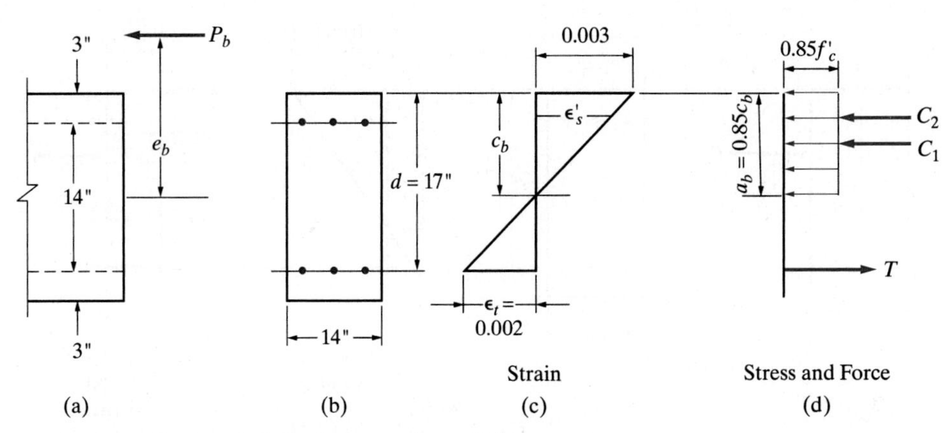

FIGURE 9-15 Example 9-5c, balanced condition.

The value of e_b may be established by summing moments about T:

$$P_b(e_b + 7) = C_1\left(d - \frac{0.85c_b}{2}\right) + C_2(14)$$

$$403(e_b + 7) = 413\left[17 - \frac{0.85(10.20)}{2}\right] + 170(14)$$

from which $e_b = 11.88$ in. Therefore, at the balanced condition,

$$\phi P_b = 0.65(403) = 262 \text{ kip}$$

$$\phi P_b e_b = \frac{262}{12}(11.88) = 259 \text{ ft.-kip}$$

d. Net tensile strain in the extreme tension steel, $\epsilon_t = 0.004$ and a corresponding compression strain of 0.003 in the compression face.

Using the strain diagram in Figure 9-16, the neutral axis depth, c is calculated as follows:

$$\frac{0.003}{c} = \frac{0.004}{17 - c}$$

from which $c = 0.429(17) = 7.29$ in.

We may then determine ϵ'_s:

$$\epsilon'_s = 0.003\,\frac{(7.29 - 3)}{7.29} = 0.00177 < \epsilon_y$$

$$f'_s = E\epsilon'_s = 29{,}000\ (0.00177) = 51.2 \text{ ksi}$$

Using similar equations from case (b), the forces in the concrete, compression steel, and tension steel are as follows:

$$C_1 = 0.85(4)(0.85)(7.29)(14) = 295 \text{ kip}$$

$$C_2 = 51.2(3) - (0.85)(4)(3) = 143 \text{ kip}$$

$$T = 60(3) = 180 \text{ kip}$$

From Σ forces $= 0$ in Figure 9-16,

$$P_n = C_1 + C_2 - T$$

$$= 295 + 143 - 180 = 258 \text{ kip}$$

From Σ moments $= 0$, taking moments about T in Figure 9-16,

$$P_n(e + 7) = C_1\left(d - \frac{0.85c}{2}\right) + C_2(14)$$

$$258(e + 7) = 295\left[d - \frac{0.85(7.29)}{2}\right] + 143(14)$$

from which $e = 16.7$ in.

For $\epsilon_t = 0.004$,

$$\phi = 0.65 + (0.004 - 0.002)\left(\frac{250}{3}\right) = 0.82$$

$$0.65 < 0.82 < 0.9 \qquad \text{(O.K.)}$$

Therefore, at $\epsilon_t = 0.004$,

$$\phi P_n = 0.82(258) = 212 \text{ kip}$$

$$\phi M_n = \phi P_n e = 212\left(\frac{16.7}{12}\right) = 295 \text{ ft.-kip}$$

e. Net tensile strain in the extreme tension steel ϵ_t is 0.005, and a corresponding compression strain of 0.003 exists in the compression face.

Using the strain diagram in Figure 9-16, the neutral axis depth, c, is calculated as follows:

$$\frac{0.003}{c} = \frac{0.005}{17 - c}$$

from which $c = 0.375(17) = 6.38$ in.

We may then determine ϵ'_s:

$$\epsilon'_s = 0.003\left(\frac{6.38 - 3}{6.38}\right) = 0.00159 < \epsilon_y$$

$$f'_s = E\epsilon'_s = 29{,}000\ (0.00159) = 46.1 \text{ ksi}$$

Using similar equations from case (b), the forces in the concrete, compression steel, and tension steel are as follows:

$$C_1 = 0.85(4)(0.85)(6.38)(14) = 258 \text{ kip}$$

$$C_2 = 46.1(3) - (0.85)(4)(3) = 128 \text{ kip}$$

$$T = 60(3) = 180 \text{ kip}$$

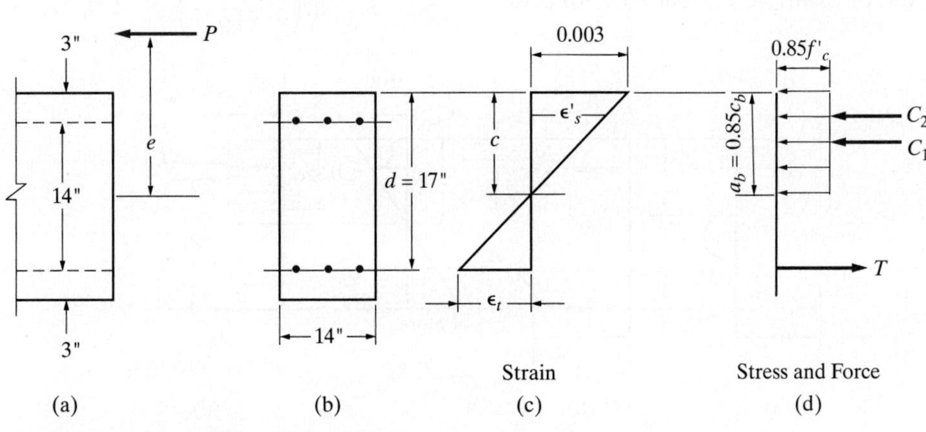

FIGURE 9-16 Examples 9-5d and 9-5e.

From Σ forces = 0 in Figure 9-16,

$$P_n = C_1 + C_2 - T$$
$$= 258 + 128 - 180 = 206 \text{ kip}$$

From Σ moments = 0, taking moments about T in Figure 9-16,

$$P_n(e + 7) = C_1\left(d - \frac{0.85c}{2}\right) + C_2(14)$$

$$206(e + 7) = 258\left[d - \frac{0.85(6.38)}{2}\right] + 128(14)$$

from which e = 19.6 in.

For $\epsilon_t \geq 0.005$, $\phi = 0.90$. Therefore, at $\epsilon_t = 0.005$,

$$\phi P_n = 0.90(206) = 185 \text{ kip}$$

$$\phi M_n = \phi P_n e = 185\left(\frac{19.6}{12}\right) = 302 \text{ ft.-kip}$$

f. The analysis of the pure moment condition is similar to the analysis of the case where eccentricity e is infinite, shown in Figure 9-17. We will find the design moment strength ϕM_n, because P_u and ϕP_n will both be zero.

With reference to Figure 9-18d, notice that for pure moment the bars on the load side of the column are in compression, whereas the bars on the side away from the load are in tension. The total tensile and compressive forces must be equal to each other. Since $A_s = A'_s$, A'_s *must* be at a stress less than yield. Assume that A_s is at yield stress. Then

C_1 = concrete compressive force
C_2 = steel compressive force
T = steel tensile force

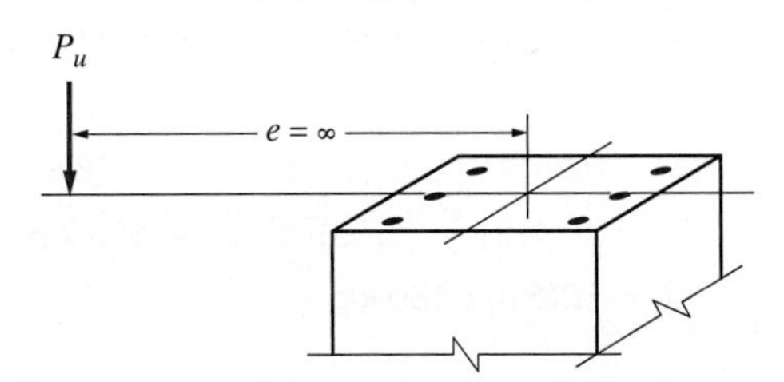

FIGURE 9-17 Column of Example 9-5 loaded with pure moment.

Referring to the compressive strain diagram portion of Figure 9-18c and noting that the basic units are kip and inches,

$$\epsilon'_s = 0.003\frac{(c - 3)}{c}$$

Because

$$f'_s = E_s\epsilon'_s$$

substituting yields

$$f'_s = 29{,}000(0.003)\frac{(c - 3)}{c}$$
$$= 87\frac{(c - 3)}{c}$$

For equilibrium in Figure 9-18d,

$$C_1 + C_2 = T$$

Substituting into the foregoing and accounting for the concrete displaced by the compression steel, we obtain

$$(0.85f'_c)(0.85c)(b) + f'_sA'_s - 0.85f'_cA'_s = f_yA_s$$

$$(0.85)(4)(0.85c)(14) + 87\left(\frac{c - 3}{c}\right)(3) - 0.85(4)(3) = 3(60)$$

Solving the preceding equation for the one unknown quantity c yields

$$c = 3.62 \text{ in.}$$

The net tensile strain in the extreme tension reinforcement may be calculated from

$$\epsilon_t = 0.003\left(\frac{d - c}{c}\right) = 0.003\left(\frac{17 - 3.62}{3.62}\right) = 0.011$$

For $\epsilon_t \geq 0.005$, the corresponding strength reduction factor ϕ is 0.90.

Therefore,

$$f'_s = 87\left(\frac{3.62 - 3}{3.62}\right) = 14.90 \text{ ksi} \qquad \text{(compression)}$$

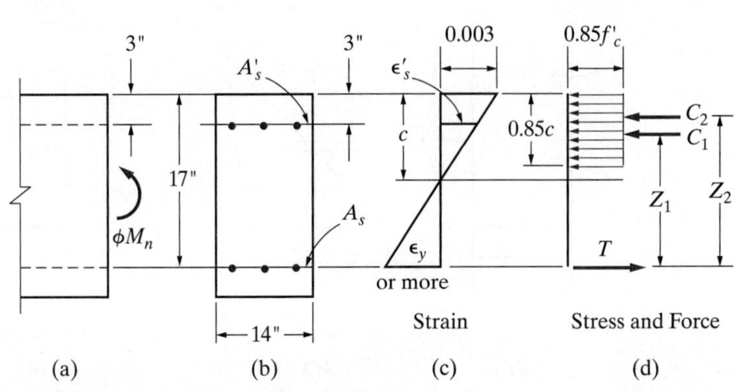

FIGURE 9-18 Example 9-5f, $e = \infty$.

Summarizing the forces,

$$C_1 = 0.85f_c'(0.85)cb = 0.85(4)(0.85)(3.62)(14) = 146.5 \text{ kip}$$

$$\text{displaced concrete} = 0.85f_c'A_s' = 0.85(4)(3) = -10.2 \text{ kip}$$

$$C_2 = f_s'A_s' = 14.90(3) = \underline{44.7 \text{ kip}}$$

$$181.0 \text{ kip}$$

$$T = f_yA_s = 60(3.0) = 180 \text{ kip}$$

The slight error between T and $(C_1 + C_2)$ will be neglected.

Summarizing the internal couples,

$$M_{n1} = C_1Z_1$$

$$= \frac{146.5}{12}\left[17 - \frac{0.85(3.62)}{2}\right]$$

$$= 188.8 \text{ ft.-kip}$$

$$M_{n2} = C_2Z_2$$

$$= (44.7 - 10.2)\left(\frac{14}{12}\right)$$

$$= 40.3 \text{ ft.-kip}$$

$$M_n = M_{n1} + M_{n2}$$

$$= 188.8 + 40.3$$

$$= 229 \text{ ft.-kip}$$

The design moment strength becomes

$$\phi M_n = 0.90(229)$$

$$= 206 \text{ ft.-kip}$$

The results of the six parts of Example 9-5 are tabulated (see Table 9-1) and plotted in Figure 9-19. All design axial load strengths are denoted ϕP_n, and all design moment strengths are denoted $\phi P_n e$. This plot is commonly called an *interaction diagram*. It applies *only* to the column analyzed, but it is a representation of *all* combinations of axial load and moment strengths for that column cross section.

In Figure 9-19, any point *on* the solid line represents an allowable combination of load and moment. Any point *within* the solid line represents a load-moment combination that is also allowable, but for which this column is *overdesigned*. Any point *outside* the solid line represents an unacceptable load-moment combination or a load-moment combination for which this column is *underdesigned*. The value of $\phi P_{n(\max)}$, which we calculated in part (a), is superimposed on the plot as the horizontal line.

Radial lines from the origin represent various eccentricities. (Actually, the slopes of the radial lines are equal to $\phi P_n/\phi P_n e$ or $1/e$.) The intersection of the $e = e_b$ line with the solid line represents the balanced condition. Any eccentricity *less* than e_b will result in compression controlling the column. For eccentricities between e_b (11.88 in.) and 19.6 in., the column will be in the transition zone ($0.002 \le \epsilon_t \le 0.005$). For eccentricities greater than 19.6 in., the column will be tension-controlled.

The calculations involved with column loads at large eccentricities are involved and tedious. The previous examples were analysis examples. Design of a cross section using the calculation approach would be a trial-and-error method and would become exceedingly tedious. Therefore, design and analysis aids have been developed that shorten the process to a great extent. These aids may be found in the form of tables and charts. A chart approach is developed in ACI Publication SP-17DA(14), *ACI Design Handbook*—Design Aids [1]. The design aids are based on the assumptions of ACI 318-14 and on the principles of static equilibrium; they are developed in a fashion similar to what was done in Example 9-5. No ϕ factors are incorporated into the diagrams. Eight interaction diagrams are included in Appendix A (Diagrams A-15 through A-22.)

The diagrams take on the general form of Figure 9-19 but are generalized to be applicable to more situations. Referring to Diagram A-15, which corresponds to our Figure 9-19, the following definitions will be useful:

$$\rho_g = \frac{A_{st}}{A_g}$$

h = column dimension perpendicular to the bending axis (see the sketch included on Diagram A-15)

γ = ratio of distance between centroids of outer rows of bars and column dimension perpendicular to the bending axis

TABLE 9-1 Column Axial Load-Moment Interaction for Example 9-5

Eccentricity, e	Net tensile strain in extreme tension steel, ϵ_t	Strength-reduction factor ϕ	Axial load strength, (ϕP_n, kip)	Moment strength, ($\phi P_n e$, ft.-kip)
Small		0.65	672	112
(i.e., 0 to 0.10 h) 5″	0.00043	0.65	476	198
11.88″ (balanced)	0.002	0.65	262	259
16.7″	0.004	0.82	212	295
19.6″	0.005	0.90	185	302
Infinite (pure moment)	>>0.005	0.90	0	206

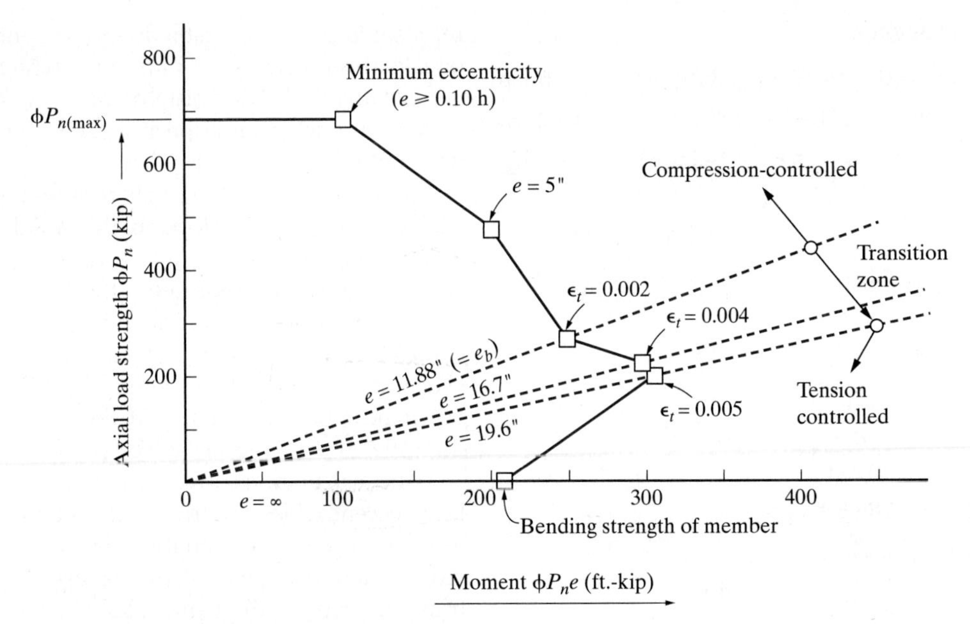

FIGURE 9-19 Column interaction diagram.

Note that the vertical axis and the horizontal axis are in general terms of K_n and R_n, where

$$K_n = \frac{P_n}{f'_c A_g}$$

$$R_n = \frac{P_n e}{f'_c A_g h}$$

Note also that P_n and $P_n e$ are nominal axial load strength and nominal moment strength. The slope of a radial line from the origin can be represented as follows:

$$\text{slope} = \frac{\text{rise}}{\text{run}} = \frac{\left(\dfrac{P_n}{f'_c A_g}\right)}{\left(\dfrac{P_n e}{f'_c A_g h}\right)} = \frac{h}{e}$$

Curves are shown for the range of allowable ρ_g values from 0.01 to 0.08. A line near the horizontal axis labeled $\epsilon_t = 0.0050$ indicates the limit for tension-controlled sections. Columns with load-moment-strength combinations below this line are tension-controlled ($\phi = 0.90$). The line labeled $f_s/f_y = 1.0$ indicates the balanced condition. Columns with load-moment-strength combinations above this line are compression-controlled ($\phi = 0.65$ for tied columns; 0.75 for spiral columns). Columns with load-moment-strength combinations between these two lines are in the transition zone. The line labeled $K_{\max}$ indicates the maximum allowable nominal load strength $[\phi P_{n(\max)}]$ for columns loaded with small eccentricities. A horizontal line drawn through the intersection of the $K_{\max}$ line and a ρ_g curve corresponds to the horizontal line near the top of the interaction diagram in Figure 9-19.

The following three examples illustrate the use of the ACI interaction diagrams for analysis and design of short reinforced concrete columns.

Example 9-6

Using the interaction diagrams of Appendix A, find the axial load strength ϕP_n and the moment strength ϕM_n for the column cross section with six No. 9 bars, as shown in Figure 9-20. Eccentricity $e = 5$ in., and use $f'_c = 4000$ psi and $f_y = 60{,}000$ psi. Compare the results with Example 9-5b.

Solution:

First, determine which interaction diagram to use, based on the type of cross section, the material strengths, and the factor γ.

$$\gamma h = 14 \text{ in.}$$

$$\gamma = \frac{14}{20} = 0.7$$

Therefore, use Interaction Diagram A-15.

$$\rho_g = \frac{6.00}{14(20)} = 0.0214$$

$$0.01 \leq 0.0214 \leq 0.08 \qquad \text{(O.K.)}$$

Next, calculate the slope of the radial line from the origin, which relates h and e:

$$\text{slope} = \frac{h}{e} = \frac{20}{5} = 4$$

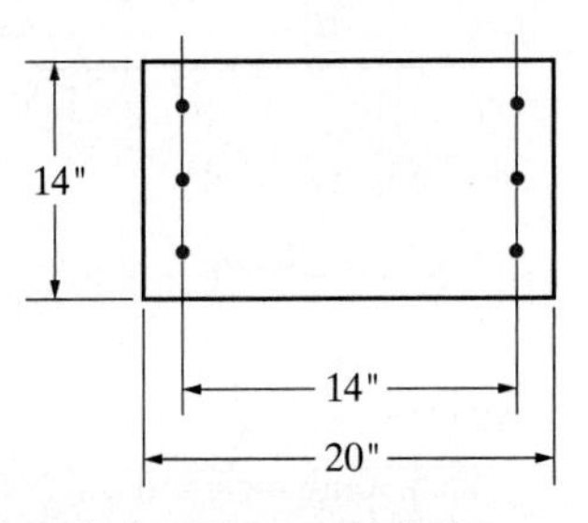

FIGURE 9-20 Sketch for Example 9-6.

A straight edge and some convenient values (e.g., $K_n = 1.0$ and $R_n = 0.25$) may be used to intersect this radial line with an estimated $\rho_g = 0.0214$ curve. At this intersection, we read $K_n \approx 0.64$ and $R_n \approx 0.16$. Because this combination of load and moment is above the $f_s/f_y = 1.0$ line, this is a compression-controlled section and $\phi = 0.65$.

$$\phi P_n = \phi K_n f_c' A_g$$
$$= 0.65\,(0.64)(4)(20)(14) = 466 \text{ kip}$$
$$\phi M_n = \phi R_n f_c' A_g h$$
$$= \frac{0.65(0.160)(4)(20)(14)(20)}{12 \text{ in./ft}} = 194 \text{ ft.-kip}$$

or

$$\phi M_n = \phi P_n e = \frac{466 \text{ ft.-kip (5 in.)}}{12 \text{ in./ft}} = 194 \text{ ft.-kip}$$

This compares reasonably well with the results of Example 9-5b: $\phi P_n = 476$ kip and $\phi M_n = 198$ ft.-kip.

Example 9-7

Design a circular spirally reinforced concrete column to support a design load $P_u = 1100$ kip and a design moment $M_u = 285$ ft.-kip. Use $f_c' = 4000$ psi and $f_y = 60{,}000$ psi.

Solution:
Estimate the column size required based on $\rho_g = 1\%$ and axial load only.

$$\text{required } A_g = \frac{P_u}{0.85\phi[0.85f_c'(1-\rho_g) + f_y\rho_g]}$$
$$= \frac{1100}{0.85\,(0.75)[0.85(4)(0.99) + 60(0.01)]}$$
$$= 435 \text{ in.}^2$$

Try a 24-in.-diameter column ($A_g = 452$ in.^{2}).

If No. 9 bars are eventually chosen (refer to Figure 9-21),

$$\gamma h = 24 - 2(1\tfrac{1}{2}) - 2\left(\frac{3}{8}\right) - 1.13 = 19.12 \text{ in.}$$

$$\gamma = \frac{19.12}{h} = \frac{19.12}{24} = 0.797$$

Therefore, use Diagram A-21 from Appendix A (ACI Interaction Diagram C4-60.8). Next, determine the required ρ_g.

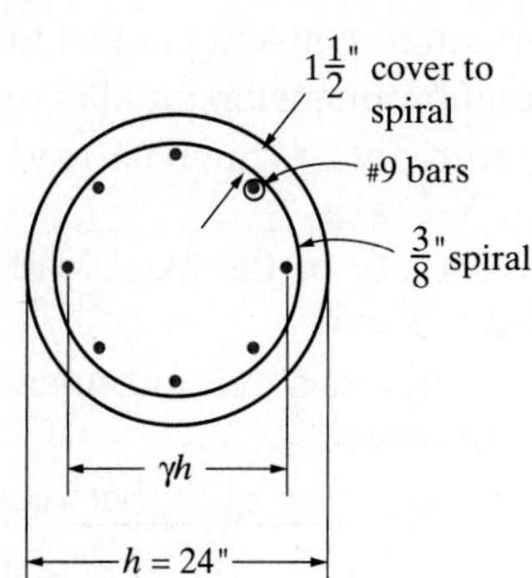

FIGURE 9-21 Sketch for Example 9-7.

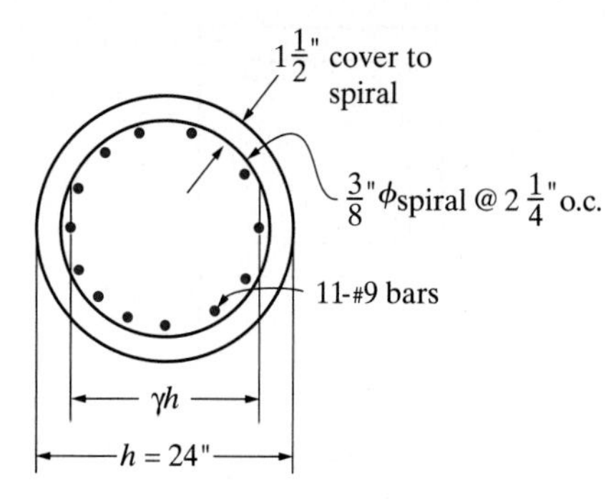

FIGURE 9-22 Design sketch for Example 9-7.

Assume that this column will be compression-controlled ($\phi = 0.75$) subject to later check.

Recognizing that required $P_n = P_u/\phi$ and required $P_n e = M_u/\phi$, we can calculate the values of required K_n and R_n:

$$\text{required } K_n = \frac{P_u}{\phi f_c' A_g} = \frac{1100}{0.75(4)(452)} = 0.811$$

$$\text{required } R_n = \frac{M_u}{\phi f_c' A_g h} = \frac{285(12)}{0.75(4)(452)(24)} = 0.105$$

From Diagram A-21, $\rho_g = 0.024$. Note that this is well above the $f_s/f_y = 1.0$ line; therefore, the column is compression-controlled and the assumption that $\phi = 0.75$ is O.K.

$$\text{required } A_s = \rho_g A_g = 0.024(452) = 10.85 \text{ in.}^2$$

Select 11 No. 9 bars ($A_s = 11.00$ in.^{2}). Check the maximum number of No. 9 bars from Table A-14: 15 (O.K.).

Design the spiral. Use a $\frac{3}{8}$-in.-diameter spiral.

The concrete core diameter is $D_{ch} = 24 - 2(1\frac{1}{2}) = 21$ in.

$$\text{required } \rho_s = 0.45\left(\frac{A_g}{A_{ch}} - 1\right)\frac{f_c'}{f_{yt}} = 0.45\left(\frac{452}{346} - 1\right)\frac{4}{60}$$
$$= 0.0092$$

$$\text{required s} = \frac{4A_{sp}}{D_{ch}\rho_s} = \frac{4(0.11)}{21(0.0092)} = 2.27$$

Use $2\frac{1}{2}$-in. spacing. The design is shown in Figure 9-22.

Example 9-8

Design a square-tied reinforced concrete column to support a design load $P_u = 1300$ kip and a design moment $M_u = 550$ ft.-kip. Use $f_c' = 4000$ psi and $f_y = 60{,}000$ psi.

Solution:
Estimate the column size required based on $\rho_g = 1\%$ and axial load only.

$$\text{required A}_g = \frac{P_u}{0.80\,\phi[0.85\,f_c'\,(1-\rho_g) + f_y\rho_g]}$$
$$= \frac{1300}{0.80\,(0.65)[0.85(4)(0.99) + 60(0.01)]}$$
$$= 631 \text{ in.}^2$$

Try a 26-in.-square column ($A_g = 676$ in.^{2}).

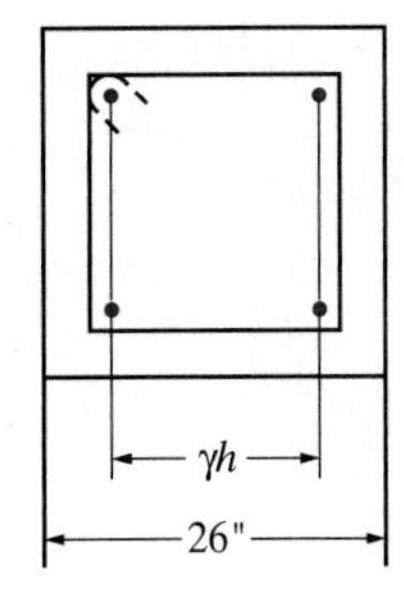

FIGURE 9-23 Sketch for Example 9-8.

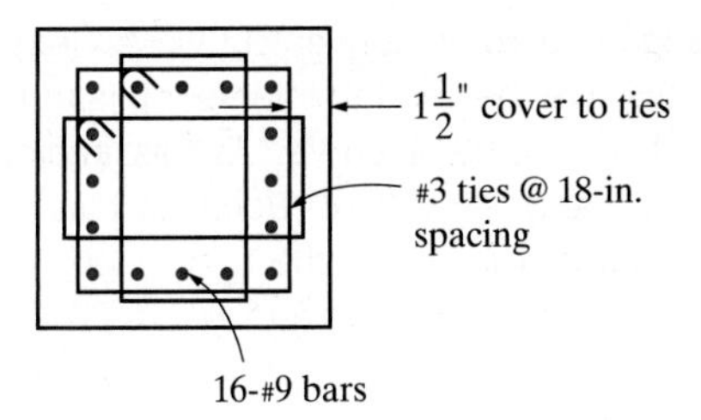

FIGURE 9-24 Design sketch for Example 9-8.

If No. 9 bars are eventually chosen (refer to Figure 9-23):

$$\gamma h = 26 - 2\left(1\tfrac{1}{2}\right) - 2\left(\frac{3}{8}\right) - 1.13 = 21.12 \text{ in.}$$

$$\gamma = \frac{21.12}{h} = \frac{21.12}{26} = 0.812$$

Therefore, use Diagram A-18 from Appendix A (ACI Interaction Diagram R4-60.8).

Next, determine the required ρ_g. Assume that this column will be compression-controlled ($\phi = 0.65$) subject to later check.

Recognizing that required $P_n = P_u/\phi$ and required $P_ue = M_u/\phi$, we can calculate the values of required K_n and R_n:

$$\text{required } K_n = \frac{P_u}{\phi f'_c A_g} = \frac{1300}{0.65(4)(676)} = 0.740$$

$$\text{required } R_n = \frac{M_u}{\phi f'_c A_g h} = \frac{550\ (12)}{0.65(4)(676)(26)} = 0.144$$

From Diagram A-18, $\rho_g \approx 0.023$. Note that this is well above the $f_s/f_y = 1.0$ line; therefore, the column is compression-controlled and the assumption that $\phi = 0.65$ is O.K.

$$\text{required } A_s = \rho_g A_g = 0.023(676) = 15.55 \text{ in.}^2$$

Select 16 No. 9 bars ($A_s = 16.00$ in.^{2}). Check the maximum number of No. 9 bars from Table A-14: 20 (O.K.).

Design the ties. Use a $\frac{3}{8}$-in.-diameter tie, because the vertical bar size (No. 9 bar) is not greater than a No. 10.

The maximum tie spacing is the smallest of the following:

$$16 \text{ (bar diameter)} = 16 \times 1.13 = 18 \text{ in.}$$

$$48 \text{ (tie diameter)} = 48 \times \frac{3}{8} = 18 \text{ in.}$$

$$\text{least column dimension} = 26 \text{ in.}$$

Therefore, use No. 3 ties at 18-in. spacing. The design is shown in Figure 9-24.

9-11 BIAXIAL BENDING

In previous sections of this chapter, we discussed the analysis and design of columns subjected to axial compression loads plus bending moment about one axis only. In buildings, columns are often subjected to axial loads plus simultaneous bending about the two orthogonal axes. Examples of such columns include corner columns and edge columns. Even interior columns may be subjected to unbalanced moments about both orthogonal axes due to unbalanced loads on the beams and girders framing into the column. The unbalanced loads and the resulting unbalanced moments may occur due to skipped or pattern live loading and/or because of differences in adjacent span lengths of the beams/girders framing into the column. There are several methods available that can be used for the design of biaxially loaded columns. In this text, a simplified approach using the equivalent eccentricity method will be used [3]. In this method, the two orthogonal moments are converted into an equivalent uni-axial moment about the weaker axis. The column is then designed for the axial load and the uniaxial moment.

The design procedure for biaxial bending in columns is as follows:

- Convert the moments $M_{u,xx}$ and $M_{u,yy}$ to an equivalent uniaxial moment, $M_{uo,yy}$.
- $M_{u,xx}$ is the factored moment about the X–X or strong axis, and $M_{u,yy}$ is the factored moment about the Y–Y or weak axis
- The eccentricity $e_x = \dfrac{M_{u,yy}}{P_u}$ and $e_y = \dfrac{M_{u,xx}}{P_u}$ (See Figure 9-25)
  - b_x = column dimension parallel to the X–X axis
  - b_y = column dimension parallel to the Y–Y axis
  - e_x = eccentricity of the axial load for bending about the Y–Y axis
  - e_y = eccentricity of the axial load for bending about the X–X axis
- The method assumes reinforcement in all four faces of the column
- The method is valid for b_x/b_y between 0.5 and 2.0
- For $\dfrac{P_u}{f'_cA_g} \leq 0.4$, $\alpha = \left(0.5 + \dfrac{P_u}{f'_cA_g}\right)\left(\dfrac{f_y + 40{,}000}{100{,}000}\right) \geq 0.60$

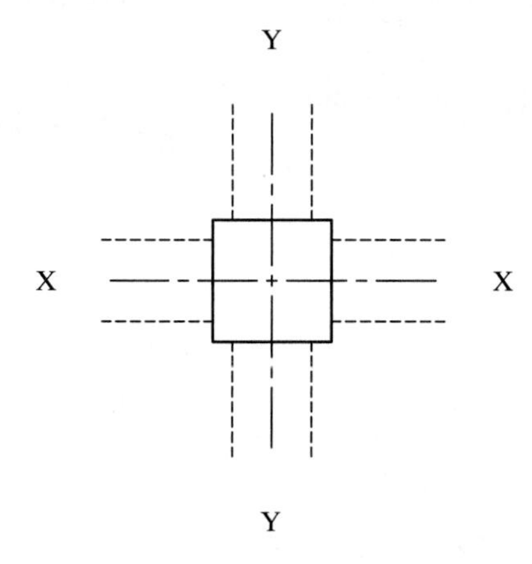

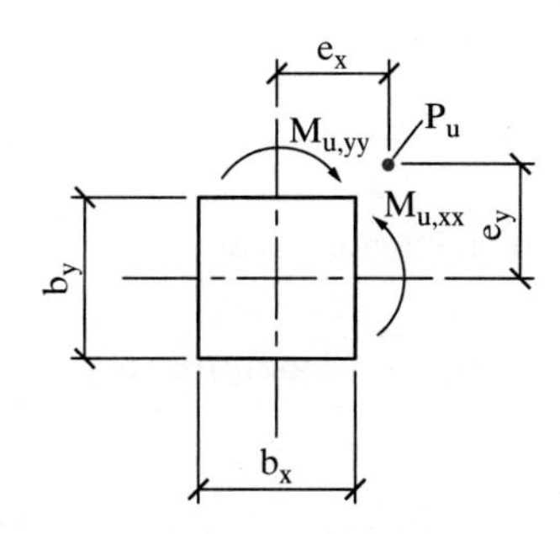

FIGURE 9-25 Biaxial moments in columns.

- For $\frac{P_u}{f'_c A_g} > 0.4$, $\alpha = \left(1.3 - \frac{P_u}{f'_c A_g}\right)\left(\frac{f_y + 40{,}000}{100{,}000}\right) \geq 0.50$

where,

f_y = rebar tensile yield strength in psi

f'_c = concrete compressive strength at 28 days

- The equivalent eccentricity for bending about Y–Y axis, $e_{ox} = e_x + \alpha\left(\frac{b_x}{b_y}\right)$
- Design the column for the axial load, P_u, and an equivalent uni-axial moment about the weaker axis, $M_{uo,yy} = P_u e_{ox}$

Example 9-9

Column Design with Biaxial Moments

Design the reinforcement for a 16 in. × 16 in. square reinforced concrete column to resist the following factored load and moments: P_u = 450 kip, M_{ux} = 70 ft-kip, M_{uy} = 45 ft-kip. The rebar tensile strength is 60,000 psi and the concrete compressive strength at 28 days is 4000, psi. Assume a short column and use the equivalent eccentricity method.

Solution:

The equivalent eccentricity method assumes reinforcement in all four faces of the column.

Calculating the eccentricities about both orthogonal axes, we obtain the eccentricity of the axial load for bending about the Y–Y axis, $e_x = \frac{M_{u,yy}}{P_u} = \frac{45 \text{ ft-kip}}{450 \text{ kip}} = 0.10 \text{ ft}$

The eccentricity of the axial load for bending about the X–X axis, $e_y = \frac{M_{u,xx}}{P_u} = \frac{70 \text{ ft-kip}}{450 \text{ kip}} = 0.156 \text{ ft}$

b_x = column dimension parallel to the X–X axis = 16 in.

b_y = column dimension parallel to the Y–Y axis = 16 in.

The method is valid for b_x/b_y between 0.5 and 2.0; $b_x/b_y = 16/16 = 1.0$ therefore O.K.

$$\frac{P_u}{f'_c A_g} = \frac{450 \text{ kip}(1000)}{(4000 \text{ psi})(16 \text{ in})(16 \text{ in.})} = 0.44 > 0.4$$

$$\alpha = \left(1.3 - \frac{P_u}{f'_c A_g}\right)\left(\frac{f_y + 40{,}000}{100{,}000}\right) \geq 0.50$$

$$= (1.3 - 0.44)\left(\frac{60{,}000 + 40{,}000}{100{,}000}\right) = 0.86 \geq 0.50,$$

Therefore, $\alpha = 0.86$

The equivalent eccentricity for bending about the Y–Y axis is

$$e_{ox} = e_x + \alpha\left(\frac{b_x}{b_y}\right) = 0.10 \text{ ft} + (0.86)(0.156 \text{ ft})\frac{16 \text{ in.}}{16 \text{ in}} = 0.234 \text{ ft}$$

Design the column for a factored axial load, P_u = 450 kip and a uni-axial factored moment about the Y–Y axis, $M_{uo,yy} = P_u e_{ox}$ = (450 kip)(0.234 ft) = 105 ft-kip.

Assume strength reduction factor ϕ = 0.65 (i.e., assuming compression controlled section)

$\gamma h = 16'' - 1.5'' \times 2$ sides − (2)(3/8 in. ties) − (1 in bar diameter) = 11.25 in., therefore, γ = (11.25/16) = 0.70.

Using the Interaction Diagram **A-17**, we obtain the interaction parameters

$$K_n = \frac{P_u}{\phi f'_c A_g} = \frac{450 \text{ kip}}{(0.65)(4 \text{ ksi})(16 \text{ in.})(16 \text{ in.})} = 0.676$$

$$R_n = \frac{M_u}{\phi f'_c A_g h} = \frac{(105 \text{ ft-kip})(12 \text{ in/ft})}{(0.65)(4 \text{ ksi})(16 \text{ in.})(16 \text{ in.})(16 \text{ in.})} = 0.118$$

From the Interaction Diagram A-17, the required gross reinforcement ratio,

$$\rho_{g,\,required} \approx 0.016$$

Therefore, A_g required = 0.016 (16 in.)(16 in.) = 4.10 in.^{2}

This falls within the compression controlled region of the interaction diagram, therefore, ϕ = 0.65 as assumed.

Select **8 # 7** vertical bars for the column (A_g = 4.80 in.^{2}). The reinforcement ratio provided is

$$\rho_{g,\,provided} = \frac{4.80 \text{ in.}^2}{(16 \text{ in.})(16 \text{ in.})} = 0.019$$

> 0.01 (minimum column vertical rebar ratio) O.K

< 0.04 (maximum column vertical rebar ratio for spliced rebars) O.K

Check the clear spacing between the column vertical reinforcement:

$$\text{clear spacing} = \frac{16\text{ in.} - (2)(1.5\text{ in side cover}) - (2)\left(\frac{3}{8}\text{ in. stirrup}\right) - \left(\frac{7}{8}\text{ in. rebar diamter}\right)}{2\text{ spaces}} - \left(\frac{7}{8}\text{ in. rebar diamter}\right)$$

$= 4.81$ in.
$> 1.5(7/8\text{ in.}) = 1.3$ in.
> 1.5 in. O.K.
< 6 in. O.K., no intermediate tie required

Design the column ties:

Select #3 ties since column vertical rebar size is #7 bar < #10 O.K.

Tie spacing $\leq$ 16 bar diameter $= (16)(7/8\text{ in.}) = 14$ in.
$\leq$ 48 tie diameter $= (48)(3/8\text{ in.}) = 18$ in.
$\leq$ least column dimension $= 16$ in.

Therefore, use #3 ties at 14 in. o.c.

See the column rebar detail in Figure 9-26.

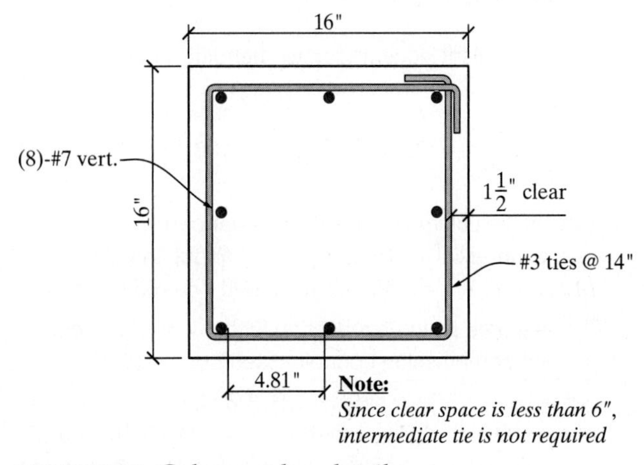

FIGURE 9-26 Column rebar detail.

The equivalent eccentricity method presented in this section is more suited for the design of biaxially loaded columns for which the factored axial load and factored moments about both orthogonal axes are known. For the analysis of biaxially loaded columns, the Bresler reciprocal load method [4] is more suitable. The reciprocal load method models the interaction between the nominal axial load capacities as shown in the following equation:

$$\frac{1}{P_{ni}} = \frac{1}{P_{nx}} + \frac{1}{P_{ny}} - \frac{1}{P_{no}}$$

Where,

The nominal concentric axial load capacity of the column with zero moments from ACI Code Equation 22.4.2.2, $P_{no} = 0.85f'_c(A_g - A_{st}) + f_yA_{st}$

P_{nx} = nominal axial load capacity of the column when subjected to the factored moment M_{ux} (with $M_{uy} = 0$)

P_{ny} = nominal axial load capacity of the column when subjected to the factored moment M_{uy} (with $M_{ux} = 0$)

P_{ni} = nominal axial load capacity of the column when subjected to the factored biaxial moments, M_{ux} and M_{uy}

Analyzing the 16 in. × 16 in. column in Example 9-9 for the given factored load and biaxial moments, we calculate the following nominal capacities:

$$P_{no} = 0.85f'_c(A_g - A_{st}) + f_yA_{st}$$
$$= 0.85(4)[(16)(16) - 4.80] + (60)(4.80)$$
$$= 1142\ kip$$

Calculate P_{nx}:

$$R_{nx} = \frac{P_{nx}e}{f'_c A_g h} = \frac{M_{ux}}{\phi f'_c A_g h} = \frac{(70\text{ ft.} - \text{kip})(12)}{0.65(4\text{ ksi})(16)(16)(16)} = 0.079$$

From the column interaction diagram (Table A-17), obtain

$$K_{nx} = 0.88 = \frac{P_{nx}}{f'_c A_g} = \frac{P_{nx}}{(4\text{ ksi})(16)(16)}$$

Therefore, $P_{nx} = 901$ kip

Calculate P_{ny}:

$$R_{ny} = \frac{P_{ny}e}{f'_c A_g h} = \frac{M_{uy}}{\phi f'_c A_g h} = \frac{(45\text{ ft.} - \text{kip})(12)}{0.65(4\text{ ksi})(16)(16)(16)} = 0.051$$

From the column interaction diagram (Table A-17), obtain

$$K_{ny} = 0.96 = \frac{P_{ny}}{f'_c A_g} = \frac{P_{ny}}{(4\text{ ksi})(16)(16)}$$

Therefore, $P_{ny} = 983$ kip

Using the Bresler reciprocal load equation yields

$$\frac{1}{P_{ni}} = \frac{1}{P_{nx}} + \frac{1}{P_{ny}} - \frac{1}{P_{no}} = \frac{1}{901} + \frac{1}{983} - \frac{1}{1142}$$

Therefore, $P_{ni} = 799$ kip

For a tied column, the strength reduction factor, $\phi = 0.65$ (i.e. assuming compression-controlled zone)

The axial load capacity for this column is $\phi P_{ni} = 0.65(799) = 519\text{ kip} > P_u = 450$ kip. O.K.

9-12 THE SLENDER COLUMN

Columns in building structures are typically braced at every floor level in both orthogonal directions. The lateral bracing of the column is provided by the beams and girders framing into the column. Thus far, our design and analysis have been limited to short columns that require no consideration of necessary strength reduction due to the possibility of buckling. All compression members will experience the buckling phenomenon as they become longer and more flexible. These are termed *slender columns* and they occur when, for aesthetic or other architectural reasons, the column is continuous over two or more floor levels. A column may be categorized as *slender* if its cross-sectional dimensions are small in comparison to its unsupported length. The degree of slenderness may be expressed in terms of the *slenderness ratio*

$$\frac{k\ell_u}{r}$$

where

k = effective length factor for compression members

ℓ_u = the unsupported length of a compression member, which shall be taken as the clear distance between floor slabs, beams, or other members capable of providing lateral support in the direction being considered (ACI Code, Section 2.2)

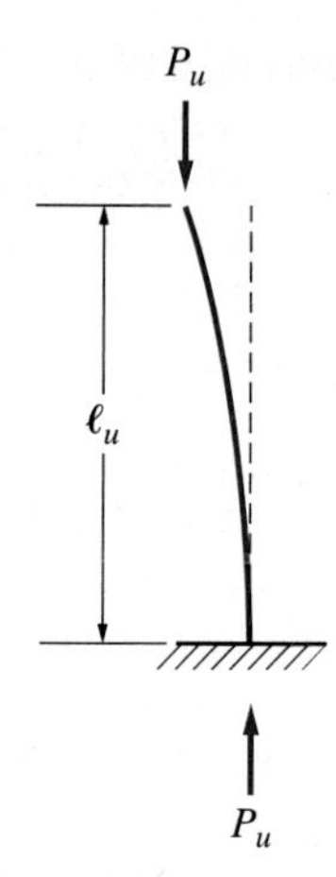

FIGURE 9-27 Fixed-free column.

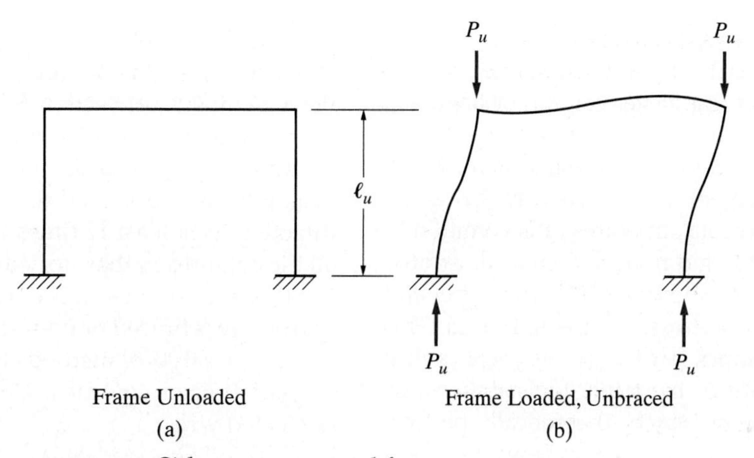

FIGURE 9-28 Sidesway on portal frame.

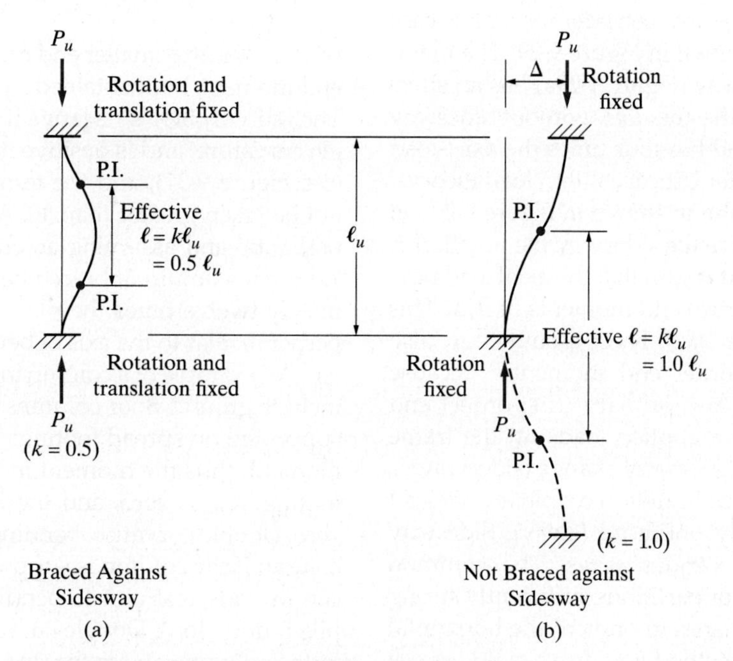

FIGURE 9-29 Sidesway and effective length.

r = radius of gyration of the cross section of the compression members, which may be taken as $0.30h$, where h is the overall dimension of a rectangular column in the direction of the moment, or $0.25D$, where D is the diameter of a circular column (ACI Code, Section 6.2.5.1)

The numerator $k\ell_u$ is termed the *effective length*. It is not only a function of the unsupported length and end conditions of the column but also a function of whether or not *sidesway* exists. Sidesway may be described as a kind of deformation whereby one end of a member moves laterally with respect to the other. Sidesway is also termed *lateral drift*.

The ACI Code, Section 6.6.4.4.3, states that for compression members braced against sidesway, k may be taken as 1.0. This is conservative. The ACI Code, Section 6.6.4.4.3, states that for compression members not braced against sidesway, the effective length must be greater than 1.0. Therefore, as a rule, compression members free to buckle in a sidesway mode are appreciably weaker than when braced against sidesway.

A simple example is a column fixed at one end and entirely free at the other (cantilever column or flagpole). Such a column will buckle, as shown in Figure 9-27. The upper end would move laterally with respect to the lower end. This lateral movement is the sidesway (or lateral drift). In reinforced concrete structures, it is common to deal with indeterminate rigid frames, such as illustrated by the simple portal frame in Figure 9-28. The upper end of the frame can move sideways as it is unbraced. This type of frame is sometimes termed a *sway frame*, and it depends on the rigidity of the joints for stability. The lower ends of the columns may be theoretically pin corrected, fully restrained, or somewhere in between.

As an example of how the effective length of a column is influenced by sidesway, consider the simple case of a single member, as shown in Figure 9-29. The member braced against sidesway (Figure 9-29a) has an effective length half that of the member without sidesway bracing (Figure 9-29b) and has four times the axial-load capacity based on the Euler critical column load theory.

If we consider the column shown in Figure 9-29b to be part of a frame and give the sidesway the notation Δ as shown in Figure 9-30, it is seen that the axial load now acts eccentrically and creates end moments of $P_u\Delta$. This is referred to as the *P-delta effect*. These moments are also referred to as "second-order end moments" because they are in addition to any primary (first-order) end moments that result from applied loads on the frame with no consideration of geometry change (sidesway).

Actual structures are rarely completely braced (nonsway) or completely unbraced (sway). Sidesway may be minimized in various ways. The common approach is to use walls or partitions sufficiently strong and rigid in their own planes to prevent the horizontal displacement. Another method is to use a rigid central core that is capable of resisting lateral loads and lateral displacements due to unsymmetrical loading conditions. ACI 318-14, Section 6.2.5 states that columns in a story may be considered braced if the total stiffness of the lateral force resisting elements within the story (e.g., total stiffness of all the shear walls in a particular direction) is at least 12 times the total lateral stiffness of all the columns in that story in the direction considered. For those cases when it is not readily apparent whether a structure is braced or unbraced, the ACI (Section 6.2.5) provides analytical methods to aid in the decision.

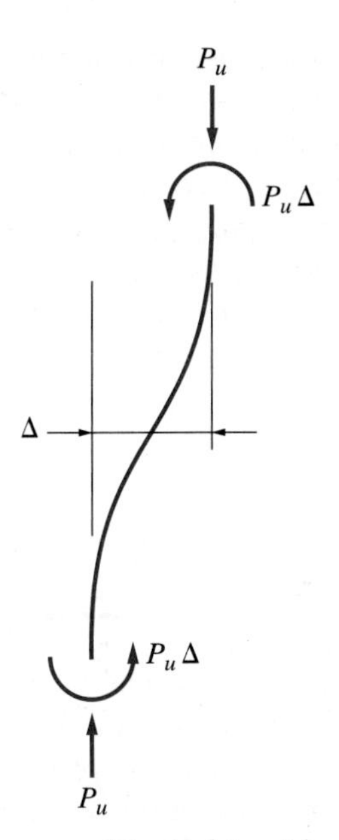

FIGURE 9-30 Column subjected to sidesway.

For braced columns, slenderness effects may be neglected when

$$\frac{k\ell_u}{r} \leq 34 + 12\left(\frac{M_1}{M_2}\right) \quad \text{[ACI Eq. (6.2.5(b))]}$$

where M_1 is the smaller end moment and M_2 is the larger end moment, both obtained by an elastic frame analysis. The ratio M_1/M_2 is negative if the column is bent in single curvature, and is positive if bent in double curvature (see Figure 9-31), and the term $[34 + 12M_1/M_2]$ shall not be taken greater than 40. At this maximum slenderness ratio and assuming an effective length factor, k, of 1.0, the maximum unbraced height ℓ_u would be approximately twelve times the plan dimension of the column perpendicular to the axis of bending (i.e., $\ell_u \leq 12h$)

An example of a column in single curvature bending include ground floor columns in braced frames that are supported on spread footings with minimal restraint to moment, thus the moment in the column at the spread footing, M_1, is zero, and therefore, the M_1/M_2 ratio is zero. Double curvature bending occurs in a ground floor column if the column is supported on a foundation that can provide restraint to bending moment (e.g., mat or pile foundation). Double curvature bending also occurs in the columns above the 2nd floor level and Table 9-2

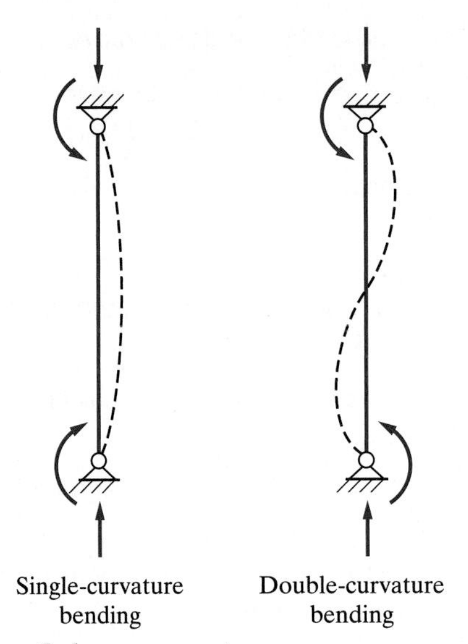

FIGURE 9-31 Column curvature.

presents some examples of the $k\ell_u/r$ limits for various M_1/M_2 ratios in a **braced column**:

For columns in sway or moment frames (i.e., in buildings where the lateral load is resisted by the bending of the beams and columns and where there are no shear walls), slenderness effects may be neglected when $k\ell_u/r$ is less than 22 (ACI 318-14, Section 6.2.5(a)). Assuming a rectangular column and an effective length factor, k, of approximately 1.2 for columns in moment frames, the preceeding requirement for columns in sway frames would indicate that the effects of slenderness can be neglected for sway or moment frames if the ℓ_u/h ratio does not exceed 5.3 (i.e., 22(0.288)/1.2).

Fortunately, for ordinary beam and column sizes and typical story heights of concrete framing systems, effects of slenderness may be neglected in more than 90% of columns in braced (nonsway) frames and in about 40% of columns in unbraced (sway) frames [2].

In cases where slenderness must be considered, the ACI Code gives the methods that can be used for the slender column design. These range from an approximate method (ACI 318-14, Section 6.6.4) in which moments are magnified to account for slenderness to a nonlinear second-order analysis. The approximate method uses a moment magnification factor that amplifies the factored moment computed from a conventional elastic analysis. In turn, the combination of the magnified factored moment and the factored axial load is used in the design of the compression member. Although the approximate analyses have been the traditional design methods of the past, more exact analyses have become possible and practical with the increased availability of sophisticated computer methods.

TABLE 9-2 Maximum Slenderness Ratio for Columns in Braced Frames

M_1/M_2 ratio	Type of curvature	Maximum $k\ell_u/r$
−1	Single curvature	22
−0.5	Single curvature	28
0	Single curvature	34
+1	Double curvature	40
+0.5	Double curvature	40

Moment Magnification for Slender Columns in Braced Frames

The effect of buckling in slender columns results in a magnification of the first-order moments, the moments that are applied to the column before buckling. These moments may be as a result of the unbalanced end moments from the beams and girders framing into the column or the from lateral loads acting on the column.

The second-order or magnified moment for a column in a braced frame is

$$M_c = \delta_{ns} M_2 \quad \text{[ACI Eq. 6.6.4.5.1]}$$

The first-order moment, M_2, is the unbalanced factored moment from the beams and girders framing into the column, but must be greater than or equal to $P_u(0.6 + 0.03h)$ about each orthogonal axis of the column (ACI 318-14, Section 6.6.4.5.4).

Where the non-sway moment magnification factor is given as

$$\delta_{ns} = C_m/(1 - [P_u/0.75P_c]) \geq 1.0 \quad \text{[ACI Eq. 6.6.4.5.2]}$$

The critical buckling load, $P_c = \pi^2(\text{EI})_{\text{eff}}/(k\ell_u)^2$ [ACI Eq. 6.6.4.4.2]

The effective length factor, k, for columns in braced frame buildings (i.e., buildings with shear walls) is usually taken as 1.0 since the column is lateral braced at both ends (ACI 318-14, Section 6.6.4.4.3).

The effective stiffness of the column, $(\text{EI})_{\text{eff}}$ is calculated using any one of the equations from the ACI 318-14, Section 6.6.4.4.4. A simplified $(\text{EI})_{\text{eff}}$ of $0.25E_cI_g$ that accounts for creep effects can be used in lieu of the other equations (ACI 318-14, Section R6.6.4.4.4).

The C_m correction factor accounts for the fact that the actual moment diagram in the column is different than an equivalent uniform moment diagram that was used to derive the moment magnification equations (ACI 318-14, Section 6.6.4.5.3).

For columns with transverse loads between supports, $C_m = 1.0$

For columns where $M_{2\text{min}} > M_2$, use $C_m = 1.0$

For all other columns, $C_m = 0.6 - 0.4(M_1/M_2)$ [ACI Eq. 6.6.4.5.3a]

M_1 and M_2 are the factored end moments in the column where M_2 is the larger end moment and is always positive. Where a column is subjected to transverse loads between supports, M_2 is the largest moment occurring within the column length.

M_1/M_2 = negative (−ve) for single curvature columns

M_1/M_2 = positive (+ve) for double curvature columns

For columns with $M_1 = M_2 = 0$, use $M_1/M_2 = -1$ which results in a value of $C_m = 1.0$.

Example 9-10

A 16 in. × 16 in. ground floor interior column in a multistory building braced with shear walls is supported on a spread footing and has an unbraced height of 15 ft. The column is subjected to a factored axial load of 600 kip and a moment about the X–X axis of 65 ft-kip. Determine if the column is slender and calculate the magnified or second-order moment, M_c. Determine the required area of the column vertical reinforcement. Assume $f'_c = 4000$ psi and $f_y = 60{,}000$ psi.

Solution:

The column unbraced length, $\ell_u = 15\text{ ft} = 180\text{ in.}$

Since the building is braced by shear walls, the slenderness factor for the column, $k = 1.0$

Since column is at the ground floor story, the recommended M_1/M_2 ratio = 0, which yields a $K\ell_u/r$ limit calculated as follows from ACI Eq. (6.2.5(b)):

$$k\ell_u/r \le 34 + 12\,(0) = 34$$

Therefore, $$(1.0)(\ell_u)/0.288h \le 34$$

Which yields $$\ell_u/h \le 9.8$$

With $\ell_u = 180$ in., this gives a minimum column size of 18.4 in. (i.e. 180/9.8) for slenderness to be neglected.

Since the column size is 16 in. which is less than 18.4 in., slenderness must be considered in the design of this column. We will now calculate the moment magnification factor, δ.

Check Minimum Moment:

$$M_{min} = P_u(0.6 + 0.03\text{ h}) = 600\text{ kip }(0.6 + 0.03[16\text{ in.}]) = 648\text{ in-kip} = 54\text{ ft-kip}$$

The applied factored moment, $M_2 = 65$ ft-kip $> M_{min}$, therefore use $M_2 = 65$ ft-kip.

(**Note:** If M_{min} were greater than M_2, then we would have used $M_2 = M_{min}$)

Since $M_1/M_2 = 0$, using ACI Eq. 6.6.4.5.3a we calculate $C_m = 0.6 - 0.4(0) = 0.6$

The effective stiffness of the column, $(EI)_{eff}$, is approximately $0.25\,E_cI_g$

$$= 0.25\,[(57{,}000\sqrt{(4000\text{ psi})}\,][(16)(16)^3/12]$$
$$= 4.9 \times 10^9\text{ lb. in.}^2 = 4.9 \times 10^6\text{ kip.in.}^2$$

The critical buckling load is calculated from ACI Eq. 6.6.4.4.2 as

$$P_c = \pi^2(EI)_{eff}/(k\ell_u)^2 = \pi^2(4.9 \times 10^6\text{ kip.in.}^2)/([1.0][180\text{ in.}])^2 = 1493\text{ kip}$$

The moment magnification factor is calculated from ACI Eq. 6.6.4.5.2 as

$$\delta_{ns} = C_m/(1 - [P_u/0.75P_c])$$
$$= 0.6/[1 - (600\text{ kip}/(0.75)(1493\text{ kip})] = 1.3$$
$$\ge 1.0, \quad \text{OK.} \quad \text{Use } 1.3$$

Therefore, the magnified moment is calculated from ACI Eq. 6.6.4.5.1 as

$$M_c = \delta_{ns}M_2 = (1.3)(65\text{ ft-kip}) = 84.5\text{ ft kip.}$$

The 16 in. × 16 in. column is then designed for a factored axial load of 600 kip and a factored magnified moment of 84.5 ft-kip.

Assuming #3 ties and #8 bars,

$$\gamma h = 16 - 2(1.5''\text{ cover}) - 2(3/8''\text{ ties}) - (\tfrac{1}{2}'') - (\tfrac{1}{2}'') = 11.25''$$

Therefore, $\gamma = 11.25/16 = 0.7$

Therefore, use the interaction diagram A-17 and calculate the following parameters:

required $K_n = P_u/\phi f'_c A_g = 600/[0.65(4)(16)(16)] = 0.9$

required $R_n = M_u/\phi f'_c A_g h$ and
$= (84.5)(12)/[0.65(4)(16)(16)(16)] = 0.095$

Therefore, entering the interaction diagram A-17, the required reinforcement ratio, $\rho = 0.024 \ge 0.01$. Therefore, A_g required $= 0.024(16)(16) = 6.14\text{ in.}^2$

Use 8 #8 rebar (A_g provided $= 6.28\text{ in.}^2 > A_g$ required)

The design of slender reinforced concrete columns in sway frames (i.e., buildings not braced with shear walls) is one of the more complex aspects of reinforced concrete design and is not within the intended scope of this book. Moreover, as discussed previously, slender columns are not common in typical reinforced concrete buildings especially in unbraced frames. For the theoretical background and applications relative to slender columns in sway frames, the reader should refer to ref [3].

9-13 CONCRETE COLUMN SCHEDULE

The column reinforcement and sizes, and the factored axial loads, are usually presented in a tabular format in a column schedule as shown in Figure 9-32. The column schedule shows, for each level, the concrete strength, the factored axial load on the column, the column cross sectional dimensions, the column vertical reinforcement and the transverse reinforcement. Other information on the schedule include the bottom elevation of the column as well as the size of the pier that transfers the column load to the foundation. Other pertinent information presented in the schedule are the column pier information and any vertical dowels.

COLUMN SCHEDULE

MARK	f'c	A-1, A-6, E-1, E-6	A-2, A-3, A-4, A-5 E-2, E-3, E-4, E-5	B-1, C-1, D-1, B-6, C-6, D-6	B-2, B-3, B-4, B-5 C-2, C-3, C-4, C-5 D-2, D-3, D-4, D-5
NO. OF COLUMNS		4	8	6	12
ROOF ELEV. = 138'-6" FACT. AXIAL LOAD PLAN SIZE VERT. REINF TIES	4000psi	100K 18"×18" (8)-#9 #3@18	180K 22"×22" (12)-#9 #3@18	190K 22"×22" (12)-#9 #3@18	350K 24"×24" (12)-#11 #4@22
3RD FLOOR = 125'-6" FACT. AXIAL LOAD PLAN SIZE VERT. REINF TIES	4000psi	215K 18"×18" (8)-#9 #3@18	380K 22"×22" (12)-#9 #3@18	400K 22"×22" (12)-#9 #3@18	350K 24"×24" (12)-#11 #4@22
2ND FLOOR = 112'-10" FACT. AXIAL LOAD PLAN SIZE VERT. REINF TIES	4000psi	330K 18"×18" (8)-#9 #3@18	580K 22"×22" (12)-#9 #3@18	610K 22"×22" (12)-#9 #3@18	750K 24"×24" (12)-#11 #4@22
1ST FLOOR = 100'-0" TOP OF PIER OR FTG = 99'-4" PLAN SIZE VERT. REINF TIES TOP OF FTG = 96'-0"	4000psi	18"×18" (8)-#9 #3@18	22"×22" (12)-#9 #3@18	22"×22" (12)-#9 #3@18	NO PIER

FIGURE 9-32 Concrete column schedule.

References

[1] *ACI Design Handbook*, "Design of Structural Reinforced Concrete Elements in Accordance with Strength Design Method of ACI 318-14," Publication SP-17(14). American Concrete Institute, P.O. Box 9094, Farmington Hills, MI 48333.

[2] Notes on ACI 318-02, Building Code Requirements for Structural Concrete, with Design Applications." Portland Cement Association, 5420 Old Orchard Road, Skokie, IL 60077-1083, 2002.

[3] James K. Wight and James G. MacGregor, *Reinforced Concrete: Mechanics and Design*, 7th ed., New York: Pearson, 2015

[4] Boris Bresler. "Design Criteria for Reinforced Columns under Axial Load and Biaxial Bending,"ACI Journal Proceedings, Vol. 57, No. 11, November 1960, pp. 481–490.

Problems

9-1. Compute the maximum design axial load strength of the tied columns shown. Assume that the columns are short. Check the tie size and spacing. Use $f'_c = 4000$ psi and $f_y = 60{,}000$ psi.

9-2. Compute the maximum design axial load strength for the tied column shown. The column is short. Check the tie size and spacing. Use $f'_c = 4000$ psi and $f_y = 60{,}000$ psi.

9-3. Find the maximum axial compressive service loads that the column of cross section shown can carry. The column is short. Assume that the service dead load and live load are equal. Check the ties. Use $f'_c = 4000$ psi and $f_y = 60{,}000$ psi.

9-4. A short, circular spiral column having a diameter of 18 in. is reinforced with eight No. 9 bars. The cover is $1\frac{1}{2}$ in., and the spiral is $\frac{3}{8}$ in. in diameter spaced 2 in. o.c. Find the maximum design axial load strength and check the spiral. Use $f'_c = 3000$ psi and $f_y = 40{,}000$ psi.

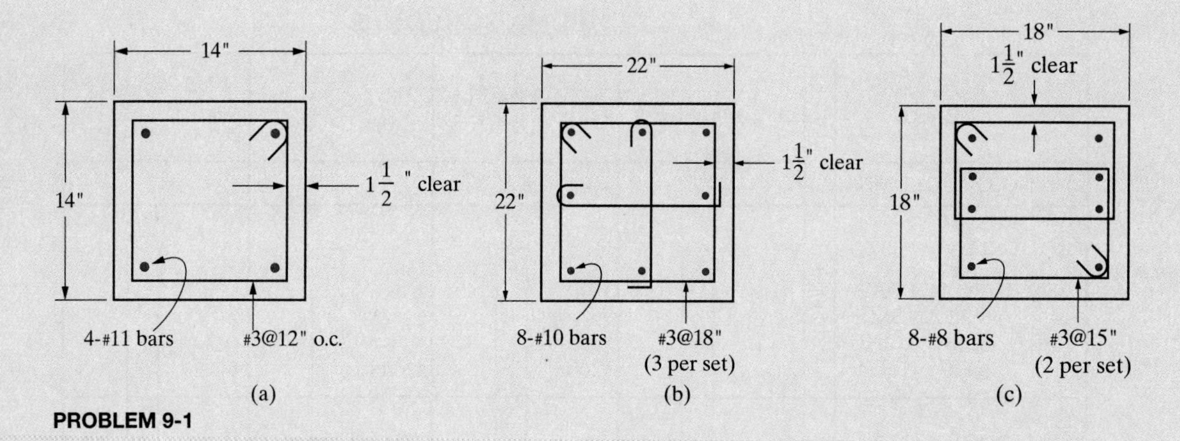

PROBLEM 9-1

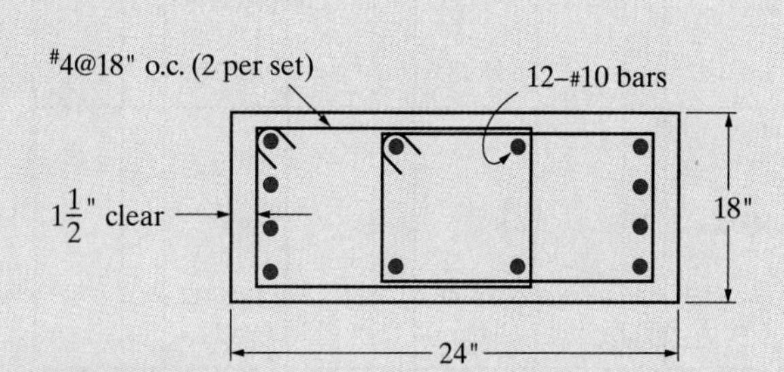

PROBLEM 9-2

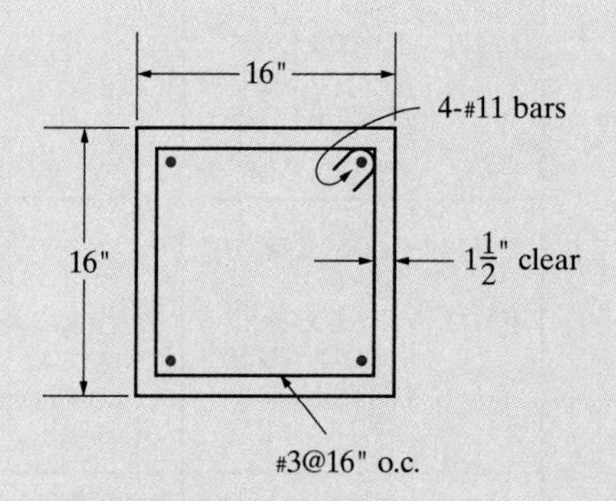

PROBLEM 9-3

9-5. Same as Problem 9-4, but $f_c' = 4000$ psi and $f_y = 60{,}000$ psi.

9-6. Compute the maximum axial compressive service live load that may be placed on the column shown. The column is short and is subjected to an axial service dead load of 200 kip. Check the ties. Use $f_c' = 3000$ psi and $f_y = 40{,}000$ psi.

9-7. Design a short, square tied column to carry a total factored design load P_u of 905 kip. Space and practical limitations require a column size of 18 in. × 18 in. Use $f_c' = 4000$ psi and $f_y = 60{,}000$ psi.

9-8. Design a short, square tied column for service loads of 205 kip dead load and 165 kip live load. Use ρ_g of about 0.04, $f_c' = 3000$ psi, and $f_y = 60{,}000$ psi. Assume that eccentricity is small.

9-9. Same as Problem 9-8, but use $f_c' = 4000$ psi, $f_y = 60{,}000$ psi, and a ρ_g of about 0.03.

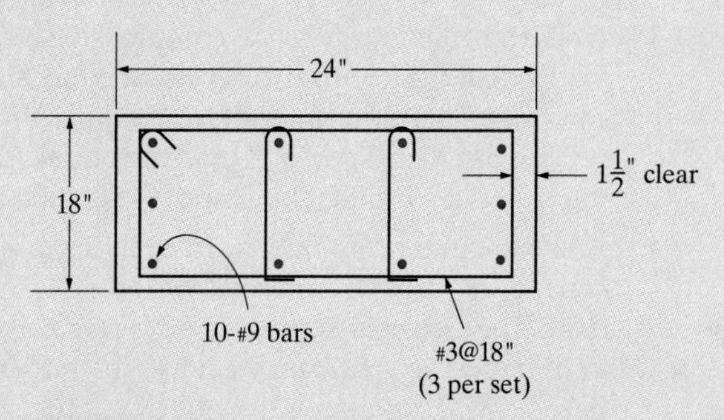

PROBLEM 9-6

9-10. Design a short, circular spiral column for service loads of 175 kip dead load and 325 kip live load. Assume that the eccentricity is small. Use $f_c' = 4000$ psi and $f_y = 60{,}000$ psi. Make ρ_g about 3%.

9-11. For the short column of cross section shown, find ϕP_n and e_b at the balanced condition using basic principles. Use $f_c' = 4000$ psi and $f_y = 60{,}000$ psi. Bending is about the strong axis.

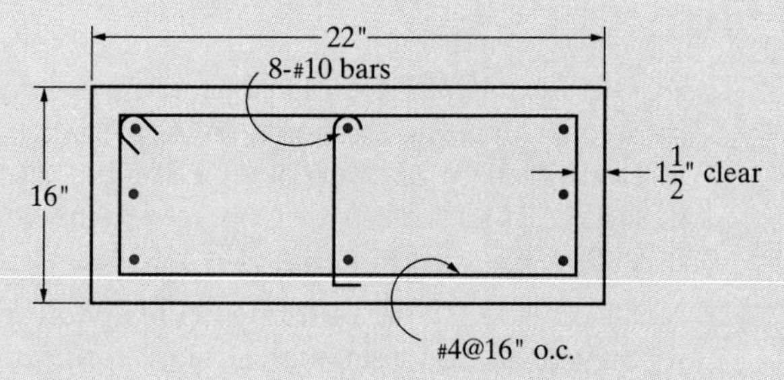

PROBLEM 9-11

9-12. Design a short square tied column to carry a factored axial design load P_u of 890 kip and a factored design moment M_u of 390 ft.-kip. Place the longitudinal reinforcing uniformly in the four faces. Use $f_c' = 4000$ psi and $f_y = 60{,}000$ psi.

9-13. Same as Problem 9-12, but design a circular spiral column.

9-14. For the short tied column of cross section shown, find the design axial load strength ϕP_n

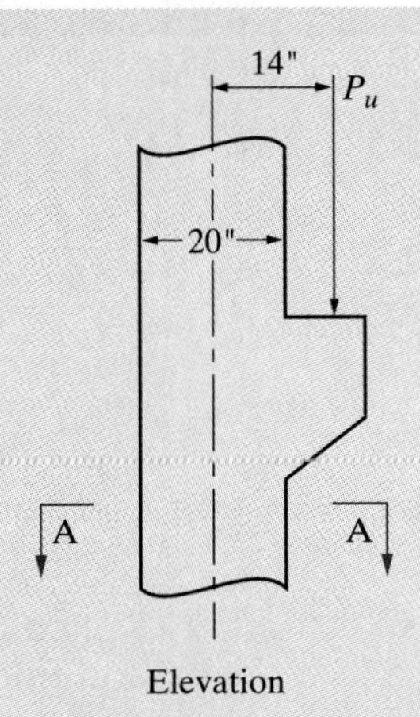

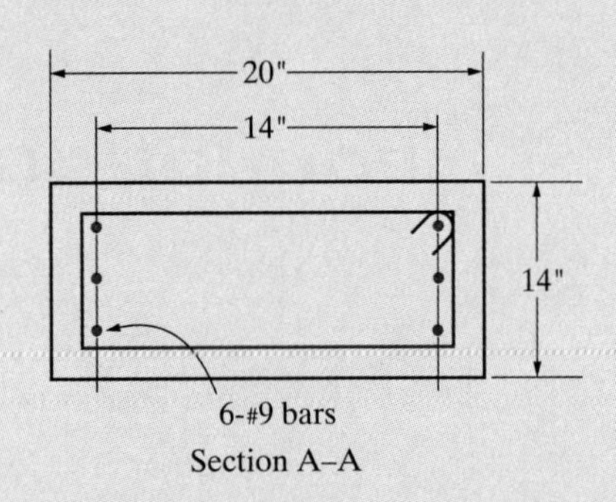

PROBLEM 9-14

for an eccentricity of 14 in. Use $f'_c = 4000$ psi and $f_y = 60{,}000$ psi. Assume that the ties and bracket design are adequate.

9-15. Refer to the Column plan detail shown. Use $f'_c = 4{,}000$ psi and $f_y = 60{,}000$ psi

a. Does the tie arrangement and vertical reinforcing ratio meet the ACI code?

b. Without using the interaction diagram, determine ϕP_n (kip) assuming minimum moments (i.e., $\phi M_n = 0.1\ \phi P_n$ h)

c. Confirm the answer from (b) using the interaction diagram; show calculations and draw appropriate lines on a copy of the appropriate interaction diagram to support your answer

d. Using the interaction diagram, determine the design bending strength, $\phi M_n(k\text{-}ft)$ when the applied axial load is $P_u = 325$k; show calculations and draw appropriate lines on the *diagram to support your answer*

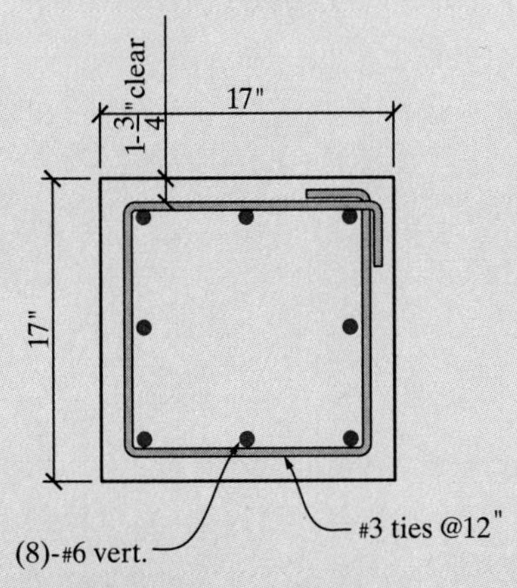

PROBLEM 9-15

9-16. Refer to the Column plan detail shown below. Use $f'_c = 4{,}000$ psi and $f_y = 60{,}000$ psi

a. Select a tie arrangement and spacing to meet ACI code.

b. Without using any interaction diagram, determine ϕP_n (kip) assuming minimum moments (i.e., $\phi M_n = 0.1\ \phi P_n$ h)

c. Confirm the answer from (b) using a copy of the appropriate interaction diagram; show calculations and draw appropriate lines on the diagram to support your answer

d. Using the same interaction diagram, determine the design axial strength, ϕP_n (k) when the applied factored moment is $M_u = 200$ k-ft; show calculations and draw appropriate lines on the diagram to support your answer

e. Identify the following regions on the diagram: $\phi = 0.65$ and $\phi = 0.9$

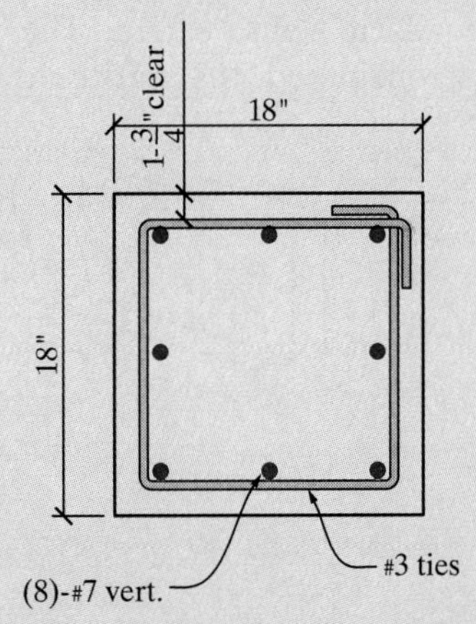

PROBLEM 9-16

9-17. A 20 in. × 20 in. ground floor interior column in a multi-story building braced with shear walls is supported on a spread footing and has an unbraced height of 20 ft. The column is subjected to a factored axial load of 650 kip and a moment about the X–X axis and the Y–Y axis of 60 ft-kip and 35 ft-kip, respectively. Determine if the column is slender about the X–X and Y–Y axis, and calculate the magnified or second-order moment, M_c about the X–X and the Y–Y axis. Determine the required area of the column vertical reinforcement. Assume $f'_c = 4000$ psi and $f_y = 60{,}000$ psi.

FOUNDATIONS

10-1 Introduction
10-2 The Geotechnical Report
10-3 Wall Footings
10-4 Wall Footings under Light Loads
10-5 Individual Reinforced Concrete Footings for Columns
10-6 Square Reinforced Concrete Footings
10-7 Rectangular Reinforced Concrete Footings
10-8 Eccentrically Loaded Footings
10-9 Combined Footings
10-10 Cantilever or Strap Footings
10-11 Analysis and Design of Mat Foundations
10-12 Deep Foundations–Piles, Drilled Shaft (Caissons), and Pile Caps
10-13 Strut-and-Tie Models for Pile Caps and Deep Beams

10-1 INTRODUCTION

The purpose of the structural portion of every building is to transmit applied loads safely from one part of the structure to another. The loads pass from their point of application into the *superstructure,* then to the *foundation,* and then into the underlying supporting material. We have discussed the superstructure and foundation walls to some extent. The foundation is generally considered the entire lowermost supporting part of the structure. Normally, a *footing* or foundation is the last structural element through which the loads pass on their path to the underlying soil or bearing strata. A footing has as its function the requirement of spreading out the superimposed load so as not to exceed the safe capacity of the underlying material, usually soil, to which it delivers the load. Additionally, the design of footings must take into account certain practical and, at times, legal considerations. Our discussion will focus on spread footings,

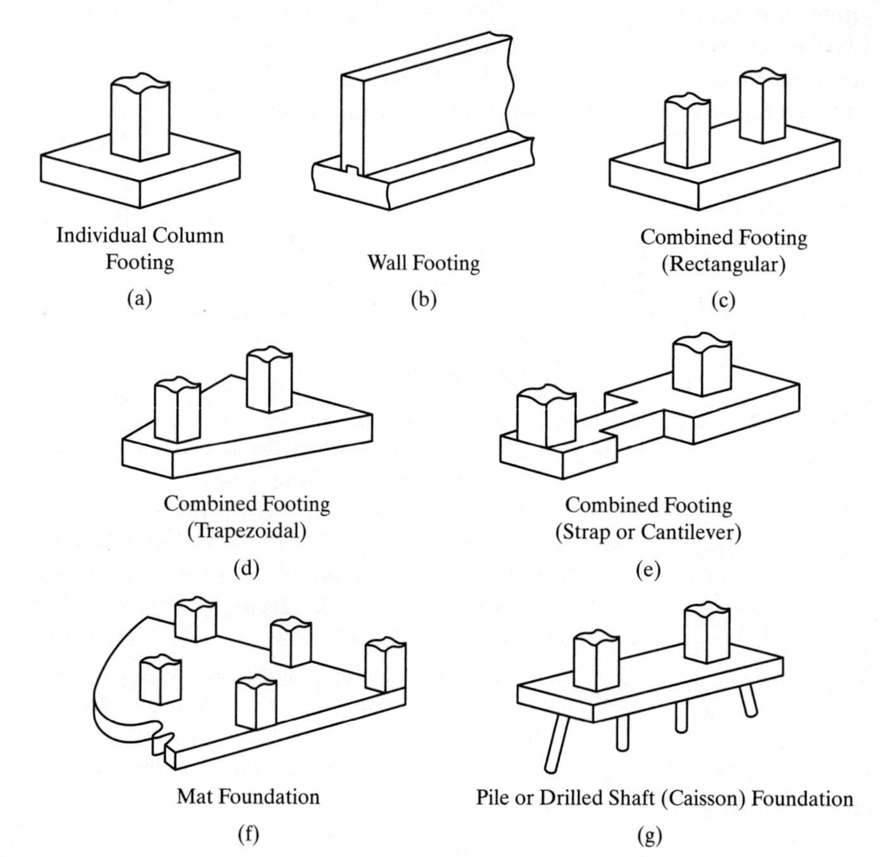

FIGURE 10-1 Types of Foundations.

mat foundations, and deep foundations, as well as pile caps and deep beams.

The more common types of foundations (see Figure 10-1) may be categorized as follows:

1. *Individual column footings* (Figure 10-1a) are often termed *isolated spread footings* and are generally square. If space limitations exist, however, the footing may be rectangular in shape.
2. *Wall footings* support walls that may be either bearing or nonbearing walls (Figure 10-1b).
3. *Combined footings* support two or more columns and may be either rectangular or trapezoidal in shape (Figure 10-1c and d).
4. If two isolated footings are joined by a *strap beam*, the footing is sometimes called a *cantilever or strap footing* (Figure 10-1e). Strap footings are used to prevent eccentric loading of individual column foundations by "strapping" an edge foundation—that will otherwise be eccentrically loaded—to a concentrically loaded interior foundation.
5. *Mat foundations* (also sometimes called raft foundations) are large continuous footings that support all columns and walls of a structure. They are commonly used where undesirable soil conditions or low allowable soil pressures prevail (Figure 10-1f) for great depths and column loads are high resulting in overlapping of individual column spread footing areas. The mat foundation can be supported directly on the soil if the soil bearing capacity is strong enough and will not result in excessive settlement; where the soil is very weak or soft which could result in excessive direct and differential settlements, the mat foundation can be supported on piles or caissons (i.e., bored reinforced concrete piles) that are in turn supported on the hard bedrock strata that could be several hundred feet beneath the structure or supported through friction between the piles and the surrounding soil. In this piled mat foundation, the mat acts like a two-way pile cap to support the column loads and in turn transmit the loads to the piles or caissons below. In piled mat foundations, load sharing between the mat foundation and the piles or caisson foundations is often considered if the soil below the mat foundation has adequate bearing capacity [5]. Mat foundations are also used for moderate height buildings with several basement levels where the foundation level is close to bedrock. For supertall buildings, a mat foundation by itself is usually not sufficient to develop the lateral and moment resistance necessary to support the high lateral loads [5]. If a "compensated" or "floating" mat foundation is desired, the required excavation can be rather deep, and the excavated area serves as the multi-level basement area. In the "compensated" mat foundation, an amount of soil approximately equal to the weight of the building is excavated, resulting in a net soil pressure (after the building is constructed) at the underside of the mat foundation that will be theoretically equal to the applied soil pressure that originally existed in the native soil at the elevation of the underside of the mat before the excavation. Thus, the weight of the excavated soil is approximately equal to the total dead weight of the building and the uniform settlement and differential settlements are reduced compared to a similar foundation located at the ground surface. The "floating" mat foundation is recommended for soils susceptible to excessive settlement and heaving during excavation (such as soft clay and loose sand), and where the elevation of the bedrock is very deep below the ground surface [5].
6. Deep foundations consist of piles, caissons (drilled shafts), and pile caps. The piles could be made of steel shapes or formed concrete piles (caissons). Pile caps *serve to transmit column loads to a group of piles or caissons, which will in turn transmit the loads to* the supporting soil through skin friction with the surrounding soil (i.e., friction piles) or through direct bearing to the underlying bedrock (Figure 10-1g). Where the piles or caissons extend down to the bedrock strata, they are usually embedded into the bedrock. For building sites with very deep bedrock bearing strata, friction piles that relies on the skin friction between the piles and the surrounding soil may be used.

The foundations for shear walls in midrise and highrise buildings usually consist of mat or pile or caisson foundations; spread footings are not used for shearwall foundations in such buildings [1].

10-2 THE GEOTECHNICAL REPORT

For most structures, a site exploration is required and it is carried out by a licensed geotechnical engineer who prepares a report that summarizes their findings. A typical geotechnical report for a project includes the following information [2]:

- The type of soil and bedrock on the site on which the structure will be built.

- The soil borings and bedrock elevation, and the elevation of the water table.
- The recommended foundation types and their net allowable bearing pressure in psf, and the pros and cons for each foundation type including estimates of the magnitude of uniform and differential settlements.
- Allowable axial (compression and tension or uplift) and lateral load capacities for steel piles and pile groups, and/or drilled cast-in-place concrete piles (caissons) if deep foundations are to be used; the minimum spacing between piles within a pile group.
- The net allowable bearing pressure of the soil and the bedrock (if the foundations are to be founded on the bedrock), and the minimum depth at which interior and perimeter footings should be founded. For footings bearing on rock, the allowable bond stress for any rock anchors.
- The site seismic classification and design properties for seismic design.
- If corrosive soils or soils with sulfates are present on the site, these information should be included in the report as this would determine the type of cement that need to be used for the foundation elements of the structure. In this situation, a recommendation for the use of epoxy-coated rebar to prevent corrosion of the rebar will also be included in the report.
- Lateral earth pressure coefficients (active, at-rest and passive pressure coefficients), the density or unit weight of the soil, coefficient of sliding friction between concrete footing and soil, and the elevation of the groundwater table for the design of basement and retaining walls.
- The modulus of subgrade reaction for the design of the slab-on-grade. The geotechnical report will also include recommendations as to whether a vapor barrier is needed or not, and if needed what type of vapor barrier should be used.
- The compaction and density requirements (i.e., modified or standard proctor density) for the native soil below the slab-on-grade, and for all the fill and backfill behind the basement and retaining walls.

Note that the allowable soil bearing pressure provided by the geotechnical engineer usually implies a factor of safety of up to 2.5 to 3.0 [3, 4], therefore, the required area of the footing or foundation is calculated using unfactored column loads, whereas the strength design of the foundation is carried out using the factored column loads. Typical values for allowable soil bearing pressures and other soil properties used for structural design are presented in Table 10-1.

10-3 WALL FOOTINGS

Wall footings (see Figure 10-2) are commonly required to support direct concentric loads. An exception to this is the footing for a retaining wall. A wall footing

TABLE 10-1 Typical Soil Properties for Structural Design

Type of soil	Allowable soil bearing pressure (psf)	Lateral bearing pressure (psf/ft below natural grade)	Lateral Sliding Resistance	
			Coefficient of friction[1]	Cohesion (psf)[2]
Crystalline bedrock	12,000	1200	0.70	-
Sedimentary and foliated rock	4000	400	0.35	-
Gravel and/or Sandy gravel	2,000	150	0.25	-
Clay, Sandy clay, Silty clay, Clayey silt, Silt, Sandy silt	1500	100	-	130

(Data from International Code Council (2014). 2015 international building code, Table 1806.2.

[1] Friction coefficient to be multiplied by dead load.

[2] Multiply cohesion value by the contact area per the limitations in IBC Section 1806.3.2.

FIGURE 10-2 Column and wall footings. The Health, Physical Education and Recreation Complex, Hudson Valley Community College, Troy, New York.
(Courtesy of George F. Limbrunner)

may be of either plain or reinforced concrete. Because it has bending in only one direction, it is generally designed in much the same manner as a one-way slab, by considering a typical 12-in.-wide strip along the length of wall. Footings supporting relatively light loads on well-drained cohesionless soil are often made of plain concrete—that is, concrete without reinforcing. This material is referred to as *structural plain concrete.*

A wall footing under concentric load behaves similarly to a cantilever beam, where the cantilever extends out from the wall and is loaded in an upward direction by the soil pressure. The flexural tensile stresses induced in the bottom of the footing are acceptable for a plain concrete footing.

From the ACI Code, Section 14.5.2, the nominal flexural design strength of a plain concrete cross section is calculated from

$$M_n = 5\lambda\sqrt{f'_c}S_m$$

if tension controls and

$$M_n = 0.85 f'_c S_m$$

if compression controls. The quantity S_m is the elastic section modulus of the section. Lamda (λ) is the modification factor reflecting the lower tensile strength of lightweight concrete relative to normal-weight concrete and is described in Chapter 1. For normal-weight concrete, $\lambda = 1.0$. In this chapter we will consider only normal-weight concrete; therefore, λ will be omitted in the examples. These formulas are based on the flexure formula,

$$f_b = \frac{Mc}{I}$$

rewritten

$$M = f_b S$$

where $5\lambda\sqrt{f'_c}$ and $0.85f'_c$ are limiting stress, f_b, values in tension and compression, respectively. Similarly, nominal shear strength for beam action for a plain concrete member is calculated from

$$V_n = \frac{4}{3}\lambda\sqrt{f'_c}bh$$

This is based on the general shear formula for a rectangular section:

$$f_v = \frac{3V}{2bh}$$

which is rewritten

$$V = \frac{2}{3}f_v bh$$

where the limiting shear stress value is $2\lambda\sqrt{f'_c}$, a familiar value from our previous discussions of shear in flexural members and $\lambda = 1.0$ for normal-weight concrete.

In each case, the basis for the design must be

$$\phi M_n \geq M_u \quad \text{and} \quad \phi V_n \geq V_u$$

as applicable. The strength-reduction factor for plain concrete is 0.60 (ACI Code, Section 21.2.1).

In a *reinforced* concrete wall footing, the behavior is identical to that just described. Reinforcing steel is placed in the bottom of the footing in a direction perpendicular to the wall, however, thereby resisting the induced flexural tension, similar to a reinforced concrete beam or slab.

In either case, the cantilever action is based on the maximum bending moment occurring at the face of the wall if the footing supports a concrete wall or at a point halfway between the middle of the wall and the face of the wall if the footing supports a masonry wall. This difference is primarily because a masonry wall is somewhat less rigid than a concrete wall.

For each type of wall, the critical section for shear in the footing may be taken at a distance from the face of the wall equal to the effective depth of the footing.

Example 10-1

Design a plain normal-weight concrete wall footing to carry a 12-in. concrete block masonry wall, as shown in Figure 10-3. The service loading may be taken as 10 kip/ft dead load (which includes the weight of the wall) and 20 kip/ft live load. Use $f'_c = 3000$ psi. The gross allowable soil pressure is 5000 psf (5.0 ksf), and the weight of earth $w_e = 100$ lb/ft^3.

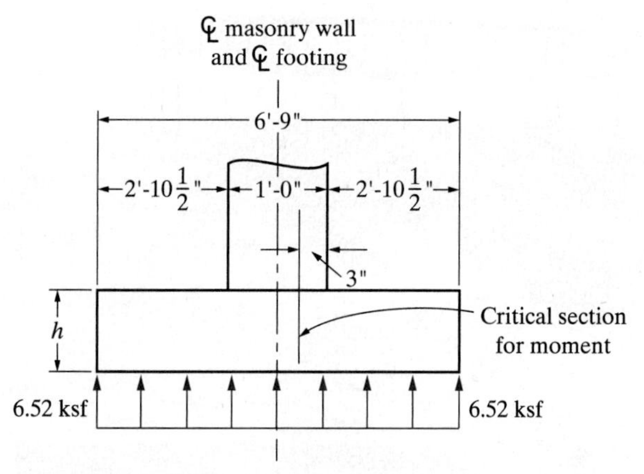

FIGURE 10-3 Plain concrete wall footing for Example 10-1 - factored soil pressure and critical section for moment.

Solution:

1. Compute the factored load:

$$\begin{aligned} w_u &= 1.2w_{DL} + 1.6w_{LL} \\ &= 1.2(10) + 1.6(20) \\ &= 44 \text{ kip/ft} \end{aligned}$$

2. Assume a footing thickness of 3 ft-0 in. This will be checked later in the design. The footing weight is

$$0.150(3) = 0.450 \text{ ksf}$$

Assuming the bottom of the footing to be 4 ft-0 in. below the finished ground line, the weight of the soil on top of the footing is

$$(1)(100) = 100 \text{ psf} = 0.100 \text{ ksf}$$

Therefore, the net allowable soil pressure for superimposed service loads is

$$5.00 - 0.45 - 0.100 = 4.45 \text{ ksf}$$

Sometimes, the geotechnical engineering report for a project may specify the net allowable soil pressure directly, thus obviating the need for the calculations in this step.

3. The maximum allowable soil pressure for strength design must now be found. It must be modified in a manner consistent with the modification of the service loads. This may be accomplished by multiplying by the ratio of the total design load (44 kip) to the total service load (30 kip). We then obtain a maximum allowable soil pressure solely for use in the strength design of the footing. This soil pressure should not be construed as an actual allowable soil pressure. Thus

$$4.45\left(\frac{44}{30}\right) = 6.53 \text{ ksf}$$

4. We may now determine the required footing width:

$$\frac{44.0}{6.53} = 6.74 \text{ ft}$$

Use 6 ft-9 in.

5. Determine the factored soil pressure to be used for the footing design if the footing width is 6 ft-9 in.:

$$\frac{44.0}{6.75} = 6.52 \text{ ksf} < 6.53 \text{ ksf} \qquad \text{(O.K.)}$$

6. With the factored soil pressure known, the bending moment in the footing may be calculated. For concrete block masonry walls, the critical section for moment should be taken at the quarter-point of wall thickness (ACI Code, Section 13.2.7.1).

   With reference to Figure 10-3, the factored moment is determined as follows:

$$M_u = \frac{6.52(3.125)^2}{2} = 31.8 \text{ ft-kip}$$

7. Find the required footing thickness based on the required moment strength. Assuming that tension controls:

$$\phi M_n = \phi 5\sqrt{f'_c}\, S_m = \phi 5\sqrt{f'_c}\frac{bh^2}{6}$$

   Setting $\phi M_n = M_u$ and considering a typical 12-in.-wide strip:

$$\phi 5\sqrt{f'_c}\frac{(12)(h^2)}{6} = 31.8 \text{ ft-kip}$$

   from which

$$\text{required } h = \frac{(31.8 \text{ ft-kip})(12 \text{ in./ft})(1000 \text{ lb/kip})(6)}{0.60(5)\sqrt{3000}\,(12 \text{ in.})}$$

$$= 34.1 \text{ in.}$$

8. It is common practice to assume that the bottom 1 or 2 in. of concrete placed against the ground may be of poor quality and, therefore, may be neglected for strength purposes. The total required footing thickness may then be determined:

$$34.1 + 2.0 = 36.1 \text{ in.}$$

   We will use $h = 38$ in. This checks closely with the assumed thickness of 36 in. No revision of the calculations is warranted.

9. Shear is generally of little significance in plain concrete footings because of the large concrete footing thickness. With reference to Figure 10-4, if the critical section for shear is considered a distance equal to the depth of the member h from the face of the wall and if the depth is taken as 34.1 in. (the required h), it may be observed that the critical section is outside the edge of the footing. Therefore, the shear check may be neglected.

10. It is common practice to use some *longitudinal* steel in continuous wall footings whether or not *transverse* steel is present. This will somewhat enhance the structural integrity by limiting differential movement between parts of the footing should transverse cracking occur. It will also lend some flexural strength in the longitudinal direction. The rationale for this stems from the many uncertainties that exist in both the supporting soil and the applied loads.

    As a guide, this longitudinal steel may be computed in a manner similar to that for temperature and shrinkage steel in a one-way slab, where

$$\text{required } A_s = 0.0018bh \text{ (using } f_y = 60{,}000 \text{ psi)}$$

    The total longitudinal steel required for the 6-ft-9-in. footing width is

$$A_s = 0.0018(6.75)(12 \text{ in./ft})(38) = 5.54 \text{ in.}^2$$

    Use 13 No. 6 bars ($A_s = 5.72$ in.^{2}). This would satisfy any longitudinal steel requirement. It is obvious, however, that this large steel requirement does not lend itself to an economical footing design, but rather leads us to conclude that we should question the use of plain concrete wall footings when superimposed loads are heavy.

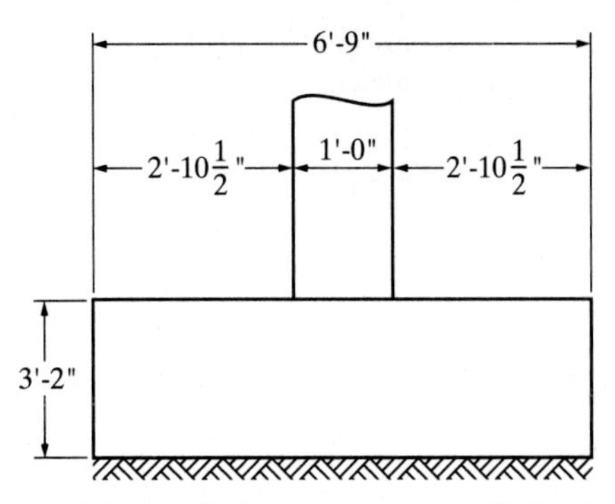

FIGURE 10-4 Detail of plain concrete wall footing for Example 10-1.

A redesign using a reinforced concrete wall footing is accomplished in the following problem.

Example 10-2

Design a normal-weight reinforced concrete wall footing to carry a 12-in. concrete block masonry wall, as shown in Figure 10-5. The service loading is 10 kip/ft dead load (which includes the weight of the wall) and 20 kip/ft live load. Use $f'_c = 3000$ psi, $f_y = 60{,}000$ psi, weight of earth = 100 lb/ft^3, and gross allowable soil pressure = 5000 psf (5.0 ksf). The bottom of the footing is to be 4 ft-0 in. below the finished ground line.

Solution:

1. Compute the factored load:

$$w_u = 1.2w_{DL} + 1.6w_{LL}$$
$$= 1.2(10) + 1.6(20)$$
$$= 44 \text{ kip/ft}$$

2. Assume a total footing thickness = 18 in. Therefore, the footing weight is

$$0.150(1.5) = 0.225 \text{ ksf}$$

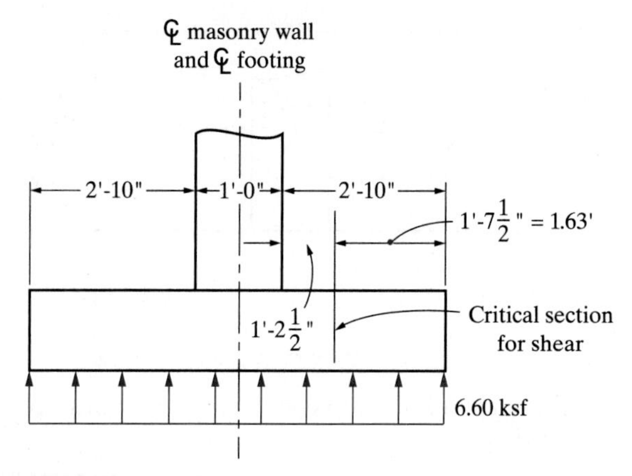

FIGURE 10-5 Reinforced concrete wall footing for Example 10-2 - factored soil pressure and critical section for shear.

Because the bottom of the footing is to be 4 ft below the finished ground line, there will be 30 in. of earth on top of the footing. This depth of earth has a weight of

$$\frac{30(100)}{12} = 250 \text{ psf} = 0.250 \text{ ksf}$$

The net allowable soil pressure for superimposed service loads is

$$5.00 - 0.225 - 0.250 = 4.53 \text{ ksf}$$

3. As in Example 10-1, we use the ratio of factored load to service load to determine the soil pressure for strength design:

$$\frac{44(4.53)}{30} = 6.64 \text{ ksf}$$

4. The required footing width is

$$\frac{44.0}{6.64} = 6.63 \text{ ft}$$

Use 6 ft-8 in.

5. The factored soil pressure to be used for the footing design is

$$\frac{44.0}{6.67} = 6.60 \text{ ksf}$$

6. The assumed effective depth d for the footing is determined by subtracting the concrete cover [see the ACI Code, Table 20.6.13.2] and one-half of the bar diameter (No. 8 assumed) from the total thickness:

$$d = 18 - 3 - 0.5 = 14.5 \text{ in.}$$

7. Because required thicknesses of reinforced concrete footings are generally controlled by shear requirements, the shear should be checked first.

With reference to Figure 10-5, as the wall footing carries shear in a manner similar to that of a one-way slab or beam, the critical section for shear will be taken at a distance equal to the effective depth of the footing (14.5 in.) from the face of the wall (ACI Code, Section 7.4.3.2 or 9.4.3.2):

$$V_u = 1.63(1)(6.60) = 10.75 \text{ kip/ft of wall}$$

The total nominal shear strength V_n is the sum of the shear strength of the concrete V_c and the shear strength of any shear reinforcing V_s.

$$V_n = V_c + V_s$$
$$\phi V_n = \phi V_c + \phi V_s$$

Assuming no shear reinforcing,

$$\phi V_n = \phi V_c$$

In footings, shear reinforcing is not required if

$$\phi V_c > V_u$$

Computing ϕV_c,

$$\begin{aligned}\phi V_c &= \phi 2\sqrt{f'_c}bd \\ &= 0.75(2)\sqrt{3000}\,(12)(14.5) \\ &= 14.30 \text{ kip/ft of wall}\end{aligned}$$

$$14.30 \text{ kip} > 10.75 \text{ kip}$$

Therefore,

$$\phi V_c > V_u$$

Thus, the assumed thickness of footing is satisfactory for shear, and no revisions are necessary with respect to footing weight.

8. With reference to Figure 10-6, the critical section for moment is taken at the quarter-point of the masonry wall thickness (ACI Code, Section 13.2.7.1). The maximum factored moment, assuming the footing to be a cantilever beam, is

$$M_u = \frac{6.60(3.08)^2}{2} = 31.3 \text{ ft-kip}$$

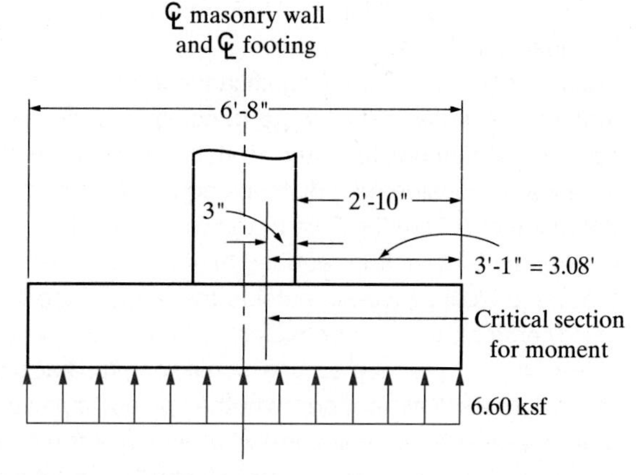

FIGURE 10-6 Reinforced concrete wall footing for Example 10-2 - factored soil pressure and critical section for moment.

9. The required area of tension steel is then determined in the normal way, using $d = 14.5$ in. and $b = 12$ in., and assuming $\phi = 0.90$:

$$\text{required } \bar{k} = \frac{M_u}{\phi bd^2} = \frac{31.3(12)}{0.9(12)(14.5)^2}$$
$$= 0.1654 \text{ ksi}$$

From Table A-8, the required $\rho = 0.0029$, and $\varepsilon_t > 0.005$; therefore, $\phi = 0.90$ (O.K.).

$$\text{required } A_s = \rho bd$$
$$= 0.0029(12)(14.5)$$
$$= 0.50 \text{ in.}^2\text{/ft of wall}$$

We will use the ACI Code minimum reinforcement requirement for beams as being applicable for footings. From Table A-5,

$$A_{s,\,\min} = 0.0033(12)(14.5)$$
$$= 0.57 \text{ in.}^2\text{/ft of wall}$$

Finally, as discussed in Chapter 8, the provisions of the ACI Code, Section 7.6.1.1, for minimum reinforcement in structural slabs of uniform thickness may be considered applicable for footings such as this that transmit vertical loads to the underlying soil. In this text, this will be used only as an absolute minimum. Checking the provision of the code for the minimum reinforcement required for grade 60 steel gives us

$$\text{required } A_s = 0.0018bh = 0.0018(12)(18)$$
$$= 0.39 \text{ in.}^2\text{/ft of wall}$$

Therefore, considering the minimum reinforcement ratio for beams as applicable, the required $A_s = 0.57$ in.^{2} per ft of wall. Therefore, use No. 6 bars at 9 in. o.c. ($A_s = 0.59$ in.^{2}).

10. The development length should be checked for the bars selected. Assume uncoated bars. This calculation for ℓ_d follows the eight-step procedure presented in Chapter 5, Section 5-2, and is summarized as follows:
    1. $K_D = 82.2$
    2. $\psi_t = 1.0$, $\psi_e = 1.0$, $\psi_s = 0.8$, $\lambda = 1.0$
    3. $\psi_t \times \psi_e = 1.0$ (O.K.)
    4. $c_b = 3.38$ in.
    5. $K_{tr} = 0$
    6. $(c_b + K_{tr})/d_b = 4.51$ in. Use 2.5 in.
    7. $K_{ER} = 0.97$
    8. $\ell_d = 19.1$ in. > 12 in. (O.K.)

    The development length provided, measured from the critical section for moment and allowing for 3-in. end cover, is 34 in. Because 34 in. > 19.1 in., the development length provided is adequate.

11. Although not specifically required in footings by the ACI Code, longitudinal steel will be provided on the same basis as for one-way slabs (Section 24.4.1). Thus

$$\text{required } A_s = 0.0018bh$$
$$= 0.0018(6.67 \text{ ft})(12 \text{ in./ft})(18 \text{ in.})$$
$$= 2.59 \text{ in.}^2$$

Use nine No. 5 bars ($A_s = 2.79$ in.^{2}) spaced equally. The footing design is shown in Figure 10-7.

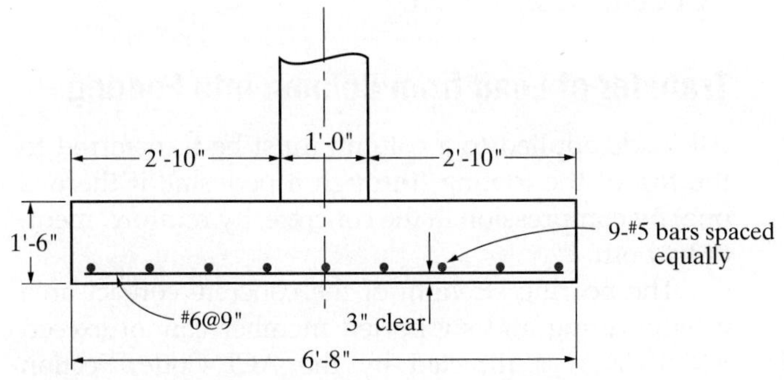

FIGURE 10-7 Design sketch for Example 10-2.

10-4 WALL FOOTINGS UNDER LIGHT LOADS

A relatively common situation is one in which a lightly loaded wall is supported on average soil. As previously indicated in this chapter, a design would result in a very small footing thickness and width.

In such a situation, experience has shown that for footings carrying plain concrete or block masonry walls, the minimum recommended dimensions shown in Figure 10-8 should be used. The minimum depth or thickness of footing should be 8 in. but not less than the wall thickness. The minimum width of footing should equal twice the wall thickness.

10-5 INDIVIDUAL REINFORCED CONCRETE FOOTINGS FOR COLUMNS

An individual reinforced concrete footing for a column, also termed an *isolated spread footing*, is probably the most common, simplest, and most economical

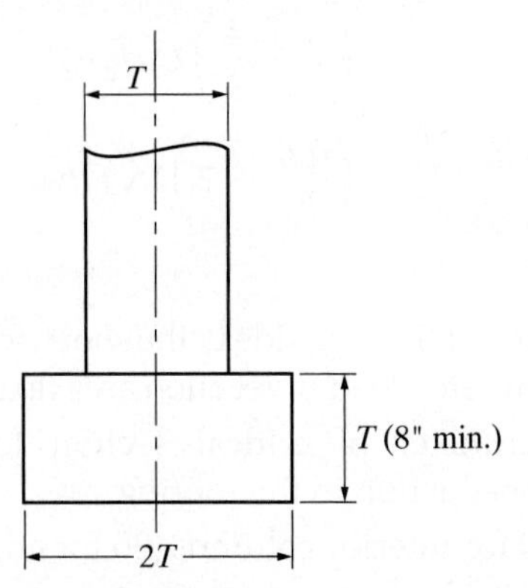

FIGURE 10-8 Recommended minimum footing dimensions for walls carrying light loads.

of the various types of footings used for structures (see Figure 10-2). Individual column footings are generally square in plan. Rectangular shapes are sometimes used where dimensional limitations exist, however. The footing is a slab that directly supports a column. At times, a pedestal or pier is placed between a column and a footing so that the base of the column need not be set below grade.

The footing behavior under concentric load is that of two-way cantilever action extending out from the column or pedestal. The footing is loaded in an upward direction by the soil pressure. Tensile stresses are induced in each direction in the bottom of the footing. Therefore, the footing is reinforced by two bottom layers of steel perpendicular to each other and parallel to the edges. The required footing-soil contact area is a function of, and determined by, the allowable soil bearing pressure and the column loads being applied to the footing.

Shear

Because the footing is subject to two-way action, two different types of shear strength must be considered: two-way shear and one-way shear. The footing thickness (depth) is generally established by the shear requirements. The two-way shear is commonly termed *punching shear*, because the column or pedestal tends to punch through the footing, inducing stresses around the perimeter of the column or pedestal. Tests have verified that, if failure occurs, the fracture takes the form of a truncated pyramid with sides sloping away from the face of the column or pedestal. The critical section for this two-way shear is taken perpendicular to the plane of the footing and located so that its perimeter, b_0, is a minimum but does not come closer to the edge of the column or pedestal than one-half the effective depth of the footing (ACI Code, Section 22.6.4.1).

The design of the footing for two-way action is based on a shear strength V_n, which is not to be taken greater than V_c unless shear reinforcement is provided. V_c may be determined from the ACI Code, Table 22.6.5.2, and shall be the smallest of

a. $$V_c = 4\lambda\sqrt{f'_c}\, b_0 d$$

b. $$V_c = \left(2 + \frac{4}{\beta_c}\right)\lambda\sqrt{f'_c}\, b_0 d$$

c. $$V_c = \left(\frac{\alpha_s d}{b_0} + 2\right)\lambda\sqrt{f'_c}\, b_0 d$$

where

β_c = ratio of the long side to the short side of the concentrated load or reaction area (loaded area)

b_0 = perimeter of critical section for two-way shear action in the footing

α_s = 40 for interior columns, 30 for edge columns, and 20 for corner columns

and V_c, f'_c, λ, and d are as previously defined. Note that the terms *interior*, *edge*, and *corner* columns in ACI Code, Section 22.6.5.3 refer to the location of the column relative to the edges of the spread footing. Therefore, *interior*, *edge*, and *corner columns* will have four-, three-, and two-sided critical sections, respectively.

The introduction of shear reinforcement in footings is impractical and undesirable purely on an economic basis. It is general practice to design spread and strip footings based solely on the shear strength of the concrete, and thus avoid the need for any shear reinforcement.

The one-way (or beam) shear may be compared with the shear in a beam or one-way slab. The critical section for this one-way shear is taken on a vertical plane extending across the entire width of the footing and located at a distance equal to the effective depth of the footing from the face of the concentrated load or reaction area (ACI Code, Section 7.4.3.2 or 9.4.3.2). As in a beam or one-way slab, the shear strength provided by the footing concrete may be taken as

$$V_c = 2\lambda\sqrt{f'_c}\, b_w d \qquad \text{[ACI Eq. (22.5.5.1)]}$$

For both one- and two-way action, if we assume no shear reinforcement, the basis for the shear design will be $\phi V_n > V_u$, where $V_n = V_c$.

Moment and Development of Bars

The size and spacing of the footing reinforcing steel is primarily a function of the bending moment induced by the net upward soil pressure. The footing behaves as a cantilever beam in two directions. It is loaded by the soil pressure. The fixed end, or critical section for the bending moment, is located as follows (ACI Code, Section 13.2.7.1):

1. At the face of the column or pedestal, for a footing supporting a concrete column or pedestal (see Figure 10-9a).
2. Halfway between the face of the column and the edge of a steel base plate, for a footing supporting a column with a steel base plate (see Figure 10-9b).

The ACI Code, Section 13.2.8.3, stipulates that the critical section for development length of footing reinforcement shall be assumed to be at the same location as the critical section for bending moment given in ACI Code, Table 13.2.7.1.

Transfer of Load from Column into Footing

All loads applied to a column must be transferred to the top of the footing (through a pedestal, if there is one) by compression in the concrete, by reinforcement, or by both.

The bearing strength of the concrete contact area of supporting and supported member cannot exceed $\phi(0.85 f'_c A_1)$ as directed by the ACI Code, Section 22.8.3.2. When the supporting surface is wider on all sides than the loaded area, the design-bearing

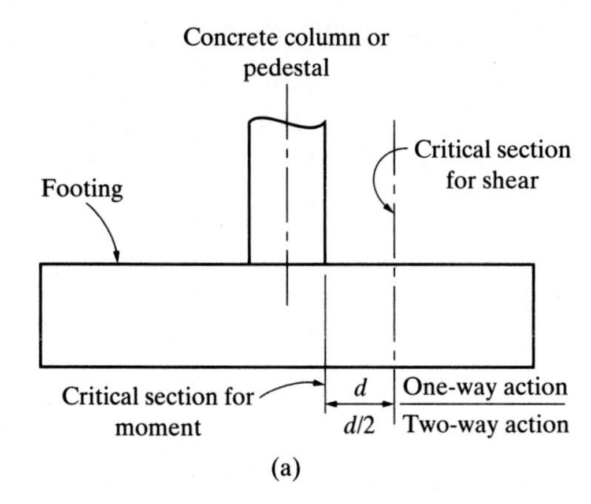

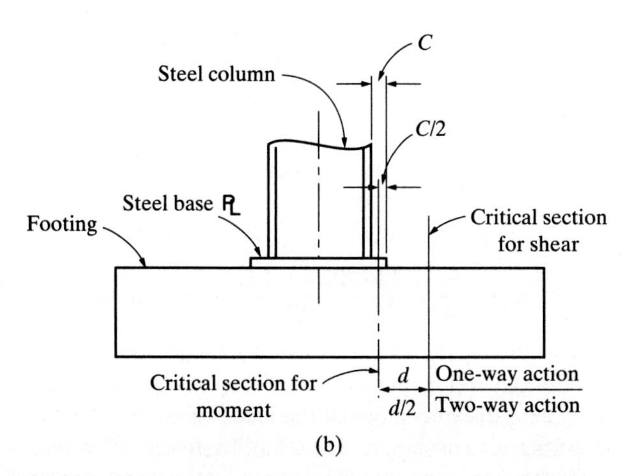

FIGURE 10-9 Critical sections for design of reinforced concrete footings supporting columns or pedestals.

strength on the loaded area may be multiplied by $\sqrt{A_2/A_1} \leq 2.0$, as discussed in Chapter 8, Section 8-4. A_2 is a concentric and geometrically similar support area that is the larger base of the frustrum of a pyramid or cone having a slope of 1 vertical to 2 horizontal and contained wholly within the supporting concrete member (e.g. footing, pier or wall) or pile cap, and A_1 is the smaller base or the loaded or reaction area (ACI Code, Section 22.8.3.2). Therefore, in no case can the design-bearing strength for the loaded area be in excess of

$$\phi(0.85f'_c A_1)(2)$$

where $\phi = 0.65$ for bearing on concrete and f'_c is as previously defined.

It is common for the footing concrete to be of a lower strength (f'_c) than the supported column concrete. This suggests that both supporting and supported members should be considered in determining load transfer.

Where a reinforced concrete column cannot transfer the load entirely by bearing, the excess load must be transferred by reinforcement where the required A_s = (excess load)/f_y. This may be accomplished by furnishing dowels, one per column bar if necessary but not larger than No. 11 (ACI Code, Sections 16.3.5.4 and 25.5.5.4).

To provide a positive connection between a reinforced concrete column and footing (whether dowels are required or not), the ACI Code, Section 16.3.4.1, requires a minimum area of reinforcement crossing the bearing surface of 0.005 of the column cross-sectional area. It is generally recommended that a minimum of four bars be used. These four bars should preferably be dowels for the four corner bars of a square column.

The development length of the dowels must be sufficient on both sides of the bearing surface to provide the necessary development length for bars in compression (see Chapter 5).

When the dowel carries excess load into the footing, it must be spliced to the column bar using the necessary compression splice. The same procedure applies where a column rests on a pedestal and where a pedestal rests on a footing.

Where structural steel columns and column base plates are used, the total load is usually transferred entirely by bearing on the concrete contact area. The design-bearing strength as stipulated previously also applies in this case. Where a column base detail is inadequate to transfer the total load, adjustments may be made as follows:

1. Increase column base plate dimensions.
2. Use higher-strength concrete (f'_c) for the pedestal or footing.
3. Increase the supporting area with respect to the base plate area until the ratio reaches the maximum allowed by the ACI Code.

In building design, it is common practice to use a concrete pedestal between the footing and the column. The pedestal, in effect, distributes the column load over a larger area of the footing, thereby contributing to a more economical footing design. Pedestals may be either plain or reinforced. If the ratio of height to least lateral dimension is in excess of 3, the member is by definition a column and must be designed and reinforced as a column (see Chapter 9). If the ratio is less than 3, it is categorized as a pedestal and theoretically may not require any reinforcement.

The cross-sectional area of a pedestal is usually established by the concrete-bearing strength as stipulated in ACI Code, Section 22.8.3.2, by the size of a steel column base plate, or by the desire to distribute the column load over a larger footing area. It is common practice to design a pedestal in a manner similar to a column using a minimum of four corner bars (for a square or rectangular cross section) anchored into the footing and extending up through the pedestal. Ties should be provided in pedestals according to the same requirements as in columns.

10-6 SQUARE REINFORCED CONCRETE FOOTINGS

In isolated square footings, the reinforcement should be uniformly distributed over the width of the footing in each direction. Because the bending moment is the same in each direction, the reinforcing bar size and spacing should be the same in each direction. In reality, the effective depth is not the same in both directions. It is common practice to use the same average effective depth for design computations for both directions, however.

It is also common practice to assume that the minimum tensile reinforcement for beams is applicable to two-way footings for each of the two directions, unless the reinforcement provided is one-third greater than required. As discussed in Section 2-8 (and in the ACI Code, Section 9.6.1.2), the minimum tensile reinforcement is determined from

$$A_{s,\min} = \frac{3\sqrt{f'_c}}{f_y} b_w d \geq \frac{200}{f_y} b_w d$$

The use of this minimum is conservative for footings. The ACI Code, Section 7.6.1.1, permits the use of a minimum reinforcement equal to that required for shrinkage and temperature steel in structural slabs of uniform thickness. This will always be somewhat less than that required by $A_{s,\min}$, but not necessarily less than that specified by the ACI Code, Section 9.6.1.3. In this book the criteria of the ACI Code, Section 7.6.1.1, will be used in isolated footing cases only as an absolute minimum of steel area to be provided.

Example 10-3

Design a square reinforced concrete footing to support an 18-in.-square tied concrete column, as shown in Figure 10-10. The column is a typical interior column in a building. Assume normal-weight concrete.

Solution:

1. The design data are as follows: service dead load = 225 kip, service live load = 175 kip, allowable soil pressure = 5000 psf (5.00 ksf), f'_c for the column = 4000 psi and for the footing = 3000 psi, f_y for all steel = 60,000 psi, and longitudinal column steel consists of No. 8 bars. The weight of earth = 100 lb/ft^3.
2. Assume a total footing thickness of 24 in. The footing weight may then be calculated as

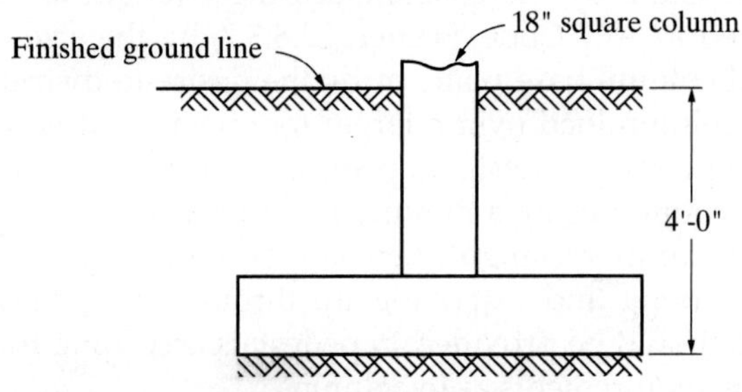

FIGURE 10-10 Sketch for Example 10-3.

$$0.150(2.0) = 0.300 \text{ ksf}$$

Because the bottom of the footing is to be 4 ft below the finished ground line, there will be 24 in. of earth on top of the footing. This depth of earth has a weight of

$$\frac{24}{12}(0.100) = 0.200 \text{ ksf}$$

Therefore, the *net allowable* soil pressure for the superimposed loads becomes

$$5.00 - 0.300 - 0.200 = 4.50 \text{ ksf}$$

The required area of footing may be determined using service loads and allowable soil pressure or by modifying both the service loads and the allowable soil pressure with the ACI load factors. Using the service loads,

$$\text{required } A = \frac{225 + 175}{4.50} = 88.9 \text{ ft}^2$$

Use a 9 ft-6 in. square footing. This furnishes an actual area A of 90.3 ft^2, which is greater than the required area.

3. The factored soil pressure from superimposed loads to be used for the footing design may now be calculated:

$$p_u = \frac{P_u}{A} = \frac{1.2(225) + 1.6(175)}{90.3} = 6.09 \text{ ksf}$$

4. The footing thickness is usually determined by shear strength requirements. Therefore, using the assumed footing thickness, check the shear strength. The thickness h was assumed to be 24 in. Therefore, the effective depth, based on a 3-in. cover for bottom steel and No. 8 bars in each direction, is

$$d = 24 - 3 - 1 = 20 \text{ in.}$$

This constitutes an average effective depth that will be used for design calculations for both directions.

5. The shear strength of individual column footings is governed by the more severe of two conditions: two-way action (punching shear) or one-way action (beam shear). The location of the critical section for each type of behavior is depicted in Figure 10-11. *For two-way action* (Figure 10-11a),

$$B = \text{column width} + \left(\frac{d}{2}\right)2$$
$$= 18 + 20 = 38 \text{ in.} = 3.17 \text{ ft}$$

The total factored shear acting on the critical section is

$$V_u = p_u(W^2 - B^2)$$
$$= 6.09(9.5^2 - 3.17^2)$$
$$= 488 \text{ kip}$$

The shear strength of the concrete ($\lambda = 1$ for normal weight concrete) is taken as the smallest of

a. $V_c = 4\sqrt{f'_c}\, b_0 d = 4\sqrt{3000}(4)(38)(20) = 666{,}000$ lb.

b. $V_c = \left(2 + \frac{4}{\beta_c}\right)\sqrt{f'_c}\, b_0 d$

$$= \left(2 + \frac{4}{1}\right)\sqrt{3000}(38)(4)(20) = 999{,}000 \text{ lb}$$

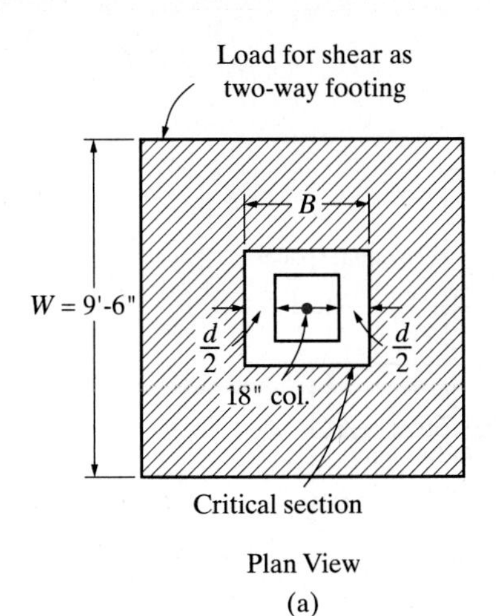

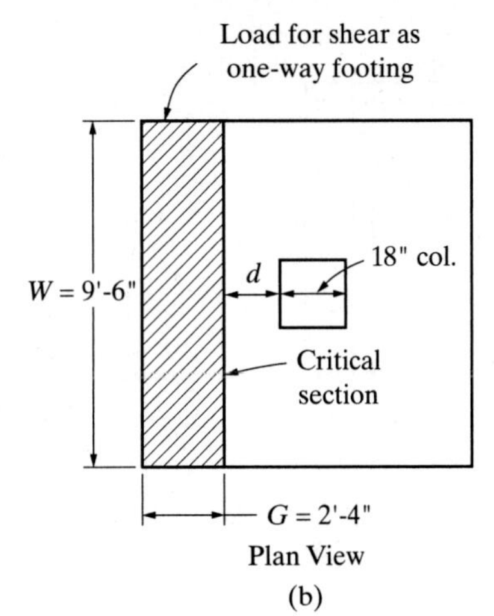

FIGURE 10-11 Footing shear analysis.

c. With $\alpha_s = 40$ for an interior column,

$$V_c = \left(\frac{\alpha_s d}{b_0} + 2\right)\sqrt{f'_c}\, b_0 d$$

$$= \left[\frac{40(20)}{38(4)} + 2\right]\sqrt{3000}(38)(4)(20) = 1{,}209{,}000 \text{ lb}$$

Thus, the lowest value governs and, $V_c = 666{,}000$ lb $= 666$ kip

$$\phi V_n = \phi V_c = 0.75(666) = 500 \text{ kip}$$

Therefore,

$$\phi V_n > V_u = 488 \text{ kip} \qquad \text{(O.K.)}$$

When V_u is relatively close to the shear strength of the concrete (ϕV_c), it indicates that the assumed footing thickness is approximately equal to that required for shear. If these two values were significantly different, the assumed footing thickness should be modified.

For one-way action, the total factored shear acting on the critical section is distance d from the face of the column:

$$V_u = p_u WG$$
$$= 6.09(9.5)(2.33)$$
$$= 134.8 \text{ kip}$$

The shear strength of the concrete is

$$V_c = 2\sqrt{f'_c}\, b_w d$$
$$= 2\sqrt{3000}(9.5)(12)(20)$$
$$= 250{,}000 \text{ lb}$$
$$= 250 \text{ kip}$$
$$\phi V_n = \phi V_c = 0.75(250) = 187.5 \text{ kip}$$

Therefore,

$$\phi V_n > V_u = 134.8 \text{ kip} \qquad \text{(O.K.)}$$

The 24-in.-deep footing is satisfactory with respect to shear. Our assumption of step 2 with regard to weight of the footing and the soil on the footing is satisfactory.

6. The critical section for bending moment may be taken at the face of the column, as depicted in Figure 10-12. Using the factored soil pressure and assuming the footing to act as a wide cantilever beam in both directions, the design moment may be computed:

$$M_u = p_u F\left(\frac{F}{2}\right)(W)$$
$$= 6.09(4)\left(\frac{4}{2}\right)(9.5)$$
$$= 463 \text{ ft-kip}$$

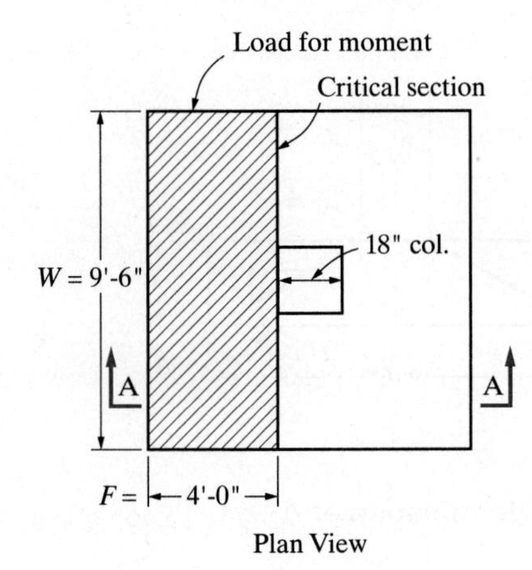

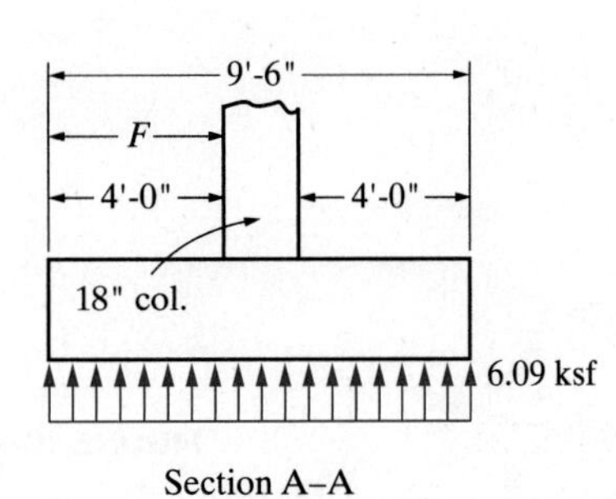

FIGURE 10-12 Footing moment analysis.

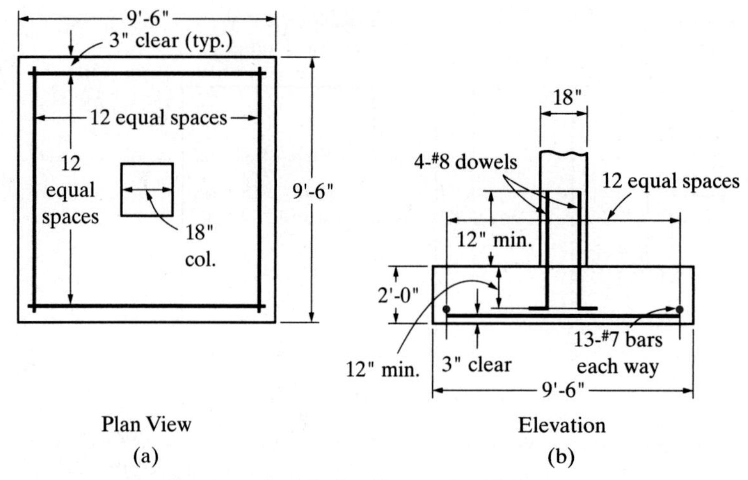

FIGURE 10-13 Design sketch for Example 10-3.

7. Assume $\phi = 0.90$ and design the tension steel at the bottom of the footing as follows:

$$\text{required } \bar{k} = \frac{M_u}{\phi bd^2} = \frac{463(12)}{0.9(9.5)(12)(20)^2} = 0.1354 \text{ ksi}$$

From Table A-8, the required $\rho = 0.0024$, $\varepsilon_t > 0.005$, and $\phi = 0.90$. Therefore,

$$\begin{aligned}\text{required } A_s &= \rho bd \\ &= 0.0024(9.5)(12)(20) \\ &= 5.47 \text{ in.}^2\end{aligned}$$

Check the ACI Code minimum reinforcement requirement. From Table A-5,

$$\begin{aligned}A_{s,\,min} &= 0.0033(9.5)(12)(20) \\ &= 7.52 \text{ in.}^2\end{aligned}$$

Of the two steel areas, the larger (7.52 in.^{2}) controls.

Because the footing is square and an average effective depth was used, the steel requirements in the other direction may be assumed to be identical. Therefore, use 13 No. 7 bars, bottom each way ($A_s = 7.80$ in.^{2} in each direction) and distribute the bars uniformly across the footing in each direction, as shown in Figure 10-13a.

Check the development length for the No. 7 bars.

This calculation for ℓ_d follows the eight-step procedure presented in Chapter 5, Section 5-2, and is summarized as follows:

1. $K_D = 82.2$
2. $\psi_t = 1.0$, $\psi_e = 1.0$, $\psi_s = 1.0$, $\lambda = 1.0$
3. $\psi_t \times \psi_e = 1.0$ (O.K.)
4. $c_b = 3.44$ in.
5. $K_{tr} = 0$
6. $(c_b + K_{tr})/d_b = 3.93$ in. Use 2.5 in.
7. $K_{ER} = 0.96$
8. $\ell_d = 27.6$ in. > 12 in. (O.K.)

Use $\ell_d = 27.6$ in. (minimum). The development length provided is 45 in., which is in excess of that required (O.K.).

8. The concrete bearing strength at the base of the column cannot exceed

$$\phi(0.85 f'_c A_1)$$

except where the supporting surface is wider on all sides than the loaded area, for which case the concrete bearing strength of the supporting surface cannot exceed

$$\phi(0.85 f'_c A_1)\sqrt{\frac{A_2}{A_1}}$$

As described in Section 8-4, A_2 is the lower base of the frustum of a pyramid having 1:2 sloping sides and fully contained within the support as shown in Figure 10-14. In this case, A_2 is the same as the area of the footing and

$$\sqrt{\frac{A_2}{A_1}} = \sqrt{\frac{90.3}{2.25}} = 6.3 > 2.0$$

Therefore, use 2.0. Then

$$\begin{aligned}\textit{footing} \text{ bearing strength} &= \phi(0.85 f'_c A_1)(2.0) \\ &= 0.65(0.85)(3.0)(18)^2(2.0) \\ &= 1074 \text{ kip}\end{aligned}$$

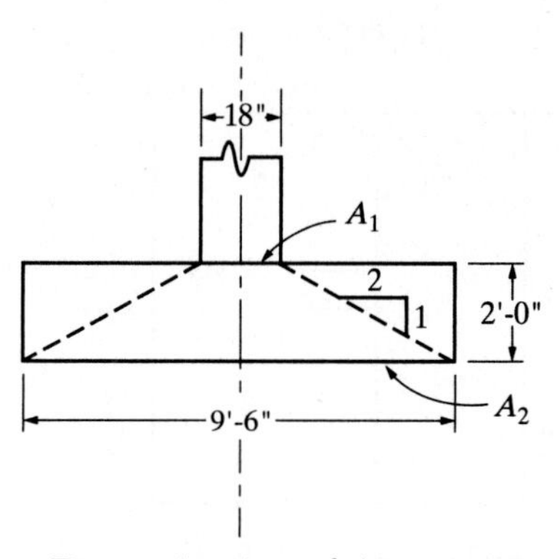

FIGURE 10-14 Determination of A_2.

The *column*-bearing strength is computed as follows:

$$\phi(0.85)f'_c A_1 = 0.65(0.85)(4.0)(18)^2 = 716 \text{ kip}$$

The *calculated* design-bearing load is

$$P_u = 1.2(225) + 1.6(175) = 550 \text{ kip}$$

Because $550 < 716 < 1074$, the entire column load can be transferred by concrete alone. The ACI Code, however, requires a minimum dowel area of

$$\text{required } A_s = 0.005A_g = 0.005(18)^2 = 1.62 \text{ in.}^2$$

Use a minimum of four bars. Four No. 6 bars, $A_s = 1.76$ in.^{2}, is satisfactory. It is general practice in a situation such as this, however, to use dowels of the same diameter as the column steel. Therefore, use four No. 8 dowels, and place one in each corner ($A_s = 3.16$ in.^{2}).

The development length for dowels into the column and footing must be adequate even though full load transfer can be made without dowels. The bars are in compression. The compression development length for the No. 8 dowels into the footings is ℓ_{dc} from Table A-12 and may be reduced by any applicable modification factors. We will use the modification factor for the case where the steel provided is in excess of the steel required:

$$\frac{\text{required } A_s}{\text{provided } A_s} = \frac{1.62}{3.16} = 0.51$$

Therefore,

$$\text{required } \ell_{dc} = 21.9(0.51) = 11.2 \text{ in.}$$

For bars in compression, ℓ_{dc} must not be less than 8 in. Therefore, use ℓ_{dc} of 12 in. into both the column and the footing, because f'_c for the column concrete is higher and the required ℓ_{dc} for the bars into the column would be less than that into the footing. The actual anchorage used may be observed in Figure 10-13b. The dowels should be placed adjacent to the corner longitudinal bars. Generally, these dowels are furnished with a 90° hook at their lower ends and are placed on top of the main footing reinforcement. This will tie the dowel in place and will reduce the possibility of the dowel being dislodged during construction. The hook cannot be considered effective as part of the required development length (ACI Code, Section 25.4.1.2).

10-7 RECTANGULAR REINFORCED CONCRETE FOOTINGS

Rectangular footings are generally used where space limitations require it. The design of these footings is very similar to that of the square column footing with the one major exception that each direction must be investigated independently. Shear is checked for two-way action in the normal way, but for one-way action, it is checked across the shorter side only. The bending moment must be considered separately for each direction. Each direction will generally require a different area of steel. The reinforcing steel running in the long direction should be placed in the bottom lower layer below the short-direction steel so that it may have the larger effective depth to carry the larger bending moments in that direction.

In rectangular footings, the *distribution* of the reinforcement is different than for square footings (ACI Code, Section 13.3.3.3). The reinforcement in the long direction should be uniformly distributed over the shorter footing width. A part of the required reinforcement in the short direction is placed in a band equal to the length of the short side of the footing centered on the column line or the pedestal (pier). The portion of the total required steel that should go into this band is

$$\frac{2}{\beta + 1}$$

where β is the ratio of the long side to the short side of the footing. The remainder of the reinforcement in the short direction is uniformly distributed in the outer portions of the footing. This distribution is depicted in Figure 10-15. Other features of the design are similar to those for the square column footing.

Example 10-4

Design a reinforced concrete footing to support an 18-in.-square tied interior concrete column, as shown in Figure 10-16. One dimension of the footing is limited to a maximum of 7 ft. Assume normal-weight concrete.

Solution:

1. The design data are as follows: service dead load = 175 kip, service live load = 175 kip, allowable soil pressure = 5000 psf (5.00 ksf), f'_c for both footing and column = 3000 psi, f_y for all steel = 60,000 psi, and longitudinal column steel consists of No. 8 bars. The weight of earth = 100 lb/ft^3.

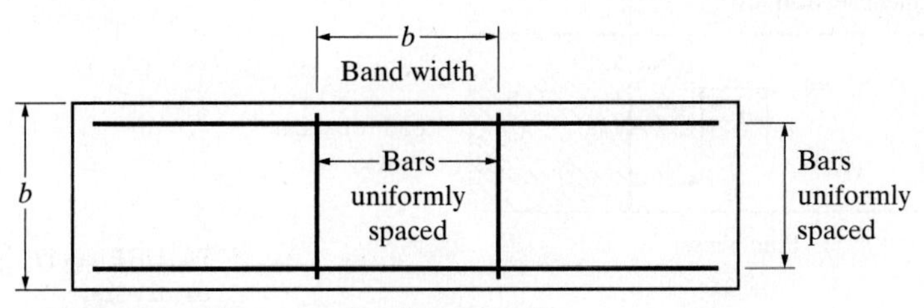

FIGURE 10-15 Rectangular footing plan.

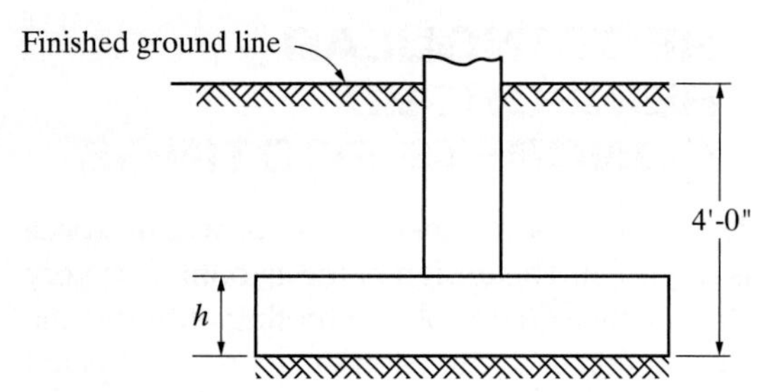

FIGURE 10-16 Sketch for Example 10-4.

2. Assume a total footing thickness h of 24 in. subject to later check. The footing weight may then be calculated as

$$0.150(2.0) = 0.300 \text{ ksf}$$

Because the bottom of the footing is to be 4 ft below the finished ground line, there will be 24 in. of earth on top of the footing. This depth of earth has a weight of

$$\frac{24}{12}(0.100) = 0.200 \text{ ksf}$$

Therefore, the *net allowable* soil pressure for the superimposed loads becomes

$$5.00 - 0.300 - 0.200 = 4.50 \text{ ksf}$$

Based on service loads, the required area of footing may be calculated:

$$\text{required } A = \frac{175 + 175}{4.50} = 77.8 \text{ ft}^2$$

Use a rectangular footing 7 ft-0 in. by 11 ft-6 in. This furnishes an actual area, A of 80.5 ft^2.

3. The factored soil pressure from superimposed loads to be used for the footing design may now be calculated:

$$p_u = \frac{P_u}{A} = \frac{1.2(175) + 1.6(175)}{80.5} = 6.09 \text{ ksf}$$

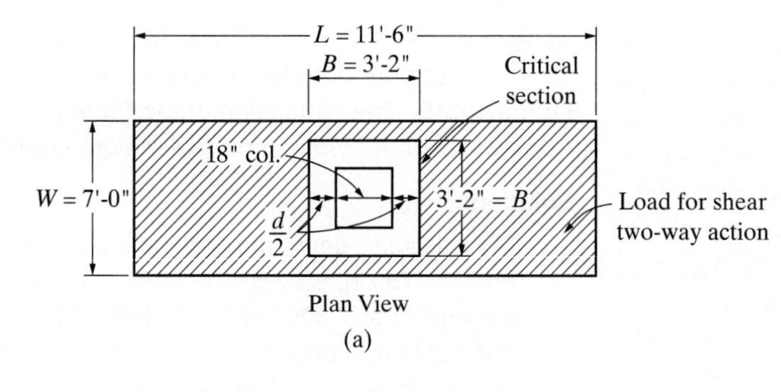

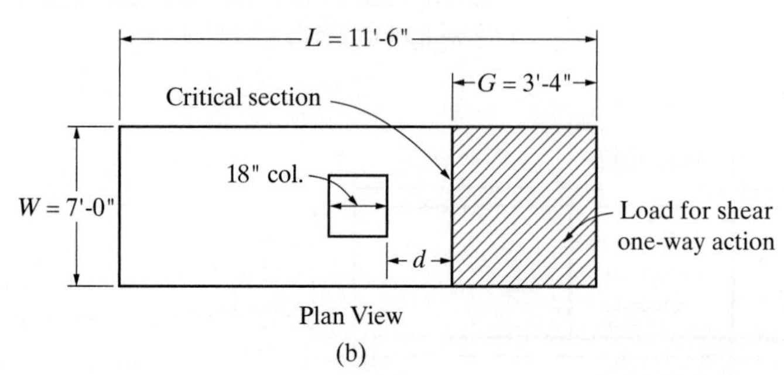

FIGURE 10-17 Footing shear analysis.

4. The footing thickness h was assumed to be 24 in. Therefore, the effective depth, based on a 3-in. cover for bottom steel and 1-in.-diameter bars in each direction, will be

$$d = 24 - 3 - 1 = 20 \text{ in.}$$

This constitutes an *average* effective depth, which will be used for design calculations for both directions.

5. Checking the shear strength *for two-way action* (with reference to Figure 10-17a),

$$B = \text{column width} + \left(\frac{d}{2}\right)2$$
$$= 18 + 20 = 38 \text{ in.} = 3.17 \text{ ft}$$

The total factored shear acting on the critical section is

$$V_u = p_u(A - B^2)$$
$$= 6.09(80.5 - 3.17^2)$$
$$= 429 \text{ kip}$$

The shear strength of the concrete is taken as the smallest of

a. $V_c = 4\sqrt{f'_c}\, b_0 d = 4\sqrt{3000}(38)(4)(20) = 666{,}000 \text{ lb}$

b. $V_c = \left(2 + \frac{4}{\beta_c}\right)\sqrt{f'_c}\, b_0 d$

$$= \left(2 + \frac{4}{1}\right)\sqrt{3000}(38)(4)(20) = 999{,}000 \text{ lb}$$

c. With $\alpha_s = 40$ for an interior column,

$$V_c = \left(\frac{\alpha_s d}{b_0} + 2\right)\sqrt{f'_c}\, b_0 d$$
$$= \left(\frac{40(20)}{38(4)} + 2\right)\sqrt{3000}(38)(4)(20) = 1{,}209{,}000 \text{ lb}$$

The lowest value governs, which means $V_c = 666{,}000$ lb $= 666$ kip. Thus

$$\phi V_n = \phi V_c = 0.75(666) = 500 \text{ kip}$$

Therefore,

$$\phi V_n > V_u = 429 \text{ kip} \qquad \text{(O.K.)}$$

For one-way action, consider shear across the short side only. The critical section is at a distance equal to the effective depth of the member from the face of the column (see Figure 10-17b).

The total factored shear acting on the critical section is

$$\begin{aligned} V_u &= p_u WG \\ &= 6.09(7.0)(3.33) \\ &= 142.0 \text{ kip} \end{aligned}$$

The nominal shear strength of the concrete is

$$\begin{aligned} V_c &= 2\sqrt{f'_c}\, b_w d \\ &= 2\sqrt{3000}(7.0)(12)(20) \\ &= 184{,}000 \text{ lb} \\ &= 184 \text{ kip} \end{aligned}$$

from which

$$\phi V_n = \phi V_c = 0.75(184) = 138.0 \text{ kip}$$

$$\phi V_n < V_u = 142 \text{ kip}$$

This is unsatisfactory. We will determine the required d based on one-way shear using $\phi V_c = V_u$.

$$\phi V_n = \phi 2\sqrt{f'_c}\, b_w d = V_u = 142.0 \text{ kip}$$

from which

$$\begin{aligned} \text{Required } d &= \frac{V_u}{\phi 2\sqrt{f'_c}\, b_w} = \frac{142.0(1000 \text{ lb/kip})}{0.75(2)\sqrt{3000}\,(7.0)(12)} \\ &= 20.6 \text{ in.} \end{aligned}$$

Therefore, the required $h = 20.6 + 3 + 1 = 24.6$ in. Use $h = 25$ in. The effect of this change on the previous calculations is small and no other revisions are considered warranted.

$$\text{new } d = 25 - 1 - 3 = 21.0 \text{ in.}$$

6. For bending moment, each direction must be considered independently, with the critical section taken at the face of the column. Using the actual factored soil pressure and assuming the footing to act as a wide cantilever beam in each direction, the design moment may be calculated. With reference to Figure 10-18b, for moment in the long direction,

$$\begin{aligned} M_u &= p_u F\left(\frac{F}{2}\right)(W) \\ &= 6.09(5)\left(\frac{5}{2}\right)(7.0) \\ &= 533 \text{ ft-kip} \end{aligned}$$

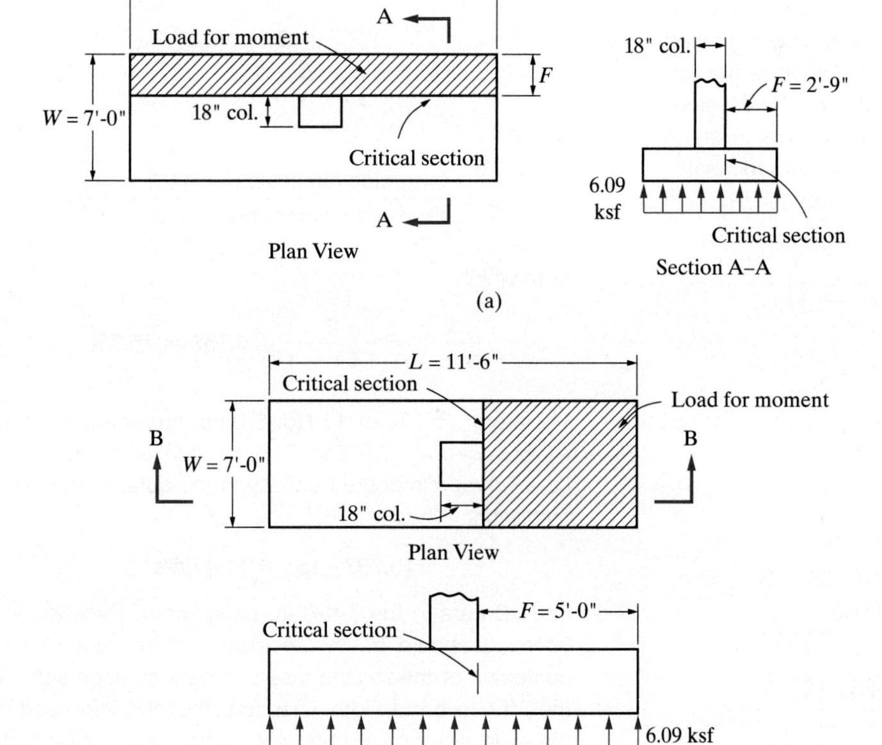

FIGURE 10-18 Footing moment analysis.

For moment in the short direction (Figure 10-18a),

$$M_u = p_u F\left(\frac{F}{2}\right)(L)$$

$$= 6.09(2.75)\left(\frac{2.75}{2}\right)(11.5)$$

$$= 265 \text{ ft-kip}$$

7. Assume $\phi = 0.90$ and design the tension steel as follows. *For the long direction*, where $M_u = 533$ ft-kip,

$$\text{required } \bar{k} = \frac{M_u}{\phi bd^2} = \frac{533(12)}{(0.9)(7)(12)(21)^2}$$

$$= 0.1918 \text{ ksi}$$

From Table A-8, the required $\rho = 0.0034$, $\varepsilon_t > 0.005$, and $\phi = 0.90$. Therefore,

$$\text{required } A_s = \rho bd$$

$$= 0.0034(7)(12)(21)$$

$$= 6.00 \text{ in.}^2$$

Check the ACI Code minimum reinforcement requirement. From Table A-5,

$$A_{s,\min} = 0.0033(7)(12)(21)$$

$$= 5.82 \text{ in.}^2$$

Of the two steel areas, the larger (6.00 in.^{2}) controls; therefore, use 10 No. 7 bars ($A_s = 6.00$ in.^{2}). These bars will run in the long direction and will be distributed uniformly across the 7 ft-0 in. width. They will be placed in the bottom lower layer where they will have the advantage of slightly greater effective depth. The development length must be checked for these bars. Assume uncoated bars. This calculation for ℓ_d follows the eight-step procedure presented in Chapter 5, Section 5-2, and is summarized as follows:

1. $K_D = 82.2$
2. $\psi_t = 1.0$, $\psi_e = 1.0$, $\psi_s = 1.0$, $\lambda = 1.0$
3. $\psi_t \times \psi_e = 1.0$ (O.K.)
4. $c_b = 3.44$ in.
5. $K_{tr} = 0$
6. $(c_b + K_{tr})/d_b = 3.93$ Use 2.5
7. $K_{ER} = 1.00$
8. $\ell_d = 28.8$ in. $>$ 12 in. (O.K.)

The development length furnished $= 60 - 3 = 57$ in., which is in excess of that required (O.K.).

For the short direction, where $M_u = 265$ ft-kip,

$$\text{required } \bar{k} = \frac{M_u}{\phi bd^2} = \frac{265(12)}{0.9(11.5)(12)(21)^2}$$

$$= 0.0581 \text{ ksi}$$

From Table A-8, the required $\rho = 0.0010$, $\varepsilon_t > 0.005$, and $\phi = 0.90$. Therefore,

$$\text{required } A_s = \rho bd$$

$$= 0.0010(11.5)(12)(21)$$

$$= 2.90 \text{ in.}^2$$

Check the ACI Code minimum reinforcement requirement. From Table A-5,

$$A_{s,\min} = 0.0033(11.5)(12)(21)$$

$$= 9.56 \text{ in.}^2$$

The $A_{s,\min}$ requirement need not be applied if the area of steel provided is at least one-third greater than that required (ACI Code, Section 9.6.1.3), however. Therefore, the required steel area may be calculated from

$$\text{required } A_s = 1.33(2.90) = 3.86 \text{ in.}^2$$

Checking further, using the ACI Code, Section 7.6.1.1, as an absolute minimum, the required steel area is

$$A_s = 0.0018bh = 0.0018(11.5)(12)(25) = 6.21 \text{ in.}^2$$

Therefore, 6.21 in.^{2} controls. Use 15 No. 6 bars ($A_s = 6.60$ in.^{2}). These bars will run in the short direction in the bottom upper layer, but will *not* be distributed uniformly across the 11 ft-6 in. side.

In rectangular footings, a portion of the total reinforcement required in the short direction is placed in a band centered on the column and having a width equal to the short side. The portion of the total required steel that goes into this band is

$$\frac{2}{\beta + 1}$$

where

$$\beta = \frac{\text{long-side dimension}}{\text{short-side dimension}} = \frac{11.5}{7.0} = 1.64$$

from which

$$\frac{2}{\beta + 1} = \frac{2}{1.64 + 1} = 0.757 = 75.7\%$$

Therefore, 75.7% of 15 No. 6 bars must be placed in a band width = 7 ft-0 in. The balance of the required bars will be distributed equally in the outer portions of the footing. Thus

$$(0.757)(15) = 11.4 \text{ bars}$$

Use 12 bars in the 7 ft-0 in. band width. Because reinforcing should be symmetrical with respect to the centerline of the footing, use two bars on each side of the 7 ft-0 in. band width: Therefore, the total steel used in the short direction will be 16 No. 6 bars ($A_s = 7.04$ in.^{2}). The bar arrangement is depicted in Figure 10-19.

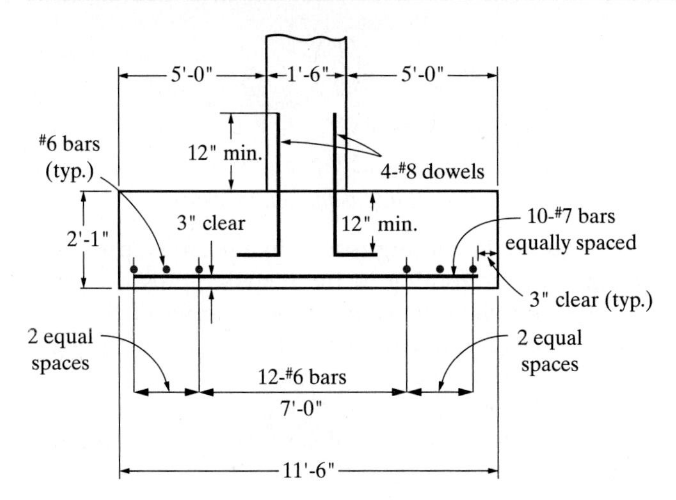

FIGURE 10-19 Design sketch for Example 10-4.

The development length will be checked for the No. 6 bars in the center band. Assume uncoated bars. This calculation for ℓ_d follows the eight-step procedure presented in Chapter 5, Section 5-2, and is summarized as follows:

1. $K_D = 82.2$
2. $\psi_t = 1.0, \psi_e = 1.0, \psi_s = 0.8, \lambda = 1.0$
3. $\psi_t \times \psi_e = 1.0$ (O.K.)
4. $c_b = 3.38$ in.
5. $K_{tr} = 0$
6. $(c_b + K_{tr})/d_b = 4.51$ Use 2.5
7. $K_{ER} = 0.882$
8. $\ell_d = 17.4$ in. > 12 in. (O.K.)

Use $\ell_d = 17.4$ in. (minimum). The development length furnished $= 33 - 3 = 30$ in., which is in excess of that required (O.K.).

8. Because the supporting surface is wider on all sides, the bearing strength for the footing may be computed as follows:

$$\text{footing bearing strength} = \phi(0.85 f'_c A_1)\sqrt{\frac{A_2}{A_1}}$$

where A_2 is calculated with reference to Figure 10-20 and is seen to be a square, 7 ft-0 in. on each side.

$$\sqrt{\frac{A_2}{A_1}} = \sqrt{\frac{49.0}{2.25}} = 4.67 > 2.0$$

Therefore, use 2.0. Then,

$$\begin{aligned}\text{footing bearing strength} &= \phi(0.85 f'_c A_1)(2.0)\\ &= 0.65(0.85)(3.0)(18)^2(2.0)\\ &= 1074 \text{ kip}\end{aligned}$$

The column bearing strength is computed as follows:

$$\phi(0.85) f'_c A_1 = 0.65(0.85)(3)(18)^2 = 537 \text{ kip}$$

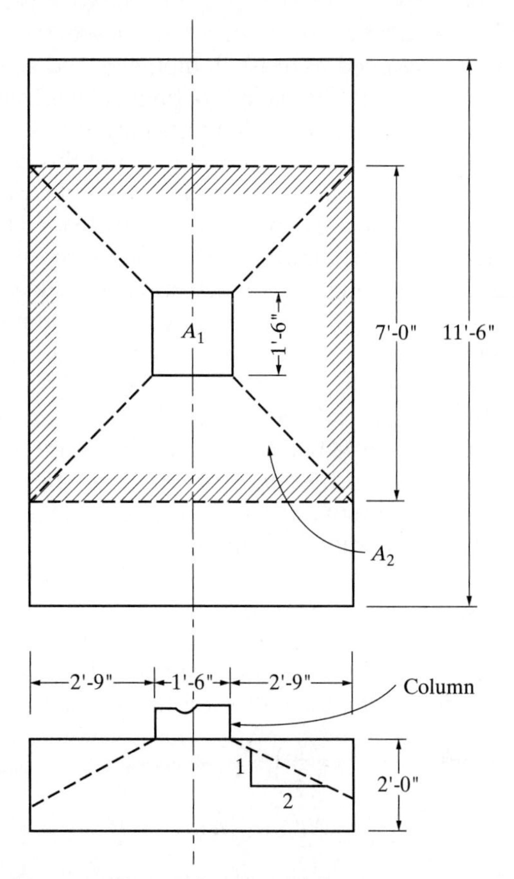

FIGURE 10-20 Determination of A_2.

The factored bearing load is

$$\begin{aligned}P_u &= 1.2(175) + 1.6(175)\\ &= 490 \text{ kip}\end{aligned}$$

Because 490 kip < 537 kip < 1074 kip, the entire column load can be transferred by concrete alone.

For proper connection between column and footing, use four No. 8 dowels (to match the column steel), one in each corner. The development length requirement for the No. 8 dowels is identical to that of Example 10-3.

10-8 ECCENTRICALLY LOADED FOOTINGS

Isolated spread footings may sometimes be subjected to uni-axial or bi-axial moments due to eccentric gravity loads and/or lateral loads. Where footings are subjected to eccentric vertical loads or to moments transmitted by the supported column, the design of the footing varies somewhat from that of the preceding sections. The soil pressure is no longer uniform across the footing width, but may be assumed to vary linearly. The resultant force should be within the middle one-third of the footing base to ensure a positive contact between the footing and the soil. With the soil pressure distribution known, the footing must be designed to resist all moments and shears, similar to concentrically loaded footings. The effects of load eccentricity on isolated footings may result in an undesirable large rotation of the footing, but this can be mitigated by connecting or strapping the eccentrically loaded footing to an adjacent concentrically loaded footing, as discussed in Section 10-10. A few examples of isolated spread footings subjected to moments and gravity loads are shown in Figure 10-21, where the moments in the footing are caused by the horizontal and vertical components of the diagonal members. The soil pressure distributions in spread footings subjected to an axial load P and moment M will be as shown in Figure 10-22.

Assuming a rigid footing and an elastic homogeneous soil, the pressure distribution under the footing will vary linearly from a minimum pressure of p_{min} to a maximum value of p_{max}. If the size of the footing is selected so that the footing is never under a negative pressure underneath the footing, the minimum and maximum soil pressures can be calculated, assuming a rigid footing, using the following stress equations:

$$p_{min} = \frac{P}{A} - \frac{M\left(\frac{L}{2}\right)}{\frac{WL^3}{12}} = \frac{P}{WL} - \frac{6M}{WL^2} \geq 0$$

$$\leq \textit{Net allowable soil pressure}$$

$$p_{max} = \frac{P}{A} + \frac{M\left(\frac{L}{2}\right)}{\frac{WL^3}{12}} = \frac{P}{WL} + \frac{6M}{WL^2}$$

$$\leq \textit{Net allowable soil pressure}$$

If the size of the footing results in a negative pressure, the area of the footing in the negative pressure zone is neglected since tension is not possible in the soil. The maximum moment on the footing to avoid negative soil pressure can be calculated by equating p_{min} to zero. Therefore, if $M \leq PL/6$ (i.e. if the load eccentricity, $e \leq L/6$), there will be no negative pressure beneath the footing. At a moment, $M = PL/6$ (i.e. the load eccentricity, $e = L/6$), the soil pressures are

$$p_{min} = 0$$

$$p_{max} = \frac{P}{WL} + \frac{6\left(\frac{PL}{6}\right)}{WL^2} = \frac{2P}{WL}$$

$$\leq \textit{Net allowable soil pressure}$$

For footings with large moments where $PL/6 < M < PL/2$ (i.e. $L/6 < e < L/2$), the modified pressure distribution will be as shown in Figure 10-22(c). Since the soil cannot resist tension, the zone of the footing with negative soil pressure is neglected. Only the footing area

FIGURE 10-21 Eccentrically loaded footings. (Courtesy of Abi O. Aghayere)

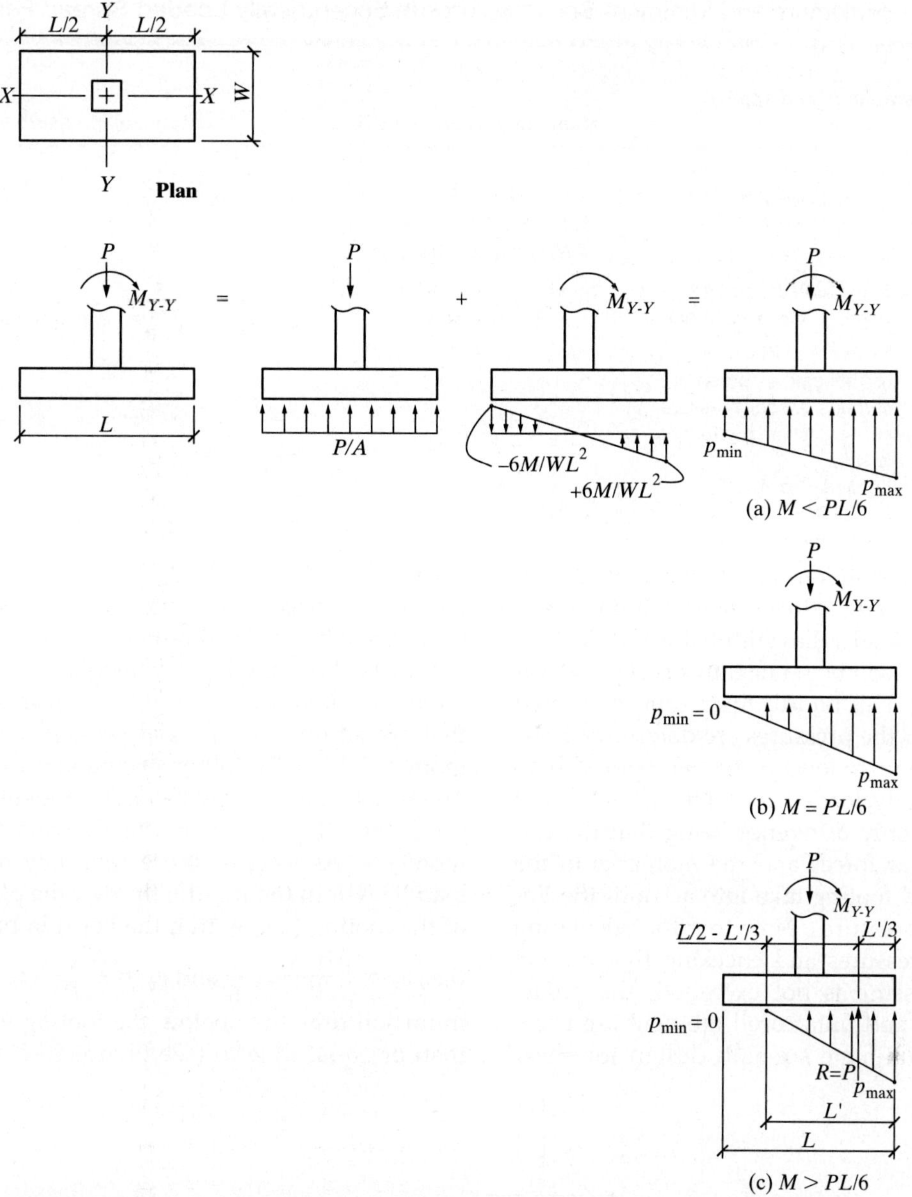

FIGURE 10-22 Pressure distributions in eccentrically loaded spread footings.

with positive soil pressure is considered. In this case the maximum soil pressure will be

$$p_{max} = \frac{2P}{WL'} \leq \textit{Net allowable soil pressure}$$

and

$$p_{min} = 0$$

where, L' is the length of the footing in contact with the soil (i.e. length of the positive soil pressure zone). Using equilibrium of forces and noting that the resultant of the reaction due to the soil pressure is equal to the applied load, P, and summing moments of this resultant reaction about the location of the gravity load, P, yields

$$P\left(\frac{L}{2} - \frac{L'}{3}\right) = M$$

Noting that the eccentricity of the vertical load, $e = M/P$, we can solve for L′ in the previous equation as follows:

$$\left(\frac{L}{2} - \frac{L'}{3}\right) = \frac{M}{P} = e$$

$$\text{Therefore, } L' = 3\left(\frac{L}{2} - e\right)$$

The maximum soil pressure for this case of large moments can be rewritten as

$$p_{max} = \frac{2P}{WL'} = \frac{2P}{3W\left(\frac{L}{2} - e\right)}$$

$$\leq \textit{Net allowable soil pressure}$$

TABLE 10-2 Maximum and Minimum Soil Pressures in Eccentrically Loaded Spread Footings

Soil pressure distribution and applied moment, *M*	Minimum soil pressure p_{min}	Maximum soil pressure p_{max}
Positive soil pressure beneath the spread footing over entire length, *L*. ($M \leq PL/6$ or $e \leq L/6$)	$\frac{P}{WL} - \frac{6M}{WL^2} \geq 0$ $\leq$ *Net allowable soil pressure*	$\frac{P}{WL} + \frac{6M}{WL^2} \leq$ *Net allowable soil pressure*
Positive soil pressure beneath the spread footing over entire length, *L*. ($M = PL/6$ or $e = L/6$)	0	$\frac{2P}{WL} \leq$ *Net allowable soil pressure*
Negative soil pressure beneath a portion of the spread footing and positive soil pressure exists over a length $L' < L$ where the footing is in contact with the soil. ($PL/6 < M < PL/2$ or $L/6 < e < L/2$)	0	$\frac{2P}{3W\left(\frac{L}{2} - e\right)} \leq$ *Net allowable soil pressure*

The pressure distributions for all the cases considered above are summarized in Table 10-2. If the eccentricity of the vertical load, e, lies within the middle third of the footing, there will be no negative soil pressures beneath the footing. This middle third zone is referred to as the kern. After the pressures are determined, the design of the footing follows a similar approach to that for concentrically loaded footings with uniform soil pressures, the only difference being that the calculation of the shear forces and the moments in the eccentrically loaded footing take into account the linearly varying soil pressures. Note that for calculating the applied soil pressures and checking that the net allowable soil pressure is not exceeded, the unfactored vertical load and unfactored moment are used. However, for the ultimate strength design for shear and flexure, the factored vertical load and factored moment are used. An isolated spread footing may also be subjected to bi-axial moments simultaneously (i.e., moments about the two orthogonal axes in plan); the moments about one axis will result in soil pressures that are additive to the soil pressures caused by the moment about the other orthogonal axis or the soil pressures from the orthogonal moments will negate each other, depending on the direction of the applied moments. As long as the eccentricity of the vertical load lie within the middle third of the plan dimension of the footing (i.e., within the kern) in both directions (i.e., $e_x = \frac{M_{Y-Y}}{P} \leq \frac{L}{6}$ and $e_y = \frac{M_{X-X}}{P} \leq \frac{W}{6}$), the minimum soil pressures below the footing will be greater than or equal to zero (see Figure 10-23). In all cases,

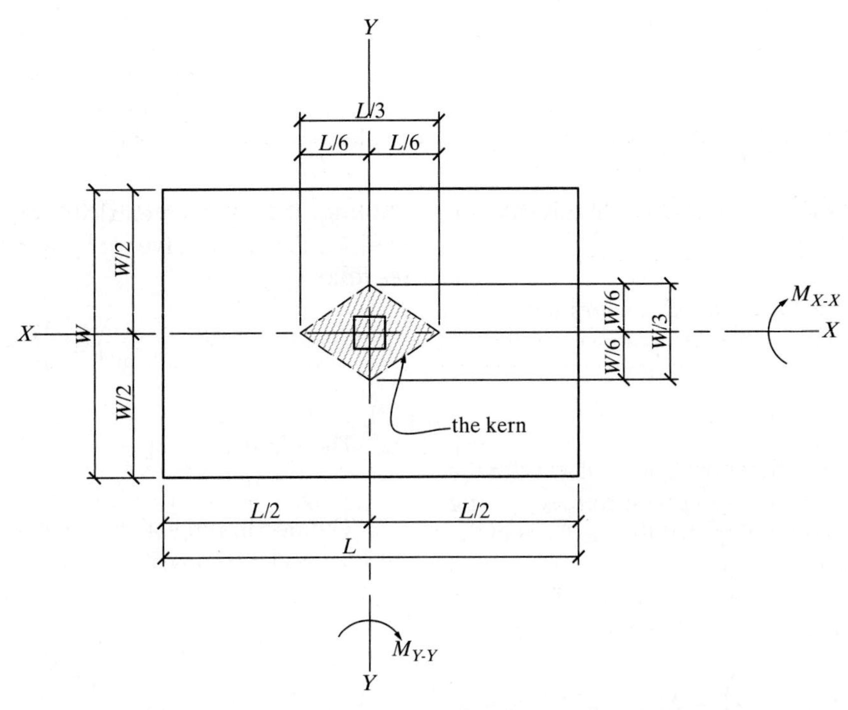

FIGURE 10-23 The kern in eccentrically loaded footings with bi-axial moments.

the designer should ensure that the size of the footing selected is such that the maximum soil pressure, p_{max}, does not exceed the net allowable soil pressure. The nonuniform pressure distribution underneath eccentrically loaded footings leads to differential settlement of the soil which could cause excessive rotation of the footing and the supported structure.

Example 10-5

Eccentrically Loaded Spread Footing

A 10 ft × 10 ft × 24″ thick footing supports an 18″ × 18″ column and is subjected to the following service loads and moments: $D = 112.5$ kip, $L = 87.5$ kip, $M_{D,\,Y\text{-}Y} = 187.5$ ft-kip, $M_{L,\,Y\text{-}Y} = 145.8$ kip. If the footing is reinforced with 14 #7 bars bottom each way (BEW), check the adequacy of the footing under the eccentric loading. The Geotechnical report specifies a net allowable soil pressure of 4500 psf. Assume f'_c of 3000 psi and f_y of 60,000 psi. Neglect the weight of the footing.

Solution:

1. Calculate the unfactored vertical load and moment, and check the applied soil pressure.

$$P_{D+L} = 1.0(112.5) + 1.0(87.5) = 200 \text{ kip}$$

$$M_{D+L,\,Y\text{-}Y} = 1.0(187.5) + 1.0(145.8) = 333.3 \text{ ft-kip}$$

$$e_{X-X} = \frac{M_{Y-Y}}{P} = \frac{333.3}{200} = 1.67 \text{ ft}$$

$$L/6 = 10 \text{ ft}/6 = 1.67 \text{ ft}$$

Since $e_{X-X} = L/6$,

$$p_{min} = 0$$

$$p_{max} = \frac{2P}{WL} = \frac{2(200 \text{ kip})}{(10 \text{ ft})(10 \text{ ft})} = 4 \text{ ksf}$$

$$\leq \textit{Net allowable soil pressure} = 4.5 \text{ ksf} \quad \text{(O.K.)}$$

For purposes of illustration, we will recalculate p_{min} and p_{max} using the general equation for the case when $e_{X-X} \leq L/6$.

$$p_{min} = \frac{P}{WL} - \frac{6M}{WL^2} = \frac{(200 \text{ kip})}{(10 \text{ ft})(10 \text{ ft})} - \frac{6(333.3 \text{ ft} - \text{kip})}{(10 \text{ ft})(10 \text{ ft})^2}$$

$$= 2 - 2 = 0 \text{ ksf}$$

$$\geq 0$$

$$\leq \textit{Net allowable soil pressure} = 4.5 \text{ ksf} \quad \text{(O.K.)}$$

$$p_{max} = \frac{P}{WL} + \frac{6M}{WL^2} = \frac{(200 \text{ kip})}{(10 \text{ ft})(10 \text{ ft})} + \frac{6(333.3 \text{ ft} - \text{kip})}{(10 \text{ ft})(10 \text{ ft})^2}$$

$$= 2 + 2 = 4 \text{ ksf}$$

$$\leq \textit{Net allowable soil pressure} = 4.5 \text{ ksf} \quad \text{(O.K.)}$$

2. Calculate the factored load and moment, and the factored soil pressure that will be used to check the footing for shear and flexure. The factored load and moment are

$$P_u = 1.2(112.5) + 1.6(87.5) = 275 \text{ kip}$$

$$M_{u,\,Y\text{-}Y} = 1.2(187.5) + 1.6(145.8) = 458.3 \text{ ft-kip}$$

$$e_{u,\,X-X} = \frac{M_{u,\,Y-Y}}{P} = \frac{458.3 \text{ ft} - \text{kip}}{275 \text{ kip}} = 1.67 \text{ ft}$$

$$L/6 = 10/6 = 1.67 \text{ ft}$$

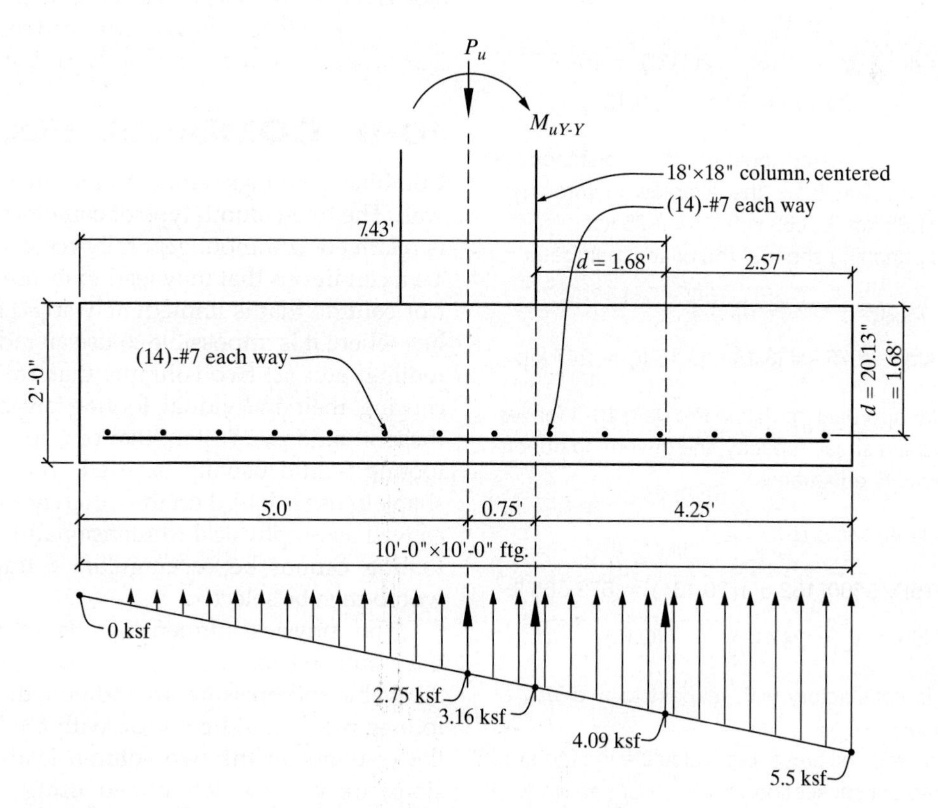

FIGURE 10-24 Factored soil pressures for Example 10-5.

Since $e_{u,\,x\text{-}x} = L/6$,

$$p_{u,\,\min} = 0$$

$$p_{u,\,\max} = \frac{2P_u}{WL} = \frac{2(275\text{ kip})}{(10\text{ ft})(10\text{ ft})} = 5.5\text{ ksf}$$

3. Check one-way shear at the critical section of "d" from the face of the column using the factored pressures (see Figure 10-24).

   Since #7 (i.e., 7/8″ diameter) bottom bars each way is specified, the average effective depth, $d = 24'' - 3''\text{ cover} - (7/8'') = 20.13''$ will be used.

   The factored shear,

$$V_u = (10')\left[4.09\text{ ksf }(2.57') + \frac{1}{2}(5.5\text{ ksf} - 4.09\text{ ksf})(2.57')\right]$$

$$= 123.2\text{ kip}$$

   The one-way shear strength of the concrete is

$$\phi V_c = 2\phi\sqrt{f'_c}\,b_w d$$
$$= 2(0.75)\sqrt{3000}(10)(12)(20.13)$$
$$= 264{,}620\text{ lb}$$
$$= 265\text{ kip}$$

$$\phi V_n = \phi V_c = 265\text{ kip} > V_u = 123.2\text{ kip} \qquad \text{(O.K.)}$$

   The footing is satisfactory with respect to one-way shear.

4. Check the two-way or punching shear at the critical perimeter of d/2 from the face of the column using the average factored pressure at the centroid of the column.

   The critical perimeter for two-way or punching shear is

$$b_o = 4(c + d) = 4(18'' + 20.13'') = 152.5''$$
$$c + d = 18'' + 20.13'' = 38.13'' = 3.18\text{ ft}$$

   At the critical perimeter located at a distance of d/2 from the column face, the average factored soil pressure = (1.88 ksf + 3.63 ksf)/2 = 2.75 ksf.

   The factored punching shear at the critical perimeter is

$$V_u = P_u - p_{avg}((c + d)(c + d)$$
$$= 275\text{ kip} - 2.75\text{ ksf }(3.18\text{ ft})(3.18\text{ ft}) = 247\text{ kip}$$

   Using a similar approach to that in Problem 10-3 for calculating punching shear capacity, the governing punching shear capacity equation is

$$\phi V_n = \phi V_c = 4\phi\sqrt{f'_c}\,b_0 d$$
$$= 4(0.75)\sqrt{3000}(152.5'')(20.13'') = 672{,}565\text{ lb}$$
$$= 673\text{ kip} > V_u = 247\text{ kip} \qquad \text{(O.K.)}$$

   The footing is satisfactory with respect to two-way or punching shear.

5. Using the factored soil pressures, calculate the factored moment at the critical section which occurs at the face of the column.

$$M_u = (10\text{ ft})\left[\frac{1}{2}(5.5\text{ ksf} - 3.16\text{ ksf})(4.25')(2/3 \times 4.25')\right.$$
$$\left. + 3.16\text{ ksf }(4.25')\left(\frac{1}{2}\right)(4.25')\right]$$
$$= 567\text{ ft-kip}$$

   Assume $\phi = 0.90$ and design the tension steel as follows:

$$\text{required } \bar{k} = \frac{M_u}{\phi b d^2} = \frac{567(12)}{0.9(10)(12)(20.13)^2}$$
$$= 0.1555\text{ ksi}$$

   From Table A-8, the required $\rho = 0.0024$, $\varepsilon_t = 0.005$, and $\phi = 0.90$. Therefore

$$\text{required } A_s = \rho b d$$
$$= 0.0027(10)(12)(20.13)$$
$$= 6.52\text{ in.}^2$$

   Check the ACI Code minimum reinforcement requirement. From Table A-5,

$$A_{s,\,\min} = 0.0033(10)(12)(20.13)$$
$$= 7.97\text{ in.}^2$$

   Of the two steel areas, the larger (7.97 in.^{2}) controls.

   Because the footing is square and an average effective depth was used, the steel requirements in the other direction may be assumed to be identical. Therefore the specified 14 No. 7 bars each way which yields an area ($A_s = 8.4\text{ in.}^2$ in each direction) is greater than the steel area required. Therefore, the footing is adequate to resist the gravity loads and moments. All other design checks such as for bond and anchorage, and bearing can be carried out similar to what was done in Example 10-3.

10-9 COMBINED FOOTINGS

Combined footings support more than one column or wall. The two-column type of combined footing, which is relatively common, generally results from necessity. Two conditions that may lead to its use are: (1) an exterior column that is immediately adjacent to a property line where it is impossible to use an individual column footing, and (2) two columns that are closely spaced, causing their individual footing areas to overlap. In these situations, a rectangular or trapezoidal combined footing would usually be used. The choice of which shape to use is based on the difference in column loads as well as on physical (dimensional) limitations. If the footing cannot be rectangular, a trapezoidal shape would then be selected.

The physical dimensions (except thickness) of the combined footing are generally established by the allowable soil pressure. In addition, the centroid of the footing area should coincide with the line of action of the resultant of the two column loads. These dimensions are usually determined using service loads in combination with the net allowable soil pressure.

Example 10-6

Determine the shape and proportions of a combined footing subject to two column loads, as shown in Figure 10-25.

Solution:

1. The design data are as follows: Service load on the footing from column A is 300 kip and from column B is 500 kip, and the allowable soil pressure is 6.00 ksf. Therefore, the unfactored resultant, $R = 300$ kip + 500 kip = 800 kip.
2. Locate the resultant column load by a summation of moments about point Z in Figure 10-25:

$$\Sigma M_z = 300(2) + 500(18) = 800(x)$$

from which

$$x = 12 \text{ ft-0 in. (measured from } Z)$$

3. Assuming a rectangular shape, establish the length of footing L so that the centroid of the footing area coincides with the line of action of resultant force R:

$$\text{required } L = 12(2) = 24 \text{ ft-0 in.}$$

4. Assume a footing thickness of 3 ft-0 in. Therefore, its weight = 0.150(3) = 0.450 ksf, and the net allowable soil pressure for superimposed loads = 6.00 − 0.450 = 5.55 ksf. This neglects any soil *on* the footing.
5. The footing area required is

$$\frac{R}{5.55} = \frac{800}{5.55} = 144.1 \text{ ft}^2$$

6. With a length = 24 ft-0 in., the footing width W required is

$$\frac{144.1}{24} = 6 \text{ ft-0 in.}$$

7. The actual uniform soil pressure is

$$\frac{800}{6(24)} + 0.450 = 6.00 \text{ ksf (O.K.)}$$

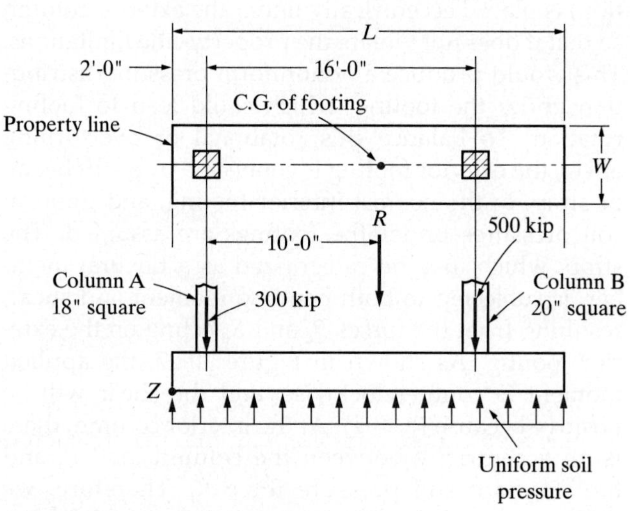

FIGURE 10-25 Sketch for Example 10-6.

Example 10-7

Using the design data from Example 10-6, determine the proportions of a combined footing if the footing length is limited to 22 ft.

Solution:

1. The resultant column load is located as in Example 10-6 at a point 12 ft-0 in. from point Z.
2. Assume a footing thickness of 3 ft-0 in. Therefore, its weight = 3(0.150) = 0.450 ksf, and the allowable soil pressure for superimposed loads is 6.00 − 0.450 = 5.55 ksf.
3. The footing area required is then

$$\frac{800}{5.55} = 144.1 \text{ ft}^2$$

4. Assume a trapezoidal shape. The area A of a trapezoid is (see Figure 10-26)

$$A = \frac{(b + b_1)L}{2}$$

from which

$$b + b_1 = \frac{2A}{L} = \frac{2(144.1)}{22} = 13.1 \text{ ft}$$

5. The center of gravity of the trapezoid and the resultant force of the column loads are to coincide. The location of the center of gravity, a distance c from point Z, may be written

$$c = \frac{L(2b + b_1)}{3(b + b_1)} = \frac{L(b + b + b_1)}{3(b + b_1)} = 12 \text{ ft}$$

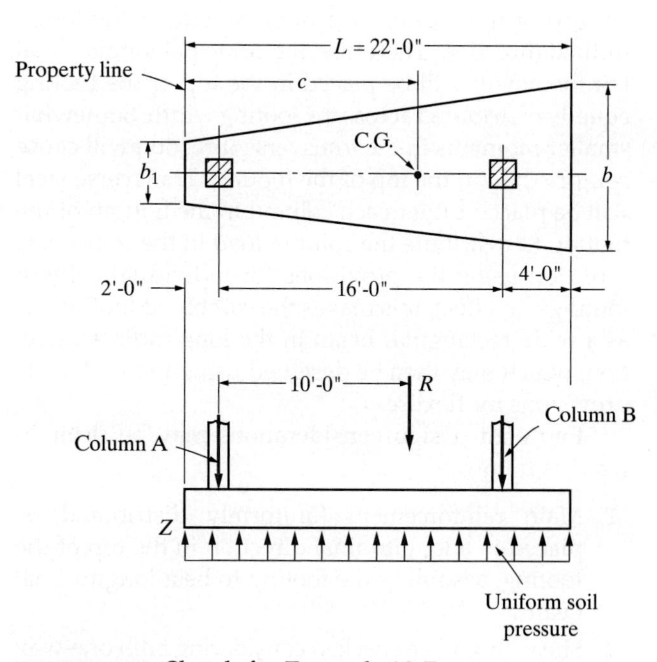

FIGURE 10-26 Sketch for Example 10-7.

6. We now have two equations that contain b and b_1. The equation of step 5 may easily be solved for b by substitution of $L = 22$ ft and $(b + b_1) = 13.1$ ft:

$$\frac{22(b + 13.1)}{3(13.1)} = 12$$

from which

$$b = \frac{12(3)(13.1)}{22} - 13.1 = 8.34 \text{ ft}$$

and

$$b_1 = 13.1 - 8.34 = 4.76 \text{ ft}$$

Use b = 8 ft-4 in. and b_1 = 4 ft-9 in. Thus

$$\text{actual } A = \frac{22(4.75 + 8.33)}{2} = 143.9 \text{ ft}^2$$

The footing is very slightly undersized. No revision is warranted.

The structural design of the rectangular and trapezoidal combined footings is generally based on a uniform soil pressure, even though loading combinations will almost always introduce some eccentricity with respect to the centroid of the footing. The determination of the footing thickness and reinforcement must be based on factored loads and soil pressure to be consistent with the ACI strength method approach. The assumed footing behavior, which is briefly described, is a generally used approach to simplify the design of the footing.

In Figure 10-26, the columns are positioned relatively close to the ends of the footing. Assuming the columns as the supports and the footing subjected to an upward, uniformly distributed load caused by the uniform soil pressure, moments that create tension in the top of the footing will predominate in the longitudinal direction. Therefore, the principal longitudinal reinforcement will be placed in the top of the footing equally distributed across the footing width. Somewhat smaller moments in the transverse direction will cause compression in the top of the footing. Transverse steel will be placed under each column in the bottom of the footing to distribute the column load in the transverse direction using the provisions for individual column footings. In effect, this makes the combined footing act as a wide rectangular beam in the longitudinal direction, which may then be designed using the ACI Code provisions for flexure.

Pertinent design considerations may be summarized as follows:

1. Main reinforcement (uniformly distributed) is placed in a longitudinal direction in the top of the footing, assuming the footing to be a longitudinal beam.
2. Shear should be checked considering both one-way shear at a distance d from the face of the column and two-way (punching) shear on a perimeter $d/2$ from the face of the column.
3. Stirrups or bent bars are frequently required to maintain an economical footing thickness. This assumes that the shear effect is uniform across the width of the footing.
4. Transverse reinforcement is generally uniformly placed in the bottom of the footing within a band having a width not greater than the column width plus twice the effective depth of the footing. The design treatment in the transverse direction is similar to the design treatment of the individual column footing, assuming dimensions equal to the band width as previously described and the transverse footing width.
5. Longitudinal steel is also placed in the bottom of the footing to tie together and position the stirrups and transverse steel. Although the required steel areas may be rather small, the effects of cantilever moments in the vicinity of the columns should be checked.

10-10 CANTILEVER OR STRAP FOOTINGS

A third type of combined footing is generally termed a *cantilever* or *strap footing*, and is often used to resist the effect of eccentric gravity loading on an isolated column; the eccentrically loaded footing or foundation is "strapped" to an interior concentrically loaded footing or foundation (see Figures 10-27 through 10-29). This is an economical type of footing when the proximity of a property line precludes the use of other types. For instance, an isolated column footing may be too large for the area available, and the nearest column is too distant to allow a rectangular or trapezoidal combined footing to be economical. The strap footing may be regarded as two individual column footings connected by a strap beam.

In Figure 10-27, the exterior footing (or foundation) is placed eccentrically under the exterior column so that it does not violate the property-line limitations. This would produce a nonuniform pressure distribution under the footing, which could lead to footing rotation. To balance this rotational or overturning effect, the exterior footing is connected by a stiff beam, or strap, to the nearest interior footing, and uniform soil pressures under the footings are assumed. The strap, which may be categorized as a flexural member, is subjected to both bending moment and shear, resulting from the forces P_e and R_e acting on the exterior footing. As shown in Figure 10-27, the applied moment is counterclockwise, and the shear will be positive because $R_e > P_e$. At the interior column, there is no eccentricity between the column load P_i and the resultant soil pressure force R_i. Therefore, we will assume that no moment is induced in the strap at the interior column.

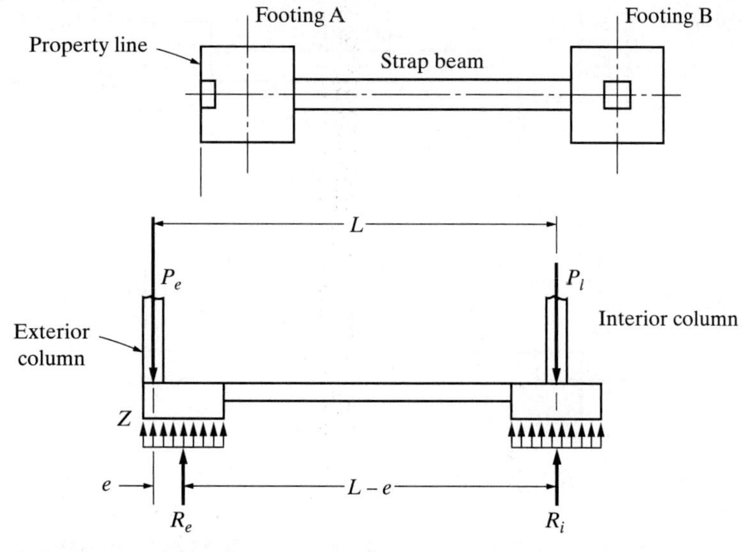

FIGURE 10-27 Cantilever footing.

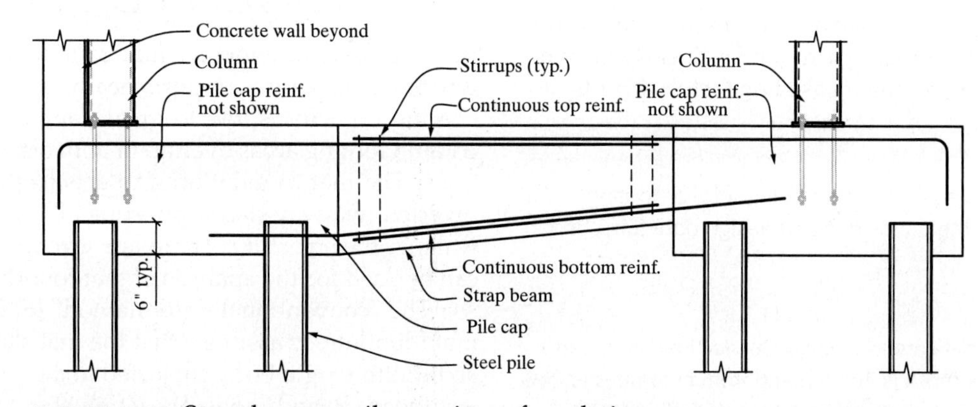

FIGURE 10-28 Strap beam at pile or caisson foundation.

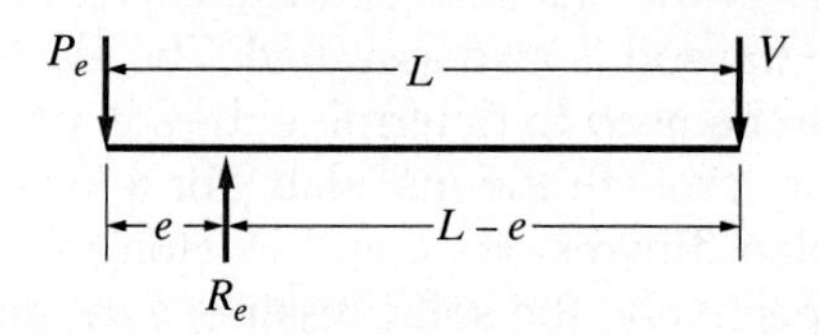

FIGURE 10-29 Strap beam free-body diagram.

We define V as the vertical shear force necessary to keep the strap in equilibrium, as shown in Figure 10-29. Then, with P_e known, V and R_e may be calculated using the principles of statics. A moment summation about R_e yields

$$P_e e = V(L - e)$$

$$V = \frac{P_e e}{L - e}$$

and from a summation of vertical forces,

$$R_e = P_e + V$$

Then, by substitution,

$$R_e = P_e + \frac{P_e e}{L - e}$$

Note that V acting downward on the strap beam also means that V is an uplift force on the interior footing. Therefore,

$$R_i = P_i - V$$

and, by substitution,

$$R_i = P_i - \frac{P_e e}{L - e}$$

In summary, R_e becomes greater than P_e by a magnitude of V, whereas R_i becomes less than P_i by a magnitude equal to V.

The footing areas required are merely the reactions R_e and R_i based on service loads divided by the net allowable soil pressure. These values are based on an assumed trial e and may have to be recomputed until the trial e and the actual e are the same.

The structural design of the interior footing is simply the design of an isolated column footing subject to a load R_i. The exterior footing is generally considered as under one-way *transverse* bending similar to a wall footing with longitudinal steel furnished by extending the strap steel into the footing. The selection of footing thickness and reinforcement should be based on factored loads to be consistent with the ACI strength design approach.

The strap beam is assumed to be a flexural member with no bearing on the soil underneath. Many

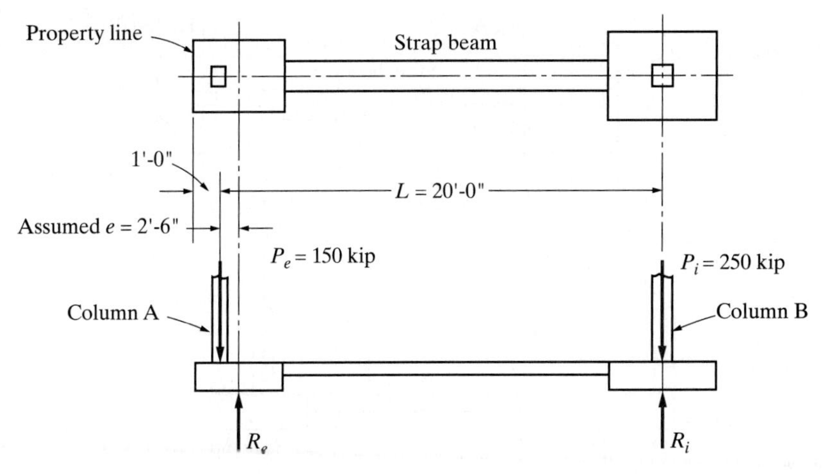

FIGURE 10-30 Sketch for Example 10-8.

designers make a further simplifying assumption that the beam weight is carried by the underlying soil; hence, the strap is designed as a rectangular beam subject to a constant shearing force and a linearly varying negative bending moment based on factored loads.

Example 10-8

Determine the size of the exterior and interior footings of a strap footing for the conditions and design data furnished in Figure 10-30.

Solution:

1. The design data are as follows: Service load on the footing from column A is 150 kip and from column B is 250 kip, and the allowable soil pressure is 4.00 ksf.
2. Assume that $e = 2$ ft-6 in. and that the footing thickness is 2 ft-0 in. Therefore, the footing weight $= 2(0.150) = 0.300$ ksf.
3. The net allowable soil pressure for superimposed loads $= 4.0 - 0.300 = 3.70$ ksf.
4. Determine the strap beam shear V:

$$V = \frac{P_e e}{L - e} = \frac{150(2.5)}{17.5} = 21.4 \text{ kip}$$

5. The footing reactions are

$$R_e = P_e + V = 150 + 21.4 = 171.4 \text{ kip}$$

$$R_i = P_i - V = 250 - 21.4 = 228.6 \text{ kip}$$

6. For the exterior footing, the required area is

$$\frac{171.4}{3.7} = 46.3 \text{ ft}^2$$

Use a footing 7 ft-0 in. by 7 ft-0 in. ($A = 49\ ft^2$). Note in Figure 10-30 that the actual e is then the same as the assumed e, and no revision of calculations is necessary. For the interior footing, the required area is

$$\frac{228.6}{3.7} = 61.8 \text{ ft}^2$$

Use a footing 8 ft-0 in. by 8 ft-0 in. ($A = 64$ ft^2).

10-11 ANALYSIS AND DESIGN OF MAT FOUNDATIONS

As discussed previously, mat foundations are used where the structure supports heavy loads and the soil bearing capacity is low to great depths such that the column footing areas overlap in both orthogonal directions. The mat foundation slab acts like a continuous inverted two-way slab supporting all the columns and walls in the structure. There are several methods that can be used for the analysis of mat foundation slabs.

The "conventional rigid method" [6] for analyzing mat foundations assumes that the mat slab behaves as an infinitely rigid body subjected to the combined total axial load, P, and moments acting about the orthogonal plan axes of the mat slab. A linear elastic stress distribution in the soil is also assumed. The elastic stress equation can be used to determine the soil pressure at any point underneath the mat slab. For a mat foundation with plan dimensions L and W along the X- and Y-axes, respectively, the soil pressure, q_i, at any point below the mat slab is given as [6]:

$$q_i = \frac{P}{WL} \pm \frac{M_{X-X}(y_i)}{\left(\frac{LW^3}{12}\right)} \pm \frac{M_{Y-Y}(x_i)}{\left(\frac{WL^3}{12}\right)}$$

$$\leq \text{net allowable soil pressure}$$

where

P = sum of all the axial dead and live loads on the mat slab from all the columns and walls

M_{X-X} is the moment about the X- axes $= Pe_y$

M_{Y-Y} is the moment about the Y- axes $= Pe_x$

x_i, y_i = the coordinates of a point on the mat along the X- and Y- axes, respectively.

Using the elastic stress equation above, the soil pressures below the mat slab can be determined for unfactored dead plus live loads. These pressures are compared to the net allowable soil pressure to ensure that the soil bearing capacity is not exceeded. The factored soil pressures are calculated using the factored load and moments. These factored pressures can then be used in the equivalent frame analysis of the mat slab where the mat is treated as an inverted slab. The mat slab is divided into design strips centered on the column lines in both orthogonal directions, and for simplicity the average factored soil pressure in the strip can be used, in lieu of the linearly varying soil pressures, for the analysis and design of the strip [6]. The analyses yield the positive and negative moments in the mat slab in both orthogonal directions from which the steel reinforcement at the top and bottom of the slab, respectively, in both orthogonal directions can be calculated using the methods presented in Chapter 2. The mat is usually reinforced with continuous top and bottom reinforcement in both orthogonal directions. Given the large depth or thickness of some mat foundations (sometimes up to 6 ft to 10 ft thick), the stability of the mat foundation reinforcement cage should be carefully addressed by the contractor since they are are responsible for the means and methods of construction. A note to this effect (see Section 14-11) should be included in the concrete specification.

An alternate approach to the analyses of mat foundations is to use one of several available finite-element based computer analysis software [7]. In this method, the mat is modeled as a flexible slab on an elastic foundation, where the slab is supported on springs in both orthogonal directions. The stiffness of the springs, k, also known as the modulus of subgrade reaction, is supplied by the geotechnical consultant. The coefficient of subgrade reaction is not constant and may vary significantly within the footprint of the mat foundation, and this variability in the modulus of subgrade reaction would influence the moments in the mat slab [8]. The conditions under which the conventional rigid method can be used are given in ref. [9]. Note that for both methods, the mat slab thickness is selected so that the punching shear capacity is not exceeded.

10-12 DEEP FOUNDATIONS—PILES, DRILLED SHAFT (CAISSONS), AND PILE CAPS

Piles/Caissons

Deep foundations, as discussed earlier, are typically used where the soil bearing capacity is low for great depths, and they are used to support structures that are subjected to high gravity and lateral loads, such as tall buildings and bridges. Common exampls of deep foundations are piles and drilled shafts - also known as caissons or drilled pier or bored piles. Piles are installed by driving or forcing them into the ground with a hammer, and this usually causes vibrations and noise. They resist loads through end bearing and/or through skin friction between the soil and the surface of the pile. Piles can be made of steel (e.g. HP steel sections or steel pipes filled with concrete) or precast or cast-in-place concrete. The cast-in-place concrete piles can be cased or uncased [10]. For piles with permanent casing, a steel shell is driven into the ground and the soil within the shell is removed and the shell is subsequently filled with concrete. For the uncased pile, the steel shell is removed gradually as the concrete is cast, starting from the bottom of the pile. Since the efficiency of a pile within a pile group is impacted by the spacing between adjacent piles, the geotechnical consultant typically specifies the minimum spacing between adjacent piles. Typical spacings commonly used in practice are 3 to 4 pile diameters between adjacent piles, but no less than 2 ft 6 in., and an edge distance of 2 pile diameters.

Drilled shaft or caissons can support loads through skin friction or through end bearing with the bottom of the caisson enlarged or "belled" to provide a larger surface area and hence a higher axial load capacity. To construct caissons—usually without permanent casing, the hole is drilled with a steel casing, and the soil within the steel shell is removed. The hole is then filled with concrete as the steel casing is gradually withdrawn. Permanent steel casings for drilled shafts or caissons may be used in areas where there is unstable soil or a debris field in the drilled shaft that may become mixed with the concrete, or there is an existing concrete debris field with large voids within the shaft which can suck away some of the fluid concrete, thus leading to the loss of a significant portion of the concrete pour. Not using permanent casings in the above situations could lead to deficient caissons which could be very costly to remediate [17]. Caissons are quieter to install compared to driven piles, and there is less vibration during the caisson installation. Caissons can be reinforced or unreinforced. Often times, a tied or spiral type reinforcement cage might be provided for the top several feet of the caisson. Because of their larger diameter compared with piles, it is possible for a human being to be lowered down to the bottom of the caisson to inspect the bearing strata to ensure that it is as sound as assumed by the geotechnical consultant. Caisson diameters typically vary from 2 ft 6 in. to 10 ft though caissons as large as 30 ft in diameter have been used in supertall buildings [16]. Since caissons are drilled rather than driven, and because they are oftentimes end bearing, there is no impact on adjacent caissons unlike driven piles where the efficiency of a pile group is impacted by the spacing between adjacent piles. The axial load capacity of a concrete caisson is the lesser of the short column capacity (since the caisson is assumed to behave as a concrete column braced all along its length by the surrounding soil) and the bearing capacity of the bearing strata [10].

Piles are laid out in groups of 2, 3, 4, 5, 6 or more piles, depending on the column or wall loads, and they are connected together at their top ends with a concrete pile cap through which the loads from the column or wall are transmitted to the group of piles (see Figure 10-31). The pile cap has to be capable of supporting the column and wall loads while spanning between the piles. Depending on the shear span-to-depth ratio (see Figure 10-39), a pile cap may act like a deep beam. Pile caps that do not fall into the deep beam category can be designed using the methods covered in previous chapters. The design of pile caps that act as deep beams is covered later in this chapter. Figure 10-32 shows a typical pile foundation plan with different pile cap layouts. More details of some typical pile cap layouts are shown in Figure 10-34, and Figure 10-33 shows a typical section at a two-pile cap. An introduction to the design of pile caps, a topic not often covered in reinforced concrete textbooks [19], is discussed later in this chapter.

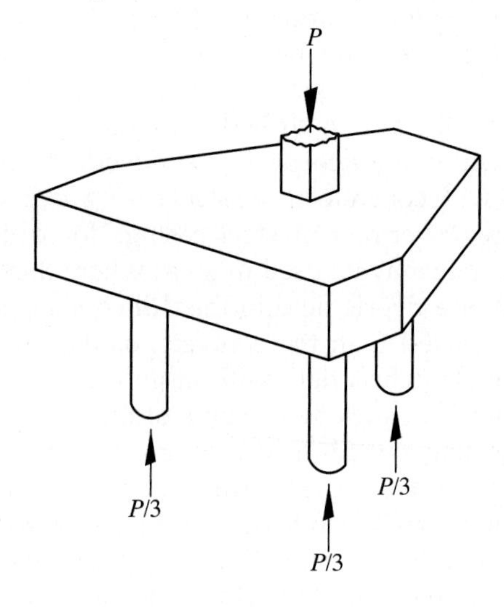

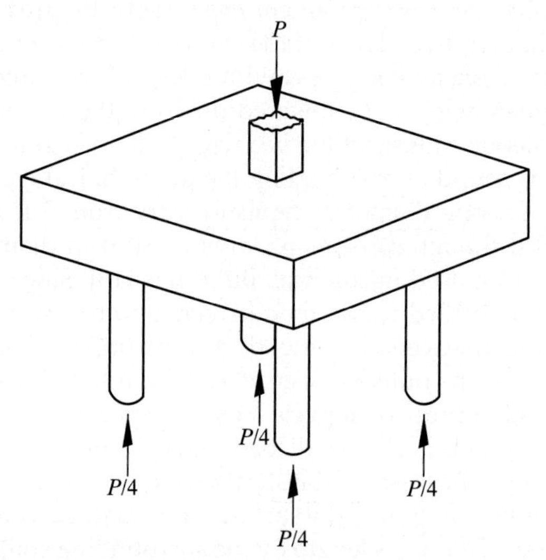

FIGURE 10-31 Piles and pile caps.

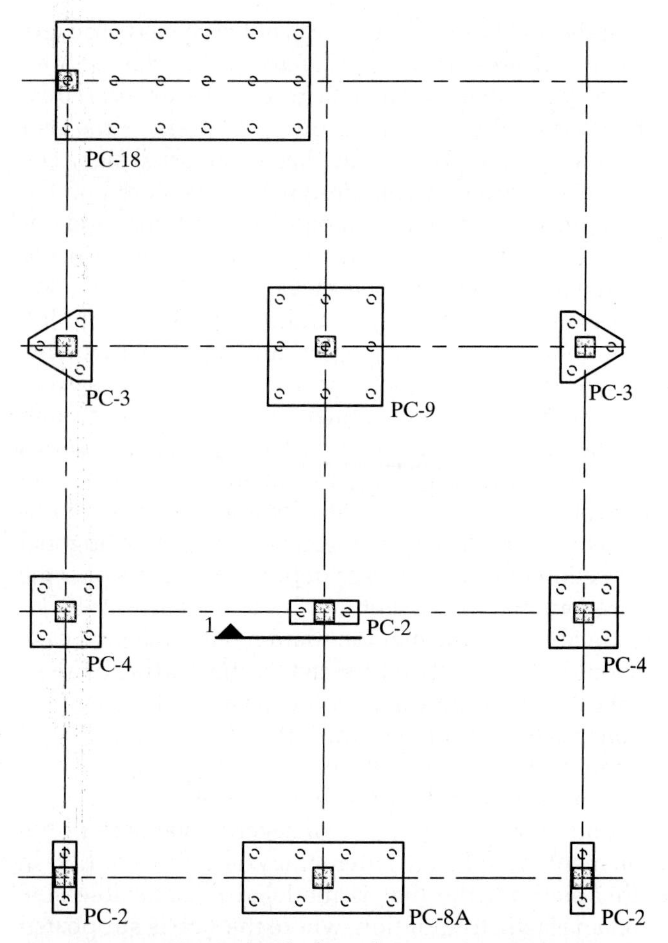

FIGURE 10-32 Typical pile foundation plan with different pile cap layouts.

The pile reactions are assumed to act at the centroid of each pile (ACI Code, Section 13.4.2.2), and the total number of piles in a pile cap layout is obtained by dividing the total unfactored total load in the column or wall by the unfactored allowable load for each pile (ACI Code, Section 13.4.1.1)—a value supplied by the Geotechnical consultant.

When the bedrock is closer to the surface and the column or wall loads are high, concrete caissons or drilled concrete shaft may make the most sense for the foundations. In this case, the load is transferred from the columns or walls through a concrete pedestal or pier which transmits the loads to the caissons, and the caisson then transfers the load to the bearing strata through end bearing. Depending on the axial and lateral loads on a caisson, they may be unreinforced, or reinforced throughout its entire length, or only in the upper portion of the caisson. Figure 10-35 shows a typical caisson (drilled shaft) foundation plan and Figure 10-36 shows a typical section at a caisson foundation. In Figure 10-35, the notations CA3, CA4 and CA5, etc., denote caissons with different diameters, and the properties (diameter and reinforcement) of these caissons are typically given in the caisson schedule. Similarly, the notations P1 and P2 denote the different pier or pedestal sizes, and their properties

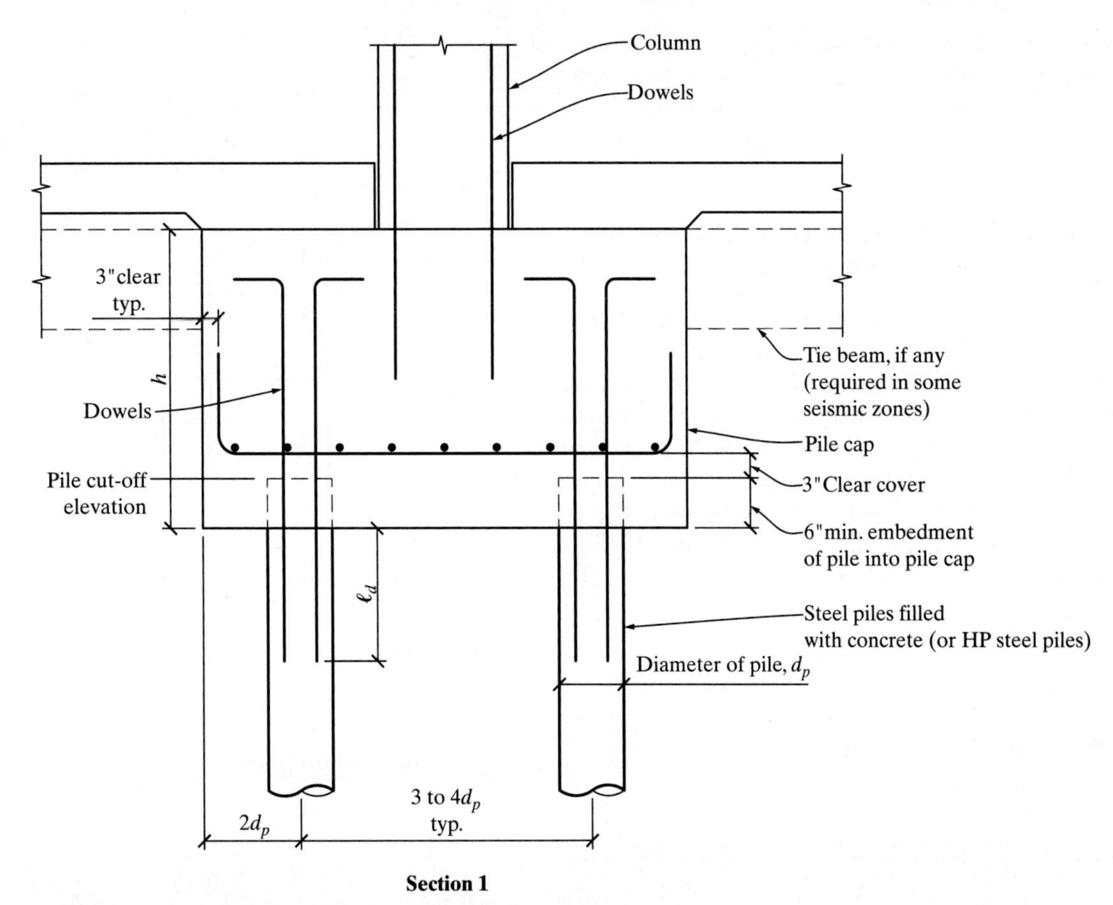

FIGURE 10-33 Typical section at two-pile cap locations.

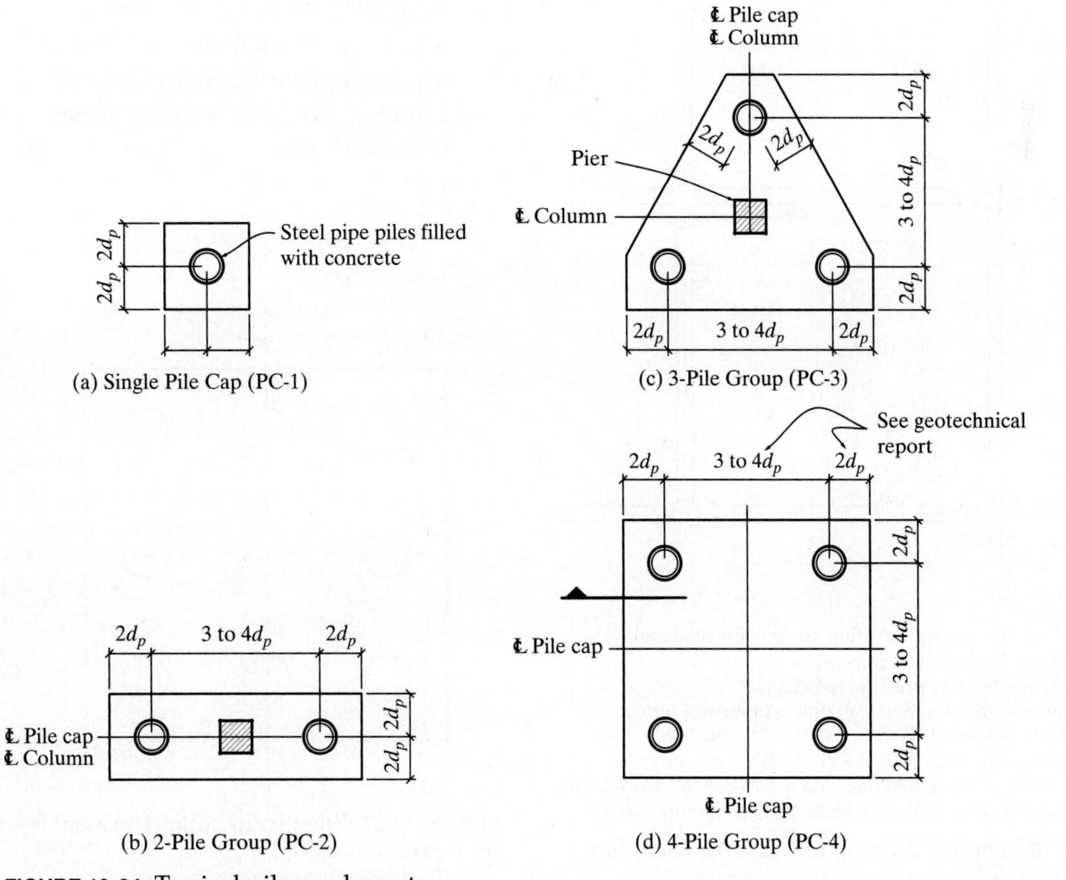

FIGURE 10-34 Typical pile cap layout.

(plan size and reinforcement) are usually indicated in the pier schedule. The notation GB1 refers to grade beams which are sometimes required to tie the caissons together.

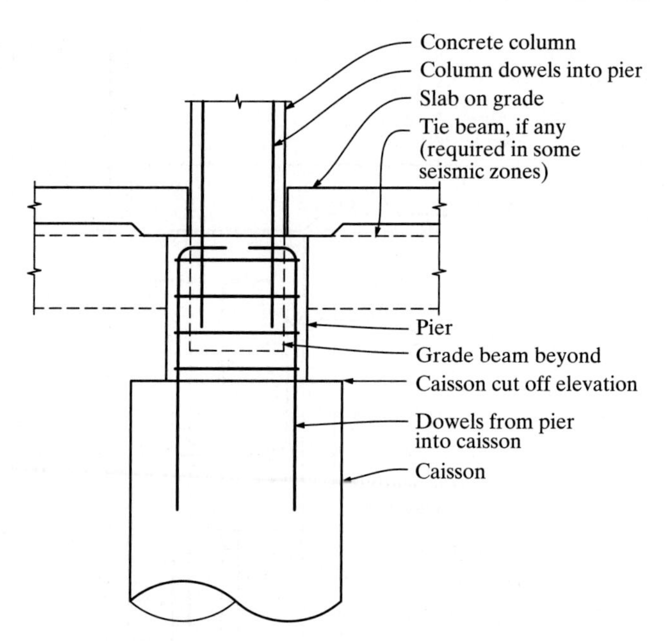

FIGURE 10-36 Typical section at Caisson foundation.

Load Distribution in Pile Caps Subject to Axial Load and Moments

Some situations occur in practice where a pile cap is subjected to moments about both orthogonal axes in addition to the gravity loads, due to the eccentricity of the column or wall load with respect to the centroid of the pile group or due to lateral loads. The maximum axial load on a pile within a pile group that is subjected to combined gravity loads and moments about both orthogonal axes of the pile cap (see Figure 10-37) can be obtained using the following elastic equation [10]:

The load on pile i,

$$P_i = \frac{P_{total}}{n} \pm \frac{M_{X-X}(y_i)}{\sum_{i=1}^{n}(y_i)^2} \pm \frac{M_{Y-Y}(x_i)}{\sum_{i=1}^{n}(x_i)^2}$$

where
n = total number of piles in the pile group
M_{X-X} = moment about the X-X axis
M_{Y-Y} = moment about the Y-Y axis
x_i and y_i are the coordinate locations of the centroid of the piles

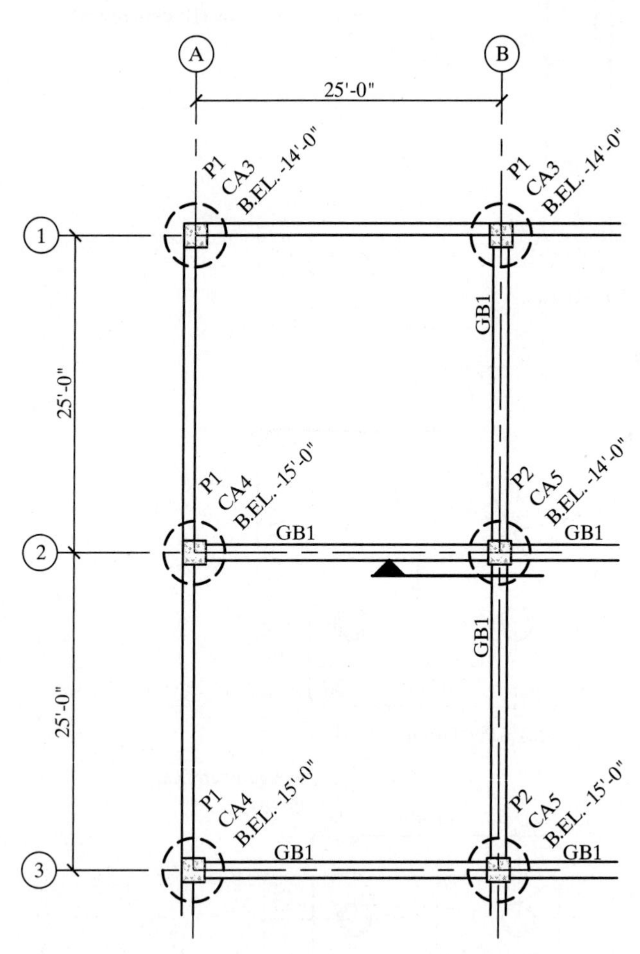

Notes:
1. B.EL. = Caisson bearing elevation–to be confirmed and approved by the geotechnical consultant
2. Caisson design bearing pressure is 12500psf
3. Center caisson on column or wall unless otherwise noted.
4. Dowels from column/wall into caisson is the same as column/wall vertical rebar (min 4-#5 dowels).
5. Slab-on-grade: 6" thick concrete with 6×6-W2.9×W2.9 WWF on 8" granular fill over 8mil vapor barrier. Finished floor elevation = 0'-0".

FIGURE 10-35 Typical Caisson foundation part plan.

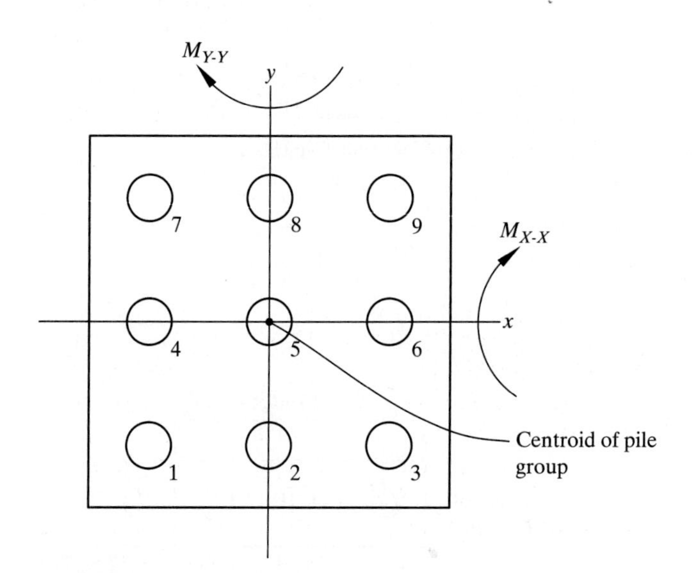

FIGURE 10-37 Pile group subject to axial load and moments.

Transfer of Lateral Base Shears in Pile and Caisson Foundations

The transfer of the lateral base shears from a building structure to the ground involves interaction between the structure and the soil, and can be accomplished through a combination of the following mechanisms [11, 12, 13]:

- Passive soil pressure resistance against the vertical face of the basement walls, pile caps, and grade beams.
- Passive resistance against the front sides of the piles or caissons. The settlement of the soil beneath the structure should be accounted for in calculating the effective depth that will be available for passive resistance.
- Bending of the caissons or pile groups as a cantilever or through some frame action due to the moment restraint provided by the pile caps or grade beams. The reduction effect of the group action of the group piles should also be accounted for.
- The use of batter piles which is quite limited in application due to the fact that many of these piles would be needed, and other batter piles will be ineffective in a direction perpendicular to the plane of inclination of the batter pile. Also, the driving of batter piles could cause major interference with the vertical piles and some of the batter piles may extend beyond the property lines.
- For pile or caisson supported structures, the friction resistance between the base of the structure and the soil below the structure is neglected because over time, the soil settlement below the base of the building eliminates this shearing resistance.

Critical Sections for Shear in Pile Caps

The critical sections for shear in pile caps are as follows [14]:

a. One-way deep beam shear located at the face of the column (see Figures 10-38b and 10-38c). The one-way shear capacity is given as

$$\phi V_c = \phi 2\sqrt{f'_c}bd \geq V_u$$

where

$\phi = 0.75$
$b =$ width of pile cap
$d =$ effective depth of pile cap bottom reinforcement
$V_u =$ the factored total reaction on the piles causing the one-way shear (e.g., in Figure 10-38, this would be the factored reaction from two piles)

b. Two-way or punching shear around the supported column, with the critical perimeter located at a distance of d/2 from the face of the column as shown in Figure 10-38a, where d is the effective depth of

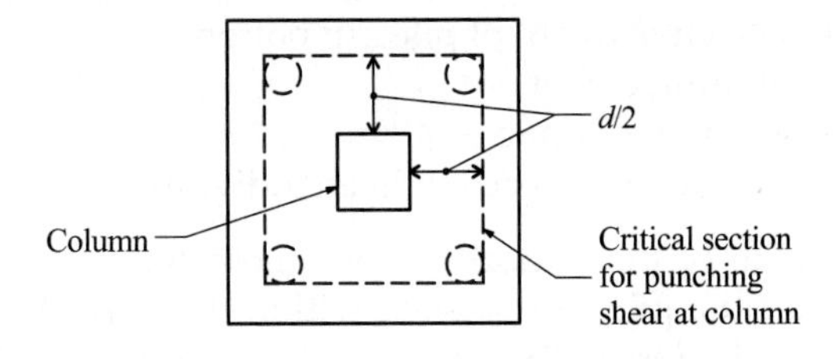

(a) Punching Shear at Column

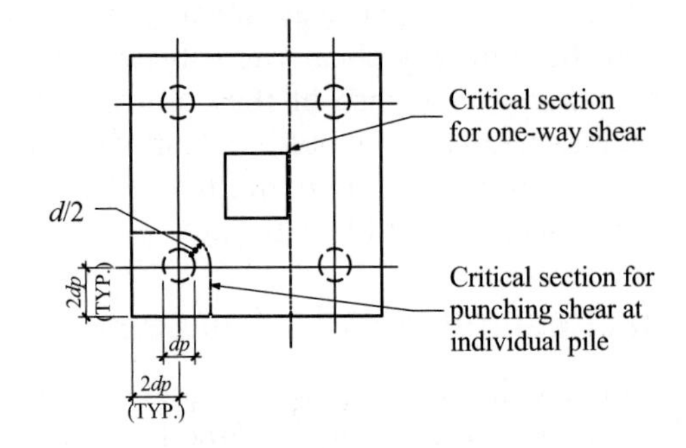

(b) Punching Shear at Individual Pile and One-Way Shear at Column Face

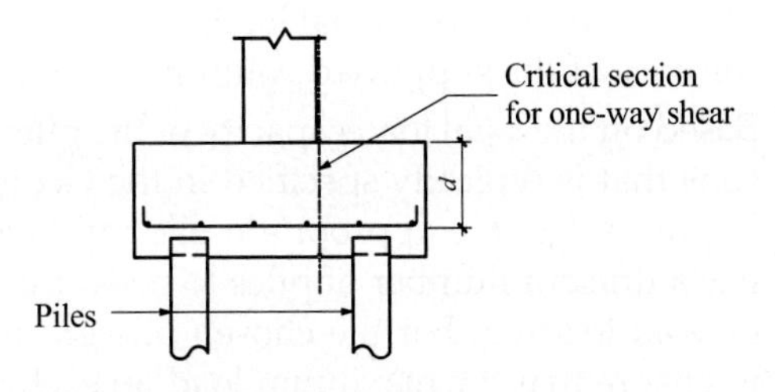

(c) Section through Pile Cap

FIGURE 10-38 Critical sections in pile caps.

the pile cap. The punching shear capacity is given as

$$\phi V_c = \phi 4\sqrt{f'_c}b_o d \geq V_u$$

where

$\phi = 0.75$
$b_o =$ perimeter of the punching shear critical section $= 2(c_1 + d + c_2 + d)$
$d =$ effective depth of pile cap bottom reinforcement
c_1 and $c_2 =$ dimensions of the column
$V_u =$ the factored load on the column

c. Punching shear around an individual edge pile with the critical perimeter located at d/2 from the face of the pile as shown in Figure 10-38b, and assuming the pile edge distance is $2d_p$.

$$\phi V_c = \phi 4\sqrt{f'_c}b_o d \geq V_u$$

where

$b_o =$ perimeter of the punching shear critical section

$$= 4d_p + \left(\frac{1}{4}\right)2\pi\left(\frac{d_p}{2} + \frac{d}{2}\right) = 4d_p + \frac{\pi}{4}(d_p + d)$$

d = effective depth of pile cap bottom reinforcement
d_p = diameter of pile
V_u = the factored reaction in an individual pile

Inspite of the critical sections indicated above, if however, the piles are located within the critical sections (see Figure 10-38a), the ACI Code sets an upper limit on the one-way shear capacity with the critical section for one-way shear taken at the face of the column. Similarly, for two-way shear, it sets an upper limit on the two-way shear capacity with the critical section taken at the face of the column (ACI Code, Section R13.4.2.5). That would indicate that for punching shear in item (b) above, the revised perimeter of the critical section would be $b_o = 2(c_1 + c_2)$.

Design Requirements for Pile Caps

Pile caps usually fall into the category of deep beams and are therefore typically designed using the strut-and-tie method that will be discussed in the next section (see ACI Code, Section 13.2.6.3). The following design parameters are required for the design of pile caps:

- The cumulative total axial load, P, and moments at the base of the supported column or wall.
- Based on the axial load capacity of the piles or caissons that is typically specified in the Geotechnical Report, select the appropriate pile cap layout that has sufficient number of piles to resist the column or wall loading. For the chosen pile group, check to ensure that the maximum load on each individual pile due to total gravity load and any moments about the two orthogonal axes do not exceed the axial load capacity of the pile (see Figure 10-37).
- Check the bearing stresses at the column-pile cap interface and at the reaction points at the pile-to pile cap interface (see Section 10-5 for the bearing capacity).
- The Geotechnical Report will specify minimum spacing and edge distance to ensure maximum efficiency for the pile group and to avoid the area of influence of one pile overlapping with other adjacent piles. In practice, a minimum spacing between piles of 3 to 4 times the pile diameter is commonly recommended. To ensure confinement in the nodal zones and provide adequate space to develop the tension tie reinforcement beyond the nodal zone, an edge distance of twice the pile diameter is commonly used and recommended [14].
- The effective depth of the bottom reinforcement in a pile cap should be at least 12 in. (ACI Code, Section 13.4.2.1).
- For pile caps designed using the strut-and-tie method, the effective strength of the strut should be calculated using $\beta_s = 0.60\lambda$ (ACI Code, Section 13.4.2.4) since confined reinforcement is not usually provided in pile caps. Confined strut reinforcement can be provided in deep beams and corbels.

10-13 STRUT-AND-TIE MODELS FOR PILE CAPS AND DEEP BEAMS

When beams or pile caps have relatively large depths with shear spans, a, such that a/h is less than 2.0 (see Figure 10-39 and ACI Code, Section 9.9.1), the load transfer is by direct strut action rather than by flexure or bending. For deep beams and pile caps, the assumptions in Chapter 2 that plane sections remain plane is no longer valid, hence the strut-and-tie method must be used for such members [18, 19, 20]. Deep beams occur as transfer beams or girders in reinforced concrete highrise buildings where they support the accumulated loads from discontinued multi-story columns or walls. Pile caps are "transitional elements" used to transfer column and wall loads from the superstructure—with its smaller construction tolerances—to a group or cluster of piles which are installed with larger construction tolerances.

The ACI Code allows the strut-and-tie method (see ACI Code Chapter 23) to be used for the design of deep beam members. The method consists of modeling the deep beam or pile cap as an idealized truss that consists of several components: the struts which are in compression, the tension tie, and the nodal zones (see Figure 10-40 for a 2-D strut-and-tie model, and Figure 10-41 for 3-D strut-and-tie models). In modeling the truss, the angle between the longitudinal axis of the strut and the horizontal must be at least 25° (ACI Code, Section 23.2.7).

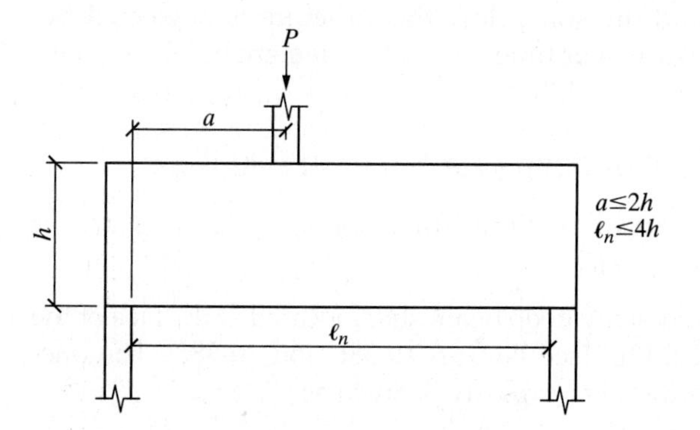

(a) Deep Beam with Single Concentrated Load

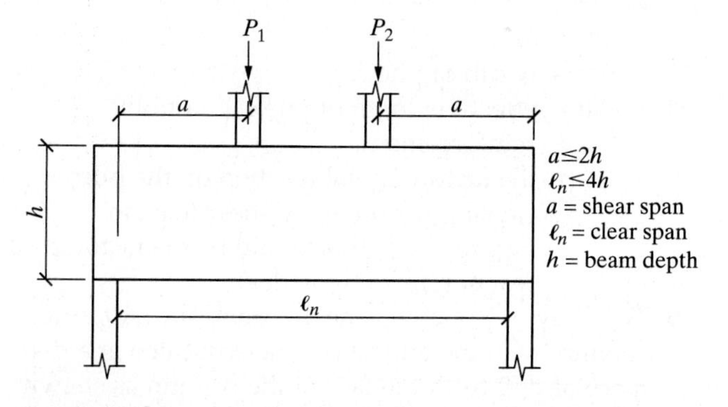

(b) Deep Beam with Single Concentrated Loads

FIGURE 10-39 Shear spans in deep beams and pile caps.

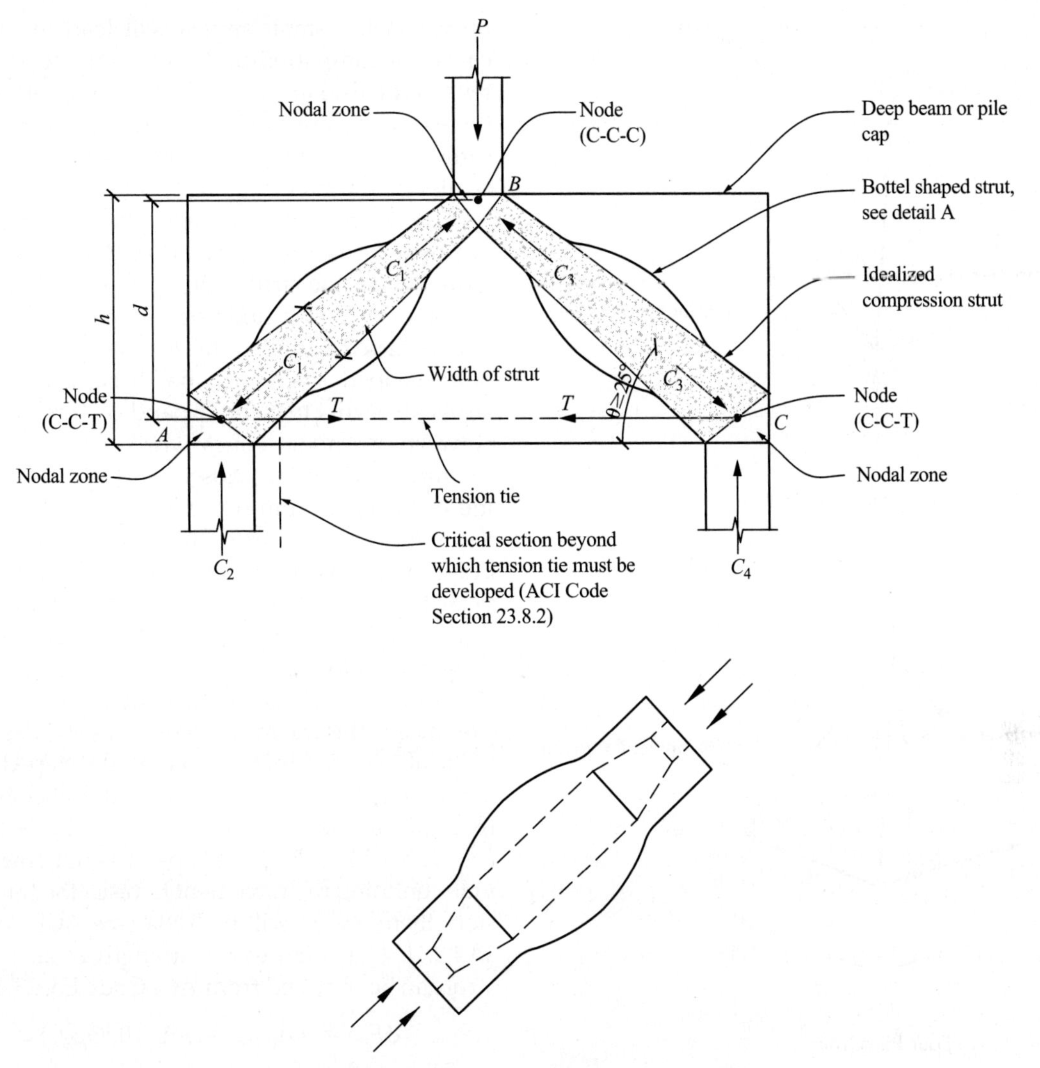

FIGURE 10-40 2-D strut-and-tie model of a deep beam or pile cap.

The struts can be idealized as a uniform prismatic member, however, where there is space for the stress in the concrete strut to spread out transversely between the nodes, this will result in the so-called bottle-shaped strut (see Figure 10-40). The transverse tension resulting from the spreading out of the stresses in the strut causes cracks parallel to the direction of the strut which leads to a reduction in the compressive strength of the strut, unless adequate reinforcement is provided to resist this transverse tension [18]. The nodes in strut-and-tie models can be classified as follows, depending on the type of members or forces meeting at the node (see Figure 10-40):

- C-C-C nodes: indicates that all the forces in all the members meeting at the node and the reaction or load induce compression on the node.
- C-C-T node: indicates that the nodal zone resists compression forces (C) on two faces, and a tension force (T) on one face.
- C-T-T node: Though not shown in Figure 10-40, this indicates that the nodal zone resists compression force (C) on one face, and tension forces (T) on the two other faces of the nodal zone.

The development of a strut-and-tie model is an iterative and graphical design process where the strut-and-tie geometry is assumed (with an angle of inclination of the strut that does not violate the Code limitations), and the forces in each strut, tie, and nodal zones are calculated and checked to ensure that their ultimate strengths are not exceeded. There are several possible solutions, but the best strut-and-tie model for a structure is one that mimics the elastic stress trajectories of the structure. The strut-and-tie method - which is based on the lower bound theorem of plasticity - results in a lower bound solution, provided the assumed stress field is in equilibrium with the external loads, and the failure or yield criterion of the struts, ties, and nodes in the model are not violated, and the member has adequate ductility to ensure any required redistribution of forces [18, 19]. The possible failure modes in the strut-and-tie model are: compression failure of the strut, tension failure of the ties, failure of the nodal zones, and bearing failure at the load and reaction points (See Section 10-5 for the calculation of bearing strength).

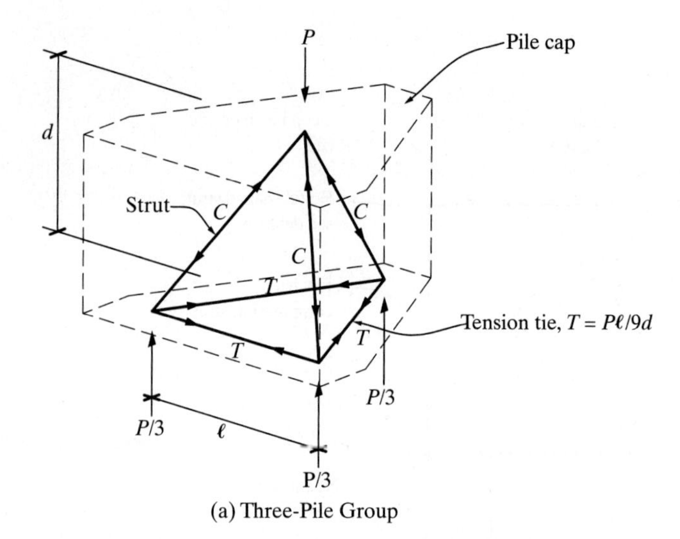

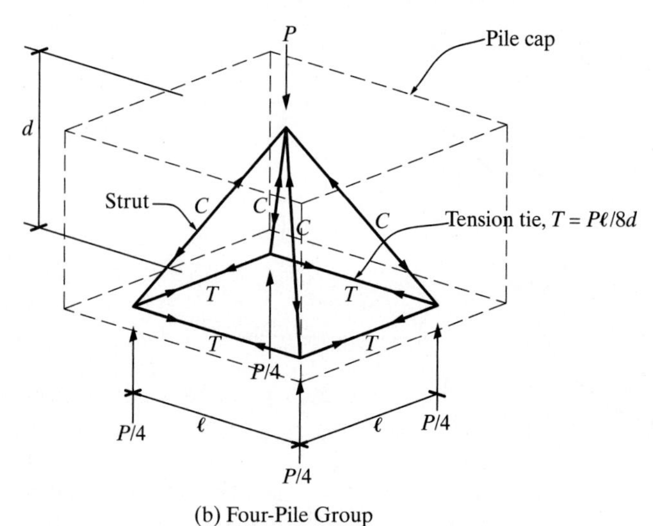

FIGURE 10-41 3-D strut-and-tie model of pile caps.

Strength of Struts

The struts are subjected to compression, and the minimum angle of inclination between the strut and a tension tie is 25° (ACI Code, Section 23.2.7) because unreasonably small angles will lead to cracking and cause "incompatibilities" between the strut and the tension tie due to the axial shortening of the strut and the elongation of the tension tie in approximately the same direction for small angle struts. The idealized struts are shown in Figure 10-40.

There are three different types of struts: prismatic or prism struts which have a constant cross-section; fan struts where the struts with different angles of inclination converge to a node or radiate away from it; and bottle-shaped struts. In this text, only the prismatic and bottle-shaped struts will be discussed. Bottle-shaped struts are struts that can expand or spread out laterally between the nodal zones when there is no confining reinforcement such as closed ties to prevent this spreading of the strut from occurring under the compression force. As is evident from the strut-and-tie model of the bottle-shaped strut in Detail "A" of Figure 10-40, transverse tension does occur as a result, and transverse reinforcement in the form of closed ties can be provided in deep beams to resist this transverse tension force. The strength of an unreinforced strut is reduced due to the presence of this transverse tension as is reflected in the β_s value of 0.60λ for unreinforced bottle-shaped struts compared with β_s of 0.75 for reinforced bottle-shaped struts that satisfies the requirements of ACI Code, Section 23.5 (see Table 10-3). For pile caps, it is not feasible to provide confining reinforcement to resist the transverse tension, therefore, β_s will be 0.60λ (see ACI Code, Section 13.4.2.4). The compressive strength of an unreinforced strut can be obtained from ACI Code Eq. 23.4.1a as

$$\phi F_{ns} = \phi A_{cs} f_{ce} = \phi A_{cs}(0.85\beta_s f_c') \geq F_{us}$$

For a strut with longitudinal reinforcement provided parallel to the longitudinal direction of the strut, the compressive strength can be obtained from ACI Code Eq. 23.4.1b as

$$\phi F_{ns} = \phi(A_{cs} f_{ce} + A_s' f_s') = \phi[A_{cs}(0.85\beta_s f_c') + A_s' f_s'] \geq F_{us}$$

TABLE 10-3 Strut Coefficient, β_s

Geometry and location of strut	Reinforcement crossing the strut	β_s
Struts with uniform cross-section along its length	N/A	1.0
Bottle-shaped struts	Reinforcement satisfies ACI Code, Section 23.5 (see equation below)	0.75[2]
Bottle-shaped struts	No reinforcement or reinforcement does not satisfy ACI Code, Section 23.5 (see equation below)	0.60λ
Struts intersecting tension ties	N/A	0.40[1]
All other cases	N/A	0.60λ

(Adapted from ACI Code Table 23.4.3)

[1]Applicable only to zones where the strut acts perpendicular to the tie

[2]Assumes that there is distributed reinforcement on the face of the deep member crossing a cracked strut that satisfies ACI Code, Eq. 23.5.3 (assuming $f_c' \leq 6000$ psi).

where

F_{us} = the factored compression force in the strut
f_{ce} = $0.85\beta_s f_c'$ is the effective concrete compression strength of a strut

β_s is the strut coefficient calculated from Table 10-3 see ACI Table 23.4.3). It accounts for the effects of cracking as well as the effects of providing crack control reinforcement across the strut. It also accounts for the effect of the spreading of the strut that results in bottle-shaped struts.

ϕ = 0.75
A_{cs} = cross-sectional area at the end of the strut
A_s' = area of longitudinal reinforcement in the strut

Since the strut dimensions may vary from node to node, the strength of the strut should be evaluated at each end of a strut and the lower value is taken as the capacity of the strut.

The longitudinal compression reinforcement in the strut must be tied similar to the ties used in columns, and the reinforcement and ties shall conform to ACI Code, Section 23.6. The distributed reinforcement per footnote 2 in Table 10-3 should satisfy the following relationship:

$$\frac{A_v}{b_s s_1}\sin\alpha_1 + \frac{A_h}{b_s s_2}\sin\alpha_2 \geq 0.003$$

where
A_v = distributed *vertical* reinforcement
s_1 = spacing of the distributed *vertical* reinforcement
α_1 = angle between the distributed *vertical* reinforcement and the centroidal axis of the strut
A_h = distributed *horizontal* distributed reinforcement
s_2 = spacing of the distributed *horizontal* reinforcement
α_2 = angle between the *distributed* horizontal reinforcement and the centroidal axis of the strut
b_s = width of the strut

Strength of Nodal Zones

The node is the intersection of the centroidal axes of the strut, the bearing or loading area and/or the tension ties. The area surrounding the nodes are known as the nodal zones. There are two types of nodal zones: hydrostatic nodal zones where the stresses on all the loaded faces of the nodes are equal, and the faces of the nodes are perpendicular to the longitudinal axes of the struts and ties intersecting at the node; in non-hydrostatic nodal zones, the face of the node is not perpendicular to the axes of the strut, and the size of the node is controlled by the face of the node with the highest stress (ACI Code, Section R23.2.6). The following nodal zone parameters are needed to establish the geometry of the strut-and-tie model: the dimensions of the loaded area (i.e. the supported columns or piers), the dimensions of the reaction points (i.e. the supporting columns or piles, and the width (i.e. vertical dimension) of the tension tie zone [18, 19]. The compressive strength of a nodal zone is obtained from ACI Code Eq 23.9.1 and rewritten as

$$\phi F_{nz} = \phi A_{nz} f_{ce} = \phi A_{nz}(0.85\beta_n f_c') \geq F_{us}$$

where
f_{ce} = $0.85\beta_n f_c'$ is the effective concrete compression strength at the face of the nodal zone,
β_n is calculated from Table 10-4 (see ACI Table 23.9.2)
ϕ = 0.75
A_{nz} = area of the face of the nodal zone perpendicular to the direction of the force, F_{us}

TABLE 10-4 Nodal Zone Coefficient, β_n

Type of nodal zone	β_n
Nodal zone with struts, bearing areas or both (e.g., C-C-C nodal zone)	1.0
Nodal zone with one tension tie (e.g., C-C-T nodal zone)	0.80
Nodal zone with two or more tension ties (e.g., C-T-T nodal zone)	0.60

(Adapted from ACI Code Table 23.9.2)

Strength of Tension Ties

Tension ties are allowed to be non-prestressed or prestressed reinforcement and must have adequate anchorage for non-prestressed reinforcement. Anchorages of tension tie reinforcement can be provided by end plates or by hooks at the ends of the deep beam or pile cap beyond the nodal zone.

The tension capacity of a tension tie is given as

$$\phi F_{nt} = \phi[A_{ts} f_y + A_{tp}(f_{se} + \Delta f_p)] \geq F_{ut}$$

where,
F_{ut} = factored tension force in the tension tie
A_{ts} = area of non-prestressed tension tie reinforcement
A_{tp} = area of tension tie prestressing steel
= 0 for non-prestressed members
f_y = yield strength of the non-prestressed reinforcement
f_{se} = effective stress in the prestressing after all prestress losses
Δf_p = 60,000 psi for bonded prestressed tension tie reinforcement
= 10,000 psi for unbonded prestressed tension tie reinforcement

Note that the centroid of the tension tie assumed in the strut-and-tie model must coincide with the centroid of the actual tension tie reinforcement selected (ACI Code, Section 23.8.1). The tension tie reinforcement must be anchored by mechanical devices such as end plates, headed bars or with standard hooks in or beyond the nodal zones (ACI Code,

Section R23.2.1). For pile caps, banded tension tie reinforcement along the pile lines are preferred and commonly used in practice rather than uniformly distributed reinforcement.

Additional Reinforcement Requirements for Deep Beams

To control the width and growth of inclined cracks in deep beams, distributed vertical and horizontal reinforcement is required in deep beams per ACI Code, Section 9.9.3.1. The distributed vertical reinforcement, A_v, on the face of the beam must satisfy the following relationship (see ACI Code, Section 9.9.3.1):

$$A_v \geq 0.0025 b_w s_1$$

where s_1 is the horizontal spacing of the vertical reinforcement, and b_w = width of beam

The distributed horizontal reinforcement, A_h, on the face of the beam should satisfy the following relationship:

$$A_h \geq 0.0025 b_w s_2$$

Where s_2 is the vertical spacing of the horizontal reinforcement, and b_w = width of beam

The maximum spacing of the distributed reinforcement is d/5 but not greater than 12 in.

Terminate the horizontal and vertical reinforcement on the face of the deep beam with standard hooks.

Design Procedure for Pile Caps and Deep Beams Using the Strut-and-Tie Method

The design of pile caps and deep beams by the strut-and-tie method is an iterative process with the procedure as follows:

- Determine the total load, P, on the beam or pile cap.
- Determine the shear span, a, and establish that the beam is a deep beam and hence can be designed using the strut-and-tie method (see Figure 10-40).
- Check the bearing stresses at the load and reaction points using the *supported* column or pier dimensions for calculating the bearing strength at the load points, and the *supporting* column or pile dimensions for calculating the bearing strength at the reaction points (see Section 10-5). These dimensions are used as the starting dimensions for the nodal zones and they are are combined with the height of the tension tie zone determined below and used to establish the geometry of the strut-and-tie model.
- Check the pile cap or deep beam for one-way shear to establish an estimate of the minimum beam depth (see ACI Code, Sections 9.9.2.1, 9.9.3.1 and 9.9.4). The size of the pile cap or deep beam should be selected so that the maximum shear in the pile cap or deep beam is such that

  $$V_u \leq \phi 10 \sqrt{f_c'}\, b_w d$$

  Where b is the width of the deep beam or pile cap. This equation is used to establish the minimum dimension for the pile cap or deep beam that helps to control cracking under service loads and prevents diagonal compression failures in deep beams (ACI Code, Section R9.9.2.1).
- For pile caps, check punching shear at the supported column and at the supporting piles.
- Lay out the struts, tension ties, and the nodal zones in the pile cap or deep beam as the idealized strut-and-tie model (see Figures 10-40 and 10-41). Ensure that the angle of inclination of the strut is not less than 25°.
- The out-of-plane thickness of the strut and nodal zones is taken as the width, b, of the deep beam and the width of the pile cap for 2-D pile cap problems. For pile caps with 3-D strut-and-tie models (e.g. pile caps with three or more piles as shown in Figure 10-41), the widths of the struts and nodal zones are not readily apparent because of the complex geometries. Initial estimates of the widths can be selected so that the ultimate strength of the struts and the bearing strength at the nodal zones exceed the factored forces in these elements. The struts and nodal zones with the assumed widths must also fit within the geometry of the pile cap. For 3-D strut-and-tie models, in lieu of checking the strength of each individual strut and node, the strengths can be deemed adequate if the bearing strength at the supporting piles and at the supported column or pier are adequate [21].
- For tension ties that are developed with hooks or straight bars beyond the nodal zone, the height of the tension tie (i.e., the in-plane dimension) can be determined by the number of layers of tension tie bars and the size of the bars. If a single layer of tension tie reinforcement is used, the minimum height, w_t, of the tension tie will be $d_b + 2c_c$ where d_b is the diameter of the tension tie bars and c_c is the clear concrete cover to the bars. If two layers of reinforcement are used and assuming a maximum clear distance of $2d_b$ between the bottom and upper layer reinforcement, the minimum height of the tension zone is $4d_b + 2c_c$. When the tension tie bars are developed by plate anchors at each end of the deep member, the height of the tension tie zone will be taken as the height of the anchor plates.
- Calculate the forces in the struts and ties, and nodal zones.
- Determine the dimensions of the struts, tension ties, and the nodal zones.
- Using the assumed dimensions above, check the struts, tension tie, and nodal zones to determine if they have adequate strength to resist the forces calculated in the previous steps. If the strengths are inadequate, revise the strut-and-tie geometry and/or the sizes of the strut, tension tie zone, and nodal zones, and repeat the process until the strengths are adequate.

- Check the minimum reinforcement required in the tension tie. For deep beams and pile caps (banded reinforcement), the minimum tension tie reinforcement is

$$A_{s,\min} = \frac{3\sqrt{f'_c}\, b_w d}{f_y} \geq \frac{200 b_w d}{f_y}$$

- Detail the reinforcement, ensuring adequate development or anchorage in or beyond the nodal zones. For pile caps, provide transverse reinforcement perpendicular to the main reinforcement, similar to temperature and shrinkage reinforcement in one-way slabs (see Chapter 2).

Example 10-9

Deep Beam Design Using the Strut-and-Tie Method

Analyze the 18 in. × 48 in. deep simply supported beam shown in Figure 10-42. For the unfactored loads shown. Ignore the self-weight of the beam. f'_c = 4000 psi and f_y = 60,000 psi. Determine the tension tie reinforcement and the additional reinforcement required in the beam.

Solution

a. Analyze the beam under factored loads:

$$P_D = 100 \text{ kip and } P_L = 160 \text{ kip}$$

Therefore, the factored load, $P_u = 1.2(100) + 1.6(160) = 376$ kip

$$M_u = \frac{P\ell}{4} = \frac{(376)(13.5)}{4} = 1269 \text{ ft.-kip}$$

$$V_u = \frac{P}{2} = \frac{376}{2} = 188 \text{ kip}$$

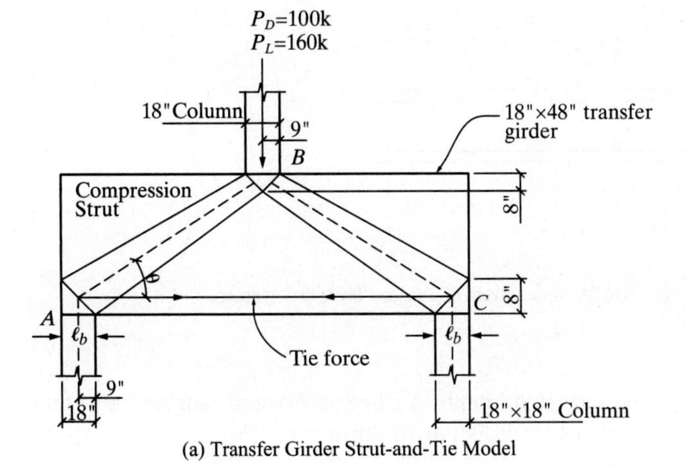

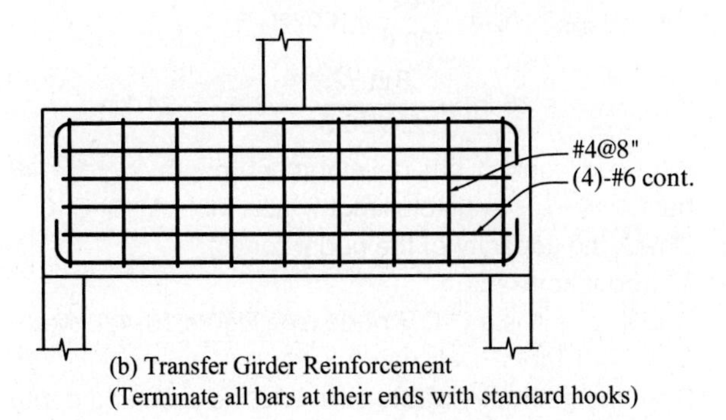

FIGURE 10-42 Transfer Girder for Example 10-9

The factored reaction, R_u = 188 kip

Check to make sure that the beam is adequate in bearing at the reaction points (A and C) for the factored reaction, R_u, and at the load point (B) for the load, P_u, using the principles presented in Section 10-5. The reader should verify that the load and reaction points are adequate in bearing.

b. Check if this is a deep beam; otherwise, analyze using the flexural methods presented in Chapters 2 and 3.
The shear span, a = 6 ft (see Figure 10-39) < 2h = 2(4 ft) = 8 ft O.K.
The clear span, ℓ_n = 12 ft < 4h = 4(4 ft) = 16 ft, O.K.
Therefore, the beam qualifies as a deep beam that can be designed or analyzed using the strut-and-tie method.

c. Check the maximum shear strength of the deep beam:
The width of the beam, b_w = 18 in. Assume the effective depth of the beam, d = 0.8h (this will be checked later. If actual the d is greater than assumed, then the initial estimate will be conservative)

$$\phi V_n = \phi 10\sqrt{f'_c} b_w d = 0.75(10)(\sqrt{4000})(18)(0.8)(48)$$
$$= 328 \text{ kip} > V_u = 188 \text{ kip O.K.}$$

d. Assume the width of the tension zone, w_t (see Figure 10-43b).
Assuming the tension tie reinforcement will be placed in two layers and assuming 2″ clear concrete cover to the tension tie reinforcement, and assuming #8 bars with 2 bar-diameters clear between the two layers of reinforcement, therefore, the width of the tension tie zone is

$$w_t = 4d_b + 2c_c = 4(1 \text{ in.}) + 2(2 \text{ in.}) = 8 \text{ in.}$$

Assume the depth of the nodal zone, w_{nz}, at point B just below the load is also 8 in, therefore, from geometry of the deep beam (see Figure 10-42(a)), the angle of inclination of the strut, θ, is calculated as follows:

$$\tan\theta = \frac{h - \frac{w_t}{2} - \frac{w_{nz}}{2}}{\frac{\ell}{2}} = \frac{48 \text{ in.} - \frac{8 \text{ in.}}{2} - \frac{8 \text{ in.}}{2}}{\frac{(13.5 \text{ ft})(12)}{2}} = 0.493$$

Therefore θ = 26.3° > 25° (see ACI 23.2.7) O.K.

e. Calculate the force in the strut, F_{us}, and check the capacity of the strut:
Equilibrium of forces at joint a requires that $F_{us}\sin(26.3°)$ = maximum reaction, R_{uA} = 188 kip.
Therefore, the compression force in the strut, F_{us} = 424 kip.
From geometry and the selected tie zone depth (see Figure 10-42(a)), the in-plane width of the strut, w_s, (see Figure 10-43) is

$$w_s = w_t\cos\theta + \ell_b\sin\theta = 8\cos(26.3°) + 18\sin(26.3°)$$
$$= 15.2 \text{ in.}$$

Assuming the strut is crossed by skin reinforcement satisfying ACI Code, Section 23.5, β_s = 0.75 (see Table 10-3). Therefore, the strut capacity is,

$$\phi F_{ns} = \phi A_{cs}(0.85\beta_s f_c) = 0.75(18 \text{ in.})(15.2 \text{ in.})[(0.85)(0.75)(4 \text{ ksi})]$$
$$= 523 \text{ kip} > F_{us} = 424 \text{ kip} \quad \text{O.K.}$$

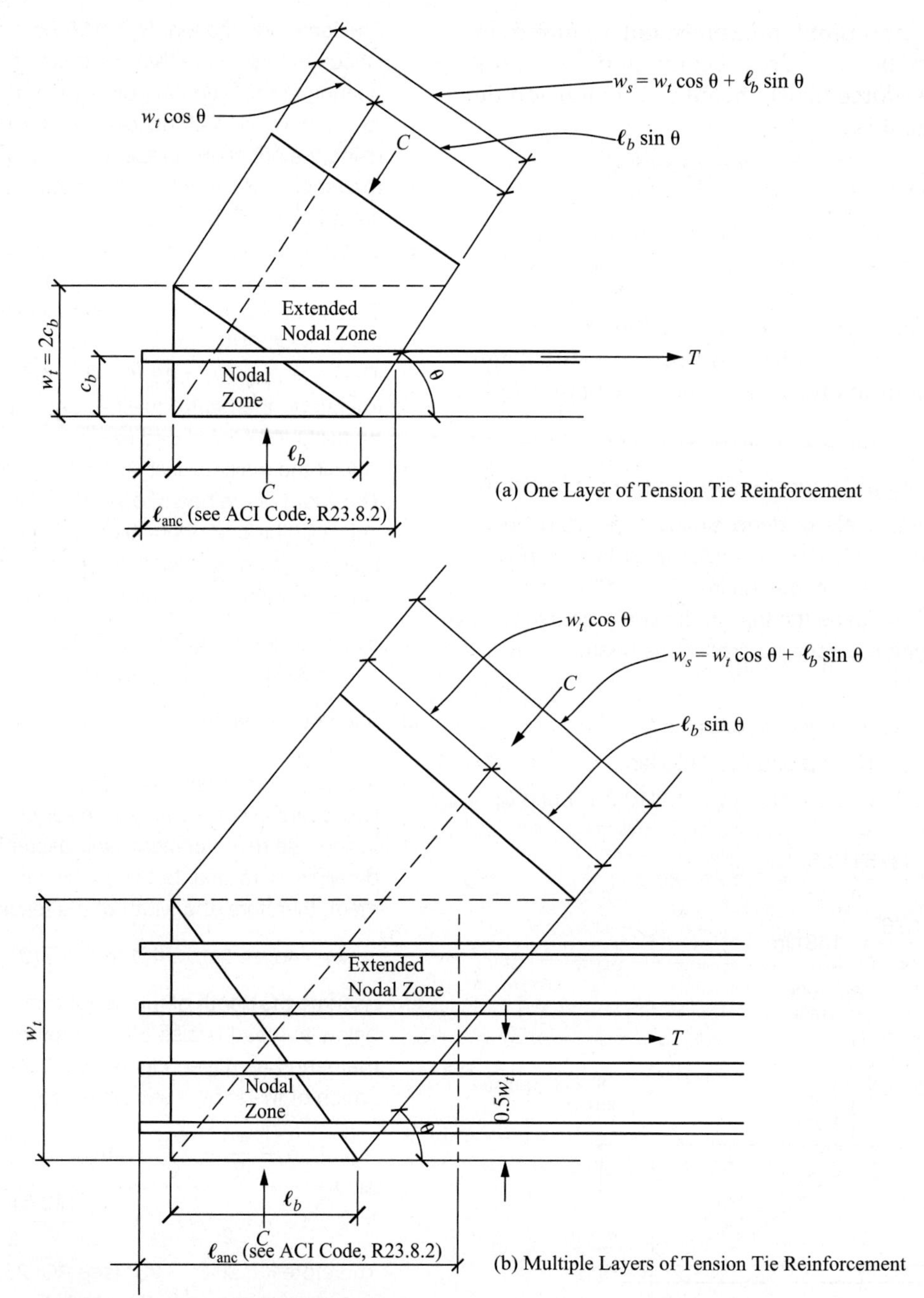

FIGURE 10-43 Size of Nodal Zones.

f. Check actual effective, d, and calculate the force in the tension tie and the area of the tension tie reinforcement:
Actual d = 48 in.–8 in./2 = 44 in. > 0.8h = 0.8(48) = 38.4 in. O.K.
Equilibrium of forces at joint A requires that $F_{us}\cos(26.3°)$ = Tension tie force, F_{ut}
Therefore, F_{ut} = 424 kip cos (26.3°) = 380 kip
Area of tension tie reinforcement required = 380 kip/[(0.75)(60 ksi)] = 8.44 in.^{2}
The minimum reinforcement required is $0.0033b_wd$ or 2.64 in^2 which is much less than the tension tie reinforcement required, therefore, O.K.
Use 2 layers of 6 #8 bars for the tension tie (A_s provided = 9.48 in.^{2} > A_s required, O.K.)
Minimum beam width to accommodate 6 #8 bars in one layer (see Table A-3) = 15 in. < 18 in. O.K.

Available length to develop hooked tension tie in two layers (see Figure 10-43) is

$$\ell_{anc} = \ell_b + \frac{w_t/2}{\tan\theta} - \text{cover} =$$

$$18\text{ in.} + \frac{8\text{ in.}/2}{\tan 26.3} - 2\text{ in.} = 24.1\text{ in.}$$

Required hooked bar development length, ℓ_{dh}, for #8 bars = 19 in. < available length, (see Table A-13) O.K

g. Check the capacity of the nodal zones:
At nodal zone A:
Nodal zone A is a C-C-T node (see Figure 10-42), therefore, from Table 10-4, $\beta_n = 0.8$
From the geometry of the strut and the assumed depth of the tension tie zone depth (see Figure 10-42(a)), the in-plane width of the strut, w_s (see Figure 10-43) is

$w_s = w_t\cos\theta + \ell_b\sin\theta = 8\cos(26.3°) + 18\sin(26.3°)$
$= 15.2\text{ in.}$

The capacity of nodal zone A is

$$\phi F_{nz} = \phi A_{nz}f_{ce} = \phi A_{nz}(0.85\beta_n f'_c)$$
$$= 0.75(18\text{ in.})(15.2\text{ in.})[(0.85)(0.8)(4\text{ ksi})]$$
$$= 558\text{ kip} > F_{us} = 424\text{ kip} \quad \text{O.K.}$$

At nodal zone B:

Nodal zone B is a C-C-C node (see Figure 10-40), therefore, from Table 10-4, $\beta_n = 1.0$. From the geometry of the strut and the assumed nodal zone depth (see Figure 10-42(a)), the in-plane width of the strut at node B, $w_s = \sqrt{(8\text{ in.})^2 + (9\text{ in.})^2} = 12.04\text{ in.}$

The capacity of nodal zone B is

$$\phi F_{nz} = \phi A_{nz}f_{ce} = \phi A_{nz}(0.85\beta_n f_c)$$
$$= 0.75(18\text{ in.})(12.04\text{ in.})[(0.85)(0.1)(4\text{ ksi})]$$
$$= 553\text{ kip} > F_{us} = 424\text{ kip} \quad \text{O.K.}$$

h. Provide additional horizontal and vertical reinforcement on the vertical surfaces of the deep beam:
The maximum spacing of the distributed reinforcement is

$$d/5 = 44\text{ in.}/5 = 8.8\text{ in. Use 8 in.}$$
$$\le 12\text{ in.}$$

Assuming #4 @ 8 in. o.c. vertical each face (VEF) and horizontal each face (HEF), therefore A_v provided = (2)(0.2 in.^{2}) = 0.4 in.^{2}

The required area of the minimum reinforcement on the vertical and horizontal surfaces of the beam are as follows:

$$A_v \ge 0.0025 b_w s_1 = 0.0025(18\text{ in.})(8\text{ in.}) = 0.36\text{ in.}^2$$
$$< A_v \text{ provided,} \quad \text{O.K.}$$
$$A_h \ge 0.0025 b_w s_2 = 0.0025(18\text{ in.})(8\text{ in.}) = 0.36\text{ in.}^2$$
$$< A_v \text{ provided,} \quad \text{O.K.}$$

where s_1 is the horizontal spacing of the vertical reinforcement, and b_w = width of beam

i. The reinforcement for the deep beam is shown in Figure 10-42(b).

Note: Though the self-weight of the deep beam was neglected in this example, we could conservatively have added the total factored self-weight of the beam to the factored concentrated load, P_u.

Example 10-10

Pile Cap Design Using the Strut-and-Tie Method

Determine the tension tie reinforcement required for a two-pile cap layout similar to that shown in Figure 10-33 that supports an 18 in. × 18 in. column with a factored load of 750 kip. The pile cap is 4 ft wide × 7 ft long × 3.5 ft deep supported on two symmetrically located 12 in. diameter concrete filled steel piles. The piles are spaced at 3 ft on centers with a 2 ft edge distance, and an embedment of the pile in the pile cap of 6 in. Ignore the self-weight of the pile cap. $f'_c = 5500\text{ psi}$ and $f_y = 60{,}000\text{ psi}$.

Solution

a. Analyze the pile cap under factored loads:

$$P_u = 750\text{ kip}$$
$$V_u = \frac{P_u}{2} = \frac{750}{2} = 375\text{ kip}$$

The factored reaction, R_u = 375 kip

b. Check if this pile cap qualifies as a deep beam; otherwise, analyze and design the cap using the flexural methods presented in Chapters 2 & 3.
The shear span, a = 12 in. (see Figure 10-39) < 2h = 2(3.5 ft) = 7 ft O.K.
The clear span,

$$\ell_n = 3\text{ ft.} - \frac{12\text{ in.}}{(2)(12)} - \frac{12\text{ in.}}{(2)(12)} = 2\text{ ft.} < 4h = 4(3.5\text{ ft})$$
$$= 14\text{ ft,} \quad \text{O.K.}$$

Therefore, the pile cap qualifies as a deep beam and should be designed or analyzed using the strut-and-tie method.

c. Check maximum shear strength of the pile cap:
The width of the pile cap, b_w = 48 in.
Assume the effective depth of the pile cap, $d = 0.8h$ (this assumption can be checked later. If actual d is greater than assumed, then the initial estimate will be conservative)

$$\phi V_n = \phi 10\sqrt{f'_c}b_w d =$$
$$= 0.75(10)(\sqrt{5500})(48\text{ in.})(0.8)(42\text{ in.})$$
$$= 897\text{ kip} > V_u = 375\text{ kip} \qquad \text{O.K.}$$

d. Check the bearing capacity of the pile cap at the piles (the reaction points) and at the load point (i.e., at the column). The bearing capacity can be obtained from Section 10.5 (ACI 22.8.3.2).
The reader can verify that the bearing capacity for the 12 in. diameter concrete-filled steel piles is 687 kip (i.e. (0.65)(113.1 in.^{2})(2)(0.85)(5.5 ksi)) which far exceeds the factored pile reaction of 375 kip, and the bearing capacity at the 18 in. × 18 in. concrete column to pile cap interface also far exceeds the factored load of 750 kip.

e. Check the pile cap for punching shear at the supporting piles and at the supported column. The reader should verify that these are adequate.

f. Assume the width of the tension zone, w_t (see Figure 10-43a).
The piles are embedded 6 in. into the pile cap. Assuming the tension tie reinforcement will be placed in one layer with a 3″ clear concrete cover to the tension tie reinforcement above the pile cut-off elevation, and assuming #8 bars
Therefore,

$$w_t = d_b + 2c_c = 1\text{ in.} + 2(3\text{ in.}) = 7\text{ in.}$$

Assume the depth of the nodal zone, w_{nz}, at the column below the load (i.e., point B) is also 7 in.
Therefore, from geometry of the pile cap (see Figure 10-40), the angle of inclination of the strut will be calculated as follows:

$$\tan\theta = \frac{h - 6\text{ in.} - \frac{w_t}{2} - \frac{w_{nz}}{2}}{\frac{\ell}{2}} = \frac{42\text{ in.} - 6\text{ in.} - \frac{7\text{ in.}}{2} - \frac{7\text{ in.}}{2}}{\frac{(3\text{ ft})(12)}{2}} = 1.611$$

Therefore $\theta = 58.2° > 25°$ (see ACI 23.2.7) O.K.

g. Calculate the force in the strut, F_{us}, and check the capacity of the strut
Equilibrium of forces at joint A requires that $F_{us}\sin(58.2°)$ = pile reaction, R_{uA} = 375 kip,
Therefore, F_{us} = 441 kip.

From the geometry of the strut and the assumed depth of the tension tie zone depth (see Figure 10-43), the in-plane width of the strut, w_s, is

$$w_s = w_t \cos\theta + \ell_b \sin\theta = (7\,\text{in.})\cos(58.2°) + (12\,\text{in.})\sin(58.2°) = 13.89\text{ in.}$$

Therefore, the cross-sectional area of the diagonal strut, $A_{cs} = (b_w)(w_s) = (48\text{ in.})(13.89\text{ in.})$

For pile caps, the strut is unreinforced, therefore, from Table 10-3, $\beta_s = 0.6$

Therefore, the strut capacity is,

$$\phi F_{ns} = \phi A_{cs}(0.85\beta_s f_c') = 0.75(48\,\text{in.})(13.89\,\text{in.})[(0.85)(0.6)(5.5\,\text{ksi})] = 1403\,\text{kip} > F_{us} = 441\,\text{kip} \quad \text{O.K.}$$

h. Calculate the force in the tension tie and the area of the tension tie reinforcement:

Equilibrium of forces at joint A requires that $F_{us}\cos(58.2°) = F_{ut}$

Therefore, $F_{ut} = 441\text{ kip}\cos(58.2°) = 232\text{ kip}$

Area of tension tie rebar $= 232\text{ kip}/[(0.75)(60\text{ ksi})] = 5.2\text{ in.}^2$

The minimum reinforcement required is $0.0033 b_w d$, therefore,

$A_{s,min} = 0.0033(48\text{ in.})(42\text{ in.} - 6\text{ in.} - 3\text{ in. cover} - 1''/2) = 5.15\text{ in}^2 <$ As required

Use one layer of 9 #7 bars for the tension tie (A_s provided $= 5.4\text{ in.}^2 > A_s$ required, O.K.).

Minimum width to accommodate 9 #7 bars in one layer (see Table A-3) $= 20$ in. $< b_w = 48$ in. O.K.

Available length to develop hooked tension tie in single layer (see Figure 10-43) is

$$\ell_{anc} = 2d_p + d_p/2 + \frac{w_t/2}{\tan\theta} - 3''\,\text{side cover}$$

$$= 2(12\,\text{in.}) + 12\,\text{in.}/2 + \frac{7\,\text{in.}/2}{\tan 58.2} - 3\,\text{in.} = 28.8\,\text{in.}$$

Required hooked bar development length, ℓ_{dh}, for #7 bars $= 16.6$ in. $<$ available length, O.K.

i. Check the capacity of the nodal zones:

At nodal zone A:

Nodal zone A is a C-C-T node (see Figure 10-40), therefore, from Table 10-4, $\beta_n = 0.8$

From the geometry of the strut and the assumed depth of the tension tie zone depth, and noting the 12 in. pile diameter, the in-plane width of the strut, w_s (see Figure 10-43) is

$$w_s = w_t \cos\theta + \ell_b \sin\theta = (7\,\text{in.})\cos(58.2°) + (12\,\text{in.})\sin(58.2°) = 13.89\,\text{in.}$$

At nodes A and C, assume the out-of-plane thickness of the nodal zone is equal to the width of the pile cap = 48 in.

The capacity of nodal zone A is

$$\phi F_{ns} = \phi A_{cs}(0.85\beta_n f_c') = 0.75(48\text{ in.})(13.89\text{ in.})[(0.85)(0.8)(5.5\text{ ksi})] = 1870\,\text{kip} > F_{us} = 441\,\text{kip} \quad \text{O.K.}$$

At nodal zone B:

Nodal zone B is a C-C-C node (see Figure 10-40), therefore, from Table 10-4, $\beta_n = 1.0$.

From the geometry of the strut and the assumed nodal zone depth (see Figure 10-42(a)), the in-plane width of the strut at node B, $w_s = \sqrt{(7\,\text{in.})^2 + (9\,\text{in.})^2} = 11.4\,\text{in.}$

At node B, assume the out of plane thickness of the nodal zone is equal to the width of the pile cap = 48 in.

The capacity of nodal zone B is

$$\phi F_{nz} = \phi A_{nz} f_{ce} = \phi A_{nz}(0.85\beta_n f_c') = 0.75(48\,\text{in.})(11.4\,\text{in.})[(0.85)(1.0)(5.5\,\text{ksi})] = 1919\text{ kip} > F_{us} = 441\,\text{kip} \quad \text{O.K.}$$

j. The reinforcement in the transverse direction = minimum reinforcement = 0.0018bh = 0.0018(12 in.)(42 in.) = 0.907 in.^{2} Use #7 @ 7½ in. on centers in the transverse direction, perpendicular to the main tension tie reinforcement.

k. The reinforcement in the pile cap will be laid out similar to the reinforcement shown in Figure 10-33. The design of this pile cap can be made more efficient by using a lower concrete strength. The reader should verify the lowest concrete strength that would be adequate for this pile cap.

An introduction to the design of pile caps has been provided in this chapter. For more detailed design of pile caps under various loading conditions, the reader is referred to Ref. [20].

References

[1] David G. Kittridge. "High Rise Shearwall Analysis, Design & Constructibility Guidelines," Proceedings of the *ASCE Structures Congress*, September 2010, pp. 381–382.

[2] Joseph A Amon. "The Geotechnical Report: What the Designer Needs to Know," *Concrete International*, September 2010, pp. 48–49.

[3] Joseph E. Bowles. *Foundation Analysis and Design*, 5th ed. McGraw-Hill, New York 2001.

[4] David Rogowsky and James K. Wight. "Load Factors Are Load Factors," *Concrete International*, Vol 32, No. 7 July 2010, p. 75.

[5] Harry G. Poulos, "Tall Building Foundations: Design Methods and Applications," *Innovative Infrastructure Solutions* Vol 1, No. 1, 2016. Springer.

[6] Braja M. Das. "Principles of Foundation Engineering," 5th Edition, Thomson, 2004.

[7] Steven C. Ball and James S. Notch. "Computer Analysis/Design of Large Mat Foundations," Journal of Structural Engineering, Vol. 110, No. 5, May 1984.

[8] Edward J. Ulrich, Jr. "Subgrade Reaction in Mat Foundation Design," Concrete International, April 1991.

[9] ACI Committee 336. "Suggested Design Procedures for Combined Footings and Mats," Journal of the American Concrete Institute, Vol. 63, No. 10, pp. 1041–1077, 1988.

[10] Cheng Liu and Jack B. Evett. "Soils and Foundations," 8th Edition, Pearson, 2014.

[11] E. Alfred Picardi. "Structural System—Standard Oil of Indiana Building," Engineering Journal, American Institute of Steel Construction, Second Quarter, 1975.

[12] S. V. DeSimone. "Distribution of Wind Load to Soil," Technical Committee No. 11, Foundation Design, State of the Art Report No. 2, International Conference on the Planning and Design of Tall Buildings, Lehigh University, Bethlehem, PA, USA, August 21–26, 1972.

[13] CPCA. "Concrete Design Handbook," Third Edition, Canadian Portland Cement Association, 2012, pp. 9-3-9-49.

[14] R. Park and W.L. Gamble. Reinforced Concrete Slabs, pp. 513–514, John Wiley & Sons, New York.

[15] Alan Williams. "Design of Reinforced Concrete Structures," Engineering Press, 1997.

[16] Dennis C. K. Poon and Torsten G. Gottlebe. "Sky High in Shenzhen," ASCE Civil Engineering, pp. 48–53, December 2017.

[17] Dominic A. Webber and Howard A. Wells. "Grant Street Pier - Drilled Shaft Repair," Structure Magazine, pp. 22–25, December 2017.

[18] Gautier Chantelot and Alexandre Mathern. "Strut-and-tie modelling of reinforced concrete pile caps," Department of Civil Engineering, Chalmers University of Technology, Goteborg, Sweden, 2010.

[19] Perry Adebar and Luke (Zongyu) Zhou. "Design of Deep Pile Caps by Strut-and-Tie-Models," ACI Structural Journal, Vol. 94, No. 4, July-August 1996.

[20] Timothy W. Mays. "CRSI Design Guide for Pile Caps," Concrete Reinforcing Steel Institute (CRSI), 1st Edition, Shaumburg, IL, 2015.

[21] Chris Willams, Dean Deschenes, and Oguzhan Bayrak. "Strut-and-Tie Model Design Examples for Bridges: Final Report," Center for Transportation Research, Report No. FHWA/TX-12/5-5253-01-1, The University of Texas, Austin, Texas, June 2012.

Problems

Note: Assume that all steel is uncoated and the soil pressures given are gross allowable soil pressures.

10-1. Design a plain concrete wall footing to carry a 12-in.-thick reinforced concrete wall. Service loads are 2.3 kip/ft dead load (includes the weight of the wall) and 2.3 kip/ft live load. Use $f'_c = 3000\ psi$ and an allowable soil pressure of 4000 psf. Assume that the bottom of the footing is to be 4 ft below grade. The weight of earth $w_e = 100\ \text{lb/ft}^3$.

10-2. Redesign the footing for the wall of Problem 10-1. Service loads are 8.0 kip/ft dead load and 8.0 kip/ft live load.

10-3. Design a reinforced concrete footing for the wall of Problem 10-1 if the service loads are 6 kip/ft dead load (includes wall weight) and 15 kip/ft live load. Use $f_y = 60{,}000$ psi.

10-4. Design a square individual column footing (reinforced concrete) to support an 18-in.-square reinforced concrete tied interior column. Service loads are 200 kip dead load and 350 kip live load. Use $f'_c = 3000$ psi and $f_y = 60{,}000$ psi. The allowable soil pressure is 3500 psf, and $w_e = 100\ \text{lb/ft}^3$. The column, reinforced with eight No. 8 bars, has $f'_c = 5000$ psi and $f_y = 60{,}000$ psi. The bottom of the footing is to be 4 ft below grade.

10-5. Design a square individual column footing (reinforced concrete) to support a 16-in.-square reinforced concrete tied interior column. Service loads are 200 kip dead load and 160 kip live load. Both column and footing have $f'_c = 4000$ psi and $f_y = 60{,}000$ psi. The allowable soil pressure is 5000 psf, and $w_e = 100\ \text{lb/ft}^3$. The bottom of the footing is to be 4 ft below grade. The column is reinforced with eight No. 7 bars.

10-6. Design for load transfer from a 14-in.-square tied column to a 13-ft-0-in.-square reinforced concrete footing. Use $P_u = 650$ kip, footing $f'_c = 3000$ psi, column $f'_c = 5000$ psi, and $f_y = 60{,}000$ psi (all steel). The column is reinforced with eight No. 8 bars.

10-7. Redesign the footing of Problem 10-5 if there is a 7-ft-0-in. restriction on the width of the footing.

10-8. Footings are to be designed for two columns A and B spaced with $D = 16$ ft as shown. There are no dimensional limitations. The service load from column A is 100 kip and from column B is 150 kip. The allowable soil pressure is 4000 psf. Determine the appropriate footing(s) size, type, and layout. Assume thickness and disregard reinforcing.

10-9. Same as Problem 10-8, except service loads are 700 kip from column A and 900 kip from column B, and $D = 14$ ft.

10-10. For the column layout shown, there is a width restriction W of 16 ft. Here $D = 14$ ft. Determine an appropriate footing size and layout if the service load from column A is 700 kip and from column B is 800 kip. Assume thickness and disregard reinforcing. The allowable soil pressure is 4000 psf.

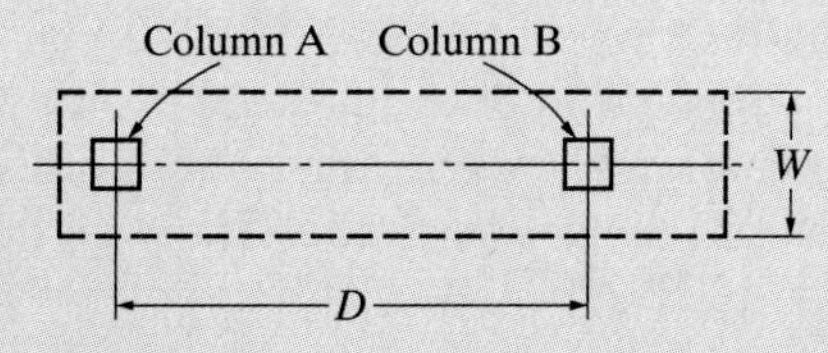

PROBLEMS 10-8 to 10-11

10-11. For the column layout shown, $D = 24$ ft. The footing for the 18-in.-square column A must be placed flush with the left side of the column. Service loads are 200 kip for column A and 300 kip for column B. The allowable soil pressure is 4500 psf. Assume a footing thickness of 2 ft-6 in. and determine an appropriate size, type, and layout for the footing.

10-12. Refer to the Footing detail shown below where the Service dead load equals the service live load. Use $f'_c = 3000$ psi and $f_y = 60{,}000$ psi. The Net allowable soil pressure is 2,000 psf (q_{net}).

a. Determine the service load 'P' that can be applied based on the net allowable soil pressure (service load).

b. Determine the service load 'P' that could be applied to the footing considering one-way shear.

c. Determine the service load 'P' that could be applied to the footing considering two-way shear.

d. Determine the service load 'P' that could be applied to the footing considering bending.

e. Based on the above, what service load could this footing be rated for?

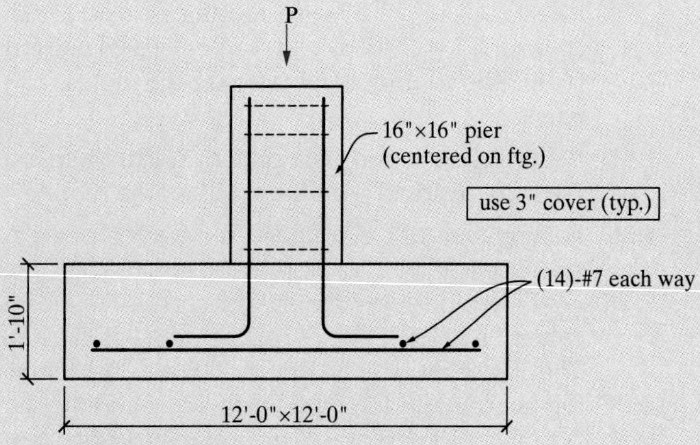

PROBLEM 10-12

10-13. Refer to the footing detail shown below. The Net allowable soil pressure is 6000 psf (q_{net}).

a. Determine the maximum column load, P, that can be applied based on the net allowable soil pressure (service load).

Use the result from (a) to complete the rest of this problem assuming the service dead load equals the service live load.

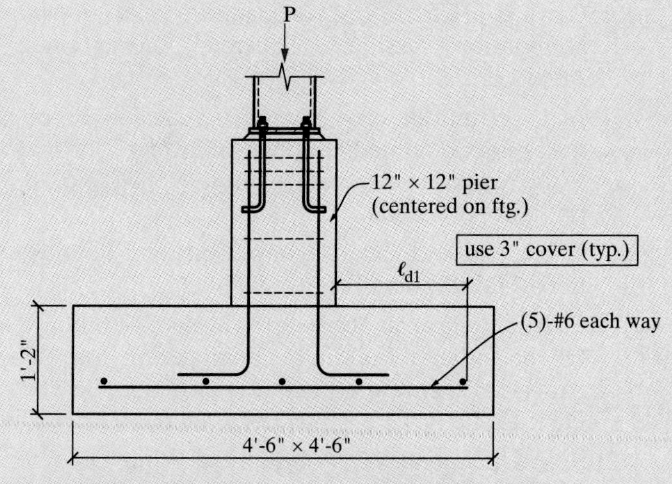

PROBLEM 10-13

b. Determine if the footing is adequate for one-way shear.

c. Determine if the footing is adequate for two-way shear.

d. Determine if the footing is adequate for bending.

e. Determine of the development length ℓ_{d1} is adequate.

10-14. Refer to the footing detail shown below. The Column service loads are: Dead = 40k, Live = 72k. The Net allowable soil pressure is 2000 psf (q_{net}).

a. Determine if the footing is adequate for one-way shear.

b. Determine if the footing is adequate for two-way shear.

c. Determine if the footing is adequate for bending.

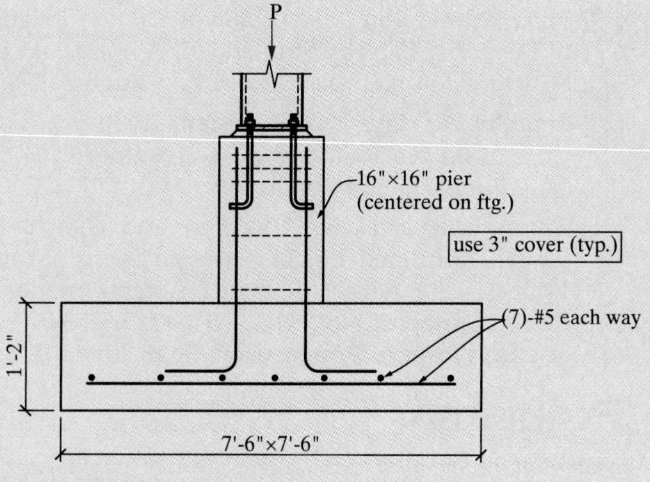

PROBLEM 10-14

PRESTRESSED CONCRETE FUNDAMENTALS

11-1 Introduction
11-2 Design Approach and Basic Concepts
11-3 Stress Patterns in Prestressed Concrete Beams
11-4 Prestressed Concrete Materials
11-5 Analysis of Rectangular Prestressed Concrete Beams
11-6 Alternative Methods of Elastic Analysis: Load Balancing Method
11-7 Flexural Strength Analysis
11-8 Notes on Prestressed Concrete Design

11-1 INTRODUCTION

According to the ACI definition, prestressed concrete is a material that has had internal stresses induced to balance out, to a desired degree, tensile stresses due to externally applied loads. Because tensile stresses are undesirable in concrete members, the object of prestressing is to create compressive stresses (prestress) at the same locations as the tensile stresses within the member so that the tensile stresses will be diminished or will disappear altogether. The diminishing or elimination of tensile stresses within the concrete will result in members that have fewer cracks or are crack-free at service load levels. This is one of the advantages of prestressed concrete over reinforced concrete, particularly in corrosive atmospheres. Prestressed concrete offers other advantages. Because beam cross sections are primarily in compression, diagonal tension stresses are reduced and the beams are stiffer at service loads. In addition, sections can be smaller, resulting in less dead weight.

Despite the advantages, we must consider the higher unit cost of stronger materials, the need for expensive accessories, the necessity for close inspection and quality control, and, in the case of precasting, a higher initial investment in plant. Figure 11-1 shows precast bridge girder components ready for post-tensioning.

11-2 DESIGN APPROACH AND BASIC CONCEPTS

As most of the advantages of prestressed concrete are at service load levels and as permissible stresses in the "green" concrete often control the amount of prestress force to be used, the major part of analysis and design calculations is made using service loads, permissible stresses, and basic assumptions as outlined in Sections 20.3.2.5.1 and 24.5 of the ACI Code. The strength requirements of the code must also be met, however. Therefore, at some point, the design must be checked using appropriate load factors and strength reduction factors.

The normal method for applying prestress force to a concrete member is through the use of steel tendons. There are two basic methods of arriving at the final prestressed member: pretensioning and post-tensioning.

Pretensioning may be defined as a method of prestressing concrete in which the tendons are tensioned before the concrete is placed. This operation, which may be performed in a casting yard, is basically a five-step process:

1. The tendons are placed in a prescribed pattern on the casting bed between two anchorages. The tendons are then tensioned to a value not to exceed 94% of the specified yield strength, but not greater than the lesser of 80% of the specified tensile strength of the tendons and the maximum value recommended by the manufacturer of the prestressing tendons or anchorages (ACI Code, Section 18.5.1). The tendons are then anchored so that the load in them is maintained.
2. If the concrete forms are not already in place, they may then be assembled around the tendons.
3. The concrete is then placed in the forms and allowed to cure. Proper quality control must be exercised, and curing may be accelerated with the use of steam or other methods. The concrete will bond to the tendons.
4. When the concrete attains a prescribed strength, normally within 24 hours or less, the tendons are cut at the anchorages. Because the tendons are now

FIGURE 11-1 Precast bridge girder components for post-tensioning. (George Limbrunner)

bonded to the concrete, as they are cut from their anchorages the high prestress force must be transferred to the concrete. As the high tensile force of the tendon creates a compressive force on the concrete section, the concrete will tend to shorten slightly. The stresses that exist once the tendons have been cut are often called the stresses at *transfer*. Because there is no external load at this stage, the stresses at transfer include only those due to prestressing forces and those due to the weight of the member.

5. The prestressed member is then removed from the forms and moved to a storage area so that the casting bed can be prepared for further use.

Pretensioned members are usually manufactured at a casting yard or plant that is somewhat removed from the job site where the members will eventually be used. In this case, they are usually delivered to the job site ready to be set in place. Where a project is of sufficient magnitude to warrant it economically, a casting yard may be built on the job site, thus decreasing transportation costs and allowing larger members to be precast without the associated transportation problems.

Figure 11-2 depicts the various stages in the manufacture of a precast, pretensioned member.

Post-tensioning may be defined as a method of prestressing concrete in which the tendons are tensioned *after* the concrete has cured. (Refer to Figure 11-3.) The operation is commonly a six-step process:

1. Concrete forms are assembled with flexible hollow tubes (metal or plastic) placed in the forms and held at specified locations.

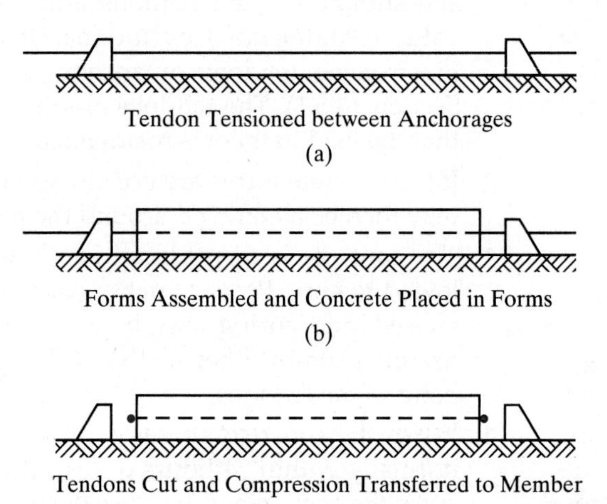

FIGURE 11-2 Pretensioned member.

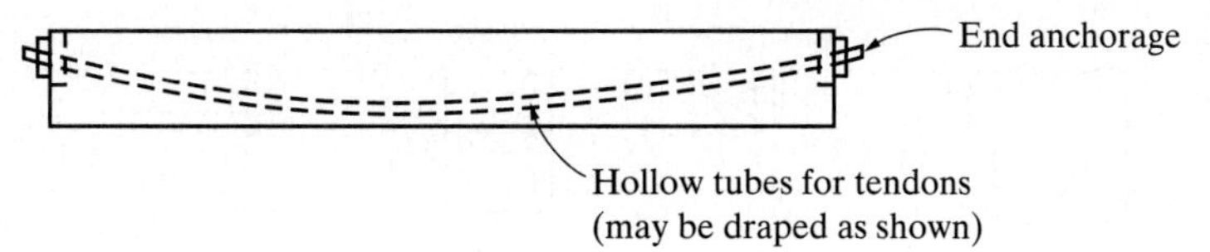

FIGURE 11-3 Post-tensioned member.

2. Concrete is then placed in the forms and allowed to cure to a prescribed strength.
3. Tendons are placed in the tubes. (In some systems, a complete tendon assembly is placed in the forms prior to the placing of the concrete.)
4. The tendons are tensioned by jacking against an anchorage device or end plate that, in some cases, has been previously embedded in the end of the member. The anchorage device will incorporate some method for gripping the tendon and holding the load.
5. If the tendons are to be bonded, the space in the tubes around the tendons may be grouted using a pumped grout. Some members use unbonded tendons.
6. The end anchorages may be covered with a protective coating.

Although post-tensioning is sometimes performed in a plant away from the project, it is most often done at the job site, particularly for units too large to be shipped assembled or for unusual applications.

The tensioning of the tendons is normally done with hydraulic jacks. Many patented devices are available to accomplish the anchoring of the tendon ends to the concrete.

11-3 STRESS PATTERNS IN PRESTRESSED CONCRETE BEAMS

The stress pattern existing on the cross section of a prestressed concrete beam may be determined by superimposing the stresses due to the loads and forces acting on the beam at any particular time. For our purposes, the following sign convention will be adopted:

Tensile stresses are positive (+).

Compressive stresses are negative (−).

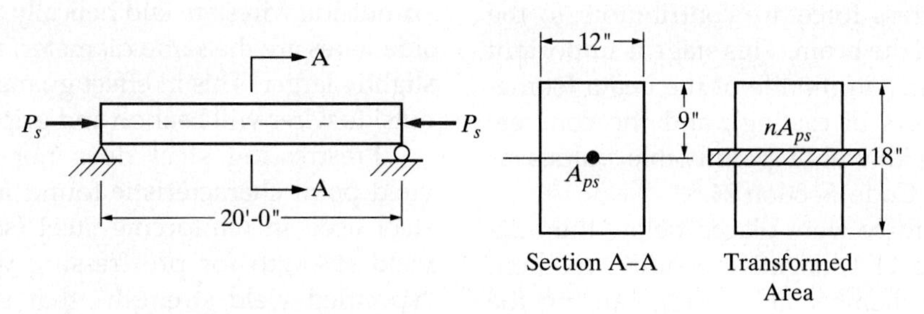

FIGURE 11-4 Sketches for Example 11-1.

Because we will assume a crack-free cross section at service load level, the entire cross section will remain effective in carrying stress. Also, the entire concrete cross section will be used in the calculation of centroid and moment of inertia.

For purposes of explanation, we will consider a rectangular shape with tendons placed at the centroid of the section, and we will investigate the induced stresses. Although the rectangular shape is used for some applications, it is often less economical than the more complex shapes.

Example 11-1

For the section shown in Figure 11-4, determine the stresses due to prestress immediately after transfer and the stresses at midspan when the member is placed on a 20-ft simple span. Use $f'_c = 5000$ psi and assume that the concrete has attained a strength of 4000 psi at the time of transfer. Use a central prestressing force of 100 kip.

Solution:

1. Compute the stress in the concrete at the time of initial prestress. With the prestressing force P_s applied at the centroid of the section and assumed acting on the gross section A_c, the concrete stress will be uniform over the entire section. Thus

$$f = \frac{P_s}{A_c} = \frac{-100}{12(18)} = -0.463 \text{ ksi}$$

2. Compute the stresses due to the beam dead load:

$$\text{Weight of beam: } w_{DL} = \frac{12(18)}{144}(0.150)$$
$$= 0.225 \text{ kip/ft}$$
$$\text{Moment due to dead load: } M_{DL} = \frac{w_{DL}\ell^2}{8} = \frac{0.225(20)^2}{8}$$
$$= 11.25 \text{ ft-kip}$$

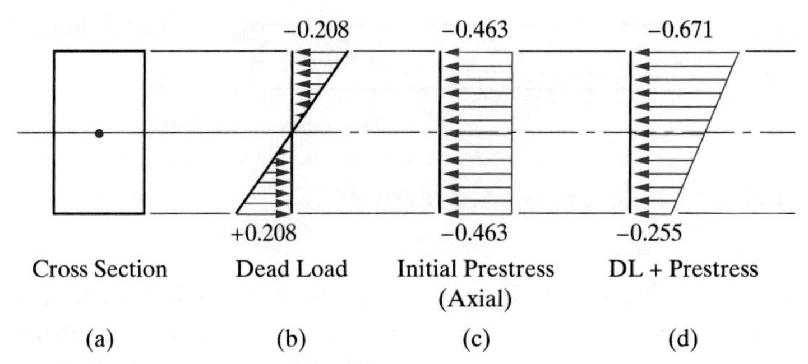

FIGURE 11-5 Midspan stresses for Example 11-1.

Compute the moment of inertia I of the beam using the gross cross section and neglecting the transformed area of the tendons (as the steel is at the centroid, it will have no effect on the moment of inertia):

$$I_g = \frac{bh^3}{12} = \frac{12(18)^3}{12} = 5832 \text{ in.}^4$$

$$\text{Dead load stresses: } f = \frac{Mc}{I} = \frac{11.25(12)(9)}{5832}$$
$$= \pm 0.208 \text{ ksi}$$

3. Compute the stresses due to prestress plus dead load:

$$f_{(\text{initial prestress+DL})} = -0.463 \pm 0.208$$
$$= -0.671 \text{ ksi} \quad (\text{compression, top})$$
$$= -0.255 \text{ ksi} \quad (\text{compression, bottom})$$

These results are shown on the stress summation diagram of Figure 11-5.

The tensile stresses due to the DL moment in the bottom of the beam have been completely canceled out and compression exists on the entire cross section. A limited additional positive moment may be carried by the beam without resulting in a net tensile stress in the bottom of the beam. This situation may be further improved on by *lowering the location of the tendon* to induce additional compressive stresses in the bottom of the beam.

Example 11-1 reflects two stages of the prestress process. The transfer stage occurs in a pretensioned member when the tendons are cut at the ends of the member and the prestress force has been transferred to the beam, as shown in Figure 11-2. When the beam in this problem is removed from the forms (picked up by its end points), dead load stresses are introduced, and in this second stage, both the beam weight (dead load) and the prestress force are contributors to the stress pattern within the beam. This stage is important because it occurs early in the life of the beam (sometimes within 24 hours of casting), and the concrete stresses must be held within permissible values as specified in the ACI Code, Section 24.5.

If the prestress force were placed below the neutral axis in Example 11-1, negative bending moment would occur in the member at transfer, causing the beam to curve upward and pick up its dead load. Hence, for a simple beam such as this, where the prestress force is eccentric, the stresses due to the initial prestress would never exist alone without the counteracting stresses from the dead load moment.

11-4 PRESTRESSED CONCRETE MATERIALS

The application of the prestress force both strains or "stretches" the tendons and at the same time induces tensile stresses in them. If the tendon strain is reduced for some reason, the stress will also be reduced. In prestressed concrete, this is known as a *loss of prestress.* This loss occurs after the prestress force has been introduced. Contributory factors to this loss are creep and shrinkage of the concrete, elastic compression of the concrete member, relaxation of the tendon stress, anchorage seating loss, and friction losses due to intended or unintended curvature in post-tensioning tendons. Estimates on the magnitudes of these losses vary. It is generally known, however, that an ordinary steel bar tensioned to its yield strength (40,000 psi) would lose its entire prestress by the time all stress losses had taken place. Therefore, for prestressed concrete applications, it is necessary to use very high-strength steels, where the previously mentioned strain losses will result in a much smaller percentage of change in the original prestress force.

The most commonly used steel for pretensioned prestressed concrete is in the form of a seven-wire, uncoated, stress-relieved strand having a minimum tensile strength (f_{pu}) of 250,000 psi or 270,000 psi, depending on grade. The seven-wire strand is made up of seven cold-drawn wires. The center wire is straight, and the six outside wires are laid helically around it. All six outside wires are the same diameter, and the center wire is slightly larger. This in effect guarantees that each of the outside wires will bear on and grip the center wire.

Prestressing steel does not exhibit the definite yield point characteristic found in the normal ductile steel used in reinforcing steel (see Figure 11-6). The yield strength for prestressing wire and strand is a "specified yield strength" that is obtained from the stress-strain diagram at 1% strain, according to the

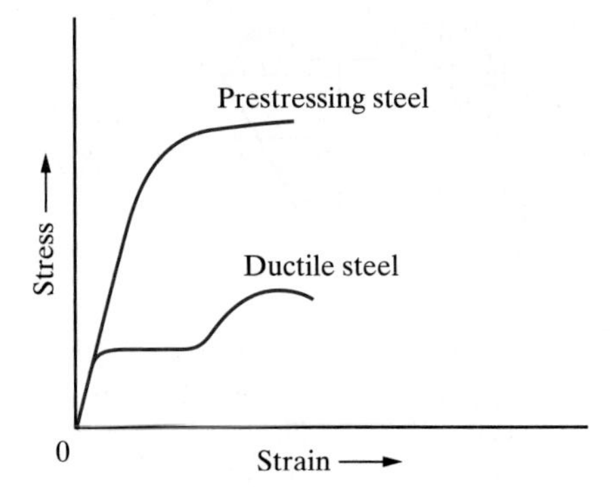

FIGURE 11-6 Comparative stress-strain curves.

American Society for Testing and Materials (ASTM). Nevertheless, the specified yield point is not as important in prestressing steel as is the yield point in the ductile steels. It is a consideration when determining the ultimate strength of a beam. For more information on prestressing wire and strand, the reader is referred to ASTM Standards A416 and A421 [1].

In normal reinforced concrete members designed to ensure tension failures, the strength of the concrete is secondary in importance to the strength of the steel in the determination of the flexural strength of the member. In prestressed applications, concrete in the range from 4000 to 6000 psi is commonly used. Some of the reasons for this are as follows: (1) volumetric changes for higher-strength concrete are smaller, which will result in smaller prestress losses; (2) bearing and development stresses are higher; and (3) higher-strength concrete is more easily obtained in precast work than in cast-in-place work because of better quality control. In addition, high early strength cement (type III) is normally used to obtain as rapid a turnover time as possible for optimum use of forms.

11-5 ANALYSIS OF RECTANGULAR PRESTRESSED CONCRETE BEAMS

The analysis of flexural stresses in a prestressed member should be performed for different stages of loading—that is, the initial service load stage, which includes dead load plus prestress before losses; the final service load stage, which includes dead load plus prestress plus live load after losses; and finally the factored load stage, which involves load and strength-reduction factors. Generally, checking of prestressed members is accomplished at the service load level based on unfactored loads. The nominal strength of a member should be checked, however, using the same strength principles as for non-prestressed reinforced concrete members.

Example 11-2

For the beam of cross section shown in Figure 11-7, analyze the flexural stresses at midspan at transfer and in service. Neglect losses. Use a prestressed steel area A_{ps} of 2.0 in.^{2}, use $f'_c = 6000$ psi, and assume that the concrete has attained a strength of 5000 psi at the time of transfer. The initial prestress force = 250 kip. The service dead load = 0.25 kip/ft, which does not include the weight of the beam. The service live load = 1.0 kip/ft. Use $n = 7$. Assume that the entire cross section is effective and use the transformed area (neglecting displaced concrete) for the moment of inertia.

Solution:

1. The beam weight is

$$w_{DL} = \frac{20(12)}{144}(0.150) = 0.25 \text{ kip/ft}$$

The moment due to the beam weight is

$$M_{DL} = \frac{w_{DL}\ell^2}{8} = \frac{0.25(30)^2}{8} = 28.1 \text{ ft-kip}$$

The moment due to superimposed loads (DL + LL) is

$$M_{DL+LL} = \frac{w_{(DL+LL)}\ell^2}{8} = \frac{1.25(30)^2}{8} = 140.6 \text{ ft-kip}$$

2. The location of the neutral axis is

$$\bar{y} = \frac{\sum(Ay)}{\sum A}$$

Using the top of the section as the reference axis,

$$\bar{y} = \frac{12(20)(10) + 14(15)}{12(20) + 14} = 10.28 \text{ in.}$$

The eccentricity e of the strands from the neutral axis is

$$15 - 10.28 = 4.72 \text{ in.}$$

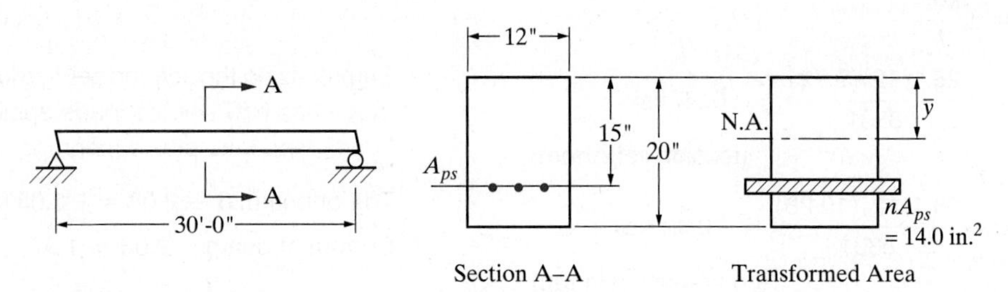

FIGURE 11-7 Sketches for Example 11-2.

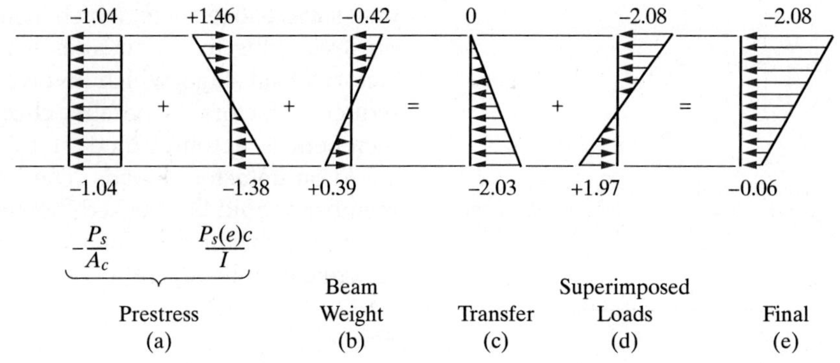

FIGURE 11-8 Midspan stress summary for Example 11-2.

The moment of inertia about the neutral axis is

$$I = \frac{12(20)^3}{12} + 12(20)\left(10.28 - \frac{20}{2}\right)^2 + 14(4.72)^2$$

$$= 8331 \text{ in.}^4$$

3. The stresses may now be calculated. These are summarized in Figure 11-8.

   a. Initial prestress: As a result of the eccentric prestressing force, the induced stress at the initial prestress stage will *not* be uniform but may be computed from

$$f = -\frac{P_s}{A_c} \pm \frac{Mc}{I}$$

$$= -\frac{P_s}{A_c} \pm \frac{P_s(e)c}{I}$$

where P_s/A_c represents the axial effect of the prestress force and $\frac{P_s(e)c}{I}$ is the eccentric, or moment, effect. Then

$$-\frac{P_s}{A_c} = -\frac{250}{12(20)} = -1.04 \text{ ksi}$$

(compression top and bottom)

$$+\frac{P_s(e)c}{I} = \frac{250(4.72)(10.28)}{8331} = +1.46 \text{ ksi}$$

(tension in top)

$$-\frac{P_s(e)c}{I} = -\frac{250(4.72)(9.72)}{8331} = -1.38 \text{ ksi}$$

(compression in bottom)

   b. The stresses due to the beam weight are

$$f = \pm\frac{Mc}{I}$$

$$= +\frac{28.1(12)(9.72)}{8331} = +0.39 \text{ ksi}$$

(tension in bottom)

$$= -\frac{28.1(12)(10.28)}{8331} = -0.42 \text{ ksi}$$

(compression in top)

Summarizing the initial service load stage at the time of transfer [prestress plus beam weight (DL)]:

Top of beam: $-1.04 + 1.46 - 0.42 = 0$
Bottom of beam: $-1.04 - 1.38 + 0.39 = -2.03$ ksi (compression)

The permissible stresses (ACI Code, Section 18.4) immediately after prestress transfer (before losses) are in terms of f'_{ci}, which is the specified compressive strength (psi) of concrete at the time of initial prestress:

$$\text{Compression} = 0.60 f'_{ci}$$

$$= 0.60(5000) = 3000 \text{ psi} = 3.0 \text{ ksi}$$

$$\text{Tension} = 3\sqrt{f'_{ci}}$$

$$= 3\sqrt{5000} = 212 \text{ psi} = 0.212 \text{ ksi}$$

Because 2.03 ksi < 3.0 ksi and 0 < 0.212 ksi, the beam is satisfactory at this stage. Should the tensile stress exceed $3\sqrt{f'_{ci}}$, additional bonded reinforcement shall be provided.

Note in Figure 11-8c that no tensile stress exists at transfer.

   c. In service, the stresses due to the superimposed loads (DL + LL) are

$$f = \pm\frac{Mc}{I}$$

$$= +\frac{140.6(12)(9.72)}{8331} = +1.97 \text{ ksi}$$

(tension in bottom)

$$= -\frac{140.6(12)(10.28)}{8331} = -2.08 \text{ ksi}$$

(compression in top)

Summarizing the second service load stage when the beam has had service loads applied (prestress plus beam dead load plus superimposed loads) gives us

Top of beam: $0 - 2.08 = -2.08$ ksi (compression)
Bottom of beam: $-2.03 + 1.97 = -0.06$ ksi (compression)

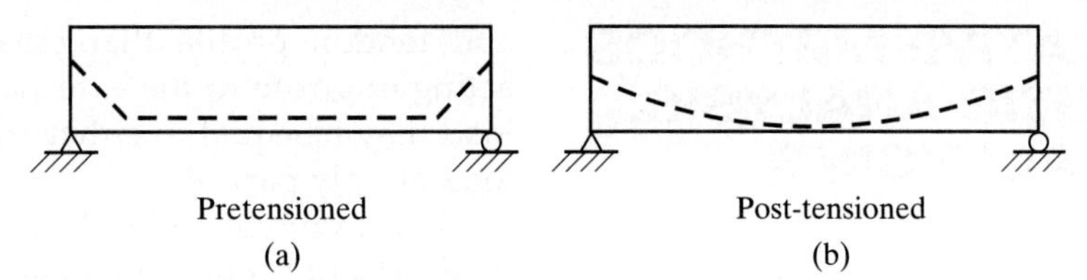

FIGURE 11-9 Draped tendons.

The permissible compressive stress in the concrete at the service load level (ACI Code, Section 24.5.4) due to prestress plus total load is 0.60 f'_c:

$$0.60(6000) = 3600 \text{ psi} = 3.60 \text{ ksi}$$

The permissible stress just given assumes that losses have been accounted for. Our analysis has neglected losses. Aside from this, the stresses in the beam would be satisfactory, because 2.08 ksi $<$ 3.60 ksi, and no tensile stress exists.

The final stress distribution (Figure 11-8e) in the beam of Example 11-2 shows that the entire beam cross section is under compression. This is the stress pattern that exists at *midspan* (as the applied moments calculated were midspan moments). It is evident that moments will decrease toward the supports of a simply supported beam and that the stress pattern will change drastically. For example, if the eccentricity of the tendon were constant in the beam in question, the net prestress (Figure 11-8a) would exist at the beam ends, where moment due to beam weight and applied loads is zero. Because the tensile stress is undesirable, the location of the tendon is changed in the area of the end of the beam, so that eccentricity is decreased. This results in a *curved or draped* tendon within the member, as shown in Figure 11-9. Post-tensioned members may have curved tendons accurately placed to satisfy design requirements, but in pretensioned members, because of the nature of the fabrication process, only approximate curves are formed by forcing the tendon up or down at a few points.

Although shear was not included in the beam analysis, the reader should be aware that the draping of the tendons, in addition to affecting the flexural stresses at the beam ends, produces a force acting vertically upward, which has the effect of reducing the shear force due to dead and live loads. Prestressed concrete flexural members are available in numerous shapes suitable for various applications. A few of the more common ones are shown in Figure 11-10.

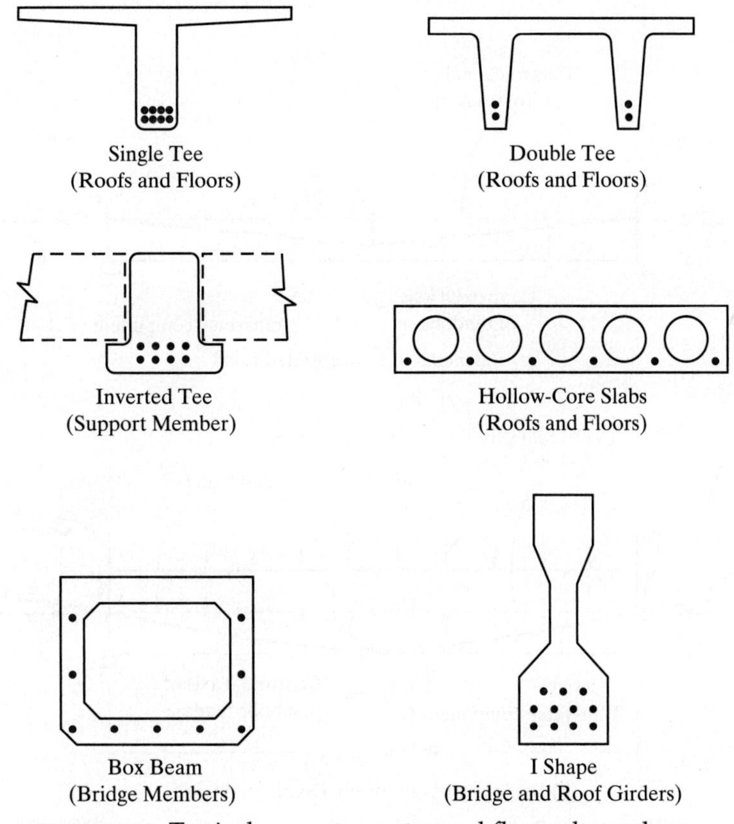

FIGURE 11-10 Typical precast, prestressed flexural members.

11-6 ALTERNATIVE METHODS OF ELASTIC ANALYSIS: LOAD BALANCING METHOD

The ACI Code requires that prestressed beams be analyzed elastically to establish whether the stresses developed by the combined action of service loads and prestressing force are within specified allowable stresses. In effect, this limit on the magnitude of stress, specifically at transfer or with service loads in place, controls cracking and subsequently prevents crushing of the concrete.

The analysis method used in Section 11-5 is often designated the *method of superposition* or the *combined loading concept*. To explain briefly, it is based on superimposing the stresses created by the applied loads with those created by the prestressing force. As previously shown, all these stresses are computed independently and then combined. This method provides the designer a complete picture of stress variation under various loading conditions.

A second method of elastic analysis is called the *load balancing method.* According to the Post-Tensioning Institute, this method is by far the most widely used method for analysis and design of post-tensioned structures. It is a technique of balancing the external load by selecting a prestressing force and tendon profile that creates a transverse load acting opposite to the external load. This transverse load may be equal to either the full external applied load or only part of it.

For example, to balance a uniformly distributed load (w) acting on a simply supported beam, a parabolic tendon profile with zero end eccentricities would be selected. The prestressing force P_s needed would be a function of the load to be balanced as well as the acceptable sag of the tendon. If the transverse load created by the tendons exactly balanced the external load, a uniform compressive stress distribution P_s/A_g will develop over the beam cross section, and the beam will remain essentially level with no deflection or camber. To balance the load, the end eccentricities should be zero; otherwise, an end moment will be developed that disturbs the uniform stress distribution.

If only a portion of the external load is balanced, a net moment in the beam at any point will develop from that portion of the load that is *not* balanced by the prestressing. It is only this net moment that must be considered in computing the bending stress. Therefore, the stress acting at any point on a cross section may be expressed as

$$f = \frac{P_s}{A_g} \pm \frac{M_{net}y}{I_g}$$

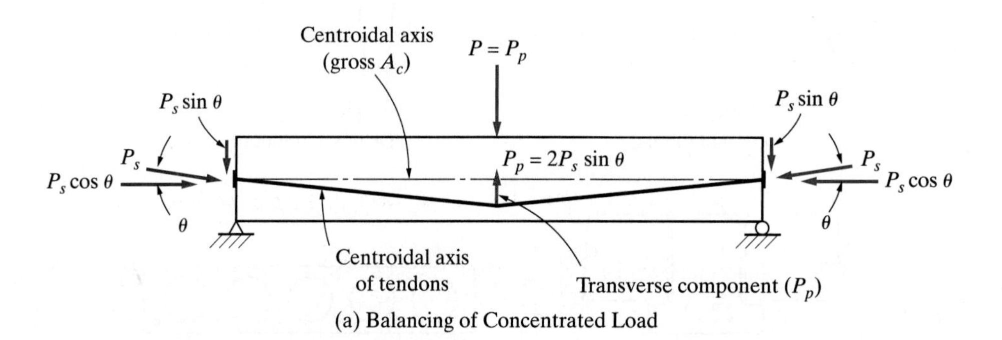

(a) Balancing of Concentrated Load

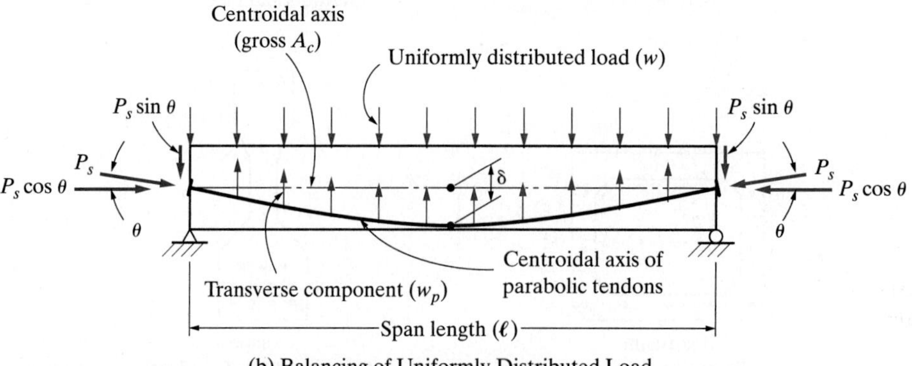

(b) Balancing of Uniformly Distributed Load

FIGURE 11-11 Load balancing.

where

P_s = prestressing force applied to the member (this is also designated T)

M_{net} = net unbalanced moment on the section

and A_g, y, and I_g are as previously defined. The computed stress due to the uniform compression and the unbalanced moment must then (as in the other methods) be compared with code-allowable stresses.

The externally applied load that will be balanced need not necessarily be uniformly distributed. It can also be a concentrated load or be a combination of both types. For a uniformly distributed load, a shallow parabolic tendon profile is generally selected, whereas a linear profile with a sharp directional change is used for a concentrated load. These are illustrated in Figure 11-11.

With respect to the balancing of a concentrated load, note the sharp directional change of the tendon under the external load at midspan. This creates an upward component

$$P_p = 2P_s \sin\theta$$

In the case shown, P_p exactly balances the applied load P. Therefore, if we neglect its own weight, the beam is not subject to net *transverse* load. At the ends of the beam, the horizontal components of P_s, which are shown as $P_s \cos\theta$ (and which are collinear with the centroidal axis of the beam), create a uniform compressive stress along the entire length of the beam. Therefore, the stress in the beam at any section may be expressed as

$$f = \frac{P_s \cos\theta}{A_g}$$

and for small values of θ,

$$f = \frac{P_s}{A_g}$$

If any additional external load is applied, the beam will act as an elastic, homogeneous concrete beam (up to the point of cracking), and a bending stress will develop, which can be evaluated by

$$f = \frac{M_{net}c}{I_g}$$

where M_{net} is the moment developed by any load applied in addition to P.

With respect to the balancing of the uniformly distributed load, note that the tendon profile is that of a parabolic curve whose upward transverse component w_p in lb/ft is given by

$$w_p = \frac{8P_s\delta}{\ell^2} \quad \textbf{(11-1)}$$

Assuming that the externally applied load w, including the weight of the beam, is exactly balanced by the transverse component w_p, there is no bending in the beam, and the beam is subject to a uniform compressive stress calculated from

$$f = \frac{P_s}{A_g}$$

If the transverse component is different from the applied external load, the bending moment developed will induce a bending stress, which can be evaluated by

$$f = \frac{M_{net}c}{I_g}$$

where M_{net} is the moment developed by a load applied over and above w.

It is questionable as to what portion of the external load should be balanced by the prestress. If too much of the applied load, such as DL plus $\frac{1}{2}$LL is to be balanced, excessive prestress may be required. The designer must exercise judgment in determining the proper amount of loading to be balanced by prestressing.

Example 11-3

The rectangular prestressed beam shown in Figure 11-12 carries uniformly distributed service loads of 1.0 kip/ft LL

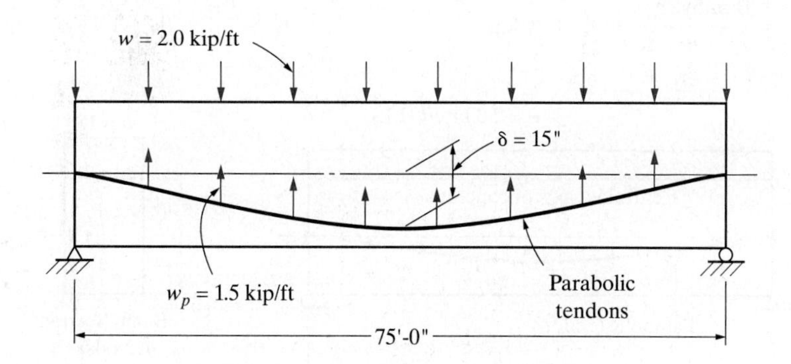

Time *Time*

FIGURE 11-12 Sketch for Example 11-3.

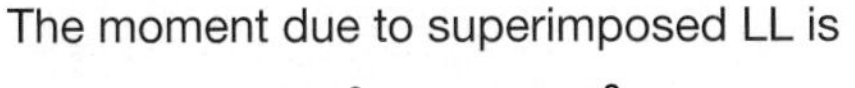

and 1.0 kip/ft DL (which includes the weight of the beam). A parabolic tendon, configured as shown, will be used. The tendon is to furnish a uniformly distributed upward balancing load of 1.5 kip/ft (DL + $\frac{1}{2}$LL). Calculate the required prestressing force and determine the net moment at midspan.

Solution:

Solving Equation (11-1) for required P_s:

$$\text{required } P_s = \frac{w_p\ell^2}{8\delta} = \frac{1.5(75)^2}{8(15/12)} = 844 \text{ kip}$$

The net moment at midspan is calculated from

$$M = \frac{w\ell^2}{8} = \frac{(2.0 - 1.5)(75)^2}{8} = 352 \text{ ft-kip}$$

Example 11-4 illustrates the two methods of elastic analysis. In this example, we will neglect the transformed area of the steel as well as any displaced concrete, and we will use the gross properties of the cross section to compute stresses. Note that this varies from the solution of Example 11-2.

Example 11-4

A rectangular prestressed concrete beam shown in Figure 11-13 is simply supported and has a span length of 25 ft-0 in. The beam carries a superimposed service LL of 2.0 kip/ft. The only service DL is the weight of the beam. The parabolic prestressing tendon is located as shown and is subjected to an effective prestress force of 300 kip (neglect any prestress losses). Determine the outer fiber flexural stresses at midspan at transfer (prestress + beam weight) and when the member is under full service load conditions (prestress + beam weight + service LL).

Solution:

a. Using the method of superposition

1. Calculate moments after determining the beam weight:

$$w_{DL} = \frac{28(18)}{144}(0.150) = 0.525 \text{ kip/ft}$$

The moment due to beam weight is

$$M_{DL} = \frac{w_{DL}\ell^2}{8} = \frac{0.525(25)^2}{8} = 41.0 \text{ ft-kip}$$

The moment due to superimposed LL is

$$M_{LL} = \frac{w_{LL}\ell^2}{8} = \frac{2.0(25)^2}{8} = 156.3 \text{ ft-kip}$$

2. Using the gross concrete section and neglecting the transformed steel area, the neutral axis coincides with the centroidal axis. Therefore, the distance from the neutral axis to the outer fiber is 14 in. The eccentricity e of the tendon is given as 6 in. The moment of inertia about the neutral axis is found from

$$I_g = \frac{18(28)^3}{12} = 32{,}928 \text{ in.}^4$$

3. Calculate the stresses.

a. At transfer, the prestressing stress is determined from

$$f = -\frac{P_s}{A_g} \pm \frac{P_s ec}{I_g}$$
$$= -\frac{300}{18(28)} \pm \frac{300(6)(14)}{32{,}928}$$
$$= -0.595 \pm 0.765$$

from which

stress at top = −0.595 + 0.765 = 0.170 ksi (tension)

stress at bottom = −0.595 − 0.765 = −1.36 ksi (compression)

The stress due to the service DL (beam weight) is calculated from

$$f = \pm\frac{Mc}{I} = \pm\frac{41.0(12)(14)}{32{,}928} = \pm 0.209 \text{ ksi}$$

Combining the preceding stresses for the transfer condition gives us

stress at top = +0.170 − 0.209 = −0.039 ksi (compression)

stress at bottom = −1.36 + 0.209 = −1.15 ksi (compression)

b. Under full service load conditions, stress due to superimposed LL is calculated from

$$f = \pm\frac{Mc}{I} = \pm\frac{156.3(12)(14)}{32{,}928} = \pm 0.797 \text{ ksi}$$

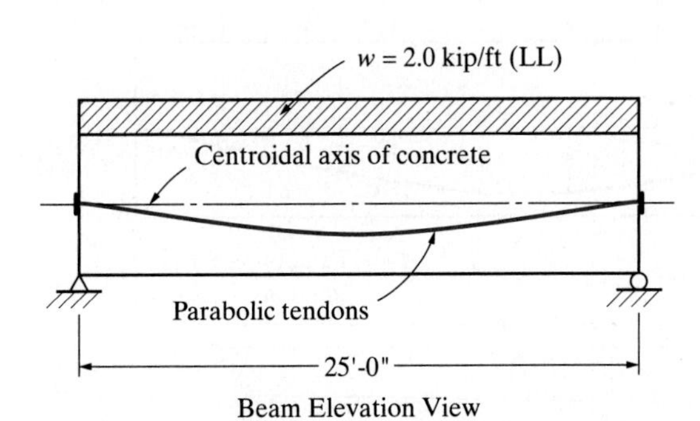

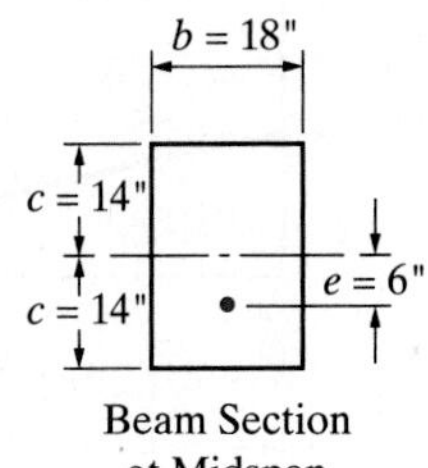

FIGURE 11-13 Sketches for Example 11-4.

Combining the preceding stresses for the full service load condition gives us

$$\text{stress at top} = -0.039 - 0.797 = -0.836 \text{ ksi (compression)}$$

$$\text{stress at bottom} = -1.15 + 0.797 = -0.353 \text{ ksi (compression)}$$

b. Using the load balancing method

1. Let $e = \delta = 6$ in. The parabolic tendons produce an upward balancing load of

$$w_p = \frac{8P_s\delta}{\ell^2} = \frac{8(300)(0.5)}{25^2} = 1.92 \text{ kip/ft}$$

a. At transfer, the unbalanced load on the beam is

$$1.92 - 0.525 = 1.395 \text{ kip/ft} \qquad \text{(upward)}$$

The net moment at midspan is

$$M_{net} = \frac{1.395(25)^2}{8} = 109.0 \text{ ft-kip} \quad \text{(tension on top)}$$

The bending stress due to the net moment is

$$f = \pm\frac{M_{net}c}{I_g} = \pm\frac{109(12)(14)}{32{,}928} = \pm 0.556 \text{ ksi}$$

The uniform compressive stress is

$$-\frac{P_s}{A_g} = -\frac{300}{18(28)} = -0.595 \text{ ksi}$$

$$\text{stress at top} = -0.595 + 0.556 = -0.039 \text{ ksi (compression)}$$

$$\text{stress at bottom} = -0.595 - 0.556 = -1.15 \text{ ksi (compression)}$$

b. Under full service load, the unbalanced load on the beam is

$$2.0 + 0.525 - 1.92 = 0.605 \text{ kip/ft (downward)}$$

The net moment at midspan is

$$M_{net} = \frac{0.605(25)^2}{8} = 47.3 \text{ ft-kip (tension on bottom)}$$

The bending stress due to the net moment is

$$f = \pm\frac{M_{net}c}{I_g} = \pm\frac{47.3(12)(14)}{32{,}928} = \pm 0.241 \text{ ksi}$$

The uniform compressive stress is

$$-\frac{P_s}{A_g} = -\frac{300}{18(28)} = -0.595 \text{ ksi}$$

$$\text{stress at top} = -0.595 - 0.241 = -0.836 \text{ ksi (compression)}$$

$$\text{stress at bottom} = -0.595 + 0.241 = -0.354 \text{ ksi (compression)}$$

Note that the results agree (with slight round-off error) using the three elastic analysis methods.

The load balancing method offers advantages of simplicity and clarity with continuous beams and slabs. It is the recommended method for those situations, particularly for preliminary designs. For simple spans, none of the two methods has any particular advantage over the other; both are equally applicable.

11-7 FLEXURAL STRENGTH ANALYSIS

As part of the design and analysis procedure of a prestressed concrete beam, the ACI Code requires that the moment due to factored service loads, M_u, not exceed the flexural design strength ϕM_n of the member. The design strength of prestressed beams may be computed using strength equations similar to those for reinforced concrete members, discussed in Chapter 2. The checking of the flexural (nominal) strength of a prestressed beam ensures that the beam is designed with an adequate factor of safety against failure.

The expression for the nominal bending strength of a rectangular-shaped prestressed beam is developed from the internal couple of an underreinforced beam at failure, as shown in Figure 11-14.

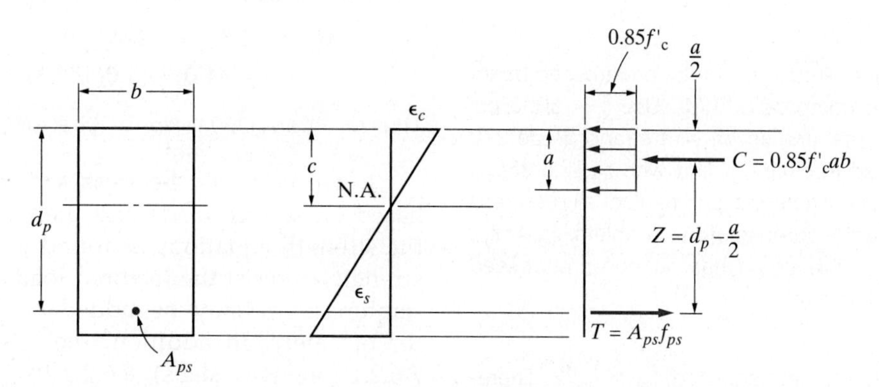

FIGURE 11-14 Equivalent stress block for strength analysis.

Assuming failure is initiated by the steel yielding, the magnitude of the internal couple is

$$M_n = A_{ps}f_{ps}\left(d_p - \frac{a}{2}\right)$$

where

A_{ps} = area of prestressed reinforcement in the tension zone

d_p = distance from extreme compression fiber to the centroid of the prestressed reinforcement

a = depth of stress block, determined by equating T and C and calculated from

$$a = \frac{A_{ps}f_{ps}}{0.85f'_c b}$$

f_{ps} = stress in the prestressed reinforcement at nominal strength

For members with bonded prestressing tendons and no non-prestressed tension or compression reinforcement (*i.e.* $\rho_s = \rho'_s = 0$), the value of f_{ps} can be obtained from

$$f_{ps} = f_{pu}\left(1 - \frac{\gamma_p}{\beta_1}\rho_p\frac{f_{pu}}{f'_c}\right) \quad \text{[ACI Eq. (20.3.2.3.1)]}$$

where

f_{pu} = ultimate tensile strength of the prestressing steel

γ_p = factor based on the type of prestressing steel: 0.55 when f_{py}/f_{pu} is not less than 0.80, 0.40 when f_{py}/f_{pu} is not less than 0.85, and 0.28 when f_{py}/f_{pu} is not less than 0.90, where f_{py} is the specified yield stress of the prestressing tendons (psi)

ρ_p = reinforcement ratio A_{ps}/bd_p

and β_1 is as defined in Chapter 2, Section 2-6. The expression for f_{ps} is valid when $f_{se} > f_{pu}/2$, where f_{se} is the effective stress in prestressing steel after losses. Here f_{ps} represents the average stress in the prestressing steel at failure. It is analogous to f_y, but as the high-strength prestressing steels do not have a well-defined yield point, it can be predicted using ACI Equation (20.3.2.3.1).

Example 11-5

Calculate the flexural strength ϕM_n of the prestressed beam of Example 11-4 and compare with M_u. Use $f'_c = 5000$ psi (normal-weight concrete) and seven-wire-strand grade 270 with $f_{pu} = 270{,}000$ psi (ordinary strand with $f_{py} = 0.85f_{pu}$). Use bonded prestressing tendons, and neglect all prestress losses. Therefore, for the purpose of this problem, $f_{se} = f_{pu}$. Use $A_{ps} = 1.224$ in.^{2}. The beam has no non-prestressed tension or compression steel.

Solution:

Because prestress losses are neglected, $f_{se} > f_{pu}/2$. Therefore, ACI Equation (20.3.2.3.1) is applicable. For the given conditions, $\gamma_p = 0.40$ and $\beta_1 = 0.80$. The reinforcement ratio is calculated from

$$\rho_p = \frac{A_{ps}}{bd_p} = \frac{1.224}{18(20)} = 0.00340$$

$$f_{ps} = f_{pu}\left(1 - \frac{\gamma_p}{\beta_1}\rho_p\frac{f_{pu}}{f'_c}\right)$$

$$= 270\left[1 - \frac{0.40}{0.80}(0.00340)\left(\frac{270}{5}\right)\right]$$

$$= 245 \text{ ksi}$$

Prestressed concrete sections are subject to the same conditions of being tension-controlled, compression-controlled, or transition sections as regular reinforced concrete sections, as discussed in Section 2-8. The appropriate values of ϕ, as previously discussed, apply. It is convenient to check the net tensile strain limit of 0.005 (for tension-controlled sections) through the use of a *reinforcement index* ω_p, where

$$\omega_p = \frac{\rho_p f_{ps}}{f'_c}$$

A reinforcement index of $0.32\beta_1$ corresponds to a net tensile stain of 0.005, the lower limit for a tension-controlled section. Therefore, checking ω_p:

$$\omega_p = \frac{\rho_p f_{ps}}{f'_c} = \frac{0.00340(245{,}000)}{5000} = 0.167$$

$$0.32\beta_1 = 0.32(0.80) = 0.256$$

Because $\omega_p < 0.256$, this is a tension-controlled section and $\phi = 0.90$.

Compute the nominal moment strength:

$$a = \frac{A_{ps}f_{ps}}{0.85f'_c b} = \frac{1.224(245)}{0.85(5)(18)} = 3.92 \text{ in.}$$

$$M_n = A_{ps}f_{ps}\left(d_p - \frac{a}{2}\right)$$

$$= 1.224(245)\left(20 - \frac{3.92}{2}\right)$$

$$= 5410 \text{ in.-kip}$$

$$= 451 \text{ ft-kip}$$

$$\phi M_n = 0.90(451) = 406 \text{ ft-kip}$$

Compute the factored service load moment:

$$M_u = 1.2M_{DL} + 1.6M_{LL}$$

$$= 1.2(41.0) + 1.6(156.3) = 299 \text{ ft-kip}$$

Thus $M_u < \phi M_n$(299 ft-kip < 406 ft-kip). Therefore, O.K.

If a prestressed beam is satisfactorily designed based on service loads and then, when checked by the strength equations, is found to have insufficient strength to resist the factored loads, non-prestressed reinforcement may be added to increase the factor of safety. In addition, the ACI Code, Sections 7.6.2.1 (one-way slabs), 8.6.2.2 (two-way slabs), and 9.6.2.1 (beams), require that the total amount of reinforcement (prestressed and non-prestressed) be sufficient to develop a factored load equal to at least 1.2 times the cracking load calculated from the modulus of rupture of the concrete. It is permissible to waive this requirement when a flexural member has shear and flexural strength at least twice that required (see ACI 318-14 Code Sections 7.6.2.2, 8.6.2.2.1, and 9.6.2.2).

11-8 NOTES ON PRESTRESSED CONCRETE DESIGN

Because the shape and dimensions of a prestressed concrete member may be established by a trial procedure or even assumed based on physical limitations, the design may be reduced to finding the prestress force, tendon profile, and the amount of the prestress steel area. The design problem then is further reduced to an analysis-type problem whereby service load stresses are checked at various stages and the flexural strength of the member is checked against the moment due to the factored service loads.

The intent of this chapter, however, is to furnish a conceptual approach to prestressed concrete members. Therefore, many significant topics normally considered in an analysis or design problem have been omitted. These include secondary moments and restraint reactions in continuous members, shear design, deflection, prestress loss, and block stresses. These topics are beyond the scope of our text. References [3] through [6] are among many other texts and publications available for further reading.

References

[1] "ASTM Standards." American Society for Testing and Materials, 100 Barr Harbor Drive, West Conshohocken, PA 19428.

[2] George F. Limbrunner and Leonard Spiegel. *Applied Statics and Strength of Materials,* 5th ed. Upper Saddle River, NJ: Prentice Hall, 2009, pp. 470–473.

[3] T. Y. Lin and Ned H. Burns. *Design of Prestressed Concrete Structures,* 3rd ed. New York: John Wiley & Sons, Inc., 1981.

[4] Post-Tensioning Institute. *Post-Tensioning Manual,* 6th ed. Phoenix, AZ: Post-Tensioning Institute, 2006.

[5] Edward G. Nawy. *Prestressed Concrete: A Fundamental Approach.* Upper Saddle River, NJ: Prentice Hall, 2009.

[6] Arthur H. Nilson. *Design of Prestressed Concrete,* 2nd ed. New York: John Wiley & Sons, Inc., 1987.

Problems

11-1. The *plain concrete* beam shown having a rectangular cross section 10-in. wide and 18-in. deep is simply supported on a single span of 20 ft. Assuming no loads other than the dead load of the beam itself, find the bending stresses that exist at midspan.

11-2. A beam identical to that of Problem 11-1 is prestressed with a force P_s of 185 kip. The tendons are located at the center of the beam. Use $A_{ps} = 1.23$ in.^{2}. Find the stresses in the beam at midspan at transfer. There is no load other than the weight of the beam. Draw complete stress diagrams.

11-3. The tendons in the beam of Problem 11-2 are placed 11 in. below the top of the beam at midspan. Assuming that no tension is allowed, determine the maximum service uniform load that the beam can carry in addition to its own weight as governed by the stresses at midspan. Assume that $f'_{ci} = 3500$ psi and $f'_c = 5000$ psi.

11-4. Same as Problem 11-3, except that the tendons are 13 in. below the top of the beam at midspan and permissible stresses are as defined in the ACI Code, Section 24.5.

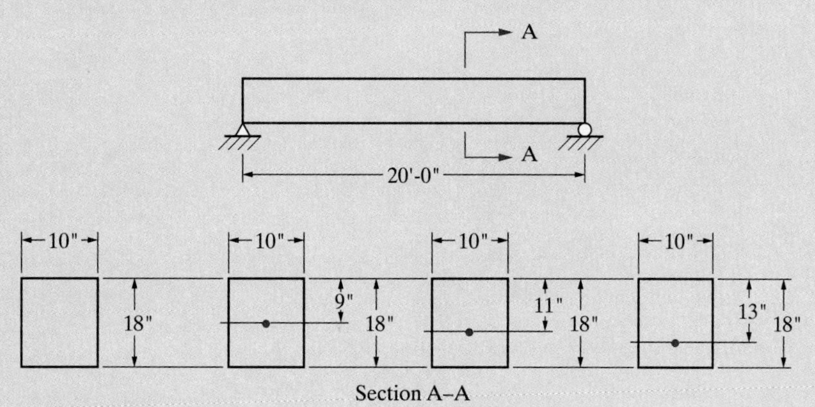

PRPBLEMS 11-1 TO 11-4

11-5. For the beam shown, calculate the outer fiber stresses at midspan and at the end supports at the following stages: (a) transfer (prestress + beam weight) and (b) under full service load. The beam is simply supported with a span length of 20 ft-0 in. and carries a superimposed live load of 2.7 kip/ft. The tendon is straight and located 9 in. below the neutral axis of the beam. The prestress force is 250 kip. Neglect prestress losses and use the gross concrete section properties. Use the method of superposition for the solution.

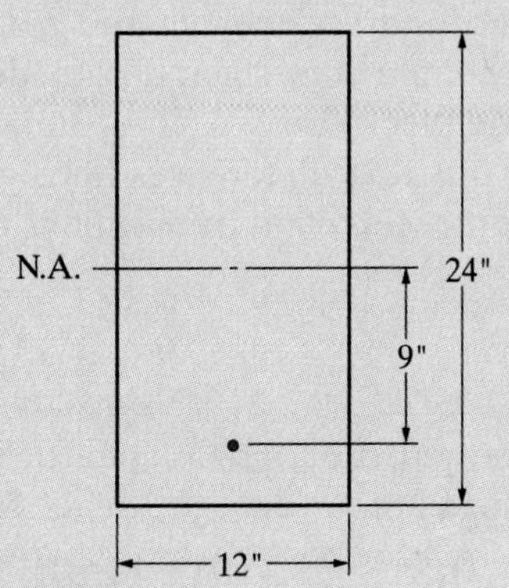

PROBLEM 11-5

11-6. The simply supported beam having a midspan cross section as shown is prestressed by a parabolic tendon with an effective prestress force of 233 kip. The beam has a span length of 30 ft and carries a total service load of 1.20 kip/ft, which includes the weight of the beam of 0.21 kip/ft. Calculate the outer fiber stresses at midspan under full service load. Use the gross properties of the cross section. Use the load balancing method.

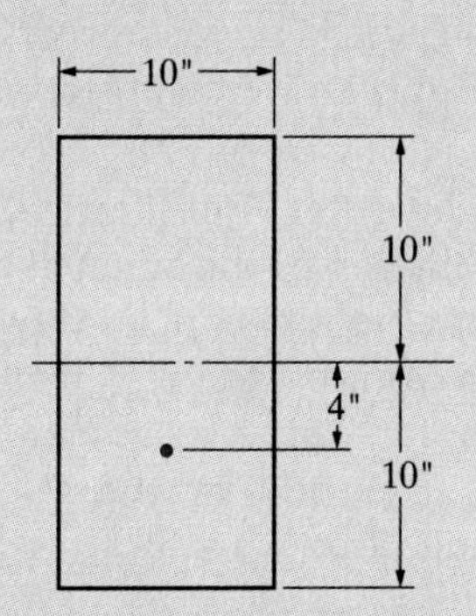

PROBLEM 11-6

11-7. For the beam of Problem 11-5, calculate the outer fiber stresses at midspan for the following stages: (a) transfer and (b) under full service load. Assume the prestress tendons are parabolically draped with a midspan eccentricity of 9 in. as shown and zero eccentricity at the end supports. Neglect prestress losses and use the gross concrete section properties. The prestress force is still 250 kip. Use the load balancing method.

11-8. Calculate the flexural strength ϕM_n of the prestressed beam shown and compare it with the factored design moment M_u. The service load moments are 125 ft-kip for dead load and 383 ft-kip for live load. Use $f'_c = 5000$ psi (normal-weight concrete) and grade 270 seven-wire strand with $f_{pu} = 270{,}000$ psi ($f_{py} = 0.85 f_{pu}$). The tendons are bonded. Neglect all prestress losses.

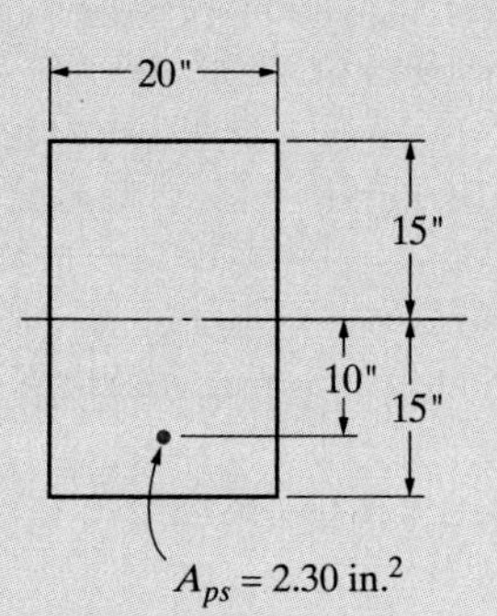

PROBLEM 11-8

CHAPTER **TWELVE**

CONCRETE FORMWORK

12-1 Introduction
12-2 Formwork Requirements
12-3 Formwork Materials and Accessories
12-4 Loads and Pressures on Forms
12-5 The Design Approach
12-6 Design of Formwork for Slabs
12-7 Design of Formwork for Beams
12-8 Wall Form Design
12-9 Forms for Columns

12-1 INTRODUCTION

Forms are temporary structures whose purpose is to provide containment for the fresh concrete and to support it until it can support itself. Forms usually must be *engineered* structures as they are required to support loads composed of the fresh concrete, construction materials, equipment, workers, impact of various kinds, and sometimes wind. The forms must support all the applied loads without collapse or excessive deflection. In addition, the forms must provide for the molding of the concrete to the desired dimensions, shape, and surface finish.

The cost of formwork is significant, generally amounting to anywhere from 40% to 60% of the cost of a concrete structure. Economy in formwork design depends partly on the ingenuity and experience of the form designer, whether a contractor or an engineer. Judgment with respect to the development of a forming system could both expedite a project and reduce costs. Although forms may be job built (design examples are presented later in this chapter), many proprietary forming systems are available. Specialty companies will design and construct forms using patented systems. In multiple-reuse situations, steel formwork has replaced wood formwork to some degree, although the use of wood is still substantial because of its availability and ease of fabrication.

Whatever the eventual type of forming may be, economy in formwork begins when the architect/engineer, during the design of the structure, examines each member and considers ease of forming and reuse of forms.

12-2 FORMWORK REQUIREMENTS

The essential requirements of good formwork may be categorized as quality, safety, and economy.

Quality in formwork requires that the forms be

1. *Accurate*: The size, shape, position, and alignment of structural elements will depend on the accuracy with which the forms are built. The designer (engineer or architect) should specify the allowable tolerances in form dimensions for form design and construction, but not make these tolerances finer than necessary, as this could add unnecessarily to formwork cost as well as delay the project.
2. *Rigid*: Forms must be sufficiently rigid to prevent movement, bulging, or sagging during the placing of concrete. Formwork must therefore be adequately propped, braced, and tied. Special consideration may have to be given to such items as corner details and the effect of any uplift pressures.
3. *Tight-jointed*: Joints that are insufficiently tight will leak cement paste. The surface of the concrete will thus be disfigured by fins of the cement paste, and honeycombing may result adjacent to the leaking joint.
4. *Properly finished*: The formwork in contact with the concrete should be so arranged and jointed as to produce a concrete surface of good appearance. Wires, nails, screws, and form surface flaws must not be allowed to disfigure the concrete surface. In some cases a provision of special form lining may be necessary to achieve the desired surface finish.

Safety in formwork requires that the forms be

1. *Strong*: To ensure the safety of the structure and the protection of the workers, it is essential that formwork be designed to carry the full load and side pressures from freshly placed concrete, together with construction traffic and any necessary equipment. On large jobs the design of formwork is usually left to a specialist, but on smaller or routine jobs it may be left to the carpenter supervisor using rule-of-thumb methods.
2. *Sound*: The materials used to construct the forms must be of the correct size and quantity, of good quality, and sufficiently durable for the job.

Economy in formwork requires that the forms be

1. *Simple*: For formwork construction to be economical, it must be designed to be simple to erect and dismantle. Modular dimensions should be used.
2. *Easily handled*: The sizes of form panels or units should be such that they are not too heavy to handle.
3. *Standardized*: Comparative ease of assembly and the possibility of reuse will lower formwork costs when sizes are standardized.
4. *Reusable*: Formwork intended for reuse should be designed for easy removal. If formwork panels have to be ripped down, the concrete may be damaged, materials will be wasted, time will be lost, and expense will be incurred in repairing and replacing damaged panels.

12-3 FORMWORK MATERIALS AND ACCESSORIES

Materials used for forms for concrete structures include lumber, plywood, hardboard, fiberglass, plastic and rubber liners, steel, paper and cardboard, aluminum, fiber forms, and plaster of Paris. Additional materials include nails, bolts, screws, form ties, form clamps, anchors, various types of inserts, and various types of form oils and compounds as well as their accessories. Forms frequently involve the use of two or more materials, such as plywood facing attached to steel frames for wall panels.

Form *lumber* generally consists of the softwoods with the species used being the type available in the local area. Various species are usually grouped together for grading and marketing purposes. Some of these groups and their included species commonly used for formwork are

1. Douglas fir–larch (Douglas fir, western larch)
2. Douglas fir–south (only Douglas fir from southern growth areas)
3. Hem–fir (Western hemlock, California red fir, grand fir, noble fir, white fir, Pacific silver fir)
4. Spruce–pine–fir (alpine fir, balsam fir, black spruce, Englemann spruce, jack pine, lodge-pole pine, red spruce, white spruce)
5. Southern pine (loblolly pine, longleaf pine, shortleaf pine, slash pine)

Form lumber is made up of standard sizes, either rough or dressed, with the grading quality of *construction grade* or *No. 2 grade* usually specified. Shoring or falsework requiring greater capacity should be of *select structural grade* lumber. Partially seasoned wood should be used because it has been found to be the most stable. Green lumber will warp and crack, whereas kiln-dried lumber will swell excessively when it becomes wet. Boards of various thicknesses may be used for sheathing when the imprint of the boards on the concrete surface is desired for architectural reasons.

Plywood is available as flat panels (4 ft × 8 ft sheets) made of thin sheets of wood (called plies) with a total thickness ranging from $\frac{1}{4}$ in. to $1\frac{1}{8}$ in. It is commonly used for sheathing or lining forms because it gives smooth concrete surfaces that require a minimum of hand finishing and because the relatively large sheets are economical and easy to use. U.S. plywood is built up of an odd number of plies with the grain of adjacent layers perpendicular. The plies are dried and joined under pressure with glues that make the joints as strong or stronger than the wood itself. The alternating direction of the grains of adjoining layers equalizes strains and thus minimizes shrinkage and warping of the plywood panels. Generally, the grain direction of the outer layers is parallel to the long dimension of the panel. For maximum strength, the direction of the grain in the face layers should be placed parallel to the span.

The plywood should be the exterior grade because of its waterproof glue. It should also be factory-treated with a form oil or parting compound. With proper care, plywood can be reused many times.

Coated plywood is occasionally used. Called *overlaid* or *plastic coated*, it is ordinary exterior plywood with a resin-impregnated fiber-facing material fused to one or both sides of the plywood sheet. The overlay covers the grain of the wood, resulting in a smoother and more durable forming surface. The resins used in overlay production are hard and resist water, chemicals, and abrasion. Two grades, high density and medium density, are available, and their difference is in the density of the surfacing material. Coated plywood is generally reused considerably more than the plain plywood form.

Nearly any exterior type of plywood can be used for concrete forming. A special panel called *plyform* is manufactured specifically for that purpose, however. Plyform is exterior-type plywood limited to certain wood species and veneer grades to ensure high performance and is manufactured in two classes, class I and class II. Class I is the strongest, stiffest, and most widely available type. Both classes have smooth, solid surfaces and can have many reuses. They are mill-oiled unless otherwise specified. Plyform is manufactured in thicknesses ranging from $\frac{15}{32}$ in. to $1\frac{1}{8}$ in., and the most commonly used thicknesses are $\frac{5}{8}$ in. and $\frac{3}{4}$ in.

Plyform can also be manufactured with a *high-density overlaid (HDO) surface* (on one side or both sides). HDO plyform has an exceptionally smooth, hard surface for the smoothest possible concrete finishes and the maximum number of reuses. The HDO surface is a hard, semiopaque resin-fiber overlay heat-fused to panel faces. It can have as many as 200 reuses, and light oiling between pours is recommended.

Structural I plyform is stronger and stiffer than plyform class I or II and is sometimes used when the face grain must be parallel to supports and there is a heavy loading against the forms. It is also available with HDO faces.

Hardboard (fiberboard) is made up of wood particles that have been impregnated, pressed, and baked. When used as a form liner, which is generally the case, hardboard must be backed with sheathing of some kind because it does not have the strength that plywood has. Tempered hardboard, which is preferred for formwork, has been impregnated with drying oil or other material that makes the board less absorbent and improves its strength. Hardboard has limited reuse capability.

Fiberglass is a glass-fiber-reinforced plastic. It is especially suitable for repetitive production of complicated shapes, particularly precast elements. The initial cost is high, but the extensive reuse of one mold without evidence of wear contributes to economy. Suitable reinforcement and backing must be provided so that the mold will not be deformed when it is handled. Molds can be used with or without form oil. Wear of the mold surface will be slightly less if an oil is used.

Plastic and rubber form liners generally come in sheets, either flexible or rigid, and are attached to the solid form sheathing. The term *form liners* includes any sheet, plate, or layer of material attached directly to the inside face of forms to improve or alter the surface texture and quality of the finished concrete. Substances that are applied by brushing, dipping, mopping, or spraying to preserve the form material and to make form stripping easier are referred to as *form coatings.* Some of these coatings are so effective as to approximate the form liner in function. Repetitive use, a great variety of textures and patterns, and ease of stripping are among the advantages of these form liners.

Various types of rigid insulation boards are used as form liners and attached to the form sheathing. The boards are generally left in place, either bonded to the concrete or held in place by form plank clips.

Steel forms have been widely used as special-purpose forms. Steel panel systems have been fabricated and used, and steel framing and bracing are important in the construction of many wood and plywood panel systems. Patented steel pans and dome components form the underside of waffle slabs and pan joists. Corrugated steel sheets serve as permanent bottom forms for decks—that is, they remain in place after the concrete has been placed. Fabricated steel form systems are extensively used in the precast and prestressed concrete industries due to their suitability for being reused extensively.

Earth as a form is used in subsurface construction where the soil is stable enough to retain the desired shape of the concrete structure. With the resulting rough surface finish, the use of earth as a form is generally limited to footings and foundations.

Fiber forms are composed of multiple layers of heavy paper bonded together and impregnated with waxes and resins to create a water-repellent cardboard. These are generally single-use, cylindrical-shape molds for columns or other applications where preformed shapes are desirable. "Sonotube" fiber forms are patented lightweight units that can be adapted to various applications.

Form accessories consist of a multitude of items necessary for form planning and construction. Types, availability, and detailed information can usually be obtained from manufacturers' catalogs as well as local concrete construction specialties distributors.

Some of the more common types of accessories are fasteners, spreaders, ties, anchors, hangers, clamps, and inserts. The most common *fastener* used is the double-headed nail, which may be withdrawn with a minimum of effort and little damage to the forms. *Spreaders* are braces that are inserted in forms to keep the faces a proper distance apart until the concrete is placed. Preferably, they are removed during or after the placement of the concrete so that they are not cast in the concrete. They may be wood blocks or proprietary metal pieces. *Ties* are tensile units adapted to holding concrete forms secure against the lateral pressure of the fresh concrete, with or without provision for spacing the forms a definite distance apart. They pass through the concrete and are fastened on each side. Various types of ties are available, many of which are patented items. *Anchors* are devices used to secure formwork to previously placed concrete of adequate strength and are normally embedded in concrete during placement. There are two basic parts, the embedded anchoring device and the external fastener, which is removed after use. *Hangers* are devices used for suspending one object from another, such as the hardware attached to a building frame to support forms. *Inserts* are of many designs and are attached to the forms in such a way that they remain in the concrete when the forms are stripped. They provide for anchorage of brick or stone veneers, pipe hangers, suspended ceilings, duct work, and any building hardware or components that must be firmly attached to the concrete. *Clamps* consist of many varied devices but serve a similar purpose to that of the tie. Beam form clamps firmly hold the beam sides and bottom together with a minimum of nailing required. Column clamps or yokes encircle column forms and hold them together securely, withstanding the lateral pressure of the freshly placed concrete. Accessories used in formwork are required to have a minimum factor of safety of 2.0 to 3.0 based on the component (see Table 4.4 of reference [5]).

12-4 LOADS AND PRESSURES ON FORMS

The concrete form is a structure. Like all structures, it must be designed and constructed with due regard to the effects of the imposed loads. A well-designed form must have adequate strength to resist failure; in addition, it must have sufficient stiffness so that deflection will not be excessive. Efficient and economical use of the material of which the forms are built is also an integral part of the design process.

The first consideration in the design of a form for concrete is the load to be supported. The loads may be considered in two categories: vertical loads and lateral loads. The lateral loads may be further categorized as externally applied load (such as wind) and internally applied pressure due to the contained fresh concrete. These loads must be carried to the ground by the formwork system or resisted by in-place construction that has adequate strength for the purpose.

Vertical loads consist of dead load and live load. The dead load consists of the weight of the formwork and the freshly placed concrete. Live load consists of the weight of workers and equipment, stored materials, and impact due to moving loads. ACI 347R-14. *Guide to Formwork for Concrete* [1] recommends that the formwork be designed for the following:

1. Minimum live load of 50 psf of horizontal projection (75 psf if motorized carts are used for delivery of the concrete).
2. Minimum combined dead and live load of 100 psf (125 psf if motorized carts are used for delivery of the concrete).

Note that these minimum design loads would be applicable to the design of structural floor and roof slabs (not to the design of slabs on ground).

Lateral pressure is exerted on all nonhorizontal surfaces that contain the fresh concrete. The pressures are imposed on wall and column forms and in essence are hydraulic loadings. The amount of lateral pressure is influenced by the weight, chemistry, and temperature of the concrete, the vertical rate of placement, size and shape of the form, height of the form, and method of consolidation (hand-spaded or mechanically vibrated). The amount and location of reinforcing also affect lateral pressure, although their effect is small and is usually neglected. ACI 347R-14 provides formulas that can be used to reasonably predict concrete pressure on wall forms and column forms. In the formulas, the following notation is used:

p = *concrete* lateral pressure (psf)
R = rate of pressure (ft/h)
T = temperature of concrete in the forms (°F)
h = height of fluid or plastic concrete above point considered (ft)
w = unit weight of fresh concrete (pcf)
C_C = chemistry coefficient
C_W = unit weight coefficient

A liquid head formula is the basis for the concrete lateral pressure calculations:

$$p = wh \qquad \textbf{[ACI 347R-14 Eq. (4.2.2.1a(a))]}$$

This formula will apply where a form is filled rapidly, before any hardening of the concrete occurs, and in situations where the conditions of Equation (4.2.2.1a(b)) and Equation (4.2.2.1a(c)) are not met. The limits stated for Equation (4.2.2.1a(b)) and Equation (4.2.2.1a(c)) do not apply to Equation (4.2.2.1a(a)).

For the special conditions of

a. concrete having a slump of 7 in. or less and
b. placed with normal internal vibration to a depth of 4 ft or less,

formwork can be designed for a lateral pressure as follows:

1. For columns (note that a column, for purposes of applying the pressure formulas, is defined as a vertical element with no plan dimension exceeding 6.5 ft),

$$p_{max} = C_W C_C \left[150 + 9000\frac{R}{T}\right]$$

[ACI 347R-14 Eq. (4.2.2.1a(b))]

with a minimum of 600 C_W lb/ft^2, but in no case greater than wh.

2. For walls, with a rate of placement of less than 7 ft/hr and a placement height not exceeding 14 ft,

$$p_{max} = C_W C_C \left[150 + \frac{9000R}{T}\right]$$

[ACI 347R-14 Eq. (4.2.2.1a(b))]

with a minimum of 600C_W lb/ft^2, but in no case greater than wh.

3. For walls with a placement rate of less than 7 ft/h, where placement height exceeds 14 ft, and for all walls with a placement rate of 7 to 15 ft/h,

$$p_{max} = C_W C_C \left[150 + \frac{43{,}400}{T} + \frac{2800R}{T}\right]$$

[ACI 347R-14 Eq. (4.2.2.1a(c))]

with a minimum of 600 C_W lb/ft ^{2}, but in no case greater than wh.

4. For all walls with a placement rate greater than 15 ft/h,

$$p_{max} = wh \qquad \textbf{[ACI 347R-14 Eq. (4.2.2.1a(a))]}$$

The unit weight coefficient C_W has the value of 1.0 for concretes having a unit weight in the range of 140 to 150 lb/ft^3. For concrete unit weights outside this range, see Table 4.2.2.1a(c) in ACI 347R-14. The chemistry coefficient C_C has the value of 1.0 for Types I, II, and III cements without retarders. For other types of cements or blends containing slag or fly ash, see Table 4.2.2.1a(b) in ACI 347R-14. For the illustrations in this text, we will assume both C_W and C_C equal to 1.0.

Formwork should also be designed to resist all foreseeable *lateral loads*, such as seismic forces, wind, cable tensions, inclined supports, dumping of concrete, and impact due to equipment. ACI 347R-14 recommends for slabs a minimum horizontal design load of 100 lb per lineal foot of floor edge or 2% of the total dead load of the floor, whichever is greater. Wall forms exposed to the elements should be designed for a minimum wind load of 15 psf, and bracing for wall forms should be designed for a lateral load of at least 100 lb per lineal foot of wall applied at the top.

In addition, formwork should be designed for any special loads likely to occur, such as unsymmetrical placement of concrete and uplift. Form designers must be alert to provide for any and all special loading conditions.

12-5 THE DESIGN APPROACH

The design of job-built forms may be considered largely as beam and post design. The bending members usually span several supports and are therefore indeterminate. Assumptions and approximations are made that simplify the calculations and facilitate the design process.

After establishing the appropriate design loads, the sheathing and supporting members are analyzed or designed in sequence. Bending members (sheathing, joists, studs, stringers, or wales) are considered uniformly loaded and supported on (1) a single span, (2) two spans, or (3) three or more spans. The uniform load assumption is common practice unless the spacing of point loads exceeds one-third to one-half of the span between supports, in which case the worst loading condition is investigated.

Each bending member should be analyzed or designed for bending moment, shear, and deflection. In addition, vertical supports (shoring) and lateral bracing (if applicable) must be analyzed or designed for either compressive or tensile loads. Bearing stresses at supports must be investigated (except for the sheathing).

In the analysis or design of the component parts, the traditional stress equations are used (see any strength of materials text). These expressions for stress are as follows.

For bending stress:

$$f_b = \frac{M}{S}$$

For shear stress:

$$f_v = \frac{1.5V}{A} \quad \text{(for rectangular wood members)}$$

$$f_v = \frac{VQ}{Ib} \quad \text{or} \quad \frac{V}{Ib/Q} \quad \text{(for plywood)}$$

For compression stress (both parallel and perpendicular to the grain):

$$f_{comp} = \frac{P}{A}$$

For tension stress:

$$f_{tens} = \frac{P}{A}$$

where

f_b = calculated unit stress in bending (psi)
f_v = calculated unit stress in shear (psi)
f_{comp} = calculated unit stress in compression (either parallel or perpendicular to the grain) (psi)
f_{tens} = calculated unit stress in tension (psi)
M = maximum bending moment (in.-lb)
S = section modulus (in.^{3})
P = concentrated load (lb)
V = maximum shear (lb)
A = cross-sectional area (in.^{2})
Ib/Q = rolling shear constant (in.^{2}/ft)

If we equate allowable unit stresses to the maximum unit stresses developed in a beam subjected to a uniformly distributed load w (lb/ft), expressions can be derived for the maximum allowable span length. Therefore, knowing the design loads and member section properties, a maximum allowable span length can be computed.

Table 12-1 contains expressions for maximum allowable span length as governed by moment, shear, and deflection. The reader may wish to derive these expressions from basic principles. The moment expressions are based on the maximum positive or negative moment. The maximum shear considered is the shear that exists at d distance from the support, where d is the depth of the member. Shear considerations for plywood vary from those accorded sawn lumber because of the cross-directional way in which the plies of the plywood are assembled. When plywood is loaded in flexure, the plies with the grain perpendicular to the direction of the span are the weakest aspect of the plywood with respect to shear. The wood fibers in these plies roll at stresses below the fiber shear strength parallel to the grain, hence the name *rolling shear*. For a discussion of the properties of plywood with respect to shear, see [2]. Deflection of forms must be limited to minimize unsightly bulges in the resulting concrete surface. The deflection limit may be specified as a fraction of span (i.e., $\ell/240$), as a limit ($\frac{1}{8}$ in.), or as the smaller of the two. The architect/engineer must decide on the deflection limits for the formwork based on bulging and sagging that can be tolerated in the surfaces of the finished structure. The casting of test panels may be warranted in some cases. Limiting deflection to $\frac{1}{360}$ of the span is acceptable in many cases where surfaces are coarse-textured and there is little reflection of light. Three design equations for deflection are given in Table 12-1. Alternatively, the designer may wish to compute the required size of the members when the design loads and span lengths are known. The same basic principles apply.

Section properties for selected thicknesses of plyform are given in Table 12-2. These section properties reflect that various species of wood used in manufacturing plywood have different stiffness and strength properties. Those species with similar properties are assigned to a species group. To simplify plywood design, the effects of using different species groups in a given panel as well as the effects of the cross-banded construction of the panel have been taken into consideration in establishing the section properties.

In calculating these section properties, all plies were transformed to properties of the face ply. As a result, the designer need not be concerned with the actual panel layup but only with the allowable stresses for the face ply and the given section properties of Table 12-2. The section properties of Table 12-2 are generally the minimums that can be expected. Hence the actual panel obtained in the marketplace will usually have a section property greater than that represented in the table.

The plyform design values presented in Table 12-2 are based on wet strength and 7-day load duration, so no further adjustment in these values is required except for the modulus of elasticity. The modulus shown is an adjusted value based on the assumption that shear deflection is computed separately from bending deflection. These values should be used for bending deflection calculations (which is the usual case). To calculate shear deflection, the modulus should be reduced to 1,500,000 psi for class I and structural plyform and to 1,300,000 for class II plyform.

Section properties for selected members of standard dressed (S4S) sawn lumber are given in Table 12-3. Typical base design values for visually graded dimension lumber are furnished in Table 12-4. This table is simplified and brief and is primarily intended as a resource to accompany the examples and problems of this text. Those who require more detailed information with respect to section properties and design values should obtain reference [4].

The base design values in Table 12-4 must be modified by applicable adjustment factors that are appropriate for the conditions under which the wood is used. Among the adjustment factors itemized in [4] are

C_D = load duration factor
C_M = wet service factor
C_t = temperature factor
C_F = size factor (not applicable to southern pine)
C_{fu} = fact use factor
C_r = repetitive member factor
C_L = beam stability factor
C_P = column stability factor
C_i = incising factor
C_b = bearing area factor

The product of the base design value (F_b for moment or F_v for shear) and any applicable adjustment factors

TABLE 12-1 Concrete Form Design Equations (to Determine Allowable Span Length)

	One span	Two spans	Three or more spans
Bending moment	$\ell = 9.8\sqrt{\frac{F_bS}{w}}$	$\ell = 9.8\sqrt{\frac{F_bS}{w}}$	$\ell = 10.95\sqrt{\frac{F_bS}{w}}$
Shear	$\ell = \frac{16F_vA}{w} + 2d$	$\ell = \frac{12.8F_vA}{w} + 2d$	$\ell = \frac{13.3F_vA}{w} + 2d$
	$\ell^a = \frac{24F_v(Ib/Q)}{w} + 2d$	$\ell^a = \frac{19.2F_v(Ib/Q)}{w} + 2d$	$\ell^a = \frac{20F_v(Ib/Q)}{w} + 2d$
Deflection $\Delta_{all.} = \frac{\ell}{240}$	$\ell = 1.57\sqrt[3]{\frac{EI}{w}}$	$\ell = 2.10\sqrt[3]{\frac{EI}{w}}$	$\ell = 1.94\sqrt[3]{\frac{EI}{w}}$
Deflection $\Delta_{all.} = \frac{\ell}{360}$	$\ell = 1.37\sqrt[3]{\frac{EI}{w}}$	$\ell = 1.83\sqrt[3]{\frac{EI}{w}}$	$\ell = 1.69\sqrt[3]{\frac{EI}{w}}$
When deflection Δ (in.) is specified	$\ell = 5.51\sqrt[4]{\frac{\Delta EI}{w}}$	$\ell = 6.86\sqrt[4]{\frac{\Delta EI}{w}}$	$\ell = 6.46\sqrt[4]{\frac{\Delta EI}{w}}$

Notes: ℓ, span length (center to center of supports) (in.); F_b, allowable bending stress (psi); S, section modulus (in.^{3}); w, uniform load (lb/ft); F_v, allowable shear stress (psi); A, cross-sectional area (in.^{2}); d, depth of member (in.); E, modulus of elasticity (psi); I, moment of inertia (in.^{4}); Ib/Q, rolling shear constant for plywood (in.^{2}/ft); Δ, deflection (in.).

[a]For plywood only.

TABLE 12-2 Section Properties and Design Values for Plyform

Thickness t (in.)	Approx. weight (psf)	Properties for stress applied parallel with face grain: Moment of inertia I (in.^{4}/ft)	Effective section modulus KS (in.^{3}/ft)	Rolling shear const. Ib/Q (in.^{2}/ft)	Properties for stress applied perpendicular to face grain: Moment of inertia I (in.^{4}/ft)	Effective section modulus KS (in.^{3}/ft)	Rolling shear const. Ib/Q (in.^{2}/ft)
Class I							
1/2	1.5	0.077	0.268	5.153	0.024	0.130	2.739
5/8	1.8	0.130	0.358	5.717	0.038	0.175	3.094
3/4	2.2	0.199	0.455	7.187	0.092	0.306	4.063
7/8	2.6	0.296	0.584	8.555	0.151	0.422	6.028
1	3.0	0.427	0.737	9.374	0.270	0.634	7.014
Class II							
1/2	1.5	0.075	0.267	4.891	0.020	0.167	2.727
5/8	1.8	0.130	0.357	5.593	0.032	0.225	3.074
3/4	2.2	0.198	0.454	6.631	0.075	0.392	4.049
7/8	2.6	0.300	0.591	7.990	0.123	0.542	5.997
1	3.0	0.421	0.754	8.614	0.220	0.812	6.987
Structural I							
1/2	1.5	0.078	0.271	4.908	0.029	0.178	2.725
5/8	1.8	0.131	0.361	5.258	0.045	0.238	3.073
3/4	2.2	0.202	0.464	6.189	0.108	0.418	4.047
7/8	2.6	0.317	0.626	7.539	0.179	0.579	5.991
1	3.0	0.479	0.827	7.978	0.321	0.870	6.981

Design values	Plyform class I	Plyform class II	Structural I plyform
Modulus of elasticity (psi)	1,650,000	1,430,000	1,650,000
Bending stress (psi)	1930	1330	1930
Rolling shear stress (psi)	72	72	102

Source: APA—The Engineered Wood Association [3].

Note: All properties adjusted to account for reduced effectiveness of plies with grain perpendicular to applied stress.

provides an allowable unit stress for use in the design and analysis of wood members. The allowable unit stress is denoted as a primed value (e.g., F'_b), which indicates that it is an adjusted design value calculated for specific conditions. For instance, for allowable shear stress,

$$F'_v = F_v C_D C_M C_t C_i$$

If no adjustment factors apply, then the allowable unit stress will be equal to the tabulated base design value (e.g., $F'_v = F_v$). One required value that is not an allowable stress is modulus of elasticity E. It will be determined in a similar fashion as the product of a base design value and applicable adjustment factors.

Determination of Allowable Bending Stress F'_b

Applicable adjustment factors for allowable bending stress are load duration, wet service, temperature, size, flat use, repetitive member, and beam stability.

The load duration factor, C_D, for 7 days or less duration of load (common for form design) is given in Table 12-4.

The wet service factor, C_M, is applicable when the moisture content of the wood is more than 19% (which would occur in a wet service situation such as when the wood is in contact with fresh concrete) and the wood is at ordinary temperature. Table 12-4 shows C_M factors.

The temperature factor, C_t, is generally not applicable in form design and may be disregarded.

The size factor, C_F, is based on member size and is shown in Table 12-5. Construction-grade-material allowable stresses are not adjusted for the size of the member. Southern pine base design values shown in Table 12-4 have size adjustments already included. The size factor C_F applies only to visually graded sawn lumber members.

TABLE 12-3 Properties of Structural Lumber

Nominal size (in.)	Standard dressed size (S4S) (in.)	Area of section, A (in.^{2})	Moment of inertia,[a] I (in.^{4})	Section Modulus,[a] S (in.^{3})	Weight[b] (lb/ft)
2 × 4	$1\frac{1}{2} \times 3\frac{1}{2}$	5.25	5.36	3.06	1.28
2 × 6	$1\frac{1}{2} \times 5\frac{1}{2}$	8.25	20.80	7.56	2.01
2 × 8	$1\frac{1}{2} \times 7\frac{1}{4}$	10.88	47.63	13.14	2.64
2 × 10	$1\frac{1}{2} \times 9\frac{1}{4}$	13.88	98.93	21.39	3.37
2 × 12	$1\frac{1}{2} \times 11\frac{1}{4}$	16.88	178.0	31.64	4.10
3 × 6	$2\frac{1}{2} \times 5\frac{1}{2}$	13.75	34.66	12.60	3.34
3 × 8	$2\frac{1}{2} \times 7\frac{1}{4}$	18.13	79.39	21.90	4.41
3 × 10	$2\frac{1}{2} \times 9\frac{1}{4}$	23.13	164.9	35.65	5.62
3 × 12	$2\frac{1}{2} \times 11\frac{1}{4}$	28.13	296.6	52.73	6.84
4 × 4	$3\frac{1}{2} \times 3\frac{1}{2}$	12.25	12.51	7.15	2.98
4 × 6	$3\frac{1}{2} \times 5\frac{1}{2}$	19.25	48.53	17.65	4.68
4 × 8	$3\frac{1}{2} \times 7\frac{1}{4}$	25.38	111.1	30.66	6.17
4 × 10	$3\frac{1}{2} \times 9\frac{1}{4}$	32.38	230.8	49.91	7.87
4 × 12	$3\frac{1}{2} \times 11\frac{1}{4}$	39.38	415.3	73.83	9.57
6 × 6	$5\frac{1}{2} \times 5\frac{1}{2}$	30.25	76.26	27.73	7.35
6 × 8	$5\frac{1}{2} \times 7\frac{1}{2}$	41.25	193.4	51.56	10.03
6 × 10	$5\frac{1}{2} \times 9\frac{1}{2}$	52.25	393.0	82.73	12.70
6 × 12	$5\frac{1}{2} \times 11\frac{1}{2}$	63.25	697.1	121.2	15.37

[a]I and S are about the strong axis.
[b]Weight in lb/ft when the unit weight of wood is 35 lb/ft^3.

The flat use factor, C_{fu}, is also shown in Table 12-5 and is applicable when dimension lumber 2 in. to 4 in. thick is loaded on the wide face.

The repetitive member factor, C_r, applies only to F_b and to members 2 in. to 4 in. thick and must meet the following requirements per reference [4]: There are at least 3 members connected by continuous sheathing and the spacing must not be greater than 24 inches.

The beam stability factor, C_L, is essentially a reduction factor for F_b where insufficient lateral restraint is furnished for the bending member. Because formwork assemblies are usually designed to meet lateral support criteria as prescribed by the *National Design Specification for Wood Construction* [4], the use of C_L is generally not applicable. For the purpose of this text, all form bending members are assumed to have adequate lateral support.

Determination of Allowable Shear Stress F'_v

Applicable adjustment factors for allowable shear stress are wet service, load duration, temperature, and incising factor.

The wet service factor C_M and the load duration factor C_D are furnished in Table 12-4. The temperature factor is generally not applicable in form design.

The incising adjustment factor C_i is a measure of the incisions that have to be made in a wood member to enhance its ability to receive pressure treatment. Thus, incising is used to increase the depth of penetration of wood preservatives in a wood member. The C_i factor is taken as 1.0 for non-incised wood, such as in the examples and problems of this chapter.

Determination of Allowable Stress for Compression Perpendicular to the Grain $F'_{c\perp}$

Applicable adjustment factors for allowable stress perpendicular to the grain are wet service, temperature, and bearing area. Note that the stress increase for short-term loading does not apply to $F_{c\perp}$.

The wet service factor C_M is furnished in Table 12-4. Again, the temperature factor is generally not applicable in form design.

The bearing area factor, C_b, is applicable where the bearing length is less than 6 in. long and at least 3 in. from the end of the member. It may be calculated from

$$C_b = \frac{\ell_b + 0.375}{\ell_b}$$

where ℓ_b is the length of the bearing (in.) measured parallel to the grain. For bearing at the end of a

TABLE 12-4 Base Design Values for Visually Graded Dimension Lumber (Normal Load Duration and Dry Service Condition)

No. 2 Grade (2″–4″ thick, 2″ and wider)

	Design values (psi)					
Species	F_b	F_t	F_v	$F_{c\perp}$	F_c	E
Douglas fir–larch	900	575	180	625	1350	1,600,000
Douglas fir–south	850	525	180	520	1350	1,200,000
Hem–fir	850	525	150	405	1300	1,300,000
Spruce–pine–fir	875	450	135	425	1150	1,400,000

Construction grade (2″–4″ thick, 2″–4″ wide)

	Design values (psi)					
Species	F_b	F_t	F_v	$F_{c\perp}$	F_c	E
Douglas fir–larch	1000	650	180	625	1650	1,500,000
Douglas fir–south	975	600	180	520	1650	1,200,000
Hem–fir	975	600	150	405	1550	1,300,000
Spruce–pine–fir	1000	500	135	425	1400	1,300,000

Southern pine (2″–4″ thick)

	Design values (psi)					
No. 2 Grade	F_b	F_t	F_v	$F_{c\perp}$	F_c	E
2″–4″ wide	1500	825	175	565	1650	1,600,000
5″–6″ wide	1250	725	175	565	1600	1,600,000
8″ wide	1200	650	175	565	1550	1,600,000
10″ wide	1050	575	175	565	1500	1,600,000
12″ wide	975	550	175	565	1450	1,600,000
Construction grade 4″ wide	1100	625	175	565	1800	1,500,000
Adjustment factors	F_b	F_t	F_v	$F_{c\perp}$	F_c	E
Load duration C_D*	1.25	1.25	1.25	—	1.25	—
Wet service C_M	0.85**	1.0	0.97	0.67	0.80***	0.90

Source: Courtesy of American Forest & Paper Association, Washington, D.C. [4].

*Values shown are typical for forming applications.

**When $F_b(C_F) \leq 1150$ psi, $C_M = 1.0$.

***When $F_c(C_F) \leq 750$ psi, $C_M = 1.0$; for southern pine, when $F_c \leq 750$ psi, $C_M = 1.0$

TABLE 12-5 Size and Flat Use Factors for Bending Stress F_b

	Size factor C_F^a		Flat use factor C_{fu}^b	
Width of lumber	2″–3″ thick	4″ thick	2″–3″ thick	4″ thick
2″–3″	1.5	1.5	1.0	—
4″	1.5	1.5	1.1	1.0
5″	1.4	1.4	1.1	1.05
6″	1.3	1.3	1.15	1.05
8″	1.2	1.3	1.15	1.05
10″	1.1	1.2	1.2	1.1
12″	1.0	1.1	1.2	1.1

Source: Courtesy of American Forest & Paper Association, Washington, D.C. [4].

[a]Applicable to No. 2 grade.

[b]Applicable to No. 2 grade and construction grade.

member and lengths of bearing equal to 6 in. or more, use $C_b = 1.0$. Note that C_b will never be less than 1.0. Therefore, it is conservative to omit it.

Determination of Modulus of Elasticity E'

The only applicable adjustment factor for modulus of elasticity in form design is the wet service C_M. It is shown in Table 12-4.

12-6 DESIGN OF FORMWORK FOR SLABS

Figure 12-1 shows a typical structural system for job-built forms for elevated slabs. The sequence of design is first to consider a strip of sheathing of the specified thickness and 12 in. in width. The maximum allowable span may then be determined based on the allowable values of bending stress, shear stress, and deflection for the sheathing. The lower of the computed values will determine the maximum spacing of the joists. This span value, usually rounded down to some lower modular value, becomes the spacing of the joists.

Based on the joist spacing used, the joist itself is analyzed to determine its maximum allowable span. Each joist must support the load from the sheathing halfway over to the adjacent joist on either side. Therefore, the *width* of the load area carried by the joist is equal to the spacing of the joists. The joist span selected becomes the spacing of the stringers. Again, a modular value is selected for stringer spacing.

Based on the selected stringer spacing, the process is repeated to determine the maximum stringer span (distance between vertical supports or shores). Notice in the design of the stringers that the joist loads are actually applied to the stringer as a series of concentrated loads at the points where the joists rest on the stringer. It is simpler and sufficiently accurate to treat the load on the stringer as a uniformly distributed load, however. Again, the width of the uniform design load applied to the stringer is equal to the stringer spacing. The calculated stringer span must next be checked against the capacity of the shores used to support the stringers. The load on each shore is equal to the shore spacing multiplied by the load per foot of stringer. Thus, the maximum shore spacing (or stringer span) is limited to the lower span length as governed by stringer strength or shore strength. In addition, it is necessary to check the bearing at the point where each joist rests on the stringer. This is done by dividing the load at this point by the bearing area and comparing the resulting stress with the allowable unit stress in

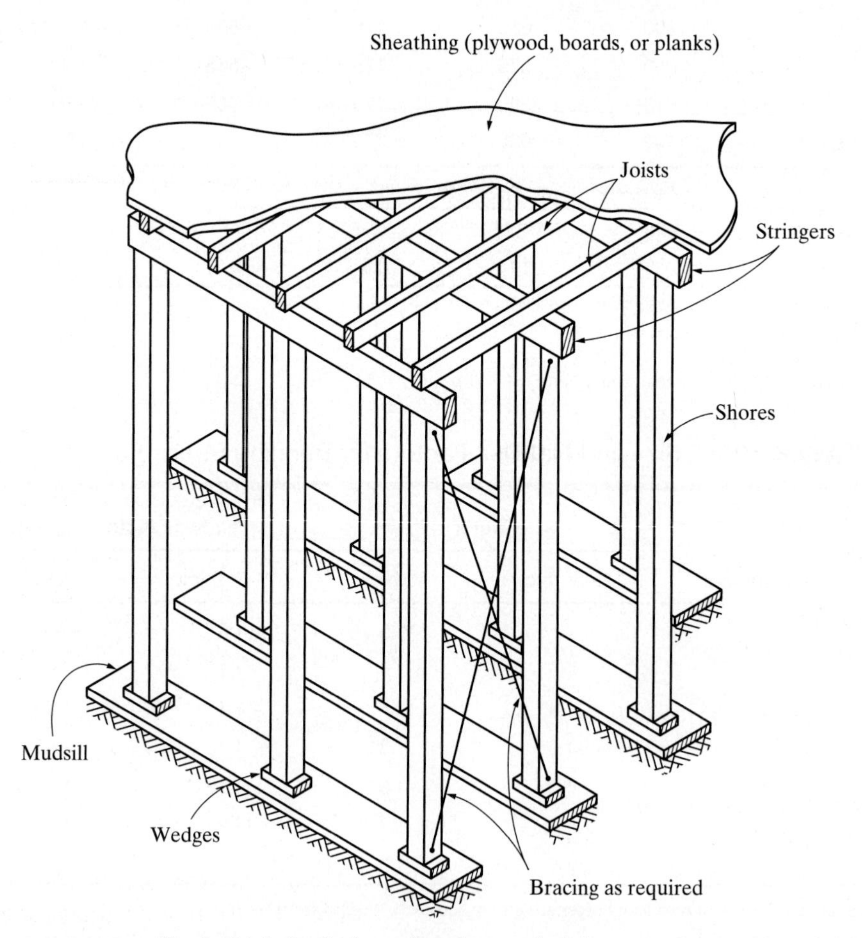

FIGURE 12-1 Typical job-built form for elevated slab.

TABLE 12-6 Effective Length Factor K_e

Buckling modes						
Theoretical K_e value	0.5	0.7	1.0	1.0	2.0	2.0
Recommended design K_e when ideal conditions approximated	0.65	0.80	1.2	1.0	2.10	2.4
End condition code	Rotation fixed, translation fixed					
	Rotation free, translation fixed					
	Rotation fixed, translation free					
	Rotation free, translation free					

Source: Courtesy of American Forest & Paper Association, Washington, D.C. [4].

compression *perpendicular* to the grain. A similar procedure is applied at the point where each stringer rests on a vertical support.

The stringers shown in Figure 12-1 are supported by solid, rectangular wood shores that are columns (which we will assume are axially loaded). As with all axially loaded columns, the allowable load is a function of the slenderness ratio ℓ/d, the ratio of the unbraced length of the member to its least lateral dimension (not to exceed 50). The slenderness ratio ℓ/d is further modified and expressed as ℓ_e/d, where ℓ_e represents an effective unbraced column length and is defined as

$$\ell_e = K_e\ell$$

where K_e is an effective length factor based on column end conditions that affect rotation and translation. Factor K_e can be obtained from Table 12-6.

The following design approach applies for the determination of the allowable stress for compression parallel to the grain for simple, solid-sawn lumber columns, as recommended in [4].

Determination of Allowable Stress for Compression Parallel to the Grain F'_c

Applicable adjustment factors for allowable stress for compression parallel to the grain are load duration, wet service, temperature, size, and column stability. This may be expressed as

$$F'_c = F_c C_D C_M C_t C_F C_P$$

where

F_c = base design value for compression parallel to the grain (psi)

C_p = column stability factor

and all other quantities have been previously defined. This allowable stress is applicable to all values of the slenderness ratio (≤ 50) and replaces the previously used short-, intermediate-, and long-column equations of earlier design specifications. Note that although the slenderness ratio ℓ_e/d for solid columns shall not exceed 50, during construction this limit is increased to 75.

Because wood shores (columns) are generally reused repeatedly, ACI Committee 347 does not recommend the use of any adjustment factor that provides increased stresses for short load duration [5]. In addition, temperature and wet service adjustments for wood shores are generally not required or considered. Therefore, only the column size factor and the column stability factor are normally considered in shore design.

The size factor, C_F, for compression parallel to the grain for the two grades indicated can be taken from Table 12-7.

The column stability factor, C_p, is a function of the effective slenderness ratio ℓ_e/d of the shore, the adjusted modulus of elasticity E', and the adjusted base value of compression parallel to the grain before C_p is applied.

The unadjusted modulus of elasticity is normally used for wood shore design as shores are rarely in the wet service condition. Therefore, $E' = E$.

TABLE 12-7 Size Factor C_F for Compression Parallel to the Grain

Width of lumber	Size factor C_F No. 2 grade	Const. grade
2″–4″	1.15	1.0
5″	1.1	—
6″	1.1	—
8″	1.05	—
10″	1.0	—
12″	1.0	—

Source: Courtesy of American Forest & Paper Association, Washington, D.C. [4].

The column stability factor C_P for solid-sawn lumber can be calculated from Equation (3.7-1) in [4]. For convenience, and to use with the limited properties of Table 12-4, this equation is modified and written as

$$C_P = \frac{1+\alpha}{1.6} - \sqrt{\left(\frac{1+\alpha}{1.6}\right)^2 - \frac{\alpha}{0.8}}$$

where

$$\alpha = \frac{0.3\,E'}{\left(\frac{\ell_e}{d}\right)^2 F_c^*}$$

and F_c^* is the adjusted base value of compression stress parallel to the grain before application of C_P.

Various tables are available to simplify the design approach and to expedite the selection of the formwork structural members [5]. These should be used with caution because of job-specific conditions and possible tabular limitations.

Example 12-1

Design the formwork for a 6-in. structural concrete floor slab. The floor system is of the type shown in Figure 12-1. Use $\frac{3}{4}$-in. class I plyform for the sheathing and No. 2 Douglas fir–larch for the rest of the lumber. The maximum deflection for the sheathing may be taken as $\frac{1}{240}$ of the span. The maximum deflection for other bending members is taken as $\frac{1}{360}$ of the span. Based on end conditions, the effective unbraced length of the shores may be taken as 10 ft. The following conditions for design have been established:

1. Joists and stringers will be designed based on adequate lateral support.
2. Joists, stringers, and shores are used under dry service conditions.

Solution:

1. *Sheathing design* (find the joist spacing): Consider a 12-in.-wide strip of sheathing perpendicular to the supporting joists. The sheathing acts as a beam and is continuous over three or more supports. Determine the maximum allowable span for the sheathing. This becomes the maximum spacing for the joists (which support the sheathing).

   a. The design values for $\frac{3}{4}$-in. class I plyform (see Table 12-2) are

   $E = 1{,}650{,}000$ psi (modulus of elasticity)
   $F_b = 1930$ psi (bending stress)
   $F_v = 72$ psi (rolling shear stress)

   The plyform properties with face grain parallel to span (perpendicular to the joists) are

   $I = 0.199\text{ in.}^4/\text{ft}$
   $S = 0.455\text{ in.}^3/\text{ft}$
   $Ib/Q = 7.187\text{ in.}^2/\text{ft}$ (rolling shear constant)
   $w_s = 2.2$ psf (weight of sheathing: for a 12 in. strip, this will be lb/ft)

   b. The loading on the sheathing is

DL (slab): (6/12)(150)	=	75 psf
LL (min.):	=	50 psf
Neglect sheathing weight		125 psf = w

   c. The maximum joist spacing based on the bending moment formula (from Table 12-1) is

   $$\ell = 10.95\sqrt{\frac{F_b S}{w}} = 10.95\sqrt{\frac{1930(0.455)}{125}} = 29\text{ in.}$$

   d. The maximum joist spacing based on shear is

   $$\ell = \frac{20F_v(Ib/Q)}{w} + 2d = \frac{20(72)(7.187)}{125} + 2(0.75) = 84.3\text{ in.}$$

   e. The maximum joist spacing based on deflection (maximum deflection is $\frac{1}{240}$ of the span) is

   $$\ell = 1.94\sqrt[3]{\frac{EI}{w}} = 1.94\sqrt[3]{\frac{1{,}650{,}000(0.199)}{125}} = 26.8\text{ in.}$$

   Deflection controls. The maximum spacing of supporting joists should not exceed 26.8 in. Use joists at 24 in. o.c.

2. *Joist design* (find the stringer spacing): Assume 2×8(S4S) joists and a 7-day maximum duration of load. Consider the joists as uniformly loaded beams continuous over three or more spans (the supports are the stringers).

a. Obtain allowable stresses using base design values from Table 12-4 and appropriate adjustment factors.

1. Bending: $F_b = 900$ psi. Adjustment factors:
   a. Load duration factor from Table 12-4: $C_D = 1.25$.
   b. Size factor from Table 12-5: $C_F = 1.2$.
   Therefore,

$$F'_b = F_b C_D C_F$$
$$= 900(1.25)(1.2) = 1350 \text{ psi}$$

2. Shear: $F_v = 180$ psi. Adjustment factor:
   Load duration factor from Table 12-4: $C_D = 1.25$.
   Therefore,

$$F'_v = F_v C_D = 180(1.25) = 225 \text{ psi}$$

3. Modulus of elasticity: $E = 1{,}600{,}000$ psi. No adjustment factors apply. Thus,

$$E' = E = 1{,}600{,}000 \text{ psi}$$

Properties for the 2 × 8(S4S) from Table 12-3 are

$$S = 13.14 \text{ in.}^3$$
$$I = 47.63 \text{ in.}^4$$
$$A = 10.88 \text{ in.}^2$$

b. Loading: Because joists support a 2-ft width of sheathing, the loading on the joist is

$$w = 125 \text{ psf } (2) = 250 \text{ lb/ft}$$

Assume the weight of the sheathing and joists to be 5 psf. Then

$$w = 250 + 5(2) = 260 \text{ lb/ft}$$

c. The maximum stringer spacing based on bending moment is

$$\ell = 10.95\sqrt{\frac{F'_b S}{w}}$$
$$= 10.95\sqrt{\frac{1350(13.14)}{260}} = 90.4 \text{ in.}$$

d. The maximum stringer spacing based on shear is

$$\ell = \frac{13.3F'_v A}{w} + 2d$$
$$= \frac{13.3(225)(10.88)}{260} + 2(7.25)$$
$$= 139.7 \text{ in.}$$

e. The maximum stringer spacing based on deflection (maximum deflection is $\frac{1}{360}$ of the span) is

$$\ell = 1.69\sqrt[3]{\frac{E'I}{w}}$$
$$= 1.69\sqrt[3]{\frac{1{,}600{,}000(47.63)}{260}}$$
$$= 112.3 \text{ in.}$$

Bending governs. Therefore, the maximum spacing of supporting stringers cannot exceed 90.4 in. Use a stringer spacing of 7 ft-0 in. (84 in.).

3. *Stringer design* (find the shore spacing): Use 4 in. × 8 in. (S4S) stringers and a 7-day maximum duration of load. Consider the stringers to be uniformly loaded beams continuous over three or more supports. The supports are the shores. The loads from the joists are concentrated loads, but for simplicity, we will assume a uniformly distributed load.

a. Obtain allowable stresses using base design values from Table 12-4 and appropriate adjustment factors.

1. Bending: $F_b = 900$ psi. Adjustment factors:
   a. Load duration factor from Table 12-4: $C_D = 1.25$.
   b. Size factor from Table 12-5: $C_F = 1.3$.
   Therefore,

$$F'_b = F_b C_D C_F$$
$$= 900(1.25)(1.3) = 1463 \text{ psi}$$

2. Shear: $F_v = 180$ psi. Adjustment factor:
   Load duration factor from Table 12-4: $C_D = 1.25$.
   Therefore,

$$F'_v = F_v C_D = 180(1.25) = 225 \text{ psi}$$

3. Modulus of elasticity: $E = 1{,}600{,}000$ psi. No adjustment factors apply. Thus

$$E' = E = 1{,}600{,}000 \text{ psi}$$

Properties for the 4 × 8 (S4S) from Table 12-3 are

$$S = 30.66 \text{ in.}^3$$
$$I = 111.1 \text{ in.}^4$$
$$A = 25.38 \text{ in.}^2$$

b. Loading: Each stringer supports a 7 ft-0 in.-wide strip of design load. Assuming a formwork weight of 5 psf, the uniformly distributed load on a stringer is calculated from

$$w = 7.0(125 + 5) = 910 \text{ lb/ft}$$

c. The maximum shore spacing based on bending (of the stringers) is

$$\ell = 10.95\sqrt{\frac{F'_b S}{w}}$$
$$= 10.95\sqrt{\frac{1463(30.66)}{910}} = 76.9 \text{ in.}$$

d. The maximum shore spacing based on shear is

$$\ell = \frac{13.3F'_v A}{w} + 2d$$
$$= \frac{13.3(225)(25.38)}{910} + 2(7.25) = 98.0 \text{ in.}$$

e. The maximum shore spacing based on deflection is

$$\ell = 1.69\sqrt[3]{\frac{EI}{w}}$$

$$= 1.69\sqrt[3]{\frac{1{,}600{,}000(111.1)}{910}} = 98.1 \text{ in.}$$

Bending governs. Therefore, the maximum spacing of supporting shores cannot exceed 76.9 in. Use a shore spacing of 6 ft-0 in. (72.0 in.).

4. *Design of shores:* The stringers are spaced at 6 ft-6 in. on center and are supported by shores at 6 ft-0 in. on center. Therefore, each shore must support a floor area of

$$7.0(6.0) = 42 \text{ ft}^2$$

Again, assuming formwork weight of 5 psf, the load per shore is calculated as

$$42(125 + 5) = 5460 \text{ lb}$$

Although commercial shores are usually readily available to support this load, we will design 4 × 4 wood shores. The effective unbraced length of the shore, ℓ_e, is 10 ft-0 in. in each direction. The capacity of the 4 × 4 (S4S) shore is calculated using the recommendations of Reference [4], as discussed previously.

a. The base design value for compression parallel to the grain from Table 12-4 is $F_c = 1350$ psi.
b. Adjustment factors:
   1. Size factor from Table 12-7: $C_F = 1.15$.
   2. For the column stability factor C_P, initially the following items must be established:
      a. For modulus of elasticity, there is no adjustment factor:

$$E' = E = 1{,}600{,}000 \text{ psi}$$

      b. Find F_c^*:

$$F_c^* = F_cC_F = 1350(1.15) = 1553 \text{ psi}$$

      c. $$\frac{\ell_e}{d} = \frac{10(12)}{3.5} = 34.3 < 50 \qquad \text{(O.K.)}$$

      d. Solve for α:

$$\alpha = \frac{0.3E'}{\left(\frac{\ell_e}{d}\right)^2 F^*_c}$$

$$= \frac{0.3(1{,}600{,}000)}{34.3^2(1553)}$$

$$= 0.263$$

   Solve for C_P:

$$C_P = \frac{1+\alpha}{1.6} - \sqrt{\left(\frac{1+\alpha}{1.6}\right)^2 - \frac{\alpha}{0.8}}$$

$$= \frac{1+0.263}{1.6} - \sqrt{\left(\frac{1+0.263}{1.6}\right)^2 - \frac{0.263}{0.8}}$$

$$= 0.247$$

c. Compute the allowable stress F_c':

$$F_c' = F_cC_FC_P = 1350(1.15)(0.247) = 383 \text{ psi}$$

Therefore, the allowable load is

$$P = F_c'A = 383(3.5)^2 = 4690 \text{ lb} \qquad \text{(N.G.)}$$

$$4690 \text{ lb} < 5460 \text{ lb}$$

There are several possible solutions. The shore size could be increased (try a 4 × 6 [S4S]), use lateral bracing (horizontal lacing) at midheight to reduce the effective length of the shore, or reduce the shore spacing. The latter choice is the simplest approach.

$$\text{new required spacing} = \frac{4690}{5460}(72 \text{ in.}) = 61.8 \text{ in.}$$

If a shore spacing of 5 ft-0 in. is used,

$$\text{load per shore} = 7.0(5.0)(125 + 5) = 4550 \text{ lb}$$

$$4550 \text{ lb} < 4690 \text{ lb} \qquad \text{(O.K.)}$$

Use 4 × 4(S4S) shores at 5 ft-0 in. on center.

5. *Bearing stresses:*

a. Where stringers bear on shores (4 × 8 stringers on 4 × 4 shores),

$$\text{contact area} = 3.5(3.5) = 12.25 \text{ in.}^2$$

$$\text{total load on shore} = 4550 \text{ lb}$$

The actual bearing stress perpendicular to the grain of the stringer is

$$\frac{4550}{12.25} = 371 \text{ psi}$$

Determine the allowable bearing stress perpendicular to the grain of the stringer. From Table 12-4, $F_{c\perp} = 625$ psi. The adjustment factor applicable in this case is C_b, since the length of bearing ℓ_b is 3.5 in.:

$$C_b = \frac{\ell_b + 0.375}{\ell_b} = \frac{3.88}{3.5} = 1.107$$

Therefore,

$$F_{c\perp}' = 625(1.107) = 692 \text{ psi} \qquad \text{(O.K.)}$$

$$371 \text{ psi} < 692 \text{ psi}$$

b. Where joists bear on stringers (2 × 8 joists on 4 × 8 stringers),

$$\text{contact area} = 1.5(3.5) = 5.25 \text{ in.}^2$$

Recalling from the design of the joists that the loading on each joist was 260 lb/ft, the load on the stringer from each joist is

$$260(7.0) = 1820 \text{ lb}$$

The actual bearing stress perpendicular to the grain is

$$\frac{1820}{5.25} = 347 \text{ psi}$$

Determine the allowable bearing stress perpendicular to the grain. Because the calculated bearing stress is low relative to the 625 psi base design value

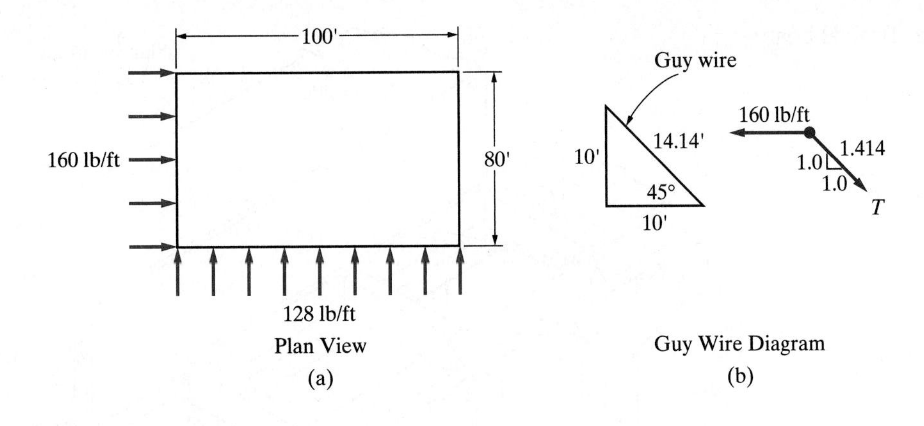

FIGURE 12-2 Floor slab lateral bracing design.

for $F'_{c\perp}$, the increase in $F'_{c\perp}$ due to C_b may be disregarded. Thus,

$$347 \text{ psi} < 625 \text{ psi} \qquad \text{(O.K.)}$$

6. *Lateral bracing:* For floor systems, the minimum load to be used in designing lateral bracing is the greater of 100 lb per lineal foot of floor edge or 2% of the total dead load of the floor. We will assume the slab to be 80 ft × 100 ft and placed in one operation. Guy wire bracing capable of carrying a load of 4000 lb each will be used on all four sides of the slab area attached at slab elevation and making a 45° angle with the ground. Guy wires can resist only tensile forces.

   Calculating lateral load H as 2% of the dead load of the floor, again assuming the formwork to be 5 psf, yields

$$H = 0.02(75 + 5)(80)(100) = 12{,}800 \text{ lb}$$

Distributing this load along the long side yields

$$\frac{12{,}800}{100} = 128 \text{ lb/ft} > 100 \text{ lb/ft}$$

and along the short side yields

$$\frac{12{,}800}{80} = 160 \text{ lb/ft} > 100 \text{ lb/ft}$$

These results are shown in Figure 12-2a. For determination of the guy wire spacing, the 160 lb/ft lateral load will be used. From Figure 12-2b, the tension in the guy wire T is calculated as

$$\frac{T}{1.414} = \frac{160}{1}$$

$$T = 226.2 \text{ lb per ft of slab being braced}$$

The maximum spacing for the guy wires is

$$\frac{4000}{226.2} = 17.7 \text{ ft}$$

Use guy wires spaced at 15 ft (max.) on center on all sides.

12-7 DESIGN OF FORMWORK FOR BEAMS

Figure 12-3 shows one of several common types of beam forms. The usual design procedure involves consideration of the vertical loads, with the following components to be designed: the beam bottom, the ledger that supports joists, and the supporting shores. Bearing stresses must also be checked.

For the deeper beams (24 in. and more), consideration should also be given to the lateral pressure produced by the fresh concrete, which must be resisted by the beam sides. The beam sides would be designed in much the same way as the sheathing in a wall form. Also of importance in Figure 12-3 are the *kickers*, which hold the beam sides in place against the pressure of the concrete, and *blocking*, which serves to transmit the slab load from the ledgers to the T-head shores.

Beam bottoms (or soffits) are usually made to the exact width of the beam. They may be composed of one or more 2-in. planks, or they may be of plywood backed by 2 × 4s. In the following example, the soffit is made of a 2 × 12, which is finished on two sides (S2S), giving it final dimensions of $1\frac{1}{2}$ in. × 12 in.

Example 12-2

Design forms to support the 12 in. × 20 in. beam shown in Figure 12-4. The beam is to support a 4-in.-thick reinforced concrete slab. Use Douglas fir–larch No. 2 grade. The maximum allowable deflection is to be $\frac{1}{360}$ of the span for bending members. The unsupported shore height will be based on an assumed floor-to-floor height of 10 ft, from which the depth of the beam will be subtracted. All bending members are to be designed based on adequate lateral support.

Solution:

1. *Beam bottom design* (compute the maximum spacing between shores): Assume a 7-day maximum duration

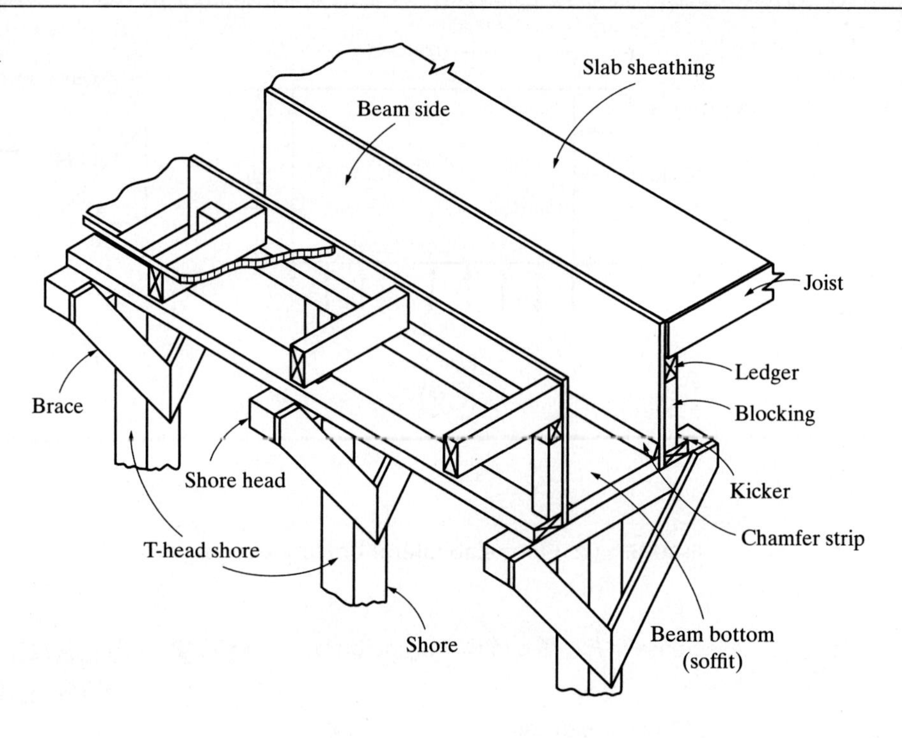

FIGURE 12-3 Typical beam formwork.

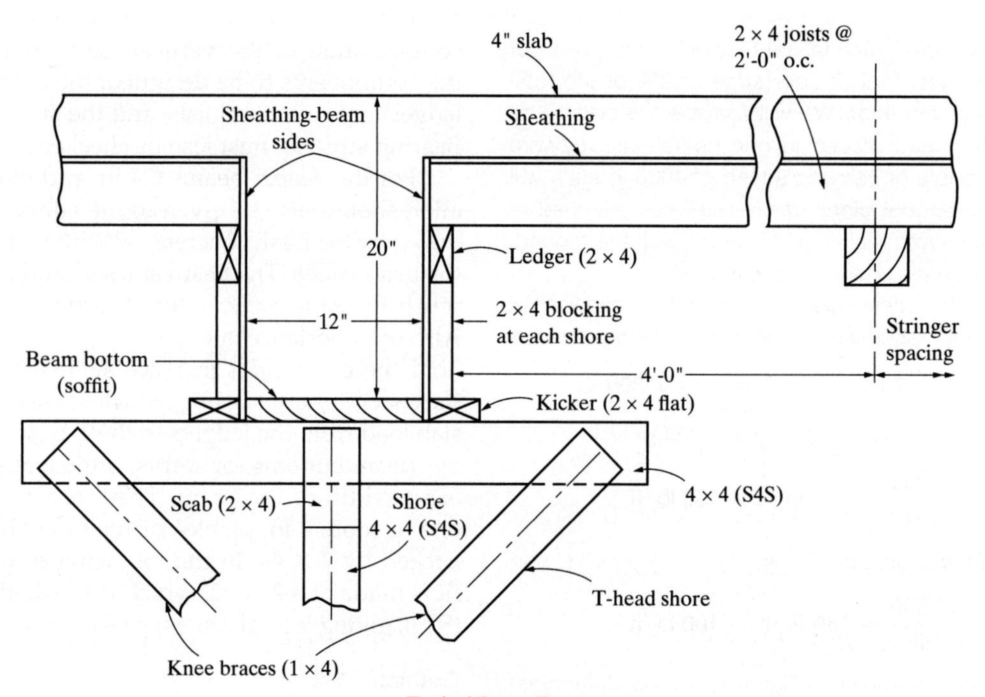

Typical Beam Forms
(a)

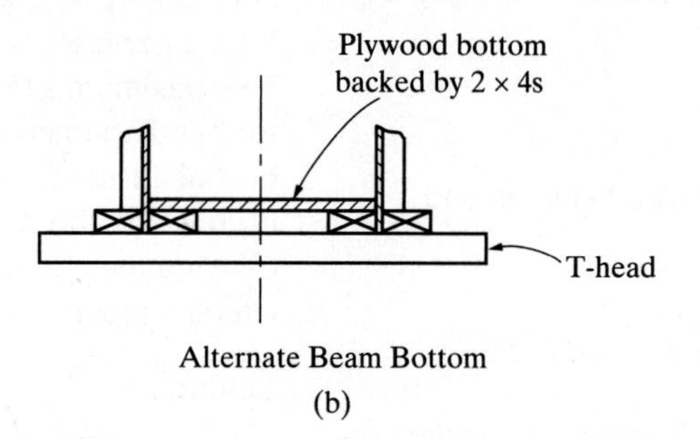

Alternate Beam Bottom
(b)

FIGURE 12-4 Beam form design.

of load. Assume the plank to be continuous over three or more supports.

a. Obtain allowable stresses using base design values from Table 12-4 and appropriate adjustment factors from Tables 12-4 and 12-5.

1. Bending: $F_b = 900$ psi. Adjustment factors: $C_D = 1.25$, $C_F = 1.0$, $C_{fu} = 1.2$. Because $F_bC_F < 1150$ psi, $C_M = 1.0$ (see footnote, Table 12-4). Therefore,

$$F'_b = F_bC_DC_FC_MC_{fu}$$
$$= 900(1.25)(1.0)(1.0)(1.2)$$
$$= 1350 \text{ psi}$$

2. Shear: $F_v = 180$ psi. Adjustment factors: $C_D = 1.25$ and $C_M = 0.97$. Therefore,

$$F'_v = F_vC_DC_M = 180(1.25)(0.97) = 218.3 \text{ psi}$$

3. Modulus of elasticity: $E = 1{,}600{,}000$ psi. Adjustment factor: $C_M = 0.9$. Therefore,

$$E' = EC_M = 1{,}600{,}000(0.9) = 1{,}440{,}000 \text{ psi}$$

Properties for the 2 × 12 (S2S) are

$$S = \frac{bh^2}{6} = \frac{12(1.5)^2}{6} = 4.5 \text{ in.}^3$$

$$I = \frac{bh^3}{12} = \frac{12(1.5)^3}{12} = 3.38 \text{ in.}^4$$

b. The loading on the beam soffit is calculated as

DL(reinforced concrete beam):

$$\frac{12(20)}{144}(150) = 250 \text{ lb/ft}$$

$$\text{LL (use 50 psf)}: \frac{12}{12}(50) = 50 \text{ lb/ft}$$

$$\text{total load} = 250 + 50 = 300 \text{ lb/ft}$$

c. The maximum shore spacing based on bending is

$$\ell = 10.95\sqrt{\frac{F'_bS}{w}}$$
$$= 10.95\sqrt{\frac{1350(4.5)}{300}}$$
$$= 49.3 \text{ in.}$$

d. The maximum shore spacing based on shear is

$$\ell = \frac{13.3F'_vA}{w} + 2d$$
$$= \frac{13.3(218.3)(12)(1.5)}{300} + 2(1.5)$$
$$= 177.2 \text{ in.}$$

e. The maximum shore spacing based on deflection is

$$\ell = 1.69\sqrt[3]{\frac{E'I}{w}}$$
$$= 1.69\sqrt[3]{\frac{1{,}440{,}000(3.38)}{300}}$$
$$= 42.8 \text{ in.}$$

Deflection governs. Try shore spacing at 42 in. o.c.

2. *Ledger design:* Use $\frac{3}{4}$-in. plyform sheathing (vertically) for the beam sides and 2 × 4 kickers as shown. The ledger is supported at each shore by a blocking piece. Because the shores are to be 42 in. o.c., the ledger will be continuous over three or more spans. Use 2 × 4s (S4S) for the ledger as shown. Compute the required spacing for the ledger supports and compare with the 42-in. spacing previously determined. Neglect the connection of the ledger to the vertical sheathing.

a. Obtain allowable stresses using base design values from Table 12-4 and appropriate adjustment factors from Tables 12-4 and 12-5.

1. Bending: $F_b = 900$ psi. Adjustment factors: $C_D = 1.25$ and $C_F = 1.5$. Therefore,

$$F'_b = F_bC_DC_F$$
$$= 900(1.25)(1.5) = 1688 \text{ psi}$$

2. Shear: $F_v = 180$ psi. Adjustment factor: $C_D = 1.25$. Therefore,

$$F'_v = F_vC_D = 180(1.25) = 225 \text{ psi}$$

3. Modulus of elasticity: $E = 1{,}600{,}000$ psi. No adjustment factors apply. Thus,

$$E' = E = 1{,}600{,}000 \text{ psi}$$

Properties for the 2 × 4 (S4S) from Table 12-3 are

$$A = 5.25 \text{ in.}^2$$
$$I = 5.36 \text{ in.}^4$$
$$S = 3.06 \text{ in.}^3$$

b. Loading: The 2 × 4 joists are supported by the ledger, as shown in Figure 12-4. Although the ledger is loaded with point loads, a uniform load will be assumed for simplicity. The loading on the slab sheathing is

$$\text{DL (slab)}: \left(\frac{4}{12}\right)(150) = 50 \text{ psf}$$
$$\text{LL(min.)} = 50 \text{ psf}$$
$$\text{assume sheathing weight} = 5 \text{ psf}$$
$$\overline{105 \text{ psf}}$$

Because the span of the joists from the ledger to the adjacent stringer is 4 ft-0 in. (see Figure 12-4), the load to the ledger is calculated as

$$w = \tfrac{1}{2}(4)(105) = 210 \text{ lb/ft}$$

c. The maximum blocking spacing based on bending is

$$\ell = 10.95\sqrt{\frac{F_b'S}{w}}$$
$$= 10.95\sqrt{\frac{1688(3.06)}{210}}$$
$$= 54.3 \text{ in.}$$

d. The maximum blocking spacing based on shear is

$$\ell = \frac{13.3F_v'A}{w}$$
$$= \frac{13.3(225)(5.25)}{210} + 2(3.5)$$
$$= 81.8 \text{ in.}$$

e. The maximum blocking spacing based on deflection is

$$\ell = 1.69\sqrt[3]{\frac{E'I}{w}}$$
$$= 1.69\sqrt[3]{\frac{1{,}600{,}000(5.36)}{210}}$$
$$= 58.2 \text{ in.}$$

Because all the three foregoing spacings *exceed* 42 in., the 2 × 4 ledgers supported by blocking at 42 in. on center are satisfactory.

3. *Design of the shores:* The shores are spaced 42 in. (or 3.5 ft) on center, and each must support a loading of

$$\text{from beam bottom: } 300(3.5) = 1050 \text{ lb}$$
$$\text{from slab forms (two sides): } 210(3.5)(2) = 1470 \text{ lb}$$
$$\text{total load per shore} = 1050 + 1470 = 2520 \text{ lb}$$

Assume 4 × 4 (S4S) wood shores. The unsupported shore height will be based on an assumed floor-to-floor height of 10 ft-0 in., from which the depth of the beam will be subtracted. The unsupported height is

$$\ell = 10(12) - 20 = 100 \text{ in.} = 8.33 \text{ ft}$$

We will assume the shores are pin-connected. From Table 12-6, $K_e = 1.0$. Therefore, the effective unbraced length is

$$\ell_e = K_e\ell = 1.0(8.33) = 8.33 \text{ ft}$$

a. The base design value for compression parallel to the grain from Table 12-4 is $F_c = 1350$ psi.

b. Adjustment factors:

1. Size factor from Table 12-7: $C_F = 1.15$.
2. For the column stability factor C_P, initially the following items must be established:

a. For modulus of elasticity, there is no adjustment factor:

$$E' = E = 1{,}600{,}000 \text{ psi}$$

b. Find F_c^*:

$$F_c^* = F_cC_F = 1350(1.15) = 1553 \text{ psi}$$

c.
$$\frac{\ell_e}{d} = \frac{8.33(12)}{3.5} = 28.6 < 50 \qquad \text{(O.K.)}$$

d. Solve for α:

$$\alpha = \frac{0.3E'}{\left(\frac{\ell_e}{d}\right)^2 F_c^*}$$
$$= \frac{0.3(1{,}600{,}000)}{28.6^2(1553)}$$
$$= 0.378$$

Solve for C_P:

$$C_p = \frac{1+\alpha}{1.6} - \sqrt{\left(\frac{1+\alpha}{1.6}\right)^2 - \frac{\alpha}{0.8}}$$
$$= \frac{1+0.378}{1.6} - \sqrt{\left(\frac{1+0.378}{1.6}\right)^2 - \frac{0.378}{0.8}}$$
$$= 0.342$$

c. Compute the allowable stress F_c':

$$F_c' = F_cC_FC_P = 1350(1.15)(0.342) = 531 \text{ psi}$$

Therefore, the allowable load is

$$P = F_c'A = 531(3.5)^2 = 6500 \text{ lb} \qquad \text{(O.K.)}$$
$$6500 \text{ lb} > 2520 \text{ lb}$$

Use 4 × 4 (S4S) shores spaced 42 in. on center.

4. *Bearing stresses:*

a. Assume 4 × 4 (S4S) T-heads on the 4 × 4 (S4S) shores. Actual bearing stress perpendicular to the grain of the T-head is calculated as

$$\frac{\text{shore load}}{\text{contact area}} = \frac{2520}{3.5^2} = 206 \text{ psi}$$

From Table 12-4, the base design value for $F_{c\perp}$ is 625 psi. We will neglect the adjustment factor for bearing area (due to $C_b > 1.0$ because $3\frac{1}{2}$ in. < 6 in.). Therefore, use $F'_{c\perp} = 625$ psi. Thus,

$$206 \text{ psi} < 625 \text{ psi} \qquad \text{(O.K.)}$$

b. Check bearing stress between the 2 × 4 ledger and the 2 × 4 blocking. The load from the ledger to the blocking is 210(3.5) = 735 lb. The actual bearing stress perpendicular to the grain of the ledger is

$$\frac{\text{load}}{\text{contact area}} = \frac{735}{1.5(3.5)} = 140 \text{ psi}$$

The allowable bearing stress, from part a, is 625 psi. Thus,

$$140 \text{ psi} < 625 \text{ psi} \qquad \text{(O.K.)}$$

c. Check bearing stress between the 2 × 4 joists and the 2 × 4 ledger. The joist loading is (105 psf)(2 ft) = 210 lb/ft, and the span of the joist is 4 ft. Therefore, the load from the joist to the ledger is (210 lb/ft)(2 ft) = 420 lb. The actual bearing stress perpendicular to the grain of the ledger (and the joists) is

$$\frac{\text{load}}{\text{contact area}} = \frac{420}{1.5(1.5)} = 186.7 \text{ psi}$$

The allowable bearing stress is determined as in part a (the bearing area adjustment factor C_b is not applicable because this is end bearing):

$$186.7 \text{ psi} < 625 \text{ psi} \qquad \text{(O.K.)}$$

d. Check the bearing of the beam soffit on the 4 × 4 (S4S) T-heads. The load from the beam soffit is (300 lb/ft)(3.5 ft) = 1050 lb. The actual bearing stress perpendicular to the grain of the T-head is

$$\frac{\text{load}}{\text{contact area}} = \frac{1050}{12(3.5)} = 25 \text{ psi}$$

For the allowable bearing stress, we will neglect the bearing area adjustment factor. The contact surface may be subjected to a wet condition, so this adjustment factor ($C_M = 0.67$ from Table 12-4) will be used:

$$F'_{c\perp} = F_{c\perp} C_M = 625(0.67) = 419 \text{ psi}$$

$$25 \text{ psi} < 419 \text{ psi} \qquad \text{(O.K.)}$$

12-8 WALL FORM DESIGN

The design procedure for wall forms is similar to that used for slab forms, substituting studs for joists, wales for stringers, and ties for shores. See Figure 12-5 for locations of these members, and see Figure 12-6 for the forms and bracing for concrete walls under construction.

The maximum lateral pressure against the sheathing must be determined first. We will assume conditions such that C_C and C_W are both 1.0. With the sheathing thickness specified, the maximum allowable span for the sheathing is computed based on bending, shear, and deflection. This will be the maximum stud spacing. (An alternative approach would be to establish the stud spacing and then calculate the required thickness of the sheathing.)

Next, the maximum allowable stud span is calculated based on stud size and loading, considering bending, shear, and deflection. This will be the maximum wale spacing. (An alternative approach would be to establish the wale spacing and then calculate the required size of the studs.)

The next step is to determine the maximum allowable spacing of wale supports (tie spacing). This is calculated based on wale size and loading. (An alternative approach would be to preselect the tie spacing and then calculate the wale size.) Double wales are commonly used to avoid the necessity of drilling wales for tie insertion.

The load supported by each tie must be computed and compared with the tie capacity. The load on each tie is calculated as the design load (psf) multiplied by the tie spacing (ft) and wale spacing (ft). If the load exceeds the tie strength, a stronger tie must be used or the tie spacing must be reduced.

Bearing stresses must also be checked where the studs bear on the wales and where the tie ends bear on the wales. Maximum bearing stress must not exceed the allowable compression stress perpendicular to the grain or crushing will result.

Finally, lateral bracing must be designed to resist any expected lateral loads, such as wind loads.

Example 12-3

Design formwork for an 8-ft-high wall. Refer to Figure 12-5. The concrete is to be placed at a rate of 4 ft/h and will be internally vibrated. Concrete temperature is expected to be 90°F. The maximum allowable deflection of bending members is to be $\frac{1}{360}$ of the span. Use $\frac{3}{4}$-in. class I plyform for the sheathing and No. 2 Douglas fir–larch for the rest of the lumber. Assume all bending members to be supported on three or more spans. The following conditions for design have been established:

1. Studs and wales are to be designed based on adequate lateral support.
2. Studs, wales, and bracing are to be used under dry service conditions.

Solution:

1. *Sheathing design* (find the stud spacing): Consider a 12-in.-wide strip of sheathing perpendicular to the supporting studs and acting as a beam continuous over three or more spans. The sheathing spans horizontally. Place the face grain perpendicular to studs.

   a. The design values for $\frac{3}{4}$-in. class I plyform (see Table 12-2) are

$$E = 1{,}650{,}000 \text{ psi}$$
$$F_b = 1930 \text{ psi}$$
$$F_v = 72 \text{ psi}$$

   Plyform properties for face grain parallel to the span are

$$I = 0.199 \text{ in.}^4$$
$$S = 0.455 \text{ in.}^3$$
$$Ib/Q = 7.187 \text{ in.}^2/\text{ft}$$

   b. Loading: The sheathing will be designed for concrete pressure, which is the lesser of $150h$ (where h will be taken as 8 ft), 2000 psf, or as determined by formula (noting that $R < 7$ ft/h and $h < 14$ ft):

$$p = 150 + \frac{9000R}{T} = 150 + \frac{9000(4)}{90}$$
$$= 550 \text{ psf} < 600 \text{ psf}$$
$$150(h) = 150(8) = 1200 \text{ psf}$$

   Use the ACI-recommended minimum of 600 psf for the sheathing design load, although the pressure could be decreased near the top of the form where

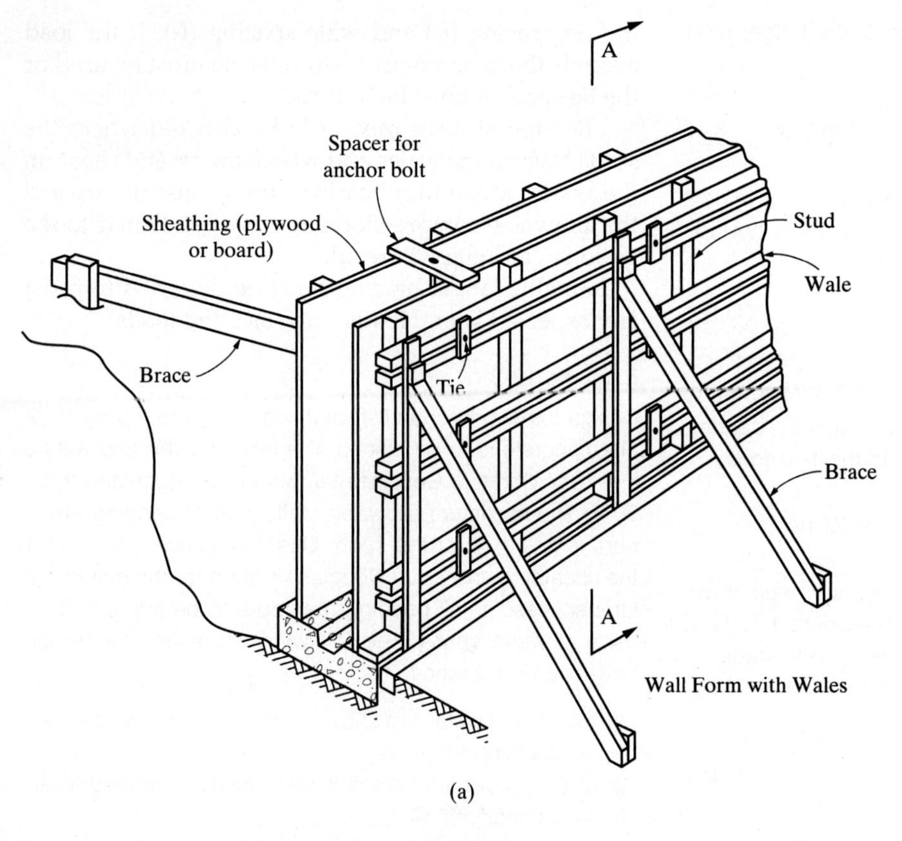

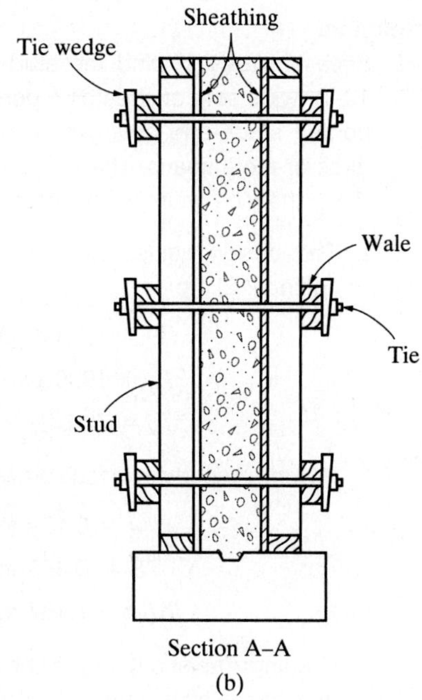

FIGURE 12-5 Typical wall forms.

150(h) controls. The length in which the decreased pressure could be used may be calculated as

$$\frac{600}{150} = 4.0 \text{ ft} \quad \text{(from the top of the form)}$$

It is conservative to design the full height for 600 psf, however.

c. The maximum stud spacing based on bending is

$$\ell = 10.95\sqrt{\frac{F_b S}{w}}$$

$$= 10.95\sqrt{\frac{1930(0.455)}{600}} = 13.25 \text{ in.}$$

d. The maximum stud spacing based on shear is

$$\ell = \frac{20F_v(Ib/Q)}{w} + 2d$$

$$= \frac{20(72)(7.187)}{600} + 2(0.75) = 18.75 \text{ in.}$$

e. The maximum stud spacing based on deflection is

$$\ell = 1.69\sqrt[3]{\frac{EI}{w}}$$

$$= 1.69\sqrt[3]{\frac{1,650,000(0.199)}{600}} = 13.82 \text{ in.}$$

Bending is critical. Use a stud spacing of 12 in. o.c.

2. *Stud design* (compute the wale spacing): Assume 2 × 4 (S4S) studs and a 7-day maximum duration of load.
   a. Obtain allowable stresses using base design values from Table 12-4 and appropriate adjustment factors from Tables 12-4 and 12-5.
      1. Bending: $F_b = 900$ psi. Adjustment factors: $C_D = 1.25$ and $C_F = 1.5$. Therefore,

$$F_b' = F_bC_DC_F$$
$$= 900(1.25)(1.5) = 1688 \text{ psi}$$

      2. Shear: $F_v = 180$ psi. Adjustment factor: $C_D = 1.25$. Therefore,

$$F_v' = F_vC_D = 180(1.25) = 225 \text{ psi}$$

      3. Modulus of elasticity: $E = 1,600,000$ psi. No adjustment factors apply. Thus,

$$E' = E = 1,600,000 \text{ psi}$$

      For the 2 × 4 lumber, from Table 12-3,

$$A = 5.25 \text{ in.}^2$$
$$I = 5.36 \text{ in.}^4$$
$$S = 3.06 \text{ in.}^3$$

   b. Loading: Since the stud spacing is 12 in. o.c., the load w will be 600 lb/ft (see step 1, part [b]).
   c. The maximum wale spacing based on bending is

$$\ell = 10.95\sqrt{\frac{F_b'S}{w}}$$

$$= 10.95\sqrt{\frac{1688(3.06)}{600}} = 32.1 \text{ in.}$$

   d. The maximum wale spacing based on shear is

$$\ell = \frac{13.3F_v'A}{w} + 2d$$

$$= \frac{13.3(225)(5.25)}{600} + 2(3.5) = 33.2 \text{ in.}$$

   e. The maximum wale spacing based on deflection is

$$\ell = 1.69\sqrt[3]{\frac{E'I}{w}}$$

$$= 1.69\sqrt[3]{\frac{1,600,000(5.36)}{600}} = 41.0 \text{ in.}$$

   Therefore, bending governs. Use a wale spacing of 24 in. o.c. (maximum).

3. *Wale design* (compute the tie spacing): Double 2 × 4 (S4S) wales will be assumed, and the allowable stresses will be adjusted for a 7-day maximum duration of load.
   a. Design values: Allowable stresses and E will be the same as for the studs. Properties for the wales will be twice those for the studs because the wales are doubled. Thus,

$$A = 10.5 \text{ in.}^2$$
$$I = 10.72 \text{ in.}^4$$
$$S = 6.12 \text{ in.}^3$$

   b. Loading: Each wale will support a strip of wall form that has a height equal to the spacing of the wales:

$$w = \frac{24}{12}(600) = 1200 \text{ lb/ft}$$

   c. The maximum tie spacing based on bending is

$$\ell = 10.95\sqrt{\frac{F_b'S}{w}}$$

$$= 10.95\sqrt{\frac{1688(6.12)}{1200}} = 32.1 \text{ in.}$$

   d. The maximum tie spacing based on shear is

$$\ell = \frac{13.3F_v'A}{w} + 2d$$

$$= \frac{13.3(225)(10.5)}{1200} + 2(3.5) = 33.2 \text{ in.}$$

   e. The maximum tie spacing based on deflection is

$$\ell = 1.69\sqrt[3]{\frac{E'I}{w}}$$

$$= 1.69\sqrt[3]{\frac{1,600,000(10.72)}{1200}} = 41.0 \text{ in.}$$

   Bending governs. A modular spacing of 24 in. would be desirable. Use tie spacing of 24 in. o.c.

4. Check the load on the ties (P_{tie}) with the capacity of the ties. Assume the tie capacity to be 3000 lb (ties of various capacities are widely available). Also assume that the ties have $1\frac{1}{2}$-in. wedges bearing on the wales. Then

$$P_{tie} = (\text{wale spacing}) \times (\text{tie spacing}) \times (\text{pressu}$$

$$= \frac{24}{12}\left(\frac{24}{12}\right)(600) = 2400 \text{ lb}$$

$2400 \text{ lb} < 3000 \text{ lb}$

Therefore, the capacity of the tie is satis

5. Check bearing stresses.
   a. Where tie wedges bear on wal
   wide),

$$P_{tie} =$$

bearing contact are

bearing stress (actual)

FIGURE 12-6 Wall forms and bracing. (George Limbrunner)

The allowable compressive stress perpendicular to the grain is $F'_{c\perp} = 625$ psi (neglect bearing area adjustment factor). Thus,

$$533 \text{ psi} < 625 \text{ psi} \qquad \text{(O.K.)}$$

b. Where studs bear on wales (double wales),

$$\text{bearing contact area} = (2)(1.5)(1.5) = 4.5 \text{ in.}^2$$

The load on the wale from the stud is

$$P = (\text{load/ft on stud}) \times (\text{wale spacing})$$

$$= 600\left(\frac{24}{12}\right) = 1200 \text{ lb}$$

$$\text{actual bearing stress} = \frac{1200}{4.5} = 267 \text{ psi}$$

As in part (a), $F'_{c\perp} = 625$ psi. Thus,

$$267 \text{ psi} < 625 \text{ psi} \qquad \text{(O.K.)}$$

6. Lateral bracing should be designed for wall forms based on the greater of wind load (using 15 psf as a minimum) or 100 lb/ft applied at the top of the wall. Calculate wind load on wall forms using the minimum 15 psf:

$$\text{wind load} = (15 \text{ psf})(1 \text{ ft})(8 \text{ ft}) = 120 \text{ lb per ft of wall}$$

This load would be considered to act at midheight of the wall, 4 ft above the base, and would create an overturning moment about the base of

$$M_{OT} = 120 \text{ lb/ft}(4 \text{ ft}) = 480 \text{ ft-lb (per ft of wall)}$$

ne equivalent force, acting at the top of the wall, that uld create the same overturning moment is

$$\text{force} = \frac{480 \text{ ft-lb}}{8 \text{ ft}} = 60 \text{ lb} \quad \text{(per ft of wall)}$$

$$60 \text{ lb/ft} < 100 \text{ lb/ft}$$

Therefore, use 100 lb/ft. This load, assumed to act at the top of the wall, can act in either direction. If *guy wires* are used, they must be placed on *both* sides of the wall. If *wooden strut bracing* is used, it can resist tension or compression and therefore *single-side* bracing may be used.

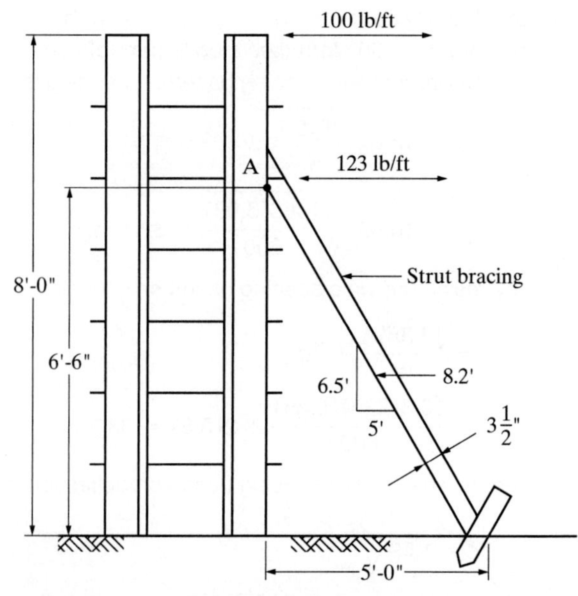

FIGURE 12-7 Lateral bracing for wall form.

In this problem use single-side strut bracing, as shown in Figure 12-7, and design for compression. The horizontal load H on the strut at point A is calculated by considering moment taken at the base of the wall:

$$H(6.5) = 100(8)$$
$$H = 123 \text{ lb}$$

The force F in the strut, using the slope triangle shown, is determined from

$$\frac{F}{8.2} = \frac{123}{5}$$

Therefore, $F = 202$ lb (per foot of wall).

Use double 2 × 4 (S4S) lumber for the strut and compute the capacity as a compression member. (This will be adequate for tension also.) For No. 2 grade Douglas fir–larch:

a. The base design value for compression parallel to the grain from Table 12-4 is $F_c = 1350$ psi.
b. Adjustment factors:
   1. Size factor from Table 12-7: $C_F = 1.15$.
   2. For the column stability factor C_P, initially the following items must be established:
      a. For modulus of elasticity, there is no adjustment factor:

$$E' = E = 1{,}600{,}000 \text{ psi}$$

      b. Find F_c^*:

$$F_c^* = F_c C_F = 1350(1.15) = 1553 \text{ psi}$$

      c. Assume that the ends are pin connected. Therefore, $K_e = 1.0$ and $\ell_e = 8.2$ ft. Then

$$\frac{\ell_e}{d} = \frac{8.2(12)}{3.0} = 32.8 < 50 \qquad \text{(O.K.)}$$

      d. Solve for α:

$$\alpha = \frac{0.3E'}{\left(\frac{\ell_e}{d}\right)^2 F_c^*} = \frac{0.3(1{,}600{,}000)}{32.8^2(1553)} = 0.287$$

      Solve for C_P:

$$C_P = \frac{1+\alpha}{1.6} - \sqrt{\left(\frac{1+\alpha}{1.6}\right)^2 - \frac{\alpha}{0.8}}$$
$$= \frac{1+0.287}{1.6} - \sqrt{\left(\frac{1+0.287}{1.6}\right)^2 - \frac{0.287}{0.8}}$$
$$= 0.267$$

c. Compute the allowable stress F_c':

$$F_c' = F_c C_F C_P = 1350(1.15)(0.267) = 415 \text{ psi}$$

Therefore, the allowable load is

$$P = F_c' A = 415(2)(5.25) = 4360 \text{ lb}$$

The maximum allowable strut spacing is calculated from

$$\frac{4360 \text{ lb}}{202 \text{ lb/ft}} = 21.6 \text{ ft}$$

Use struts at 21 ft-0 in. on center.

Single 2 × 4 struts could have been used if an intermediate brace were used to reduce the unbraced length.

12-9 FORMS FOR COLUMNS

Concrete columns are usually one of five shapes: square, rectangular, L-shaped, octagonal, or round. Forms for the first four shapes are generally made of sheathing, consisting of vertical planks or plywood, with wood yokes and steel bolts, patented steel clamps, or steel bands used to resist the concrete pressure acting on the sheathing. Forms for round columns may be wood, steel plate, or patented fiber tubes.

Because forms for columns are usually filled rapidly, frequently in less than 60 min, the pressure on the sheathing will be high, especially for tall columns. The ACI recommendations for lateral concrete pressure in column forms were discussed in Section 12-4. AC1 347R-14 (Equation [4.2.2.1a(b)]) is applicable for concrete having a slump of 7 in. or less and placed with normal internal vibration to a depth of 4 ft or less. However, assuming normal-weight concrete (150 pcf), the pressure should not be taken as greater than $150h$ (psf), where h is the depth in feet below the upper surface of the freshly placed concrete. Thus, the *maximum* pressure at the bottom of a form 10-ft high should be taken as $150(10) = 1500$ psf regardless of the rate of filling the form or concrete temperature. It is suggested that the pressure be conservatively calculated using the equation

$$p = 150h$$

Figure 12-8a illustrates typical construction of a column form using plywood sheathing backed by vertical stiffening members and clamped with adjustable metal column clamps. The sheathing must be selected to span between the stiffening members using the concrete pressure that exists at the bottom of the column form. The vertical stiffening members must span between the column clamps, the spacing of which can be increased as the pressure decreases toward the top of the form.

Figure 12-8b illustrates a method suitable for forming smaller columns where no vertical stiffening members are required and the plywood sheathing is backed directly by battens that are part of a wood and bolt column yoke. Column clamps can be used in this

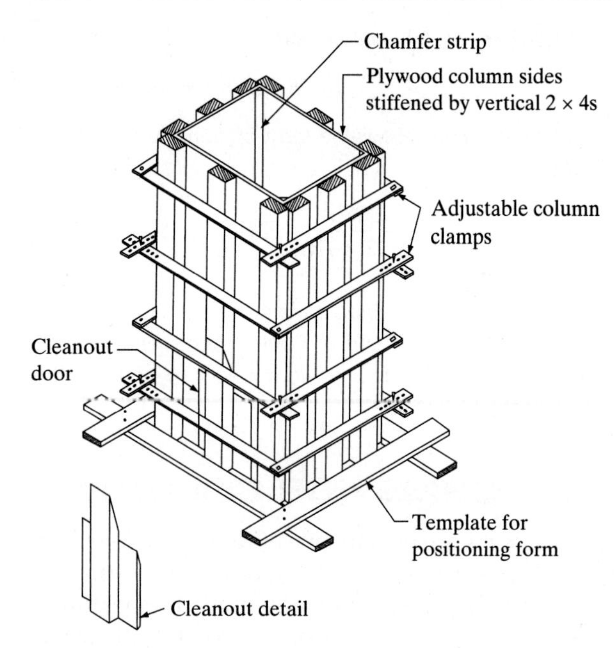

Typical Construction for Larger Column Forms
(a)

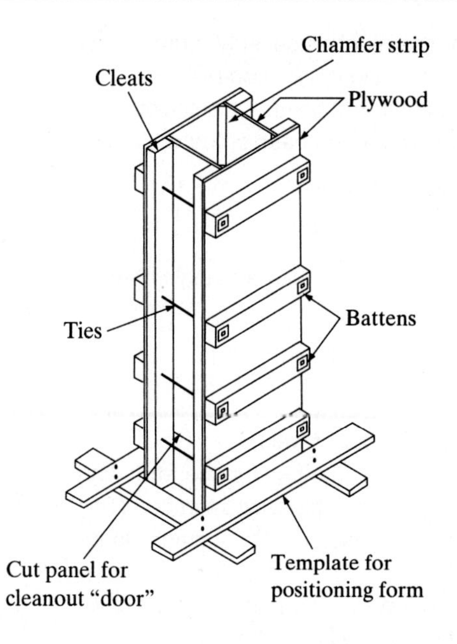

Typical Construction for Smaller Column Forms
(b)

FIGURE 12-8 Typical column forms. (*Source:* Courtesy of the American Concrete Institute)

situation as well. If the thickness of the sheathing is selected, the design consists of determining the maximum safe spacing of the column clamps considering the pressure from the concrete as well as the permissible deflection, allowable bending stress, and allowable shearing stress.

The sheathing span length ℓ may be calculated for moment, shear, and deflection, with the shortest of these span lengths being the controlling value.

Because the pressure against the forms varies with height, however, the determination of the optimum clamp spacing becomes laborious. As a result, tables have been developed that aid in quick determination of support (clamp) spacing. Table 12-8 is an example of one such table specifically set up for plywood sheathing. The tabular values are based on the assumption that the lateral pressure is uniform between clamps and of an intensity equal to that at the lower clamp.

Clamps must also be investigated to determine if they can resist the applied loads. The manufacturer usually has recommended capacities for steel column clamps. Wood-yoke-type clamps with tie rods must be designed.

Generally, the type of forming system is a function of column size and height. As column size increases, either the thickness of the sheathing must be increased or vertical stiffeners must be added to prevent sheathing deflection. If vertical supports or stiffeners are used (see Figure 12-8) in combination with a plywood sheathing, the sheathing should span between the vertical supports and the plywood face grain should be horizontal (in the direction of the span) for maximum strength. The clamp spacing is then a function of the vertical support member strength. If plywood sheathing spans between clamps (without vertical supports), the face grain should be vertical (in the direction of the span) for maximum strength.

Example 12-4

Determine a clamp spacing pattern for column form sheathing made up of $\frac{3}{4}$-in.-thick plywood. The column height is to be 12 ft-0 in. Assume the sheathing continuous over four or more supports and its face grain parallel to the span (vertical). Use class I plyform design values of

$$F_b = 1930 \text{ psi}$$
$$F_v = 72 \text{ psi}$$
$$E = 1{,}650{,}000 \text{ psi}$$

with allowable deflection of span/360 but not greater than $\frac{1}{16}$ in.

Solution:

1. Table 12-8 is used to determine the maximum span of the plywood between clamps. This depends on the pressure on the form, which is determined from

$$p = wh$$

where

w = unit weight of concrete (pcf)
h = depth of fresh concrete (ft)

2. Denoting the vertical distance from the bottom of the form as y (ft), the pressure is determined from

$$p = wh = 150(12 - y)$$

TABLE 12-8 Safe Span in Inches for Class I Plyform, Continuous over Four or More Supports

	Stress parallel to gain				Stress perpendicular to gain			
Pressure (psf)	1/2 in.	5/8 in.	3/4 in.	1 in.	1/2 in.	5/8 in.	3/4 in.	1 in.
75	20	24	26	32	14	16	21	28
100	18	22	24	30	12	14	19	26
125	17	20	23	28	12	13	18	25
150	16	19	22	27	11	13	17	24
175	15	18	21	26	10	12	16	23
200	15	17	20	25	10	11	15	22
300	13	15	17	22	9	10	13	19
400	12	14	16	20	8	9	12	18
500	11	13	15	18	7	8	11	16
600	10	12	13	17	7	8	11	15
700	9	11	12	16	6	8	10	14
800	9	10	11	15	6	7	9	14
900	8	10	11	14	5	6	8	13
1000	8	9	10	13	5	6	7	12
1100	8	9	10	12	5	5	7	11
1200	7	8	9	12	4	5	6	10
1300	7	8	9	11	4	5	6	10
1400	6	7	9	11	4	4	6	9
1500	6	7	8	11	4	4	5	9
1600	6	6	8	10	—	4	5	8
1700	5	6	8	10	—	4	5	8
1800	5	6	7	9	—	4	5	8
1900	5	6	7	9	—	4	5	7
2000	5	5	7	9	—	—	4	7
2200	4	5	6	8	—	—	4	7
2400	4	5	6	8	—	—	4	6
2600	4	4	5	7	—	—	4	6
2800	4	4	5	7	—	—	4	6
3000	—	4	5	6	—	—	—	5

Notes: $F_b = 1930$ psi; $F_v = 72$ psi; $E = 1{,}650{,}000$ psi; allowable deflection = span/360 but not greater than $\frac{1}{16}$ in. Safe spans less than 4 in. are not shown. Tabulated spans are rounded to the nearest inch.

TABLE 12-9 Clamp Spacing Determination

y^a (ft)	Pressure (psf)	Maximum clamp spacing (in.)[b]
0	1800	7
3	1350	9
6	900	11
9	450	15
11.5	75	26

[a]Quantity y is measured upward from the bottom of the form.
[b]From Table 12-8.

For some arbitrary values of y, the calculated pressures and the maximum spans (clamp spacings) from Table 12-8 are shown in Table 12-9.

3. A plot of maximum clamp spacing as a function of distance above top-of-footing is shown in Figure 12-9. The final clamp layout, also shown in Figure 12-9, is determined by trial and error. This procedure is similar to stirrup design (Chapter 4). One should attempt to minimize the number of clamps without having too many different-size spacings.

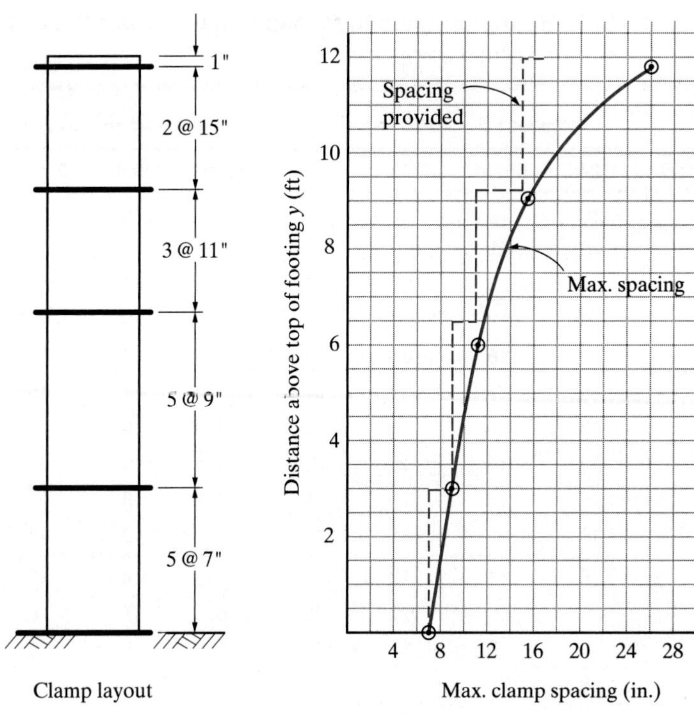

FIGURE 12-9 Sketch for Example 12-4.

References

[1] *Guide to Formwork for Concrete* (ACI 347R-14). American Concrete Institute, 38800 Country Club Drive, Farmington Hills, MI 48331.

[2] *Plywood Design Specification*. APA—The Engineered Wood Association, P.O. Box 11700, Tacoma, WA 98411-0700, January 1997.

[3] *Concrete Forming*. APA—The Engineered Wood Association, P.O. Box 11700, Tacoma, WA 98411-0700, 2004.

[4] *National Design Specification for Wood Construction* (ANSI/AF&PA NDS-2015), with supplements and commentary. American Forest & Paper Association/American Wood Council, 1111 19th Street, N.W., Suite 800, Washington, D.C. 20036, 2015.

[5] M. K. Hurd, *Formwork for Concrete*, SP-4, 8th ed. American Concrete Institute, 38800 Country Club Drive, Farmington Hills, MI 48331, 2014.

Problems

Where applicable in these problems, and unless otherwise noted, *assume:*

1. The vertical live load is to be 75 psf (motorized carts).
2. Bending members are continuous over three or more spans and are assumed to have adequate lateral support.
3. There is a 7-day duration of load.
4. Lumber is No. 2 grade hem-fir.
5. The forming weight is 5 psf; neglect sheathing weight.
6. The maximum deflection is $\frac{1}{360}$ of the span.
7. All lumber is (S4S).
8. Plywood is placed with the face grain perpendicular to the supports.
9. No available information exists with respect to wood splits, checks, and shakes. Assume $C_i = 1.0$, $C_C = 1.0$, and $C_W = 1.0$.

12-1. Using basic principles, derive the equations of Table 12-1 for allowable span length ℓ as governed by moment, shear, and deflection. Handbooks (such as the AISC *Manual*) may be helpful for moment, shear, and deflection equations. Show load diagrams.

a. One span

b. Two spans

c. Three spans

12-2. A slab form is to be built for an 8-in.-thick concrete slab. The plywood sheathing face grain is perpendicular to the joists. The maximum allowed deflection is to be the smaller of $\frac{1}{360}$ of the span or $\frac{1}{8}$ in. Compare the maximum allowable joist spacing for class II plyform $\frac{1}{2}$ in. thick and 1 in. thick. Draw a sectional view showing slab, sheathing, and joists.

12-3. A 6-in.-thick concrete slab is to be formed using plywood supported on 2 × 8 (S4S) joists that are spaced 1 ft-6 in. on center. Find the maximum allowable stringer spacing if maximum allowable deflection is 1/240 of the span. Draw a sketch of the formwork.

12-4. For the slab of Problem 12-3, assume that double 2 × 10 stringers spaced 5 ft-0 in. o.c. will support the joists. Determine the maximum shore spacing. Draw a sketch of the formwork. (Neglect shore capacity.)

12-5. In Problem 12-4, if the shores were to be spaced 6 ft-0 in. o.c., select a new stringer size. The stringer may be either doubled 2-in.-thick planks or a single wood beam.

12-6. Compare the capacities of 6 ft-0 in. and 12 ft-0 in. 4 × 4 (S4S) wooden shores. What are the capacities if the shores are full nominal size (4 in. × 4 in.)?

12-7. Design a soffit (beam bottom) for an $11\frac{1}{4}$ in. × 24 in. reinforced concrete beam form. Use a 2 × 12. Determine the maximum shore spacing. Draw a sectional view through the beam.

12-8. Design a plywood beam bottom form shown as an alternative in Figure 12-4. The beam will be 14 in. × 24 in. First design the class I plyform to span the clear span between 2 × 4s (single span, maximum deflection of $\frac{1}{16}$ in.). Then determine maximum shore spacing assuming that the weight of the beam is supported by the 2 × 4s, which back the beam bottom.

12-9. In Problem 12-8, replace the 2 × 4s with 2 × 6s and determine the new maximum shore spacing.

12-10. A 12-ft-high concrete wall is to be placed at a rate of 6 ft/h, and the temperature is expected to be 90°F. What is the *maximum* lateral pressure due to fresh concrete for which the wall forms must be designed? Draw a diagram showing lateral pressure versus distance from the bottom of the wall for the full 12-ft height of wall.

12-11. Design the formwork for an 8-in.-thick reinforced concrete floor slab. Use $\frac{3}{4}$-in. class II plyform and No. 2 grade Douglas fir–larch for the other lumber. Use 2 × 8 joists and double 2 × 8 stringers. Use 4 × 4 shores with an unsupported height of 10 ft. Assume that the slab will be 120 ft × 120 ft in plan. Use guy wire bracing (at 45° to the 10-ft height) that has a tensile capacity of 4400 lb.

12-12. Design forms to support reinforced concrete beams as shown in Figure 12-4a with the following changes. The beam will be $11\frac{1}{4}$ in. × 22 in., joists are 1 ft-8 in. o.c., and stringers are spaced 7 ft-0 in. o.c. Shores will be full nominal size 4×4s with an unsupported height of 10 ft. All lumber will be No. 2 grade Southern pine.

12-13. Design the formwork for a 12-ft-high reinforced concrete wall. Concrete will be placed at a rate not to exceed 5 ft/h and will be internally vibrated. Temperature is expected to be 80°F. Use $\frac{3}{4}$-in. class I plyform for the sheathing and No. 2 grade Douglas fir–larch for the rest of the lumber. Use 2×4s for studs and doubled 2×6s for wales. Ties will have a capacity of 5000 lb (2-in.-wide wedges). For lateral bracing design, assume the wind to be 15 psf and use a guy wire (at 45° to the 10-ft height) that has a tensile capacity of 4400 lb.

12-14. Determine a clamp spacing pattern for a column form that has sheathing of 1-in.-thick plywood. The column height is to be 14 ft-0 in. Use Table 12-8. Assume that $F_b = 1930$ psi, $F_v = 72$ psi, and $E = 1{,}650{,}000$ and that the face grain is parallel to the span between clamps.

CHAPTER **THIRTEEN**

DETAILING REINFORCED CONCRETE STRUCTURES

13-1 Introduction
13-2 Placing or Shop Drawings
13-3 Marking Systems and Bar Marks
13-4 Schedules
13-5 Fabricating Standards
13-6 Bar Lists
13-7 Extras
13-8 Bar Supports and Bar Placement
13-9 Computer Detailing

13-1 INTRODUCTION

The contract documents package for a typical building as developed by an architect/engineer's office commonly includes both drawings and specifications. The drawings typically concern the following areas: site, architectural, structural, mechanical, and electrical. The specifications supplement and amplify the drawings. The contract documents package is the product that results from what may be categorized as the planning and design phase of a project.

The next sequential phase may be categorized as the construction phase. It includes many subcategories for reinforced concrete structures, two of which are detailing and fabricating of the reinforcing steel. As described in the *ACI Detailing Manual-2004* [SP-66(04)] [1], detailing consists of the preparation of placing drawings, reinforcing bar details, and bar lists that are used for the fabrication and placement of the reinforcement in a structure. Fabricating consists of the actual shopwork required for the reinforcing steel, such as cutting, bending, bundling, and tagging. Figure 13-1 shows the reinforcement for a mat foundation prior to the concrete pour.

Most bar fabricators not only supply the reinforcing steel but also prepare the placing drawings and bar lists, fabricate the bars, and deliver to the project site.

FIGURE 13-1 Reactor containment foundation mat. Seabrook Station, New Hampshire. (Courtesy of George Limbrunner)

In some cases, the bar fabricator may also act as the placing subcontractor.

It is general practice in the United States for all reinforced concrete used in building projects to be designed, detailed, and fabricated in accordance with the latest ACI Code. In addition, the Concrete Reinforcing Steel Institute regularly publishes its *Manual of Standard Practice* [2], which contains the latest recommendations of the reinforcing steel industry for standardization of materials and practices.

Techniques have also been developed that make use of electronic computers and other data-processing equipment to facilitate the generation of bar lists and other components of the detailing process. This not only aids in standardization and accuracy of the documents produced but also can be readily incorporated into the stock control system and the shopwork planning of the reinforcing steel fabricator.

13-2 PLACING OR SHOP DRAWINGS

The placing drawing (commonly called a shop drawing) consists of a plan view with sufficient sections to clarify and define bar placement. As such, it is the guide that the ironworkers will use as they place the reinforcing steel on the job. In addition, the placing drawing will contain typical views of beams, girders, joists, columns, and other members as necessary. Frequently, tabulations called "schedules" are used to list similar members, which vary in size, shape, and reinforcement details. A bar list, bending details, or both may or may not be shown on the placing drawing, as some fabricators not only have their own preferred format but prefer the list to be prepared as a separate entity.

The preparation of the placing drawing is based on the complete set of contract documents and generally contains only the information necessary for bar fabrication and placing. Building dimensions are not shown unless they are necessary to locate the steel properly.

The structural drawing (Figure 13-2), which is a part of the structural plans of the contract documents, is the drawing on which the placing drawing is based. Figure 13-3 shows a placing drawing of the same system shown in Figure 13-2. This building is an example of a framing system that uses girders between columns to support beams, which in turn support one-way slabs. The girders support only two-thirds of the beams; the remainder frame directly into the columns.

The structural drawing, as may be observed, shows the floor plan view locating and identifying the structural elements, along with views of typical beams, girders, and slabs and their accompanying schedules. The placing drawing supplements the structural drawing by furnishing all the information necessary for bar fabrication and placement. The precise size, shape, dimensions, and location of each bar are furnished, using a marking system discussed later in this chapter. Information relative to bar supports may also be included. Figure 13-3 illustrates a placing drawing that includes placing data for bar supports, as well as a bending details schedule.

Placing drawings, in addition to controlling the placement of the steel in the forms, serve as the basis for ordering the steel. Therefore, a proper interpretation of the contract documents by the fabricator is absolutely essential. Generally, all placing drawings are submitted to the architect or engineer for checking and review for conformance to the specifications and contract documents before shop fabrication begins.

13-3 MARKING SYSTEMS AND BAR MARKS

With respect to buildings, two identification systems are required. The first involves the identification of the various structural members, and the second involves the identification of the individual bars within the members. The marking system for the structural members may consist of an alphabetical-numerical identification for each beam, girder, and slab, with the columns designated numerically as in Figures 13-2 and 13-3. Also used is a system of alphabetical and numerical coordinates in which the centerlines of columns are numbered consecutively in one direction and lettered consecutively in the other. A coordinate system may be observed in the foundation-engineering drawing of Figure 13-4, where a column may have a coordinate designation such as B2 or C3. The system is generally established on the architectural and structural drawings and adopted by the detailer, unless the detailer requires a more precise identification system.

Footings, as may be observed in Figure 13-4, are generally designated with an F prefix followed by a number, such as F1 and F3, without regard to a coordinate system. Footing piers or pedestals may be identified using the coordinate system, such as B2 or D4, or may be designated with a P prefix followed by a number, such as P1 or P3. Beams, joists, girders, lintels, slabs, and walls are generally given designations that indicate the specific floor in the building, the type of member, and an identifying number. For example, 1G2 indicates a first-floor girder numbered 2, and RB4 indicates a roof beam numbered 4.

In some situations, the floor designation is omitted, as shown in Figures 13-2 and 13-3. The beams and girders are then designated with a prefix B or G, respectively, followed by a number (e.g., B4 or G2). In some cases, suffixes have been added, such as G2A, indicating that there is a difference in the member.

Along with a marking system established for the structural members, a system of identifying and

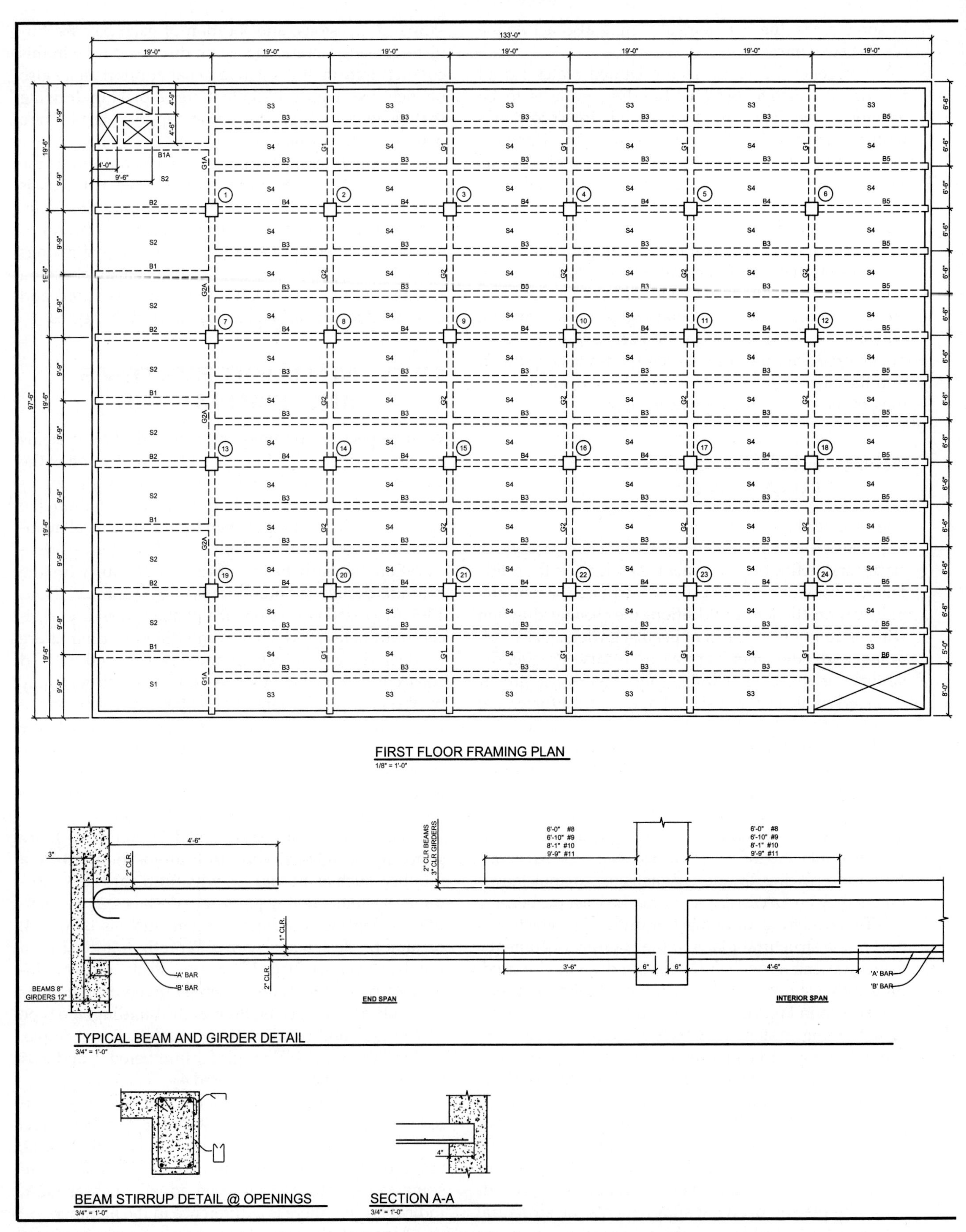

FIGURE 13-2 Structural drawing for beam and girder framing. (Courtesy of the American Concrete Institute)

BEAM AND GIRDER SCHEDULE

MARK	SIZE		BOTTOM		TOP	STIRRUPS	
	B	D	'A' BARS	'B' BARS		NO. - SIZE	SPACING FROM FACE OF SUPPORT
G1	14	32	3 - #9	3 - #8 ★	3 - #9 @ NON CONT. ENDS 3 - #9 } @ COLS 1 TO 6 & 3 - #8 } 19 TO 24 (2 LAYERS, #9 BARS IN TOP LAYER	22 - #3	1@2, 3@6, 3@9, 2@10, 2@12
G1A	SAME AS G1 U.O.N., ADD 3 - #4 E.F.					26 - #4	1@2, 3@6, 3@5, BAL @10
G2	14	32	3 - #8	2 - #6 ★	3 - #11 AT COLS 7 TO 18	16 - #3	1@2, 7@12
G2A	SAME AS G2 U.O.N., ADD 3 - #4 E.F.					22 - #4	1@2, 10@10
B1	12	22	3 - #9	3 - #7 ★	2 - #8 NON CONT. END 4 - #9 CONT. END	16 - #3	1@2, 2@5, 5@6
B2	12	22	3 - #7	2 - #6 ★	2 - #8 NON CONT. END 4 - #9 CONT. END	16 - #3	1@2, 2@5, 5@6
B3	10	22	2 - #7	2 - #6	2 - #7 EACH END	10 - #3	1@2, 4@6
B4	10	22	2 - #6	2 - #6	2 - #10 EACH END EXCEPT @ END CONT. WITH B2, B3, B5 (SEE B2, B3, B5 FOR THIS REINF.)	10 - #3	1@2, 4@6
B5	10	22	2 - #7	2 - #7 ★	2 - #11 CONT. END 2 - #8 NON-CONT. END	10 - #3	1@2, 4@6
B6	10	16	2 - #7	2 - #7	2 - #6 CONT	10 - #3	1@2, 4@6
B7	10	18	2 - #5		2 - #6 CONT	10 - #3	1@2, 4@7
B8	8	8	2 - #5			NONE	
L1	8	12	2 - #5			NONE	

★ UPPER LEVEL

SLAB SCHEDULE

MARK	DEPTH	REINFORCING		
		BOTTOM	TOP	TEMP.
S1	5"	#4@ 6 1/2"	#4@12" NON CONT., END #4@7" CONT., END	#3@11"
S2	5"	#4@ 9 1/2"	#4@7" CONT., END	#3@11"
S3	4"	#3@12"	#3@12" NON CONT., END #4@12" CONT., END	#3@14"
S4	4"	#3@12"	#4@12" CONT., END	#3@14"

NOTES:

1. ALL CONCRETE WORK SHALL CONFORM TO ACI 318-11.
2. REINFORCING STEEL SHALL CONFORM TO ASTM A615, GRADE 60.
3. f'c = 4000psi AT 28-DAYS MAXIMUM AGGREGATE SIZE IS 3/4".
4. PLACE MAIN REINFORCING STEEL SO THAT BOTTOM OF STEEL IS 2" ABOVE FORMS IN BEAMS AND 3/4" IN SLABS.
5. WHERE BEAM OR GIRDER IS PARALLEL TO MAIN SLAB REINFORCING, PLACE #4x5'-0"@12" IN TOP OF SLAB OVER GIRDER AND AT RIGHT ANGLES TO SAID MEMBER.
6. STIRRUPS TO HAVE 2 - #3 SUPPORT BARS AS REQUIRED.
7. LINTELS TO BEAR 8" ON EACH SIDE OF OPENING.
8. PROVIDE 2 - #3 BARS TOP AND BOTTOM AT ALL OPENINGS AND EXTEND 1'-9" BEYOND OPENING.
9. PROVIDE #3 BARS AT RIGHT ANGLES TO MAIN REINFORCING STEEL AT OPENINGS IN SLAB AS SHOWN ON PLAN.
10. LAP ALL TEMPERATURE BARS 16".
11. PLACE E/W TOP GIRDER BARS AT COLUMNS FIRST BELOW N/S BEAM TOP BARS.
12. BAR SUPPORTS TO BE CLASS 3 (NO PROTECTION).
13. PROVIDE DETAILS IN ACCORDANCE WITH ACI-315 DETAILING MANUAL.

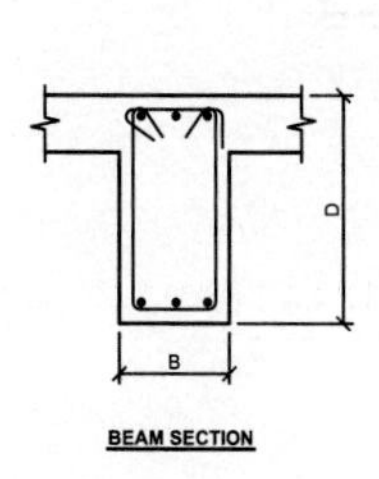

BEAM SECTION

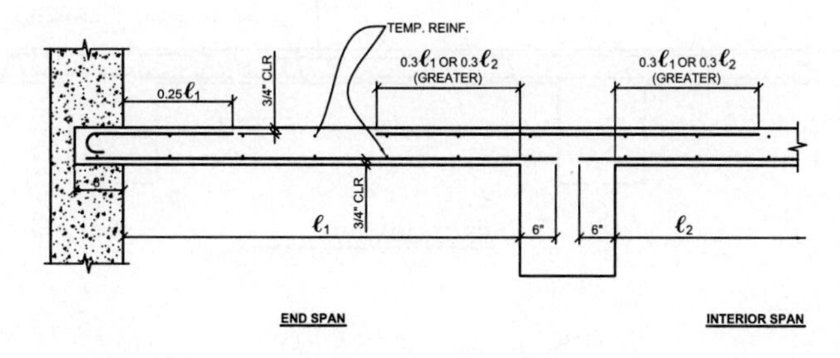

END SPAN **INTERIOR SPAN**

TYPICAL SLAB DETAIL

3/4" = 1'-0"

PROJECT: NEW OFFICE BUILDING

LOCATION:

TITLE: FIRST FLOOR FRAMING PLAN

ENGINEER:

REVISIONS	
BY	DATE

DRAWN BY: CAD

DATE: 01-01-12

JOB. NO.: ##

DWG. NO.: S6

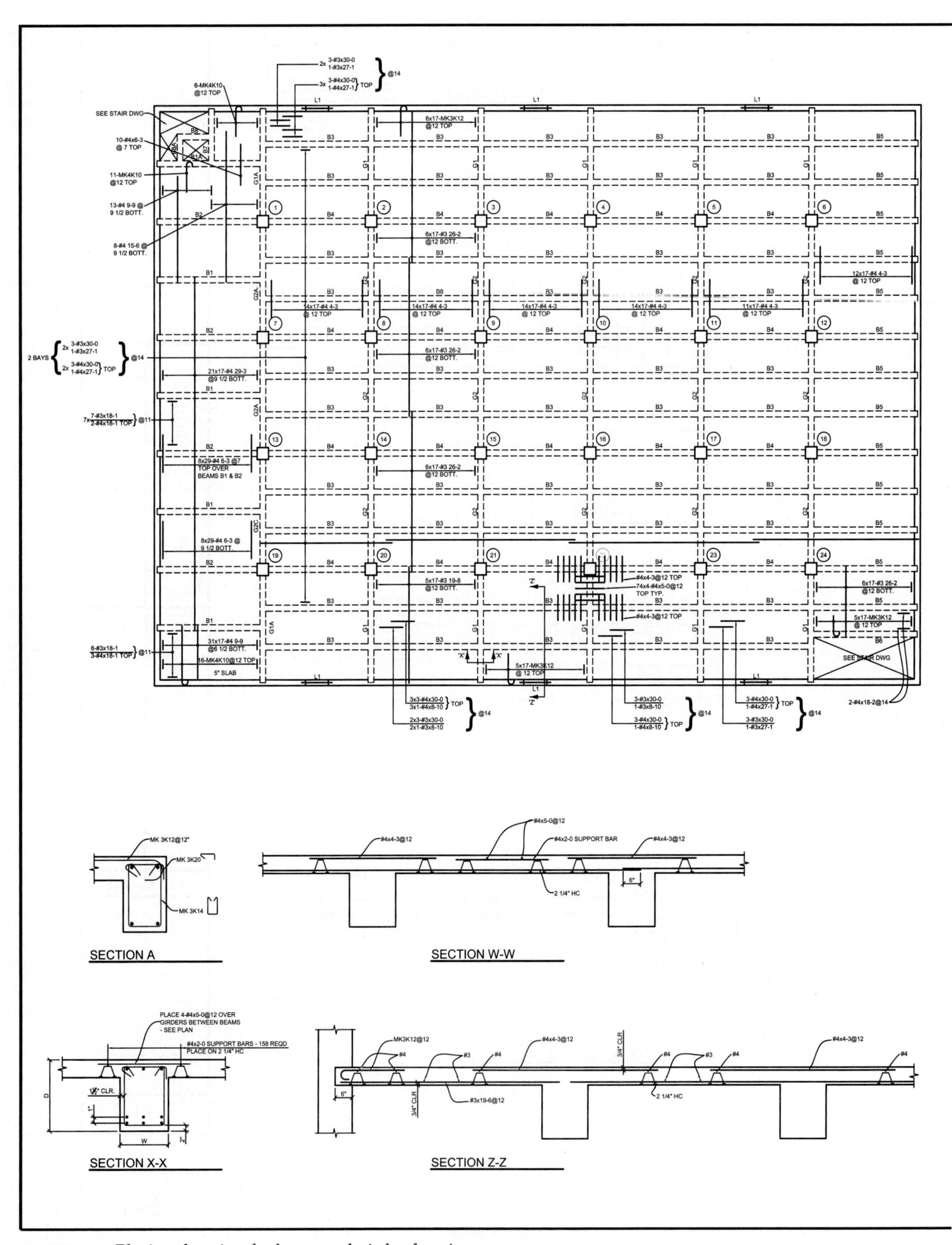

FIGURE 13-3 Placing drawing for beam and girder framing. (Courtesy of the American Concrete Institute)

BENDING DETAILS

MARK	SIZE	LGTH	TYPE	A	B	C	D	E	F	G	J		H	K
9K1	9	6-7	1	1-3	5-4						11 3/4			
3K2	3	6-3	S3	4	2-4	11	2-4			4				
3K3	3	1-7	T9	4	11					4				
10K4	10	10-9	3		8-1	6-3							4-5	4-5
8K6	8	5-11	1	11	5-0						8			
3K8	3	4-5	S3	4	1-7	7	1-7			4				
5K9	5	13-6	1	7	12-11						5			
4K10	4	3-2	1	6	2-8						4			
4K11	4	4-0	3		2-0	2-0							1-5	1-5
3K12	3	2-4	1	3	1-11						3			
3K13	3	3-5	S3	4	1-1	7	1-1			4				
3K14	3	3-9	S3	4	1-3	7	1-3			4				
3K15	3	4-7	S3	4	1-7	9	1-7			4				
4K16	4	4-0	1	6	3-6						4			
4K17	4	4-5	1	6	3-11						4			
4K18	4	2-7	1	6	2-1						4			
8K19	8	22-5	1	11	21-6						8			
3K20	3	1-3	T9	4	7					4				
4K21	4	6-4	S3	4 1/2	2-4	11	2-4			4 1/2				
4K22	4	1-8	T9	4 1/2	11					4 1/2				
3K23	3	1-5	T9	4	9					4				

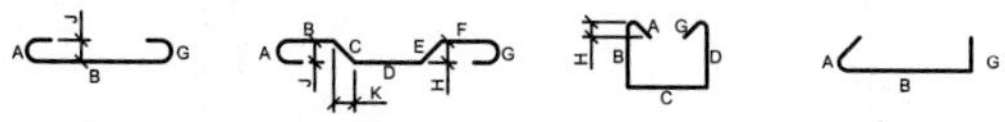

TYPE 1 TYPE 3 TYPE S3 TYPE T9

BAR SUPPORTS (CLASS 3)

490 PCS. 2" BEAM BOLSTERS x 5'-0" (BB)
63 PCS. 1" BEAM BOLSTERS UPPER x 5'-0" (BBU)
402 PCS. 3/4" SLAB BOLSTERS x 10'-0" (SB)
149 - 3 1/4" HC SPACE @ 3'-0" FOR 5" SLAB
1522 - 2 1/4" HC SPACE @ 3'-0" FOR 4" SLAB

NOTES:
ALL SLABS ARE 4" THICK UNLESS OTHERWISE SPECIFIED
MAIN SLAB BARS IN BOTTOM TO BE SUPPORTED IN EACH INTERIOR BAY BY TWO ROWS OF 3/4" SLAB BOLSTERS
MAIN SLAB BARS IN BOTTOM TO BE SUPPORTED IN EACH EXTERIOR BAY & 5" SLABS BY THREE ROWS OF 3/4" SLAB BOLSTERS
FOR BEAM & GIRDER DETAIL SEE ARCH'S DWG. 3 OF 12
SLAB BARS TO BE CUT IN FIELD AT OPENINGS
REINFORCING - ASTM A615 GRADE 60
SEE ACI 315 DETAILING MANUAL FOR TYPICAL BAR BENDS

BEAM AND GIRDER SCHEDULE

NO.	MARK	SIZE		REINFORCING				STIRRUPS SPACE EACH END	STIRRUP SPACERS
				BOTTOM		TOP *			
		W	D	'A' BARS	'B' BARS	NON-CONT. END	CONT. END		
9	G1	14	32	3 - #9x19-0	3 - #8 x 16-3 * EXTEND INTO EXTERIOR SUPP.	3 - MK 9K1	3-#9x17-8 & 3-#8x12-0 COLS. 2 TO 6 & 20 TO 23 2 LAYERS, #9 BARS INTO TOP LAYER	22 MK 3K2 22 MK 3K3 (1@2", 3@6" 3@9", 2@10", 2@12")	2-#3x7-10 LAP 12" WITH TOP BARS
2	G1A	14	32	3 - #9x19-0	3 - #8 x 16-3 * EXTEND INTO EXTERIOR SUPP.	3 - MK 9K1	3-#9x17-8 & 3-#8x12-0 COLS. 1 & 19 2 LAYERS, #9 BARS INTO TOP LAYER	25 MK 4K21 25 MK 4K22 (1@2", 3@6", 3@9", BAL@10")	2-#4x7-10 TOP LAP 12" 4-#4x19-0 2 EF
1	G1B	14	32	3 - #9x19-0	3 - #8 x 16-3 EXTEND INTO EXTERIOR SUPP.	3 - MK 9K1 SEE PLAN	3-#9x17-8 & 3-#8x12-0 COL. 24 #9 BAR IN TOP LAYER FOR DOWELS TO STAIR	25 MK 4K21 25 MK 4K22 (1@2", 3@6", 3@9", BAL@10")	2-#4x7-10 TOP LAP 12" WITH TOP BARS 4-#4x19-0 2 EF
10	G2	14	32	3 - #8x19-0	2 - #6 x 13-6 * PLACE SYM.		3-#11x19-6 COLS. 8-12, 14-18	16 MK 3K2 16 MK 3K3 (1@2", 7@12")	
2	G2A	14	32	3 - #8x19-0	2 - #6 x 13-6 * PLACE SYM.		3-#11x19-6 COLS. 7 & 13	22 MK 4K21 22 MK 4K22 (1@2", 10@10")	2-#4x3-0 TOP 4-#4x19-0 2 EF
5	G2B	14	32	3 - #8x19-0	2 - #6 x 13-6 PLACE SYM.		SEE G1, G1B & G2 FOR TOP STEEL	16 MK 3K2 16 MK 3K3 (1@2", 7@12")	
1	G2C	14	32	3 - #8x19-0	2 - #6 x 13-6 PLACE SYM.		SEE G1A & G2A FOR TOP STEEL	22 MK 4K21 22 MK 4K22 (1@2", 10@10")	2-#4x3-0 TOP 4-#4x19-0 2 EF
5	B1	12	22	3 - #8x18-0	3 - #7 x 15-5 * EXTEND INTO EXTERIOR SUPP.	2 - MK 8K6	4 - #9x15-0	16 MK 3K15 16 MK 3K23 (1@2", 2@5", 5@8")	
4	B2	12	22	3 - #7x18-4	2 - #8 x 15-5 * EXTEND INTO EXTERIOR SUPP.	2 - MK 8K6	4 - #9x17-8	16 MK 3K15 16 MK 3K23 (1@2", 2@5", 5@6")	
10	B3	10	22	3 - #7x18-10	2 - #6 x 13-0 PLACE SYM.	2 - #10x14-4	2 - #10x16-2	10 MK 3K8 10 MK 3K20 (1@2", 4@6")	
1	B3A	10	22	2 - #7x18-10	2 - #6 x 15-10 EXTEND INTO G1B	2 - MK 10K4 @ STAIR	SEE B4	10 MK 3K8 10 MK 3K20 (1@2", 4@6")	
46	B4	10	22	2 - #6x18-10	2 - #6 x 13-0 PLACE SYM.		2 - #10x16-2 SEE B2, B3, B5	10 MK 3K8 10 MK 3K20 (1@2", 4@6")	
13	B4A	10	22	2 - #6x18-10	2 - #6 x 13-0 PLACE SYM.	SEE B3, B4, B5 FOR TOP STEEL		10 MK 3K8 10 MK 3K20 (1@2", 4@6")	
13	B5	10	22	2 - #7x18-4	2 - #7 x 15-5 * EXTEND INTO EXTERIOR SUPP.	2 - MK 8K6	2 - #11x19-6	10 MK 3K8 10 MK 3K20 (1@2", 4@6")	
1	B6	10	16	2 - #6x18-4	2 - #8 x 15-5 * EXTEND INTO EXTERIOR SUPP.	2 - MK 8K19 EXT. 3-8 PAST @ G1B		10 MK 3K8 10 MK 3K20 (1@2", 4@6")	
1	B7	10	18	2 - #5x9-9		2 - MK 5K9 CONT. EXT. 2-8 @ B1		10 MK 3K14 10 MK 3K20 (1@2", 4@7")	
1	B8	8	8	2 - #5 x 7-0					
1	B8A	8	8	2 - #5 x 5-4					
6	L1	8	12	2 - #5 x 5-0					

* NOTE - PLACE GIRDER TOP BARS FIRST UPPER LAYER

REINFORCING STEEL PLACING DRAWING

USE THIS DRAWING IN CONJUNCTION WITH THE ARCHITECTURAL & STRUCTURAL DRAWINGS

ELEVATIONS & DIMENSIONS SHOWN ON THIS DRAWING ARE FOR PURPOSES OF PLACING REINFORCING BARS ONLY AND ARE NOT TO BE USED FOR CONSTRUCTION UNLESS VERIFIED BY THE CONTRACTOR.

REV. NO.	DATE	DESCRIPTION

DWG.: FIRST FLOOR FRAMING PLAN
JOB: NEW OFFICE BUILDING
LOCATION:
ENGINEER:
CONTRACTOR:
DATE: DWG. BY: FILE NO.:
CHK. BY: DWG NO.: P-6

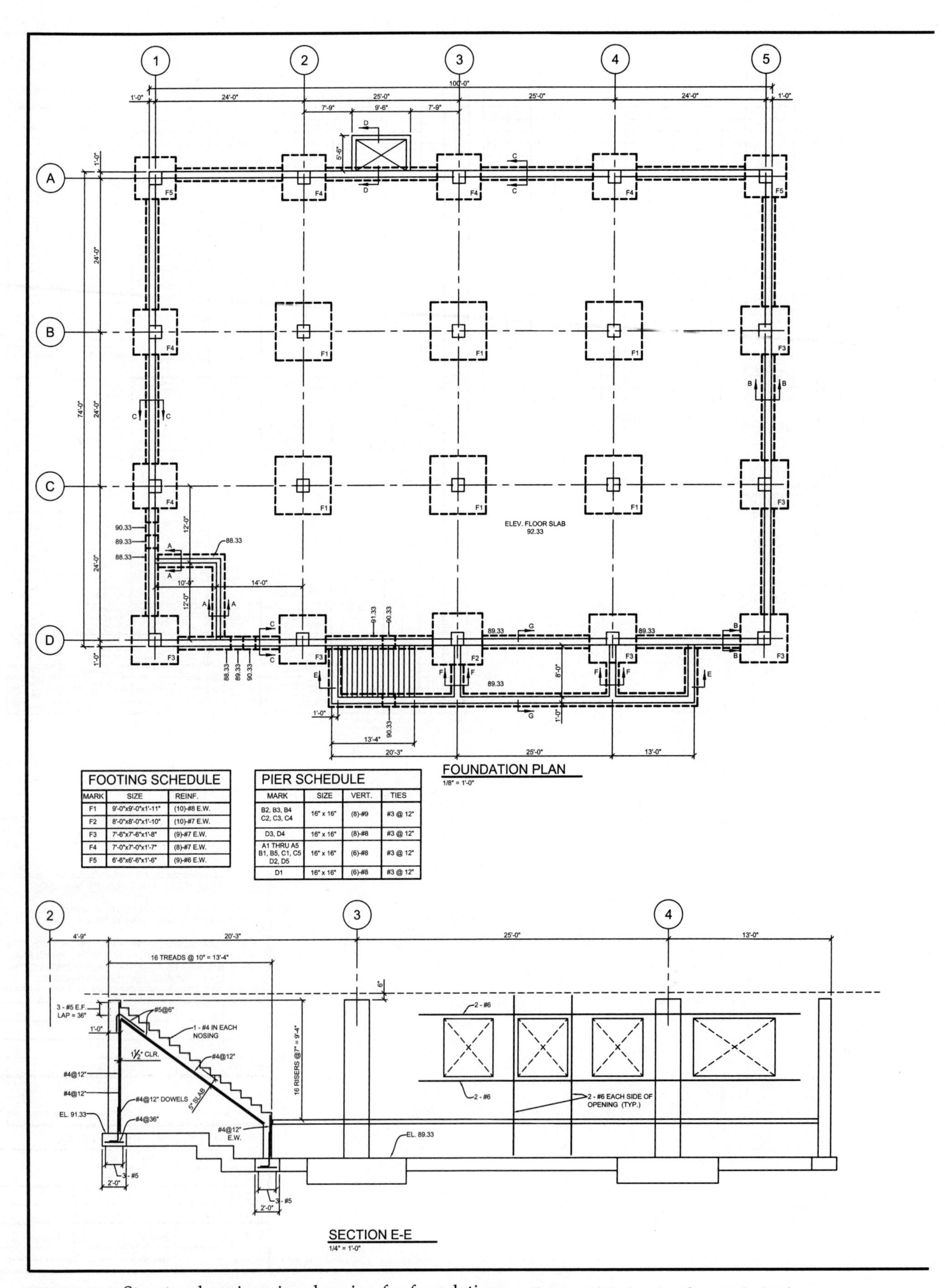

FOOTING SCHEDULE

MARK	SIZE	REINF.
F1	9'-0"x9'-0"x1'-11"	(10)-#8 E.W.
F2	8'-0"x8'-0"x1'-10"	(10)-#7 E.W.
F3	7'-6"x7'-6"x1'-8"	(9)-#7 E.W.
F4	7'-0"x7'-0"x1'-7"	(8)-#7 E.W.
F5	6'-6"x6'-6"x1'-6"	(9)-#6 E.W.

PIER SCHEDULE

MARK	SIZE	VERT.	TIES
B2, B3, B4 C2, C3, C4	16" x 16"	(8)-#9	#3 @ 12"
D3, D4	16" x 16"	(8)-#8	#3 @ 12"
A1 THRU A5 B1, B5, C1, C5 D2, D5	16" x 16"	(6)-#8	#3 @ 12"
D1	16" x 16"	(6)-#8	#3 @ 12"

FIGURE 13-4 Structural engineering drawing for foundations. (Courtesy of the American Concrete Institute)

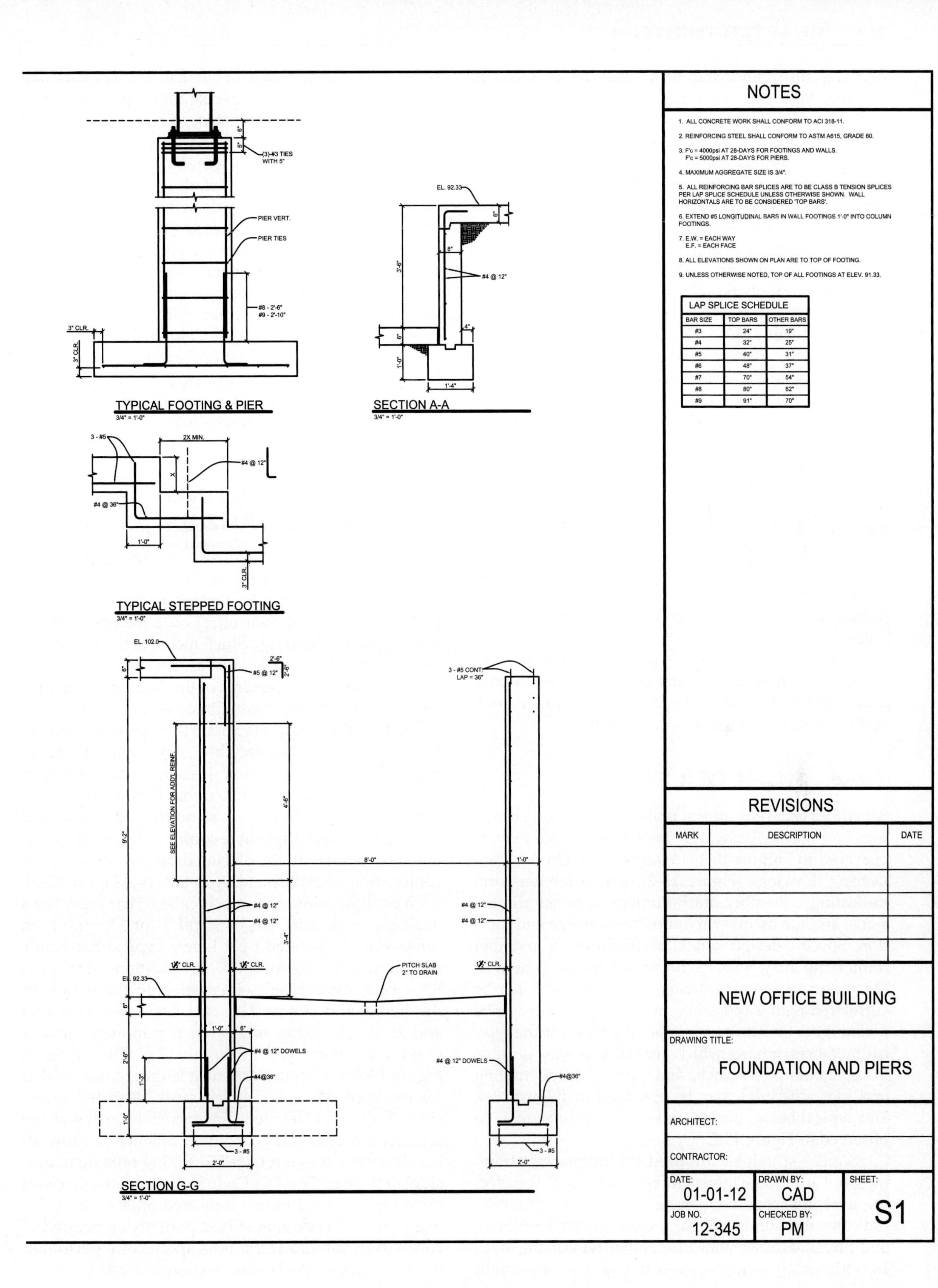

LAP SPLICE SCHEDULE

BAR SIZE	TOP BARS	OTHER BARS
#3	24"	19"
#4	32"	25"
#5	40"	31"
#6	48"	37"
#7	70"	54"
#8	80"	62"
#9	91"	70"

marking the reinforcing bars must be established. In buildings, only bent bars are furnished with a mark number or designation. The straight bar has its own identification by virtue of its size and length. Numerous systems are in use throughout the industry, with the system choice generally a function of the building type, size, and complexity as well as the standards of each fabricator.

One common system is the use throughout the project of an arbitrary letter such as K followed by consecutive numbers, without regard to bar location or shape. This system may be observed in the placing drawing of Figure 13-3. The letter is prefixed with the size of the bar. For example, 8K19 represents a No. 8 bar whose shape and dimensions may be observed in the bending details schedule and whose location, in this case, may be established by the beam and girder schedule as being the top reinforcing at the noncontinuous end of the B6 members.

An alternative system is to use many letters rather than one arbitrary letter. Column bars may be designated with a C and footing bars with an F. For example, a 7F5 would be a No. 7 footing bar, whose shape and dimensions would be established in a bending details schedule and whose location would be observed in a footing schedule, typical footing details, or the foundation plan.

Other acceptable systems are currently being used. Of primary importance in any system that is chosen is that it should be simple, logical, easy to understand, and not lead to ambiguity or confusion.

13-4 SCHEDULES

Schedules generally appear both on engineering drawings and placing drawings. Typical schedules may be observed in Figures 13-2, 13-3, and 13-4. On the engineering drawings (Figure 13-2), it is a tabular form indicating a member mark number, concrete dimensions, and the member reinforcing steel size and location. Specific design details pertinent to the member reinforcing may also be furnished in the schedule. On the engineering drawing, the schedule must be correlated with a typical section and plan view to be meaningful. Schedules are generally used for the typical members, among which are slabs, beams, girders, joists, columns, footings, and piers. Typical footing and pier schedules may be observed in Figure 13-4, and typical beam, girder, and slab schedules may be observed in Figure 13-2.

Similar schedules are used on the placing drawings, but additional information is furnished. The placing drawing schedule is more detailed and generally indicates the number of bars, member mark number, and physical dimensions of member reinforcing steel. In addition, it indicates size and length of straight bars, mark numbers (if bent bars), location of bars, and spacing, along with all specific notes and comments relative to the reinforcing bars. This schedule must also be worked together with typical sections and plan views. The schedules are generally accompanied by a bar list, indicating bending details that may or may not be presented on the placing drawing. There is no standard format for either engineering drawing or placing drawing schedules. They are merely a convenient technique of presenting information for a group of similar items, such as groups of beams, girders, columns, and footings. The schedule format will vary somewhat to conform to the requirements of a particular job.

A schedule for all structural components may not be necessary on a placing drawing. In Figure 13-3 the slab reinforcement is shown directly on the plan, so there is no need for a slab schedule.

13-5 FABRICATING STANDARDS

The fabrication process consists of cutting, bending, bundling, and tagging the reinforcing steel. Our discussion will primarily be limited to the cutting and bending because of their effect on a member's structural capacity. Bending, which includes the making of standard hooks, is generally accomplished in accordance with the requirements of the ACI Code, the provisions of which have been discussed in Chapter 5.

In the fabricating shop, bars to be bent are first cut to length as stipulated and then sent to a special bending department, where they are bent as designated in the bending details schedule or bar lists. The common types of bent bars have been standardized throughout the industry, and applicable configurations are generally incorporated in the placing drawing or bar lists in conjunction with the bending details (see Figure 13-3). Each configuration, sometimes called a bar type, has a designation such as 7, 8, or 9 and S1 or T1, with each dimension designated by a letter. Typical bar bends are shown in Figure 13-5. In addition, standards have been established with respect to the details of the hooks and bends. The ACI Code (Sections 25.3 and 26.6) establishes minimum requirements and is graphically portrayed in Figure 13-6. The tables in Figure 13-6 also shows the extra length of bar needed for the hook (A or G), which must be added to the sum of all other detailed dimensions to arrive at the total length of bar. It is common practice to show all bar dimensions as out to out (meaning outside to outside) of the bar. The ACI Code also stipulates that bars must be bent cold unless indicated otherwise by the engineer. Field bending of bars partially embedded in concrete is not allowed unless specifically permitted by the engineer (ACI Code, Section 26.6.3.1b).

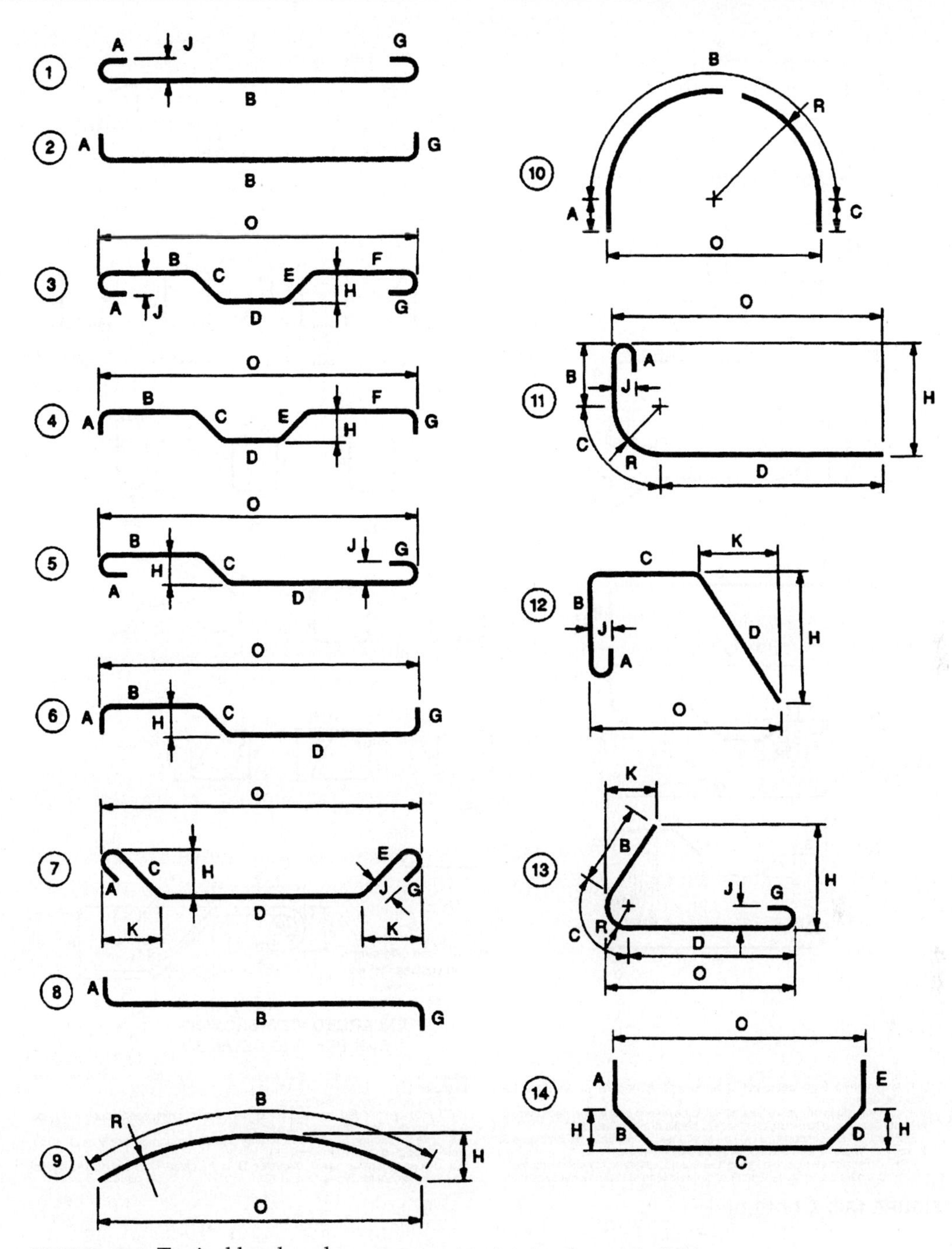

FIGURE 13-5 Typical bar bends. (Courtesy of the American Concrete Institute)

Straight bars are cut to the prescribed length from longer stock-length bars, which are received in the fabricating shop from the mills. Tolerances in fabrication of reinforcing steel are generally standardized and are given in Figure 13-7. For instance, the cutting tolerance for straight bars is the specified length ±1 in., unless special tolerances are called for. Due consideration for these tolerances must be made by both the engineer and contractor in the design and construction phases.

13-6 BAR LISTS

The bar list serves several purposes. It is used for fabrication, including cutting, bending, and shipping, as well as placement and inspection. It represents a bill of materials indicating complete descriptions of the various bar items. The information in a bar list is obtained from the placing drawing or while the placing drawing is being prepared. A typical bar list form is shown in Figure 13-8; its similarity to the bending details schedule of Figure 13-3 is apparent.

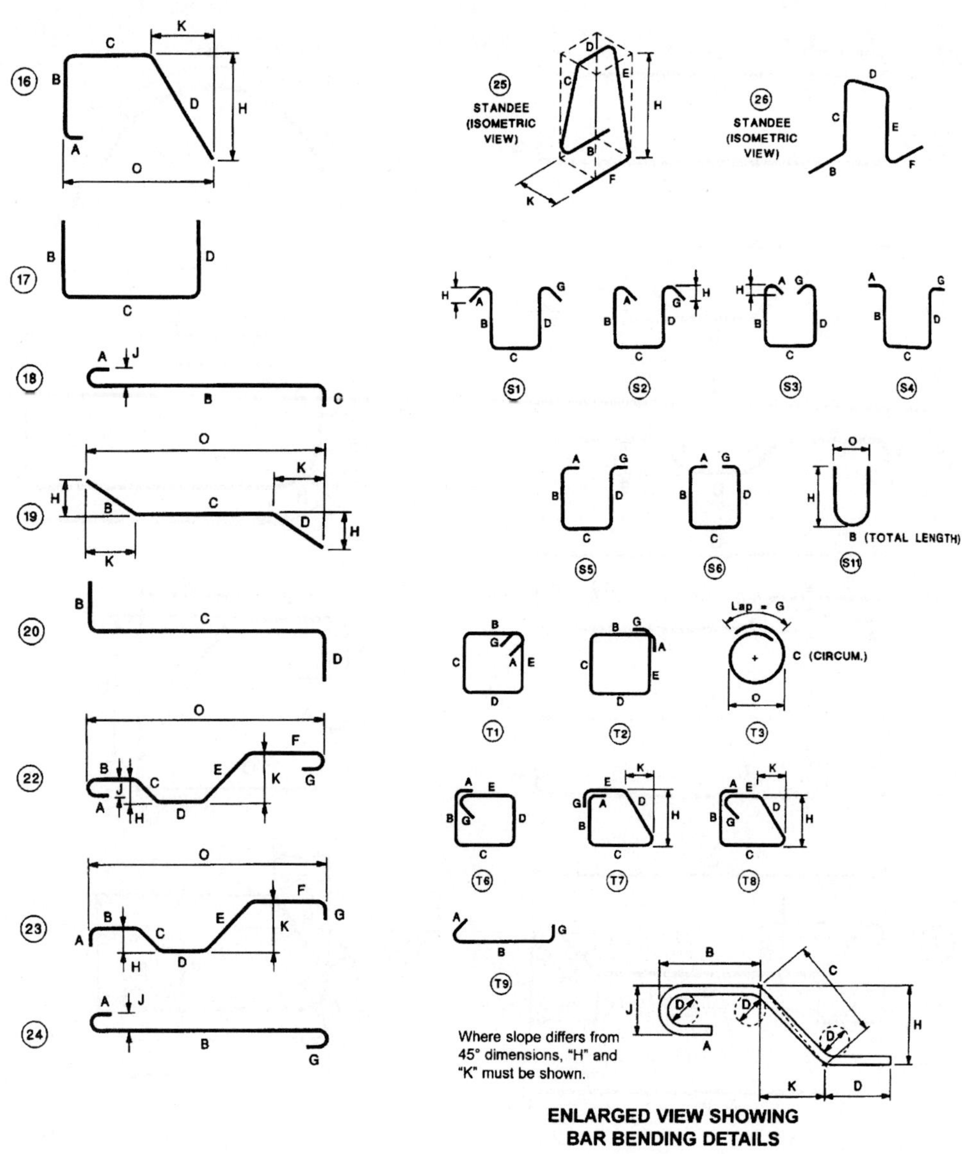

Notes:
1. All dimensions are out-to-out of bar except "A" and "G" on standard 180 and 135 degree hooks.
2. "J" dimensions on 180 degree hooks to be shown only where necessary to restrict hook size, otherwise ACI standard hooks are to be used.
3. Where "J" is not shown, "J" will be kept equal or less than "H" on Types 3, 5, and 22. Where "J" can exceed "H," it should be shown.
4. "H" dimension stirrups to be shown where necessary to fit within concrete.
5. Where bars are to be bent more accurately than standard fabricating tolerances, bending dimensions that require closer fabrication should have limits indicated.
6. Figures in circles show types.
7. For recommended diameter "D" of bends and hooks, see Section 3.7.1; for recommended hook dimensions, see Table 1.
8. Type S1 through S6, S11, T1 through T3, T6 through T9: apply to bar sizes No. 3 through 8 (No. 10 through 25).
9. Unless otherwise noted, diameter "D" is the same for all bends and hooks on a bar (except for Types 11 and 13).

FIGURE 13-5 Continued

The bar list generally includes both straight and bent bars and indicates all bar dimensions and bends as well as the grade of steel and the number of pieces. A bar list of this type may be used in addition to a bending detail schedule on the placing drawing. Some fabricators, however, prefer not to use a bending detail schedule. With this system, the detailers will use sketches until the drawing is complete and then transfer the information to separate bar lists. Two bar lists are prepared, one for straight bars and one for bent bars. Dimensions less than 12 in. are given in inches; over 12 in. they are given in feet and inches. In some cases, when the dimension is less than one inch, a digit is added at the end to indicate the fractional length in increments of one-eight of an inch (see Figure 13-9).

13-7 EXTRAS

Bars are sold on the basis of weight. To a base price are added various extra charges (extras) dependent principally on the amount of effort required to produce the final product. Among these extras are bending extras and special fabrication extras.

Bent bars are generally classified as heavy bending or light bending. Extra charges are made for all shop bending, with the charge a function of the

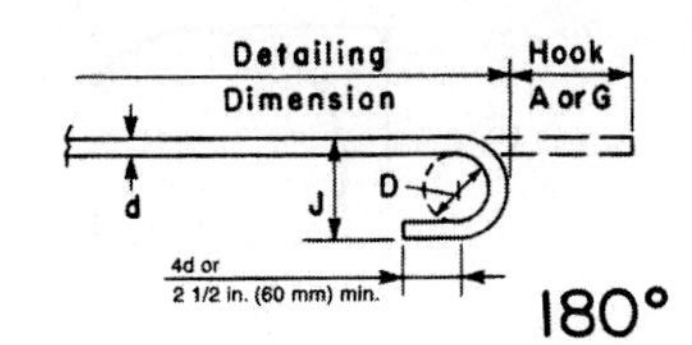

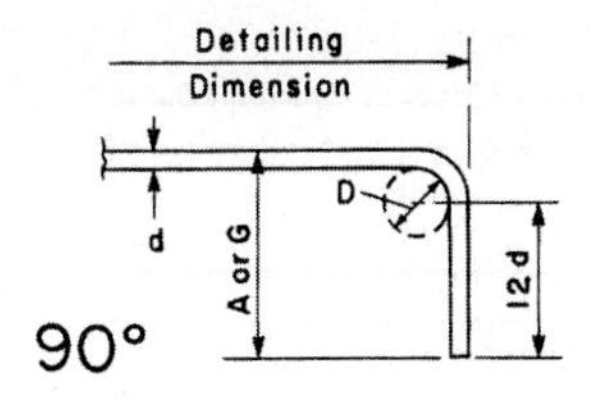

RECOMMENDED END HOOKS
All grades
D = Finished bend diameters

Bar size, No.	D,* in (mm)	180 degree hook		90 degree hook
		A or G, ft-in (mm)	J, ft-in. (mm)	A or G, ft-in. (mm)
3 (10)	2 1/4 (60)	5 (125)	3 (80)	6 (155)
4 (13)	3 (80)	6 (155)	4 (105)	8 (200)
5 (16)	3 3/4 (95)	7 (180)	5 (130)	10 (250)
6 (19)	4 1/2 (115)	8 (205)	6 (155)	1-0 (300)
7 (22)	5 1/4 (135)	10 (250)	7 (175)	1-2 (375)
8 (25)	6 (155)	11 (275)	8 (205)	1-4 (425)
9 (29)	9 1/2 (240)	1-3 (375)	11 3/4 (300)	1-7 (475)
10 (32)	10 3/4 (275)	1-5 (425)	1-1 1/4 (335)	1-10 (550)
11 (36)	12 (305)	1-7 (475)	1-2 3/4 (375)	2-0 (600)
14 (43)	18 1/4 (465)	2-3 (675)	1-9 3/4 (550)	2-7 (775)
18 (57)	24 (610)	3-0 (925)	2-4 1/2 (725)	3-5 (1050)

Table 1(cont.)—Standard hooks: All specific sizes recommended meet minimum requirements of ACI 318

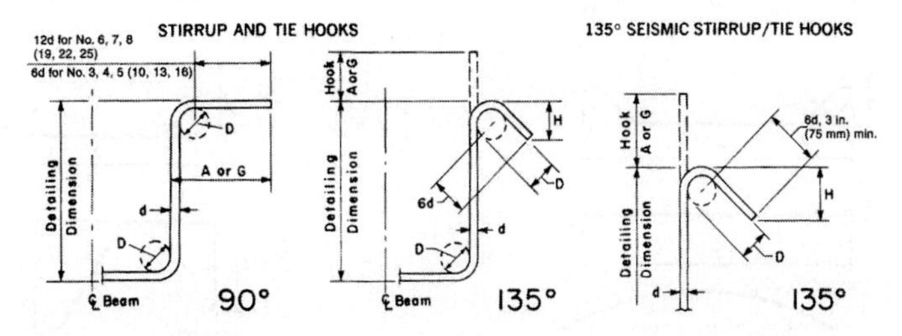

STIRRUP
(TIES SIMILAR)
STIRRUP AND TIE HOOK DIMENSIONS
ALL GRADES

Bar size, No.	D,* in. (mm)	90 degree hook	135 degree hook	
		Hook A or G, ft-in. (mm)	Hook A or G, ft-in. (mm)	H approx., ft-in. (mm)
3 (10)	1 1/2 (40)	4 (105)	4 (105)	2 1/2 (65)
4 (13)	2 (50)	4 1/2 (115)	4 1/2 (115)	3 (80)
5 (16)	2 1/2 (65)	6 (155)	5 1/2 (140)	3 3/4 (95)
6 (19)	4 1/2 (115)	1-0 (305)	8 (205)	4 1/2 (115)
7 (22)	5 1/4 (135)	1-2 (355)	9 (230)	5 1/4 (135)
8 (25)	6 (155)	1-4 (410)	10 1/2 (270)	6 (155)

135 DEGREE SEISMIC STIRRUP/TIE
HOOK DIMENSIONS
ALL GRADES

Bar size, No.	D,* in. (mm)	135 degree hook	
		Hook A or G, ft-in. (mm)	H approx., ft-in. (mm)
3 (10)	1 1/2 (40)	4 1/4 (110)	3 (80)
4 (13)	2 (50)	4 1/2 (115)	3 (80)
5 (16)	2 1/2 (65)	5 1/2 (140)	3 3/4 (95)
6 (19)	4 1/2 (115)	8 (205)	4 1/2 (115)
7 (23)	5 1/4 (135)	9 (230)	5 1/4 (135)
8 (25)	6 (155)	10 1/2 (270)	6 (155)

*Finished bend diameters include "spring back" effect when bars straighten out slightly after being bent and are slightly larger than minimum bend diameters in 3.7.2.

FIGURE 13-6 Standard hook details. (Courtesy of the American Concrete Institute)

classification. Due to the increased amount of handling and number of bends per pound of steel, light bending charges per pound are appreciably more than the charges for heavy bending. According to the ACI, heavy bending is defined as bar sizes No. 4 through No. 18 that are bent at not more than six points, radius bent to one radius, and bending not otherwise defined. Light bending includes all No. 3 bars, all stirrups and column ties, and all bars No. 4 through No. 18 that are bent at more than six points, bent in more than one plane, radius bent with more than one radius in any one bar, or a combination of radius and other bending. Special fabrication includes fabrication of bars specially suited to conditions for a given project. This may include special tolerances and variations from minimum standards as well as unusual bends and spirals.

13-8 BAR SUPPORTS AND BAR PLACEMENT

Bar supports are used to hold the bars firmly at their designated locations before and during the placing of concrete. The ACI Code Section 26.6.2.2 requires that all reinforcing be in place and secured prior to pouring concrete, which precludes the placing or positioning or reinforcing while the concrete is wet (i.e., "wet-setting"). These supports may be of metal, plastic, precast concrete, or other approved materials. The most commonly used bar supports are factory-made wire bar supports, which are available in various sizes and types and which may be provided some corrosion resistance by having exposed parts covered or capped with plastic or being made wholly or in part of galvanized or stainless steel. The Concrete Reinforcing Steel Institute

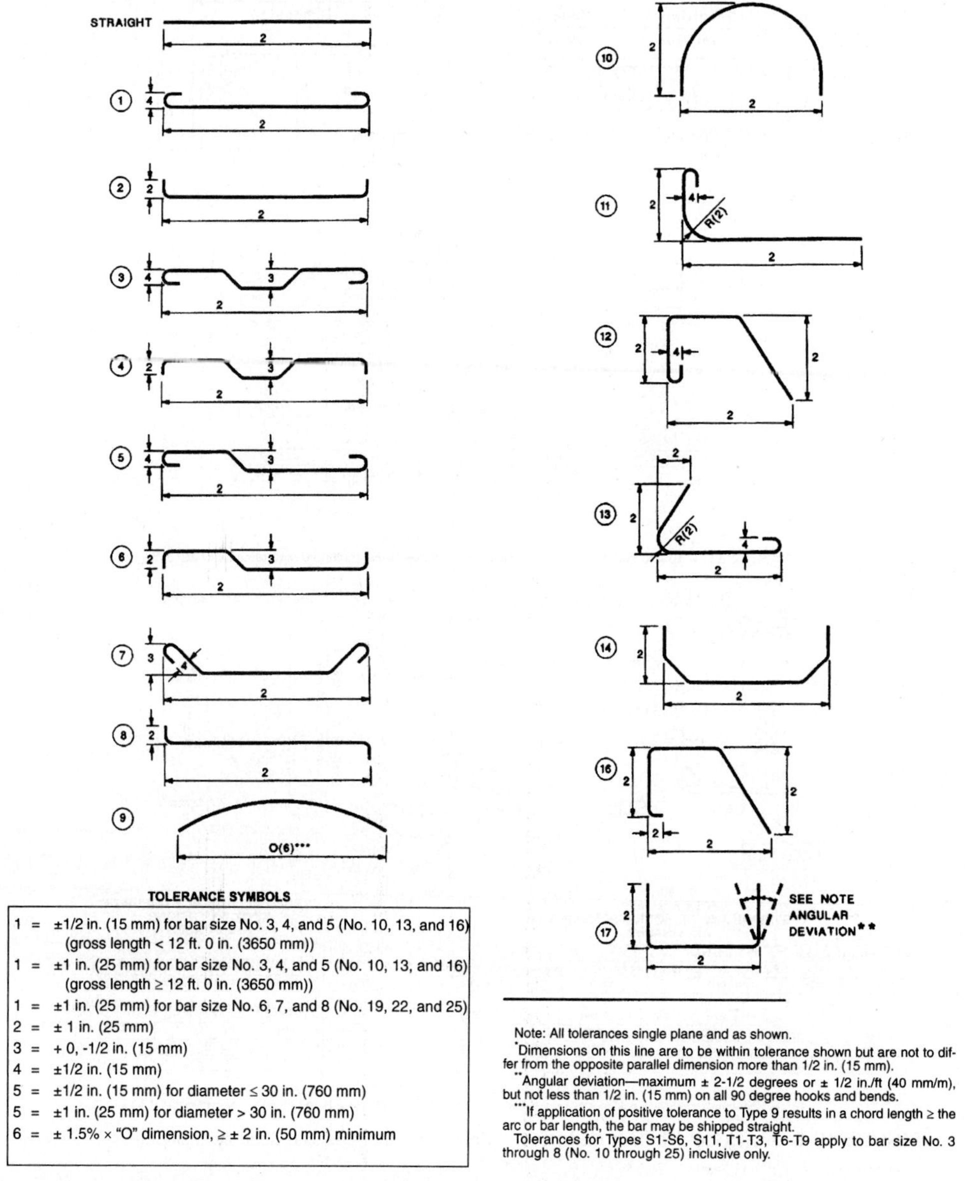

FIGURE 13-7 Standard fabricating tolerances for bar sizes No. 3 through No. 11 (No. 10 through No. 36). (Courtesy of the American Concrete Institute)

(CRSI) publishes information intended to serve as a guide for the selection and utilization of steel wire bar supports used to position reinforcing steel. It is a general practice that unless the engineers' drawings or specifications show otherwise, bar supports will be furnished in accordance with CRSI standards. More detailed information relative to bar supports may be found in the CRSI publications *Placing Reinforcing Bars* [3] and *Manual of Standard Practice* [2].

13-9 COMPUTER DETAILING

The term *computer detailing* is somewhat of a misnomer. Although computers and other electronic data-processing equipment have been used in the bar fabricating industry for many years, it has not been until recently that the actual detailing has been done with computers. The generation of the placing drawings is well within the capability of currently available computer-aided design and drafting (CADD) software, but only the larger steel bar companies, which can afford to dedicate staff to this function, use CADD in this way. The most beneficial aspect of the use of computers in the bar fabrication business concerns the handling and manipulation of data and fabrication management in the shop. Several commercially available programs perform at various levels of sophistication, and bar fabricators sometimes create their own in-house software to perform unique functions geared to their own needs. Each package and each in-house

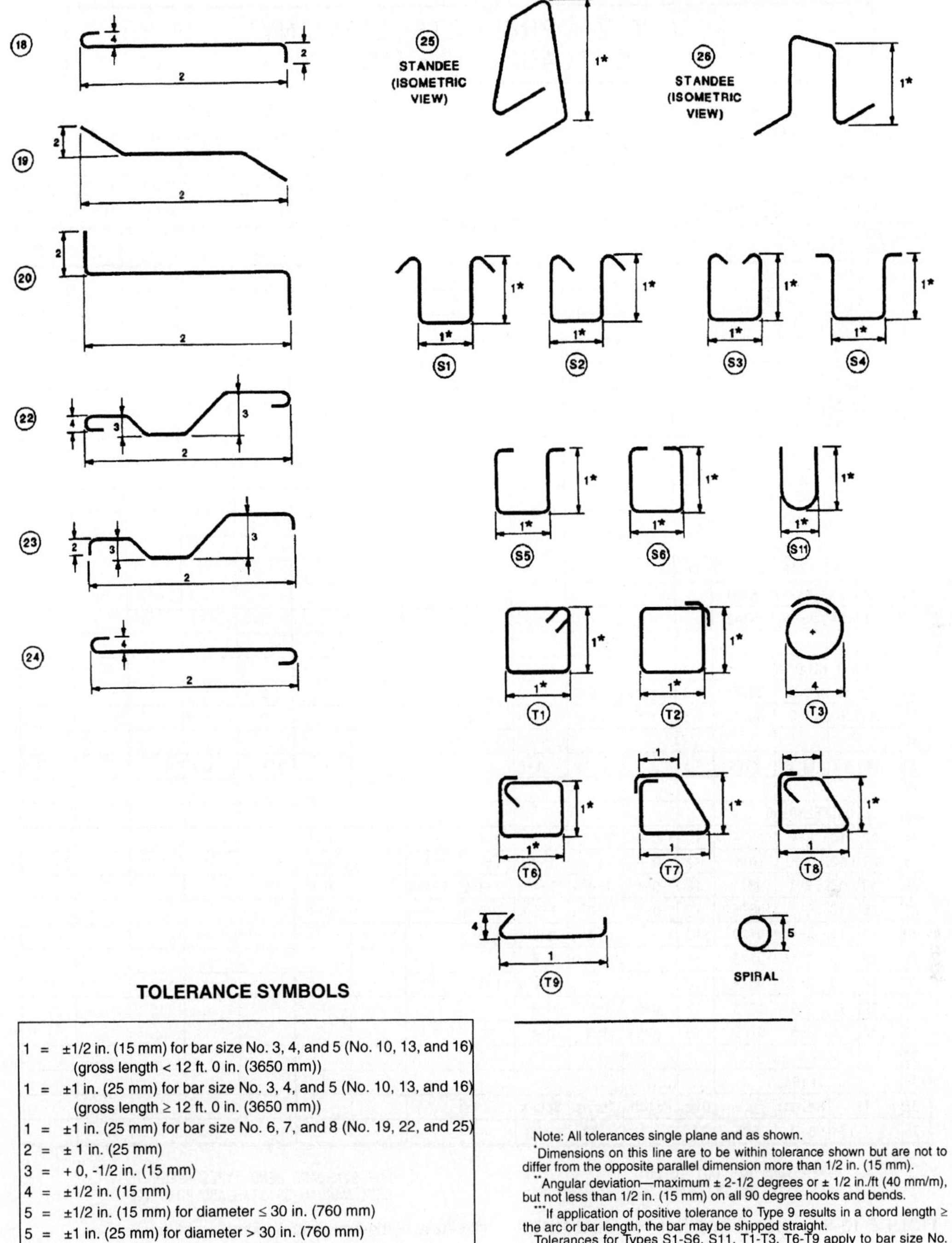

1	=	±1/2 in. (15 mm) for bar size No. 3, 4, and 5 (No. 10, 13, and 16) (gross length < 12 ft. 0 in. (3650 mm))
1	=	±1 in. (25 mm) for bar size No. 3, 4, and 5 (No. 10, 13, and 16) (gross length ≥ 12 ft. 0 in. (3650 mm))
1	=	±1 in. (25 mm) for bar size No. 6, 7, and 8 (No. 19, 22, and 25)
2	=	± 1 in. (25 mm)
3	=	+ 0, -1/2 in. (15 mm)
4	=	±1/2 in. (15 mm)
5	=	±1/2 in. (15 mm) for diameter ≤ 30 in. (760 mm)
5	=	±1 in. (25 mm) for diameter > 30 in. (760 mm)
6	=	± 1.5% × "O" dimension, ≥ ± 2 in. (50 mm) minimum

Note: All tolerances single plane and as shown.

*Dimensions on this line are to be within tolerance shown but are not to differ from the opposite parallel dimension more than 1/2 in. (15 mm).

**Angular deviation—maximum ± 2-1/2 degrees or ± 1/2 in./ft (40 mm/m), but not less than 1/2 in. (15 mm) on all 90 degree hooks and bends.

***If application of positive tolerance to Type 9 results in a chord length ≥ the arc or bar length, the bar may be shipped straight.

Tolerances for Types S1-S6, S11, T1-T3, T6-T9 apply to bar size No. 3 through 8 (No. 10 through 25) inclusive only.

FIGURE 13-7 Continued

program is different, and the person entering the field can expect to receive training on the particular equipment and software that the company is using. The development and availability of increasingly sophisticated software can be expected.

The functions most widely performed by computers in the bar fabricating business involve the generation and printing of notes and labels, schedules (beam, column, slab, pier, footing, joist, and the like), bending details, bar tags, and weight summaries. The software will also optimize the cutting schedule to minimize waste (which is very important in the high-volume, low-margin rebar fabricating business). In addition, information can be generated that will limit bar bundles based on maximum weight and configuration.

The process begins with the preparation of the placing drawing (see Figure 13-3), whether by CADD or manual drafting. For ease of use, it is preferable to furnish the labels on the placing drawing. If the drawing is complex, the detailer will label a reinforcing bar or a group of bars with a single letter or number, which is then keyed to a label list and becomes part of

X Y Z PRODUCTS COMPANY
CHICAGO, ILLINOIS

PROJECT NO. 27693
CUSTOMER: JONES BROS. CONST. CO. DRAWING NO. Figs.18-5a,18-6a
PROJECT: FIELDCREST APT. BLDG. SHEET 1 of 2
LOCATION: SMITHVILLE, N.C. DATE 9/15/97 REVISED 9/19/97
MATERIAL FOR: PARTIAL BASEMENT COLUMNS DRAWN BY H.N.H.

ITEM	NO. PCS.	SZ	LENGTH	BAR MARK	TYPE	A	B	C	D	E	F	G	H	J	K	O	R
1	STRAIGHT																
2	4	57	23-11														
3	4	57	18-11														
4	12	57	8-11														
5																	
6	8	43	23-11														
7	4	43	8-11														
8																	
9	12	29	12-8														
10	6	29	10-8														
11	6	29	4-1														
12																	
13	STRAIGHT (SAW CUT BOTH ENDS)																
14	8	57	23-11	57W1													
15	8	57	11-5	57W2													
16																	
17	HEAVY BENDING																
18	4	36	20-0	36BC5	3		16-0	1-0	3-0				0-3	1-0			
19	64	36	13-6	36C16	3		3-0	1-3	9-3				0-4	1-3			
20																	
21	18	29	12-8	29C4	3		1-11	1-8	9-1				0-4 1/2	1-8			
22																	
23	LIGHT BENDING																
24	22	10	8-4	10T6	T2	0-4	2-1	1-9	2-1	1-9		0-4					
25	22	10	7-8	10T9	T2	0-4	1-11	1-7	1-11	1-7		0-4					
26	50	10	6-3	10BT1	T2	0-4	1-4 3/4	1-4 3/4	1-4 3/4	1-4 3/4		0-4					
27	26	10	6-3	10BT3	T2	0-4	1-0 3/4	1-8 3/4	1-0 3/4	1-8 3/4		0-4					
28	44	10	3-4	10T10	S10		1-3 1/2	0-9	1-3 1/2								
29	22	10	2-10	10T23	T5	0-5	2-1	0-4									
30	52	10	2-10	10BT4	S10		1-0 1/4	0-9 1/4	1-0 1/4								
31	22	10	2-8	10T8	T5	0-5	1-11	0-4									
32	22	10	2-6	10T20	T5	0-5	1-9	0-4									
33																	
34			SPIRALS														
35			Height		Dia.	Pitch	Turns	Spcrs									
36	4	13	8-9	SP5	21"	3"	38	3									

ALL DIMENSIONS ARE OUT TO OUT
ALL BARS ASTM A615M GRADE 420

FOR STANDARD BEND TYPES REFER TO
CRSI MANUAL OF STANDARD PRACTICE

FIGURE 13-8 Typical bar list for buildings. (Courtesy of the American Concrete Institute)

the placing drawing. Data are keyed directly into the computer or recorded on a standard input form, which is then followed by the keyboarding operation. Older systems used punched tape or punched cards.

The data become part of a database, which then provides information for the various other operations. A bill of material, or bar list, is commonly generated on which the bars are segregated by grade and size in descending order. Some programs allow the merging of rebar requirements for several jobs; lower-level (and less expensive) programs may not have this capability. It is not difficult to imagine the complexities involved. The bar shop must keep orderly track of bars that vary in grade (40 or 60), type (uncoated, epoxy-coated, galvanized), shape (see Figure 13-5), length, and size.

Several illustrations provide examples of the types of documents that are commonly used in the bar fabricating business. Figure 13-3 shows a placing drawing and has been previously discussed. A computer program can also generate a rebar shearing schedule and a rebar shearing schedule summary (weight summary). The cutting of the 60-ft-long bars is optimized to minimize the scrap.

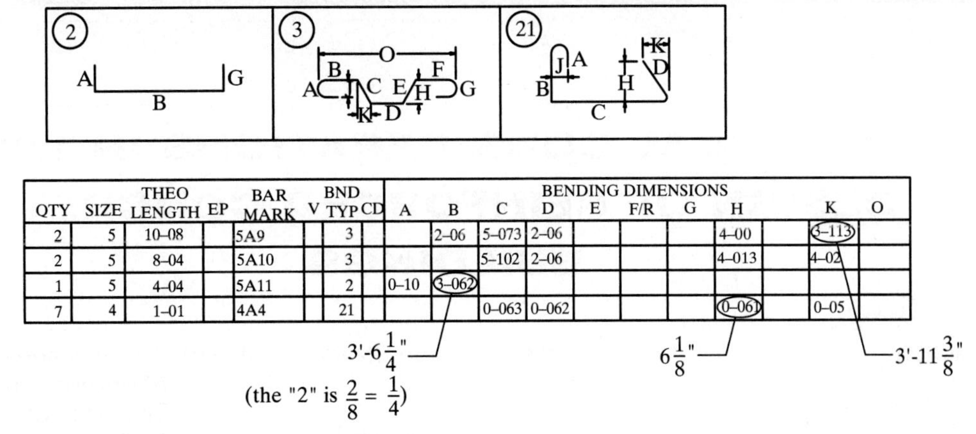

QTY	SIZE	THEO LENGTH	EP	BAR MARK	V	BND TYP	CD	A	B	C	D	E	F/R	G	H		K	O
2	5	10–08		5A9		3			2–06	5–073	2–06				4–00		3–113	
2	5	8–04		5A10		3				5–102	2–06				4–013		4–02	
1	5	4–04		5A11		2		0–10	3–062									
7	4	1–01		4A4		21				0–063	0–062				0–061		0–05	

FIGURE 13-9 Rebar bend details.

Computers have become an indispensable tool in the bar fabricating business. They allow for more productive use of the detailer's talents, freeing the detailer from tedious and repetitive clerical tasks.

References

[1] *ACI Detailing Manual* [SP-66(04)], includes "Details and Detailing of Concrete Reinforcement" (ACI 315-99), "Manual of Structural and Placing Drawings for Reinforced Concrete Structures" (ACI 315R-04), and "Supporting Reference Data." American Concrete Institute, 38800 Country Club Drive, Farmington Hills, MI 48331, 2004.

[2] *Manual of Standard Practice*, 28th ed. Concrete Reinforcing Steel Institute, 933 North Plum Grove Road, Schaumburg, IL 60173, 2009.

[3] *Placing Reinforcing Bars*, 9th ed. Concrete Reinforcing Steel Institute, 933 North Plum Grove Road, Schaumburg, IL 60173, 2011.

CHAPTER **FOURTEEN**

PRACTICAL CONSIDERATIONS IN THE DESIGN OF REINFORCED CONCRETE BUILDINGS

14-1 Introduction

14-2 Rules of Thumb and Practical Considerations for Reinforced Concrete Design

14-3 Approximate Moments and Shears in Continuous Girders

14-4 Strengthening and Rehabilitation of Existing Reinforced Concrete Structures

14-5 Diaphragms, Drag Struts, and Chords

14-6 One-way Slabs Subjected to Concentrated Loads

14-7 Load Testing of Structures

14-8 Closure or Pour Strips in Reinforced Concrete Floors

14-9 Fire Resistance of Concrete Structural Elements

14-10 Analysis and Design of Edge-supported Two-Way Slabs on Stiff Supports

14-11 Cast-In Place Concrete Specifications

14-12 Student Design Projects

14-1 INTRODUCTION

In this chapter, several additional topics of importance to the structural engineer for the proper design and construction of reinforced concrete structures are covered. The topics covered range from the rules of thumb and practical means for designing reinforced concrete structures, to the retrofitting of existing structures and the load testing of structures. The topic of diaphragms, drag struts and chords is introduced, a topic often neglected and obscure to the typical student in a concrete design course, but important enough that the ACI Code devotes an entire chapter to it. The reader is also introduced to concrete slabs subjected to concentrated loads, construction pour gaps that are needed in the casting of large concrete floor areas, fire resistance of concrete elements, edge-supported two-way slabs on stiff supports, and concrete specifications. The chapter concludes with the introduction to two reinforced concrete building design projects. These projects, which mimic realworld building projects, will enable the reader to futher hone their skills in the design of individual structural members presented in previous chapters, and enable them learn how each of these structural elements fit together holistically within the context of an entire concrete building structure.

14-2 RULES OF THUMB AND PRACTICAL CONSIDERATIONS FOR REINFORCED CONCRETE DESIGN

In reinforced concrete construction, simplicity and repetitiveness are the keys to economy [1,2,3] since formwork is a major cost component in concrete construction. In beam and girder framing schemes with one-way slabs, the following should be considered:

1. When laying out reinforced concrete beams and girders in one-way slab systems, the girders should span in the shorter direction along the column lines and the beams should span in the longer direction. See Figure 3-2 for typical one-way slab framing schemes. For greater economy, bay sizes are typically 20–35 ft maximum for beam and girder framing schemes with one-way slabs in residential and office buildings [3,4]. The beam spacing determines the span of the one-way slabs and typically ranges from 10 ft to 16 ft maximum in order to keep the slab depth to a reasonable size.
2. The depths of reinforced concrete beams and one-way slabs should be determined using the minimum depths specified in Table 6-1.

The depths of beams or girders should be varied in 1 in. increments and the depths of slabs should be varied in ½-in. increments. *Even* sizes for the beam width and depth are more commonly used. The width of the beam or girder should be no less than (⅓)h to (½)h, where h is the overall beam depth, but for economy and simplicity of formwork, a beam or girder width equal to or greater than the column width is recommended [29]. For beam and girder design, the beam stem width, b_w, and effective depth, d, should also satisfy the following relationship [29]:

$$b_w d^2 \geq 20 M_u,$$

where b_w and d are in inches and M_u is in ft-kip.

The overall depth of the beam, $h \approx d + 2.5''$ for beams and girders with single layer rebar and, $h \approx d + 3.5''$ for beams and girders with double layer rebar.

3. The concrete columns should be laid out in a regular orthogonal grid, as much as possible, preferably with square bays, where possible. In one-way slab construction, the columns are usually spaced a maximum of approximately 35 ft apart for economy.
4. Since concrete formwork constitutes a major component of the cost of reinforced concrete, it is more cost-effective to use the same depth for the beams and girders throughout the height of the building, where feasible, and vary the beam width and/or reinforcement where necessary. Similarly, it is more economical to keep the size of continuous beams constant from span to span, and vary the reinforcement as required [29].
5. Where possible, column sizes should be kept constant for multiple stories of the building, varying the concrete strength and/or the amount of reinforcing steel where higher capacity is required. Where changes in column sizes need to be made, it is cost-effective to change only one dimension at a time and to limit the size change to increments of 2 in. [5]. Consider using rectangular columns instead of square or circular columns to help in hiding large portions of the columns within walls.
6. Using the charts in Reference [5], the preliminary column size (square column) in inches for normal-height columns (less than or equal to 15 ft story height) can be calculated as a function of the sum of the tributary areas supported by the column as follows:

$$h = 0.0025\, \Sigma A_t + 5$$

where,

h = square column cross-sectional dimension in inches

ΣA_t = sum of the total tributary floor/roof area in ft^2 supported by the column.

7. A concrete building structure is considered "braced" if the stiffness of the shearwalls is at least equal to or greater than *twelve* times the sum of the gross stiffness of the columns in a given direction within a story (ACI Code, Section 6.2.5). Since in reinforced concrete buildings the elevator or stair walls typically function as shearwalls, most of the small to moderate height buildings are usually considered "braced" by the shearwalls.

   For columns in "braced" frames bent in *double curvature* with approximately equal end moments (i.e., $M_1/M_2 = +1$) which would apply to columns above the 2nd floor level, the maximum clear story heights for which slenderness can be neglected in the design of the columns is approximately $12h$; for ground floor columns supported on spread footings which provide minimal restraint to bending at the bottom of the column, the maximum clear story heights is approximately $10h$; the values for maximum clear story height for different column sizes in braced frames are given in Table 14-1 [3]:
8. Visualize the construction sequence for the structure and its elements and how the reinforcement will be placed. This will help ensure constructability.
9. Provide adequate spacing between adjacent reinforcing bars to ensure free passage of concrete between the rebars and to prevent voids from forming.
10. Draw to scale the detail of reinforcing in slabs, beams, girders, and columns in order to catch any potential congestion of the reinforcing steel.

TABLE 14-1 Maximum Clear Story Heights for Columns in Braced Frames for which Slenderness Can Be Neglected

	Ground floor columns $\left(\frac{\ell_u}{h} \approx 10\right)$	Columns on the 2nd floor and above $\left(\frac{\ell_u}{h} \approx 12\right)$
Column size, h	**Maximum clear story height, ℓ_u (ft)**	**Maximum clear story height, ℓ_u (ft)**
12	10	12
14	11.67	14
16	13.33	16
18	15	18
20	16.67	20
22	18.33	22
24	20	24

Source: David B. Fanella and S. K. Ghosh. Simplified Design—Reinforced Concrete Buildings of Moderate Size and Height. Skokie, IL: Portland Cement Association, 1993.

11. Preferable to use 90° hook instead of 180° hooks in concrete slabs because of the difficulty in placing bars with 180° hook around the edge reinforcement in the slab in the perpendicular direction. When using 90° hooks in reinforced concrete slabs, using reinforcement with a small-enough diameter that results in a hook length no greater than 80% of the slab thickness will ensure that the hook will fit within the thickness of the slab. If the length of the hook is greater than the slab thickness, the 90° hook would need to be oriented at an angle in order to fit within the slab thickness.
12. Though the ACI Code allows the maximum reinforcement in columns of up to 8% of the gross area (which amounts to 4% when lap splices are used), use between 1% and 2% vertical reinforcing steel in reinforced concrete columns for economy and to avoid congestion of reinforcement, especially at beam–column connections [3, 4]. Instead of using more than 2% reinforcement in a column, consider increasing the concrete strength and/or the cross-sectional size of the column.
13. Instead of using bundled vertical bars in columns, increase the size of the column to obviate the need for bundled bars.
14. Coordinate placement of mechanical ducts and piping to avoid interference with location of reinforcement in slabs, beams, girders, and columns.
15. For typical concrete buildings, a 7-day floor-to-floor cycle (i.e., 7 days between casting the concrete floors) will ensure adequate strength development, but the use of admixtures can reduce this time to as low as three days.
16. As previously mentioned, the most economical formwork is achieved when the beam or girder has the same width as the column [3]. Care should be taken when the beam or girder width is much larger than the column width because only a portion of the girder top reinforcement will be developed within the column width. If in the modeling and analysis of such a girder–column connection, a full moment connection was assumed, this assumption would be incorrect because the connection will have less than a full moment capacity since all the girder top reinforcement cannot be fully developed within the column width.

Example 14-1

Preliminary Sizing of Columns

Determine the preliminary size of the first floor typical interior column for a 10-story building with 30-ft square bay dimensions, considering axial loads only. The concrete and steel strengths are 4000 psi and 60,000 psi, respectively. Assume a dead load of 100 psf and a live load of 50 psf.

Solution:

The factored floor load, $w_u = 1.2\,(100\text{ psf}) + 1.6\,(50\text{ psf}) = 200\text{ psf}$

The tributary floor area of a typical interior column $= (30\text{-ft})(30\text{ ft}) = 900\text{ ft}^2$

Sum of tributary area supported by the first floor interior column $= (10\text{ floors})(900\text{ ft}^2) = 9000\text{ ft}^2$

The preliminary square column cross-sectional size, $h = (0.0025)(9000) + 5 = 27.5\text{ in.}$

Therefore, use $28'' \times 28''$ column at the ground floor level.

Total factored axial load on the typical interior ground floor column, $P_u = (9000\text{ ft}^2)(200\text{ psf}) = 1800\text{ kip}$

It should be noted that the self-weight of the column has been neglected in this example, and the tributary area method for calculating the column load above neglects to account for the self-weight of the beam and girder stems below the soffit of the slab. The self-weight of the beam and girder stems will add, approximately, an additional 5% to 10% to the factored load calculated above. Furthermore, the first interior columns will also support a higher load than the typical interior columns. An additional 7.5% should be added to the factored load for the first interior columns to account for the higher shears in the end-span beams and girders at the first interior columns (see the ACI shear coefficients for beams in Figure 6-1).

In the example just cited, the load-reducing effect of live load reduction permitted by the building codes has not been taken into account. Taking advantage of the live load reduction will result in a smaller column size.

14-3 APPROXIMATE MOMENTS AND SHEARS IN CONTINUOUS GIRDERS

Girders usually frame into columns and support the concentrated reactions from the infill beams, in addition to any other uniformly distributed loads from the self-weight of the girder stem and cladding loads. Consequently, the ACI moment and shear coefficients for continuous beams and one-way slabs, introduced in Chapter 6, are not applicable to continuous girders. The approximate moments and shears in continuous girders can be obtained from Tables 14-2A and 14-2B, respectively [6].

The moment for which the columns are designed, in addition to the axial loads, can be obtained by calculating the unbalanced moment in the column due to the factored vertical loads on the girders. The factored live loads on the girder should

TABLE 14-2A Approximate Factored Moment for Continuous Girders [6]

Positive Moment

At end spans:

$$M_u^+ = \frac{w_u \ell_n^2}{14} + \frac{\ell_n}{6} \sum P_u$$

At interior spans:

$$M_u^+ = \frac{w_u \ell_n^2}{16} + \frac{\ell_n}{7} \sum P_u$$

Negative Moment at Supports

At interior face of external column or perpendicular structural wall:

$$M_u^- = \frac{w_u \ell_n^2}{16} + \frac{\ell_n}{10} \sum P_u$$

At exterior face of first internal column or perpendicular structural wall, only two spans:

$$M_u^- = \frac{w_u \ell_n^2}{9} + \frac{\ell_n}{6} \sum P_u$$

At faces of internal columns or perpendicular structural walls, more than two spans:

$$M_u^- = \frac{w_u \ell_n^2}{10} + \frac{\ell_n}{6.5} \sum P_u$$

At faces of structural walls parallel to the plane of the frame:

$$M_u^- = \frac{w_u \ell_n^2}{12} + \frac{\ell_n}{7} \sum P_u$$

At support of girder cantilevers:

$$M_u^- = \frac{3 w_u \ell_n^2}{4} + \ell_n \sum P_u$$

Source: American Concrete Institute (2002). Essential Requirements for Reinforced Concrete Buildings (International Publication Series IPS-1). Reproduced with permission from the American Concrete Institute.

TABLE 14-2B Approximate Factored Shear for Continuous Girders [6]

At exterior face of first interior column:

$$V_u = 1.15 \frac{w_u \ell_n}{2} + 0.80 \Sigma P_u$$

At faces of all other columns:

$$V_u = \frac{w_u \ell_n}{2} + 0.75 \Sigma P_u$$

At supports of girder cantilevers:

$$V_u = w_u \ell_n + \Sigma P_u$$

where,

w_u = factored uniformly distributed load on girder (kip/ft)

ΣP_u = sum of all the factored concentrated loads acting on the girder span (kip)

ℓ_n = clear span of the girder (ft)

Source: American Concrete Institute (2002). Essential Requirements for Reinforced Concrete Buildings (International Publication Series IPS-1). Reproduced with permission from the American Concrete Institute.

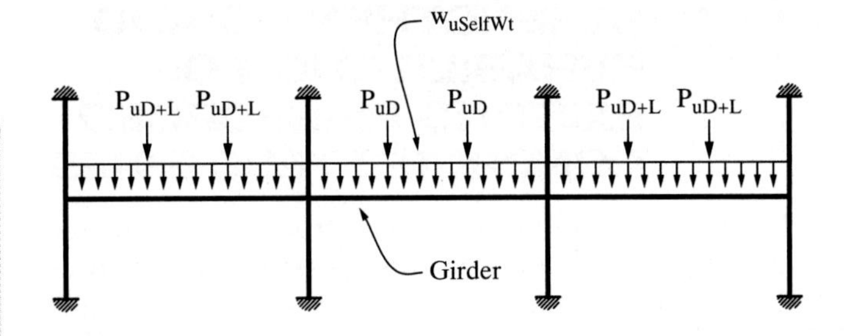

a) Case 1

$P_{uD+L} = 1.2P_D + 1.6P_L$

$P_{uD} = 1.2P_D$

$w_{uSelfWt}$ = 1.2 x (weight of girder stem and any wall load directly supported by the girder)

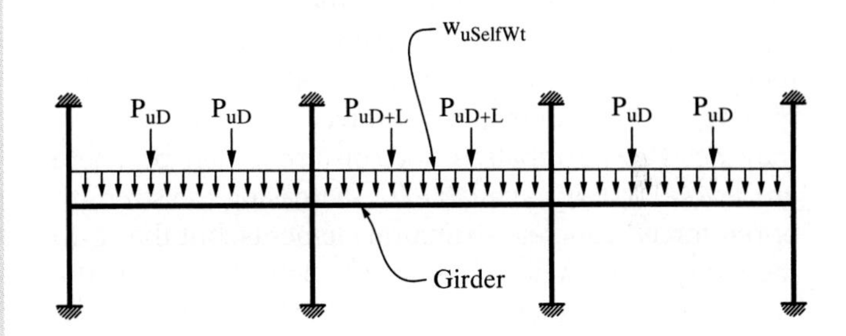

b) Case 2

FIGURE 14-1 Pattern Live Load analysis for girders.

be alternated to create the maximum value for the unbalanced moments in the column at the column–girder connection. The two cases to be considered are as follows: in the first case, the *odd* spans of the girder should be loaded with the factored live loads (i.e., the beam factored live load reactions), in addition to the factored uniform dead load and the beam's factored dead load reaction on all girder spans; in the second case, the *even* spans of the girder should be loaded with the factored live loads (i.e., the beam's factored live load reactions), in addition to the factored uniform dead load and the beam's factored dead load reaction on all girder spans (see Figure 14-1). The unbalanced moment, ΔM_u, in the column at a girder–column connection is the maximum difference between the factored moments in the girder on both sides of the column. This unbalanced moment is distributed to the column above and below the floor level under consideration in proportion to the stiffness ratios of the columns above and below the floor (i.e., $I_{c,\text{above}}/H_{\text{above}}$: $I_{c,\text{below}}/H_{\text{below}}$), where I_c is the moment of inertia of the column for bending in the plane of the girder and H is the floor-to-floor height of the column. For roof girders, the top floor columns resist all the unbalanced moment, ΔM_u.

14-4 STRENGTHENING AND REHABILITATION OF EXISTING REINFORCED CONCRETE STRUCTURES

The use of a reinforced concrete structure may change during the life of the structure necessitating modifications to the structure in cases where the change results in increased loads. Converting a building from office use to a library or for storage, for example, will result in a significant increase in the applied loads on the structure. A floor opening may need to be cut in the slabs of an existing building to provide natural lighting or a new stair, which may reduce the capacity of the existing slab, beams, or girders. Also, the structural members in a building may be degraded due to exposure to a corrosive environment or damaged due to a seismic event, or they may need to be upgraded to meet new code requirements. Consequently, the existing slabs, beams, girders, and columns in a reinforced concrete building may need to be repaired and/or strengthened. There are several options available for increasing the strength of reinforced concrete structural elements, but the common goal is to achieve composite action between the existing concrete member and the added reinforcement:

1. For reinforced concrete slabs, Carbon Fiber Reinforced Polymer (CFRP) strips could be applied to the top or bottom surface of the slab to add flexural capacity to an existing slab [7]. Steel plates with drilled epoxy anchors installed at the bottom and top surfaces of a reinforced concrete slab can also be used to increase the moment capacity of the slab [8]. For slabs with large openings, steel beam framing at the underside of the slab at the edges of the floor opening could be used to increase the load-carrying capacity of the slab. In addition to reinforcing and strengthening the edges of the slab opening, the adjacent slabs should be evaluated for redistribution of moments, and the adjacent reinforced concrete beams should also be evaluated for torsion and a reduced moment capacity due to the floor opening.

   The punching shear capacity of two-way slabs can be increased by providing a steel angle collar around the column at the underside of the flat plate or flat slab. The angles are through-bolted to the column and epoxy anchored into the existing slab. The horizontal leg of the angle helps to increase the effective perimeter of the critical section for punching shear, b_o, thus leading to a reduction in the punching shear stress. The punching shear capacity of a flat plate can also be increased by casting a concrete collar or jacket around the column at the underside of the existing slab. The jacket or collar is attached to the existing columns with through-bolts or dowels, and attached to the slab with epoxy anchored dowels.
2. For reinforced concrete beams and girders, the following options are possible:
   a. Add a steel plate bonded with epoxy to the underside of an existing concrete beam or girder.
   b. Provide CFRP reinforcement on the sides, top, or bottom of existing concrete beams or girders to resist bending and shear. For shear, the CFRP strips are placed on the two vertical sides of the beam. CFRPs are preferred to bonded steel plates because no corrosion protection is required and they possess good fatigue properties. CFRPs are also lighter than steel plates and therefore more easily handled on site.
   c. Externally post-tension the concrete beam or girder using parabolic cables to relieve the beam/girder of some of the existing dead loads.
   d. Enlarge the size of the beam or girder by casting additional concrete around the existing member. The surfaces of the existing member would need to be roughened and dowels drilled and epoxy anchored into the existing concrete member in order to bond the new concrete to the existing concrete.
3. For reinforced concrete columns, the methods of strengthening or repair include the use of CFRP wraps around existing columns. Fiber Reinforced Polymer (FRP) laminates are also used where the FRP acts like a jacket around the existing column and the annular gap between the laminate and the face of the existing column is filled with resin or grout [9]. Concrete columns can also be strengthened using steel jackets around the existing column. The annular space between the steel jacket and the existing column is filled with grout or epoxy resin. These wrap-around help to confine the concrete, thus increasing the compressive strength of the column.

   Reinforced concrete columns can also be strengthened or repaired by concrete jacketing or encasement of the existing column. To assure composite action between an existing reinforced concrete member and a new concrete jacket, it is recommended that the surface of the existing concrete be roughened, with a bonding agent applied.
4. Other considerations include the strength of the retrofitted member during a fire. CFRP reinforcement are typically exposed and will lose their strength in a fire. Before deciding which methodology to use, the engineer should determine the cause of the structural deficiency and choose a repair method accordingly that will correct the problems and minimize the impact on building operations. Other factors to consider include the available amount of space for the retrofit, the cost of the retrofit, aesthetics, fire protection, and the local contractors' experience with the proposed retrofit methods.

Examples of structural details for the repair and strengthening of damaged or degraded reinforced concrete columns, beams, and girders [10, 11] are shown in Figures 14-2 and 14-3.

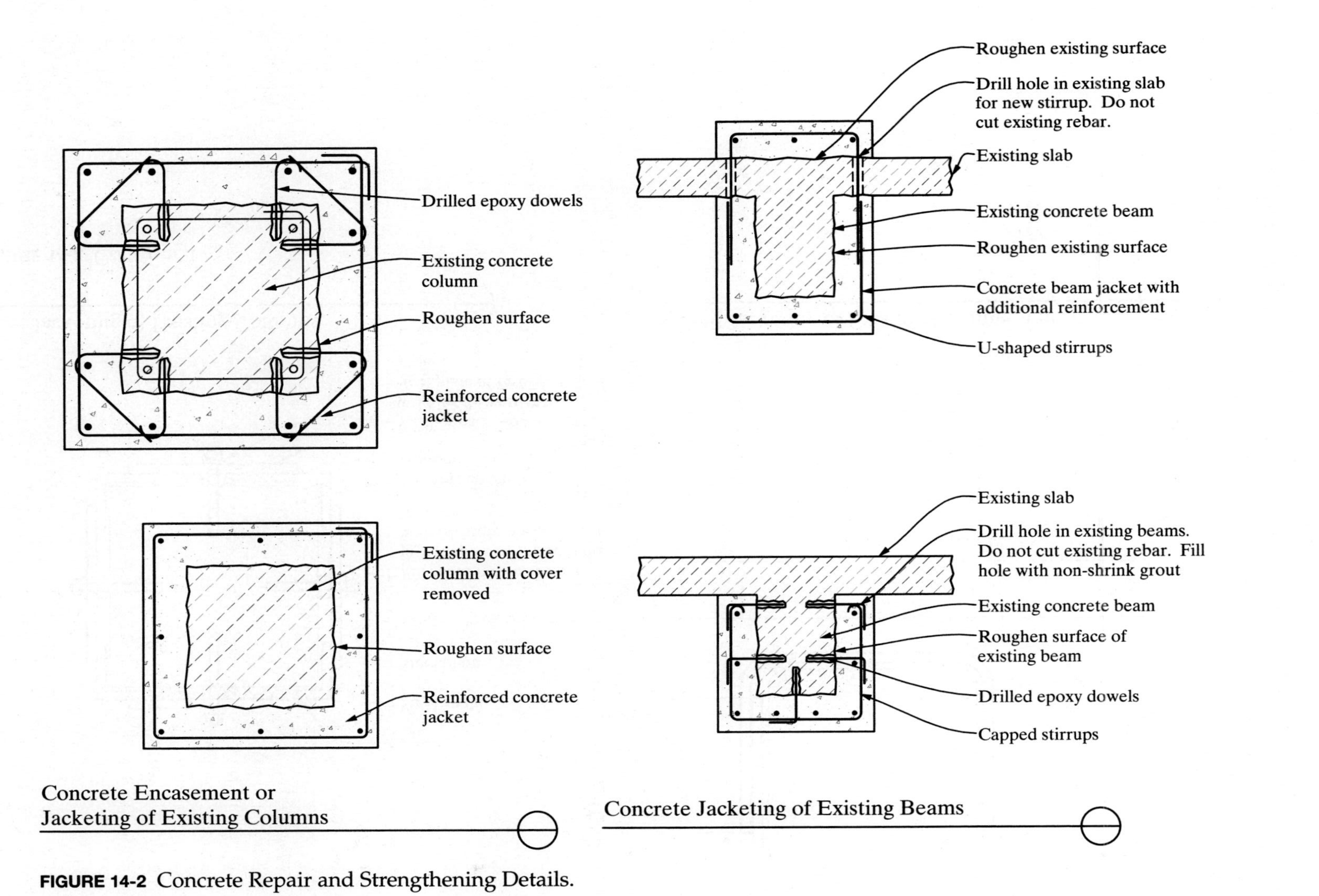

FIGURE 14-2 Concrete Repair and Strengthening Details.

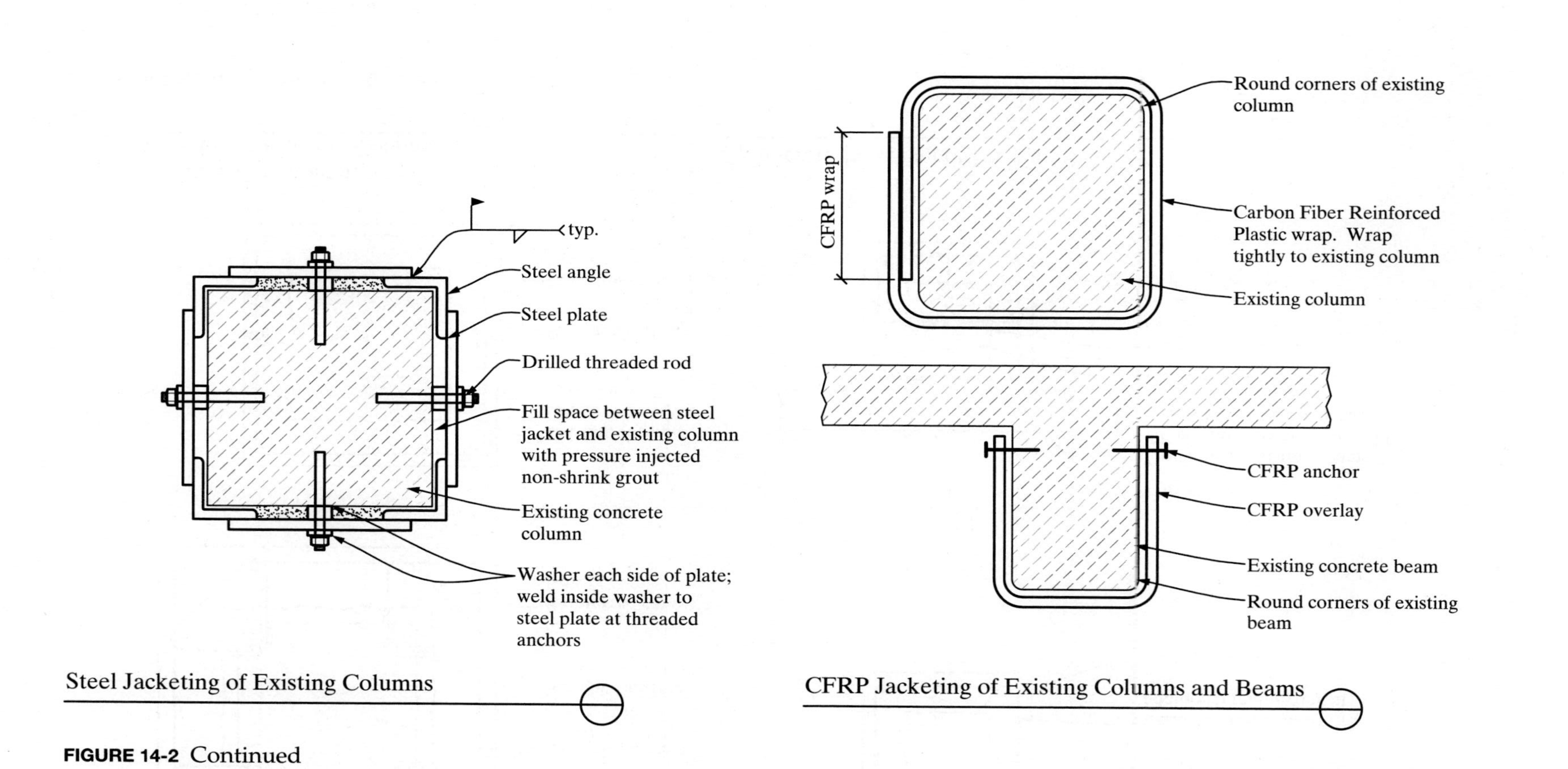

FIGURE 14-2 Continued

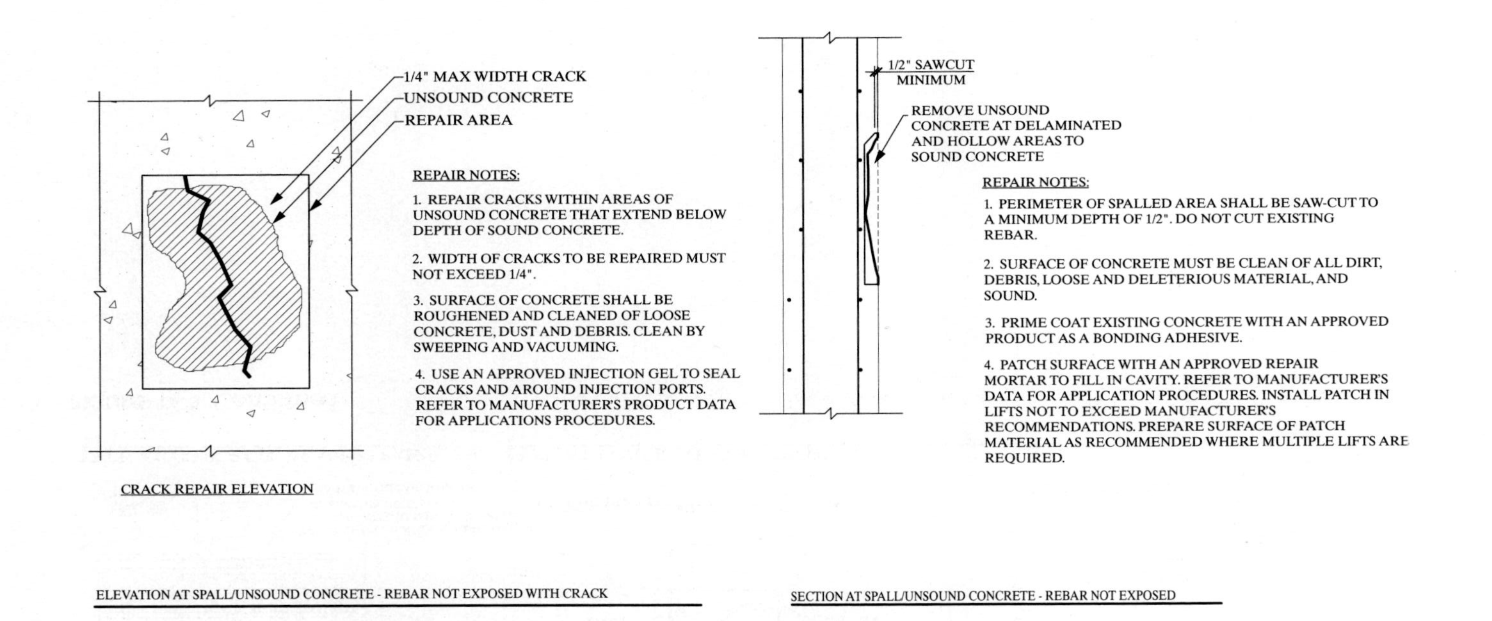
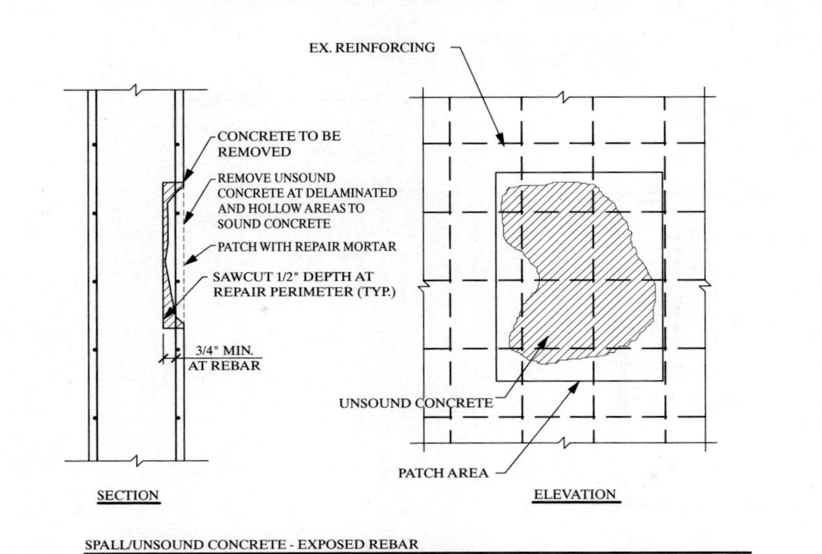

FIGURE 14-3 Concrete Repair Details.

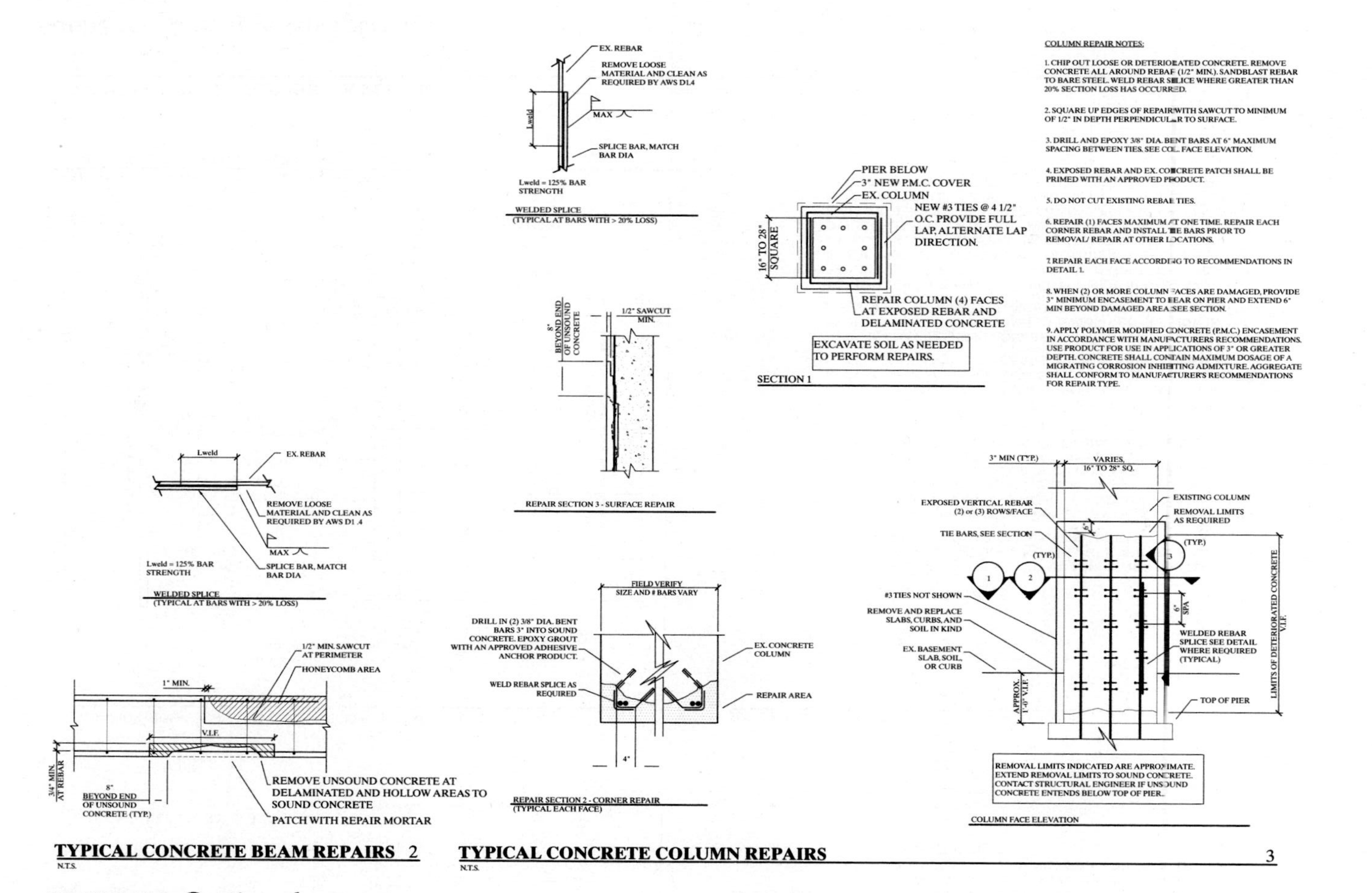

FIGURE 14-3 Continued

14-5 DIAPHRAGMS, DRAG STRUTS, AND CHORDS

In building structures, the floor and the roof slabs, in addition to supporting gravity loads, act as deep horizontal plates or beams in transferring the in-plane lateral shear forces to the lateral force resisting systems (i.e., the shear walls or moment frames). When floor and roof slabs act in this capacity, they are referred to as diaphragms. The ACI Code, Chapter 12, has introduced a new chapter on diaphragms for structures in seismic design category (SDC) C and lower. In the author's experience, diaphragms are sometimes either ignored or taken for granted in the structural design of concrete buildings, especially in non-seismic regions, so it is instructive to note that a chapter has been dedicated to this all-important structural element in the ACI 318-14 Code, in addition to the already existing provisions for diaphragms in Section 18.12 of Chapter 18 (Earthquake Resistant Structures) of the ACI Code. In high-rise buildings, the ground floor or podium concrete slab may be designed to act as a "super" diaphragm to transfer all or a portion of the lateral shears in the shear walls into the perimeter basement or slurry walls. In this and all other similar situations, the necessary reinforcement required in the floor diaphragm to facilitate this transfer of lateral shears must be provided, in addition to the reinforcement required in the slab to support the gravity loads. See Figures 14-4 through 14-7.

There are three types of diaphragms—rigid diaphragms, flexible diaphragms, and semi-rigid diaphragms:

Rigid diaphragms:

- These are floor or roof slabs that are relatively stiffer in the horizontal plane when compared to the lateral stiffness of the shear walls or moment frames (i.e., the LFRS).
- The lateral forces in structures with rigid diaphragms are distributed to the lateral force resisting systems (LFRS) in proportion to their lateral stiffnesses.

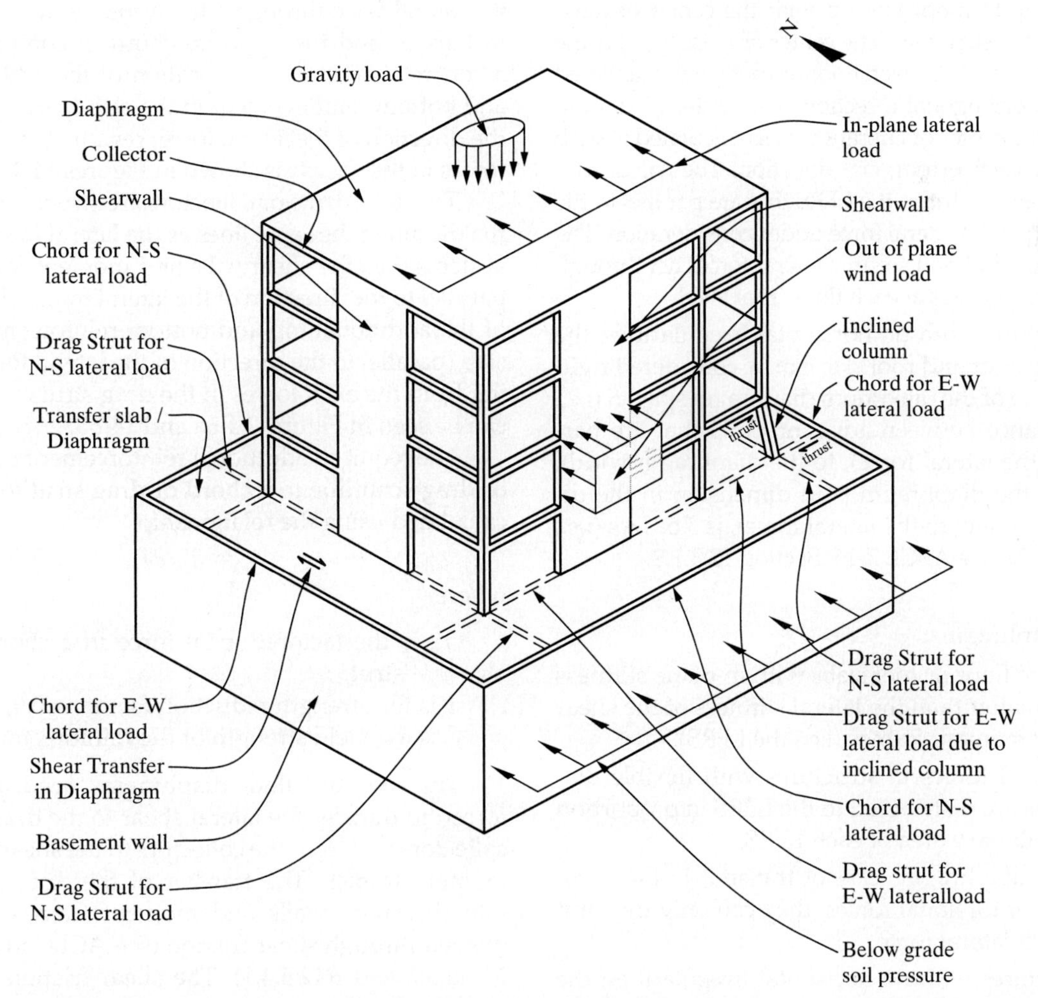

FIGURE 14-4 Diaphragms, drag struts, and chords: Multi-story building (adapted from ref. [18]).

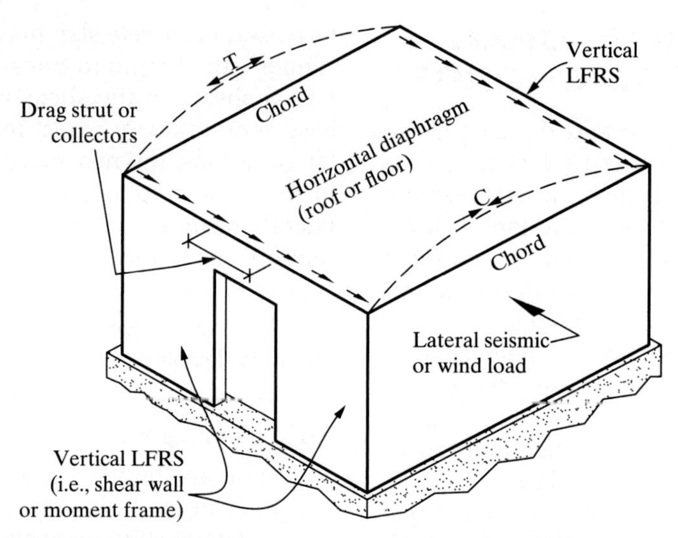

FIGURE 14-5 Diaphragms, drag struts, and chords: One-story building.

- Rigid diaphragms can transmit in-plane rotational or torsional forces arising from the lateral force acting on the structure. These in-plane rotational forces are due to the non-symmetrical placement of the LFRS; that is, the center of rigidity (CR) of all the lateral force resisting systems in the structure at a floor or roof level does not line up with the center of mass (CM) of the structure. The center of mass (CM) is the center of gravity at each floor or roof level, calculated for each orthogonal direction. The center of rigidity (CR) is the center of lateral stiffness calculated at each level for each orthogonal direction. The calculation for CR only includes the LFRS that are parallel to the direction of the lateral force under consideration. The resultant wind and seismic lateral forces act through the center of mass at each floor/roof level.
- In structures with no horizontal irregularities, the concrete floor and roof slab can be considered rigid if the ratio of the maximum diaphragm span, S (i.e., the distance between adjacent LFRS that are parallel to the lateral force), to the diaphragm width, D_e (i.e., the diaphragm plan dimension in the direction parallel to the lateral force), is 3 or less (i.e., $S/D_e \leq 3$). See ASCE 7-16, Section 12.3.1.2.

Flexible diaphragms:

- These are floor or roof slabs with in-plane stiffness much smaller than the lateral stiffness of the shear walls or moment frames (i.e., the LFRS).
- The lateral forces in structures with flexible diaphragms are distributed to the LFRS in proportion to the tributary area of each LFRS.
- Flexible diaphragms cannot transmit in-plane rotational or torsional forces; they can only transmit the direct lateral force.
- In structures with no horizontal irregularities, the concrete floor and roof slab can be considered flexible if $S/D_e > 3$ (See ASCE 7-16, Section 12.3.1.3.)

Diaphragms that do not qualify as rigid or flexible diaphragms are called semi-rigid diaphragms. Floor and roof diaphragms consist of *drag struts* and *chords*. The chords are the edge members in the diaphragm perpendicular to the direction of the lateral force, and they resist the in-plane moment on the diaphragm caused by the lateral force through a force couple, with one chord in tension and the opposite chord in compression. In a concrete structure, the location of the additional top and bottom reinforcement in the slab (perpendicular to the direction of the lateral force) resulting from the axial forces in the chords is shown in Figures 14-4 and 14-7.

The drag struts or collectors are structural elements that lie along the same lines as the lateral force resisting systems (i.e., the shearwalls and moment frames), and parallel to the direction of the lateral force. The location of the additional top and bottom reinforcement in the slab (parallel to the direction of the lateral force) resulting from the axial forces in the drag struts or collectors can be seen in Figures 14-6a and 14-6b.

The required additional reinforcement in the chord or drag strut due to a chord or drag strut force can be calculated using the relationship

$$A_s = T_u/\phi f_y,$$

where,

T_u is the factored axial force in a chord or drag strut,
ϕ is the strength reduction factor $= 0.9$, and
f_y is the yield strength of the reinforcement.

The roof and floor diaphragms need to be reinforced to transfer the lateral shear to the drag struts or collectors and from the collectors to the shear walls and moment frames. The transfer of the drag strut force into the shear walls and moment frames is accomplished through shear friction (see ACI Code, Sections R12.5.3.7 and R12.5.4.1). The shear friction reinforcement is placed perpendicular to the drag strut tension/compression reinforcement (see Figure 14-6b).

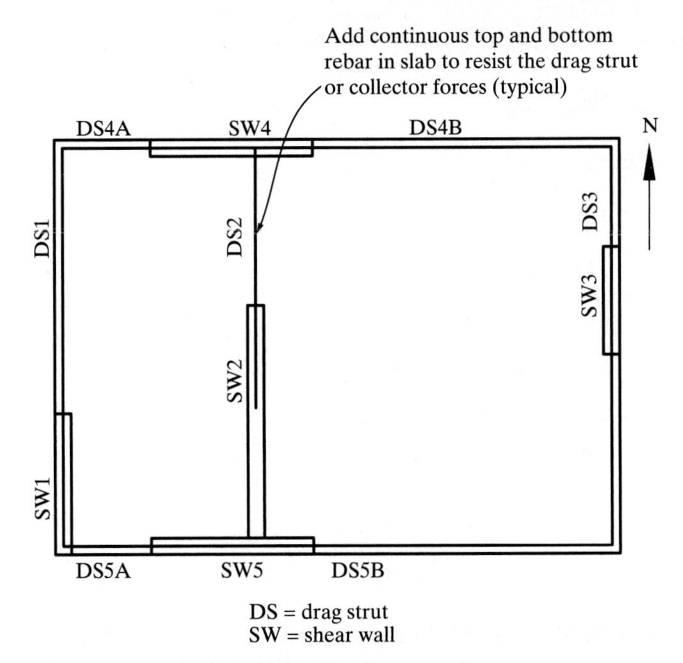

FIGURE 14-6a Plan view of floor or roof slab showing drag struts and additional reinforcement.

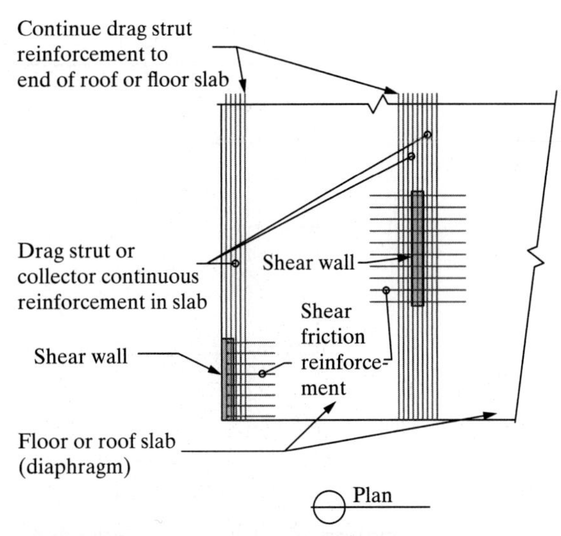

FIGURE 14-6b Drag strut and Shear friction reinforcement (Adapted from ACI Code Figures R12.5.3.7 and R12.5.4.1)

14-6 ONE-WAY SLABS SUBJECTED TO CONCENTRATED LOADS

There are a number of situations in practice where one-way slabs may be subjected to concentrated loads. Examples include slabs that support wheel loads due to trucks or forklift or heavy safes, or slabs that support concentrated loads from hung equipment. The issue of what effective width to use in the design of one-way slabs in these situations is not readily apparent.

For hollowcore slabs in bending, the Spancrete Manufacturers Association [20–22] recommends for bending moment calculations an effective width at the mid-span of 0.50ℓ for an interior concentrated load or line load, and 0.25ℓ at the mid-span when the concentrated or line load is close to an unsupported edge of the slab, where ℓ is the span of the hollowcore plank.

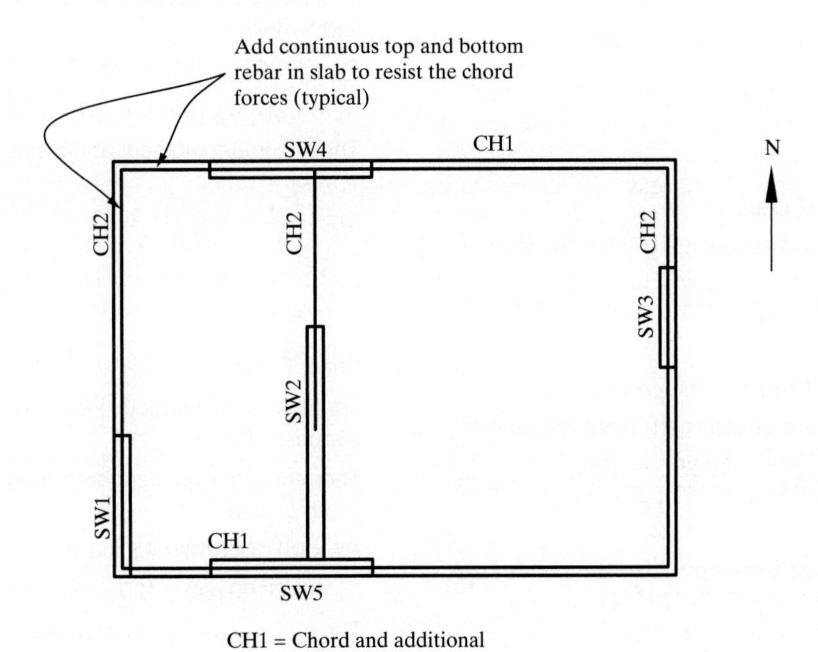

FIGURE 14-7 Plan view of floor or roof slab showing chords and additional reinforcement.

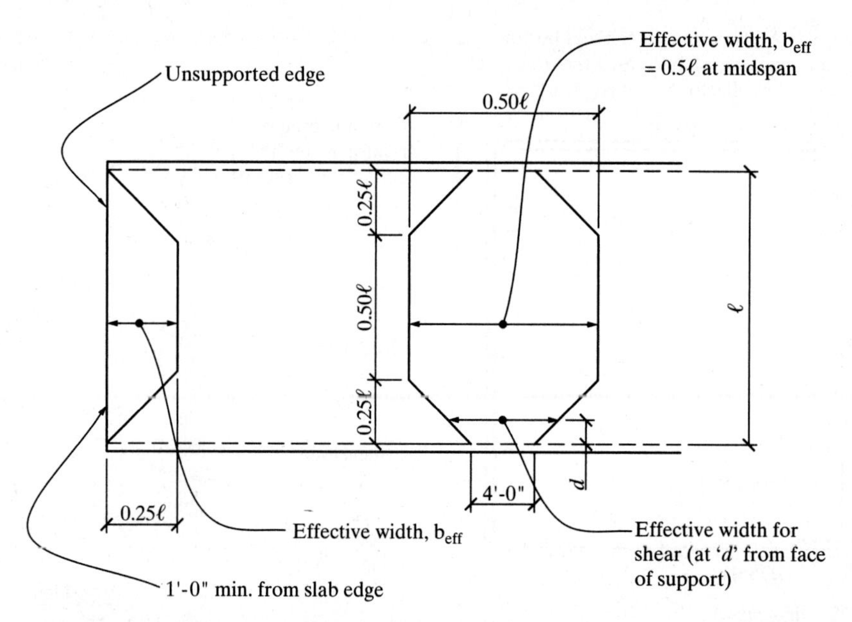

FIGURE 14-8 Effective widths for one-way slabs subject to concentrated loads (Adapted from Ref. 21 and 22)

For shear design, the effective width varies from 4 ft to a maximum of 0.50ℓ at a distance of 0.25ℓ from the support face for interior loads. For loads close to an unsupported edge, the effective width for shear design varies from 1.0 ft to a maximum of 0.25ℓ at a distance of 0.25ℓ from the support face (see Figure 14-8). In summary, the effective widths are as follows:

Bending Moment:

Interior Concentrated Loads at slab midspan–
Effective width, $b_{eff} = 0.50\ell$

Concentrated Loads at slab midspan Close to Unsupported Slab Edge –
Effective width, $b_{eff} = 0.25\ell$

Shear:

Interior Concentrated Loads –
Effective width at a distance x (ft) from the face of

support is: $b_{eff} = 4.0\text{ ft} + \left[\dfrac{0.50\ell - 4.0\text{ ft}}{0.25\ell}\right]x \leq 0.5\ell$

Concentrated Loads Close to Unsupported Slab Edge –
Effective width at a distance x (ft) from the face of

support is: $b_{eff} = 1.0\text{ ft} + \left[\dfrac{0.25\ell - 1.0\text{ ft}}{0.25\ell}\right]x \leq 0.25\ell$

Effective width equations similar to the above have been recommended by others [23, 24].

EXAMPLE 14-2

One way slabs under concentrated loads

A one way slab spanning 12 ft between simple supports is subjected to a uniformly distributed load of 100 psf (includes its self-weight and superimposed dead load) and supports a concentrated live load of 12 kip at the midspan. Calculate the maximum moment in ft-kip/ft width of slab and the maximum shear in kip/ft.

Solution:

Bending Moment:

For the concentrated load, the effective width at midspan = 0.5L = 0.5 (12 ft) = 6 ft

Therefore, the concentrated load is 12 kip/6 ft = 2 kip/ft width of slab

$P_u = 1.6\ (2\text{ kip/ft}) = 3.2\text{ kip/ft}$

$w_u = 1.2(100\text{ psf}) = 120\text{ psf} = 0.12\text{ ksf}$

The maximum moment at the slab midspan is calculated as

$$M = \frac{P_u\ell}{4} + w_u\left(\frac{\ell^2}{8}\right) = \frac{(2\text{ kip/ft})(12\text{ ft})}{4} + (0.12\text{ ksf})\left(\frac{(12\text{ ft.})^2}{8}\right)$$

$$= 8.16\text{ ft} - \text{kip/ft width of slab}$$

Shear Force:

For the concentrated load, the effective width at the support = 4 ft

Therefore, the concentrated load is 12 kip/4 ft = 3 kip/ft width of slab

$P_u = 1.6\ (3\text{ kip/ft}) = 4.8\text{ kip/ft}$

$w_u = 1.2(100\text{ psf}) = 120\text{ psf} = 0.12\text{ ksf}$

The maximum shear is calculated as

$$V = \frac{P_u}{2} + w_u\left(\frac{\ell}{2}\right) = \frac{4.8\text{ kip/ft}}{2} + (0.12\text{ ksf})\left(\frac{(12\text{ ft.})}{2}\right)$$

$$= 3.12\text{ kip/ft width of slab}$$

14-7 LOAD TESTING OF STRUCTURES

The load testing of an existing structure may become necessary for several reasons: (1) there is an increase in the occupancy load and the load capacity of an existing concrete structure cannot be determined analytically because there are no available structural drawings, and the size and location of the existing reinforcement cannot be determined; or (2) the existing structure has been damaged due to corrosion, fire, or other causes, and the load capacity is desired; or (3) it is necessary to determine the load capacity of repaired or retrofitted structures; or (4) it is necessary to determine the effects of construction or design errors such as misplacement or omission of reinforcement or the effect of lower-than-specified concrete strength on the load carrying capacity of a structure. There are two available methods for conducting these load tests on existing structures [12–19]:

1. The static and monotonic 24-hour uniformly distributed load test; this is the most commonly used load test.
2. The cyclic load test (this is a more rapid load test). For a more detailed discussion of this load test, the reader should refer to Ref. [12].

The static and monotonic load tests have been in use since the 1800s and the ACI Code, Chapter 27, provides the procedure for evaluating the strength of existing structures using the static and monotonic load test. Loading materials that can be used for this load test include non-hydraulic methods such as sand bags, steel plates, and water containers (with a water meter). These materials can be stacked to achieve the desired load. Care should be taken if these materials are placed on the structure using fork lifts or other equipment whose own weights could negatively affect the safety of the structure [16, 17]. These non-hydraulic loading methods can only be used for static and monotonic load tests. The lack of ability to control or promptly reduce the non-hydraulic loads in case of imminent failure of the structure is one of the disadvantages of these types of loads. On the other hand, hydraulic loading can be used for both static and cyclic load tests and it has the obvious advantage that the applied load can be controlled, thus enhancing safety during a load test.

Static and Monotonic Uniformly Distributed Load Test

According to Section 27.4 of the ACI 318-14, the requirements and specifications for the strength evaluation of an existing structure using a static and monotonic load test are as follows (see Figure 14-9):

1. The minimum age of the structure at the time of the load test is 56 days. This requirement can be waived and the structure tested at a younger age if all the stakeholders involved with the structure including the owner, the contractor, and the structural engineer are in agreement.
2. The test duration when the total test load is on the structure is 24 hours.
3. The total test load is applied in a minimum of four equal increments (ACI 318-14, Section 27.4.3.1).
4. Record the maximum deflections and other load effects at each load increment. Within 1 hour before the start of the load test, the initial deflection of the structural member, Δ_{initial}, should be measured (ACI 318-14, Section 27.4.4.2).
5. Check for cracking or spalling of concrete at each load stage.
6. The test load arrangement and location should be selected so as to maximize the deflection and load effects on the member that is being evaluated (ACI 318-14, Section 27.4.2.1). The load combinations for the load test (ACI 318-14, Section 27.4.2.2) are as follows:
   a. $\text{TTL} = 1.15\text{D} + 1.5\text{L} + 0.4(\text{L}_r \text{ or S or R})$
   b. $\text{TTL} = 1.15\text{D} + 0.9\text{L} + 1.5(\text{L}_r \text{ or S or R})$
   c. $\text{TTL} = 1.3\text{D}$

   where,

   TTL = Uniform Total Load on the structure (includes the dead load that is already in place)
   TL = Uniform Test Load = TTL − Existing dead load that is already in place on the structure

   The test load, TL, is applied in four equal increments as shown in Figure 14-9 (ACI 318-14, Section 27.4.3).
7. The structure is deemed capable of supporting the applied test load if Criterion (a) *or* (b), *and* Criterion (c) below are satisfied:.

   a. $\Delta_1 - \Delta_{initial} \leq \dfrac{\ell_t^2}{20{,}000\, h}$

   b. $\Delta_r = \Delta_2 - \Delta_{initial} \leq \dfrac{\Delta_1 - \Delta_{initial}}{4}$ (that is 75% or more deflection recovery after 48 hours)

   c. There should be no crushing or spalling of concrete or cracks; the presence of these would indicate imminent shear failure (ACI 318-14, Sections 27.4.5.1 and 27.4.5.2).

   where,

   ℓ_t = span of the concrete member in inches (for 2-way slabs, use the shorter span).
   h = overall depth of the member in inches.

The equations above were derived based on simple-span members. A safety plan is required for any load test to prevent injury or death in the event of a premature failure, and to prevent

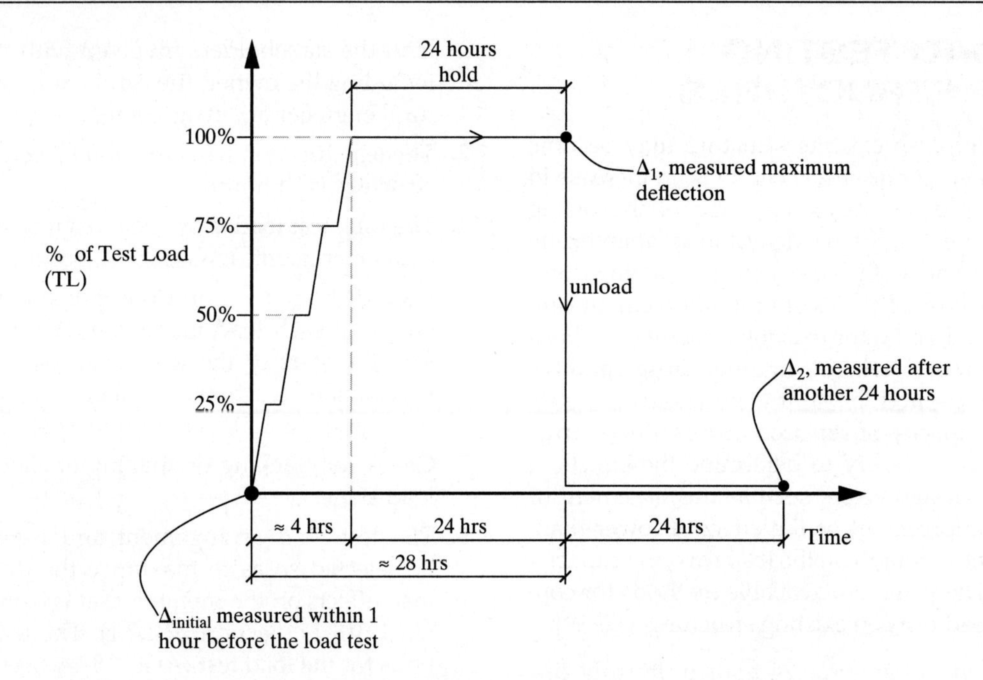

FIGURE 14-9 Load-time graph for initial load test.

structural damage. The safety plan should include adequate shoring below the structure that is to be load tested in order to "catch" the structural member in case of imminent failure and thus prevent progressive collapse of the entire structure.

Second or Repeated Load Test

If the structure does not satisfy the first two criteria of the first load test for the load rating considered, a second load test can be carried out after at least 72 hours after the removal of the test load from the first load test (see Figure 14-10). The structure shall be deemed acceptable for the second load test if there are no concrete cracks or spalling, and if the following acceptance criterion (ACI 318-14, Section 27.4.5.7) is satisfied:

$$\Delta_{r\,(repeated)} = \Delta_{2\,(repeated)} - \Delta_{initial\,(repeated)} \leq \frac{\Delta_{1\,(repeated)}}{5}$$

If a structure fails to satisfy the acceptance criterion for the second or repeated load test or the first two acceptance criteria for the initial load test, a reduced maximum load rating can be calculated for the structure or structural member using these acceptance criteria.

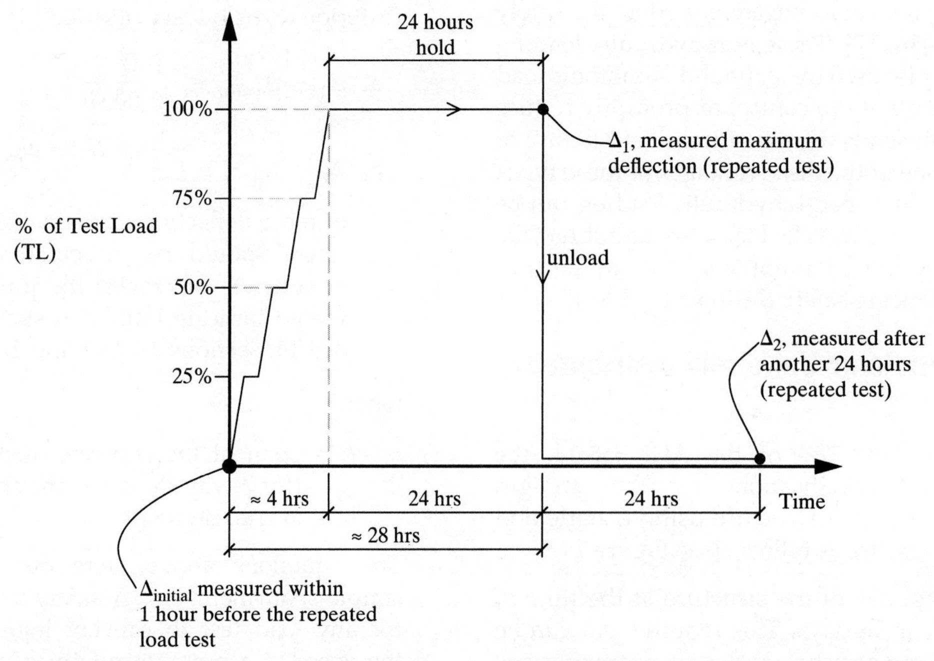

FIGURE 14-10 Load-time graph for second or repeated load test.

14-8 CLOSURE OR POUR STRIPS IN REINFORCED CONCRETE FLOORS

As concrete cures, it shrinks and this movement, if restrained, leads to shrinkage cracks. The larger the concrete pour, the greater the shrinkage movement-unless strategies are adopted to reduce the movement due to shrinkage as well as minimize the restraint to shrinkage, and thus control the distribution and size of the shrinkage cracks. For reinforced concrete floor or roof plans larger than approximately 200 ft to 250 ft, the movement due to shrinkage can be mitigated through the use of expansion joints or through the use of closure or pour strips. Shrinkage compensating cement can are also be used in the concrete mix. Closure strips are also called shrinkage strips or construction or pour gaps. They are often used in multistory building construction and involves pouring adjacent portions of the concrete floor, and leaving a gap between two adjacent concrete pours to allow the two portions of the concrete floor to shrink independently of each other without restraint from each portion. The pour strip is then filled in with concrete at a later date. Closure strips can be provided in floors in two orthogonal directions so that no concrete pour area is larger than 200 ft × 200 ft in plan. The width of the closure strip is typically 3 ft to 5 ft and should be large enough to accommodate a tension lap splice for the slab or beam reinforcement, since the slab or beam reinforcement on either side of the pour gap must be continuous through the closure strip, but "physically independent of each other" until the pour gap is filled in. The closure strip or pour gap is subsequently filled in after 14 to 90 days [29]. In some structures where shrinkage cracks must be greatly minimized, the length of time it takes to fill in the closure strip might be much longer and could be as long as 120 days. Before the closure strip is filled in, the primary shoring for the floor slab and beams near the closure strip and the adjacent floor bays must be kept in place, and fall prevention measures at the location of the closure strip should be installed by the contractor. The use of closure strips and the length of time that the strip is unfilled can greatly impact the construction cost and construction schedule of a project. Recently, "lockable dowels" have been used to *obviate* the need for shoring near the closure strips [30].

Typical reinforcement details for closure strips are shown in Figure 14-11.

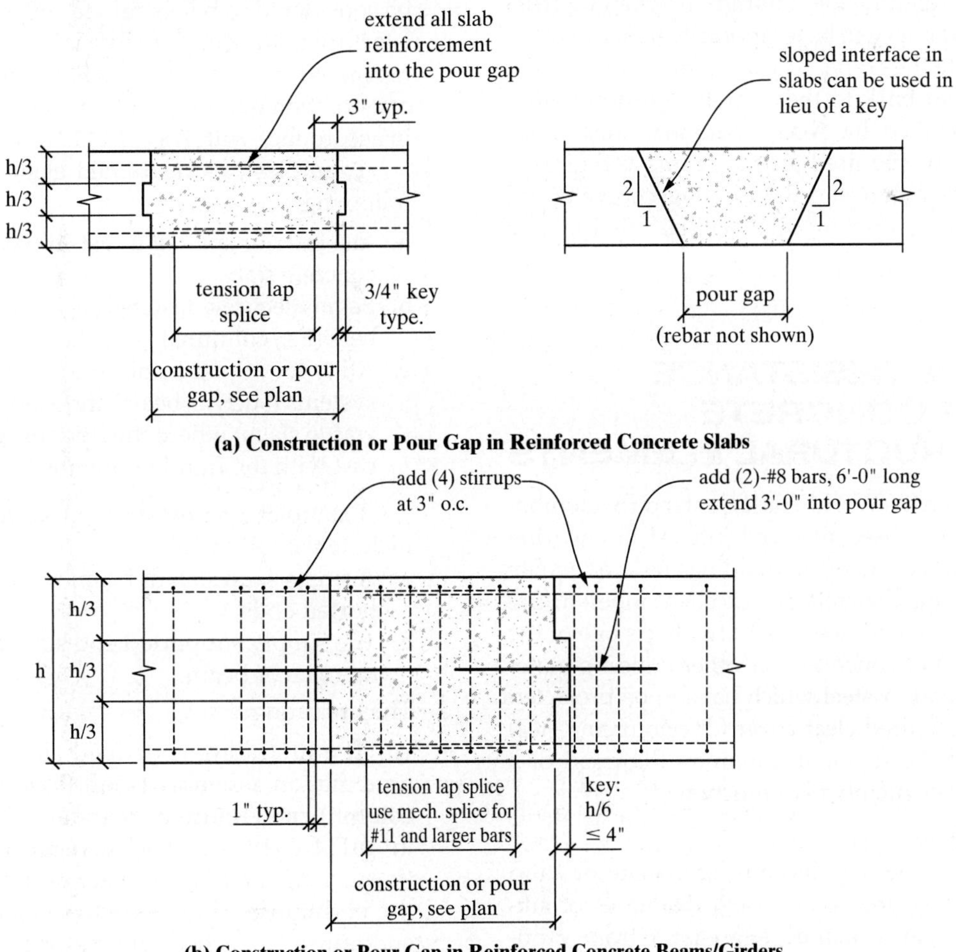

FIGURE 14-11 Closure or pour strip detail.

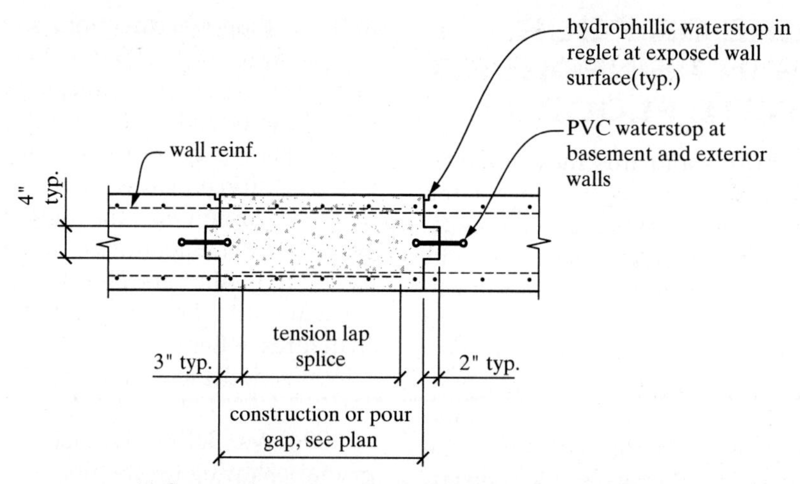

(c) Construction or Pour Gap in Reinforced Concrete Walls

FIGURE 14-11 (Continued).

Similar to the pour gap detail in Figure 14-11 that is specifically used to control shrinkage, temporary openings are sometimes required in floor/roof slabs and beams to accommodate construction cranes that may need to be located within the building footprint because of existing adjacent properties. The detail used for such cases will be similar to Figure 14-11 except that the reinforcement within the opening (that will be filled in later) will be temporarily bent upwards to accommodate the crane. After the crane is removed, the bars are bent back to the straight position before the opening is filled in. Also, additional shear reinforcement may be required for the beams and girders near the interface between the existing concrete and the concrete in the opening that will be filled in at a later date.

14-9 FIRE RESISTANCE OF CONCRETE STRUCTURAL ELEMENTS

Since the attack on the World Trade Center on September 11, 2001, there has been renewed interest in ensuring that the structural elements in buildings have adequate fire resistance [25]. Concrete has excellent inherent fire resistive properties and does not burn; this is one of the major advantage of concrete over other building materials such as wood or steel which have to be protected from fire. The required clear cover for reinforcing steel in concrete and the required minimum thickness of a structural concrete member is a function of:

1. The type of aggregate used.

   That is, whether siliceous, carbonate or sand lightweight aggregate are used. Examples of siliceous aggregates include aggregates from granite and sandstone while examples of carbonate aggregates include aggregates from limestone and dolomite.

2. Whether the structural member is *restrained* or *unrestrained*.

   According to Section 703.2.3 of the International Building Code (IBC), "fire-resistance-rated assemblies tested under ASTM E119 or UL 263 shall not be considered to be restrained unless evidence satisfactory to the building official is furnished by the registered design professional showing that the construction qualifies for a restrained classification in accordance with ASTM E119 or UL 263" [27].

   Examples of a restrained element include the following:

   a. The interior span of a cast-in-place continuous concrete slab
   b. Beams securely fastened to the framing members (i.e., columns)
   c. All types of cast-in-place concrete floor or roof systems (such as beam-and-slab, flat slabs, and waffle slabs) where the floor or roof system is cast with the framing members

   Examples of unrestrained structural element include the following:

   a. A single span slab or beam with simply supported ends
   b. The simply supported end span of a continuous slab or beam.

3. The fire rating.

   The fire rating is defined as "the duration of time that an assembly (roof, floor, beam, wall, or column) can endure a "standard fire" as defined in ASTM E 119" [25, 28]. There are four criteria defined in ASTM E119 that act as a measure of the fire endurance of the structural element. The fire endurance is reached when any of those four criteria is reached first. One of those conditions is the time in hours it takes for the structural element or member to lose its strength and collapse under the applied service load when subjected to a fire.

Concrete with siliceous aggregates are more susceptible to the deleterious effects of fire and require more clear cover and higher thicknesses to achieve the same fire rating as concrete with carbonate or sand-lightweight aggregates. Also, a restrained structural member (e.g., the interior span of a continuous slab or beam) requires smaller clear cover and minimum thickness than an unrestrained structural member (e.g., the end span of a continuous slab or beam, or a simply supported slab or beam). The minimum concrete covers specified in ACI 318-14, Section 20.6.1.3.1 sometimes exceeds or is equal to the prescriptive values specified in ACI 216 or Section 721 of the International Building Code (IBC) [25–28], but for fire ratings greater than or equal to 2 hours, this is not necessarily the case. Therefore, it is imperative that the minimum cover and minimum thickness requirements needed to satisfy the fire resistance ratings in ACI 216 or the IBC not be overlooked during the design phase because ignoring these requirements could lead to costly repairs. For the higher fire ratings (e.g., 3 hours or 4 hours), the values of minimum concrete cover tabulated in ACI 216 are more stringent than the minimum covers prescribed in ACI 318. Since most ready-mix concrete contractors generally use aggregates that are indigenous to the area where the project is located, the minimum concrete covers and minimum thicknesses given below from ACI 216 and the IBC [25–27] are for the worst-case scenario; that is, for concrete with *siliceous* aggregates and assuming *unrestrained* structural members:

- Minimum cover to main rebar in *floor and roof slabs*:
  - 1-hour and 1.5-hour fire rating: Use the concrete covers specified in ACI 318-14, Table 20.6.1.3.1
  - 2-hour fire rating: Minimum cover for floor and roof slabs = 1 in.
  - 3-hour fire rating: Minimum cover for floor and roof slabs = $1\frac{1}{4}$ in.
  - 4-hour fire rating: Minimum cover for floor and roof slabs = 1-5/8 in.
- Minimum cover to main rebar in *beams and girders*:
  - Less than 4-hour fire rating: Minimum cover to main rebar = values in ACI 318-14, Table 20.6.1.3.1.
  - 4-hour fire rating: For beams that are 10 in wide or greater, minimum clear cover to main rebar = $1\frac{3}{4}$ in.
- Minimum cover to main rebar in *columns*:
  - 1-hour and 1.5-hour fire rating, minimum clear cover = values in ACI 318-14, Table 20.6.1.3.1
  - 2-hour, 3-hour, or 4-hour fire rating, minimum clear cover = 2 in.
- Minimum thickness of *floor and roof slabs* (practical minimum slab thickness = 4 in.):
  - 1.5-hour fire rating, minimum slab thickness = 4.3 in.
  - 2-hour fire rating, minimum slab thickness = 5.0 in.
  - 3-hour fire rating, minimum slab thickness = 6.2 in.
  - 4-hour fire rating, minimum slab thickness = 7.0 in.
- Minimum concrete column dimensions (practical minimum column size = 12 in.):
  - 3-hour fire rating, minimum column size = 12 in.
  - 4-hour fire rating, minimum column size = 14 in.

14-10 ANALYSIS AND DESIGN OF EDGE-SUPPORTED TWO-WAY SLABS ON STIFF SUPPORTS

The analysis of two-way slabs is a complex and indeterminate system. In this section, the analysis of rectangular slab panels supported on stiff supports (e.g. walls and stiff beams) along all four edges is presented. The slab bends in two directions with curvature in the two orthogonal directions and a portion of the load is carried by bending in the short direction while the remaining portion is carried by bending in the long direction of the slab. An approximate equation for distributing the uniform load on the slab can be derived by assuming equal deflection of simply supported strips at the center of the slab in both orthogonal directions, and neglecting the effects of the twisting moments at the corners of the slab.

The maximum deflection of the center strip in the long (ℓ_1) direction is

$$\delta_1 = \frac{5w_1\ell_1^4}{384EI}$$

The maximum deflection of the center strip in the short (ℓ_2) direction is

$$\delta_2 = \frac{5w_2\ell_2^4}{384EI}$$

For compatibility, $\delta_1 = \delta_2$ and noting that $w_1 + w_2 = w$ yield the following expressions for the approximate distribution of the uniform load in the two orthogonal directions of the slab:

$$w_1 = w\left(\frac{\ell_2^4}{\ell_1^4 + \ell_2^4}\right) \qquad w_2 = w\left(\frac{\ell_1^4}{\ell_1^4 + \ell_2^4}\right)$$

Where,

w_1 = the portion of the total uniformly distributed load carried by bending in the ℓ_1 direction
w_2 = the portion of the total uniformly distributed load carried by bending in the ℓ_2 direction
w = total uniformly distributed load supported by the slab

Note that in deriving the above equations, though the corners are usually restrained, the torsional moments at the corners of the slab have been neglected for simplicity. The load distribution above is very conservative when compared to the elastic moment distribution in a two-way edge supported slab [34].

In the preceding approximate method, the complex behavior of the 2-way edge-supported concrete slab is reduced to the behavior of only the two intersecting center strips, and the inelastic moment redistribution that occurs in the concrete slab is also ignored. The actual behavior of the 2-way edge supported slab will involve both bending and twisting moments in both orthogonal directions and it also involves the inelastic redistribution of moments, resulting in a higher load carrying capacity for the slab, and smaller bending moments than would be obtained using the approximate load distribution derived previously.

A more accurate method for calculating the moments in edge-supported two-way slabs, which has been in use for many decades, is the moment coefficient method presented in Ref. [35]. The moment coefficients in both orthogonal directions are tabulated for various slab aspect ratios, $\frac{\ell_2}{\ell_1}$, and edge support conditions–simply supported or continuous. The minimum slab thickness for edge-supported two-way slabs can be calculated as follows [35]:

- h_{min} = (perimeter of slab panel)/140 $\geq$ 4 in. for simply supported two-way slab panels
- h_{min} = (perimeter of slab panel)/160 $\geq$ 4 in. for continuous two-way slab panels

Ref. [6] also provides moment coefficients that can be used for calculating the design moments for continuous edge supported two-way slabs on stiff supports.

Example 14-3

A slab is supported on walls on all four sides. The clear spans of the long side of the slab is 27 ft and the short side is 21 ft. The slab supports a total unfactored dead load of 150 psf (includes the selfweight of the 10 in. thick slab, 15 psf partition load, and 10 psf for floor finishes, ceiling, and mechanical and electrical equipment) and a live load of 50 psf. Calculate the factored moment in ft.-kip/ft in the long direction (i.e. M_1) and in the short direction (i.e. M_2) using the moment coefficient method. Calculate the required area of steel using the methods presented in Chapter 2. Assume f'_c is 4000 psi and f_y is 60,000 psi.

Solution:

$$w_u = 1.2\,(150\text{ psf}) + 1.6(50\text{ psf}) = 260\text{ psf} = 0.26\text{ ksf}$$

$$\ell_1 = 27\text{ ft}$$

$$\ell_2 = 21\text{ ft}$$

$$\frac{\ell_1}{\ell_2} = \frac{27\text{ ft}}{21\text{ ft}} = 1.29 < 2.0\text{, therefore, this is a two-way slab.}$$

The slab panel aspect ratio, $\frac{\ell_2}{\ell_1} = \frac{21\text{ ft}}{27\text{ ft}} = 0.78$

$$h_{min} = 2(21\text{ ft.} + 27\text{ ft.})(12\text{ in./ft})/140 = 8.2\text{ in.}$$

Also, use Table 6-4 to check the minimum slab thickness as follows:

$$h_{min} = \frac{\ell_n}{33} = \frac{\ell_{n1}}{33} = \frac{(27\text{ ft})(12)}{33} = 9.8\text{ in. 10 in. slab. O.K.}$$

For the slab panel aspect ratio of 0.78, the maximum design moment coefficients can be obtained by linear interpolation from the table of moment coefficients in Ref. [35] as follows:

Long direction: $C_a = 0.021;\ M_1 = 0.021 w_u(\ell_1)^2 = 0.021(0.26\text{ ksf})(27)^2 = 4\text{ ft.-k/ft width}$

Short direction: $C_a = 0.059;\ M_2 = 0.059 w_u(\ell_2)^2 = 0.059(0.26\text{ ksf})(21)^2 = 6.8\text{ ft.-k/ft width}$

Note that these are the moments in the middle strips of the slab panel (i.e. the middle 10.5 ft for long direction bending and the middle 13.5 ft for short direction bending). Though the moments can be assumed to vary from a maximum at the edge of the middle strips to 1/3 of the maximum value at the non-deflecting edge supports, for convenience and ease of rebar placement, some engineers for practical reasons assume a uniform maximum moment across the slab width, resulting in a uniform layout of the slab reinforcement.

Using the methods presented in Chapter 2, the slab reinforcement can be determined. For the M_2 moment, use an effective depth, $d = 9''$ (i.e. $10'' - ¾''$ cover $- 0.5''/2$) in the short direction assuming #4 reinforcement in the bottom lower layer – BLL; for the moment, M_1, use an effective depth, $d = 8.5''$ (i.e. $10'' - ¾''$ cover $- 0.5'' - 0.5''/2$) in the long direction (i.e. bottom upper layer – BUL). Note that the bottom reinforcement will be uniformly distributed in the slab.

In the short direction,

$$\bar{k} = \frac{M_{u2}}{\phi b d^2} = \frac{6.8(12)}{(0.9)(12\text{ in.})(9\text{ in.})^2} = 0.093$$

From Table A-10, we get $\rho = 0.00157 = \frac{A_s}{bd}$

Therefore, the required $A_s = 0.00157(12\text{ in.})(9\text{ in.}) = 0.17\text{ in.}^2 < A_{s,\,min}$

Therefore, use $A_{s,\,min}$.

$$A_{s,\,min} = 0.0018bh = 0.0018(12)(10) = 0.22\text{ in}^2\text{/ft width}$$

Use #4 @ 11″ o.c. BLL (see Table A-4)

In the long direction,

$$\bar{k} = \frac{M_{u1}}{\phi b d^2} = \frac{4(12)}{(0.9)(12\text{ in.})(8.5\text{ in.})^2} = 0.0615$$

From Table A-10, we get $\rho = 0.0011 = \frac{A_s}{bd}$

Therefore, the required $A_s = 0.0011(12\text{ in.})(8.5\text{ in.}) = 0.11\text{ in.}^2 < A_{s,\,min}$

Therefore, use $A_{s,\,min} = 0.22\text{ in}^2/\text{ft width}$

Use #4 @ 11″ o.c. BUL (see Table A-4)

Maximum spacing of main reinforcement in the slab = 2h = 2(10 in.) = 20 in. > 11 in., O.K.

To resist the twisting moments at the corners of the slab, provide reinforcement at the top of the slab at the four corners in accordance with Section 6.4-11 and Figure 6-36. Since the slab is assumed to be simply supported at all four edges, minimum top reinforcement (hooked at the wall supports) should also be provided in the slab at the edges in the two orthogonal directions.

14-11 CAST-IN PLACE CONCRETE SPECIFICATIONS

The contract documents for building structures include the contract drawings and the written specifications for the materials used in the structure. There is much information regarding the performance requirements for the concrete materials and accessories used in the structure that cannot all be placed on the structural or architectural drawings. Therefore, more in-depth performance requirements are usually provided in the written specifications. In essence, the structural drawings are the graphic representation, and the specifications are the written representation, of the design intent of the structural engineer of record (SER). Both documents are meant to be complementary to each other and there is sometimes some overlap between the drawings and the specifications, but conflicts between both documents should be avoided. In practice, it is common to find a note on the drawings or in the specifications that states that the more stringent specification requirements shall govern, as determined by the engineer of record when there is a conflict between the specifications and the drawings. The concrete specifications, and indeed all material specifications, follow the standard master format prescribed by the Construction Specifications Institute (CSI). There are 50 divisions in the CSI master specifications: the concrete specifications are in division 030000, and the cast-in-place concrete specifications are in section 033000 of division 030000; the masonry specification is in division 040000 while the metals specification is in division 050000 [31, 32].

The typical cast-in-place concrete specification includes three parts with many subsections: Part 1—General, Part 2—Products, and Part 3—Execution [32]. The subsections for each of the three parts include specifications for items that are pertinent to the proper functioning of the concrete structure. Some pertinent areas that are covered in the concrete specifications are listed below [32]:

- Concrete mixtures for each of the different structural elements are listed in this section. A sample proportion of normal weight concrete is as follows:
  - Minimum Compressive Strength: 4500 psi at 28 days
  - Maximum Water-Cementitious Materials Ratio: 0.45
  - Slump Limit: 4 inches plus or minus 1 inch. If admixtures are used to improve workability, the maximum slump limits may be relaxed with Engineer's approval
  - Air Content: 6 percent at point of delivery
  - Course Aggregate: 1-in. nominal maximum aggregate size"
- Construction joints—The locations of construction joints are specified in this section and an example statement in the specification is as follows: "Place joints perpendicular to main reinforcement. Continue reinforcement across construction joints unless otherwise indicated. Do not continue reinforcement through sides of strip placements of floors and slabs; Locate joints for beams, slabs, joists, and girders in the middle third of spans. Offset joints in girders a minimum distance of twice the beam width from a beam-girder intersection; Locate horizontal joints in walls and columns at underside of floors, slabs, beams, and girders and at the top of footings or floor slabs."
- Concrete placement—In this section, the performance requirements for placing the wet concrete are specified (e.g., "Do not add water to concrete during delivery, at Project site, or during placement unless approved by Architect; Deposit concrete continuously in one layer or in horizontal layers of such thickness that no new concrete will be placed on concrete that has hardened enough to cause seams or planes of weakness. If a section cannot be placed continuously, provide construction joints as indicated. Deposit concrete to avoid segregation").
- Cold-weather placement—In this section, the performance requirements for concrete placement during cold weather are covered. The specification requirements include the following instructions: "Comply with ACI 306.1. Protect concrete work from physical damage or reduced strength that could be caused by frost, freezing actions, or low temperatures. When average high and low temperature is expected to fall below 40°F for three successive days, maintain delivered concrete

mixture temperature within the temperature range required by ACI 301. Do not use frozen materials or materials containing ice or snow. Do not place concrete on frozen subgrade or on subgrade containing frozen materials. Do not use calcium chloride, salt, or other materials containing antifreeze agents or chemical accelerators unless otherwise specified and approved in mixture designs."

- Hot-weather placement—In this section, the performance requirements for concrete placement during hot weather are covered. The specification requirements include the following instructions:

"Comply with ACI 301 and as follows: Maintain concrete temperature below 90°F at time of placement. Chilled mixing water or chopped ice may be used to control temperature, provided water equivalent of ice is calculated and included in the total amount of mixing water. Using liquid nitrogen to cool concrete is Contractor's option. Fog-spray forms, steel reinforcement, and subgrade just before placing concrete. Keep subgrade uniformly moist without standing water, soft spots, or dry areas."

- Concrete protection and curing—Instructions are given in this section for protecting freshly poured concrete from premature drying, and the curing of formed and unformed surfaces per ACI 308.1. One or a combination of the several curing methods listed—moisture curing, moisture-retaining-cover curing, or curing compound—can be used. Slab-on-grade are moist cured for a minimum of 10 days. Examples of specification items in this section include the following:
- "Moisture curing—Keep surfaces continuously moist for not less than seven days with the following materials: Water, Continuous water-fog spray, Absorptive cover, water saturated, and kept continuously wet. Cover concrete surfaces and edges with 12-in. lap over adjacent absorptive covers; Moisture-Retaining-Cover Curing: Cover concrete surfaces with moisture-retaining cover for curing concrete, placed in widest practicable width, with sides and ends lapped at least 12 in., and sealed by waterproof tape or adhesive. Cure for not less than seven days. Immediately repair any holes or tears during curing period using cover material and waterproof tape. Moisture cure or use moisture-retaining covers to cure concrete surfaces to receive floor coverings. Moisture cure or use moisture-retaining covers to cure concrete surfaces to receive penetrating liquid floor treatments. Cure concrete surfaces to receive floor coverings with either a moisture-retaining cover or a curing compound that the manufacturer certifies will not interfere with bonding of floor covering used on the Project."

Additional Specifications for Thick Concrete Slabs

For thick concrete slabs such as mat or raft foundations that could be classified as mass concrete, the heat of hydration given off during the concrete curing process could result in excessive shrinkage which leads to excessive cracking of the mass concrete. Temperatures of up to 170–200°F have been measured in the cores of mass concrete foundations [33]. To mitigate the effects of high temperatures on thick concrete elements, additional specification items are required. The reader should refer to ACI 301 and ACI 207—Mass Concrete for further guidance. In one highrise building supported on a 6 ft thick mat foundation, 6,000 cubic yards (or 162,000 ft^3) of concrete was poured continuously (i.e. a monolithic pour) using several pumps, and it took approximately 28 hours to complete the casting of the mat foundation. This was done so as to avoid any construction joints in such a thick slab especially since the area where the highrise building is located has a high water table. In fact, the water table was so high on the site that, in order to prevent uplift of the mat foundation, the water table was lowered and monitored until the 10th floor slabs and beams were cast, thus ensuring enough dead load to resist the buoyancy forces from the hydrostatic uplift pressures at the base of the mat foundation. To further control the heat of hydration, the 28 day strength of the concrete was specified as 3000 psi with a 120-day strength of 4500 psi. Prequalification trial mixtures submitted by three concrete suppliers were used to cast 6 ft thick trial blocks of concrete that were tested by the owner's materials consultant prior to the mat foundation concrete pour; and the concrete supplier with the trial mix that best met the performance specifications was selected. Examples of additional specification items for concrete mat foundations that were successfully used on the project are as follows:

- Cementitious material: Moderate heat of hydration portland cement (Type II) with flyash (Type F) and cementitious hydraulic slag (Type H).
- A portion of the water in the concrete mix should be in the form of crushed ice with temperatures below 32°F in order to achieve the specified placing temperature for the concrete.
- To further reduce the shrinkage in thick mat foundations, a maximum aggregate size of $1\frac{1}{2}$ inches should be specified.
- The concrete mix for the mass concrete should be selected to achieve the specified placing temperature and to ensure that the temperature in the concrete and the maximum temperature differential do not exceed the specified limits.

- To measure the temperatures, thermocouples should be placed in the concrete. The delivery temperature of the mat foundation concrete should be between 41°F and 50°F, and the maximum internal temperature of the concrete should not exceed 130°F.

The preceding requirements were included in Part 2 of the cast-in-place concrete specifications for the project. In part 3 of the cast-in-place concrete specifications, the following additional items were added to the specification:

- Provide a 3 inch thick mud slab or skim coat with f'_c of 750 psi at the underside of the mat foundation
- Ensure the stability of the mat top reinforcement at all times and during the concrete pour
- The construction of the concrete mat foundation should be accomplished in one continuous pour to ensure that no cold joints develop between successive lifts or sections of concrete placement.
- The maximum temperature differential should not exceed 68°F. To measure the maximum temperature and the maximum temperature differential, thermocouples should be inserted in the concrete at 2 inches from the bottom and top surfaces, and at the mid-depth of the mat foundation. The thermocouples should be spaced longitudinally and transversely a maximum distance of 100 ft. Protective covers and insulation placed over the mat should be kept in place until the temperature differential is within limits, and the temperatures should be regularly monitored to check compliance.

14-12 STUDENT DESIGN PROJECTS

In this section, two reinforced concrete building design projects are presented, and any one of them could be assigned as a term project in a reinforced concrete design course. These design projects are very similar to projects found in consulting structural engineering practice. The design brief for each project is as follows:

1. **Building type:**
   a. **Residential Building:** A two-story reinforced concrete residential building with plan dimensions of 135 ft by 57 ft. The floor-to-floor height is 15 ft (see Figures 14-12 and 14-13).
   b. **Hospital Building:** A two-story reinforced concrete building with plan dimensions of 150 ft by 100 ft. The floor-to-floor height varies (see Figures 14-14 and 14-15).

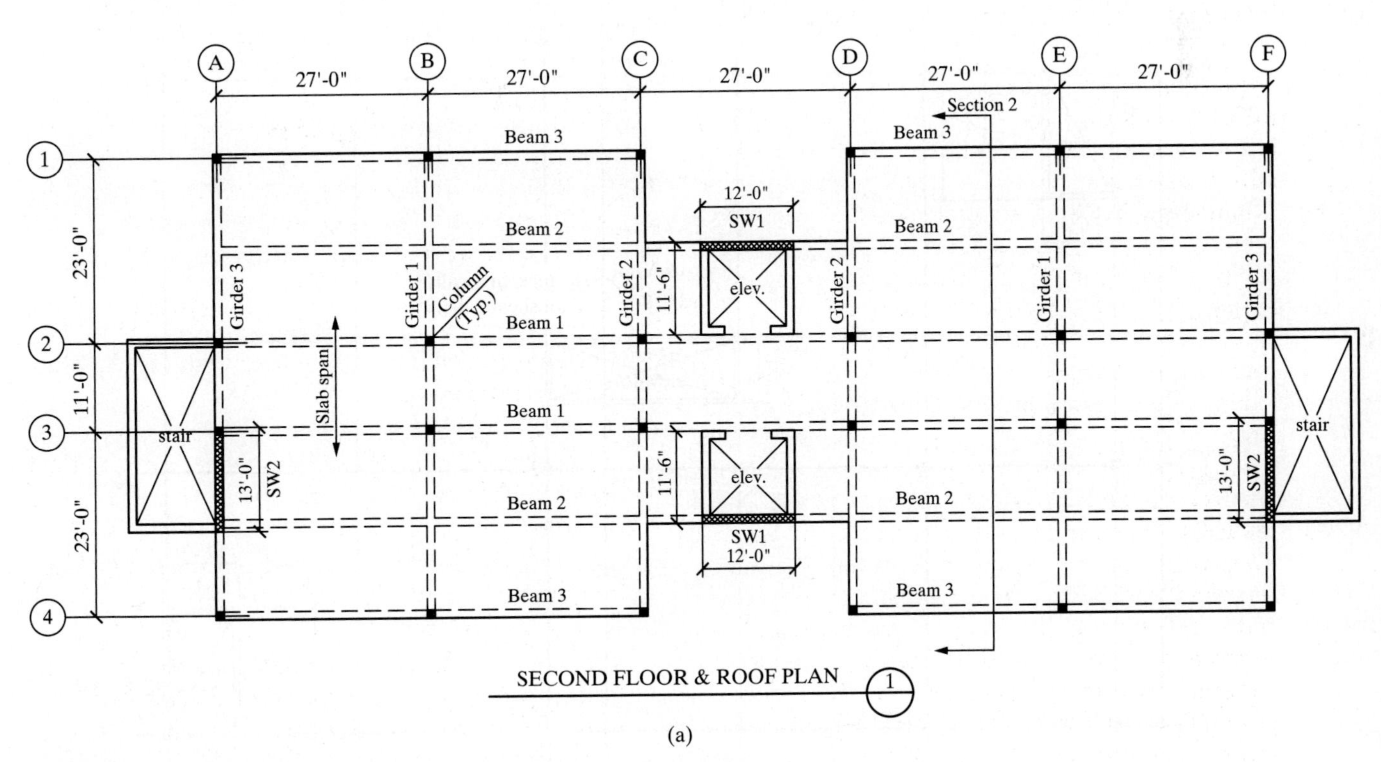

FIGURE 14-12 Residential building - Typical floor and roof plan.

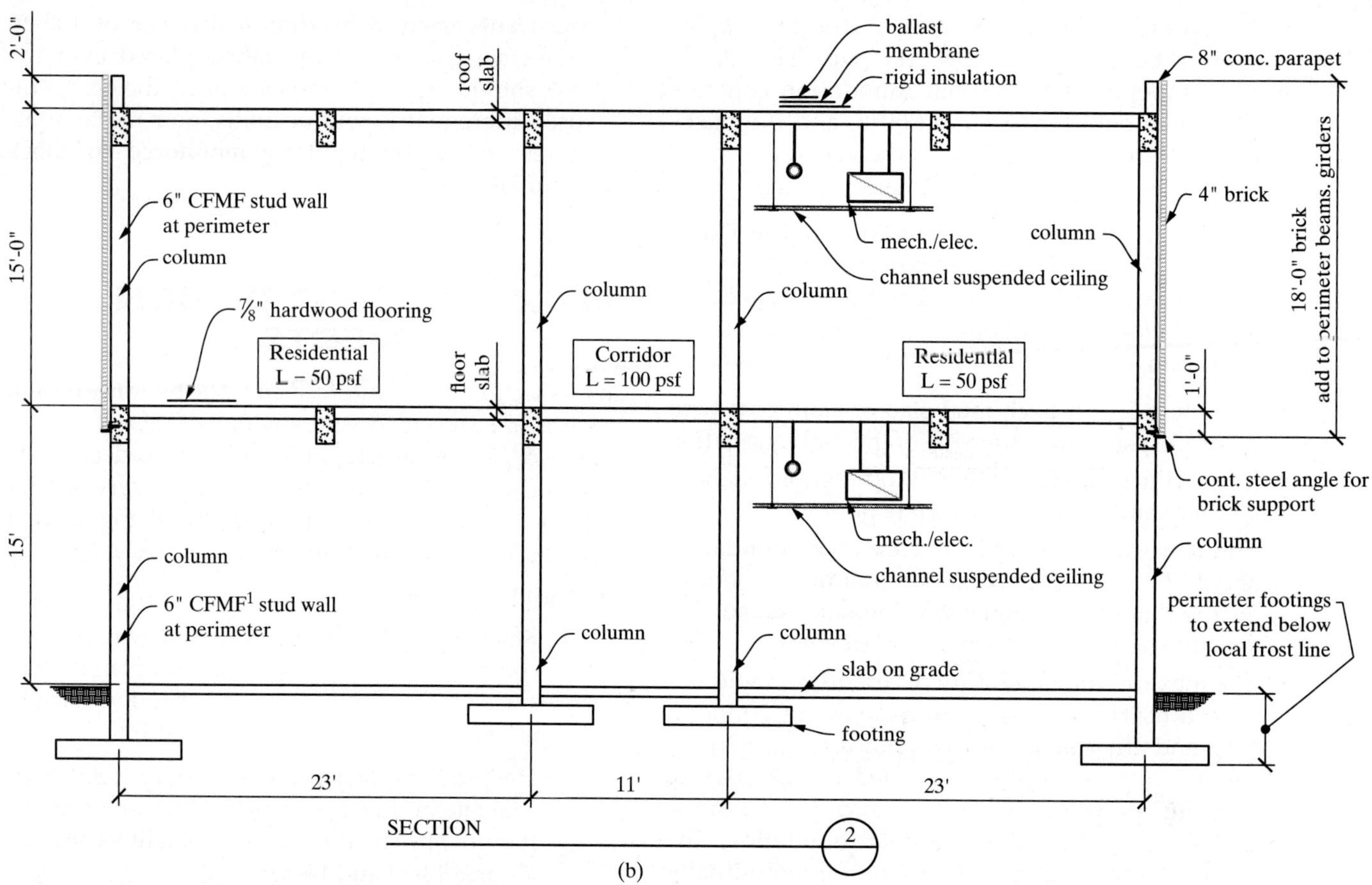

FIGURE 14-13 Residential building - section.
[1]CFMF is cold-formed metal framing

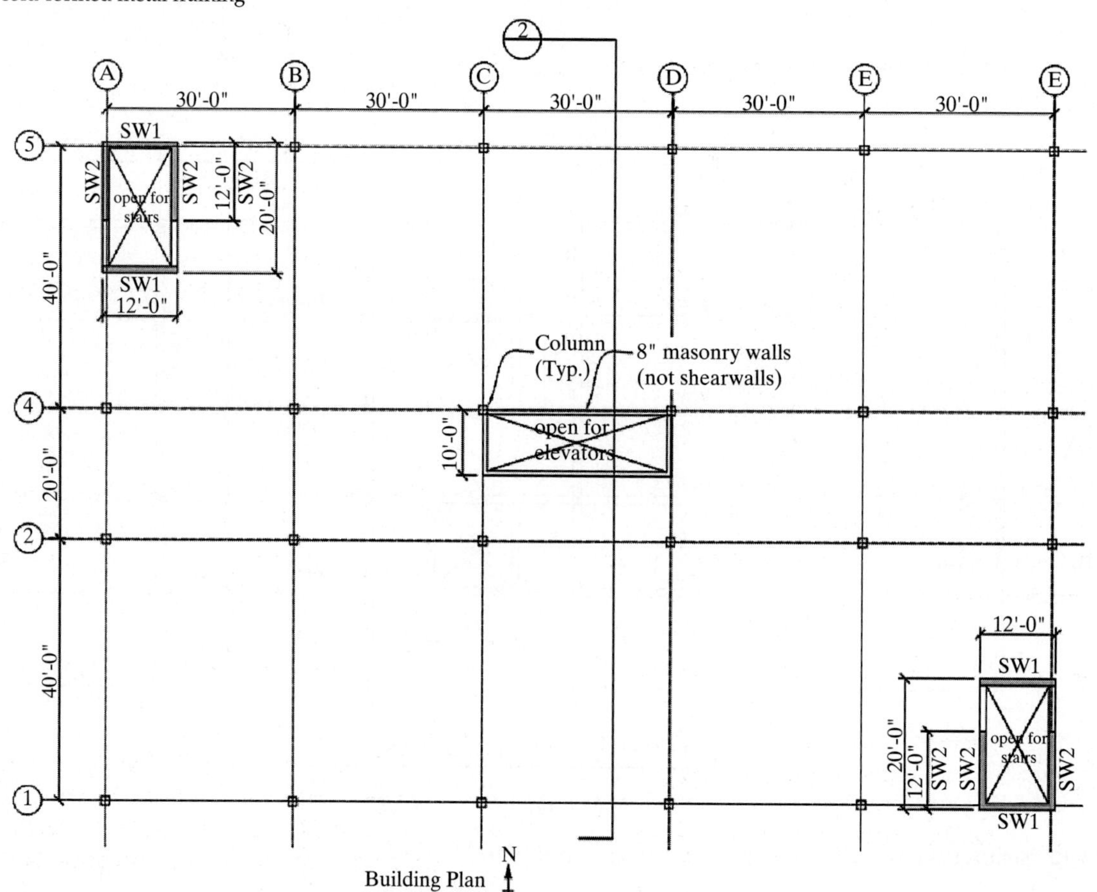

FIGURE 14-14 Hospital building - Typical floor and roof plan.

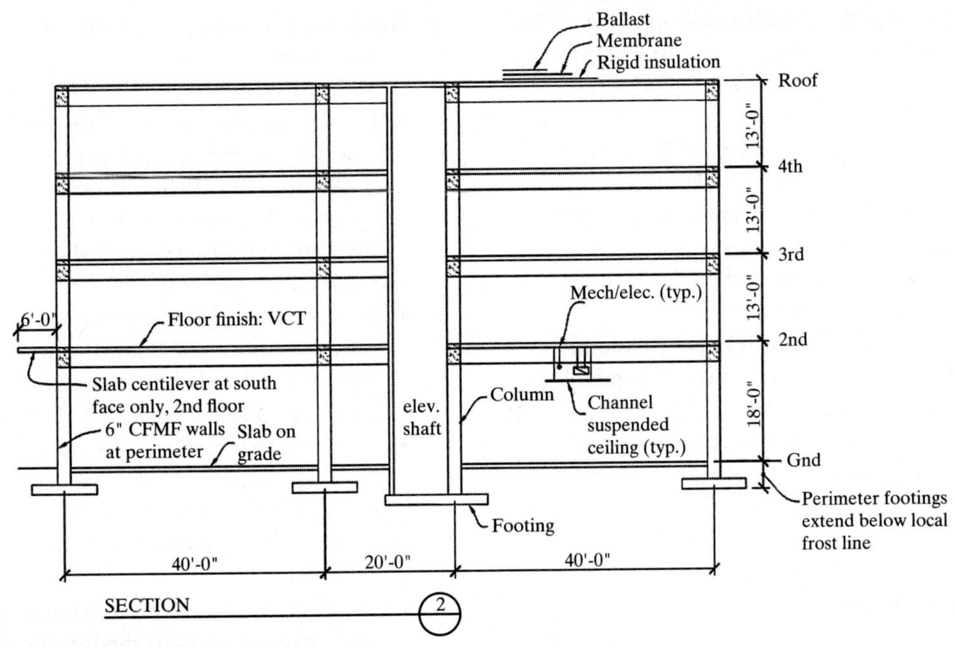

FIGURE 14-15 Hospital building - section.

- Each Building is located in Philadelphia, PA.
- The elevated slabs are supported on beams and girders.
- Lateral loads are to be resisted by shearwalls that are symmetrically located as indicated.
- The ACI 318 specification and the ASCE-7 load standard should be used.
- Assume that the stairs will be designed by others.
- Determine the critical lateral loading—seismic or wind—and use this to design the lateral force resisting systems or shearwalls.
- *Member Properties:* Use $f'_c = 4000$ psi and $f_y = 60$ ksi.

2. **Drawings and details:** May be either large size, foldout size, or on 8-1/2″ × 11″ and should include the following:
   - *Framing Floor and Roof Plans* that show the floor and roof slabs, floor and roof beams and girders, columns, and shearwalls.
   - Provide appropriate marks for all of the framing members (e.g., B-1, B-2, G-1, etc.)
   - *Foundation Plans* that show the slab-on-grade and foundations.
   - Show control joint locations in the slab-on-grade.
   - *Schedules*: Slab, beam, girder, column, and footing schedules.
   - *Sections, Details, and Elevations* that show typical beams, slabs, columns, footings, and shearwalls with reinforcing sizes and placement, as well as cover requirements and lap splice details.
3. **Loads:**
   - Calculate the dead and live loads (floor live loads and roof live loads).
   - Live loads:
     (a) Residential Building—Use a design live load of 50 psf for the second floor except at the corridor between grids 2 and 3 where the live load of 100 psf is required.
     (b) Hospital Building—Use a design live load of 80 psf for all floors
   - Use at least 15 psf for partitions (if applicable) and 10 psf for mechanical and electrical fixtures.
   - Follow the ASCE-7 load standard to calculate the wind load, snow load, and seismic load.
   - Assume a net allowable soil bearing capacity of 3000 psf.
   - Perimeter Cladding:
     (a) Residential Building—Assume the perimeter cladding (4″ brick wall with steel stud backing), which weighs 40 psf, will be supported by the perimeter (spandrel) beams and girders only.
     (b) Hospital Building—Perimeter is curtain wall and weighs 20 psf supported at each floor.
4. **Checklist of Design Items:**
   - *Provide the following at the front of the project submission:* cover page, an executive summary, design assumptions and a table of contents.
   - *Gravity and Lateral Loads:* Provide a load summation table for the floor and roof loads (gravity) and the wind and seismic loads (lateral). Provide a column load summation table.

- *Floor and Roof Slabs:* Continuous one-way slabs, check deflections against ACI requirements on the floors; framing plans should show typical slab reinforcing.
- *Floor Beams:* Design the interior "T" beams and girders, and the exterior "L" beams and girders. Draw the shear force diagrams for each and check deflections against Code requirements; show final beam and girder sizes in a beam schedule.
- *Columns:* Show the column loads in a load summation table and show final column sizes in a column schedule.
- *Foundations:* Design column footings, wall footings for shearwalls, and exterior wall footings and show final footing sizes in a footing schedule.
- *Shearwalls:* Design for lateral loads in both orthogonal directions and show reinforcing details.
- Check the development lengths for all flexural reinforcements.
- Provide quantities of rebar and concrete and cost estimates of each (installed cost).

5. **Analysis and Design considerations:**
   - For simplicity, the box-type behavior of the stair and elevator walls is ignored.
   - For the analysis of the slabs, beams, and girders, use a computer analysis to more accurately determine the shears, moments, and deflections, and compare the results with the approximate moments and shears obtained from the ACI coefficients
   - Provide line diagrams of the slabs, beams, and girders showing the loads and other input data used for either hand calculations or the computer analysis.
   - The following deflection limits are recommended: Total Loads: L/240, Live Load: L/360, Brick plus Live Load: L/600 or 0.3 inch (use the smaller value) for members supporting brick cladding.
   - Shearwalls should be modeled with lateral and gravity loads to design the footings. The footings are designed as a continuous wall footing and should be checked for overturning.

A project checklist is presented in Table 14-3.

TABLE 14-3 Project Checklist

Design Item	Description
Project Description	Title page, Table of contents, Executive summary, Design assumptions
Design Loads	Dead, Live, Snow, Wind
Slab Design	Continuous design, shear/moment diagrams, Bending, strain, shear, main steel, temperature steel
Beams	Line diagrams, Continuous design, shear/moment diagrams, Bending, strain, shear, deflection
Girders	Line diagrams, Continuous design, shear/moment diagrams, Bending, strain, shear, deflection
Columns	Load summation tables, compression check, ties and vertical bars
Footings (spread & wall ftgs.)	1- & 2-way shear, bending
Shearwalls	Load diagrams (lateral & gravity), shear/moment diagrams, shear, bending
Slab on Grade	Joints, reinforcing
Bond/Anchorage	lap splices & dowels: columns, walls; dev length: footings; slabs and beams
Schedules	slab, beams, girders, columns, footings
Drawings & Details	**Notes:** Material properties and assumptions **Plans:** marks for beams, columns, footings **Slabs:** rebar layout in plan & section **Beams:** rebar & stirrup layout in an elevation & section **Shearwalls:** plans, sections, detail **Columns:** section **Slab on grade:** Joints & reinforcing; **Footings:** sections **Schedules:** as needed for slabs, beams, columns and footings All details are to be to scale and include all appropriate dimensions and call-outs
Spreadsheets/ computer analysis	Line diagrams for analysis input, shear/moment diagrams for output; show beam marks where appropriate; clearly identify variables
Quantities	cubic yards of concrete, tons of reinforcing

References

[1] Jim Delahay and Brad Christopher. "Current Trends in Economical Concrete Construction," *STRUCTURE Magazine*, October 2007, pp. 20–21.

[2] Clifford W. Schwinger. "Tips for Designing Constructible Concrete Structures." *STRUCTURE Magazine*, February 2011, pp. 42–43.

[3] David B. Fanella and S. K. Ghosh. *Simplified Design—Reinforced Concrete Buildings of Moderate Size and Height*." Skokie, IL: Portland Cement Association, 1993.

[4] R. E. Shaeffer. *Reinforced Concrete—Preliminary Design for Architects and Builders*. Blacklick, OH: McGraw Hill, 1992, 196 pp.

[5] Edward Allen and Joseph Iano. *The Architect's Studio Companion—Rules of Thumb for Preliminary Design*, 5th ed. John Wiley, Hoboken, New Jersey, 2011.

[6] ACI 314-16. "Guide to Simplified Design for Reinforced Concrete Buildings," American Concrete Institute, P.O. Box 9094, Farmington Hills, MI 48333-9094, 2016.

[7] B. Keith Brenner. "Innovative Reinforcing Gives Old Structure New 'Light.'" *STRUCTURE Magazine*, January 2012, pp. 26–27.

[8] T. Alkhrdaji and J. Thomas. "Keys to Success: Structural Repair and Strengthening Techniques for Concrete Facilities." Structural Engineer, V. 5, No. 4, May 2004, pp. 24–27.

[9] Mo Ehsani, Majid Farahani, and Eric Raatz. "Repair of Columns with FRP Laminates." *STRUCTURE Magazine*, January 2012, pp. 35–37.

[10] FEMA-547. "Techniques for the Seismic Rehabilitation of Existing Buildings." Federal Emergency Management Agency, Washington, D.C., 2006.

[11] Institute for Research in Construction. "Guidelines for Seismic Upgrading of Building Structures." National Research Council of Canada, 1995, pp. 16.

[12] M. K. ElBatanouny, Antonio Nanni, Paul H. Ziel, and Fabio Matta. "Condition Assessment of Prestressed Concrete Beams Using Cyclic and Monotonic Load Tests. "*ACI Structural Journal*, V. 112, No. 1, January-February 2015, pp. 81–89.

[13] Subhash Kulkarni. "Forensic Study of a Partial Collapse." *ACI Concrete International*, October 2000, pp. 45–50.

[14] Alexander Newman. "Structural Renovation of Buildings. McGraw Hill, September 2000.

[15] Dov Kaminestzky. "Construction and Design Failures: Lessons from Forensic Investigations." McGraw-Hill, January 1991.

[16] Gustavo Tumialan, Nestore Galati and Antonio Nanni. "In-Situ Load Testing of Concrete Structures: Part 1 – Rational, Objectives and Execution. "*STRUCTURE Magazine*," April 2014, pp. 10–12.

[17] Gustavo Tumialan, Nestore Galati and Antonio Nanni. "Load Testing of Concrete Structures: Part 2 – Test Protocols and Case Studies. "*STRUCTURE Magazine*," June 2014, pp. 16–18.

[18] ACI 318-14. "Building Code Requirements for Structural Concrete." American Concrete Institute (ACI), Farmington Hills, Michigan, 2014.

[19] ACI 437-12. "Code Requirements for Load Testing of Existing Concrete Structures and Commentary." American Concrete Institute (ACI), Farmington Hills, Michigan, 2012.

[20] *PCI Design Handbook – Precast and Prestressed Concrete*, 7th Edition, Precast/Prestressed Institute, Chicago, IL, 2010.

[21] PCI Design Manual for the Design of Hollow Core Slabs, Precast/Prestressed Institute, Chicago, IL, 1998.

[22] Alex Aswad and Francis J. Jacques, "Behavior of Hollow-Core Slabs Subject to Edge Loads," PCI Journal, March-April 1992.

[23] AS 3600-2009, Australian Standard, Concrete Structures, Section 9.6, Council of Standards Australia, Sydney, Australia, 2009.

[24] Eva O. L. Lantsought, Cor van der Veen, and Joost C. Walraven, "Shear in One-Way Slabs under Concentrated Load Close to Support," *ACI Structural Journal*, V 110, No. 2, March-April 2013.

[25] David N Bilow and Mahmoud E. Kamara. *Fire and Concrete Structures*, 2008 ASCE Structures Congress, Vancouver, B.C., Canada. April 24–26, 2008.

[26] ACI. Standard Method for Determining Fire Resistance of Concrete and Masonry Construction Assemblies, ACI 216.1, American Concrete Institute, Farmington Hills, MI, 2007.

[27] IBC. International Building Code, International Code Council, Falls Church, VA, 2016.

[28] ASTM E 119-00a, "Standard Test Methods for Fire Tests of Building Construction and Materials." ASTM International, West Conshohocken, PA, 2000, p. 21.

[29] David A. Fanella "Time-saving Design Aids for Reinforced Concrete." *Structural Engineer*, pp. 39-41, August 2001.

[30] Andre Brault, Neil Hoult, Tom Greenough, Ian Trudeau, and Barry Charnish. "Closure Strip Strategies," *ACI Concrete International*, pp. 29–35, July 2017.

[31] Madan Mehta, Walter Scarborough, and Diane Armpriest. "Building Construction: Principles, Materials, and Systems," Pearson, 2010.

[32] CSI MasterFormat®, "Master List of Numbers and Titles for the Construction Industry." Construction Specifications Institute (CSI), 2012.

[33] Abdol R. Chini, Larry C. Muszynski, Lucky Acquaye, and Sophia Tarkhan, "Determination of the maximum placement and curing temperature of mass concrete to avoid durability problems and DEF." Final Report to Florida Department of Transportation (Contract No. BC 354-29), February 2003.

[34] Stephen P. Timoshenko and Woinowsky-Krieger. "Theory of Plates and Shells," Second Edition, McGraw-Hill, 1959, New York.

[35] CAC. "Concrete Design Handbook," Cement Association of Canada (CAC), 3rd Edition, 2006, Ottawa.

APPENDIX A

TABLES AND DIAGRAMS

A-1 Reinforcing Steel

A-2 Areas of Multiples of Reinforcing Bars

A-3 Minimum Required Beam Widths

A-4 Areas of Reinforcing Bars per Foot of Slab

A-5 Design Constants

A-6 Properties and Constants for Normal-Weight Concrete

A-7 Through A-11 Coefficient of Resistance ($\bar{k}$) versus Reinforcement Ratio (ρ)

A-8 Coefficient of Resistance ($\bar{k}$) versus Reinforcement Ratio (ρ)

A-9 Coefficient of Resistance ($\bar{k}$) versus Reinforcement Ratio (ρ)

A-10 Coefficient of Resistance ($\bar{k}$) versus Reinforcement Ratio (ρ)

A-11 Coefficient of Resistance ($\bar{k}$) versus Reinforcement Ratio (ρ)

A-12 Development Length for Compression Bars

A-13 Development Length for Hooked Bars

A-14 Preferred Maximum Number of Column Bars

A-15 Through A-22 Column Interaction Diagrams

TABLE A-1 Reinforcing Steel

Type of steel and ASTM specification number	Bar sizes	Grade	Minimum tensile strength (psi)	Minimum yield strength f_y (psi)	Yield strain ϵ_y
Billet Steel A615	Nos. 3–6	40	70,000	40,000	0.00138
	Nos. 3–18	60	90,000	60,000	0.00207
	Nos. 6–18	75	100,000	75,000	0.00259
Low-Alloy Steel A706	Nos. 3–18	60	80,000 (Min.: 1.25 f_y)	60,000 (Max.: 78,000)	0.00207

Bar number	3	4	5	6	7	8	9	10	11	14	18
Unit weight per foot (lb)	0.376	0.668	1.043	1.502	2.044	2.670	3.400	4.303	5.313	7.65	13.60
Diameter[a] (in.)	0.375	0.500	0.625	0.750	0.875	1.000	1.128	1.270	1.410	1.693	2.257
Area (in.^{2})	0.11	0.20	0.31	0.44	0.60	0.79	1.00	1.27	1.56	2.25	4.00

[a]The nominal dimensions of a deformed bar (diameter and area) are equivalent to those of a plain round bar having the same weight per foot as the deformed bar.

TABLE A-2 Areas of Multiples of Reinforcing Bars (in.^{2})

Number of bars	Bar number								
	#3	#4	#5	#6	#7	#8	#9	#10	#11
1	0.11	0.20	0.31	0.44	0.60	0.79	1.00	1.27	1.56
2	0.22	0.40	0.62	0.88	1.20	1.58	2.00	2.54	3.12
3	0.33	0.60	0.93	1.32	1.80	2.37	3.00	3.81	4.68
4	0.44	0.80	1.24	1.76	2.40	3.16	4.00	5.08	6.24
5	0.55	1.00	1.55	2.20	3.00	3.93	5.00	6.35	7.80
6	0.66	1.20	1.86	2.64	3.60	4.74	6.00	7.62	9.36
7	0.77	1.40	2.17	3.08	4.20	5.53	7.00	8.89	10.9
8	0.88	1.60	2.48	3.52	4.80	6.32	8.00	10.2	12.5
9	0.99	1.80	2.79	3.96	5.40	7.11	9.00	11.4	14.0
10	1.10	2.00	3.10	4.40	6.00	7.90	10.0	12.7	15.6
11	1.21	2.20	3.41	4.84	6.60	8.69	11.0	14.0	17.2
12	1.32	2.40	3.72	5.28	7.20	9.48	12.0	15.2	18.7
13	1.43	2.60	4.03	5.72	7.80	10.3	13.0	16.5	20.3
14	1.54	2.80	4.34	6.16	8.40	11.1	14.0	17.8	21.8
15	1.65	3.00	4.65	6.60	9.00	11.8	15.0	19.0	23.4
16	1.76	3.20	4.96	7.04	9.60	12.6	16.0	20.3	25.0
17	1.87	3.40	5.27	7.48	10.2	13.4	17.0	21.6	26.5
18	1.98	3.60	5.58	7.92	10.8	14.2	18.0	22.9	28.1
19	2.09	3.80	5.89	8.36	11.4	15.0	19.0	24.1	29.6
20	2.20	4.00	6.20	8.80	12.0	15.8	20.0	25.4	31.2

TABLE A-3 Minimum Required Beam Widths (in.)

Number of bars in one layer	Bar number							
	#3 and #4	#5	#6	#7	#8	#9	#10	#11
2	6.0	6.0	6.5	6.5	7.0	7.5	8.0	8.0
3	7.5	8.0	8.0	8.5	9.0	9.5	10.5	11.0
4	9.0	9.5	10.0	10.5	11.0	12.0	13.0	14.0
5	10.5	11.0	11.5	12.5	13.0	14.0	15.5	16.5
6	12.0	12.5	13.5	14.0	15.0	16.5	18.0	19.5
7	13.5	14.5	15.0	16.0	17.0	18.5	20.5	22.5
8	15.0	16.0	17.0	18.0	19.0	21.0	23.0	25.0
9	16.5	17.5	18.5	20.0	21.0	23.0	25.5	28.0
10	18.0	19.0	20.5	21.5	23.0	25.5	28.0	31.0

Note: Tabulated values based on No. 3 stirrups, minimum clear distance of 1 in., and a $1\frac{1}{2}$ in. cover.

TABLE A-4 Areas of Reinforcing Bars per Foot of Slab (in.^{2})

Bar spacing (in.)	Bar number								
	#3	#4	#5	#6	#7	#8	#9	#10	#11
2	0.66	1.20	1.86						
$2\frac{1}{2}$	0.53	0.96	1.49	2.11					
3	0.44	0.80	1.24	1.76	2.40	3.16	4.00		
$3\frac{1}{2}$	0.38	0.69	1.06	1.51	2.06	2.71	3.43	4.35	
4	0.33	0.60	0.93	1.32	1.80	2.37	3.00	3.81	4.68
$4\frac{1}{2}$	0.29	0.53	0.83	1.17	1.60	2.11	2.67	3.39	4.16
5	0.26	0.48	0.74	1.06	1.44	1.90	2.40	3.05	3.74
$5\frac{1}{2}$	0.24	0.44	0.68	0.96	1.31	1.72	2.18	2.77	3.40
6	0.22	0.40	0.62	0.88	1.20	1.58	2.00	2.54	3.12
$6\frac{1}{2}$	0.20	0.37	0.57	0.81	1.11	1.46	1.85	2.34	2.88
7	0.19	0.34	0.53	0.75	1.03	1.35	1.71	2.18	2.67
$7\frac{1}{2}$	0.18	0.32	0.50	0.70	0.96	1.26	1.60	2.03	2.50
8	0.16	0.30	0.46	0.66	0.90	1.18	1.50	1.90	2.34
9	0.15	0.27	0.41	0.59	0.80	1.05	1.33	1.69	2.08
10	0.13	0.24	0.37	0.53	0.72	0.95	1.20	1.52	1.87
11	0.12	0.22	0.34	0.48	0.65	0.86	1.09	1.39	1.70
12	0.11	0.20	0.31	0.44	0.60	0.79	1.00	1.27	1.56
13	0.10	0.18	0.29	0.41	0.55	0.73	0.92	1.17	1.44
14	0.09	0.17	0.27	0.38	0.51	0.68	0.86	1.09	1.34
15	0.09	0.16	0.25	0.35	0.48	0.64	0.80	1.02	1.25
16	0.08	0.15	0.23	0.33	0.45	0.59	0.75	0.95	1.17
17	0.08	0.14	0.22	0.31	0.42	0.56	0.71	0.90	1.10
18	0.07	0.13	0.21	0.29	0.40	0.53	0.67	0.85	1.04

TABLE A-5 Design Constants

f'_c (psi)	$\left[\frac{3\sqrt{f'_c}}{f} \geq \frac{200}{f}\right]$	Recommended design values	
		ρ	$\bar{k}$(ksi)
	$f_y = 40{,}000$ psi		
3000	0.0050	0.0135	0.4828
4000	0.0050	0.0180	0.6438
5000	0.0053	0.0225	0.8047
6000	0.0058	0.0270	0.9657
	$f_y = 50{,}000$ psi		
3000	0.0040	0.0108	0.4828
4000	0.0040	0.0144	0.6438
5000	0.0042	0.0180	0.8047
6000	0.0046	0.0216	0.9657
	$f_y = 60{,}000$ psi		
3000	0.0033	0.0090	0.4828
4000	0.0033	0.0120	0.6438
5000	0.0035	0.0150	0.8047
6000	0.0039	0.0180	0.9657
	$f_y = 75{,}000$ psi		
3000	0.0027	0.0072	0.4828
4000	0.0027	0.0096	0.6438
5000	0.0028	0.0120	0.8047
6000	0.0031	0.0144	0.9657

[a]Does not apply to T-beams with flanges in tension (see Section 3-2). To compute $A_{s,min}$, see Section 2-8.

TABLE A-6 Properties and Constants for Normal-Weight Concrete

	f'_c (psi)			
	3000	3500	4000	5000
E_c (psi)[a]	3,120,000	3,370,000	3,605,000	4,030,000
n[b]	9	9	8	7
$7.5\sqrt{f'_c}$ (ksi)[c]	0.411	0.444	0.474	0.530

[a]E_c for normal-weight concrete $= 57{,}000\sqrt{f'_c}$.
[b]Nearest whole number.
[c]Modulus of rupture (f_r).

TABLE A-7 Coefficient of Resistance ($\bar{k}$) versus Reinforcement Ratio (ρ)
($f'_c = 3000$ psi; $f_y = 40{,}000$ psi; units of $\bar{k}$ are ksi)

ρ	$\bar{k}$	ρ	$\bar{k}$	ρ	$\bar{k}$
0.0010	0.0397	0.0054	0.2069	0.0098	0.3619
0.0011	0.0436	0.0055	0.2105	0.0099	0.3653
0.0012	0.0476	0.0056	0.2142	0.0100	0.3686
0.0013	0.0515	0.0057	0.2178	0.0101	0.3720
0.0014	0.0554	0.0058	0.2214	0.0102	0.3754
0.0015	0.0593	0.0059	0.2251	0.0103	0.3787
0.0016	0.0632	0.0060	0.2287	0.0104	0.3821
0.0017	0.0671	0.0061	0.2323	0.0105	0.3854
0.0018	0.0710	0.0062	0.2359	0.0106	0.3887
0.0019	0.0749	0.0063	0.2395	0.0107	0.3921
0.0020	0.0788	0.0064	0.2431	0.0108	0.3954
0.0021	0.0826	0.0065	0.2467	0.0109	0.3987
0.0022	0.0865	0.0066	0.2503	0.0110	0.4020
0.0023	0.0903	0.0067	0.2539	0.0111	0.4053
0.0024	0.0942	0.0068	0.2575	0.0112	0.4086
0.0025	0.0980	0.0069	0.2611	0.0113	0.4119
0.0026	0.1019	0.0070	0.2646	0.0114	0.4152
0.0027	0.1057	0.0071	0.2682	0.0115	0.4185
0.0028	0.1095	0.0072	0.2717	0.0116	0.4218
0.0029	0.1134	0.0073	0.2753	0.0117	0.4251
0.0030	0.1172	0.0074	0.2788	0.0118	0.4283
0.0031	0.1210	0.0075	0.2824	0.0119	0.4316
0.0032	0.1248	0.0076	0.2859	0.0120	0.4348
0.0033	0.1286	0.0077	0.2894	0.0121	0.4381
0.0034	0.1324	0.0078	0.2929	0.0122	0.4413
0.0035	0.1362	0.0079	0.2964	0.0123	0.4445
0.0036	0.1399	0.0080	0.2999	0.0124	0.4478
0.0037	0.1437	0.0081	0.3034	0.0125	0.4510
0.0038	0.1475	0.0082	0.3069	0.0126	0.4542
0.0039	0.1512	0.0083	0.3104	0.0127	0.4574
0.0040	0.1550	0.0084	0.3139	0.0128	0.4606
0.0041	0.1587	0.0085	0.3173	0.0129	0.4638
0.0042	0.1625	0.0086	0.3208	0.0130	0.4670
0.0043	0.1662	0.0087	0.3243	0.0131	0.4702
0.0044	0.1699	0.0088	0.3277	0.0132	0.4733
0.0045	0.1736	0.0089	0.3311	0.0133	0.4765
0.0046	0.1774	0.0090	0.3346	0.0134	0.4797
0.0047	0.1811	0.0091	0.3380	0.0135	0.4828
0.0048	0.1848	0.0092	0.3414	0.0136	0.4860
0.0049	0.1885	0.0093	0.3449	0.0137	0.4891
0.0050	0.1922	0.0094	0.3483	0.0138	0.4923
0.0051	0.1958	0.0095	0.3517	0.0139	0.4954
0.0052	0.1995	0.0096	0.3551	0.0140	0.4985
0.0053	0.2032	0.0097	0.3585	0.0141	0.5016

TABLE A-7 Continued

ρ	$\bar{k}$	ρ	$\bar{k}$	ρ	$\bar{k}$	ϵ^*_t
0.0142	0.5047	0.0173	0.5981	**0.02033**	**0.6836**	**0.00500**
0.0143	0.5078	0.0174	0.6011	0.0204	0.6855	0.00497
0.0144	0.5109	0.0175	0.6040	0.0205	0.6882	0.00493
0.0145	0.5140	0.0176	0.6069	0.0206	0.6909	0.00489
0.0146	0.5171	0.0177	0.6098	0.0207	0.6936	0.00485
0.0147	0.5202	0.0178	0.6126	0.0208	0.6963	0.00482
0.0148	0.5233	0.0179	0.6155	0.0209	0.6990	0.00478
0.0149	0.5264	0.0180	0.6184	0.0210	0.7017	0.00474
0.0150	0.5294	0.0181	0.6213	0.0211	0.7044	0.00470
0.0151	0.5325	0.0182	0.6241	0.0212	0.7071	0.00467
0.0152	0.5355	0.0183	0.6270	0.0213	0.7097	0.00463
0.0153	0.5386	0.0184	0.6298	0.0214	0.7124	0.00460
0.0154	0.5416	0.0185	0.6327	0.0215	0.7150	0.00456
0.0155	0.5447	0.0186	0.6355	0.0216	0.7177	0.00453
0.0156	0.5477	0.0187	0.6383	0.0217	0.7203	0.00449
0.0157	0.5507	0.0188	0.6412	0.0218	0.7230	0.00446
0.0158	0.5537	0.0189	0.6440	0.0219	0.7256	0.00442
0.0159	0.5567	0.0190	0.6468	0.0220	0.7282	0.00439
0.0160	0.5597	0.0191	0.6496	0.0221	0.7308	0.00436
0.0161	0.5627	0.0192	0.6524	0.0222	0.7334	0.00432
0.0162	0.5657	0.0193	0.6552	0.0223	0.7360	0.00429
0.0163	0.5687	0.0194	0.6580	0.0224	0.7386	0.00426
0.0164	0.5717	0.0195	0.6608	0.0225	0.7412	0.00423
0.0165	0.5746	0.0196	0.6635	0.0226	0.7438	0.00419
0.0166	0.5776	0.0197	0.6663	0.0227	0.7464	0.00416
0.0167	0.5805	0.0198	0.6691	0.0228	0.7490	0.00413
0.0168	0.5835	0.0199	0.6718	0.0229	0.7515	0.00410
0.0169	0.5864	0.0200	0.6746	0.0230	0.7541	0.00407
0.0170	0.5894	0.0201	0.6773	0.0231	0.7567	0.00404
0.0171	0.5923	0.0202	0.6800	0.0232	0.7592	0.00401
0.0172	0.5952	0.0203	0.6828	**0.02323**	**0.7600**	**0.00400**

*$d = d_t$.

TABLE A-8 Coefficient of Resistance ($\bar{k}$) versus Reinforcement Ratio (ρ) ($f'_c = 3000$ psi; $f_y = 60{,}000$ psi; units of $\bar{k}$ are ksi)

ρ	$\bar{k}$	ρ	$\bar{k}$	ρ	$\bar{k}$	ϵ^*_t
0.0010	0.0593	0.0059	0.3294	0.0108	0.5657	
0.0011	0.0651	0.0060	0.3346	0.0109	0.5702	
0.0012	0.0710	0.0061	0.3397	0.0110	0.5746	
0.0013	0.0768	0.0062	0.3449	0.0111	0.5791	
0.0014	0.0826	0.0063	0.3500	0.0112	0.5835	
0.0015	0.0884	0.0064	0.3551	0.0113	0.5879	
0.0016	0.0942	0.0065	0.3602	0.0114	0.5923	
0.0017	0.1000	0.0066	0.3653	0.0115	0.5967	
0.0018	0.1057	0.0067	0.3703	0.0116	0.6011	
0.0019	0.1115	0.0068	0.3754	0.0117	0.6054	
0.0020	0.1172	0.0069	0.3804	0.0118	0.6098	
0.0021	0.1229	0.0070	0.3854	0.0119	0.6141	
0.0022	0.1286	0.0071	0.3904	0.0120	0.6184	
0.0023	0.1343	0.0072	0.3954	0.0121	0.6227	
0.0024	0.1399	0.0073	0.4004	0.0122	0.6270	
0.0025	0.1456	0.0074	0.4054	0.0123	0.6312	
0.0026	0.1512	0.0075	0.4103	0.0124	0.6355	
0.0027	0.1569	0.0076	0.4152	0.0125	0.6398	
0.0028	0.1625	0.0077	0.4202	0.0126	0.6440	
0.0029	0.1681	0.0078	0.4251	0.0127	0.6482	
0.0030	0.1736	0.0079	0.4300	0.0128	0.6524	
0.0031	0.1792	0.0080	0.4348	0.0129	0.6566	
0.0032	0.1848	0.0081	0.4397	0.0130	0.6608	
0.0033	0.1903	0.0082	0.4446	0.0131	0.6649	
0.0034	0.1958	0.0083	0.4494	0.0132	0.6691	
0.0035	0.2014	0.0084	0.4542	0.0133	0.6732	
0.0036	0.2069	0.0085	0.4590	0.0134	0.6773	
0.0037	0.2123	0.0086	0.4638	0.0135	0.6814	
0.0038	0.2178	0.0087	0.4686	**0.01355**	**0.6835**	**0.00500**
0.0039	0.2233	0.0088	0.4734	0.0136	0.6855	0.00497
0.0040	0.2287	0.0089	0.4781	0.0137	0.6896	0.00491
0.0041	0.2341	0.0090	0.4828	0.0138	0.6936	0.00485
0.0042	0.2396	0.0091	0.4876	0.0139	0.6977	0.00480
0.0043	0.2450	0.0092	0.4923	0.0140	0.7017	0.00474
0.0044	0.2503	0.0093	0.4970	0.0141	0.7057	0.00469
0.0045	0.2557	0.0094	0.5017	0.0142	0.7097	0.00463
0.0046	0.2611	0.0095	0.5063	0.0143	0.7137	0.00458
0.0047	0.2664	0.0096	0.5110	0.0144	0.7177	0.00453
0.0048	0.2717	0.0097	0.5156	0.0145	0.7216	0.00447
0.0049	0.2771	0.0098	0.5202	0.0146	0.7256	0.00442
0.0050	0.2824	0.0099	0.5248	0.0147	0.7295	0.00437
0.0051	0.2876	0.0100	0.5294	0.0148	0.7334	0.00432
0.0052	0.2929	0.0101	0.5340	0.0149	0.7373	0.00427
0.0053	0.2982	0.0102	0.5386	0.0150	0.7412	0.00423
0.0054	0.3034	0.0103	0.5431	0.0151	0.7451	0.00418
0.0055	0.3087	0.0104	0.5477	0.0152	0.7490	0.00413
0.0056	0.3139	0.0105	0.5522	0.0153	0.7528	0.00408
0.0057	0.3191	0.0106	0.5567	0.0154	0.7567	0.00404
0.0058	0.3243	0.0107	0.5612	**0.01548**	**0.7597**	**0.00400**

*$d = d_t$.

TABLE A-9 Coefficient of Resistance ($\overline{k}$) versus Reinforcement Ratio (ρ) ($f'_c = 4000$ psi; $f_y = 40{,}000$ psi; units of ($\overline{k}$) are ksi)

ρ	$\overline{k}$	ρ	$\overline{k}$	ρ	$\overline{k}$	ρ	$\overline{k}$
0.0010	0.0398	0.0054	0.2091	0.0098	0.3694	0.0142	0.5206
0.0011	0.0437	0.0055	0.2129	0.0099	0.3729	0.0143	0.5239
0.0012	0.0477	0.0056	0.2166	0.0100	0.3765	0.0144	0.5272
0.0013	0.0516	0.0057	0.2204	0.0101	0.3800	0.0145	0.5305
0.0014	0.0555	0.0058	0.2241	0.0102	0.3835	0.0146	0.5338
0.0015	0.0595	0.0059	0.2278	0.0103	0.3870	0.0147	0.5372
0.0016	0.0634	0.0060	0.2315	0.0104	0.3906	0.0148	0.5405
0.0017	0.0673	0.0061	0.2352	0.0105	0.3941	0.0149	0.5438
0.0018	0.0712	0.0062	0.2390	0.0106	0.3976	0.0150	0.5471
0.0019	0.0752	0.0063	0.2427	0.0107	0.4011	0.0151	0.5504
0.0020	0.0791	0.0064	0.2464	0.0108	0.4046	0.0152	0.5536
0.0021	0.0830	0.0065	0.2501	0.0109	0.4080	0.0153	0.5569
0.0022	0.0869	0.0066	0.2538	0.0110	0.4115	0.0154	0.5602
0.0023	0.0908	0.0067	0.2574	0.0111	0.4150	0.0155	0.5635
0.0024	0.0946	0.0068	0.2611	0.0112	0.4185	0.0156	0.5667
0.0025	0.0985	0.0069	0.2648	0.0113	0.4220	0.0157	0.5700
0.0026	0.1024	0.0070	0.2685	0.0114	0.4254	0.0158	0.5733
0.0027	0.1063	0.0071	0.2721	0.0115	0.4289	0.0159	0.5765
0.0028	0.1102	0.0072	0.2758	0.0116	0.4323	0.0160	0.5798
0.0029	0.1140	0.0073	0.2795	0.0117	0.4358	0.0161	0.5830
0.0030	0.1179	0.0074	0.2831	0.0118	0.4392	0.0162	0.5863
0.0031	0.1217	0.0075	0.2868	0.0119	0.4427	0.0163	0.5895
0.0032	0.1256	0.0076	0.2904	0.0120	0.4461	0.0164	0.5927
0.0033	0.1294	0.0077	0.2941	0.0121	0.4495	0.0165	0.5959
0.0034	0.1333	0.0078	0.2977	0.0122	0.4530	0.0166	0.5992
0.0035	0.1371	0.0079	0.3013	0.0123	0.4564	0.0167	0.6024
0.0036	0.1410	0.0080	0.3049	0.0124	0.4598	0.0168	0.6056
0.0037	0.1448	0.0081	0.3086	0.0125	0.4632	0.0169	0.6088
0.0038	0.1486	0.0082	0.3122	0.0126	0.4666	0.0170	0.6120
0.0039	0.1524	0.0083	0.3158	0.0127	0.4701	0.0171	0.6152
0.0040	0.1562	0.0084	0.3194	0.0128	0.4735	0.0172	0.6184
0.0041	0.1600	0.0085	0.3230	0.0129	0.4768	0.0173	0.6216
0.0042	0.1638	0.0086	0.3266	0.0130	0.4802	0.0174	0.6248
0.0043	0.1676	0.0087	0.3302	0.0131	0.4836	0.0175	0.6279
0.0044	0.1714	0.0088	0.3338	0.0132	0.4870	0.0176	0.6311
0.0045	0.1752	0.0089	0.3374	0.0133	0.4904	0.0177	0.6343
0.0046	0.1790	0.0090	0.3409	0.0134	0.4938	0.0178	0.6375
0.0047	0.1828	0.0091	0.3445	0.0135	0.4971	0.0179	0.6406
0.0048	0.1866	0.0092	0.3481	0.0136	0.5005	0.0180	0.6438
0.0049	0.1904	0.0093	0.3517	0.0137	0.5038	0.0181	0.6469
0.0050	0.1941	0.0094	0.3552	0.0138	0.5072	0.0182	0.6501
0.0051	0.1979	0.0095	0.3588	0.0139	0.5105	0.0183	0.6532
0.0052	0.2016	0.0096	0.3623	0.0140	0.5139	0.0184	0.6563
0.0053	0.2054	0.0097	0.3659	0.0141	0.5172	0.0185	0.6595

TABLE A-9 Continued

ρ	$\bar{k}$	ρ	$\bar{k}$	ρ	$\bar{k}$	ϵ^*_t
0.0186	0.6626	0.0229	0.7927	**0.0271**	**0.9113**	**0.00500**
0.0187	0.6657	0.0230	0.7956	0.0272	0.9140	0.00497
0.0188	0.6688	0.0231	0.7985	0.0273	0.9167	0.00494
0.0189	0.6720	0.0232	0.8014	0.0274	0.9194	0.00491
0.0190	0.6751	0.0233	0.8043	0.0275	0.9221	0.00488
0.0191	0.6782	0.0234	0.8072	0.0276	0.9248	0.00485
0.0192	0.6813	0.0235	0.8101	0.0277	0.9275	0.00482
0.0193	0.6844	0.0236	0.8130	0.0278	0.9302	0.00480
0.0194	0.6875	0.0237	0.8159	0.0279	0.9329	0.00477
0.0195	0.6905	0.0238	0.8188	0.0280	0.9356	0.00474
0.0196	0.6936	0.0239	0.8217	0.0281	0.9383	0.00471
0.0197	0.6967	0.0240	0.8245	0.0282	0.9410	0.00469
0.0198	0.6998	0.0241	0.8274	0.0283	0.9436	0.00466
0.0199	0.7029	0.0242	0.8303	0.0284	0.9463	0.00463
0.0200	0.7059	0.0243	0.8331	0.0285	0.9490	0.00461
0.0201	0.7090	0.0244	0.8360	0.0286	0.9516	0.00458
0.0202	0.7120	0.0245	0.8388	0.0287	0.9543	0.00455
0.0203	0.7151	0.0246	0.8417	0.0288	0.9569	0.00453
0.0204	0.7181	0.0247	0.8445	0.0289	0.9596	0.00450
0.0205	0.7212	0.0248	0.8473	0.0290	0.9622	0.00447
0.0206	0.7242	0.0249	0.8502	0.0291	0.9648	0.00445
0.0207	0.7272	0.0250	0.8530	0.0292	0.9675	0.00442
0.0208	0.7302	0.0251	0.8558	0.0293	0.9701	0.00440
0.0209	0.7333	0.0252	0.8586	0.0294	0.9727	0.00437
0.0210	0.7363	0.0253	0.8615	0.0295	0.9753	0.00435
0.0211	0.7393	0.0254	0.8643	0.0296	0.9779	0.00432
0.0212	0.7423	0.0255	0.8671	0.0297	0.9805	0.00430
0.0213	0.7453	0.0256	0.8699	0.0298	0.9831	0.00427
0.0214	0.7483	0.0257	0.8727	0.0299	0.9857	0.00425
0.0215	0.7513	0.0258	0.8754	0.0300	0.9883	0.00423
0.0216	0.7543	0.0259	0.8782	0.0301	0.9909	0.00420
0.0217	0.7572	0.0260	0.8810	0.0302	0.9935	0.00418
0.0218	0.7602	0.0261	0.8838	0.0303	0.9961	0.00415
0.0219	0.7632	0.0262	0.8865	0.0304	0.9986	0.00413
0.0220	0.7662	0.0263	0.8893	0.0305	1.0012	0.00411
0.0221	0.7691	0.0264	0.8921	0.0306	1.0038	0.00408
0.0222	0.7721	0.0265	0.8948	0.0307	1.0063	0.00406
0.0223	0.7750	0.0266	0.8976	0.0308	1.0089	0.00404
0.0224	0.7780	0.0267	0.9003	0.0309	1.0114	0.00401
0.0225	0.7809	0.0268	0.9031	**0.03096**	**1.0130**	**0.00400**
0.0226	0.7839	0.0269	0.9058			
0.0227	0.7868	0.0270	0.9085			
0.0228	0.7897					

*$d = d_t$.

TABLE A-10 Coefficient of Resistance ($\bar{k}$) versus Reinforcement Ratio (ρ) ($f'_c = 4000$ psi; $f_y = 60{,}000$ psi; units of $\bar{k}$ are ksi)

ρ	$\bar{k}$	ρ	$\bar{k}$	ρ	$\bar{k}$	ρ	$\bar{k}$
0.0010	0.0595	0.0039	0.2259	0.0068	0.3835	0.0097	0.5322
0.0011	0.0654	0.0040	0.2315	0.0069	0.3888	0.0098	0.5372
0.0012	0.0712	0.0041	0.2371	0.0070	0.3941	0.0099	0.5421
0.0013	0.0771	0.0042	0.2427	0.0071	0.3993	0.0100	0.5471
0.0014	0.0830	0.0043	0.2482	0.0072	0.4046	0.0101	0.5520
0.0015	0.0889	0.0044	0.2538	0.0073	0.4098	0.0102	0.5569
0.0016	0.0946	0.0045	0.2593	0.0074	0.4150	0.0103	0.5618
0.0017	0.1005	0.0046	0.2648	0.0075	0.4202	0.0104	0.5667
0.0018	0.1063	0.0047	0.2703	0.0076	0.4254	0.0105	0.5716
0.0019	0.1121	0.0048	0.2758	0.0077	0.4306	0.0106	0.5765
0.0020	0.1179	0.0049	0.2813	0.0078	0.4358	0.0107	0.5814
0.0021	0.1237	0.0050	0.2868	0.0079	0.4410	0.0108	0.5862
0.0022	0.1294	0.0051	0.2922	0.0080	0.4461	0.0109	0.5911
0.0023	0.1352	0.0052	0.2977	0.0081	0.4513	0.0110	0.5959
0.0024	0.1410	0.0053	0.3031	0.0082	0.4564	0.0111	0.6008
0.0025	0.1467	0.0054	0.3086	0.0083	0.4615	0.0112	0.6056
0.0026	0.1524	0.0055	0.3140	0.0084	0.4666	0.0113	0.6104
0.0027	0.1581	0.0056	0.3194	0.0085	0.4718	0.0114	0.6152
0.0028	0.1638	0.0057	0.3248	0.0086	0.4768	0.0115	0.6200
0.0029	0.1695	0.0058	0.3302	0.0087	0.4819	0.0116	0.6248
0.0030	0.1752	0.0059	0.3356	0.0088	0.4870	0.0117	0.6296
0.0031	0.1809	0.0060	0.3409	0.0089	0.4921	0.0118	0.6343
0.0032	0.1866	0.0061	0.3463	0.0090	0.4971	0.0119	0.6391
0.0033	0.1922	0.0062	0.3516	0.0091	0.5022	0.0120	0.6438
0.0034	0.1979	0.0063	0.3570	0.0092	0.5072	0.0121	0.6485
0.0035	0.2035	0.0064	0.3623	0.0093	0.5122	0.0122	0.6532
0.0036	0.2091	0.0065	0.3676	0.0094	0.5172	0.0123	0.6579
0.0037	0.2148	0.0066	0.3729	0.0095	0.5222	0.0124	0.6626
0.0038	0.2204	0.0067	0.3782	0.0096	0.5272	0.0125	0.6673

TABLE A-10 Continued

ρ	$\bar{k}$	ρ	$\bar{k}$	ρ	$\bar{k}$	ϵ_t^*
0.0126	0.6720	0.0154	0.7985	**0.01806**	**0.9110**	**0.00500**
0.0127	0.6766	0.0155	0.8029	0.0181	0.9126	0.00498
0.0128	0.6813	0.0156	0.8072	0.0182	0.9167	0.00494
0.0129	0.6859	0.0157	0.8116	0.0183	0.9208	0.00490
0.0130	0.6906	0.0158	0.8159	0.0184	0.9248	0.00485
0.0131	0.6952	0.0159	0.8202	0.0185	0.9289	0.00481
0.0132	0.6998	0.0160	0.8245	0.0186	0.9329	0.00477
0.0133	0.7044	0.0161	0.8288	0.0187	0.9369	0.00473
0.0134	0.7090	0.0162	0.8331	0.0188	0.9410	0.00469
0.0135	0.7136	0.0163	0.8374	0.0189	0.9450	0.00465
0.0136	0.7181	0.0164	0.8417	0.0190	0.9490	0.00461
0.0137	0.7227	0.0165	0.8459	0.0191	0.9529	0.00457
0.0138	0.7272	0.0166	0.8502	0.0192	0.9569	0.00453
0.0139	0.7318	0.0167	0.8544	0.0193	0.9609	0.00449
0.0140	0.7363	0.0168	0.8586	0.0194	0.9648	0.00445
0.0141	0.7408	0.0169	0.8629	0.0195	0.9688	0.00441
0.0142	0.7453	0.0170	0.8671	0.0196	0.9727	0.00437
0.0143	0.7498	0.0171	0.8713	0.0197	0.9766	0.00434
0.0144	0.7543	0.0172	0.8754	0.0198	0.9805	0.00430
0.0145	0.7587	0.0173	0.8796	0.0199	0.9844	0.00426
0.0146	0.7632	0.0174	0.8838	0.0200	0.9883	0.00422
0.0147	0.7676	0.0175	0.8879	0.0201	0.9922	0.00419
0.0148	0.7721	0.0176	0.8921	0.0202	0.9961	0.00415
0.0149	0.7765	0.0177	0.8962	0.0203	0.9999	0.00412
0.0150	0.7809	0.0178	0.9003	0.0204	1.0038	0.00408
0.0151	0.7853	0.0179	0.9044	0.0205	1.0076	0.00405
0.0152	0.7897	0.0180	0.9085	0.0206	1.0114	0.00401
0.0153	0.7941			**0.02063**	**1.0126**	**0.00400**

$^*d = d_t$.

TABLE A-11 Coefficient of Resistance ($\bar{k}$) versus Reinforcement Ratio (ρ) (f'_c = 5000 psi; f_y = 60,000 psi; units of $\bar{k}$ are ksi)

ρ	$\bar{k}$	ρ	$\bar{k}$	ρ	$\bar{k}$	ρ	$\bar{k}$
0.0010	0.0596	0.0048	0.2782	0.0086	0.4847	0.0124	0.6789
0.0011	0.0655	0.0049	0.2838	0.0087	0.4899	0.0125	0.6838
0.0012	0.0714	0.0050	0.2894	0.0088	0.4952	0.0126	0.6888
0.0013	0.0773	0.0051	0.2950	0.0089	0.5005	0.0127	0.6937
0.0014	0.0832	0.0052	0.3005	0.0090	0.5057	0.0128	0.6986
0.0015	0.0890	0.0053	0.3061	0.0091	0.5109	0.0129	0.7035
0.0016	0.0949	0.0054	0.3117	0.0092	0.5162	0.0130	0.7084
0.0017	0.1008	0.0055	0.3172	0.0093	0.5214	0.0131	0.7133
0.0018	0.1066	0.0056	0.3227	0.0094	0.5266	0.0132	0.7182
0.0019	0.1125	0.0057	0.3282	0.0095	0.5318	0.0133	0.7231
0.0020	0.1183	0.0058	0.3338	0.0096	0.5370	0.0134	0.7280
0.0021	0.1241	0.0059	0.3393	0.0097	0.5422	0.0135	0.7328
0.0022	0.1300	0.0060	0.3448	0.0098	0.5473	0.0136	0.7377
0.0023	0.1358	0.0061	0.3502	0.0099	0.5525	0.0137	0.7425
0.0024	0.1416	0.0062	0.3557	0.0100	0.5576	0.0138	0.7473
0.0025	0.1474	0.0063	0.3612	0.0101	0.5628	0.0139	0.7522
0.0026	0.1531	0.0064	0.3667	0.0102	0.5679	0.0140	0.7570
0.0027	0.1589	0.0065	0.3721	0.0103	0.5731	0.0141	0.7618
0.0028	0.1647	0.0066	0.3776	0.0104	0.5782	0.0142	0.7666
0.0029	0.1704	0.0067	0.3830	0.0105	0.5833	0.0143	0.7714
0.0030	0.1762	0.0068	0.3884	0.0106	0.5884	0.0144	0.7762
0.0031	0.1819	0.0069	0.3938	0.0107	0.5935	0.0145	0.7810
0.0032	0.1877	0.0070	0.3992	0.0108	0.5986	0.0146	0.7857
0.0033	0.1934	0.0071	0.4047	0.0109	0.6037	0.0147	0.7905
0.0034	0.1991	0.0072	0.4100	0.0110	0.6088	0.0148	0.7952
0.0035	0.2048	0.0073	0.4154	0.0111	0.6138	0.0149	0.8000
0.0036	0.2105	0.0074	0.4208	0.0112	0.6189	0.0150	0.8047
0.0037	0.2162	0.0075	0.4262	0.0113	0.6239	0.0151	0.8094
0.0038	0.2219	0.0076	0.4315	0.0114	0.6290	0.0152	0.8142
0.0039	0.2276	0.0077	0.4369	0.0115	0.6340	0.0153	0.8189
0.0040	0.2332	0.0078	0.4422	0.0116	0.6390	0.0154	0.8236
0.0041	0.2389	0.0079	0.4476	0.0117	0.6440	0.0155	0.8283
0.0042	0.2445	0.0080	0.4529	0.0118	0.6490	0.0156	0.8329
0.0043	0.2502	0.0081	0.4582	0.0119	0.6540	0.0157	0.8376
0.0044	0.2558	0.0082	0.4635	0.0120	0.6590	0.0158	0.8423
0.0045	0.2614	0.0083	0.4688	0.0121	0.6640	0.0159	0.8469
0.0046	0.2670	0.0084	0.4741	0.0122	0.6690	0.0160	0.8516
0.0047	0.2726	0.0085	0.4794	0.0123	0.6739	0.0161	0.8562

TABLE A-11 Continued

ρ	$\bar{k}$	ρ	$\bar{k}$	ρ	$\bar{k}$	ϵ^*_t
0.0162	0.8609	0.0194	1.0047	**0.02125**	**1.0838**	**0.00500**
0.0163	0.8655	0.0195	1.0090	0.0213	1.0859	0.00498
0.0164	0.8701	0.0196	1.0134	0.0214	1.0901	0.00494
0.0165	0.8747	0.0197	1.0177	0.0215	1.0943	0.00491
0.0166	0.8793	0.0198	1.0220	0.0216	1.0985	0.00487
0.0167	0.8839	0.0199	1.0263	0.0217	1.1026	0.00483
0.0168	0.8885	0.0200	1.0307	0.0218	1.1068	0.00480
0.0169	0.8930	0.0201	1.0350	0.0219	1.1110	0.00476
0.0170	0.8976	0.0202	1.0393	0.0220	1.1151	0.00473
0.0171	0.9022	0.0203	1.0435	0.0221	1.1192	0.00469
0.0172	0.9067	0.0204	1.0478	0.0222	1.1234	0.00466
0.0173	0.9112	0.0205	1.0521	0.0223	1.1275	0.00462
0.0174	0.9158	0.0206	1.0563	0.0224	1.1316	0.00459
0.0175	0.9203	0.0207	1.0606	0.0225	1.1357	0.00456
0.0176	0.9248	0.0208	1.0648	0.0226	1.1398	0.00452
0.0177	0.9293	0.0209	1.0691	0.0227	1.1438	0.00449
0.0178	0.9338	0.0210	1.0733	0.0228	1.1479	0.00446
0.0179	0.9383	0.0211	1.0775	0.0229	1.1520	0.00442
0.0180	0.9428	0.0212	1.0817	0.0230	1.1560	0.00439
0.0181	0.9473			0.0231	1.1601	0.00436
0.0182	0.9517			0.0232	1.1641	0.00433
0.0183	0.9562			0.0233	1.1682	0.00430
0.0184	0.9606			0.0234	1.1722	0.00426
0.0185	0.9651			0.0235	1.1762	0.00423
0.0186	0.9695			0.0236	1.1802	0.00420
0.0187	0.9739			0.0237	1.1842	0.00417
0.0188	0.9783			0.0238	1.882	0.00414
0.0189	0.9827			0.0239	1.1922	0.00411
0.0190	0.9872			0.0240	1.1961	0.00408
0.0191	0.9916			0.0241	1.2001	0.00405
0.0192	0.9959			0.0242	1.2041	0.00402
0.0193	1.0003			**0.02429**	1.2076	**0.00400**

*$d = d_t$.

TABLE A-12 Development Length (ℓ_{dc}) for Compression Bars with $f_y = 60{,}000$ psi(in.)

Bar Size	f'_c (normal-weight concrete), psi			
	3000	4000	5000	6000
3	8.2	7.1	6.8	6.8
4	11.0	9.5	9.0	9.0
5	13.7	11.9	11.3	11.3
6	16.4	14.2	13.5	13.5
7	19.2	16.6	15.8	15.8
8	21.9	19.0	18.0	18.0
9	24.7	21.4	20.3	20.3
10	27.8	24.1	22.9	22.9
11	30.9	26.8	25.4	25.4
14	37.1	32.1	30.5	30.5
18	49.4	42.8	40.6	40.6

Note: See Chapter 5 for calculation of development length for compression bars.

TABLE A-13 Development Length (ℓ_{dh}) for Hooked Bars with $f_y = 60{,}000$ psi(in.)

Bar Number	f'_c (psi)			
	3000	4000	5000	6000
3	8.2	7.1	6.4	5.8
4	11.0	9.5	8.5	7.7
5	13.7	11.9	10.6	9.7
6	16.4	14.2	12.7	11.6
7	19.2	16.6	14.8	13.6
8	21.9	19.0	17.0	15.5
9	24.7	21.4	19.1	17.5
10	27.8	24.1	21.6	19.7
11	30.9	26.8	23.9	21.8

Note: Modification factors may apply.
$\ell_{dh} \times$ applicable modification factors $\geq$ larger of $8d_b$ or 6 in. See Chapter 5 for calculation of development length for hooked bars.

TABLE A-14 Preferred Maximum Total Number of Column Bars

$1\frac{1}{2}$ bar diameters or $1\frac{1}{2}$ in.

Cover

Cover

Recommended spiral or tie bar number	Core size (in) = column size Size − 2 × cover	Circular area (in.^{2})	Bar number							Square area (in.^{2})	Bar number						
			#5	#6	#7	#8	#9	#10	#11		#5	#6	#7	#8	#9	#10	#11[a]
3[a]	9	63.6	8	7	7	6	—	—	—	81	8	8	8	8	4	4	4
	10	78.5	10	9	8	7	6	—	—	100	12	8	8	8	8	4	4
	11	95.0	11	10	9	8	7	6	—	121	12	12	8	8	8	8	4
	12	113.1	12	11	10	9	8	7	6	144	12	12	12	8	8	8	8
	13	132.7	13	12	11	10	8	7	6	169	16	12	12	12	8	8	8
	14	153.9	14	13	12	11	9	8	7	196	16	16	12	12	12	8	8
	15	176.7	15	14	13	12	10	9	8	225	16	16	16	12	12	12	8
4	16	201.1	16	15	14	12	11	9	8	256	20	16	16	16	12	12	8
	17	227.0	18	16	15	13	12	10	9	289	20	20	16	16	12	12	8
	18	254.5	19	17	15	14	12	11	10	324	20	20	16	16	16	12	12
	19	283.5	20	18	16	15	13	11	10	361	24	20	20	16	16	12	12
	20	314.2	21	19	17	16	14	12	11	400	24	24	20	20	16	12	12
	21	346.4	22	20	18	17	15	13	11	441	28	24	20	20	16	16	12
	22	380.1	23	21	19	18	15	14	12	484	28	24	24	20	20	16	12
5	23	415.5	24	22	21	19	16	14	13	529	28	28	24	24	20	16	16
	24	452.4	25	23	21	20	17	15	13	576	32	28	24	24	20	16	16
	25	490.9	26	24	22	20	18	16	14	625	32	28	28	24	20	20	16
	26	530.9	28	25	23	21	19	16	14	676	32	32	28	24	24	20	16
	27	572.6	29	26	24	22	19	17	15	729	36	32	28	28	24	20	16

[a] No. 4 tie required for No. 11 or larger longitudinal reinforcement (ACI, Section 7.10.5.1).

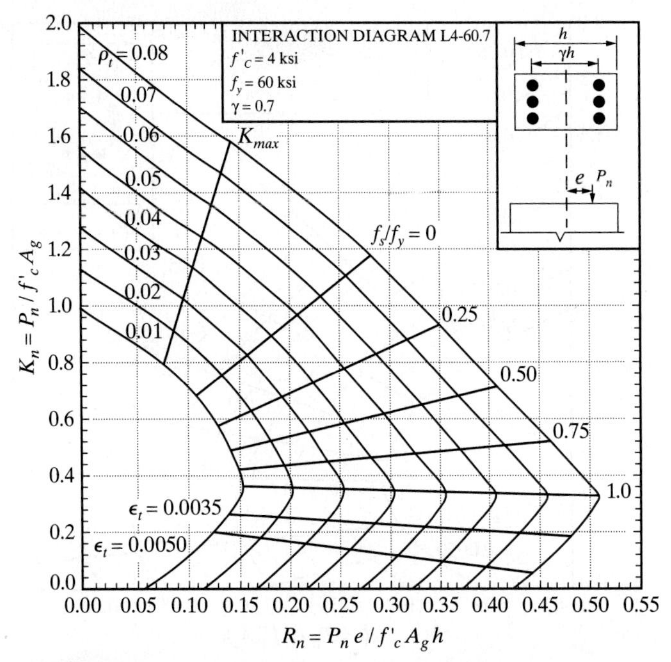

FIGURE A-15

Source: Diagrams A-15 through A-22 are from the ACI Reinforced Concrete Design Handbook Design Aid—Analysis Tables (ACI SP-17DA(14)) and are reprinted here with the permission of the American Concrete Institute.

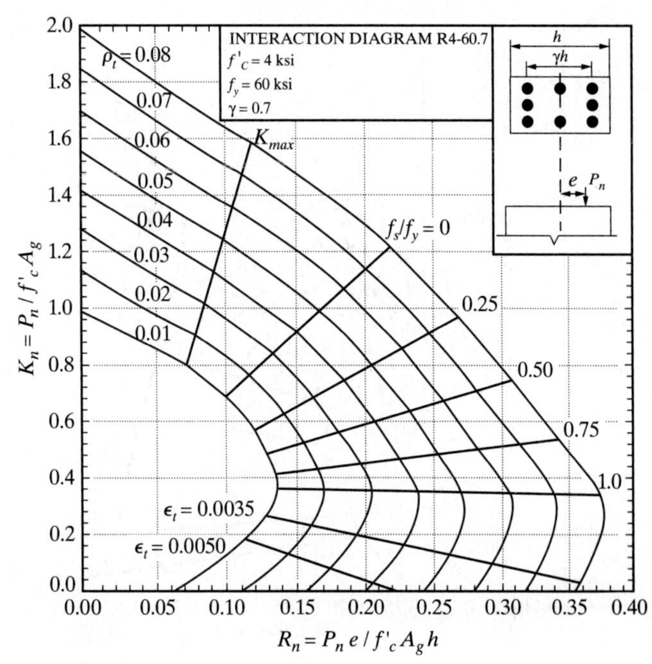

FIGURE A-17

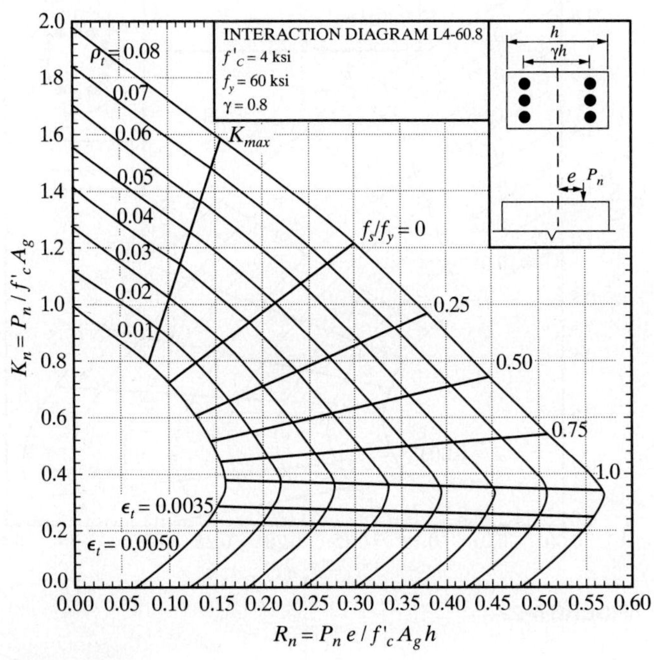

FIGURE A-16

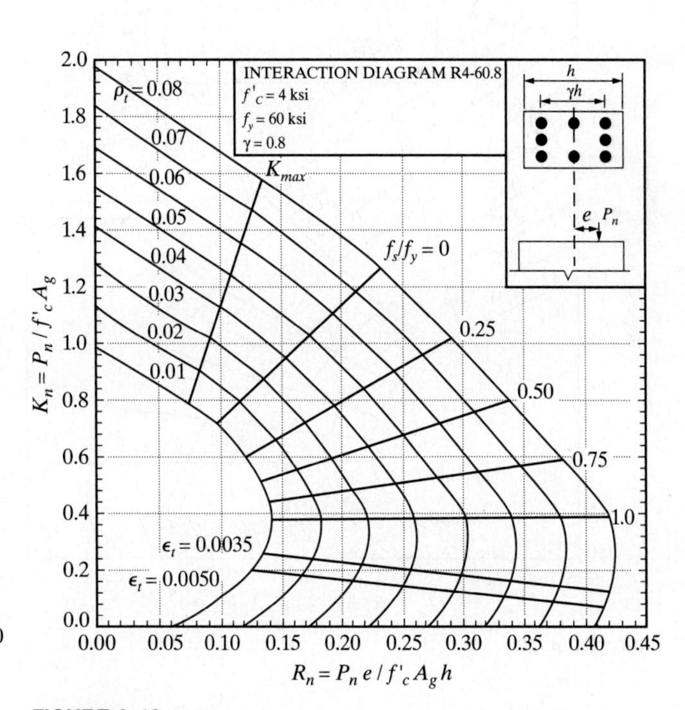

FIGURE A-18

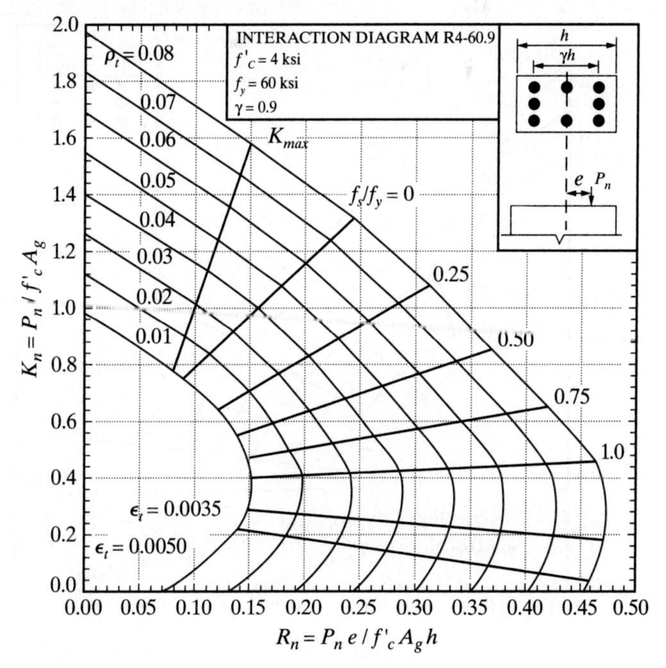

FIGURE A-19

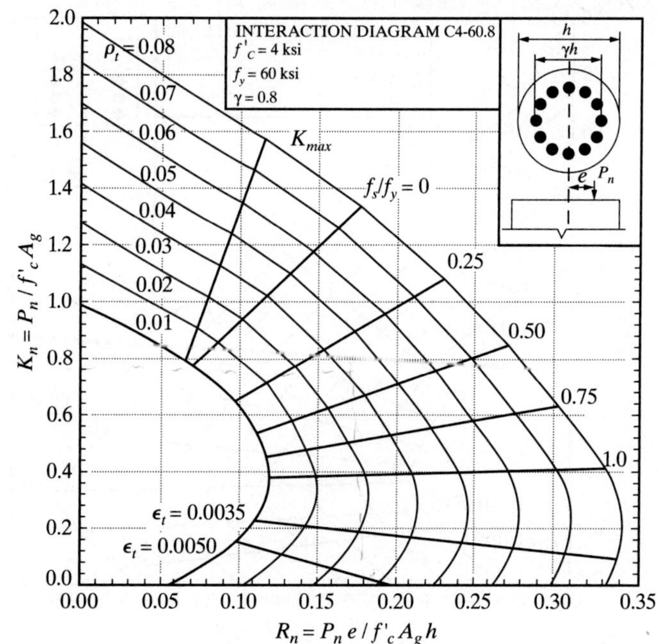

FIGURE A-21

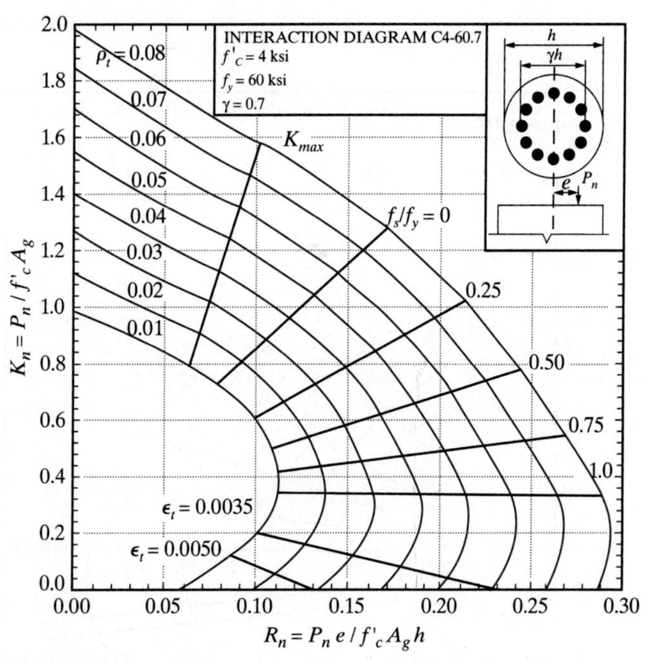

FIGURE A-20

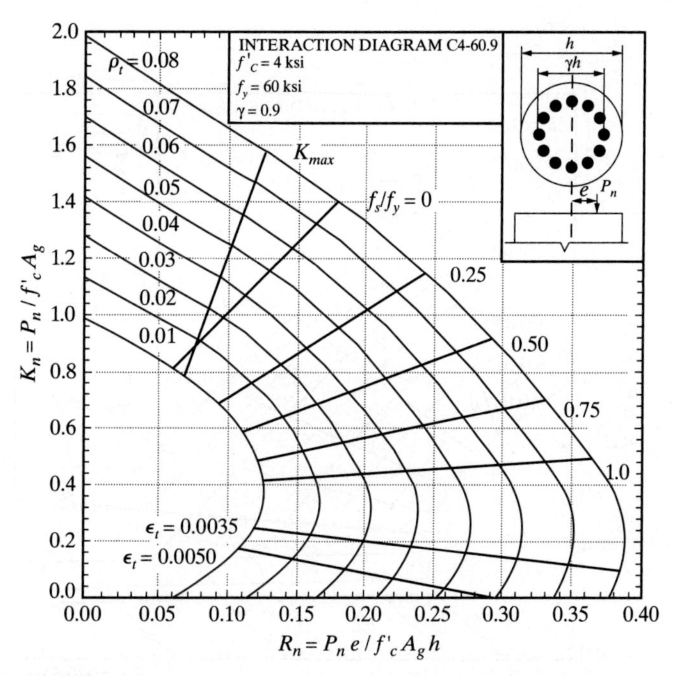

FIGURE A-22

SUPPLEMENTARY AIDS AND GUIDELINES

B-1 Accuracy for Computations for Reinforced Concrete
B-2 Flow Diagrams

B-1 ACCURACY FOR COMPUTATIONS FOR REINFORCED CONCRETE

The widespread availability and use of electronic calculators and computers for even the simplest of calculations have led to the use of numbers that represent a very high order of accuracy. For instance, a calculator having an eight-digit display will yield the following:

$$\frac{8}{0.7} = 11.428571$$

It should be recognized that the numerator and denominator, each of one-figure accuracy, resulted in a number that indicates eight-figure accuracy. Because the quotient cannot be expected to be more accurate than the numbers that produced it, the result as represented may lead one into a false sense of security associated with numbers of very high accuracy. For instance, it is illogical to calculate a required steel area to four-figure accuracy when the loads were to two-figure accuracy and the bars to be chosen have areas tabulated to two- (sometimes three-)figure accuracy. Likewise, the involved mathematical expressions developed in this book and those presented by the ACI Code should be thought of in a similar light. They deal with a material, concrete, that

1. Is made on site or plant-made, is subject to varying amounts of quality control, and will vary from the design strength.
2. Is placed in forms that may or may not produce the design dimensions.
3. Contains reinforcing steel of a specified minimum strength but that may vary above that strength.

In addition, the reinforced concrete member has reinforcing steel that may or may not be placed at the design location, and the design itself is generally based on loads that may be only "best estimates."

With the foregoing in mind, the following has been suggested by the Concrete Reinforcing Steel Institute as a rough guide for numerical accuracy in reinforced concrete calculations:

1. Loads to the nearest 1 psf; 10 lb/ft; 100 lb concentration
2. Span lengths to about 0.1 ft
3. Total loads and reactions to 0.1 kip or three-figure accuracy
4. Moments to the nearest 0.1 ft-kip or three-figure accuracy
5. Individual bar areas to 0.01 in.^{2}
6. Concrete sizes to $\frac{1}{2}$ in.
7. Effective beam depth of 0.1 in.
8. Column loads to the nearest 1.0 kip

In general (admittedly, not always), the reader will find that in this text we have represented numbers used in calculations to an accuracy of three significant digits. If the number begins with 1, then four significant digits are shown. We round intermediate and final numerical solutions in accordance with this rule of thumb. When working on a calculator, however, one will normally maintain all digits and round only the final answer. For this reason, the reader may frequently obtain numerical results that are slightly different from those printed in the text. This should not cause undue concern.

B-2 FLOW DIAGRAMS

The step-by-step procedures for the analysis and design of reinforced concrete members may be presented in the form of flow diagrams (see Figures B-1 to B-4). To aid the reader in grasping the overall calculation approach, which may sometimes include cycling steps, flow diagrams for the analysis and design of rectangular beams and T-beams are presented here. These flow diagrams represent, on an elementary level, the type of organization required to develop computer programs to aid in analysis and design calculations.

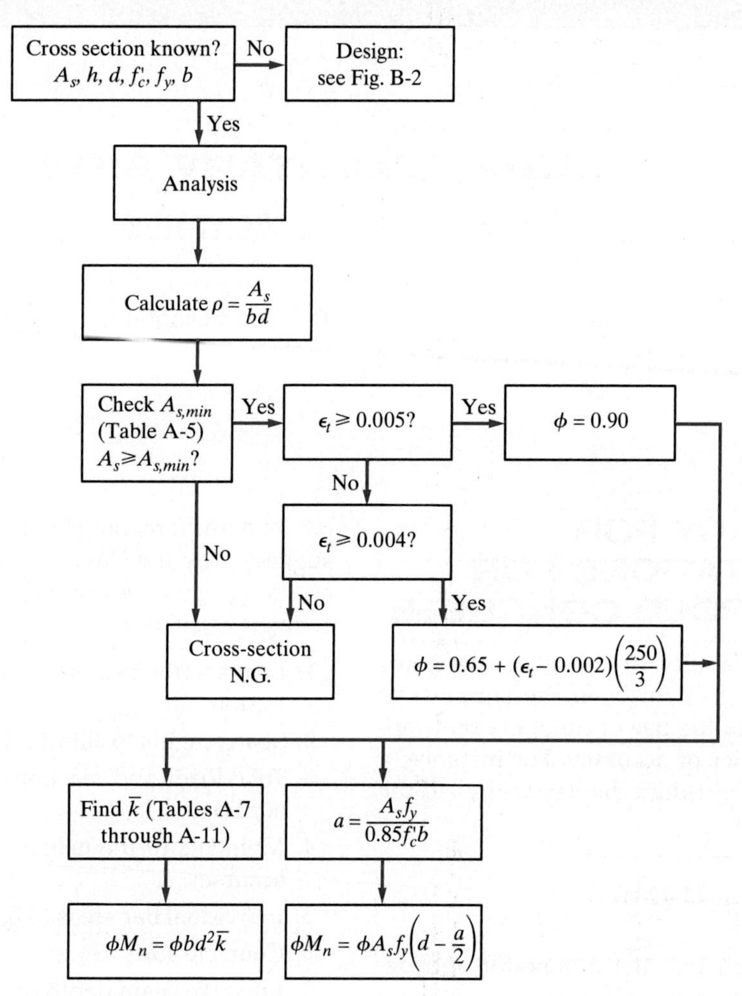

FIGURE B-1 Rectangular beam analysis for moment (tension steel only).

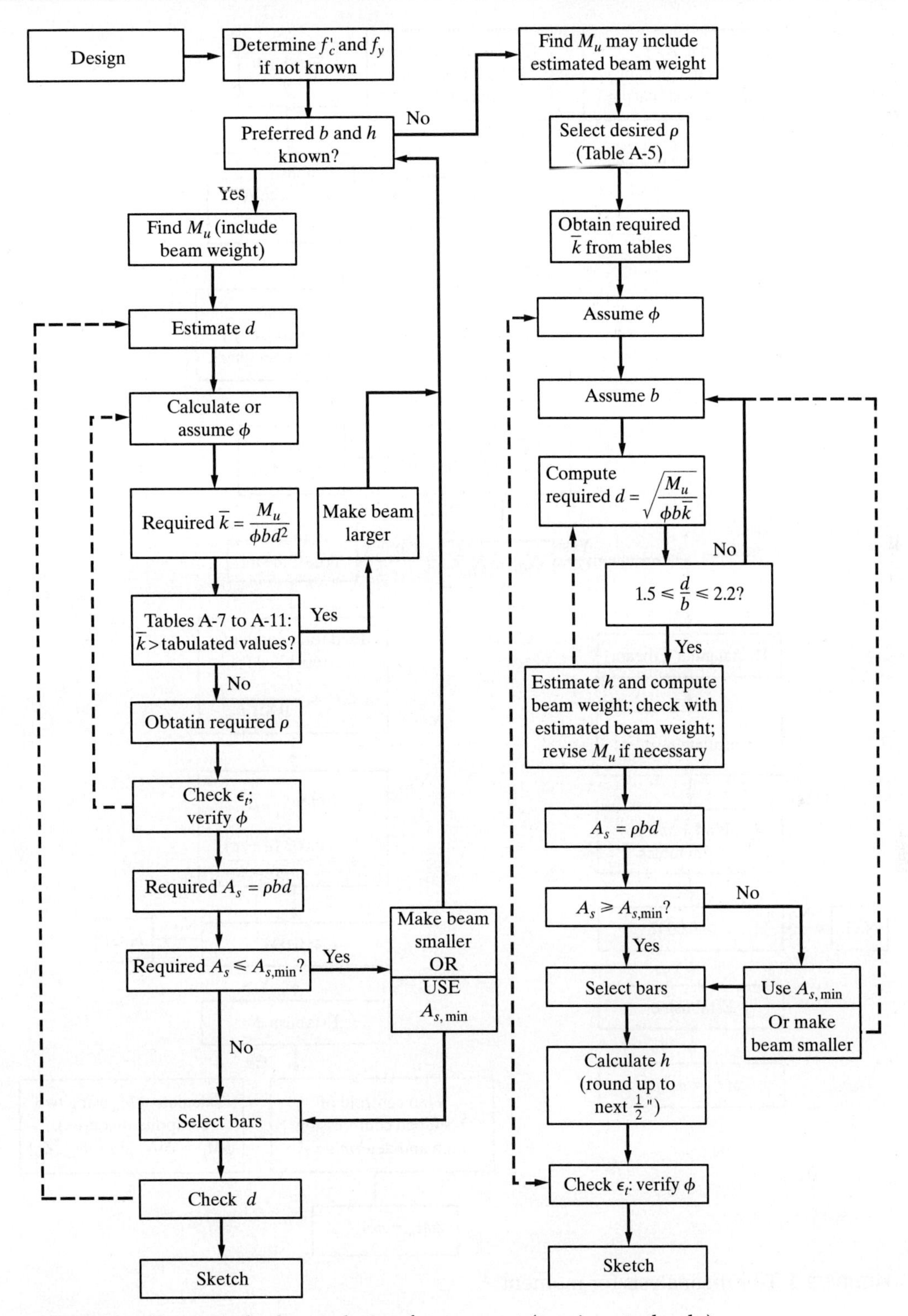

FIGURE B-2 Rectangular beam design for moment (tension steel only).

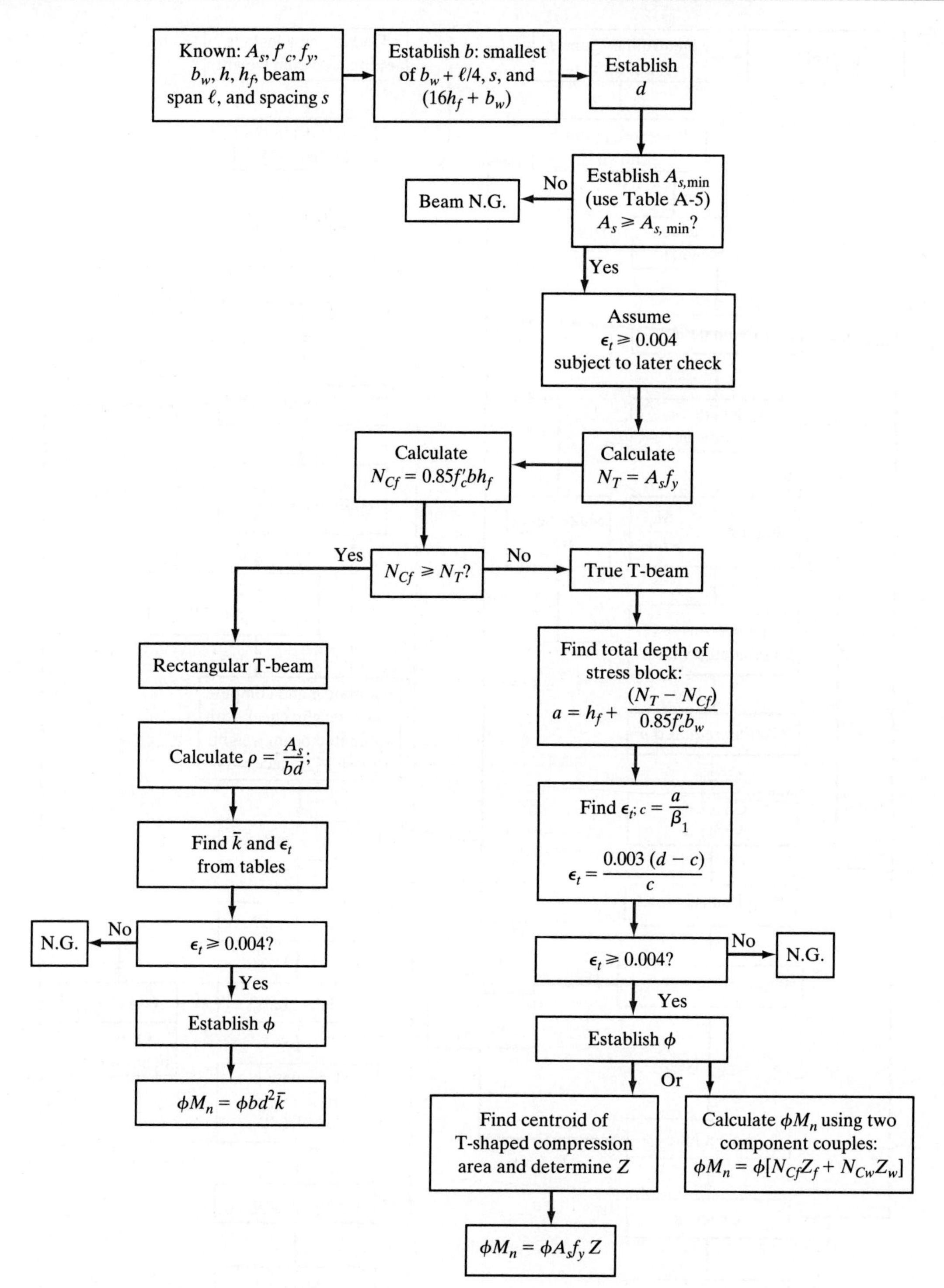

FIGURE B-3 T-beam analysis for moment.

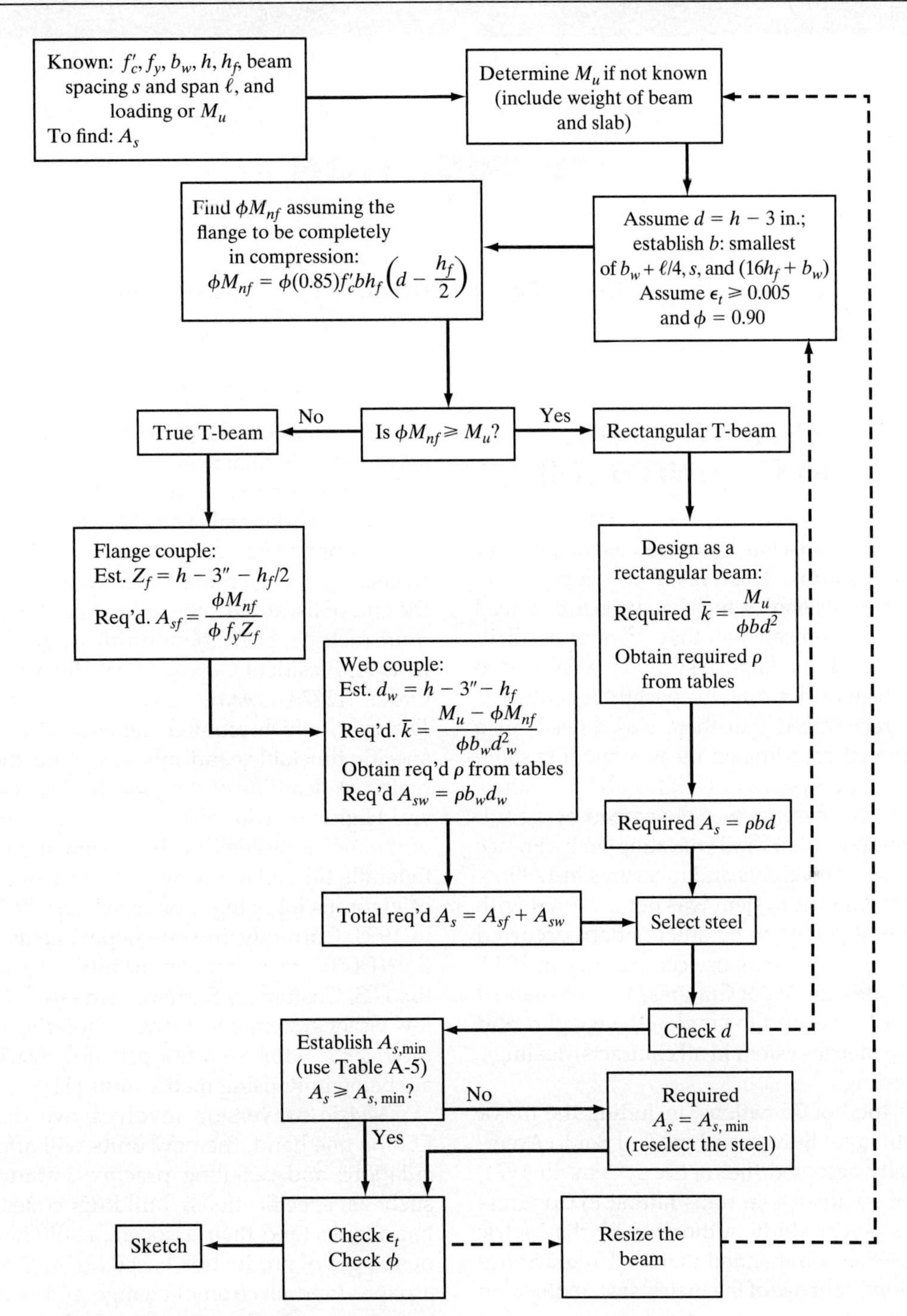

FIGURE B-4 T-beam design for moment.

METRICATION

C-1 The International System of Units (SI)
C-2 SI Style and Usage
C-3 Conversion Factors

C-1 THE INTERNATIONAL SYSTEM OF UNITS (SI)

The U.S. Customary System (or "inch-pound system") of weights and measures has been used as the primary unit system in this book. This system developed from the English system (British), which had been introduced in the original 13 colonies when they were under British rule. Even though the English system was spread to many parts of the world during the past three centuries, it was widely recognized that there was a need for a single international coordinated measurement system. As a result, a second system of weights and measures, known as the metric system, was developed by a commission of French scientists and was adopted by France as the legal system of weights and measures in 1799.

Although the metric system was not accepted with enthusiasm at first, adoption by other nations occurred steadily after France made its use compulsory in 1840. In the United States, an Act of Congress in 1866 made it lawful throughout the land to employ the weights and measures of the metric system in all contracts, dealings, or court proceedings.

By 1900, a total of 35 nations, including the major nations of continental Europe and most of South America, had officially accepted the metric system. In 1971, the secretary of commerce, in transmitting to Congress the results of a 3-year study authorized by the Metric Study Act of 1968, recommended that the United States change to predominant use of the metric system through a coordinated national program. Congress responded by enacting the Metric Conversion Act of 1975 that established a U.S. Metric Board to carry out the planning, coordination, and public education that would facilitate a voluntary conversion to a modern metric system. Today, with the exception of a few small countries, the entire world is using the metric system or is changing to its use. Because of the many versions of the metric system that developed, an International General Conference on Weights and Measures in 1960 adopted an extensive revision and simplification of the system. The name *Le Système International d'Unités* (International System of Units), with the international abbreviation SI, was adopted for this modernized metric system.

The American Society of Civil Engineers (ASCE) resolved in 1970 to actively support conversion to SI and to adopt the revised edition of the American Society for Testing and Materials (ASTM) *Metric Practice* guide. In 1988, Congress passed the Omnibus Trade and Competitiveness Act that, among other things, mandated that by the end of fiscal 1992, the federal government would require metric specifications on all the goods it purchased. In 1991, President George H. W. Bush signed Executive Order 12770, *Metric Usage in Federal Government Programs*, which required federal agencies to develop specific timetables and milestones for the transition to metric. A deadline of the year 2000 was set by the Federal Highway Administration for state implementation of the metric system for the design and construction of federally funded highway projects. Under the influence of vigorous lobbying, Congress cancelled this deadline in 1998. Currently, in state departments of transportation (DOTs), there is no uniformity of systems. Some use the U.S. Customary System, some use SI, and some allow either system. Most federal building construction is metricated, while very few private construction projects are being built using metric units [1].

Metric conversion involves two distinct aspects. On the one hand, the new units will affect design calculations and detailing practices. Many publications such as specifications, building codes, and design handbooks (and their associated software) are in various stages of production based on SI. This is essentially a paper (and electronic) change and is called a "soft" conversion. On the other hand, the physical sizes of some products will be affected (for instance, plywood will change from a 4-ft width to a slightly smaller 1200-mm width); this is called a "hard" conversion.

Metric reinforcing steel sizes have been standardized based on a soft conversion rather than a hard conversion. That is, the physical sizes of the 11 (Nos. 3–11, No. 14, and No. 18) inch-pound bars remain the same, but the designations change and the dimensions (diameter, area, and so on) are specified in metric units. The metric bar number is the nominal diameter rounded to the nearest mm. Virtually all new reinforcing steel is labeled in metric units (see Figure C-1). Table C-1 provides in inch-pound bar sizes the soft metric bar sizes and soft metric bar data.

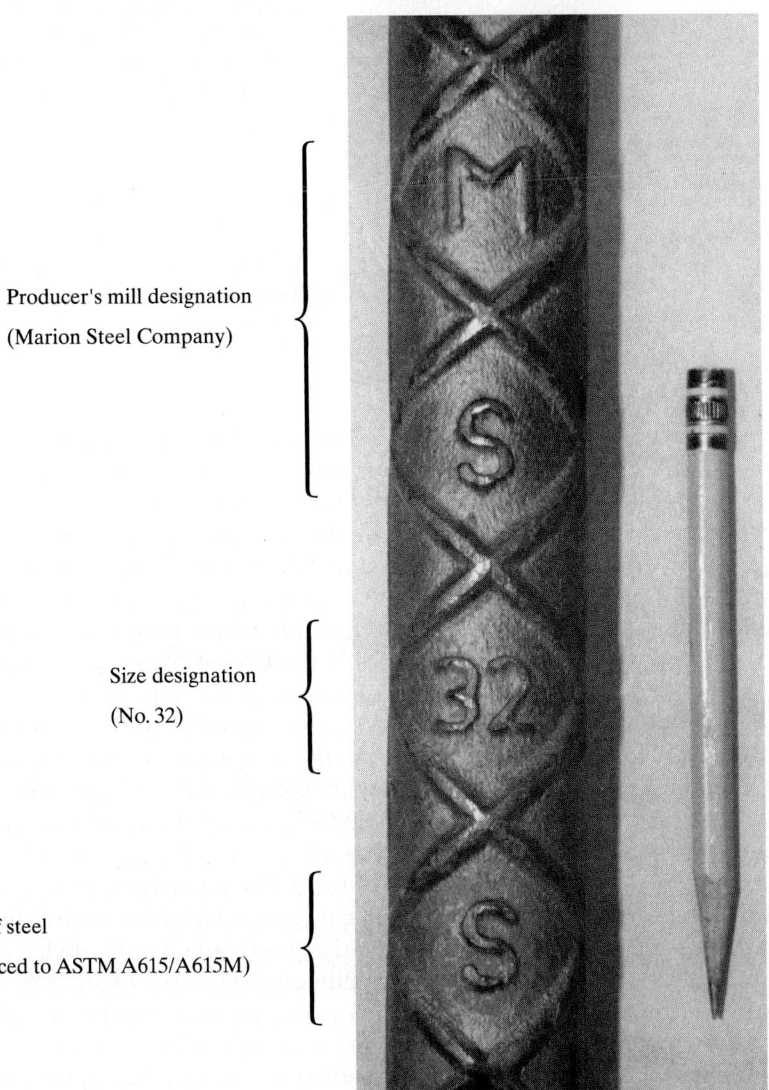

FIGURE C-1 Identification marks. This bar has a single longitudinal grade line (indicating grade 420 steel) on its opposite side (hidden in this photo).

TABLE C-1 Inch-Pound Bar Designation versus Soft Metric Bar Designation (with Soft Metric Bar Data)

	Soft metric		
Inch-pound bar designation	**Bar designation**	**Nominal diameter (mm)**	**Nominal area (mm^2)**
#3	#10	9.5	71
#4	#13	12.7	129
#5	#16	15.9	199
#6	#19	19.1	284
#7	#22	22.2	387
#8	#25	25.4	510
#9	#29	28.7	645
#10	#32	32.3	819
#11	#36	35.8	1006
#14	#43	43.0	1452
#18	#57	57.3	2581

TABLE C-2 Minimum Yield Strengths (ASTM A615/A615M-04b)

Metric (MPa)	Inch-pound equivalent (psi)
520 (grade 520)	75,400 (grade 75)
420 (grade 420)	60,900 (grade 60)
280 (grade 280)	40,600 (grade 40)

TABLE C-3 SI Units and Symbols

Quantity	Unit	SI symbol
Length	meter	m
Mass	kilogram	kg
Time	second	s
Angle[a]	radian	rad

[a]It is also permissible to use the arc degree *and its decimal submultiples* when the radian is not convenient.

Minimum yield strengths have been established as 420 MPa (grade 420) and 520 MPa (grade 520). These values are intended to be equivalent to inch-pound grade 60 and grade 75 materials, respectively. ASTM A615/A615M-04b also includes a grade 280 (280 MPa), which is equivalent to inch-pound grade 40. These minimum yield strengths are summarized in Table C-2.

C-2 SI STYLE AND USAGE

The SI consists of a limited number of *base* units that establish fundamental quantities and a large number of *derived* units, which come from the base units, to describe other quantities.

The SI base units pertinent to reinforced concrete design are listed in Table C-3, and the SI-derived units pertinent to reinforced concrete design are listed in Table C-4.

Because the orders of magnitude of many quantities cover wide ranges of numerical values, SI *prefixes* have been established to deal with decimal-point placement. SI prefixes representing steps of 1000 are recommended to indicate orders of magnitude. Those recommended for use in reinforced concrete design are listed in Table C-5. From Table C-5, it is clear that there is a choice of ways to present numbers. It is preferable to use numbers between 1 and 1000, whenever possible, by selecting the appropriate prefix. For example, 18 m is preferred to 0.018 km or 18,000 mm.

A brief note must be made at this point concerning the presentation of numbers with many digits. It is common practice in the United States to separate digits into groups of three by means of commas. To avoid confusion with the widespread European practice of using a comma on the line as the decimal marker, the method of setting off groups of three digits with a gap, as shown in Table C-5, is recommended international practice. Note that this method is used on both the left and right sides of the decimal marker for any string of five or more digits. A group of four digits on either side of the decimal marker need not be separated.

A significant difference between SI and other measurement systems is the use of explicit and distinctly separate units for mass and force. The SI base unit kilogram (kg) denotes the base unit of mass, which is the quantity of matter of an object. This is a constant quantity that is independent of gravitational attraction. The derived SI unit newton (N) denotes the absolute de-

TABLE C-4 SI-Derived Units

Quantity	Unit	SI symbol	Formula
Acceleration	meter per second squared	—	m/s^2
Area	square meter	—	m^2
Density (mass per unit volume)	kilogram per cubic meter	—	kg/m^3
Force	Newton	N	$kg \cdot m/s^2$
Pressure or stress	pascal	Pa	N/m^2
Volume	cubic meter	—	m^3
Section modulus	meter to third power	—	m^3
Moment of inertia	meter to fourth power	—	m^4
Moment of force, torque	newton meter	—	$N \cdot m$
Force per unit length	newton per meter	—	N/m
Mass per unit length	kilogram per meter	—	kg/m
Mass per unit area	kilogram per square meter	—	kg/m^2

TABLE C-5 SI Prefixes

Prefix	SI symbol	Factor
mega	M	1,000,000 = 10^6
kilo	K	1000 = 10^3
milli	M	0.001 = 10^{-3}
micro	μ	0.000 001 = 10^{-6}

rived unit of force (mass times acceleration: kg·m/s²). The term *weight* should be avoided since it is confused with mass and because it describes a particular force that is related solely to gravitational acceleration, which varies on the surface of the earth. For the conversion of mass to force, the recommended value for the acceleration of gravity in the United States may be taken as $g = 9.81\ \text{m/s}^2$.

As an example, we will consider the mass of reinforced concrete, which, in the design of bending members, must be considered as a load or force per unit length of span. In the U.S. Customary System, reinforced concrete weighs 150 lb/ft³. This is equivalent in the SI to

$$150\ \frac{\text{lb}}{\text{ft}^3} \times \left(\frac{3.2808\ \text{ft}}{1\ \text{m}}\right)^3 \times \frac{1\ \text{kg}}{2.2046\ \text{lb}} \approx 2400\ \frac{\text{kg}}{\text{m}^3}$$

This represents a mass per unit volume (density), where the unit volume is 1 cubic meter, m³. To use the density of the concrete to obtain a force per cubic meter, Newton's law must be applied:

$$\begin{aligned} F &= \text{mass times acceleration of gravity} \\ &= mg \\ &= 2400(9.81) \\ &= 23{,}500\frac{\text{kg}\cdot\text{m}}{\text{s}^2\text{m}^3} \\ &= 23{,}500\ \text{N/m}^3 \\ &= 23.5\ \text{kN/m}^3 \end{aligned}$$

The dead load per unit length of a beam of dimensions $b = 500$ mm and $h = 1000$ mm can then be determined:

$$\left(\frac{500}{1000}\right)\left(\frac{1000}{1000}\right)(23.5) = 11.75\ \text{kN/m}$$

Thus, the load or force per unit length (1 meter) equals 11.75 kN.

C-3 CONVERSION FACTORS

Table C-6 contains conversion factors for the conversion of the U.S. Customary System units to SI units for quantities frequently used in reinforced concrete design.

TABLE C-6 Conversion Factors: U.S. Customary to SI Units

	Multiply		By		To Obtain
Length	inches	×	25.4	=	millimeters
	feet	×	0.3048	=	meters
	yards	×	0.9144	=	meters
	miles (statute)	×	1.609	=	kilometers
Area	square inches	×	645.2	=	square millimeters
	square feet	×	0.0929	=	square meters
	square yards	×	0.8361	=	square meters
Volume	cubic inches	×	16,387.	=	cubic millimeters
	cubic feet	×	0.028 32	=	cubic meters
	cubic yards	×	0.7646	=	cubic meters
	gallons (U.S. liquid)	×	0.003 785	=	cubic meters
Force	pounds	×	4.448	=	newtons
	kip	×	4,448.	=	newtons
Force per unit length	pounds per foot	×	14.594	=	newtons per meter
	kip per foot	×	14,594.	=	newtons per meter
Load per unit volume	pounds per	×	0.157 14	=	kilonewtons per
	cubic foot	×	0.0283		cubic meter
Bending moment or torque	inch-pounds	×	0.1130	=	newton meters
	foot-pounds	×	1.356	=	newton meters
	inch-kip	×	113.0	=	newton meters

TABLE C-6 Continued

	Multiply		By		To Obtain
	foot-kip	×	1356.	=	newton meters
	inch-kip	×	0.1130	=	kilonewton meters
	foot-kip	×	1.356	=	kilonewton meters
Stress, pressure, loading (force per unit area)	pounds per square inch	×	6895.	=	pascals
	pounds per square inch	×	6.895	=	kilopascals
	pounds per square inch	×	0.006 895	=	megapascals
	kip per square inch	×	6.895	=	megapascals
	pounds per square foot	×	47.88	=	pascals
	pounds per square foot	×	0.047 88	=	kilopascals
	kip per square foot	×	47.88	=	kilopascals
	kip per square foot	×	0.047 88	=	megapascals
Mass	pounds	×	0.454	=	kilograms
Mass per unit volume (density)	pounds per cubic foot	×	16.02	=	kilograms per cubic meter
	pounds per cubic yard	×	0.5933	=	kilograms per cubic meter
Moment of inertia	inches^4	×	416,231.	=	millimeters^4
Mass per unit length	pounds per foot	×	1.488	=	kilograms per meter
Mass per unit area	pounds per square foot	×	4.882	=	kilograms per square meter

Although specified in the SI, the pascal is not universally accepted as the unit of stress. Because section dimensions and properties are generally in millimeters, it is more convenient to express stress in newtons per square millimeter ($1\,\text{N/mm}^2 = 1\,\text{MPa}$).

Reference is made to the metric version of the code, ACI 318M-14 [2]. The metric version of the code furnishes equivalents for equations and data necessary for use in the SI. Other reference sources that contain treatment of the many aspects of metrication in the design and construction field are listed at the end of this appendix [3,4].

Example C-1

Find ϕM_n for the beam of cross section shown in Figure C-2. The steel is grade 420, and $f'_c = 20\ \text{N/mm}^2$.

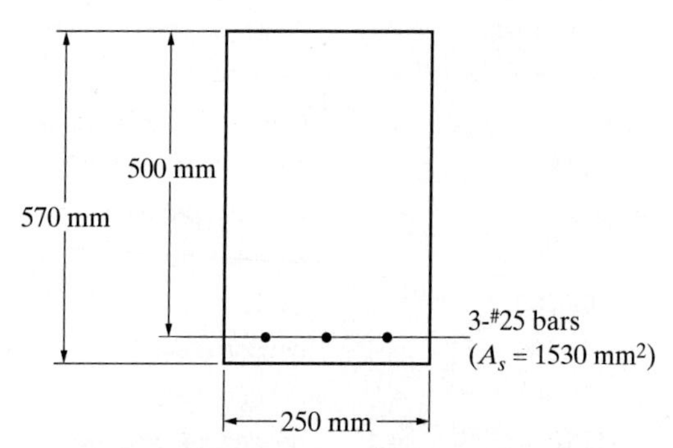

FIGURE C-2 Cross section for Example C-1.

Solution:

1.
$$f_y = 420\ \text{N/mm}^2$$
$$f_c = 20\ \text{N/mm}^2$$
$$b = 250\ \text{mm},\quad d = 500\ \text{mm}$$
$$h = 570\ \text{mm},\quad A_s = 1530\ \text{mm}^2$$

2. To be found: ϕM_n.

3.
$$\rho = \frac{A_s}{bd} = \frac{1530}{250(500)} = 0.0122$$

4. Check $A_{s,\min}$ (ACI 318M-14, section 9.6):

$$A_{s,\min} = \frac{0.25\sqrt{f'_c}}{f_y} b_w d \geq \frac{1.4}{f_y} b_w d$$

$$\frac{0.25\sqrt{f'_c}}{f_y} = \frac{0.25\sqrt{20}}{(420)} = 0.00266$$

$$\frac{1.4}{f_y} = \frac{1.4}{420} = 0.00333$$

Therefore, use 0.00333. Then

$$A_{s,\min} = 0.00333(250\ \text{mm})\ (500\ \text{mm}) = 416\ \text{mm}^2$$
$$1530\ \text{mm}^2 > 416\ \text{mm}^2 \qquad \text{(O.K.)}$$

5. Determine ϵ_t from Equation (2-2) (Chapter 2):

$$\epsilon_t = \frac{0.00255\ f'_c \beta_1}{\rho f_y} - 0.003$$
$$= \frac{0.00255(20\ \text{N/mm}^2)(0.85)}{0.0122(420\ \text{N/mm}^2)} - 0.003$$
$$= 0.00546 > 0.005$$

Therefore, $\phi = 0.90$

6. $$a = \frac{A_s f_y}{0.85 f'_c b} = \frac{1530(420)}{0.85(20.0)(250)} = 151.2 \text{ mm}$$

$$Z = d - \frac{a}{2} = 500 - \frac{151.2}{2} = 424 \text{ mm}$$

$$M_n = A_s f_y Z$$

$$\phi M_n = 0.9(A_s f_y Z)$$

All the quantities needed for the ϕM_n calculation have been determined, but some conversion is required to make prefixes compatible. Rather than set up prefix conversion for each lengthy calculation, it is suggested for situations such as this that quantities be substituted in units of meters and newtons. The results will be in the same units. This method also lends itself to the use of numerical values expressed in powers-of-10 notation. For the ϕM_n calculation,

$$A_s = 1.530 \times 10^{-3} \text{m}^2, \; f_y = 420 \times 10^6 \text{ N/m}^2,$$
$$Z = 0.424 \text{ m}$$

from which

$$\phi M_n = 0.9(1.530 \times 10^{-3})(420 \times 10^6)(0.424)$$
$$= 245 \times 10^3 \text{N} \cdot \text{m}$$
$$= 245 \text{ kN} \cdot \text{m}$$

Note that the final ϕM_n is changed to kN · m. The kilo prefix is the most appropriate prefix for the majority of flexural problems that are presented in this book using the U.S. Customary System of units (see "Bending moment or torque" in Table C-6 of this text for comparison with ft-kip).

Example C-2

Design a rectangular reinforced concrete beam for a simple span of 10 m to carry service loads of 17.1 kN/m dead load (does not include the dead load of the beam) and 31.0 kN/m live load. The maximum width of beam desired is 400 mm. Use $f'_c = 20.0 \text{ N/mm}^2$ and $f_y = 420 \text{ N/mm}^2$. Assume a No. 10 stirrup and sketch the design. (See Chapter 2 for design procedure.)

Solution

1. $$w_u = 1.2\, w_{DL} + 1.6 w_{LL}$$
$$= 1.2(17.1) + 1.6(31.0)$$
$$= 70.1 \text{ kN/m}$$
$$M_u = \frac{w_u \ell^2}{8} = \frac{70.1(10)^2}{8} = 876 \text{ kN} \cdot \text{m}$$

2. Assume that $\rho = 0.0090$ (Table A-5). Note that the given f'_c and f_y correspond approximately to 3000 psi and 60,000 psi, respectively.
3. From Table A-5 (or A-8)

$$\text{required } \bar{k} = 0.4828 \text{ ksi}$$

Converting to SI (see Table C-6),

$$\bar{k} = 0.4828(6.895) = 3.329 \text{ N/mm}^2$$

4. Assume that $b = 400$ mm:

$$\text{required } d = \sqrt{\frac{M_u}{\phi b \bar{k}}}$$

Substituting quantities in terms of meters and newtons yields

$$\text{required } d = \sqrt{\frac{876 \times 10^3}{0.9(0.400)(3.329 \times 10^6)}} = 0.855 \text{ m}$$
$$= 855 \text{ mm}$$
$$\frac{d}{b} \text{ ratio} = \frac{855}{400} = 2.14 \qquad \text{(O.K.)}$$

5. Estimate the total beam depth for purposes of determining the beam dead load. Assume a No. 36 main bar, a No. 10 stirrup, and a minimum cover of 40 mm. Then

$$h = 855 + 35.8/2 + 9.5 + 40 = 922 \text{ mm}$$

Use a beam depth of 950 mm. The density of reinforced concrete is 23.5 kN/m^3; therefore, the dead load of the beam per meter length is

$$0.400(0.950)(23.5) = 8.93 \text{ kN/m}$$

6. The additional M_u due to the beam dead load is

$$M_u = \frac{1.4 w_{DL} \ell^2}{8} = \frac{1.4(8.93)(10)^2}{8}$$
$$= 156.3 \text{ kN} \cdot \text{m}$$
$$\text{total } M_u = 876 + 156.3 = 1032 \text{ kN} \cdot \text{m}$$

7. Using ρ, $\bar{k}$, and b as before, the effective depth required is

$$\text{required d} = \sqrt{\frac{1.032 \times 10^6}{0.9(0.400)(3.329 \times 10^6)}} = 0.928 \text{ m}$$
$$= 928 \text{ mm}$$
$$\frac{d}{b} \text{ ratio} = \frac{928}{400} = 2.32 \qquad \text{(O.K.)}$$

8. $$\text{required } A_s = \rho bd$$
$$= 0.0090(400)(928)$$
$$= 3340 \text{ mm}^2$$

Check $A_{s,\min}$ (use Table A-5 because the given values of f'_c and f_y correspond approximately to 3000 psi and 60,000 psi):

$$A_{s,\min} = 0.0033 b_w d$$
$$= 0.0033(400 \text{ mm})(928 \text{ mm})$$
$$= 1225 \text{ mm}^2$$
$$3340 \text{ mm}^2 > 1225 \text{ mm}^2$$

9. Use four No. 36 bars:

$$A_s = 1006(4) = 4024 \text{ mm}^2 \qquad \text{(O.K.)}$$

The minimum beam width for four No. 36 bars may be determined (closely) from Table A-3, noting that a No. 36 bar is approximately equivalent to a No. 11 bar. Thus,

$$\text{minimum } b = 14(25.4) = 356 \text{ mm} < 400 \text{ mm (O.K.)}$$

10. The total beam depth h may be taken as the effective depth required, plus minimum concrete cover, plus stirrup diameter, plus one-half the main steel diameter:

$$\text{required } h = 928 + 40 + 9.5 + 35.8/2 = 995 \text{ mm}$$

Use $h = 1000$ mm.

11. Check ϵ_t by calculation. Due to rounding of h, $d = 928 + 5 = 933$ mm. The final ρ is

$$\rho = \frac{A_s}{bd} = \frac{4024 \text{ mm}^2}{400 \text{ mm}(933 \text{ mm})} = 0.01078$$

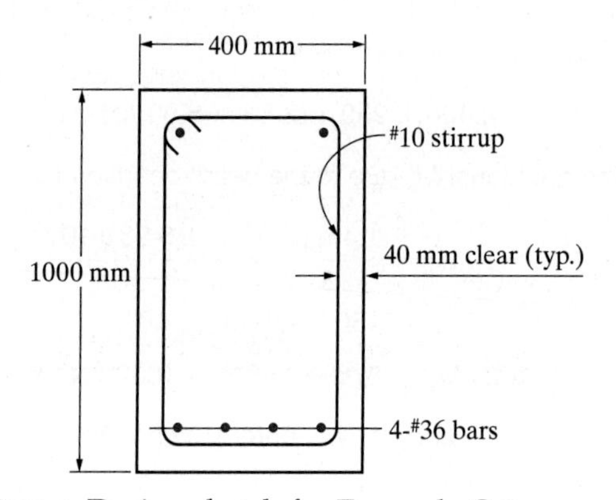

FIGURE C-3 Design sketch for Example C-2.

From Equation (2-2) (Chapter 2):

$$\epsilon_t = \frac{0.00255\, f'_c \beta_t}{\rho f_y} - 0.003$$

$$= \frac{0.00255(20 \text{ N/mm}^2)(0.85)}{0.01078(420 \text{ N/mm})^2} - 0.003$$

$$= 0.0066 > 0.005$$

Therefore, $\phi = 0.90$ as assumed.

The design sketch is shown in Figure C-3.

Example C-3

A simply supported, rectangular, reinforced concrete beam 300 mm wide and having an effective depth of 500 mm carries a total factored load w_u of 70 kN/m on a 9.0-m clear span. (The given load includes the dead load of the beam.) Design the web reinforcement (stirrups). The steel is grade 280, and $f'_c = 20$ N/mm^2.

Solution

1. Draw the shear force V_u diagram (see Figure C-4):

$$V_u = \frac{w_u \ell}{2} = \frac{70(9)}{2} = 315 \text{ kN}$$

At the critical section

$$V_u^* = 315 - \frac{500}{1000}(70) = 280 \text{ kN}$$

2. Determine if stirrups are required:

$$V_c = 0.17\sqrt{f'_c}\, b_w d$$

$$= 0.17\sqrt{20}(300)(500) = 114 \times 10^3 \text{ N} = 114 \text{ kN}$$

$$\phi V_c = 0.75(114) = 85.5 \text{ kN}$$

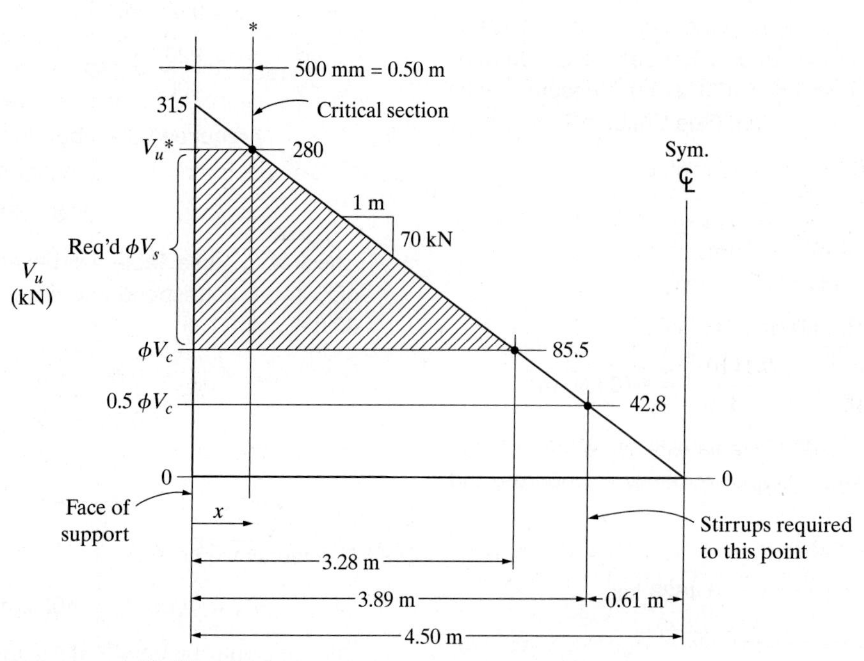

FIGURE C-4 V_u diagram.

Stirrups must be provided if $V_u > 0.5\phi V_c$:

$$0.5\ \phi V_c = 0.5(85.5) = 42.8\text{ kN}$$

Stirrups *are* required, as 280 kN > 42.8 kN.

3. Find the length of span over which stirrups are required, referencing from the face-of-support:

$$\frac{315 - 42.8}{70} = 3.89\text{ m}$$

Note this location on the V_u diagram as well as the location where $V_u = \phi V_c = 85.5$ kN. This location is obtained from

$$\frac{315 - 85.5}{70} = 3.28\text{ m}$$

4. Designate "Req'd ϕV_s" on the V_u diagram:

$$\begin{aligned}\text{required } \phi V_s &= max.V_u - \phi V_c - mx\\ &= 315 - 85.5 - 70x\\ &= 230 - 70x\end{aligned}$$

This applies in the range

$$500\text{ mm} \leq x \leq 3280\text{ mm}$$

5. Assume a No. 10 stirrup ($A_v = 2A_s = 142\text{ mm}^2$) and compute the spacing requirement at the critical section based on the required ϕV_s^*. At this location, the stirrups will be most closely spaced. From ACI Equation (22.5.10.5.3),

$$\text{required } s^* = \frac{A_v f_{yt} d}{V_s^*} = \frac{\phi A_v f_{yt} d}{\text{required } \phi V_s^*}$$

where the denominator is determined with reference to Figure C-4:

$$\text{required } \phi V_s^* = V_u^* - \phi V_c$$

Using basic units of meters and newtons,

$$\begin{aligned}\text{required } s^* &= \frac{0.75(142 \times 10^{-6})(280 \times 10^6)(0.500)}{(280 - 85.5) \times 10^3}\\ &= 0.077\text{ m} = 77\text{ mm}\end{aligned}$$

This is less than our 100-mm (4-in.) minimum spacing rule of thumb. Therefore, increase the stirrup size to a No. 13 bar ($A_v = 258\text{ mm}^2$). Then

$$\begin{aligned}\text{required } s^* &= \frac{0.75(258 \times 10^{-6})(280 \times 10^6)(0.500)}{(280 - 85.5) \times 10^3}\\ &= 0.139\text{ m} = 139\text{ mm}\end{aligned}$$

We will use a 130-mm spacing as the stirrup spacing between the face-of-support and the critical section, subject to further checks.

6. Establish the ACI Code maximum spacing requirements. From the ACI Code (318M-14), Section 9.7.6.2.2, if V_s is less than $0.33\sqrt{f_c'}b_w d$, the maximum spacing is $d/2$ or 600 mm, whichever is smaller; otherwise, the maximum spacing will be the smaller of $d/4$ or 300 mm.

$$\begin{aligned}0.33\sqrt{f_c'}b_w d &= 0.33\sqrt{20.0}(300)(500)\\ &= 221 \times 10^3\text{ N}\\ &= 221\text{ kN}\end{aligned}$$

At the critical section, the required V_s is

$$\begin{aligned}V_s^* &= \frac{\phi V_s}{\phi} = \frac{V_u^* - \phi V_c}{\phi}\\ &= \frac{280 - 85.5}{0.75} = 259\text{ kN}\end{aligned}$$

Because 259 kN > 221 kN, the maximum spacing will be $d/4$ or 300 mm, whichever is smaller, from the face-of-support out to where the required V_s drops below 221 kN. This maximum spacing is

$$\frac{d}{4} = \frac{500}{4} = 125\text{ mm}$$

125 mm is less than 300; therefore, use 125 mm. Next, determine where $V_s = 221$ kN, which is where the maximum spacing can be increased to the smaller of $d/2$ or 600 mm.

$$\frac{d}{2} = \frac{500}{2} = 250\text{ mm}$$

250 mm < 600 mm; therefore, use 250 mm.

$$V_s = \frac{\phi V_s}{\phi} = \frac{V_u - \phi V_c}{\phi} = \frac{(315 - 70x) - \phi V_c}{\phi} = 221\text{ kN}$$

from which

$$x = \frac{\phi 221 + \phi V_c - 315}{-70} = 0.911\text{ m}$$

Therefore, the maximum spacing allowed increases to 250 mm at 0.911 m from the face-of-support.

A second criterion for maximum spacing is based on the code minimum area requirement (ACI 318M-14, Section 9.6.3.3). The governing equation may be rewritten in the form

$$s_{max} \leq \frac{A_v f_{yt}}{0.062\sqrt{f_c'}b_w} = \frac{258(280)}{0.062\sqrt{20}(300)} = 868\text{ mm}$$

Check the upper limit:

$$s_{max} = \frac{A_v f_{yt}}{0.35 b_w} = \frac{258(280)}{0.35(300)} = 688\text{ mm}$$

Of the foregoing maximum spacing criteria, the smallest value will control. Maximum spacing requirements are summarized in Figure C-5.

7. Next, determine the spacing requirements based on shear strength. At the critical section, the required spacing is 130 mm. The maximum spacing is 125 mm to 0.911 m from the face-of-support and 250 mm thereafter. At other points along the span (x meters from the

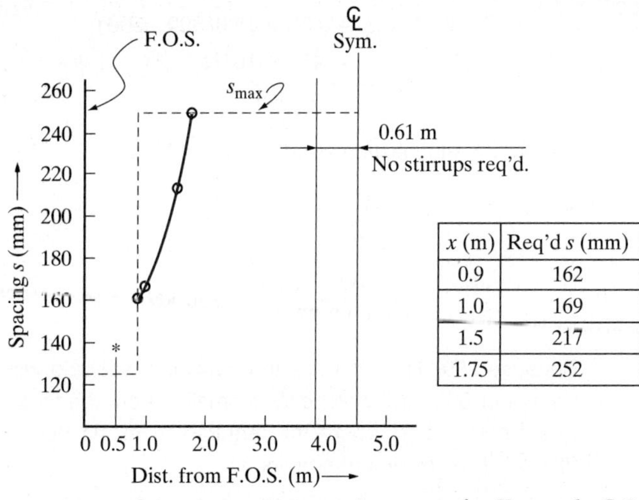

x (m)	Req'd s (mm)
0.9	162
1.0	169
1.5	217
1.75	252

FIGURE C-5 Stirrup spacing requirements for Example C-3.

face-of-support), the required spacing may be determined as follows:

$$\text{required } s = \frac{A_v f_{yt} d}{\text{required } V_s} = \frac{\phi A_v f_{yt} d}{\text{required } \phi V_s}$$

where the denominator can be determined from the expression given in step 4, where

$$\text{required } \phi V_s = 230 - 70x$$

Using basic units of meters and newtons, the calculation for required spacing results in

$$\text{required } s = \frac{0.75(258 \times 10^{-6})(280 \times 10^{6})(0.500)}{230 \times 10^{3} - (70 \times 10^{3})x}$$

$$= \frac{27.1 \times 10^{3}}{230 \times 10^{3} - (70 \times 10^{3})x}$$

$$= \frac{27.1}{230 - 70x}$$

where the resulting spacing is in meters.

The results for several arbitrary values of x are shown tabulated and plotted in Figure C-5. As an example, compute the required stirrup spacing at a distance of 1 m from the face-of-support ($x = 1$ m):

$$\text{required s} = \frac{27.1}{230 - 70(1)} = 0.169 \text{ m} = 169 \text{ mm}$$

Similarly, the required spacing may be found at other points along the beam (see Figure C-5).

8. Using Figure C-5, the stirrup pattern shown in Figure C-6 may be developed. Stirrups have been placed the full length of the span, which is a common, conservative practice.

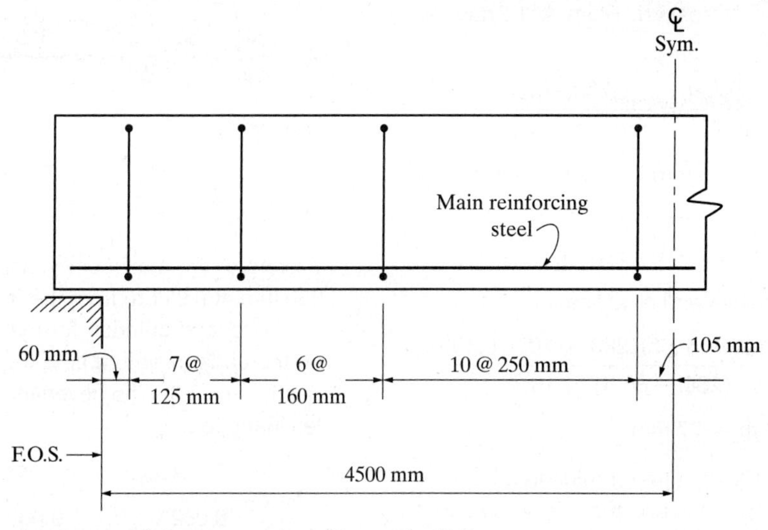

FIGURE C-6 Stirrup spacing, Example C-3.

References

[1] *Construction Metrication.* Construction Metrication Council of the National Institute of Building Sciences, 1090 Vermont Ave. NW, Washington, D.C. 20005-4905, V. 9, No. 4 (4th Qtr.), 2000.

[2] *Building Code Requirements for Structural Concrete* (ACI 318M-14). American Concrete Institute. 38800 Country Club Drive, Farmington Hills, MI 48331, 2014.

[3] ISO 1000/AMD1:1998. *SI Units and Recommendations for the Use of Their Multiples and of Certain Other Units,* 3rd ed., American National Standards Institute, Inc., 1819 L Street, NW, Washington, D.C. 20036, 1992 (amended 1998).

[4] *Using Soft-Metric Reinforcing Bars in Non-Metric Construction Projects,* Engineering Data Report No. 42. Concrete Reinforcing Steel Institute, 933 N. Plum Grove Road, Schaumburg, IL 60173-4758, 1997.

ANSWERS TO SELECTED PROBLEMS

Chapter 1

1-1. **a.** $w = 467$ lb/ft
b. $w = 488$ lb/ft
1-3. $f_r = 0.356$ ksi; ACI $f_r = 0.411$ ksi
1-5. $f_{top} = 0.396$ ksi
1-9. $M_{cr} = 531$ in.-kip

Chapter 2

2-1. **a.** $M_n = 421$ ft-kip
b. $M_n = 515$ ft-kip (+22%); (A_s: +27%)
c. $M_n = 501$ ft-kip (+19%); (d: +16.7%)
d. $M_n = 436$ ft-kip (+3.6%); (f'_c: +33.3%)
2-3. **a.** $\phi M_n = 213$ ft-kip
b. $\phi M_n = 310$ ft-kip (+45.5%); (f_y:+50%)
2-5. $M_u = 939$ ft-kip
a. $\phi M_n = 894$ ft-kip (N.G.)
b. $\phi M_n = 1072$ ft-kip (O.K.)
2-7. $M_u = 164.5$ ft-kip
$\phi M_n = 169$ ft-kip (O.K.)
2-9. $\phi M_n = 350$ ft-kip $>$ 304 ft-kip (O.K.)
2-11. $\phi M_n = 41.4$ ft-kip $>$ 35.5 ft-kip (O.K.)
2-13. As designed, $\phi M_n = 19.32$ ft-kip; as built, $\phi M_n = 11.96$ ft-kip (−38.1%)
2-15. 4 No. 9
2-17. 3 No. 11 $\phi M_n = 417$ ft-kip
2-19. 6 No. 9 (two layers, 1-in. clear)
$\phi M_n = 540$ ft-kip
2-21. $b = 16$ in., $h = 32$ in., 5 No. 9
2-23. $b = 18$ in., $h = 35$ in., 6 No. 9
2-25. $b = 16$ in., $h = 34$ in., 5 No. 10
2-27. $b = 12$ in., $h = 27$ in., 3 No. 11
2-29. (+Moment): $b = 11$ in., $h = 24$ in., and 2 No. 9
(−Moment): same b and h, and 2 No. 6
2-31. **a.** 6 in. slab, No. 5 @ 15 in. o.c. main steel, No. 4 @ 18 in. o.c. shrinkage and temperature steel
b. 4 in. slab, No. 6 @ 12 in. o.c. main steel, No. 3 @ 14 in. o.c. shrinkage and temperature steel

Chapter 3

3-1. $\phi M_n = 303$ ft-kip
3-3. $\phi M_n = 439$ ft-kip
3-5. **a.** $\phi M_n = 1527$ ft-kip
b. $A_s = 12.43$ in.^{2}
3-7. $\phi M_n = 856$ ft-kip
3-9. $\phi M_n = 1120$ ft-kip
3-11. $\phi M_n = 422$ ft-kip
3-13. 6 No. 10
3-15. **a.** 4 No. 9 bars (two layers)
b. 4 No. 8 bars (one layer)
3-17. **a.** $w_{LL} = w_{DL} = 1.87$ kip/ft
b. **1.** $\phi M_n = 341$ ft-kip
2. $\phi M_n = 344$ ft-kip
3-19. **a.** $\phi M_n = 582$ ft-kip
b. with 4 No. 8 compression bars, $\phi M_n = 696$ ft-kip
3-21. 3 No. 8 (compression steel), 5 No. 11 (tension steel 2 layers, 2 up, 3 down)
3-23. 2 No. 9 bars (compression steel), 4 No. 9 bars (two layers, tension steel)
3-25. 2 No. 10 bars (compression steel), 5 No. 9 bars (tension steel, in one layer)

Chapter 4

4-1. 7150 lb $<$ 9000 lb (N.G. in shear)
4-3. Max. $V_u = 35.6$ kip
4-5. $s = 7.76$-in. spacing
4-7. No. 3 stirrups (from F.O.S.): 3 in., 8 sp @ 6 in., 11 sp @ 10 in.
4-9. Double-loop No. 3 stirrups (from F.O.S.): 4 in., 11 sp @ 9 in., 6 sp @ 20 in.
4-11. No. 3 stirrups (from F.O.S.): 2 in., 7 sp @ 9 in., 7 sp @ 16 in.
4-13. No. 3 stirrups (from F.O.S.): 2 in., 6 sp @ 6 in., 5 sp @ 8 in., 3 sp @ 12 in.
4-15. No. 3 stirrups (from F.O.S.): 4 in., 14 sp @ 13 in.
4-17. $T_{cr} = 24.3$ ft-k: torsion may not be neglected

Chapter 5

5-1. 55.4 in.
5-3. 32.7 in. Use 2-in. (min.) side cover.
5-5. No. 4
5-7. $\ell_d = 69.6$ in. $>$ 49.5 in. (N.G.); 180° hook: $\ell_{dh} = 16.9$ in.
5-9. At A, req'd lap = 32.2 in.; at B: req'd lap = 54.3 in.
5-11. Req'd lap = 25.4 in.
5-13. **a.** Cut 2 No. 9 @ 13 ft-0 in. from centerline.
b. No. 3 stirrups (from F.O.S.): 3 in., 3 sp @ 15 in., 6 sp @ 5 in., 8 sp @ 15 in.

Chapter 6

6-1. $M_u = -3.03$ ft-kip, +5.19 ft-kip, −7.26 ft-kip, +4.54 ft-kip, −6.60 ft-kip; $V_u = 3.30$ kip, 3.8 kip

6-3. $b = 12$ in., $h = 24$ in.; end span: 2 No. 6 bars for −M @ end support (with 180° hook), 2 No. 8 bars for +M: interior span: 3 No. 8 bars for −M @ interior support, 2 No. 8 bars for +M

Chapter 7

7-1. **a.** $\bar{y} = 11.61$ in., $I_{cr} = 20{,}350$ in.^{4}
b. $\bar{y} = 1.68$ in., $I_{cr} = 72.7$ in.^{4}
c. $\bar{y} = 7.63$ in., $I_{cr} = 6159$ in.^{4}

7-3. **a.** $\Delta = 0.36$ in.
b. $\Delta = 0.43$ in.

7-5. $\Delta = 0.34$ in. < 0.40 in. (O.K.)

7-7. **a.** $s = 3.3$ in.; max. $s = 10.31$ in. (O.K.)
b. $s = 8.0$ in.; max. $s = 12$ in. (O.K.)

Chapter 8

8-1. **a.** Total $H = 7.01$ kip/ft
b. Total $H = 6.9$ kip/ft
c. Total $H = 8.0$ kip/ft
d. Total $H = 12.5$ kip/ft

8-3. Overturning F.S. = 3.29, sliding F.S. = 1.35, $p_{max} = 0.80$ ksf, $p_{min} = 0.55$ ksf

8-5. No. 7 @ 8 in. o.c.; use a 90° standard hook

8-9. No. 4 HEF @ 18 in. o.c., No. 4 VEF @18 in. o.c., 6 No. 9 vertical bars each end of wall spread over an end zone length of 12 in. (required minimum end zone length is 6 in.)

Chapter 9

9-1. **a.** $\phi P_{n(\max)} = 530$ kip, required ties are No. 4 @ 14 in. o.c.
b. $\phi P_{n(\max)} = 1156$ kip, required ties are No. 3 @ 18 in. o.c. (three per set)
c. $\phi P_{n(\max)} = 759$ kip, required ties are No. 3 @ 16 in o.c. (two per set)

9-3. $\phi P_{n(\max)} = 636$ kip, required ties = No. 4 @ 16 in. o.c., $P_{DL} = P_{LL} = 227$ kip

9-5. $\phi P_{n(\max)} = 840$ kip, required spiral = $\frac{3}{8}$ in. diam. @ 2 in. o.c.

9-7. Use 12 No. 9 bars (four per face), required ties are No. 3 @ 18 in. o.c. (three per set)

9-9. Use a column 14 in. × 14 in., 6 No. 9 bars, No. 3 ties @ 14 in. o.c.

9-11. $e_b = 13.29$ in., $\phi P_n = 341$ kip

9-13. Use a column 24 in. diameter; 10 No. 10 bars, $\frac{3}{8}$ in. diameter spiral at $2\frac{1}{4}$ in. o.c.

Chapter 10

10-1. Width = 24 in., depth = 12 in., longitudinal steel: 3 No. 4 bars

10-3. Width = 6 ft-0 in depth = 1′-3″, transverse steel: No. 6 @ 11 in. o.c.; longitudinal steel: 7 No. 5 bars

10-5. 9 ft-0 in. square, depth = 2 ft-0 in., 11 No. 7 bars each way

10-7. Rectangular footing, 7 ft-0 in × 11 ft.-6 in., depth = 2 ft-0 in., 9 No. 8 bars (long direction), 15 No. 6 bars (short direction)

10-9. Rectangular footing, 19 ft-6 in. × 23 ft-9 in.

10-11. Footing for column A: 7 ft-6 in. square; footing for column B: 8 ft-2 in. square

Chapter 11

11-1. $f = \pm 0.209$ ksi

11-3. $w = 1.34$ kip/ft

11-5. Midspan (transfer): $f_{top} = 0.929$ ksi (tens), $f_{bott} = 2.66$ ksi (comp); midspan (full load): $f_{top} = 0.477$ ksi (comp), $f_{bott} = 1.259$ ksi (comp); at end supports (only prestressing stresses exist): $f_{top} = 1.085$ ksi (tension), $f_{bott} = 2.82$ ksi (comp)

11-7. $f_{top} = 2.20$ ksi (comp), $f_{bott} = 0.132$ ksi (comp)

11-9. $\phi M_n = 889$ ft-kip, $M_u = 763$ ft-kip (O.K.)

Chapter 12

12-3. Max. stringer spacing = 92.9 in.

12-5. Use 4 × 10(S4S)

12-7. Max. shore spacing = 37.0 in.

12-9. Max. shore spacing = 35.4 in.

12-11. Joists @ 16 in. o.c., stringers @ 5 ft-9 in. o.c., 4 × 4 (S4S) shores @ 4 ft-6 in. o.c., guy wire bracing @ 12 ft (max.) o.c. on all sides

12-13. Studs @ 12 in. o.c., wales @ 28 in. o.c., max. tie spacing = 32 in. o.c., guy wire bracing @ 25 ft (max.) o.c. on each side of the wall

INDEX

A

Accelerators, 3
Accuracy of computations. *See* Reinforced concrete
ACI Building Code (318-11), 1
Admixtures, 3
Aggregates for concrete, 1–2
Air entrainment, 2
Allowable bending stress, 9, 323–324
Allowable soil pressure, 264, 273, 278–279
Allowable stress
  design, 22
  for form lumber, 324, 327
Anchorage of bars. *See* Development length
Anchors, 319
Answers to selected problems, 417
Approximate moments in continuous ¡girders, 363
Approximate shears in continuous girders, 363
ASTM, 3, 6, 386
ASTM A767, 7
A_v defined, 77
Axial load on columns defined, 236

B

Balanced beam. *See* Beams, reinforced concrete
Balanced failure mode, 27–28
Bar
  fabricator, 344–345
  lists, 345, 353–354
  marks, 345–352
  moment and development of bars, 268
  sizes, 6
  supports, 356
  tags, 357
  typical bends, 352
Bar cutoffs, 118
  in continuous spans, 132
  at simple supports, 122
  in tension zones, 121
Bars, areas of
  for multiples of bars, 387
  per foot of slab, 36, 388
Basement walls, 221
Base shears
  in caisson transfer, 291
  in pile, 291
Base SI units, 410
Basic development length, 111, 399
Beams, reinforced concrete, 21
  analysis and design methods, 21
  analysis for moment
    doubly reinforced, 62
    flow diagram, 404
    rectangular, 34
  balanced, 27
  behavior under load, 22
  crack control, 194
  depths of, 361
  design of (rectangular)
    depth-width ratios, 39
    recommended ρ values, 39, 389
  doubly reinforced, 62
    additional code requirements, 70
    compression steel not yielding, 65
    design for moment, 68
    summary of analysis procedure, 67
    summary of design procedure, 69
  equivalent stress distribution, 25
  failure of, 24
  flow diagram, 406
  formwork design example, 331
  irregular cross-sections, 56
  methods of analysis and design, 21
  minimum steel requirement, 30
  practical moment strength, 31
  rectangular (tension steel only)
    analysis for moment, 31
    design for moment, 37
    flexural strength, 25
    free design, 39
    sketch for, 39
    summary of analysis procedure, 34
    summary of design procedure, 41
  shear in, 75
  strength requirements, 30
  T-beam, 53
  torsion design procedure, 91
Bearing lug, 216
Bearing strength. *See* Footing
Bearing walls, 201, 219, 220
Bending, heavy/light, 355
Bends, typical, 353
Billet steel, 6
Bottle-shaped struts, 293, 294
"Braced" frames, 361
Brackets, 94–99
  with anchor bar, 95
  at expansion joints, 95
  with steel angle anchor, 96
Brittle failure mode, 27
Buildings, reinforced concrete, 360–384
Bundled bars. *See* Development length
Buttress wall, 201
β_1, value of, 25

C

Caissons, 287–292
  transfer of lateral base shears in, 291
Cantilever (strap) footing, 284–286
Cantilever retaining walls
  design of, procedure, 204
    additional details, 214
    footing base shear key, 216
    heel, 208
    soil pressures, 207
    stability analysis, 206
    stem, 211
    stem bar cutoffs, 211
    toe, 204
  rules of thumb for proportioning, 204
Carbon Fiber Reinforced Polymer (CFRP) strips, 364
Cast-in place concrete specifications, 379–381
Chords, 369–371
Closure strips, 375–376
Code. *See* ACI Building Code (318-11)
Coefficient
  of active earth pressure, 202
  of passive earth pressure, 202
Coefficient of resistance, 32
  tabulated values for, 390–398
Coefficients for shear and moment, 132
Cold joints, formation of, 3
Cold-weather placement, 379–380
Column capital, 150
Columns
  analysis of, short, 238
  bi-axial bending, 250
  capital, 34
  code requirements for, 236
  composite, 235
  cover requirements (*See* Cover requirements)
  design of, short, 239
  effective length, 253
  footings for, 268
  forms for, 339
  interaction diagram, 247
  large eccentricity, 242
    analysis, 243
    design, 247
  load-moment relationship, 241
  minimum spiral steel reinforcement ratio, 237
  ϕ factor considerations, 242
  preliminary sizing, 362
  reinforcing details, 236
  schedule, 256
  short, 235
  slender, 252
  stability factor, 328
  strength of axially loaded, small eccentricity, 236
  typical tie arrangements for, 237
Column offset in two-way slabs, 147, 165
Column strips, 148
  widths, 147
Compatibility torsion, 87
Composite columns. *See* Columns
Compression-controlled section, defined, 28
Compression splices, 117
Compression steel, 63
Compressive strength of concrete. *See* Concrete
Computation accuracy, 403
Computer detailing, 356
Concrete, 1, 8
  aggregates for, 1
  in compression, 3
  compressive strength of, 4
  cover, 8
  creep in, 5, 186
  design constants, 389
  fire protection requirements (*See also* Fire resistance of concrete), 8
  mix design, 2
  mixtures, 379
  modulus of elasticity, 4
  modulus of rupture, 5
  plain (*See* Plain concrete)
  protecting and curing, 380
  reinforcement in, 8, 365–366
  repair, 367–368
  sections, gross and cracked section properties of, 197–198
  specifications, 379–381
  strength-time relationship, 5
  stress-strain curve, 4
  in tension, 5
  weight of, 4

Concrete column schedule, 256
Concrete mixes
accelerators and retarders, 3
ACI Code, 2
air entrainment, 2
components of, 2
corrosion-inhibiting admixtures, 3
design and selection of, 2
water reducers, 2–3
Concrete slab systems
gravity load distribution in, 14–16
one-way slab, 13, 131, 146
two-way slab, 13, 145, 165
Construction Specifications Institute (CSI), 379
Continuous beams and slabs, deflection of simply supported and, 189–190
Continuous construction, 132
coefficients for shear and moment, 132
Continuous floor systems, 133
Continuous girders, 190–191
Continuous one-way floor systems, 133
Continuous-span bar cutoffs, 132–133
Continuous two-way slabs
advantages of, 146
analysis and design of, 145
column and middle strips, 148
column capital, 150
combined shear and unbalanced moment transfer in slab-column connections, 159–162
drop panels, 150–151
effective depth of, 162–163
effect of transverse edge beams, 165–167
elastic behavior of, 146
equivalent frame method, 176–180
longitudinal distribution, 155
minimum thickness of, 148–149
one-way shear in, 155–157
openings in, 165
punching shear in, 157–159
relative stiffness of beam and slab, 152–153
shear cap, 152
slab systems with beams, 167–168
solutions for, 146
static longitudinal moment, 150
structural integrity reinforcement in, 163-165
transverse distribution, 155
wall-supported, 162
Contract documents, 344
Conventional rigid method (mat foundations), 286
Conversion factors (U.S. Customary to SI), 411
Corbels, 94–99
with anchor bar, 95
at existing columns, 95
at expansion joints, 95
at offset columns, 95
with steel angle anchor, 95
Corrosion-inhibiting admixtures, 3
Corrosion of reinforcing steel, 7
Counterfort wall, 201
Cover, concrete, definition of, 8
Cover requirements
beams and girders, 37
columns, 237
slabs, 36
Crack control, 194
in deep beams, 194
Cracking moment, 5, 9, 186
Creep. *See* Concrete
Critical section for shear, 81

D

Dead loads, 14
Deep beams
design of, 296
reinforcement requirements for, 296
strut-and-tie models for, 292–296
Deflections, 183
control measures, 193
continuous beams and slabs, 189–190
continuous girders, 190–191
cracking moment M_{cr}, 186
immediate, 186
long term, 187
Deformed bars, 5
Derived SI units, 410
Design loads, 30
Detailing concrete structures, 344
Development length, 104
bundled bars, 109
compression bars, 111
hooks, 112
at simple supports, 122
tension bars, 107
of web reinforcement, 115
Diagonal tension, 76
Diaphragms, 369–371
Direct Design Method, 146, 147
Doubly reinforced beams. *See* Beams, reinforced concrete
Dowels, 216, 269, 273
Drag struts (collectors), 369–371
Drilled shaft. *See* Caissons
Drop panel, 34, 150–152
thickness, 145
Ductile failure mode, 28
Ductility requirements, 28

E

Earth pressure. *See* Walls, lateral forces on
Eccentrically loaded columns, 242
Eccentrically loaded footing, 277
Effective depth, 23
Effective length for formwork shores, 327
Elastic design of beams, 21
Elastic stress equation, 286–287
E, modulus of elasticity. *See* Concrete
End anchorage, 105, 305
Epoxy-coated reinforcing, 7, 108
Equilibrium torsion, 87
Equivalent fluid pressure, 203
Equivalent fluid weight, 203
Equivalent frame method, 146, 176–180
Equivalent stress distribution. *See* Beams
ϵ_t defined, 43
Excess reinforcement factor, 109
Extreme tension steel, defined, 28

F

ϕ. *See* Strength reduction factor
F^*_c defined, 328
Fabricating standards, 352
Fabricating tolerances, 356
Fiber Reinforced Polymer (FRP), 364
Fins, 317
Fire resistance of concrete, 376
Flat plates, 34
practical range of sizes for, 146
Flat slabs, 34, 145
practical range of sizes for, 146
Flexible diaphragms, 371
Flexural bond, 104
Flexure, 21
formula, 9
Floor dead load, 15
components of, 14
Floor slab, 14
formwork design example, 328–331
Floor system
defined, 196
frequency equation for, 196
maximum velocity of, 196
types of, 13
Floor vibrations, 195–197
damping, 195
frequency, 195
harmonic, 195
period, 195
types of, 196
Flow diagrams
rectangular beam analysis, 404
rectangular beam design, 405
T-beam analysis, 406
T-beam design, 407
Footing. *See also* Foundations
base shear key, 216
bearing strength, 269, 272
cantilever (strap), 284–286
combined, 282
design of, 268
eccentrically loaded, 277
individual column, 267
rectangular, 273
shear reinforcement in, 268
square, 269–270
steel distribution in, 273, 276
transfer of load from column into, 268–269
trapezoidal, 282
types of, 261
unreinforced concrete, 63
wall, 262
under light loads, 267
Form liners, 319
Formwork, 317
for beams, 331
for columns, 339
deflections in, 321
design approach for, 321
design equations for, 322
guy wire design, 331
loads and pressures on, 319
lumber for, 318
allowable stresses, 323
base design values for, 325
properties of, 324
materials and accessories, 318
plywood for, 318
requirements for, 317
for slabs, 326
for walls, 335
wood shore design, 327, 330
Foundations
drilled shafts (*See* Caissons)
mat foundation, 261, 286–287
pile caps, 287–292
pile foundations, 287–292
Four-pile group, 288, 294

G

Galvanized reinforcing, 7
Geotechnical report, 261–262
Girders, 37, 51, 362
continuous, 363
depths of, 361
pattern live load analysis for, 363
Grade line, 6, 409
Gravity load distribution in concrete slab systems, 14
dead loads, 14
spandrel beam loads, 16
tributary width and areas, 14–16
typical interior beam loads, 16

H

Hangers, 319
Hardboard, 319
Heavy bending, 355

High-density overlaid (HDO) plyform, 318
Hollowcore plank, 371
Hollowcore slabs, 371
Hook details, standard, 355
Hooked bars, 112, 399
Hooks, 112
Hot-weather placement, 380
Human-induced floor vibrations, 197
Human-induced vibrations, 196
Hydration, 1

I

I. See Moment of Inertia
Idealized stress-strain diagram for steel, 6
Immediate deflection, 186
Inch-pound system, 408
Incising factor, 324
Inserts, 319
Interior footing, 285
Internal couple method, 9
International Building Code (IBC), 377
Inverted T-girder, 57
Irregular cross sections, 56
Isolated spread footings, 261

K

k. See Coefficient of resistance
K_e defined, 327
Keys in walls, 215
Kickers, 331

L

l_a defined, 122
Lambda (λ) defined, 5
Lateral drift, 254
L-beams, 51
  analysis, 53
  design for moment, 57
  distribution of tension steel in flange, 140
  effective flange width, 53
  minimum steel, 53
  rectangular, 54
l_d. *See* Development length
l_{db}. *See* Basic development length
l_{dh} basic development length for hooks, 112
l_e defined, 327
Light bending, 355
Load factors, 30
Loads on formwork, 319
Load testing of structures, 373–374
Long-term deflections, 189
Loop stirrup, 37
Lumber for formwork. *See* Formwork, lumber for

M

Marking systems, 345
Mass, 410
Mat foundations, analysis and design of, 286–287
Maximum number of column bars in one row, 400
M_{cr}, 9, 186
Mesh. *See* Welded wire reinforcing
Metrication, 408
Middle strips, 148
  widths, 147
Minimum area of shear reinforcement, 77
Minimum flexural tension steel area
  one-way slabs, 34
  rectangular beams, 30
  T-beams, 53
Minimum required beam widths, 387
Minimum thickness for slabs, 36, 42
Modular ratio, 184
Modulus of elasticity
  for concrete (*See* Concrete)
  for reinforcing steel, 6
Modulus of rupture for concrete. *See* Concrete
Moisture curing, 380
Moment arm, 9
Moment frames, concrete, 222–223
Moment of inertia
  cracked section, 183, 184
  effective, 184
  gross section, 13, 184
Monotonic uniformly distributed load test, 373–374
Multiple-loop stirrups, 79
Multi-story columns, 292

N

n. See Modular ratio
Net tensile strain, 26, 28
Neutral axis, 9, 23, 28
  transformed section, 186
Nodal zones, strength of tension ties, 295
Nominal moment strength, 25

O

One-way shear, 155–157
One-way slabs, 13, 371
  analysis for moment, 34
  design for moment, 42
  load calculations for, 15
  load distribution in, 13
  moments and shears in, 146
  shear and moment for, 131
  subjected to concentrated loads, 371–372
Overlaid plywood, 318
Overturning factor of safety, 206

P

Pavement, 3
P-delta effect, 254
Pedestal, 234, 269
Pile caps, 287–292
  critical section for shear in, 291–292
  design of, 292, 296
  layouts, 288
  load distribution in, 290
  strut-and-tie models for, 292–296
Piles, 287–290
  lateral base shears in, 291
  transfer of lateral base shears in, 291
Placing drawings, 345
Plain concrete, 9, 263
Plastic coated plywood, 318
Plyform, 318
  safe span, table, 339
  section properties and design values, 323
Plywood, 318
Portland cement, 1
Post-tensioning, 304
Pour strips, 375–376
Practical moment strength of beams. *See* Beams, reinforced concrete
Pressure on formwork, 319
Prestressed concrete, 303
  cracks in, 303
  design notes, 315
  draped tendons, 309
  flexural strength analysis, 315
  load balancing method, 310
  loading stages, 307
  materials, 306
  method of superposition, 310
  rectangular beam analysis, 307
  typical precast members, 309
Prestress losses, 309
Pretensioning, 303
Principal planes, 75
Principal stresses, 75
Project for students, reinforced concrete buildings, 381–384
Proportional limit, 21
Punching shear, 268

R

Rectangular beams. *See* Beams
Rectangular combined footings, 284
Rectangular reinforced concrete footings, 273
Rectangular T-beam, 54
Reinforced concrete, 2, 21
  accuracy of computations, 403
  buildings, practical considerations, 360–384
    approximate moments and shears in girders, 362–363
    cast-in place concrete specifications, 379–381
    closure or pour strips in, 375–376
    columns, preliminary sizing of, 364
    diaphragms, drag struts, and chords, 369–371
    fire resistance of concrete structural elements, 376–377
    girders, 359
    load testing of structures, 373–374
    one-way slabs, 371
    project for the students, 381–384
    rules of thumb, 360
    strengthening and rehabilitation, 364–368
    student design projects, 381
    two-way slabs on stiff supports, 377–378
  footings, 269–270, 273
  wall footing, 264
  weight, 17
Reinforced concrete columns, 364
Reinforced concrete structure
  use of, 364
Reinforcement
  of concrete structures, 193
  ratio, defined, 31
  in shear walls, 226
  for torsion, 90
Reinforcing bars, 6
  anchorage (*See* Development length)
  areas of (*See* Bars, areas of)
  cover required (*See* Cover requirements)
  distribution in tension flanges, 140
  lengths available, 6, 117
  metric, 408
  properties and specifications, 386
  selection of, for beams, 38
  tolerances, 353
  typical bends, 353
Relative stiffness of beam and slab, 152–153
Repair of concrete structures. *See* Strengthening of concrete structures
Repeated load test, 374
Resisting moment, 9
Retaining walls, 201
  lateral forces on, 201
Retarders, 3
Rolling shear, 321
Roof dead load
  components of, 14
Roof slab, 14
ρ. See Reinforcement, ratio defined
Rules of thumb, 360

S

Schedules, 352
Second load test, 374
Sensitive equipment vibrations, 197
Serviceability, 183
Service loads, 22, 30, 183

Shear
analysis procedure, 78
design for
in beams (*See* Beams, reinforced concrete)
in footings, 265, 266, 268
in slabs, 77, 136
friction, 216
in girders, 362
and moment equations, 131
reinforcement, design requirements, 76
strength in footings, 268
stresses, 146
Shear cap, 152
Shearing schedule, 358
Shear span, 76
in deep beams, 292
Shear walls, 223
design considerations, 226
design example, 228
layout, 223
strength of, 228
Shop drawing, 345
Shores. *See* Formwork, wood shore design
Short column, defined, 235
Shrinkage and temperature steel. *See* Slabs
SI, 408
prefixes, 410
style and usage, 410
Sidesway, 254
Simply supported beams, 190
Slab-column connections, 159–162
Slabs, 34
cover requirements, 36
effective widths for, 372
on grade, 43
loads, 15
minimum reinforcement, 36
minimum thickness, 35
one-way, 35
analysis for moment, 36
design for moment, 43
shrinkage and temperature steel, 35
thickness selection, 43
two-way, 34
Slab systems
with beams, 167–168
one-way, 13
two-way, 13
Slab thickening, 145
Slender columns. *See* Columns
Slenderness ratio, 254
Sliding factor of safety, 206
Spacing requirements for bars, 37
in slabs, 35
Spancrete Manufacturers Association, 371
Spandrel beam loads, 16
Span length of beams and slabs, 42
Special fabrication, 355
Spiral columns, 234
Spirals, 234, 236
Splices, 117
Split-cylinder test, 5
Splitting tensile strength, 5
Spreaders, 319
Square reinforced concrete Footing, 269–270
Static longitudinal moment, 150
Static uniformly distributed load test, 373–374
Steel ratio. *See* Reinforcement, ratio defined
Stirrup design
notes on, 79
procedure, 79
requirements at bar cutoffs, 121
Strain in concrete, maximum, 24
Strand, tensile strength, 306
Strap beam, 261, 284–285
Strap footing. *See* Footing, cantilever (strap)
Strength design method assumptions, 23
Strengthening of concrete structures, 362
Strength of columns, 235
Strength reduction factor (ϕ), 30
Strength requirements for beams. *See* Beams, reinforced concrete
Stress-strain curve
for concrete (*See* Concrete)
for reinforcing steel, 6
Structural drawings, 345
Structural engineer of record (SER), 379
Structural integrity reinforcement
for beams, 125–126
negative (-ve) moment regions, 126
positive (+ve) moment regions, 125–126
for two-way slabs, 163–165
Structural lumber. *See* Formwork, lumber for
Structural plain concrete, 263
Structural steel
on bearing walls, 219
columns on footings, 267
composite columns, 235
Strut-and-tie method, 292–293
C-C-C nodal zone, 293
C-C-T nodal zone, 293
C-T-T nodal zone, 293
failure modes in, 293
pile caps and deep beams using, 296
strength of nodal zones, 295
strength of tension ties, 295–296
Strut coefficient, 294–295
Student design project problem, 360–362
Student design projects, 381
Summary of procedure
calculation of tension l_d, 109
doubly reinforced beam analysis, 67
doubly reinforced beam design, 69
L-beam design for moment, 61–62
L-beams analysis for moment, 60–61
one-way slab design, 43
rectangular beam analysis, 34
rectangular beam design, 41
short column analysis and design, 241
stirrup design, 79
T-beam analysis for moment, 60–61
T-beam design for moment, 61–62
torsion design, 91
Surcharge, 203
Sustained live load, 187
Sway frame, 254

T

T-beams, 51
analysis, 53
design for moment, 57
distribution of tension steel in flange, 140
effective flange width, 53
minimum steel, 53
rectangular, 54
Tendons, 309
Tensile strength of concrete, 5
Tensile stresses, 268
Tension-controlled section, defined, 28
Tension ties, tension capacity of, 295–296
Three-pile group, 288, 294
Tied columns, 234
Ties
in columns, 238
in forms, 319
Torsion, 86
design procedure, 91
Transfer stage, 304, 306
Transformed concrete cross-section, 185, 305
Transition region, 28
Transverse bending, 285
Transverse distribution, 155
Transverse edge beams, 165–167
Trapezoidal combined footings, 284
Tributary area, 14–15
Tributary width (TW), 14–16
True T-beam, 54
Truss bars, 118
Two-pile cap, 289
Two-way slab systems, 13, 34
advantages of, 145
analysis of, 377–378
continuous (*See* continuous two-way slabs)
load distribution in, 13
maximum reinforcement spacing in, 165
minimum reinforcement in, 155–157, 164–165
one-way shear in, 155–157
openings in, 165
shear capacity of, 364
on stiff supports, 377–378
Typical interior beam loads, 16

V

Vibrations. (*See* Floor Vibrations)

U

Ultimate load, 22
Ultimate strength design, 22
Unreinforced strut, 294
Unrestrained structural members, 377

W

Wall footings, 262
under light loads, 267
longitudinal steel in, 265, 267
Walls
form design, 335
lateral forces on, 201
retaining (*See* Retaining walls)
types of, 200
Water-cementitious material ratio, 1
Water reducers, 2–3
Web reinforcement, 77
development of, 115
Weight
of concrete, 2
vs. mass, 411
of reinforced concrete (*See* Reinforced concrete, weight)
Welded wire reinforcing, 7
Workability, degree of, 3
Workability of concrete, 1
Working stress design (WSD), 22

Y

Yield stress, 6